1 MONTH OF
FREE
READING

at
www.ForgottenBooks.com

ISBN 978-0-332-51239-6
PIBN 10566634

SCIENCE-GOSSIP

AN ILLUSTRATED MONTHLY RECORD OF

NATURE AND COUNTRY-LORE

AND

APPLIED SCIENCE

EDITED BY

JOHN T. CARRINGTON

ASSISTED BY

F. WINSTONE

VOL. VII.—NEW SERIES

LONDON

110 STRAND, w.c.

WHOLESALE AGENTS—Horace Marshall & Son

BERLIN : R. Friedländer & Sohn

25-98651- 72b9

OUR ANNUAL GREETING.

WE cannot forbear from asking our readers to felicitate with us upon the success of our past volume. The excellence of many of the contributions has obtained for the magazine congratulations from widely-distributed sources. Added to this has been the satisfaction of recording in our pages descriptions of several animals new to Science.

With the conclusion of the volume come changes among our departmental editors through the retirement of Messrs. James Quick, Edward A. Martin, and Harold M. Read, who have so kindly conducted the respective departments of Physics, Geology and Chemistry. To these gentlemen we tender our sincere thanks for their past assistance. Mr. C. A. Mitchell, B.A. (Oxon.), F.I.C., has undertaken the last department, and we believe that our readers will greatly appreciate the varied experience that he will bring to bear upon that section of Science. We hope at an early date to be able to announce that conductors have volunteered for the other departments.

We have long felt the desirability of paying more attention to the Botanical section of Science, and we are glad to state that two gentlemen, well known as botanists, have been good enough to undertake the honorary departmental editorship. Mr. James Saunders, A.L.S., of Luton, will conduct Field Botany, and Mr. Harold A. Haig, of University College, London, will supervise Structural Botany. We trust that our readers may help to make their columns of universal interest.

Dr. G. H. Bryan, F.R.S., has generously undertaken to assist with our mathematical columns.

We have again to thank our contributors for their valuable articles and notes, and to express our regret that limitation of space has precluded the consideration of numerous other manuscripts. We have also to tender gratitude to our old and to the many new subscribers for the support they have rendered, without which it would be impossible to continue our efforts to popularise Science.

110 *Strand, London.*
May 1901.

Editor.

CONTENTS.

VOLUME VII.—NEW SERIES.

ARTICLES, NOTES, ETC.

CONTENTS.

CONTENTS.

SCIENCE-GOSSIP.

◆•◆

BIRDS AT LYNMOUTH.

By Thomas H. Mead-Briggs, M.A., F.E.S.

THE village of Lynmouth, in North Devon, lies at the mouth of the two small rivers, the East and West Lyn, that meet there near the sea. It is from this circumstance that the place receives its name. Lynmouth is situated in Latitude

Lynmouth. The hills are higher inland than on the coast. Hollardy Hill, behind Lynton, rises to nearly 800 feet, looking down upon that place more than 300 feet below its summit. Summer House Hill, formerly called Lyn Cliff, is situated

From Photo by]   HAUNT OF THE WATER-OUZELS, EAST LYN.   [John T. Carrington.

51° 13·50′ N., and Longitude 5° 49·40′ W., and is on the south coast of the Bristol Channel, nearly opposite to Swansea, on the Welsh coast, from which it is distant about thirty miles. Our village is surrounded by high hills, only broken by the valleys of the two rivers. The adjoining village of Lynton lies in a hollow, about 450 feet above

between the two valleys of the Lyn, reaches to quite that altitude, and is our highest hill. There is, however, a hill on the east side of the village of Countisbury, distant about two miles from this place, that rises to 1,125 feet above the sea, into which a portion juts, forming a bold headland. Chapman Burrows, five miles inland, reaches an altitude of 1,575 feet, being the highest in North Devon. These hills are chiefly of sandstone on the coast, with some beds of coarse slate a little distance inland. The cliffs on the coast are in many places prettily coloured by peroxide of iron, which causes various shades of red, yellow, and grey, often curiously intermixed. Portions of the cliffs along the coast have from time to time fallen into the sea, but from some cause the rocks on the shore are only covered with algae in isolated places, the rest being smooth and bare.

Lynmouth, owing to its northern aspect and high surrounding hills, suffers from an absence of sunshine in the winter as regards many of the houses. The inhabitants of some of them in the valley of the East Lyn do not see the sun on their residences from the end of October until February 14th, a date eagerly looked forward to by them. Owing, however, to the sheltered position of our valley, myrtles bloom and fruit freely, fuchsias and geraniums live in the open gardens through the winter, and *Euonymus japonica* also bears fruit in most years.

The absence of marshes and sea-sands in the immediate neighbourhood of this place is the reason that the number of species of birds I have seen here are comparatively so few. With the exception of an occasional flock of geese or ducks, flying too high in the air to distinguish the species, and a common sandpiper now and then seen flying along the rocks on the shore, the large groups of birds to which they belong are absent. The following list includes those birds I have seen in the Lynmouth district during the last three years :—

Turdus viscivorus Lin. Missel-thrush. Common here at all times of the year. The young birds assemble in flocks in the autumn.

Turdus musicus Lin. Song-thrush. Very common.

Turdus iliacus Lin. Redwing. My brother (C. A. Briggs) told me that he saw a flock of these birds at Countisbury on September 7th, 1899. I saw a few others at times all through the winter of 1899-1900, which has been the longest and most severe winter since we have been here. I have not noticed this bird in the other winters.

Turdus pilaris Lin. Fieldfare. In 1897 Mr. E. B. Jeune, my brother, and I saw three fieldfares flying on the high ground between Countisbury and Glenthorne. In 1899 I saw a few on October 17th from the window of the railway train travelling between Lynton and Barnstaple. I have seen one or two others at different times, but it is not a common winter visitor.

Turdus merula Lin. Blackbird. Resident and very common.

Saxicola oenanthe Lin. Wheat-ear. On March 25th, 1897, I saw a great number of these birds in the Valley of Rocks, but they were apparently on migration, as only a few subsequently remained to breed. On May 28th there appeared to be only

one there. Next year, on March 22nd, there were several in the Valley of Rocks, and I saw a few on May 23rd. On April 11th, 1900, there was a male in full plumage at Countisbury, and another, later, on the Tors above East Lyn.

Pratincola rubetra Lin. Whinchat. May 23rd, 1898, several were in the Valley of Rocks, where the bird is often common.

Pratincola rubicola Lin. Stonechat. Resident. On March 22nd, 1898, I saw two in the Valley of Rocks, and one on May 23rd in the same place. This was a male bird, and the female probably had her nest near by, as the male bird would not be driven away.

Ruticilla phoenicurus Lin. Redstart. Common here, especially in 1898, when I knew of three nests up the valley of the East Lyn, between Lynmouth and Rockford.

Erithacus rubecula Lin. Redbreast. Resident and very common.

Sylvia cinerea Bechstein. White-throat. On April 5th, 1897, I saw one sitting on a telegraph wire, on the Barnstaple road. This was the only occasion of my seeing one of these birds until a male and a female appeared in our garden on May 5th, 1900.

Phylloscopus rufus Bechstein. Chiff-chaff. Not uncommon. April 4th, 1899, one heard near the Countisbury road; April 8th, 1898, one heard; April 15th, 1899, one heard. These are the first dates of hearing this bird in the respective years; but it is not common on the coast of North Devon.

Accentor modularis Lin. Hedge-sparrow. Resident and common.

Cinclus aquaticus Bechstein. Dipper, or Water-ouzel. This is one of our most interesting birds, and one that gives endless pleasure in observation, on account of its quick movements and rapid flight. It is not uncommon, and resident, though never abundant, in the valleys of the East and West Lyn rivers. These birds chiefly frequent the rocky portions of the streams, though they are often to be seen on rocks below the point of junction, and where the river Lyn passes through Lynmouth. They even descend to the tidal region below our house, and hunt about among stones covered by *Zostera* and other sea-weeds. Our dippers can hardly be described as shy birds, as they take little notice of people walking on the paths by the riverside. In breeding plumage the *C. aquaticus* of the Lyn have not the lower portion of the breast of dark chestnut-brown, as described by Mr. Howard Saunders in his manual of British birds. There is a variety named *melanogaster*, a Scandinavian form, that has been reported from some parts of Britain, to which I imagine our water-ouzels belong. These birds breed regularly by both our streams. The accompanying illustrations of the "Haunt of the Water-ouzels," and of a nest with young, were taken on May 11th last, the latter from a nest found by my brother and the Editor of this magazine. When photographed, the nest apparently contained four fully-fledged young ones; but as not more than three could at the same time show themselves through the entrance to the dome-shaped nest, it was difficult to know the number. We have to thank our friend the Vicar of Lynmouth, the Rev. Albert H. Hockley, for photographing this nest under very difficult circumstances. The open beak of one of the young ones is visible in the picture. The nest was constructed entirely of moss, and was situated about five feet above the summer level of the water. The illustration on the first page, of the haunt of

Parus palustris Lin. Marsh-tit. Resident, rather more common than *P. ater*, and in the same localities.

Parus coeruleus Lin. Blue tit. Resident and common.

Troglodytes parvulus K. L. Koch. Wren. Resident and common. Several are always to be seen in our garden, where they nest.

Certhia familiaris Lin. Tree-creeper. Resident. I have only seen this bird here occasionally. Once or twice one was climbing round the poles of our verandah.

Motacilla lugubris Temminck. Pied wagtail.

From a Photo by]   *[Rev. A. H. Hockley.*
NEST AND YOUNG OF WATER-OUZEL.

the water-ouzels and site of the nest, is from a photograph by Mr. John T. Carrington. The lady shown in the picture is pointing to the rock, just below the fissure containing the nest. This view shows the sweetly pretty character of one haunt of dippers on the East Lyn.

Parus major Lin. Great tit. Resident and common.

Parus ater Lin.. Cole-tit. Resident. A few seen every year in our garden, where every winter they come for food placed for them in a cocoanut shell suspended from the verandah.

Resident and common. I have often seen the young birds in the summer.

Motacilla alba Lin. White wagtail. I saw a pair of these birds on the wall of the esplanade here on April 19th, 1898. They had the mantle of a pure French grey, the same colour as that of an adult *Larus argentatus* (the herring-gull). I have looked at numbers of *M. lugubris*, in the hope of being able to distinguish this form or species (*M. alba*), but these two birds were strikingly distinct from any others I have seen. I have not observed this bird since that occasion.

B 2

Motacilla melanope Pallus. Grey wagtail. Resident and common on the banks of the East Lyn. It also comes into our garden. The male bird in breeding plumage is a very handsome object on account of the triangular black patch under the throat.

Anthus pratensis Lin. Meadow-pipit. Resident and common, especially a little inland.

Anthus obscurus La. Rock-pipit. Resident and very common on the coast. The male in breeding plumage is decidedly vinous-tinted on the breast.

Muscicapa grisola Lin. Spotted fly-catcher. Resident. I have found the nest here.

Hirundo rustica Lin. Swallow. Common; one was seen here on February 13th, 1897.

Chelidon urbica Lin. Martin. Common here on arrival and before departure, but they seem to move inland to breed.

Cotile riparia Lin. Sand-martin. Common, but seen most plentifully on arrival and departure in spring and autumn. I have not observed it breeding on the coast.

Ligurinus chloris Lin. Greenfinch. Resident and common.

Carduelis elegans Stephens. Goldfinch. Scarce. I caught a young bird in our stable in the summer of 1899, and let it fly again.

Carduelis spinus Lin. Siskin. I saw one on November 7th, 1896, flying amongst some fir-trees on the road between Lynton and Lynmouth. This was the only example I have yet seen here.

Passer domesticus Lin. House-sparrow. Resident, but not abundant here. The chaffinches appear to take the place of these birds about the houses in Lynmouth.

Passer montanus Lin. Tree-sparrow. I saw a pair on May 3rd, 1898, near the "Cottage Inn," on Barnstaple road, near Lynton. I have not seen any others in North Devon.

Fringilla coelebs Lin. Chaffinch. Resident and very abundant. The males are exceedingly highly coloured in the breeding season.

Fringilla montifringilla Lin. Brambling. I saw one male in our garden on June 18th, 1899.

Pyrrhula europaea Vieillot. Bullfinch. Resident and not uncommon.

Emberiza citrinella Lin. Yellow bunting. Resident and common.

Emberiza cirlus Lin. Cirl bunting. Resident and fairly common.

Sturnus vulgaris Lin. Starling. Resident and common.

Pyrrhocorax graculus Lin. Chough. Resident, rare. Occurs in a locality some distance from here, where it breeds annually.

Garrulus glandarius Lin. Jay. Resident and common. Each autumn these birds come to our plantation adjoining the garden for the acorns of the evergreen oaks.

Pica rustica Scopoli. Magpie. Resident and common, especially a little distance inland. I have often seen the nests near here.

(To be concluded.)

ANIMALS AS WEATHER PROPHETS.

BY ARTHUR H. BELL.

OF all the proposed methods for forecasting the weather, it is doubtful if any of them are so curious as the one that suggests that future climatic conditions may be foretold by observing the movements of the common leech. One observer states that if a leech is kept in a bottle of water it will be seen to disport itself in ways that closely synchronise with changes in the weather, sinking to the bottom of the bottle when the weather is going to be stormy, and rising to the top when conditions are more promising. Now, so much faith was put in the leech as a weather-prophet that an admirer of its performances proposed to attach a small chain to the creature and cause it to make electric contact whenever it travelled about its jar in response to supposed coming changes in the weather. Enamoured, moreover, of his suggestion, it was further proposed that the electric current thus created should be utilised to set the largest bell in the parish in motion, and so give warning to all whom it might concern that changes in the weather were coming. It cannot, however, be said that any great success has

attended this proposition, and the attempt to elevate the leech to the front rank of weather-prophets has not met with general approval among serious students of meteorology. The leech indeed, like so many other creatures, has nowadays to submit to a rigorous cross-examination; and when dealing with the supposed powers of insects, birds, and animals to forecast the weather it is customary to compare such pretensions with the weather maps and synchronous charts upon which modern meteorology may be said to be based. Popular prognostics, founded on the movements of birds and animals, there are in great numbers; but it is now possible to say why it is that these prognostics are successful, and how it is that they sometimes fail.

Meteorologists compile charts on which are plotted the barometer readings taken at a large number of stations at the same hour over a large tract of country. Lines are then drawn through all places having the same barometer reading, such lines being called isobars, or lines of equal atmospheric pressure. An examination of one of these isobaric charts, as they are sometimes called, quickly reveals the fact that the shapes assumed by these isobars vary with the type of weather. Thus during stormy, rainy weather the isobars are crowded together, the barometer readings being lowest at the centre of the disturbance, the conditions being called "cyclonic." On the other hand, when the weather is fine and the wind slight in force, the isobars are far apart, and the highest barometer readings are to be found at the centre, the conditions being then called "anti-cyclonic." Roughly speaking, therefore, all types of weather are either cyclonic or anti-cyclonic; so in looking at a weather map it will be noticed that the cyclones occupy one part of the country and the anti-cyclones another, and the kind of weather to be expected depends on the relative positions of these two systems to one another. In forecasting the weather, therefore, by the aid of such charts, it is as if one had to foretell the future positions of the pieces on a chess-board; for the cyclones and anti-cyclones move at varying rates, and it is not surprising, therefore, that in compiling his forecasts the weather-prophet sometimes makes mistakes. A further important fact is that each quadrant in a cyclone, and in an anti-cyclone also, has certain definite kinds of weather associated with it; and it is this circumstance which explains the success of many popular weather prognostics based on the movements of birds and animals, and on such things as haloes and sunset colours. The front of a cyclone or depression, as it is sometimes termed, is always associated with a rising temperature and damp weather; while as the storm departs, temperature falls and the wind flies round to a dry quarter. It is therefore not surprising that as the cyclone advances, the dampness of the atmosphere causes certain birds, quadrupeds, insects, and flowers to exhibit an unusual amount of restlessness; the closing of petals on the part of flowers, and unwonted migrations on the part of animals, being solely due to the hygrometric condition of the atmosphere produced by a coming storm. There is therefore reason to expect bad weather when these movements are observed, since they simply mean that the front part of a cyclone has arrived, and the usual sequence of rainy weather may be expected to follow. Most people, again, are familiar with the feathery wisp-like cloud called cirrus, which is invariably associated with the forefront of a cyclone. When the weather-prophet sees this cloud he is sure that stormy weather is near at hand. Cirrus, the highest of all clouds, is probably composed of ice crystals, and it is these that act like prisms, causing the haloes which during bad weather form round the sun and moon.

Time out of mind haloes have been popular prognostics for stormy weather, and receive sanction from the weather charts, for they simply mean that cirrus cloud is forming, or, in other words, that the front of a cyclone has arrived. A similar comparison shows that when people forecast rainy weather from the flaring of lamps and candles, from rheumatic pains, shooting corns, and creaking furniture, the indications are the same; for all these phenomena are to be set down to the credit of the high temperature and increased moisture which are associated with the front portion of a cyclonic system. In the same way the bawling of peacocks, the braying of asses, the occasionally excessive quacking of ducks, and an unusual activity and brilliance of glowworms, which are all popular prognostics for rainy weather, probably have gained their popularity from the fact that they manifest themselves at such times as atmospheric pressure is decreasing and moisture increasing.

As already mentioned, fine-weather conditions are called anti-cyclonic, and the prominent features of this kind of weather are dew and mist in the morning, brilliant midday sunshine, with mist again during the evening hours. At such times, also, the barometer stands high, and it is not surprising that under the exhilarating influences of a blue sky and gentle breezes birds fly high and go far afield, and animals and insects show an abnormal activity. Bird-catchers know that when larks fly high and swallows soar aloft in the empyrean a continuance of fine weather may be expected. Owls hooting in the stilly night are not pleasant to those suffering from insomnia; but there is consolation in the thought that such hootings betoken the formation of an anti-cyclone and subsequent fine weather. Indeed, concerning anti-cyclonic conditions there are a very large number of popular prognostics which have reference to fine weather, and some of these proverbs and weather saws have been in use during many centuries. Modern meteorology, moreover, does not seek to discountenance the use of this folk-lore, since it can be shown, as above, that it owes its success to the fact that these prognostics agree with the conclusions arrived at by more scientific methods. The meaning, therefore, of all these fine-weather prognostics is that the birds, insects, and other animals are active and lively because the atmospheric conditions are anti-cyclonic.

Cyclones and anti-cyclones, however, are not always of the same size and shape; and as they drift through the atmosphere they get squeezed and distorted, spreading out in unexpected directions. Two cyclones, for instance, may be separated by a very narrow anti-cyclone. As this "anti-cyclonic wedge," as it is sometimes called, passes over the country the birds or other animals commence all those movements which the weather-wise associate with fine weather, and the prophets say that fine weather is coming. Yet, since this wedge of high pressure is so narrow, the fine weather lasts only an hour or so, and cyclonic and stormy weather quickly follows. This is an instance where the popular prognostics fail, and it is by looking at a weather chart whereon the cyclones and anti-cyclones and their relative positions are set forth that the explanation of the failure is revealed. Moreover, in an ordinary way the isobars are circular, but at times they run in straight lines like a series of railway lines, and it is no uncommon thing for one storm area to be divided from another by these straight isobars. It is at such times as these that distant sounds are plainly heard, such as those made by railway trains, canal boats, and waterfalls. Distant objects are also plainly seen, and this unusual audibility and visibility have always been a popular prognostic for bad weather.

As we can understand, the success of this . prognostic depends on the fact that a cyclone is ⁻

following closely behind what meteorologists call some "straight isobars." The cyclone at times, however, shows a disposition to move in unexpected directions, or to disperse and fill up; and it is when either of these contingencies happens that the prognostics fail. Not only, therefore, do the modern methods of dealing with weather-problems show how it is that birds and animals, and such creatures as leeches, provide dependable weather-prognostics, but they also show why it is that they often fail. It is sometimes said that animals prognosticate changes in the weather; but the philosophically minded attribute no such powers to them, and the weather charts and isobaric maps demonstrate that at best they are but prognostics.⁻

London, S.W., May 1900.

GEOLOGY AROUND BARMOUTH.

By JOHN H. COOKE, F.G.S., F.L.S.

FEW districts offer so many attractions. to the geologist as Merionethshire. The diversity of its rocks gives rise to a great variety of landscape, from the brown moorland enveloped in gorse and heather, enclosuring pastures of the softest and richest verdure, to the wooded heights of Lylfaen and the rugged barrenness of Cader and the Arrans. It is an ideal region for a holiday, comprising within a small and easily accessible area a rare combination of barbaric wildness and cultivated repose. The region around the Mawddach is undoubtedly the finest in Wales. In picturesqueness it rivals the Bosphorus, to which it bears a striking resemblance; and nothing short of a panoramic view can give any adequate idea of its wonderful variety. In the autumn of 1898 I paid my first visit to this classic region, and in my intervals of leisure I jotted down a few notes on its geological phenomena. These I have now gathered together and embodied in the following short paper, in the hope that they may prove of some interest to future visitors.

Merionethshire consists, for the most part, of a broad, oval-shaped mountain boss of coarse quartzose and greenish-grey grits of Cambrian age. Barmouth is situated at the south-western extremity of the boss, and on the line of junction of the Upper Cambrian and the *Lingula* beds of the Lower Silurian. The strata of Cambrian age which are exposed in the hillsides overlooking Barmouth consist of an alternating series of coarse and fine grits and purple shales dipping from 50° to 60° E.S.E., with intrusive masses of quartz porphyries and of greenstone dikes.

The grits are exceedingly compact, and so coarse that many of the angular fragments of quartz which they enclose measure upwards of a quarter of an inch in diameter. They consist generally of

fragmental quartz and felspar with an appreciable quantity of greenish chlorite united in a siliceous cement; but they are very variable in constitution and structure within even limited areas. The coarse grits alternate rapidly with conglomerates and fine-grained sandstones, these again with slates and shales. The conglomeratic masses are irregular and pockety, and seldom exceed 3 feet in depth. On the pathway at the back of Moss Bank, overlooking the Barmouth Morfa, the fine-grained grits and sandstones enclose isolated rounded boulders of coarse grit varying from 1 inch to 18 inches in diameter. The variability of the formation is well shown in this section, and also in a huge boulder which lies on the shore-line immediately opposite the bridge which crosses the railway at Llanaber. The face of the boulder, measuring 3 feet 8 inches in width, indicates seven periods of depositions, each of which possesses marked characteristics that differentiate it from the others.

The Barmouth grits do not afford the fossil-hunter many opportunities, as the only evidences of organic life that have hitherto been forthcoming are the tracks and borings of annelids, and, in the direction of Harlech, two species of *Oldhamia*. The petrographer, however, will find the strata full of interesting points and problems. Two species of felspar occur in the rocks in considerable masses, while the quartz exhibits well-defined, hair-like crystals of rutile as well as liquid cavities containing bubbles. A vein of auriferous quartz dipping 50° E.S.E., and attaining 3 feet in thickness, passes through Cellfechan Farm at Barmouth, but though it has been worked for some years it barely pays its way. The manganese mines on the slopes around the new church at Barmouth are well worth a visit. The grits at this point have been contorted, and one of the mines has been opened out in a small anticlinal fold. The deposits are limited in extent, and the mines are therefore of no great size. The manganic ore varies in quality, from an impure whitish-grey carbonate, with nearly 20 per cent. of siliceous matter, to a black hydrated oxide which is comparatively rich in the metal. The black oxide is a result of the oxidation of the carbonate, and it contains from 32 to 35 per cent. of metallic manganese.

. An interesting line of investigation is afforded to the wayfarer in determining the number, character, and position of the dike intrusions of the district. The heights that rise behind . Craig Abermaw are crossed by considerable numbers of igneous ·dikes, which do not appear to have received at the hands of geologists the attention they merit. Owing to the conditions of exposure it is not always possible to trace their horizontal

streamlet are the barren slopes of the Cambrian grits; on the other side, the many-hued *Lingula* flags with their well-wooded sides chequered here and there with grassy knolls and patches of the exquisitely blended colourings of gorse and heather. The exposures of the *Lingula* flags and slates may be traced from Aberamfra to the fifth milestone on the Dolgelly road. They exhibit a series of banded flags, formed by the repetition of numerous wavy felspathic and siliceous layers of a light bluish-grey colour, often so felspathic that the finer-grained layers present the appearance of being formed of spathose dust and ashes. Many of the flags offer marked evidences of current-bedding. The colourings of these rocks, due· to the decomposition of the chlorite and iron pyrites, are exceedingly fine. They appear to the best advantage on a wet day. The *Lingula* beds in the

FIG. 1. GLACIAL DEBRIS AT LLANABER, NEAR BARMOUTH.

extensions, but where this has been done they have been found to be very persistent and of great length. In most cases there are distinct evidences of the meeting of the boundary faces, and many opportunities are therefore afforded for studying the varying phases of metamorphism. Proceeding eastwards from Barmouth down the steep descent of Aberamfra, the junction of the grits and the *Lingula* beds is well marked by an ice-cold streamlet which flows beneath the roadway into the Mawddach.

The characteristic surface features which mark the line of demarcation between the Cambrian and the Ordovician at this point serve as a useful index in tracing out the line of junction, which runs almost due north. On the western side of the

neighbourhood of Dolgelly are fairly fossiliferous, but around Barmouth very few organic remains have been forthcoming. This is probably due to the fact that the latter have not been so thoroughly overhauled as the former; but Mr. Salter's discovery of *Lingulae* at Banc-y-frain shows further research would probably be well repaid. Further east the flags have yielded a rich series, including *Lingula davisii*, the·crustacean *Hymenocaris vermicauda*, and the trilobite *Olenus micrurus*. There is, therefore, ample work in this charming district both for the fossil-hunter and for the petrologist.

The glacialist, too, will find much to occupy his energies in tracing out the directions of flow of the various ice sheets which enveloped the district during the Quaternary period. All of the hills and valleys around Barmouth, and much farther afield, afford conspicuous evidences characteristic of ice moulding by glaciers of great size. The slates, flags, quartzose, sandstones, and conglomerates, all alike are crossed and recrossed by polished and striated grooves, and many of the hill-tops have been moulded into well-formed *roches moutonnées*. A series of upwards of twenty of these bosses and domes lie on the summit of Abermaw and Celfawr, immediately overlooking Barmouth town, at a height of about 800 feet O.D. Many other similar and equally well developed examples occur on the sides and summits of most of the neighbouring hills. From the top of Abermaw may be seen, dotted here and there on slope and crest, large perched blocks that have been lowered gently into their present positions as the glaciers that once carried them melted away. One of these blocks, lying above Ceilwart, measures 18 feet 8 inches by 7 feet 6 inches. It demonstrates admirably, in the polished and striated surfaces of its under sides, the manner in which it had been used in the grip of the glacier as a gouge to form the grooves and striae on the hillside around it. None of these perched blocks are "travellers," in the glacial acceptation of the term. All are of local origin, and may be traced to the Cambrian strata in an area of a few miles. I was told of a "granite" boulder having been seen on one of the hills in the vicinity of Barmouth, but I was not successful in finding it. I am inclined to think that it was perhaps one of the many blocks of felspar porphyry, which abound in the neighbourhood, that had their origin either in the Arrans or Cader. In the course of my rambles around Barmouth I saw no granites either on the hills or in the moraines; but I met with many felspathic porphyritic rocks which, at first sight, one might be excused for mistaking for granite.

The lateral moraines of the glaciers that descended the Mawddach from the Arrans and Cader are a conspicuous feature around Barmouth. These glaciers, after forcing their way through the Mawddach estuary, spread out fan-shaped, and extended right and left along what is now the shore line of Cardigan Bay. Most of the walls and farm buildings between Barmouth and Harlech have been built of the rounded and striated boulders obtained from the morainal accumulations that flank the hillsides and shore line between the two places. Sections of the right moraine are well shown at Llanaber with the debris and boulders (Fig. 1). This portion of the moraine extends from Llanaber to the Barmouth Morfa, where it disappears beneath a stretch of sand dunes; but its continuation may be traced in the boulder-beach which extends along the Barmouth sea-front. It varies in thickness from 40 feet at Llanaber to a thin stratum of stiff yellow clay at the Morfa. An examination of the boulders shows that the accumulations near Barmouth offer a greater variety of

rocks than those near Llanaber. In all parts of the deposit the Cambrian grits and shales predominate; but the beds off the Morfa also contain considerable numbers of boulders of pyritous shale from the Ordovician, felspar porphyries such as are to be found at Arran Mowddy, dolerites from the Cambrian dikes, calcareous ash, and rubbly vesicular conglomerates similar to those found at Cader. These interesting beds would repay a careful and patient investigation; and to such as care to undertake it I would suggest as the most favourable points of attack the Llanaber and Morfa sections, and also the left moraine which lies on the southern side of the Mawddach.

In the middle of the Mawddach estuary are numerous islets, all of which show in their polished, rounded, and striated surfaces that they are typical examples of *roches moutonnées* formed by the old Arran *mer de glace*.

(*To be continued.*)

BRITISH COCKCHAFERS.

By E. J. BURGESS SOPP, F.E.S.

(*Concluded from Vol. VI. page 355.*)

AS perfect insects cockchafers, which belong to the Phyllophaga, or "leaf-cutters," are wholly phytophagous and almost wholly crepuscular or nocturnal in their habits. They shun the sunlight during the day, when they may often be found clinging to the under side of twigs and leaves, which, as a rule, they do not quit until the approach of twilight, when they commence to buzz about amongst the branches of trees and shrubs to feed upon the foliage. Their flight is clumsy and heavy, which causes them to fall an easy prey to bats, owls, night-jars, and other nocturnal creatures. Nevertheless, there are usually enough left to gladden the heart of the coleopterist and more than sufficient to satisfy the horticulturist whose apple-trees, roses, strawberries, asparagus, and other garden produce is often badly injured by these destructive chafers, in either the larva or imago form. Orchard trees that have been seriously attacked are said not to bear satisfactorily for several seasons afterwards. Julius Pollux, who wrote towards the close of the second century, seems to have noted some connection between the two, as shown by the following passage: "The Melolonthe is a winged animal which they also call Melolanthe, either from the bloom of apples, or its occurring with their bloom" (Bk. 9, ch. 7).

As in the case of many other insects, the great strength of the cockchafer becomes apparent to anyone who takes the trouble to test it. In an interesting series of experiments carried out by Plateau he found that the ratio of weight lifted to weight of body (·940 grammes) in one of these insects was 14·3; but, as we all know, muscular strength is not everything in this world, and when we come to compare the proportion of brain to body we find that whereas in the worker bee it is 1 to 175, in the cockchafer it amounts to but 1 in 3,500 (Clodd). The only consolation we can offer the May-bug is that it is in the convolutions rather than in the volume of brain wherein the true measure of capacity lies.

The uses in one form or another to which cockchafers after death have been put are numerous and varied, but in no way do they suffice to act as a set-off to the mischief occasioned by them during life. They form an excellent food for pigs, game, and poultry, all of which are exceedingly fond of them. By boiling them in water, Hungarians have succeeded in extracting an oil which has been largely used as a grease for carriage wheels (Farkas); whilst, in addition to their contributing to the manufacture of a useful artists' colour (Mulsant), Lesser (l. ii. 173) claims that they form an infallible panacea both for the bite of a mad dog and an attack of the plague.

Our two species of Melolontha are very similar in appearance, but may be separated as follows; whilst, if knowing the locality whence one's specimens come, the coleopterist will often find it of material service in naming his captures.

Melolontha vulgaris is a large oblong chafer, measuring from an inch to an inch and an eighth in length. The head and thorax are usually black, although the latter not infrequently partakes of a rich dark brown coloration, a hue also exhibited in a more constant and brighter tint in the clypeus and antennae. Both the head and thorax are covered with long yellowish hairs extending in unrubbed specimens over the whole surface of the insect, but on the elytra and exposed dorsal portions of the abdomen it is both whiter and shorter than on the thorax. The antennae are 10-jointed, and the scutellum is large, black, and conspicuous. The general ground-colour of the elytra is reddish-brown, but the close white pubescence with which freshly emerged beetles are thickly " dusted " gives them a warm grey appearance. Each elytron bears four slightly raised longitudinal lines which in rubbed specimens often stand out with great distinctness. The under side of the cockchafer is also covered with whitish pubescence becoming thicker and yellower towards the anterior portion of the abdomen and thorax. One of the most conspicuous features of the genus is the prolongation of the last dorsal segment of the abdomen, or pygidium, into a broadly elongated point, gradually narrowing to the apex, giving to the chafers a somewhat curious appearance. Another characteristic is the series of triangular white marks so much in evidence when the insects are viewed sideways. The sexes are easily distinguished. In the male the sides of the thorax are closely covered with long erect hairs, and the club of the antenna is longer than the whole of the remaining joints; whereas in the female the pubescence on the thorax

is of a more downy nature, and the club is plainly shorter than the remainder of the antenna. In shape the clubs also differ in themselves, as in the male cockchafer they are composed of seven lamellae or leaves, and in the female of only six. These numbers will at once serve to separate the true Melolonthae from the closely allied genus *Ithizotrogus*, or "summer chafer," in which the club of the antenna is composed of but three similar leaf-like processes.

M. vulgaris is common and generally distributed over the greater part of England and Wales, although it becomes decidedly scarcer towards the north, and everywhere varies considerably in abundance during a sequence of years. In Scotland it occurs locally in several districts, and is present in many parts of Ireland, more especially in the South, where in some seasons it occurs in considerable profusion.

Melolontha hippocastani differs from our last-described beetle in being both darker and smaller, rarely attaining to more than an inch in length. The thorax is generally red or reddish-brown, and the pygidium less elongate and slightly more broadened at the apex, which gives it the appearance of being more abruptly truncated than in *M. vulgaris*. The scutellum, too, is often red. The insect is closely pubescent all over, the pubescence being thicker, shorter, and of slightly finer texture than in the preceding species. In the male "the third joint of the antennae is thickened at the apex and armed in front with a sharp tooth, and in the female the first lamella of the club is shorter than the remainder" (Fowler, " Col. Brit. Islands," vol. iv. p. 53). Figuier mentions ("Insect World," p. 456) that one of the chief points of difference between our two species of *Melolontha* lies in the fact that *M. hippocastani* has black legs, but this is undoubtedly an error; for though, as also with the antennae, the examination of a series will show that they run evidently darker than in our commoner cockchafer, yet the colour cannot be said to depart from some shade of dark brown.

M. hippocastani usually appears slightly earlier in the spring than does *M. vulgaris,* to which beetle, in both the larva and imago forms, it is very closely allied in habits. It is always a local insect in Britain, and in England does not appear to have occurred south of Westmorland and Durham. It has been recorded from the Dublin district, and probably occurs in other Irish localities, whilst in Scotland it has been taken in the " Clyde, Forth, Tay, and Moray districts," and appears " partly to take the place of the preceding species " (Fowler).

Rhizotrogus solstitialis, the " summer chafer," differs from the typical cockchafers in not having the abdomen produced to anything like the same extent as in the Melolonthae; by having only nine in lieu of ten joints to the antennae; and, as before stated, by having but three lamellae in the

B 3

composition of the club. It is, moreover, a smaller, less robust-looking insect than either of the others, and measures about three-quarters of an inch in length. The triangular white patches which are so effectively arranged along the sides, and form such a noticeable feature in the general appearance of its larger allies, are also distinctly traceable in the present species. although not developed in so remarkable a degree. In form *R. solstitialis* may be described as long-oval, and in colour reddish-brown, with the thorax and scutellum thickly covered with long yellowish hairs. The elytra, upon which can be traced very faintly raised longitudinal lines, are but slightly punctured in comparison with the thorax, and sparsely pubescent. The legs are long and red. In their internal anatomy the beetles of this genus are remarkable for their great concentration of nerve-centres.

Summer chafers appear towards the middle or end of June, and are said to be most destructive in those years when cockchafers are abundant. as they follow up their ravages (Dr. E. Hofmann). The larvae differ little in habit or appearance—except in being slightly smaller—from those of the Melolonthae ; and the imagos sally forth towards nightfall, as do our other cockchafers. and may be seen flying over grass-land or about trees. shrubs, hedges, and similar situations. *R. solstitialis* is a decidedly local chafer, and confined to the southern portions of England and Wales. It has been recorded from the Isle of Wight, Southampton, Dover, Hastings, Blandford, Swansea, and one or two other places. Although generally a scarce beetle, yet where occurring it is occasionally found in numbers. At Ferndown (Dorset) in some seasons these chafers appear in the greatest profusion, much to the delight and benefit of the local poultry, which run them down and consume them with avidity and evident relish.

    " Saxholme," Hoylake,
        Cheshire.

BUTTERFLIES OF THE PALAEARCTIC REGION.

BY HENRY CHARLES LANG, M.D., M.R.C.S., L.R.O.P. LOND.

(Continued from Vol. VI. page 358[1].)

PIERIS (continued).

8. **P. shawii** Bates. Henderson and Hume, "Lahore to Yarkand," p. 305 (1873). (*Mesapia shawii.*)

35—37 mm.

Wings white, ♂ f.w. with a black spot at end of disc. cell ; midway between this and apex a black spot touching the costa, and another just beneath it. Along the ou. marg. a row of three or four black spots. H.w. base dusky, a very narrow black spot at end of disc. cell. ♀ has the costal spots continued to form a broken band reaching to in. marg.; along ou. marg. a row of five spots. H.w. with an ante-marg. row of dusky spots, and an indistinct marginal spot at the end of each nervule ; ground colour somewhat dusky, owing to the presence of finely scattered dark scales. U.s. fringes of all the wings light brownish-pink. H.w. in ♀ with the ground colour tinged with the above, and sprinkled with black scales; otherwise the markings of the upper side are repeated beneath in both sexes.

HAB. South Trans-alai, Pamir, Central Asia. At great elevations. VI—VIIe.

9. **P. deota** Niceville. Roborowskyi Alph. R. H. p. 121.

51—56 mm.

Wings white throughout their whole area. ♂

F.w. with apex black as in *P. brassicae*, the black extending along ou. marg. and deeply indentated internally. ♂ has a narrow black spot between second and third nervules. H.w. with a marginal row of black spots more or less triangular in shape. ♀ All the markings stronger and larger—three conspicuous spots on f.w. H.w. with a large spot on costa, and another below the disc. cell not seen in any of the allied species. U.s. h.w. and apices of f.w. ochreous-brown, with a slight purplish tinge, and sprinkled with black.

HAB. Eastern Pamir, Lob-noor (R. H.), Kurag Tag. At great elevations. VI—VIIe.

a. var. *ceris* Stgr. 1st generation. Smaller and with darker markings than in the later brood. HAB. Amür (Ask. Vlad. etc.). V,

10. **P. cheiranthi** Hüb. Samml. Exot. Schmett. (1816).

♂ 56—62. ♀ 60—64 mm.

The upper side of ♂ resembles *P. brassicae*, but the black marking at the apex f.w. is more decided. There is a short black line or elongated spot near centre of wing. I have never seen a specimen without at least a trace of this ; sometimes it is well marked. H.w. have the costal spot somewhat darker than in *P. brassicae.* ♀ differs conspicuously from that of *P. brassicae*, first in the extension of the black markings. The three black spots seen on the f.w. in that species are enlarged and coalescent, forming a conspicuous black patch.

H.w. strongly tinged with ochreous-yellow. U.s. ♂ and ♀, f.w. with a large black irregular spot reaching from the level of disc. cell to in. marg. and crossed by white nervures. Apices yellow. H.w. bright ochre-yellow, dusted with black.

HAB. Canary Islands, coast of Teneriffe (Mrs. Holt White, in "Butt. and Moths of Teneriffe," p. 29), Orotava (R. H.). I have received many specimens from the last locality taken by Don Ramon Gomez. IV—IX.

LARVA. "The larva is smooth, has a ground colour of grey, finely dotted over with black spots. There is a yellow stripe on the back and along each side. It feeds on the nasturtium (*Tropaeolum*) gregariously."—"Butt. and Moths of Ten." loc. cit.

This very beautiful and striking *Pieris* has been thought by some to be merely an insular development of *P. brassicae*, the common large-white butterfly of the British Isles. It is very difficult to speak with certainty with regard to insular species, and especially so as *P. brassicae* does not occur in the Canary Islands, where the form *cheiranthi* is

Pieris hippia.

common. There is another species, the one whose description follows, inhabiting the Canaries and Madeira, that certainly more resembles this than it does *P. brassicae.*

11. **P. wollastoni** Staint. R.H. p. 120.
50—60 mm.

Closely allied to *P. cheiranthi*, but smaller. ♂ with the black spot on f.w. similar to that seen in the last species. ♀ has the spots as in *P. cheiranthi*, though not so strongly marked. H.w. tinged with yellow above. U.s. of ♂ and ♀ has the yellow colouring on f. and h.w. replaced by grey, finely powdered with black scales.

HAB. Madeira, also Teneriffe.

12. **P. brassicae** L. Syst. Nat. x. 467. Lg. B. E. p. 28, pl. VI. fig. 2. "The large-white."
50—60 mm.

Wings white in both sexes, bases dusky, f.w. tipped with black. ♀ with three black spots, two of them nearly circular, placed one above the other, nearly midway between the centre and in. marg. ; the third is triangular, with its apex inwards, placed beneath the other two, and generally touching the lower one. But the three spots are never coalescent, as in the last two species; neither is there ever in the ♂ a black spot in centre of f.w. H.w. with a black spot in the middle of the costal border in both sexes. U.s. ground colour of h.w. and apices of f.w. greenish, and not ochreous-yellow (as in *P. cheiranthi*). H.w. dusted with black.

HAB. The entire region; except the Polar portion, the Canaries, and Madeira. Usually a very common species, sometimes occurring in abundance, and often migratory. V—IX.

(*To be continued.*)

ON THE NATURE OF LIFE.

BY GEOFFREY MARTIN.

I TRUST I may be permitted to answer the courteous criticism of my paper on "Life Under Other Conditions" (S.-G., N.S., vol. vi. pp. 291 and 326), by Dr. Allen in the May number (vol. vi. p. 365) of SCIENCE-GOSSIP. I substantially agree with Dr. Allen concerning the functions of carbon and nitrogen in living matter, but it appears to me that he devotes too much attention to the nitrogen atom and neglects the carbon. Dr. Allen holds that nitrogen in living matter is the "central or linking element." That it may be a "linking" element in the sense of the "floating linkages" (') of Knarr, I am willing to admit ; but that nitrogen ever forms the "central" element in living matter is more than open to doubt.

Certainly the chemical facts do not justify this view. The products of the breakdown of living matter seldom, if ever, give evidence of the "central" position of the nitrogen atom. Indeed, in all such products the "kernel," or "core," appears to consist of carbon atoms, and to this "core" the nitrogen is attached. This may be seen by glancing at the formulæ given below of some simple substances produced by the breakdown of living matter :—

$$
\begin{array}{ll}
\text{NH—C—NH} & \\
\text{CO} \quad \text{C—NH} \end{array} \text{CO} \qquad \text{NH}_2 \\
\text{NH—CO} \qquad\qquad \text{NH} = \text{C} \\
\qquad\qquad\qquad\qquad\qquad\qquad \text{N.(CH}_3\text{).CH}_2\text{.CO}_2\text{H}
$$

Uric Acid.   *Creatine.*

$$
\text{(CH}_3\text{)}_2\text{—CH—CH}_2\text{—C—CO}_2\text{H} \\
\qquad\qquad\qquad\qquad\quad \text{H}
$$

$$
\text{CH.(NH}_2\text{).CO}_2\text{H} \\
\text{CH} \\
\text{CH}_2\text{.CO}_2\text{H}
$$

Leucine.   *Glutaminic Acid.*

If such products result from the breakdown of living matter, it is reasonable to suppose that in

(1) Vide Richter's "Organic Chemistry," vol. i. p. 56.

B 4

the enormously more complicated protoplasm whence they are derived the atomic systems have an internal atomic construction identical with that they possess on passing out of the organism. This is certainly the case in all other branches of organic chemistry, and I see no necessity for making an arbitrary assumption that it is otherwise in living matter. One may, therefore, assume that protoplasm consists of a number of atomic systems or groupings—we may call them "units," each with a "kernel," or "core," of carbon atoms, and attached to this core a number of nitrogen atoms. These nitrogen atoms may serve to connect up system with system in an unstable way, so that system is eternally breaking away from system, and a kind of circulating internal movement of these atomic groupings results. In the way suggested above, nitrogen may be the "linking" element—the element that is eternally oscillating between adjacent groups; but, so far as I can discover, there is no chemical evidence that nitrogen ever forms the "core" of a complete atomic grouping.

Animals are internally connected up by a complex circulating system, which serves to convey nourishment to every part of the body.

I would extend the principle further, and assume that in every fragment of living protoplasm there is an analogous stream of atoms continually circulating throughout all its parts; that, in fact, a fragment of living matter is a fragment in a state of organised motion. All inert bodies—for example, a crystal—are eternally quivering with motion, but their motion is not *directed.* I would assume that the difference between a living fragment and a crystal is simply that, in the one case, the motion of the atoms is *directed* so as to produce continual motion, whereas in the other case the motion is not thus directed.

If carbon is truly the "core" of each protoplasmic unit, the internal cohesion of every such unit must depend upon the central attractive force of this "core." The force with which the nitrogen is bound to the carbon core must be a junction of the attractive strengths, both of carbon and nitrogen. At the temperature of living matter these forces are, according to my theory, almost balanced. Consequently there results a chemical union of extreme instability, so unstable that the ever-varying influence of the neighbouring atoms is sufficient to cause a continual breakdown of the system, thus giving rise to the phenomenon of vitality.

It will be noticed that the fundamental characteristic of the carbon compounds is their plasticity or liquidity. Even those compounds of carbon which are solid at ordinary temperatures melt at quite moderate temperatures. When the whole range of compounds of an element are stamped with a common property, we may assume that it is conferred upon them by the properties

of the central atom. I would therefore suggest that at normal temperatures the degree of motion of the atoms in carbon compounds is such that, when set in opposition to the attractive strength of the carbon atoms, a balance is attained, and the carbon atom has but little external energy left over wherewith to cause such an intense internal cohesion of the neighbouring molecules as is characteristic of the silicon compounds. Such a plasticity or fluidity is of the utmost importance in order to ensure a continuous internal movement in the living structure. The nitrogen compounds have by no means the monopoly of unstability. Indeed, an eminent chemist [2] has declared that the chemistry of the carbon compounds is peculiarly the region of unstable compounds. One cannot fail to be struck with the numerous and deep-seated changes that are continually occurring among carbon compounds.

The slightest elevation of temperature effects chemical changes in countless numbers of purely carbon compounds. In this respect the carbon compounds are quite as unstable as the nitrogen compounds, although free from that rapidity of change that characterises the latter. The decompositions of the carbon compounds are slow, and it is this "time element" that makes them appear so much more stable than the nitrogen compounds.

The unstability of the carbon compounds is an intramolecular unstability, and takes the form of isomeric changes. A slight increase of temperature usually suffices to transform a compound into its isomeric modification.

I take it that such a fundamental characteristic throughout all the range of its compounds indicates that the carbon atom is unable at ordinary temperatures to fully control the motion of the attached atoms, and consequently, when this motion is increased, intramolecular rearrangement takes place.

I therefore claim for carbon the property of being an element in its "critical" state, although Dr. Allen's view, that nitrogen is also such an element, cannot be contested.

Dr. Allen justly remarks that in order to effect important changes among inorganic substances such as the silicates, we require the employment of strong chemical reagents, such as the caustic alkalies and the like, as well as a high temperature; whereas in a living being the range of temperature during which the processes of growth, secretion, or excretion goes on is restricted to a few degrees. I maintain that were the silicates at a sufficiently high temperature, such smooth isomeric changes that characterise the carbon compounds would certainly become manifest. In contradistinction to these imperceptible intramolecular transformations of the carbon compounds, we may contrast the violent, abrupt, and

(2) Nernst. "Theoretical Chemistry," p. 570. Ed. 1895.

sudden changes that characterise the nitrogen compounds. Indeed, it appears to me that the slow, continuous changes of the former are far more analogous to the chemical changes that take place within the living organism than the awful swiftness and violence of the changes that accompany the molecular transformations of the latter. It would be well to remember that the majority of the so-called "nitrogen" compounds have an equal right to be regarded as carbon compounds, for in most cases they arise from the substitution of nitrogen atoms in a carbon compound. Their excessive unstability may be due to a very large extent to the fact that the carbon is not quite capable of controlling the motion of the nitrogen atom, and which, therefore, is always liable to "run amock" among the other atoms of the compound.

Dr. Allen maintains " we have yet to find a class of silicon compounds which would behave like carbon compounds in the storing of energy. Silicon seems to perform a passive, not a dynamic function in the life of this world." Applied to ordinary temperatures this statement is undoubtedly correct, but applied to a temperature at which the compounds of silica are fluid, this ceases to be so. Silicon, even to a greater degree than carbon, possesses an affinity for oxygen. This is especially exemplified in the enormous number of oxy-compounds to which silicon gives rise.

The difference between the compounds of carbon and the compounds of silicon is mainly a difference in temperature. Nothing appears more probable than at such a temperature that the internal cohesion of the siliceous bodies is reduced to the same degree that at ordinary temperatures characterises the carbon compounds, that silicon would possess energy-storing properties to an even greater degree than carbon; but the storage of energy by the silicon would not be brought about by the addition of hydrogen, as appears to be the case with carbon, in ordinary life. Some other element, perhaps oxygen itself, would perform this function in life at the temperature I am considering.

The idea that carbon gradually displaced the silicon in living matter with the falling temperature appears to offer a satisfactory explanation of the otherwise unaccountable presence of silica in all living matter. Indeed, it will be noticed that the earlier and more rudimentary forms of life often contain a comparatively large amount of silica; for example, the sponges. In still earlier types of life, now long since extinct, silica appeared to play an even more important part than it does at present. To give one example, there is used for polishing purposes, under the name "Tripoli," a collection of siliceous skeletons of the lowest microscopical infusoria, which are sometimes found in considerable layers in the form of a sandy mass. The microscopic remains of the infusoria have a

pointed, though not angular shape, and it is for this reason they can be used for polishing without scratching.

On Dr. Allen's theory no explanation can be given of the existence of such siliceous forms of life. The presence of silica in living matter bears witness to the slow process of evolution, much in the same way that the gills on the neck of the embryonic babe bear witness to an aqueous origin.

Dr. Allen doubts whether the temperature of the world's surface was ever sensibly higher than it is at present. Equally doubtful is the hypothetical meteoric origin of the earth. The nebular theory of cosmic origin, in spite of certain mathematical difficulties, is a far more probable theory than that supported by Dr. Allen. Indeed, we have in Jupiter and the other giant planets of the solar system actual examples of planetary bodies so hot as to be in a fluid state, and this in spite of their immense gravitational force. As Jupiter now is, so may the earth have been. This world, being of much smaller mass, would cool far more rapidly than Jupiter. The spectroscope, too, teaches us divers facts concerning the evolution of the heavenly bodies. It shows how there exist in space vast masses of incandescent gas—nebulæ. Further, it indicates that there are suns white-hot, suns yellow-hot, suns red-hot, and suns but barely visible and fast cooling down into everlasting night. Such facts as these must be taken into account in framing any hypothesis concerning the making of worlds.

My critic's argument that the earth's crust does not bear any geological evidence of ever having been at a sensibly higher temperature than at present cannot be taken seriously. For a great space of time the temperature of the earth's surface has undoubtedly not been very greatly different from that which now reigns, and this accounts for the complexity of life as it at present exists upon the earth. This space of time, however, bears no more ratio to the time I am considering than does a drop of rain in a thunderstorm bear to the total number of falling drops. The forces of disintegration would have long since swept out any traces of such a period.

In conclusion, I might state that I am but appealing to the universal law of evolution, the law that appears to run throughout all Nature.

The conception that life originated in a sea of white-hot fluid is immensely superior, from a purely chemical point of view, to that advocated by Dr. Allen. Under such conditions all the elements could freely intermingle together, and life could thus take form, substance, and nutriment out of the complex mixture that such a fluid must have been. The original elements that composed early living matter could not, however, have been those of which it is now formed.

13 *Hampton Road, Bristol,* May 21, 1900.

AN INTRODUCTION TO BRITISH SPIDERS.

By Frank Percy Smith.

(Continued from Vol. VI. page 361.¹)

GENUS *CHIRACANTHIUM* C. KOCH.

The spiders in this genus may be distinguished from those of *Clubiona* by the greater length of the legs. Their relative length is also different, for whereas in *Clubiona* the fourth pair is the longest, in *Chiracanthium* the longest is the first pair.

Chiracanthium carnifex C. Koch.

Length. Male 8.4 mm., female 9 mm.
Abdomen of a dull yellowish-green shade, with a dark red band along its upper side. This is a very pretty spider, and appears to be generally distributed.

Chiracanthium erraticum Wlk. (*Clubiona erratica* Bl.)

This spider is very similar to the last, but may be distinguished from it by the form of the radial apophysis, which in this species ends in a simple point, whereas in *C. carnifex* C. Koch it is distinctly notched. This species is not very common.

Chiracanthium pennyi Cambr.

This is a rare spider, and can only be satisfactorily distinguished from the preceding by the form of the digital joint.

Chiracanthium lapidicolens Sim. (*C. nutrix* Westr. and "Spiders of Dorset.")

Length. Male 8.5 mm., female 9 mm.
Cephalo-thorax devoid of markings. Abdomen similar to *C. carnifex* C. Koch, but the central band is restricted to its fore part. This spider is not common.

GENUS *ANYPHAENA* SUND.

The spiders of this genus may be distinguished from *Clubiona* by their possessing a curious transverse fold in the integument of the under side of the abdomen, also by the first leg being the longest.

Anyphaena accentuata Wlk. (*Clubiona accentuata* Bl.)

Length. Male 6 mm., female 6.5 mm.
Cephalo-thorax light brown with a black band on each side. Legs light brown spotted with black. Abdomen brownish-yellow marked with black. This is a rather common spider, and seems to be generally distributed.

GENUS *AGROECA* WESTR.

In this genus the eyes are disposed in two rows, of which the anterior is the shorter and less curved. The clypeus is very narrow. The fourth leg is the longest.

(1) This series of articles on British Spiders commenced in Science-Gossip, No. 67, December 1899.

Agroeca brunnea Bl. (*Agelena brunnea* Bl.)

Length. Male 6.5 mm., female 8 mm.
Cephalo-thorax reddish-brown with some black lateral markings. Legs dull yellow. Abdomen bright yellowish-brown, with a blackish band along its fore part followed by a series of angular markings.

Agroeca proxima Cambr.

Length. Male 5 mm., female 6 mm.
This spider is similar to the last, but the markings are less distinct. The radial apophysis of the male is also shorter.

Agroeca gracilipes Bl. (*Agelena gracilipes* Bl. ; *Liocranum gracilipes* in "Spiders of Dorset.")

Length. Male 3 mm., female 3.5 mm.
Cephalo-thorax reddish-yellow, with a black marginal marking. Legs reddish-yellow, with some of the joints of a dark brown colour. Abdomen yellow, darker in front and with some dark brown markings on the hinder part. This spider is not common.

Agroeca celer Cambr. (*Liocranum celer* in "Spiders of Dorset.")

This rare spider is similar to the last, but may be distinguished by the greater distance between the lateral eyes.

Agroeca celans Bl. (*Agelena celans* Bl. ; *Liocranum celans* in "Spiders of Dorset.")

Length. Male 4 mm., female 4.5 mm.
Cephalo-thorax dark brown with yellow margins, and a central marking of the same colour. Legs reddish- or yellowish-brown. Abdomen brown with some yellow markings, which are brightest towards the spinners. This spider is rather rare.

Agroeca inopina Cambr.

Somewhat similar to *A. brunnea* Bl. and *A. proxima* Cambr. It may be distinguished by its smaller size and by the legs being more or less distinctly annulated. Its general colour is not nearly so bright as in these two species. Very rare and local.

GENUS *LIOCRANUM* L. KOCH.

In this genus the eyes are arranged in two curved rows, the distance between the individual eyes being not very great. The convexity of the curve is directed backwards. The legs are very long, their relative length being 4' 1, 2, 3.

Liocranum domesticum Wid. (*Clubiona domestica* Bl.)

Length. Male 6.5 mm., female 8.5 mm.
Cephalo-thorax yellow with four brownish-black

longitudinal bands. Legs yellowish-brown with some indistinct annulations. Abdomen brownish-yellow, with numerous black markings. This spider is uncommon and local.

GENUS *MICARIA* C. Koch.

The spiders included in this genus are more or less of an ant-like form, and of the most brilliant colours. The eyes are very small, and are arranged in two nearly equal parallel rows, slightly curved, with the convexities of the rows directed backwards.

Micaria pulicaria Sund. (*Drassus nitens* Bl.)
Length. Male 3.5 mm., female 4 mm.
This is one of the prettiest spiders found in Britain, and cannot be mistaken for any other, except perhaps *M. scintillans* Cambr. The general ground colour is black, but the spider is covered with peculiar scale-like hairs which reflect metallic tints. This spider is not uncommon and is widely distributed. I have taken it plentifully between Bexhill and St. Leonards, where it lives amongst the ants with which the shore is infested. The similarity in the appearance of the ant and the spider is very striking.

Micaria scintillans Cambr.
Length. Male 5 mm., female 6 mm.
This spider is similar to the last, but is decidedly larger, and is not so brilliantly coloured. It is less common, and very local.

GENUS *MICARIOSOMA* Simon.

In this genus the eyes are arranged in two short, slightly curved, concentric rows. The maxillae are strong and very broad at the base.

Micariosoma festivum C. Koch. (*Drassus propinquus* Bl. *Phrurolithus festivus* in "Spiders of Dorset.")
Length. Male 3 mm., female 3.5 mm.
Cephalo-thorax brown and hairy. Legs yellowish-brown, with the femora of the first two pairs much darker. Abdomen brown with some white and yellow markings. This species is not uncommon.

GENUS *ZORA* C. KOCH.

In this genus the convexities of the two rows of eyes are directed towards one another, the curve of the posterior row being very considerable.

Zora spinimana Sund. (*Hecaerge maculata* Bl. and "Spiders of Dorset.")
Length. Male 5 mm., female 6.5 mm.
Cephalo-thorax dull yellow, with two broad, brown, longitudinal bands. Legs yellow, with some brown markings. Abdomen yellowish-brown with distinct markings of a blackish tint. This species is not uncommon, but is liable, on account of the position of the eyes, to be confounded with some of the Lycosidae. It may readily be distinguished from these, however, by its possessing only two tarsal claws.

Zora nemoralis Bl. (*Hecaerge nemoralis* Bl.)
Closely allied to the last. The abdomen is clothed with long hairs, which impart to it a smooth silky appearance. This spider is very rare.

FAMILY SPARASSIDAE.

In this family the eyes are arranged in two transverse rows. The anterior row is strongly curved, and has its convexity directed forwards; whilst the posterior row is almost straight, with its slight convexity directed backwards. In many respects the spiders included in the Sparassidae are allied to the Clubionidae, between which family and the Thomisidae they seem to form a connecting link. A single genus only, containing but one species, has hitherto been found in Britain.

GENUS *MICROMMATA* LATR.

Maxillae long and straight. Labium semicircular. Legs 4, 2, 1, 3. Tarsal claws 2 in number.
Micrommata virescens Clk. (*Sparassus smaragdulus* Bl.)
Length. Male 8.5 mm., female 13 mm.
The cephalo-thorax of the male is of a greenish-yellow tint, the colour of the legs being very similar. The abdomen is of a bright yellow hue above, the sides shading off to a dull green. On the upper surface are three bands of brilliant scarlet. The female is wholly of a bright green colour. This is one of our most beautiful spiders, but unfortunately it is rather uncommon. It is generally distributed over the Southern Counties of England, and it has been taken in the South of Ireland. It frequents low plants in open spaces in woods, and, although the male will be distinguished at once by its bright colours, the female so closely assimilates the tint of the leaves that it can only be detected by practised eyes.

FAMILY THOMISIDAE.

In this family the eyes are usually small, and are arranged in a semicircle or crescent. The terminal tarsal claws are two in number, and the spinners short. The falces are not prominent. The spiders contained in this family, in many cases, bear a striking resemblance to crabs, both in respect to their general appearance and their method of progression. The Thomisidae are found in very varied situations, such as among reeds and heather, under stones, and in the blossoms of plants. They spin no snare, but lie in wait for their prey.

GENUS *TIBELLUS* SIM.

The lateral eyes of the posterior row are somewhat the largest of the eight, and are widely separated from the other eyes. The cephalo-thorax is longer than broad, and the abdomen rather elongated.
Tibellus oblongus Wlk. (*Philodromus oblongus* Bl.)
Length. Male 7.5 mm., female 9.5 mm.
The cephalo-thorax is dull yellow, with three dark

longitudinal bands : one near each margin, and a central one somewhat divided towards the eyes. The relative length of the legs is 2, 4, 1, 3 ; their colour being similar to that of the cephalo-thorax, but usually paler, and they are speckled with minute brown spots. The abdomen is long and narrow, and of a dull creamy-yellow hue, with a central longi-

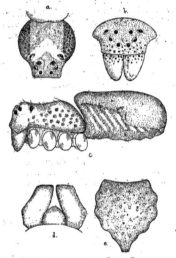

FIG. XI. CHARACTERISTICS OF GENUS PHILODROMUS.

a. Cephalo-thorax viewed from above ; *b.* eyes and falces seen from in front ; *c.* side view of spider, legs and palpi truncated ; *d.* maxillae and labium ; *e.* sternum.

tudinal brown band and a lateral mark of the same colour. On the upper side, towards the posterior extremity, are two distinct red-brown spots. This spider is not rare, but it seems to be rather local. I have received it in large numbers from Norwich.

GENUS *THANATUS* C. L. KOCH.

In this genus the anterior eyes are rather closely grouped, the laterals being larger than the centrals. The relative length of the legs is 4, 2, 1, 3. Cephalo-thorax as broad as long.

Thanatus striatus C. L. Koch. (*Thanatus hirsutus* Cambr.)

Length. Male 4 mm., female 4.3 mm.

The cephalo-thorax is of a dull yellow-brown colour with three longitudinal dark brown bands, and is thinly covered with distinct bristles. The legs are a little paler than the cephalo-thorax, and are furnished with a few fine spines. The abdomen is of a creamy-yellow colour with some long black bristles ; and upon its upper side are five dark brown longitudinal bands. This spider is rather rare and local, but has been found in large numbers in the Fens of Cambridgeshire.

GENUS *PHILODROMUS* WLK.

The eyes in this genus are small and almost equal in size. The anterior row is much the shortest and, with the centrals of the posterior row, form a regular hexagonal figure. The cephalo-thorax is broader than long. Relative length of legs 2, 1, 4, 3 ; or 2, 1, 3, 4.

Philodromus margaritatus Clk. (*P. pallidus* Bl.)

Length. Male 5 mm., female 6.3 mm.

The cephalo-thorax is grey with some brown markings. The legs are very long, of a yellowish-grey colour with some dark brown spots and streaks. The colour of the abdomen is similar to that of the legs, but it usually has a greenish tint ; there are some brown and black markings on its upper side. This spider is not rare on Scotch firs, with the lichen-covered bark of which its colours harmonise. There is a curious variety known as *jejunus* Panz, which is of a white colour with a few black markings. This form is found on apple-trees, its colours coinciding with the lichen found in similar situations.

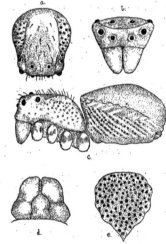

FIG. XII. CHARACTERISTICS OF GENUS XYSTICUS.

a. Cephalo-thorax viewed from above ; *b.* eyes and falces seen from in front ; *c.* side view of spider, legs and palpi truncated ; *d.* maxillae and labium ; *e.* sternum.

Philodromus dispar Wlk.

Length. Male 5 mm., female 6 mm.

The cephalo-thorax is of a dull yellow colour with a rather indistinct central band. The legs are long, and of a dull greenish-yellow hue. The abdomen is reddish-brown with a dark marking in the centre of its anterior part, followed by a number of dark angular bars. In the female these markings are usually more prominent than in the male. This spider is fairly common and generally distributed.

(To be continued.)

ON COLOURING OF BIRDS' EGGS.

By Reginald J. Hughes.

MR. WHELDON in his last article (SCIENCE-GOSSIP, vol. vi., N.S., p. 362) shows very clearly the difficulties and anomalies which arise in attempting to account for the markings on birds' eggs, either by the nature of their food or the theory of protection, and I agree to a considerable extent with many of his remarks.

With permission, I will attempt to review the present state of our knowledge of this subject as impartially as possible. I consider Mr. Wheldon has proved that the nature of the food can have no effect in the case of, at any rate, those pigments whose principal basis is carbon, and which do not contain iron. Carbon is an important constituent of all organic substances, and it can form colouring matters of many tints, so that animals with similar diets could produce differently coloured pigments, and *vice versâ*. Thus we have on many eggs, besides the blue oocyan, brown, red, and yellow pigments—named by Mr. Sorby "rufous ooxanthine, lichen ooxanthine, and yellow ooxanthine "—which are carbonaceous and contain no iron. Again, the dark skin of negroes is caused by melanin, a compound of carbon, oxygen, hydrogen, and nitrogen, and hence cannot be affected by their diet, since all food-substances contain these elements. The case of iron seems to me somewhat different. It will hardly be disputed that some foods contain a larger proportion of iron than others. A herbivorous animal must consume less of it than a carnivorous one, and a granivorous bird gets even less from the seeds it eats than it would from a diet of, for instance, grass. Yet, as Mr. Wheldon points out, the amount in the haemoglobin of all warm-blooded animals is approximately the same. It is evident that the surplus iron is got rid of in some manner, and the simplest explanation of iron-containing pigments on feathers or eggs seems to be that some is carried by the blood as impurities into the radii of each feather while young, and into the glands that secrete the shell-staining matter. I do not think that every radius has a special gland to extract the pigment; the deposit is chiefly mechanical, aided probably by the action of light. Glands, however, no doubt, do take an important part in the formation of the colour destined for the eggs, and modify it to a great extent. There are many instances of the deposit in extremities of the body of impurities in the blood. A botanical analogy is the secretion of superabundant silica in the edges of blades of grass, and the readiness of blood to perform this function is shown by the liability of poisons to affect the tint of the feathers. It should be noticed that carbonate of iron does not act like poisons, as, instead of discolouring, as in the case of hemp-seed, it intensifies the natural colours. It is known that the blood of birds contains more iron at one time than at others, especially before the egg-laying period, as is well shown by the reddening of fowls' combs. The primary reason is, no doubt, that an extra quantity is required to furnish the contents of the eggs, the yolk especially containing much iron. In spring birds obtain a more abundant supply of food furnishing this element. The surplus appears in the male in the brighter colours of his spring dress; in the hen it is used among granivorous birds entirely for the contents of the eggs; but I believe, in the case of insectivorous species and others that obtain a large supply of iron, some of that still remaining unused is deposited on the shell.

It seems to me that any argument on the relation of diet to the colour of eggs must apply equally well to that of the feathers; so, although the pigment of birds marked principally by melanin and other carbonaceous substances may not be the same as that on their eggs, and neither may bear any relation to the food—yet the majority with feathers containing zoonerythrin, the pigment which most probably contains iron, should, if my theory is correct, have eggs coloured chiefly by oorhodine and should feed on matter containing comparatively much iron. That this is the case we have at present insufficient evidence, and some apparent exceptions, though there are certainly a great many instances of it, and I hope to return to this subject at some future time. Possibly there is no general rule for explaining the colours of birds' eggs, and the best way to do so will turn out to be by examining the case of each species in detail, and considering all the causes—such as habits, surroundings, heredity, and food—which may have influenced the result.

Norman Court, Southsea.

HONOURS FOR MEN OF SCIENCE.

AMONG the honours commemorating her Majesty Queen Victoria's birthday in 1900 several men of learning and science are included. Professor Richard Calverhouse Jebb, Litt.D., D.C.L., LL.D., Regius Professor of Greek, and Conservative Member of Parliament for Cambridge University since 1891, has received a knighthood. Sir Richard Jebb was born in 1841 at Dundee. Dr. David Gill, C.B., J.P., LL.D., F.R.S., H.M. Astronomer Royal at the Cape of Good Hope since 1879, becomes a K.C.B. Sir David Gill was born in 1843 in Aberdeenshire. Professor Thomas Edward Thorpe, Ph.D., D.Sc., LL.D., F.R.S., Director of Government Laboratories, London, is created a C B. He is President of the Chemical Society. Dr. Thorpe was educated at Owens College, Manchester and the Universities of Heidelberg and Bonn. Dr. Thorpe was born near Manchester in 1845. Major-General Thomas Fraser, C.B., C.M.G., Commandant of the School of Military Engineering at Chatham since 1896, and General Officer commanding Thames District, is promoted to K.C.B. He was born at Chilham Castle, Kent, in 1840.

BRITISH FRESHWATER MITES.

By CHARLES D. SOAR, F.R.M.S.

(Continued from Vol. VI., page 365.)

4. *Atax intermedius* Koenike.

BODY.—Female about 0.68 mm. long. Breadth about 0.56 mm. Colour dirty yellow, with dark brown markings on the dorsal surface. Oval in shape. In some specimens the marginal line bends a little inwards on the posterior portion. It is without the two papillae we find in *Atax crassipes*.

LEGS.—First pair about 0.80 mm. Fourth pair about 1.0 mm. All of the legs are fairly thick, not the first pair only as in *A. crassipes*. All are well supplied with hairs, par-

FIG. 1. *Atax intermedius.* Tarsus.

ticularly the fourth pair. The first pair have the usual sword hairs common to members of this genus, but they are not so strong and powerful as in *A. crassipes*. One of the best points of identity in this species is the double claw (fig. 1). There are two claws to each of the tarsi, though I have only drawn one, so that its structure can be easily seen.

EPIMERA.—In four groups.

PALPI.—About 0.28 mm. long. The first segments are very thick. On the last segment but one are two or three small pegs, but very insignificant when compared with those we find on the same segment of *A. crassipes*.

GENITAL AREA.—Composed of two pairs of plates, the upper pair having each two discs, the lower three discs (fig. 2). The plates are very

FIG. 2. *A. intermedius.* Genital plates of female?

similar to those of *Atax taverneri*, except the arrangement of the three discs on each lower plate, which in the case of this mite are placed in the form of an angle.

MALE.—Is a little smaller than female, but showing similar structure, except that in the genital area we find two plates instead of the four. In the female each plate has five discs.

LOCALITIES.—I have as yet only taken this mite in the Norfolk Broads. I found two females and one male, also several nymphs in the gills of fresh-

water mussels; but no doubt it is to be found at other places where *Anodonta* is common.

5. *Atax bonzi* Claparède, 1868.

This mite is very much like the preceding species. In fact, so much so that it does not require any particular description, so long as the differences between the two are pointed out for identification. It is about the same size, but the legs are much thinner and with less hairs. Although it has two claws to each foot, each claw is single, not double, as we found in *A. intermedius*. The genital area is also much the same in appearance, excepting that the three discs on the lower plates of the female are arranged more in a row, and do not form such an angle as shown on fig. 2.

LOCALITIES.—I found several females of this species also in the gills of some freshwater mussels in the Norfolk Broads. There appears to be two species of these mussels in the Broads: one, *Anodonta cygnea*, and another; but this mite appears to favour both, for I found some in each kind.

6. *Atax ypsilophorus* Bonz.

This mite has a number of discs on each of the genital plates, which are situated at the extreme margins of the body.

LOCALITIES.—This species was taken by myself in the Norfolk Broads; but unfortunately I did not make any drawing of it, therefore fuller description must await further captures, unless some of our readers are kind enough to send me specimens. It is parasitic on freshwater mussels, like the two preceding species.

GENUS *COCHLEOPHORUS* PIERSIG.

The characteristics of this genus differ very little from the preceding one. The palpi are small and thin, also less in width than the first pair of legs. The genital plates are nearer the epimera, and not so much on the extreme margin of the body as we found in *Atax*. A number of discs are found on each plate, though there is one species of *Atax* which also has a number of discs on each genital plate, viz. *Atax ypsilophorus* Bonz.

1. *Cochleophorus spinipes* Müller.

BODY.—Female about 1.20 mm. in length. Width about 1.0 mm. Colour a pale greenish-yellow, with brown markings on the dorsal surface, in the centre of which is a yellow T-shaped mark. The dermal glands on the dorsal surface are also very prominent. Eyes very conspicuous, and near margin of body.

LEGS.—First pair about 1.40 mm. long ; second pair longer than third, being about 1.60 mm. Fourth pair about same length as third, though sometimes a little longer. Colour of legs a dirty green. They are all very much the same thick-

FIG. 3. *Cochleophorus spinipes.* Genital area of female.

ness, and not like the first pair of *Atax crassipes,* which are so much thicker than the others. The first two pairs are fitted with strong stiff spines, some of which have very much the appearance of twisted glass. The fourth pair have a number of very beautiful feathered hairs on the inner edge towards the body.

EPIMERA.—In four groups, which cover a great deal of the ventral surface.

PALPI.—Small and thin, about 0.34 mm. in length.

GENITAL AREA.—Composed of two plates. Each plate is covered with a large number of discs (fig. 3).

MALE.—Very much the same appearance as the

FIG. 4. *C. spinipes.* Ventral surface of male.

female. About 1.0 mm. in length. Fig. 4 is of the ventral surface, showing the plates which are quite different to those of the female.

LOCALITIES. — Sunningdale on the Thames ; Snaresbrook, Essex ; Enfield, Middlesex ; Oxshott, Surrey ; Suffolk ; and Dr. George has found it in Lincolnshire.

2. *Cochleophorus vernalis* Koch.

BODY.—Female. Length about 1.22 mm., width about 1.0 mm. Yellow in colour, with very dark markings on the dorsal surface, and a bright red mark in the centre, which is very conspicuous and brilliant.

LEGS.—First pair about 1.20 mm., fourth pair about 1.62 mm. First two pairs are fitted with the spines I mentioned as found on *C. spinipes.* The fourth pair are also well supplied on the inner edge with the feathered hairs (fig. 5).

EPIMERA.—In four groups, covering a great part of the ventral surface (fig. 5).

PALPI.—Small, about 0.25 mm. This has also

FIG. 5. *C. vernalis.* Ventral surface of female.

about five or six feathered hairs on the upper edge.

GENITAL AREA.—Composed of two plates with a number of discs (fig. 6).

FIG. 6. *C. vernalis.* Genital area of female.

MALE.—A little smaller than female, but it has a very remarkable feature about the structure of the fourth pair of legs which I have not noticed in the males of any other mite. The first four seg-

FIG. 7. *C. vernalis.* Ventral surface of male.

ments of the leg are very thick and strong, particularly the third segment, as seen in fig. 7.

LOCALITIES.—Not very common. Dr. George has found it in Lincolnshire. I have taken it at Woking, in Surrey.

(To be continued.)

THE NEW F.R.S.

WE understand that the usual fifteen candidates selected by the Council of the Royal Society for election for the year 1900 are as follows:—GEORGE JAMES BURCH, M.A., Lecturer at University Extension College, Reading ; Chemistry and Physics. T. W. EDGWORTH DAVID, B.A., F.G.S., Professor of Geology in University of Sydney, N.S.W. ; Geology. JOHN BRETLAND FARMER, M.A., F.L.S., Professor of Botany, Royal College of Science, London ; Botany. LEONARD HILL, M.B., Lecturer London Hospital Medical College ; Physiology. JOHN HORNE, F.G.S., F.R.S.E., Staff of Geological Survey of Scotland ; Geology. JOSEPH JACKSON LISTER, M.A., F.Z.S., Demonstrator, Comparative Anatomy, University of Cambridge ; Zoology. JAMES GORDON MACGREGOR, D.Sc., M.A., Professor of Physics, Halifax, N.S. ; Physics. PATRICK MANSON, C.M.G., M.D., F.R.C.P., LL.D. ; Parasitology and Tropical Medicine. THOMAS MUIR, LL.D., M.A., F.R.S.E., Superintendent-General Education, Cape Colony ; Mathematics. ARTHUR ALCOCK RAMBAUT, M.A., Sc.D., Radcliffe Observer ; Astronomy. WILLIAM JAMES SELL, M.A., Demonstrator of Chemistry, University of Cambridge ; Chemistry. W. BALDWIN SPENCER, B.A., M.A., Professor of Biology, Melbourne University ; Zoology and Comparative Anatomy. JAMES WALKER, D.Sc., Ph.D., Professor of Chemistry, University College, Dundee ; Chemistry. PHILIP WATTS, Director, War-Shipbuilding Department, Armstrong, Whitworth & Co. ; Naval Architecture. CHARLES THOMSON REES WILSON, M.A., B.Sc., Investigations on Atmospheric Electricity ; Physics.

EDITORIAL.—Dr. Bryan's article on " Desmids" unavoidably stands over until next month. Mr. Read's Chemical Department will also commence in the next number.

A NATIONAL REPOSITORY FOR SCIENCE AND ART.—The remarkable paper read on May 16th by Professor Flinders Petrie before the Society of Arts on this subject has created considerable discussion. We hope to refer to the subject on a future occasion.

PROTECTION OF RARE BIRDS.—We are informed that the British Ornithologists' Union has passed a stringent resolution forbidding the members from directly, or indirectly, destroying in Britain the nests, eggs, young or parent birds of chough, golden oriole, hoopoe, osprey, kite, white-tailed eagle, honey buzzard, common buzzard, bittern, and ruff. We are glad the Union has taken this course, as it is a sufficient example to others, and an answer to those who accuse naturalists of exterminating rare animals.

SCIENTIFIC GLASS-BLOWING.—Some time since we referred to the advantage of an amateur worker, requiring special designs of scientific glass apparatus, being able to make his own. For this purpose the Camera Construction Company, of 38 Eagle Street, London, W.C., have designed a complete outfit necessary for glass-blowing, and at a price within the reach of most students. As there are also full instructions for practising the art, this forms a good opportunity for commencing new and interesting home experiments in art-glass work.

MOLLUSCA

CONDUCTED BY WILFRED MARK WEBB, F.L.S.

NOTE.—When this column was inaugurated it was hoped that our readers would send in contributions from month to month rather than that it should simply consist of a series of editorial notes. As this may not have been made quite clear, the present opportunity is taken of saying that the Editor would be glad to receive notes upon all subjects connected with molluscs, and that they may be sent to the office of SCIENCE-GOSSIP. It often happens that notes upon apparently common incidents bring forth interesting discussions.—ED. S.-G.

DURATION OF LIFE IN HELIX POMATIA.—In the current number of the " Essex Naturalist " Mr. Benjamin Cole gives some interesting data with regard to the longevity of our largest snail. He says that on June 3rd, 1894, when in company with Professor Meldola, he found two living specimens of the mollusc in question at Newlands Corner, near Gomshall, Surrey. The snails were quite full grown, but there was no evidence forthcoming of their age when taken. They were subsequently kept as pets, and when active were fed exclusively upon garden lettuce. They hybernated every year, closing up the mouth of the shell with the secreted epiphragm, and remaining shut in from October to about the end of March or the beginning of April. One of the specimens was accidentally killed at the end of two years, but its companion was still alive when Mr. Cole wrote his note, and apparently quite healthy, judging from its weight—being then still enjoying its winter sleep. Presuming therefore that this *Helix pomatia* was two years old when found, it is now at least seven and a half years old, and Mr. Cole hopes that it will live much longer. This is an interesting fact, confirming Miss Armitage's views on this subject, given at page 323 of the last volume of SCIENCE-GOSSIP.

VARIETAL NAMES.—"Nature," some weeks ago, pointed out in a few words how open to criticism are those who persist in using " varietal " names for what are mere variations. "Variety-mongers"—as Mr. B. B. Woodward, I think it was, long ago dubbed the collectors in question—must feel the comparison very odious which "Nature" draws between a " bird man " who rationally talks of " sports " and the " shell man " who records a white form of a snail under the title of *Helix marmorata* var. *alba* nov. A dozen years or more ago the writer expressed himself strongly in these pages and elsewhere, as did also some others, with regard to what is really a variety. The crusade may have done some good. but " varieties " still flourish. The importance of recording " variations " has been advanced as a reason for recording them all by name ; but perhaps some of our readers could try to bring forward some real convincing reason for the process, which fails to impress the general biologist as necessary. —*Wilfred Mark Webb.*

NOTICES BY JOHN T. CARRINGTON.

Micro-organisms and Fermentation. By ALFRED JÖRGENSEN. Translated by ALEX. K. MILLER, Ph.D., F.I.C., and A. E. LENNHOLM. Third edition. xiii + 318 pp., with 83 illustrations. (London and New York: Macmillans, 1900.) 10s. net.

This—the third—edition has been completely revised, and a large portion of the book has been entirely re-written, with considerable additions, so as to bring the text up to the most recent knowledge of this branch of the very progressive science of bacteriology. There are several new features in the book, including a summary of observations on the behaviour of high-fermentation yeasts in use in brewery-factories. It includes also descriptions of some interesting yeast species discovered in recent years. The organisms occurring in milk and the pure cultures of lactic-acid bacteria in dairies and distilleries are illustrated with new figures. To those who are connected with breweries, distilleries, and dairies, the work will be invaluable, and the general reader who desires a knowledge of the remarkable organisms of fermentation will find this work of great help to a better understanding of the subject. Added to the book is a voluminous bibliography, extending to no less than forty-two pages. The absence of an index is not convenient. The first chapter will be useful to microscopists who have not studied these curious organisms, as it contains directions for procedure, and particulars of the apparatus necessary for their culture, microscopical preparation, staining, &c. There are also valuable chapters on bacteria generally and mould-fungi. This latter will be of service to cryptogamic-botany students, as it really forms a modern treatise on moulds that appear among the products of fermentation and alcoholic distillation. These beautiful vegetable organisms are easily obtained, and form pleasing objects for the microscope.

Letters of Berzelius and Schönbein. Edited by GEORG W. A. KAHLBAUM, translated by Francis V. Darbishire, Ph.D., and N. V. Sedgwick. 112 pp. 7⅞ in. × 5 in. (London, Edinburgh, and Oxford: Williams & Norgate. 1900.) 3s.

Christian Friedrich Schönbein will ever be known as the discoverer of ozone and the inventor of gun-cotton. He was at one time Professor of Chemistry and Physics at the University of Bâle, and the centenary of his birth occurred last year. His correspondent, Baron Jöns Jakob Berzelius, was twenty years the elder of the two, and he had a wider range of study than Schönbein, who concentrated his attention on a narrowly contracted field of chemical and electrical energy. These letters began in 1836, and continued during a period of great activity among chemists and physicists. In them are frequent references to the work of Faraday and others whose names are familar to every educated person; they are thus of more than passing interest, on account of the personal views expressed by the two eminent savants who conducted the correspondence.

The Temple Encyclopaedic Primers. AN INTRODUCTION TO SCIENCE. By ALEXANDER HILL, M.D. 139 pp., 6 in. × 4 in., with 6 portraits. ETHNOLOGY, by D. MICHAEL HABERLANDT. viii + 169 pp., 6 in. × 4 in., with 56 illustrations. (London: J. M. Dent & Co., 1900.) 1s. each.

The publishers of these short educational works have long been known for the issue of artistic and well-chosen books. This new series will be found handy little volumes to carry in one's pocket when travelling or for reading at odd intervals. They are well printed, and written by people who know their respective subjects, and will doubtless find a large sale among the public generally.

The Flowering Plant. By J. R. AINSWORTH DAVIES, M.A., F.C.P. Third Edition, xv + 195 pp., 8 in. + 5¼ in., with 70 illustrations. (London: Charles Griffin & Co., Limited, 1900.) 3s. 6d.

The intention of this successful work is to illustrate the first principles of botany. It will be found useful as a class-book, as it is well arranged and contains an appendix on practical work; also a second appendix, giving a series of examination questions founded on the South Kensington and London Matriculation systems.

Common Objects of the Microscope. By J. G. WOOD, M.A., F.L.S. viii + 186 pp., 7⅝ in. × 5 in., with 14 coloured or other plates, and 16 illustrations in text. Second Edition, revised by E. C. Bousfield, L.R.C.P. (London: George Routledge & Sons, Limited, 1900.) 3s. 6d.; with plain plates, 1s.

There are few books more suited to the beginner, or more likely to inculcate in him a desire to examine for himself Nature's handiwork, than the late Rev. J. G. Wood's well-known "Common Objects of the Microscope." We welcome therefore the new edition that Dr. Bousfield has revised and largely re-written for Messrs. Routledge. The familiar plates by the late Tuffen West appear as before, but to them have been added two excellent plates of infusoria, rotifera, worms, etc., drawn by Dr. Bousfield. The bulk of the text has wisely been left much as the Rev. J. G. Wood wrote it; but the reviser has contributed an entirely new chapter on pond-life and a popular account of marine life, a new chapter on the preparation and mounting of objects for the microscope, also an introductory chapter on the practical manipulation of the instrument. This last, however, is somewhat inadequate, even with due regard to the scope and aims of the work. Whilst thoroughly sympathising with the reviser's desire to leave intact the author's original text, we think it is misleading and unnecessary to illustrate or recommend to the notice of beginners—into whose hands this volume will largely fall—such an antiquated and defective microscope as is figured on page 14. The making of microscopes has taken large strides since this book was first written, and a good microscope costs no more than a bad one In all other respects Dr. Bousfield has done his work excellently. The book has been well produced by the publishers, as the print, paper, and binding are all good. The work is what it professes to be, a revision and not merely a reprint.—*F. S. S.*

THE total amount realised at the sale by auction of the Samuel Stevens collection, referred to in our last number (page 379), was £835. This included a few scientific books and the cabinets.

MADRAS has founded a new literary and scientific monthly journal entitled "The Indian Review." We have received a specimen copy, and find it well arranged and brightly conducted.

WE are requested by the principal of the Burlington Classes to notify that candidates for scientific and other degrees can be prepared by correspondence. On application to 27 Chancery Lane, London, W.C., particulars of the system will be forwarded to applicants.

A SPECIAL number of the "Photogram," No. 77, was issued last month. It is remarkable as being written entirely by men of the city of Birmingham. The contents are of much technical value to photographers. The forty-two pages are beautifully illustrated with thirty illustrations.

THE "Photogram" Publishing Company are about to issue by subscription "An Index of Standard Photograms." Particulars may be obtained from Messrs. Dawbarn & Ward, Limited, Farringdon Avenue, London, E.C. This new work of reference promises to be of much value.

WE regret to record the death, on May 4th, of Dr. Edmund Atkinson, one of the older generation of men of science, and at one time assistant to the late Sir Edward Frankland in the laboratory of Owens College, Manchester. Dr. Atkinson was perhaps best known as the translator of Ganot's "Physics."

Two county lists, useful to entomologists, are being issued by Mr. Claude Morley, F.E.S., of 16 Bath Street, Ipswich. They are the Coleoptera of Suffolk, and the Hymenoptera Aculeata of the same county. The names of species are accompanied by full notes, and exact localities are given for the rarer insects.

ANOTHER report from the distributor to the Botanical Exchange Club of the British Isles has been issued. It is dated March, 1900, but refers to the distribution for 1898. The number of specimens sent in for exchange reached 3,273, and were received from twenty-one contributors. The report contains thirty pages of notes from members upon the plants circulated for exchange. Many of these remarks are of value.

IN connection with "The Scientific Roll," Mr. A. Ramsay has entered upon a systematisation of the literature, no matter how scattered or fragmentary, relating to bacteria. Mr. Ramsay appeals for obscure references, and desires subscribers to the three volumes, each of about 500 pages. The work will be issued in parts at the rate of one volume per year. The first part is to be issued in September, 1900. Mr. Ramsay's address is 3 Cowper Road, Acton, London, W.

OWING to the absence of so many astronomers who have gone abroad in the hope of observing the total eclipse of the sun, the annual visitation to Greenwich Observatory has been postponed to the end of June.

WE are pleased to find from the balance-sheet issued by the honorary treasurer of the North London Natural History Society that the prosperous condition was fully maintained during last year. There is a substantial balance in favour of the society, which has fully kept up to a high standard the work done at its meetings.

AS not infrequently our readers have applied to us for advice with regard to obtaining protection for inventions, we may refer them and others to Messrs. Rayner & Co., 37 Chancery Lane, London, who pay especial attention to patents of a scientific character. The experience of the firm is useful in such cases.

THE International Committee, which has been sitting at the Foreign Office, London, for the past few weeks to consider the question of International agreement for the protection of wild animals, birds, and fishes in South Africa, signed, on May 19th, a joint convention aiming at the desired protection. The exact terms of the agreement are to be announced at a later date.

IT has been suggested that there are no means of enforcing a "close time" for big game in Central Africa. We do not agree with this opinion. If a similar system to that in being in the North-West of Canada is founded, there will be little difficulty. There the informer gets a large portion of the fine, and the native Indians are excellent informers. This would equally apply to the negroes.

THE Millport Marine Biological Station has issued its annual report for 1899. Considerable progress has been made during the year. This station is under the patronage of the University of Glasgow and a number of other learned and municipal Scotch bodies. It is managed by an influential committee. The station is now prepared to receive students to work at the tables, or will supply material to those unable to visit the institution.

THE Egyptian Ministry of Finance have forwarded to us the annual report for 1899 of the Ghizeh Zoological Gardens, near Cairo, of which Mr. Stanley S. Flower, F.Z.S., son of the late Sir William Flower, F.R.S., is the director. During the year 43,567 persons visited the gardens On October 6th, 1899, there were on view 473 animals, representing 132 species, exclusive of large numbers of wild birds and other animals that are encouraged to take up their abode in the gardens. A list of these appears in the report.

THE South London Entomological and Natural History Society has issued its Proceedings for 1899. The annual volume continues to increase in importance, this issue containing 136 pages. The report of the council and the balance-sheet indicate a prosperous condition of the Society. Fourteen papers were read at the meetings, several of them, with the presidential address, appearing in the volume before us. It is embellished by a plate of some antennae of Psychides. Much useful information may be gained from the reports of the bi-monthly meetings.

PLANTS OF SOUTH HANTS AND DORSET.—A Flora of the Bournemouth district has been compiled by the Rev. E. F. Linton, M.A. Mr. Linton's position as a botanist is a guarantee for the correctness of the information therein contained.

A LONDON FIELD-BOTANY CLUB.—Several of our readers have from time to time suggested that a Field-Botanists' Club should be formed in connection with SCIENCE-GOSSIP. The Editor will be pleased to hear from any persons who would care to join such a club. Only a nominal subscription of One Shilling per Annum would be required, and meetings might be held either once or twice a month. We are not aware of any place in London where those interested in the study of Field-Botany can meet wholly for the purpose of discussing plants, their characteristics, habits, and identification. Another object of the Club will be to facilitate exchange of specimens between the members ; also to assist and encourage those who are commencing the study of field or structural botany. The idea is to give facilities for chatty intercommunication between botanists residing or visiting London. On receiving a sufficient number of names, the Editor of SCIENCE-GOSSIP will call a meeting to discuss the proposal.— Send name and address to *John T. Carrington,* 110 *Strand, London, W.C.*

LUNGWORT IN HANTS.—*Pulmonaria angustifolia* was, at Easter time, in fine blossom, both in a coppice and some hedgerows a little to the south-west of Sway station, on the Bournemouth direct line in Hampshire. Apparently those plants that grew in the hedgerows, bordering some moist meadows, were relics of the flora of the woodlands that formerly covered this district.—*James Saunders, A.L.S., Luton.*

POTAMOGETON RUTILUS IN SUSSEX.—Among the rarer plants found by me in Sussex is *Potamogeton rutilus* from near Rye. It is probably the only station now remaining in Britain where it is known certainly to occur. Mr. A. Bennett recently referred to my finding this plant in the "Journal of Botany."—*Thomas Hilton, Brighton.*

ABNORMAL COWSLIP FLOWER.—I send you herewith a very curious aborted specimen of cowslip (*Primula veris*), gathered here yesterday in rough pasture ground. There are about a dozen flowers in the "umbel," two being very large. In every case both calyx and corolla appear to be represented by green leaflets, among which the ill-developed pistils or stamens can be seen. There is not the slightest trace of the yellow colour so conspicuous in the normal cowslip flower.—*Frank Sich, junior, Niton, Isle of Wight, May 18th, 1900.*

PRUNUS CERASIFERA, NEW TO BRITAIN.— The new Prunus discovered near Hemel Hempstead in Hertfordshire, and about which an article appeared in your issue of June 1899 (SCIENCE-GOSSIP, N.S., vol. vi. p. 14), proves to be *Prunus cerasifera.*

I am just in receipt of a letter from Mr. George Nicholson, Curator of the Royal Gardens, Kew, saying that it is probably a native of the Caucasus. Last year some specimens were removed to Kew Gardens for observation. It is now in full bloom, and I shall be pleased to indicate the spots to any of your readers who may wish to procure specimens. It blossoms some weeks earlier than the "sloe," and is from this fact alone an interesting addition to our flora.—*B. Piffard, Hemel Hempstead, April 20th, 1900*

SALICORNIA APPRESSA IN SUSSEX.—This plant, hitherto only recorded as occurring as British in the county of Kent, I found last year growing on the Sussex coast, a long distance from Kent. It is a most interesting find, as this addition to our county flora is also an indication that its range is wider in the South of England than hitherto supposed. I have to thank the Rev. E. S. Marshall, joint author of the "Flora of Kent," for checking my identification. He says: "Yes, they are fine specimens of *Salicornia appressa*, and exactly similar to my Kent plants."—*Thomas Hilton, 16 Kensington Place, Brighton, April 3rd, 1900.*

SILENE ITALICA IN SUSSEX.—On several occasions during last summer I found flowering and fruiting plants of the rare *Silene italica*, or Italian catchfly, on the Sussex downs in the neighbourhood of Newmarket Hill. At the time they were overlooked for *Silene nutans*, but I have since heard from my friend Mr. Thomas Hilton, of Brighton, that he has found specimens in two other stations on the downs, also east of Brighton. As the plants are not uncommon and well established, far distant from any influence likely to have caused their introduction, the *Silene italica* growing on the Sussex downs may be looked upon as truly native. We should, I think, now consider this species as a native British plant ; it has not thus generally been hitherto ranked. If any botanist will write to me on the subject, I shall be pleased to indicate the habitat.—*John T. Carrington, 110 Strand, London, W.C., May 1900.*

CERASTIUM ARCTICUM, VAR. EDMONDSTONII.—In the annual report of the Botanical Exchange Club is a note upon the variety *edmondstonii* Beeby of *Cerastium arcticum*, Lange. It is from the pen of Mr. W. H. Beeby, who says :—"When I first gathered this plant in 1886 I brought home roots, and, being very desirous of growing it, also a bag of its native soil. Under these conditions it maintained the dark purplish-copper colour of its foliage fairly well, until the plants were lost in a removal some few years later. In 1897 and 1898 I brought home seeds and roots, and have the plants growing this time, not in their native soil, but in a mixture of Surrey soils. These plants have entirely lost their original colour, and have become completely green ; so that it appears that the only character which separates this variety from the type is merely temporary, and due to habitat. The serpentine gravels of Unst contain a number of minerals, notably chromate of iron, and the colour of the leaves may probably be due to the influence of one of them. The *Cerastium* is by no means the only plant growing on these hills which is affected in this way. J. M. Norman's ' *C. latifolium*' is, of course, *C. arcticum* (*C. latifolium* proper not being known in Scandinavia or other boreal countries), consequently his reference of *C arcticum* to a hybrid *C. alpinum × C. latifolium* is mythical."

NATIONAL PHYSICAL LABORATORY.—A question upon the exact position of the building about to be erected for the National Physical Laboratory was asked in the House of Commons on May 7th. Mr. Akers-Douglas, in reply, stated that the new laboratory would be situated quite outside the Kew Gardens, on Crown lands. The building will be so placed as not to interfere with the views, from the Gardens, of the Old Deer Park.

FASCIATED GROWTH OF ASH.—I am sending with this a curious specimen of fasciated ash which I thought might be of interest to you. It was found in my woods here by my son, and I do not remember seeing anything quite so peculiar in an ash before, though I suppose fasciated specimens are not really uncommon.—*Dora Twopeny, Woodstock, Sittingbourne.*

[The specimen is one of those curious cases of fasciation that sometimes occur in shoots from tree-roots when the leading stem has been cut. A similar case was figured in SCIENCE-GOSSIP, vol. ii. N.S., page 6.—ED. S.-G.]

BARBASTELLE BAT IN HAMPSHIRE.—One of our school children has to-day brought me a living specimen of this rare bat, taken in a barn in the parish. I recognised it at once by the strange way in which the ears meet over the nose and by the very dark fur. I have never found it before, but Dr. Laver of Colchester told me that Dr. J. E. Taylor knew of its occurrence in Hampshire, so that I have always reckoned it among our eleven species of bats.—*(Rev.) J. E. Kelsall, Milton, Lymington, April 6th, 1900:*

BAT SWIMMING.—Until a few days ago I was not aware that bats possessed the power of swimming. On the Saturday before Easter last I was walking with a relative, and crossing the Nun's Bridge at Thetford, when we noticed something unusual swimming across the river. We watched until it landed, and found to our surprise it was a bat. The river Thet is here about 30 feet wide, overhung with trees covered with ivy. The bat was in mid-stream when we first saw it. On the side for which it was aiming a wall rises out of the water, and when it came to this obstruction at first attempted to climb up, but failing, swam along until it reached a landing-place. I secured the little animal in a very wet condition and placed him up among some ivy to dry. I believe it was a long-eared bat, but am not sufficiently acquainted with bats to be certain. It had long ears, and was larger than others I have handled.—*J. S. Warburton, Methwold, Norfolk.*

AN EXHIBITION OF LIVING MOLLUSCS.—In connection with the collections to be arranged by the Society of Experimental Fish Culture, a series is to be made of living examples of our British Freshwater Molluscs, which will be on view at the Crystal Palace. These should prove of considerable general interest, and those of our readers who can put their hands upon exceptionally fine specimens might do worse than communicate with the undersigned or the curator, Mr. Edgar Shrubsole, at the Crystal Palace, London, S.E.—*Wilfred Mark Webb, 2 The Broadway, Hammersmith.*

THE LATE SPRING.—In reading through your note on the late spring of 1900 (vol. vi. 369), I see that you mention that "wild flowering plants are generally backward." I have noticed this lateness especially in the common *Arum maculatum*, of which I have not yet found a spathe fully expanded. This is, I think, very backward, for they generally commence blooming in the middle of April.—*S. Albert Webb, 41 Rothesay Road, Luton, May 4th, 1900.*

GUINEA-PIGS AND RATS.—I have seen it stated that rats bear an antipathy towards guinea-pigs, and that the presence of the latter will drive away rats from a house. As I have also seen this idea refuted, the following incident may be of interest. Not long ago I gave away guinea-pigs which I had in my possession. The house to which I sent them was infested with rats. I now hear that the rats have disappeared since the arrival of the guinea-pigs. The disappearance of the rats may be due to some other cause, but if so the coincidence would be curious. Could any of your readers inform me if the truth as to this point has been ascertained, and what reasons have been assigned?—*McTaggart Cowan, jnr., 53 Ashton Terrace, Glasgow, W.*

SOUTH-EASTERN UNION OF SCIENTIFIC SOCIETIES.—The Annual Congress of this body will be held on June 7th, 8th, and 9th at the Pavilion, Brighton, under the presidency of Mr. W. Whitaker, F.R.S., President of the Geological Society; the Honorary Secretaries being G. Abbott, M.R.C.S., Tunbridge Wells, and Mr. E. Alloway Pankhurst, of 3 Clifton Road, Brighton. The arrangements include (Thursday, June 7th, at 3.30 p.m.) opening of an Exhibition of Photographs and Photographic Apparatus. On the same evening, at 8 p.m., the Mayor of Brighton will receive the members of the Congress at the Pavilion, when the President-Elect, Professor Howes, LL.D., F.R.S., will deliver the annual address. On Friday, June 8th, at 10 a.m., will be held the Council Meeting, and afterwards, from 11 a.m. to 1 p.m., papers will be read on "The Skin of Liquids," by C. H. Draper, B.A., D.Sc.; "The Structure of the Lower Greensand near Folkestone," by D. H. C. Sorby, F.R.S.; "On Instincts which in Some Insects produce Results corresponding to those of the Moral Sense in Man," by F. Merrifield, F.E.S. On Friday, 3 p.m. to 5 p.m., "Dust: its Living and Dead Constituents" (lantern illustrations), by H. Garbett, M.D.; "Science at the End of the Eighteenth Century," by Arthur W. Brackett, F.S.I.; "On Colouring of Pupae in Relation to their Surroundings," by F. Merrifield, F.E.S. On Friday evening, 8 p.m., reception of members of the Congress by the Mayor of Hove, at Hove Town Hall. On Saturday, 9.45 a.m., Council Meeting; at 10.30 a.m., Delegates' Meeting; 12 noon, paper on "The Brighton Raised Beaches and their Microscopical Contents" (lantern illustrations), by Mr. F. Chapman, A.L.S. On Saturday there will be visits to the Brighton Museum (2.30. p.m.), Brighton Aquarium (3 p.m.), and the Booth Bird Museum at 3 p.m. At the last, Mr. Allcbin, of Maidstone, will read a paper on "Protection of Birds in the South-Eastern Counties."

GEOLOGY

CONDUCTED BY EDWARD A. MARTIN, F.G.S.

A PRIMITIVE ORNAMENT.—The little globular hydrozoan which is so common in the Upper Chalk, known as *Porosphaera globularis* (*Millepora, Coscinopora*), was probably used as the earliest form of ornament by our Palaeolithic ancestors. In association with the implements that occur in the High-Level Gravels of St. Acheul, near Amiens, large numbers of these rounded bodies were found by Prestwich, suggesting the idea that they had been carefully collected from the chalk, and then strung together, being perforated in such a way as to resemble beads. At the present time we see the avidity with which the naked savage seizes upon ornamental beads for the purpose of adorning his person, and we seem to see, in this association of these bodies with palaeoliths, the germ of that artistic faculty which afterwards showed itself in the etchings of deer and other animals that have been found on slabs and antlers of the Cave period.

THE CRAG DEPOSITS.—Mr. F. W. Harmer, F.G.S., read an important paper before the Geological Society on May 9th on "The Crag of Essex, and its relation to that of Suffolk and Norfolk." He pointed out that the term "Red Crag," including, as it does, beds differing considerably in age, is vague, and, when we attempt to correlate the East Anglian deposits with those of other countries, inconvenient. It will be remembered that Prestwich divided the Coralline Crag into seven zones. Mr. Harmer now divides the Red Crag, the Norwich Crag, the Chillesford Beds, the Weybourn Crag, and the Cromer Forest Bed series, into a series of ten zones—that is to say, all the beds between the Coralline Crag and the Arctic Freshwater Bed (Clement Reid) of Suffolk. The line separating the Older and Newer Pliocene is now drawn by the author between the Lenham Beds, containing *Arca diluvii* and other characteristic Miocene species of the North Sea or of the Italian Pliocene, and the Coralline Crag, the latter being considered as the oldest member of a more or less continuous and closely connected series of Newer Pliocene age. The palaeontological difference between the Coralline and Walton Crags is shown to be less than has hitherto been supposed. The Norwich Crag occupies an area entirely distinct from that of the Red Crag, no instance being known where the one overlies the other in vertical section: the fauna of the former is, moreover, more boreal and comparatively poor in species. This crag thickens rapidly towards the north and the east, and is believed to form part of the great delta formation of the Rhine. The mammalian remains found at the base of the different horizons of the crag in a *remanié* bed containing material derived from various sources, are probably derivative from deposits older than the Coralline Crag, formerly existing to the south. The Chillesford (estuarine) and Weybourn (marine) deposits—the latter charac-

terised by the sudden appearance in the Crag basin in prodigious abundance of *Tellina balthica*—represent separate stages in the continued refiguration of East Anglia during the Pliocene period; but the Cromer Forest Bed (fresh-water and estuarine), with its southern mammalia and its flora—similar to that of Norfolk at the present day—clearly indicates a return to more temperate conditions, and must therefore be separated alike from the Weybourn Crag on the one hand, and from the *Leda myalis* sands and the Arctic Fresh-Water Bed on the other. The two latter seem naturally to group themselves together, and with the Glacial deposits.

THE GLOPPA GLACIAL DEPOSITS.—About two miles north-east of Oswestry is a small farm named Gloppa. In the Glacial Gravels about there have been found many species of shells. As these occur at a height of 1,120 feet above sea-level, they have been regarded, together with the shells at similar heights at Moel Tryfaen (1,330 to 1,360) and at Macclesfield (1,150), as evidence of the submergence of the land to these depths at the time of deposition of the gravels. The gravels and sands are said by Mr. A. C. Nicholson, F.G.S., to be spread out around Gloppa, the main mass being comprised in a ridge of eskers about 1,000 yards long. About 60 feet of material was exposed at Gloppa between 1888 and 1891. The gravel consists of an agglomeration of erratics of many kinds, Silurian grit and argillite, felspathic traps, granites, etc., and bears a close resemblance to those at Moel Tryfaen; the larger ones being striated. Although the bulk of the shells are in fairly good condition, many are broken, rotted, and fragmentary. A portion of an elephant's tusk was also discovered. The shells found include some not now living in British seas, but proper to Arctic and Scandinavian waters. Such are *Leda pernula, Astarte borealis, Dentalium abyssorum, Natica affinis, Cardium groenlandicum, Trophon clathratus,* and others. The majority are now living in British seas, and include *Pecten opercularis, Mytilus edulis, Pectunculus glycimeris, Cardium edule, Venus casina, Littorina littorea,* etc. Besides eighty-five living species, seventeen derived fossils have been found, of Silurian, Carboniferous Limestone, Coal-Measure Sandstone, Lias, Gault, and Chalk age. What are we to think of such a deposit? What is the origin of this High-Level Glacial gravel? Mr. Nicholson doubted whether it could have been derived from a Boulder Clay, the nearest deposit of this in the neighbourhood being at a level of only 700 feet. Whatever its origin, the remarkable state of preservation of the shells appears to show that these were contemporaneous in age with the deposit in which found. Professor Hull thought that they lived in a sea which contained rafts of ice or small icebergs, and these deposited the boulders amongst the beds of sand. As submergence of the land took place, the molluscs would live at a successively higher level up the sides of the sunken mountains, whilst the ice-foot with derived Lias and other fossils would also gradually mount, until the fossils were left intermingled with the remains of a contemporaneous fauna. The deposit may indeed have been part of the great terminal moraine of Professor Carvell-Lewis, which accumulated where the melting of ice was in progress, and which he drew across England, approaching very near to this spot.

ASTRONOMY

CONDUCTED BY F. C. DENNETT.

	1900	Rises.	Sets.	R.A.	Dec.
		Position at Noon.			
		h.m.	h.m.		° ′
Sun	.. 9 ..	3.46 a.m. ..	8.12 p.m. ..	5.9 ..	22.55 N.
	19 ..	3,44	.. 8.18	.. 5.50 ..	23.26
	29 ..	3,47	.. 8.19	.. 6.32 ..	23,15

		Rises.	Souths.	Sets.	Age at Noon.
	June	h.m.	h.m.	h.m.	d. h.m.
Moon ..	9 ..	4.36 p.m. ..	9.14 p.m. ..	1.17 a.m. ..	11 21.10
	19 ..	11.35	.. 5.14 a.m.	..11.31 a.m. ..	21 21.10
	29 ..	6.32 a.m. ..	2.5 p.m. ..	9.25 p.m. ..	2 10.23

		Souths.	Semi-	R.A.	Dec.
				Position at Noon.	
	June	h.m.	Diameter.	h.m.	° ′
Mercury	.. 9 ..	0.52 p.m. ..	2.7″ ..	6.1 ..	25.18 N.
	19 ..	1.32	.. 3.1″ ..	7.21 ..	24.3
	29 ..	1.52	.. 3.6″ ..	8.21 ..	20.23
Venus	.. 9 ..	2.31 p.m. ..	21.1″ ..	7.40 ..	23.10 N.
	19 ..	1.54	.. 24.6″ ..	7.43 ..	21.21
	29 ..	1.0	.. 27.8″ ..	7.29 ..	19.37
Mars	.. 19 ..	9.39 a.m. ..	2.2″ ..	3.28 ..	18.31 N.
Jupiter	.. 19 ..	10.15 p.m. ..	20.7″ ..	16.6 ..	20.0 S.
Saturn	.. 19 ..	0.22 a.m. ..	8.5″ ..	18.9 ..	22.26 S.
Uranus	.. 19 ..	10.42 p.m. ..	1.9″ ..	16.32 ..	21.53 S.
Neptune	.. 19 ..	11.57 a.m. ..	1.2″ ..	5.46 ..	22.13 N.

MOON'S PHASES.

		h.m.		h.m.
1st Qr. ..	June 5 ..	6.39 a.m.	Full .. June 13 ..	3.38 a.m.
3rd Qr. ..	„ 20 ..	0.57 a.m.	New .. „ 27 ..	1.27 a.m.

In apogee June 5th at 9 p.m. ; and in perigee on 19th at 2 a.m.

METEORS.

				h.m.
May 29–June 4 ..	η Pegasids ..	Radiant R.A. 22.12	Dec. 27° N.	
June 10–28	.. δ Cepheids ..	„	„ 22.20	„ 57° N.
„ 13–July 7	.. Vulpeculids..	„	„ 20.8	„ 24° N.

CONJUNCTIONS OF PLANETS WITH THE MOON.

					° ′
June 11	..	Jupiter	.. 8 p.m. ..	Planet 1.29 N.	
„ 13	..	Saturn	.. 11 p.m. ..	„ 0.56 S.	
„ 24	..	Mars°	.. 8 a.m. ..	„ 1.31 S.	
„ 28	..	Venus°	.. 9 a.m. ..	„ 1.29 N.	
„ 29	..	Mercury°	.. 7 a.m. ..	„ 5.9 N.	

* Daylight.   All are above the English horizon.

OCCULTATIONS.

		Magni-tude.	Dis-appears. h.m.	Angle from Vertex.	Re-appears. h.m.	Angle from Vertex.
June	Object.			°		°
2..	κ Cancri	.. 5.0 ..	8.33 p.m. ..	82 ..	9.35 p.m. ..	247
13..	Saturn	.. — ..	9.40 p.m. ..	116 ..	10.52 p.m. ..	283

THE SUN should be watched, as spots are appearing more frequently on the disc. Summer is said to commence at 10 p.m., June 21st, when the sun enters the sign, not the constellation, of Cancer.

MERCURY is an evening star all the month. Starting from close proximity to the sun, it is in conjunction with Neptune at 3 p.m. on June 7th, being 2° 54′ to the north. At 7 p.m. on 13th it is only 3′ south of ε Geminorum. At 10 a.m. on 22nd it is in conjunction with Venus, being 2° 19′ to the north.

VENUS is also an evening star all the month, being at the point of its greatest brilliancy at 6 a.m. on 1st.

MARS is a morning star, rising less than two hours before the sun all the month.

JUPITER is well situated for the observer, save for its great south declination. It rises closely east of β Scorpii at 7.17 p.m. at the beginning, and 5.10 p.m. at the end of the month.

SATURN comes into opposition at 5 p.m. on June 23rd, and so is best situated for observation this month, although English observers will find its great south declination detrimental to good definition. It is a magnificent object. The major axis of the outer ring is, on June 19th, 42·61″, and the minor axis 18·87″ ; so that the southern pole of the planet will be hidden behind the rings, and the northern pole will be apparently lying upon the rings. It will be near μ Sagittari.

URANUS is in opposition at 11 a.m. on June 1st, and so is in the best position for observation. It appears as a 5·5 magnitude star to the naked eye, a little east-south-east of the 4·6 magnitude star ω Ophiuchi.

NEPTUNE, being in conjunction at 11 a.m. on June 18th, is too near to the sun for observation.

ECLIPSE OF THE MOON, June 13th.—A very slight partial eclipse commencing with :—

		h. m.	Angle from N. point.
First contact with penumbra ..	..	1 16·2 a.m.	
„ „ „ shadow	..	3 24·2 a.m.	176°
Middle of eclipse ..	..	3 27·6 a.m.	
Last contact with shadow	..	3 31·0 a.m.	180°

Moon sets 3h. 54m. The eclipse is so slight that the shadow only grazes the limb.

THE BRUCE GOLD MEDAL of the Astronomical Society of the Pacific has been awarded to David Gill, C.B., LL.D., F.R.S., Her Majesty's Astronomer at the Cape of Good Hope.

MINOR PLANETS.—According to Professor Bauschinger, the total number which have been certainly discovered hitherto now reaches 451.

THE PARIS EXHIBITION TELESCOPE, which is being belauded before the public as if it were *the* greatest telescope of the age, is only so by reason of its focal length. The great telescope of the Paris Observatory should be its equal in definition and light-grasping power, whilst Dr. A. A. Common's five-foot silvered-glass Newtonian ought to be a long way its superior in both these respects. The announcements recently published about the observations on the sun were peculiar reading. It is hardly necessary to say that the spectroscope is necessary to see the prominences on the limb.

COMETS.—Mr. W. F. Denning has been drawing attention to the remarkable fact that during the present century large comets have appeared at intervals of about nineteen and a half years: 1823, 1843, 1862, 1881. The next period falls at the end of the present year ; so far no comet has appeared to keep up the succession. Giacobini's comet has been a very small one, its total light being equal to between 12 and 13 magnitudes. At the end of June, when its distance will be 128 millions of miles, its R.A. will be about 23h. 47m. and its North Declination 41° 31′, near the centre of the triangle formed by β Andromedaea, β Cassiopeia and β Pegasi, and having a motion towards the west.

CHAPTERS FOR YOUNG ASTRONOMERS.

By Frank C. Dennett.

VENUS.

(Continued from Vol. VI. page 377.)

MANY observers never see spots on Venus which they can delineate, whilst others have drawn a great amount of detail. That markings really do exist there is not any real doubt ; but whether the markings are of a stable character or only atmospheric is questionable. Observers for the past 230 years have from time to time seen and drawn such spots. Yet among the drawings there is little, if any, of that similarity which one meets with in the case of Mars. Most of those who have studied the planet agree that whilst they have no doubt of the reality of the markings, there is a strange want of definiteness which makes them difficult to secure. The size of the telescope seems to be less necessary in this work than is usually the case. A good, rather than a large, telescope appears to be the important item, combined with persistent observation, so that the transient times may not be missed when our own atmosphere is in good condition and the gaseous envelope of Venus itself equally transparent. Excellent observations have been made when the planet has been seen through a cloud of Aurora, the detail being then remarkably visible.

The spots were carefully observed and charted by Bianchini at Rome in 1726 and 1727 with a non-achromatic telescope 66 feet long and just over 2½ inches aperture, bearing a power of 112. He, however, made the mistake of supposing that the rotation period was 24d. 8h. More than a hundred years later, 1839 to 1841, De Vico and six assistants, with the 6¼-inch achromatic instrument then at the Roman observatory, confirmed the work of Bianchini so far as the accuracy of his chart is concerned, save that he added one new spot. An important fact noted was that those who were most successful in seeing the spots were those astronomers who were least successful in detecting faint close companions of brilliant stars. Amongst the dark markings brilliant spots and patches have sometimes been seen. Huggins repeatedly saw a round bright spot. With, and Key, in April 1868, observed a small brilliant spot on the limb which appeared as a projection. Browning has also seen a bright patch on the disc. With frequently observed with an unsilvered glass-mirror, which reduced the glare of the planet. A similar effect may also be obtained by using a solar diagonal on an achromatic. In using a diagonal, solar or star, it must be kept in mind that the object viewed is inverted, but not reversed.

· The accompanying drawings of Venus were made by the writer in February 1881. The one marked A was made on the 11th at 6.45 p.m. The most striking object was the little oval "sea" near the limb, reminding one of the Mare Crisium on the moon when of similar phase. At 7.5 p.m. it had moved perceptibly nearer to the limb. On the 15th, at 6.12 p.m. and 7.27 p.m., the drawings B and C were made, the brightish patch near the centre of the disc having much increased in size in the latter. The instrument employed was a 9½-inch Calver Newtonian.

In 1890 the astronomical world was startled by an announcement by Schiaparelli that the revolution of Venus, instead of taking just less than a day, really occupied between six and nine months. Observations made at Lowell's Flagstaff Observatory, Arizona, confirmed this opinion. Drawings of the planet which were made give it the appearance almost of a cart wheel. The markings were said to be very evident, though not with the large instruments, but with the smaller, and best of all with the 3-inch. There has been much discussion on the subject, and tried observers, such as Niesten of Brussels, Stuyvaert, and Trouvelot, have considered their observations and supported the short-time rotation period. Now, apparently, the matter has been set at rest by the spectroscope. Dr. Belopolsky of Moscow finds that the amount of displacement of the lines indicates a short period of rotation. There can be little doubt that the surface we see consists of something analogous to clouds, and is not the real surface of the planet.

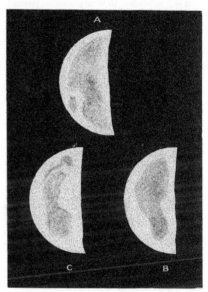

VENUS.—As seen on February 11th and 15th, 1881.

The spectroscope makes it evident that aqueous vapour is certainly présent in its atmosphere.

Respecting the transits of Venus across the sun, little need be said, as anyone now living is never likely to see another. The next transit will occur on the morning of June 8th. 2004, followed by one on June 5th, 2012. The "black drop," due to irradiation, is a very striking object at the internal contacts, when the planet, instead of appearing round, becomes lengthened out into the shape of a pear. Venus has been seen projected on the bright, yet invisible. corona of the sun. When on the disc many observers have seen this planet surrounded by a bright aureole, and having one or two bright spots near its centre.

(To be continued.)

MICROSCOPY

CONDUCTED BY F. SHILLINGTON SCALES, F.R.M.S.

PROTHALLIUM OF LYCOPODIUM CLAVATUM.—
A recent number of the "Journal" of the Royal
Microscopical Society summarises a contribution
by Mr. W. H. Lang in the "Annals of Botany," 1899,
pp. 279-317, in which he describes and illustrates
the hitherto unknown prothallium of this club-
moss. It is a nearly flat plate of tissue, with
numerous rhizoids, its structure closely resembling
that of *L. annotinum.* Of seven prothallia examined,
six were female and one male. No archegones
were found on the latter, but on two of the female
prothallia a few antherids were seen. The
archegones and antherids were both confined to
the upper surface of the prothallium; each organ
developed from a single cell. In the young plant
no structure was recognised comparable to the
protocorm of *L. cernum.* The large foot persists
for a considerable time after the prothallium has
disappeared. The cells of the prothallium are infested by an endophylic fungus. The author then
discusses the comparative structure of the prothallia of species of *Lycopodium* at present known,
and dissents from the view of Bruchmann that they
should, on this ground, be split up into a number
of distinct genera.

BECK'S NEW $\frac{1}{10}$-INCH IMMERSION OBJECTIVE.—
Messrs. R. & J. Beck, Limited, have sent for our
inspection a new $\frac{1}{10}$-inch homogeneous oil-immersion objective which will, we think, prove specially
useful to histological students and those workers
who do not require a high aperture. The N.A. of
this lens, like Messrs. Beck's cheaper $\frac{1}{12}$-inch, is
only 1·0; but its definition is very good, whilst its
price, inclusive of oil, is only £3. We think it
should be a popular lens, as it is much superior to
the water-lenses often used of this power, though
a water-lens has facilities and advantages attending its use that make it of service to many workers,
especially when dealing with mounts made rapidly
for temporary examination only. The objective
has a good working distance, and is corrected for
the short tube length.

SECTION OF LIMESTONE.—Mr. Mason, of Park
Road, Clapham, has recently sent us an exceptionally interesting microscopical slide of a section
of Carboniferous Limestone from Llanynynech,
Montgomeryshire. This section is peculiarly rich
in organic remains of unusual variety, and has, we
believe, called forth some diversity of opinion
amongst those interested in the subject. Apart
from this, the slide is a really beautiful one, and
we recommend it to the notice of those of our
readers who may be concerned with such matters.

SWIFT'S NEW BACTERIOLOGICAL MICROSCOPE.—
Messrs. James Swift & Son have recently brought
out and exhibited before the Royal Microscopical
Society a new microscope for bacteriological purposes that differs in some ways from the model
hitherto adopted by this firm. The coarse adjustment rack works in a circular groove instead of the
usual angular one, this being the arrangement
largely used abroad, whilst the fine adjustment is
also of the Continental triangular bar form. The
drawback to this type of fine adjustment is that it
necessarily bears the whole weight of the body of
the instrument, and Messrs. Swift have therefore
improved on the Continental form hitherto made
by the use of their patent differential screw. The
milled head of this last is so divided that one
division equals ·003 millimètre, and the index can be
shifted to zero to allow of direct readings being made.
The draw-tube is graduated in millimètres, and
varies the tube-length from 160 mm. to 220 mm. The
stage is covered with vulcanite, and is of specially
ample size to allow of the use of the largest petric
dish, without any risk of the plate slipping off the
stage. The right-hand side of the stage is divided

BACTERIOLOGICAL MICROSCOPE.

into squares as a finder, which enables the observer to record the position of a slide for future
reference. A full-sized improved Abbé condenser
with iris diaphragm beneath the stage is focussed
by means of a special spiral focussing adjustment
that has proved both convenient and easy of manipulation. The base of the stand is Messrs. Swift's
well-known claw-tripod, giving freedom of manipulation combined with stability; and this form of
base is greatly to be preferred to the ordinary
Continental horseshoe and pillar form. The price
of the microscope as described and illustrated,
with one eyepiece, Abbé condenser with iris dia-
phragm, dust-proof triple nose-piece, $\frac{2}{3}$ inch, $\frac{1}{6}$, and
$\frac{1}{12}$ oil immersion—all of Messrs. Swift's well-known
and excellent "Pan-aplanatic" series—and case is
£16, or without the $\frac{1}{12}$ immersion £11. This microscope has been submitted to our personal inspection, and we can recommend it to the notice of
our readers.

MICROSCOPY FOR BEGINNERS.

BY F. SHILLINGTON SCALES, F.R.M.S.

(Continued from Vol. VI. page 375.)

OBJECTIVES should be carefully treated, and it should be borne in mind that they are delicate pieces of apparatus. Dew on a lens should be allowed to evaporate; dust on the back lens should be removed with a soft camel-hair or sable brush; and if the lenses really require cleaning, a specially soft piece of chamois-leather or cambric should be kept free from dust and used for that purpose only. The lenses should n_{ever} be unscrewed: that is a matter for a first-rate optician only, and the maker is the proper man. An oil-immersion lens should be carefully and gently wiped immediately after use, and if by any chance any of the oil should have dried on the lens it is best to put another drop of oil on it, and to leave it for a time before wiping clean. Under any circumstances use as little pressure and friction in cleaning as possible. When objectives are in use, but temporarily removed from the microscope, they should be laid end upwards on the table to keep out the dust. For this reason it is well also to keep one of the eyepieces habitually in the tube of the microscope.

A glass shade is preferable to the ordinary wooden case, except, of course, for travelling, as the microscope is apt to get jarred or knocked about through being constantly taken out of and put into its case. It is well, however, to remember that a microscope should not be allowed to stand in direct sunlight, if for no other reason than that the heat might prove injurious to the balsam or cement connecting the lenses of the objectives.

On p. 249 of the last volume of SCIENCE-GOSSIP we gave instructions in the use of Beale's camera lucida, and what we said then applies very largely to the other forms of camerae lucidae. The type known as "Wollaston's" is not now much used, as it has been superseded by other forms more easy to use. It consists of a small prism placed over the eyepiece, which reflects the microscopic image into the eye. The microscope is inclined horizontally, and the eye must be so placed that one-half of the pupil is covered by this prism, and the other half looks directly at the paper placed beneath. The difficulty is in keeping this position, as any displacement causes an unequal illumination of either the image or the paper. Beale's camera lucida has not this objection, and is, perhaps, at present the most popular. A makeshift camera on this principle can be made by means of a piece of cork, a cover-glass, and a couple of pins; all that is necessary being to adjust the cover-glass in front of, but at an angle of 45° with, the eye-lens of the eyepiece. The cover-glass gives, however, a somewhat troublesome double reflection; and this is obviated in Beale's arrangement, as supplied by the opticians, by the use of tinted glass. The Abbé form of camera lucida is considered by many workers to be the best in the market, but it is also the most costly. An arm projecting from the camera carries a plane mirror, which reflects the image of paper and pencil into a silvered prism placed above the eyepiece, and so into the eye. The paper must lie in the same plane as the object; and if the microscope is therefore to be used in any but the vertical position, the drawing board must be sloped accordingly. In both this and the Wollaston form provision is made for adjusting the light by means of tinted glass. Lately a new form of combined eyepiece and camera lucida, made by Swift, Leitz, and other makers, has found great favour on account of its simplicity. It is made in two forms—one for use with the microscope in a vertical position, and another in an inclined position. A prism placed above the eye-lens projects the pencil and paper clearly into the field. Some little practice is also required with this form to get the pupil of the eye placed in such a position over the prism that neither the image of the object nor of the paper overpowers the other. The beginner will find that in all forms of camerae lucidae the secret of success, as we have already pointed out (see SCIENCE-GOSSIP, vol. vi., N.S., p. 249), lies in the proper adjustment of the illumination for both microscope and paper. It is here that the value of an independent lamp for the paper makes itself felt. With low powers the illumination in the microscope is the stronger, and the lamp-flame must be adjusted accordingly, or even a piece of white paper may be placed over the mirror when that is used. With high powers the paper is generally the brighter, and tinted screens must be used, or the light modified. The usual standard for distance between eyepiece and table is ten inches, and this should be adhered to approximately. Any variation will alter the size of the drawing. It may not be superfluous to add that short-sighted people will require to use their spectacles if they are to see the paper and pencil clearly. The pencil should have a sharp point, and the lines should not be drawn too heavily in the first place. With all forms of drawing apparatus the paper must lie in the position for which the camera lucida is designed, as detailed above, or the result will be an elliptical image.

The use of the stage-micrometer in connection with the camera lucida will suggest itself to any-one. It is only necessary to replace the object on the stage by the micrometer, taking care not to alter the other adjustments of the microscope, and to note the measurements thus shown upon the drawing. Supposing the portion of the drawing to be measured corresponded with one-hundredth of an inch, as shown on the stage-micrometer, and with one inch when measured with an ordinary rule, the actual magnification is one hundred diameters. If the micrometer be a millimetre scale, it will be necessary to provide oneself with a rule divided in millimetres—or to convert the English measurements accordingly, either by reference to a table or by calculation. For rough purposes the English inch may be taken as 25·4 mm.

The use of the stage-micrometer in conjunction with the eyepiece micrometer has been dealt with on p. 248 in the last volume. In making measurements by this method when using high powers, difficulty is often encountered in causing the object on the stage or the stage-micrometer to come into exact alignment with the lines in the eyepiece. To obviate this a mechanical stage is a great convenience, or the form of micrometer designed by Mr. Jackson, with a slight adjustment to the scale by means of a screw. The most perfect form of micrometer eyepiece is the screw-micrometer, containing one fixed and one travelling wire; the movement of the latter being accurately recorded by means of a drum, whilst each revolution of the drum corresponds to one of many serrated teeth in the field of view.

(To be continued.)

· CONDUCTED BY JAMES QUICK.

WIRELESS TELEGRAPHY.—It is to be hoped that the removal of the wireless telegraph installation at the South Foreland lighthouse and the Goodwin lightship is not a permanent one. It has certainly proved of great practical value, and has demonstrated to a marked degree the utility of the system for signalling to lighthouses and lightships. About a year ago the equipment was installed upon an experimental basis ; but as, apparently, the Trinity House authorities had no funds at their disposal to purchase the instruments, and the latter are wanted elsewhere by the Wireless Telegraph Company, they were removed some weeks since.

RELATIONS BETWEEN ELECTRICITY AND ENGINEERING.—This formed the subject of this year's "James Forrest" lecture, delivered by Sir William Preece, C.B., F.R.S., before the Institution of Civil Engineers. Feeling keenly that the subject was far too wide a one to be discussed from all points, the lecturer confined himself almost exclusively to the miscellaneous uses to which electricity had been, and was being, put to obtain engineering results. Of the extensiveness and ease of applied electricity Sir William Preece holds as high opinions as anyone. "No magician or poet," he said, "ever conceived so potent a power within the easy reach of man." In treating of the transmission of electricity over great distances he said : "Sitting on the shores of the Atlantic in Ireland, one can manipulate a magnetic field in Newfoundland so as to record simultaneously on paper in conventional characters slowly written words. Thus we have bridged the ocean and annihilated space." If Sir William Preece's opinion upon the commercial success of wireless telegraphy methods is accepted as an authoritative one, possibly the ardour of some commercial men has been damped by a remark made by the lecturer that there was "no commercial business in it."

NEW TYPEWRITING TELEGRAPH.—This instrument, the invention of Mr. W. S. Steljes, is shortly to be put upon the market by the Typewriter Telegraph Corporation, Limited. It is of simple construction, and requires no battery power, the electrical energy being generated by a magneto-instrument. In working the instrument a record is made at both ends of the line in the form of a printed copy of the message sent, and, as it can be used in conjunction with telephone lines, both verbal and printed messages can be sent.

NINETEENTH-CENTURY CLOUDS OVER PHYSICS. The Friday evening lectures at the Royal Institution were resumed, after the Easter vacation, on April 27th, Lord Kelvin commencing the present series with a discourse upon "Nineteenth-Century Clouds over the Dynamical Theory of Heat and Light." It could not be considered a brilliant one from the experimental and popular points of view ; nevertheless, it is a very valuable one to physicists, coming as it did from such an authority as Lord Kelvin. The first cloud referred to was that which came into evidence with the undulatory theory of light. If the ether be assumed to be an elastic solid, as required by this theory, it is difficult to conceive how the earth and other bodies can move so freely through the ether. In spite, however, of the investigations of Young, Michelson, Morley, Lodge, and others, on the relation between ether and matter, leading to the result from which Lord Kelvin saw no possibility of escape, viz. that there is no motion of the ether relative to matter, he persisted in his opinion that matter does not move freely through the ether. The second great undissolved cloud over the dynamical theory lay in the Maxwell-Boltzmann doctrine of the partition of kinetic energy. This doctrine also Lord Kelvin, from many recent calculations, is induced to say is not true. At the conclusion of the discourse Lord Kelvin brought forward some considerations respecting the structure of the atom and the ether, pointing out that the ether must be truly imponderable, and quite outside the law of universal gravitation.

SOCIETY OF ARTS.—At a special meeting of the Council of the Society of Arts, held at Marlborough House on May 8th, the Prince of Wales (President) presented the Albert Medal of the Society to Sir W. Crookes, F.R.S., "for his extensive and laborious researches in chemistry and physics, researches which have in many instances developed into useful, practical applications in the arts and manufactures."

THE TELEPHONOGRAPH.—No one can deny the advantages of being provided with a telephone in one's office or rooms, in spite of the constant irritation of the call-bell and the shortcomings of the telephone companies. The telephonograph, however, is to prove an additional boon, in that telephonic messages can be received and recorded in one's absence. This instrument, which is a modification of the phonograph, is provided with a steel band which replaces the wax cylinder of the Edison phonograph ; a magnet, controlled by a telephone, being also substituted for the ordinary phonographic style. Currents transmitted by the telephone pass through an electro-magnet and produce consequent poles on the steel band, a somewhat converse operation being employed for reproducing the sound. A long line can, of course, intervene between the transmitting telephone and the phonograph itself.

THE ROYAL SOCIETY.—The gentlemen's conversazione of the Royal Society took place on Wednesday evening, May 9th. Among the electrical exhibits shown was one by Professor S. P. Thompson, illustrating the converse to De la Rive's experiment with a floating battery. Little floating magnets enclosed in glass tubes took the place of the floating battery. An immersed hollow coil carrying an electric current provided the necessary magnetic field, which determined the movements of the magnets. This experiment should be a most useful and simple one to lecturers upon electricity, as it would certainly assist students in grasping the important principles underlying. Mr. P. E. Shaw exhibited his electrical micrometer, a description of which appeared in this column for last month. Among the candidates recently selected by the Council of the Royal Society for election into the Society are two distinguished physicists, Professor G. J. Burch, M.A., and Dr. James Gordon MacGregor.

ROYAL METEOROLOGICAL SOCIETY.—The first afternoon meeting of the present session was held on Wednesday, May 16th, at the Society's rooms, 70 Victoria Street, Westminster, Dr. C. Theodore Williams, President, in the chair. A most interesting paper on "The Wiltshire Whirlwind of October 1st, 1899," was read which had been prepared by the late Mr. G. J. Symons, F.R.S., a few days before he was stricken down with paralysis. This whirlwind occurred between 2 and 3 P.M., commencing near Middle Winterslow and travelling in a north-north-easterly direction. The length of the damage was nearly twenty miles, but the average breadth was only about 100 yards. In this narrow track, however, buildings were blown down, trees were uprooted, and objects were lifted and carried by the wind a considerable distance before they were deposited on the ground. Fortunately the greater part of the district over which the whirlwind passed was open down; otherwise the damage, and perhaps loss of life, would have been considerable. At Old Lodge, Salisbury, the lifting power of the whirlwind was strikingly shown by some wooden buildings being raised and dropped several feet north-west of their original position. At a place eighteen miles from its origin the whirlwind came upon a rick of oats, a considerable portion of which it carried right over the village of Ham and deposited in a field more than a mile and a half away. A paper by Dr. Nils Ekholm, of Stockholm, was also read on "The Variations of the Climate of the Geological and Historical Past and their Causes." In this the author attempts to apply the results of physical, astronomical, and meteorological research in order to explain the secular changes of climate unveiled by geology and history.—*William Marriott, Assistant Secretary.*

SOUTH LONDON ENTOMOLOGICAL AND NATURAL HISTORY SOCIETY.—MARCH 8th, Mr. W. J. Lucas, B.A., F.E.S., President, in the chair. Mr. Harwood exhibited a species of *Blatta* from the Eastern Counties, which was apparently new to Britain. Mr. Adkin, a bred series of *Eugonia autumnaria* from Bournemouth. Mr. Colthrup, a specimen of *Euchelia jacobaeae*, with the red areas unusually pale, a very beautifully marked variety of *Eurrhypara urticata*, and very small examples of *Pieris rapae*, including a yellow variety. Mr. Lucas exhibited living specimens of the immature stage of *Blatta australasia* from Kew, and a case containing preserved examples of the whole of the British Cockroaches, with drawings of several species. Mr. Main, living specimens of *Blatta americana* from Silvertown. Mr. Edwards, living specimens of *Phyllodromia germanica*, male, female, and immature. Mr. Moore, numerous exotic species of Cockroaches. Mr. Tutt, a long and varied series of *Epunda lutulenta*, taken at Mucking, Essex, by Rev. E. Burrows in 1898-99, and gave notes as to the occurrence and variation of the species. Mr. Lucas read a paper entitled "Cockroaches: Natives

and Aliens," illustrating it with numerous lantern slides. MARCH 22nd, Mr. W. J. Lucas, B.A., F.E.S., President, in the chair. Mr. Montgomery exhibited specimens of a second generation and a partial third brood of *Coremia designata*, and gave notes on their life-history and variation. Mr. F. N. B. Carr, a varied series of *Hibernia leucophearia* from Lee. Some very beautiful lantern slides on Ornithological subjects were then exhibited under the auspices of the Society for the Protection of Birds, including copies of a number of plates from Lord Lilford's "British Birds," and numerous studies of the nests and haunts of birds by Mr. E. B. Lodge. APRIL 12th, Mr. F. Noad Clark in the chair. Mr. Edwards exhibited a living specimen of *Scorpio europaeus*, sent by Dr. Chapman from Cannes. It fed readily upon cockroaches. Mr. Sich, living larvae and cases of *Coleophora lineolea* from Chiswick. Mr. Noad Clark, photomicrographs of the ova of (1) *Eugonia fuscantaria*, showing clearly the serrated edges; (2) *Geometra vernaria*, in piles as deposited; and (3) *Neuronia popularis*. Mr. Colthrup, specimens of *Bombyx quercûs* var. *callunae*; and Mr. Tutt gave an interesting account of the Lasiocampid moths, to which he had recently been devoting his attention. He showed that they formed a clearly definable section, and contained numerous easily distinguishable, although closely allied, sub-sections and genera. The various points of view of ovum, larva, pupa, and imago were taken into consideration, and contrasted and compared with allied groups as well as among themselves. APRIL 26th, Mr. W. J. Lucas, B.A., F.E.S., President, in the chair. Mr. Buckstoné, specimens of *Triphaena fimbria* bred from ova. The larvae had been fed exclusively on cabbage. Mr. Turner, Longicorn Coleoptera—(1) *Saperda populnea*, taken by Mr. Day at Carlisle; (2) *Rhagium bifasciatum* from the New Forest; (3) *Clytus mysticus* from Brockley; (4) *C. arietis* from Lewisham; together with larvae of (i) *Callimorpha dominula* from Deal, where they were comparatively scarce; (ii) *Bombyx quercûs* from Deal, on garden-rose; (iii) *Pericallia syringaria* from Bexley. Mr. Lucas, specimens of the snake's-head plant, *Fritillaria meleagris*, including a white variation from Oxford. Mr. Moore showed a Kaffir necklace made of the so-called "eggs" of the white ant, *Termes bellicosus*, but he found they were really the encysted pupae of a species of Coccid of the genus *Margarodes*, having subterranean habits. Mr. Lucas, a specimen of the dragon-fly, *Sympetrum vulgatum*, a male taken by Mr. Hamm, of Oxford, at Torquay, on August 15th, 1899. This is the second authenticated British specimen, the other being in the collection of Mr. C. A. Briggs, of Lynmouth. Mr. Adkin, a fine bred series of *Eugonia fuscantaria*, reared from Lewes ova. He stated it was easy to breed when sleeved. Mr. Clark reported that he had received ova of *Gonepteryx rhamni* which had been found deposited on the stems of the buckthorn. Mr. Harrison reported having seen a dragon-fly, *Libellula quadrimaculata*, on wing at Easter. Mr. Step exhibited a considerable number of lantern-slides, made by himself, of "Wild Flowers at Home," and described their characteristics and surroundings at some length. A discussion ensued upon this branch of natural history study. The use of the photographic camera was recommended as an exact record of observations on natural objects.—*Hy. J. Turner, Hon. Report Sec.*

CORRESPONDENCE.

WE have pleasure in inviting any readers who desire to raise discussions on scientific subjects, to address their letters to the Editor, at 110 Strand, London, W.C. Our only restriction will be, in case the correspondence exceeds the bounds of courtesy ; which we trust is a matter of great improbability. These letters may be anonymous. In that case they must be accompanied by the full name and address of the writer, not for publication, but as an earnest of good faith. The Editor does not hold himself responsible for the opinions of the correspondents.—Ed. S.-G.

NATURE PICTURES OF LEPIDOPTERA.

To the Editor of SCIENCE-GOSSIP.

SIR,—Can any of your readers tell me, through your columns, the process for stamping off on paper the scales of butterflies' wings, so as to get a counterfeit representation of the species ?

ROGER VERITY.

Florence, Italy.

PHONETICS AND ETHNOLOGY.

To the Editor of SCIENCE-GOSSIP.

SIR,—Does there exist any systematic treatment of the connection between these two sciences ? Has any divergence in the structure of the vocal organs or of the speech and auditory centres been noticed among various races, especially among the inhabitants of the once-Roman Provinces, who have now developed Neo-Latin into seven principal forms? Have any of their conflicting and contradictory phonetic tendencies corresponded with such physiological and racial divergences ? To what climatic, or any other non-biological, causes can be attributed the tendency to divergent alteration of pronunciation ? I should be obliged for any references, especially recent ones. to these points.   CHARLES G. STUART-MENTEATH.

23 Upper Bedford Place,
London, W.C., May 11th, 1900.

NOTICES OF SOCIETIES.

*Ordinary meetings are marked †, excursions * ; names of persons following excursions are of Conductors. § Lantern Illustrations.*

GEOLOGISTS' ASSOCIATION OF LONDON.
June 1.—† " Our Older Sea Margins." Sir A. Geikie, F.R.S. (In Theatre of Museum of Practical Geology, Jermyn Street, S.W. 8 p.m.)
  „ 2-5.—* Malvern and District. Prof. Theodore T. Groom, M.D., D.Sc., F.G.S.
  „ 16.—* Caterham, Godstone, and Tilburstow. W. Whitaker, B.A., F.R.S.
  „ 23.—* Guildford. A. K. Coomara-Swamy, F.G.S.
  „ 30.—* Silchester. J. H. Blake, F.G.S.

SOUTH-EASTERN UNION OF SCIENTIFIC SOCIETIES.
June 7-9.—Annual Congress at the Pavilion. Brighton.
  E. Alloway Pankhurst, 3 Clifton Road, Brighton, Hon. Local Sec.

NOTTINGHAM NATURAL SCIENCE RAMBLING CLUB.
June 2.—° The Hemlock Stone. Geology. J. Shepman, F.G.S.
  „ 16.—° Hucknall and High Park Wood. Botany. W. Stafford.

NORTH LONDON NATURAL HISTORY SOCIETY.
June 1-4.—* New Forest. C. Nicholson, F.E.S.
  „ 4.—° Westerham. L. B. Prout.
  „ 7.—† " Reptiles." F. W. Jones.
  „ 23.—* Cycle Excursion from Blackheath Station.
  „ 24.—† Debate. Sexual Selection.
  „ 30.—° Harefield. R. W. Robins.

LAMBETH FIELD CLUB.
June 4.—° Broxbourne and Wormley Woods. C. S. Cooper.
  „ 11.—† Annual Meeting.
  „ 16.—* Woking and Chobham. W. Wright.

SOUTH LONDON NATURAL HISTORY SOCIETY.
June 4.—New Forest.
  „ 16.—Chipstead.
July 7.—E. Horsley.
Sept. 22.—Paul's Cray Common.

NOTICES TO CORRESPONDENTS.

TO CORRESPONDENTS AND EXCHANGERS.—SCIENCE-GOSSIP is published on the 25th of each month. All notes or other communications should reach us not later than the 18th of the month for insertion in the following number. No communications can be inserted or noticed without full name and address of writer. Notices of changes of address admitted free.

BUSINESS COMMUNICATIONS.—All Business communications relating to SCIENCE-GOSSIP must be addressed to the Proprietor of SCIENCE-GOSSIP, 110 Strand, London.

SUBSCRIPTIONS.—The volumes of SCIENCE-GOSSIP begin with the June numbers, but Subscriptions may commence with any number, at the rate of 6s. 6d. for twelve months (including postage), and should be remitted to the Office, 110 Strand, London, W.C.

EDITORIAL COMMUNICATIONS, articles, books for review, instruments for notice, specimens for identification, &c., to be addressed to JOHN T. CARRINGTON, 110 Strand, London, W.C.

NOTICE.—Contributors are requested to strictly observe the following rules. All contributions must be *clearly* written on one side of the paper only. Words intended to be printed in *italics* should be marked under with a single line. Generic names must be given in full, excepting where used immediately before. Capitals may only be used for generic, and not specific names. Scientific names and names of places to be written in round hand.

THE Editor will be pleased to answer questions and name specimens through the Correspondence column of the magazine. Specimens, in good condition, of not more than three species to be sent at one time, *carriage paid.* Duplicates only to be sent. which will not be returned. The specimens must have identifying numbers attached, together with locality, date and particulars of capture.

THE Editor is not responsible for unused MSS., neither can be undertake to return them unless accompanied with stamps for return postage.

NOTICE.

SUBSCRIPTIONS (6s. 6d. per annum) may be paid at any time. The postage of SCIENCE-GOSSIP is really one penny, but only half that rate is charged to subscribers.

ANSWERS TO CORRESPONDENTS.

A. S. H. (Surrey).—Yes, the dipterous fly is *Bombylius major.*

A. M. (Hayling).—Yes, the nodules on the roots of *Vicia sativa* are the product of bacteria, and known as nitrogen nodules.

EXCHANGES.

NOTICE.—Exchanges extending to thirty words (including name and address) admitted free, but additional words must be prepaid at the rate of threepence for every seven words or less.

WANTED, to exchange this season's good stuffed specimens. Sheldrakes, golden-eyes. redshank, &c., for others.—H. Stevens, 23 Mandella Terrace, Heaton, Newcastle-on-Tyne.

WANTED, grains of so-called " ginger-beer plant," in exchange for rare algae or shells.—Miss Stanley, Heathfield, Sussex.

WANTED.—London Cat. B. Plants, 9th edition, Nos. 23, 35. 117, 159, 179, 304, 305, 314, 1849. &c. Offered 8, 191, 194, 208, 391, 1047, 1368, 1553, &c.—Rev. G. H. Waddell, Saintfield, co. Down, Ireland.

WANTED.—British birds' eggs and a large egg cabinet. Offered, foreign stamps, books on geology, engineering, and many for Civil Service candidates.—E. Pull, 5 New Terrace, Chesham Bois, Bucks.

OFFERED.—*Rhopalocera* of Sikkim (Himalaya). Wanted, European (British excepted), Tropical American, and other *Rhopalocera.* List of species and other information.— Roger Verity, 1 Via Leone Decimo, Firenze, Italy.

MOSSES.—I wish to exchange mosses. Please send lists of duplicates and desiderata.—(Rev.) Hunt Painter, Stirchley Rectory, Shifnal, Shropshire.

GEOLOGICAL NOTES IN ORANGE RIVER COLONY.

By Major B. M. Skinner, R.A.M.C.

1. From Enslin to Bloemfontein.

A FEW remarks on the geology of the narrow tract of country lying between Enslin and Bloemfontein may perhaps prove of some interest, even though the notes from which they are taken were jotted down often on stray scraps of paper. The country observed was strictly confined to the line of march, the collection of rock specimens being an impossibility, owing to want of any means of carriage.

Enslin is on the railway between the Modder and Orange rivers, and consists of a small railway station in a red plain. Out of the red plain rise small rounded hills, all of them covered with the same loose rounded boulders; those on the east of the rail at this spot being celebrated for the battle which occurred on them on November 25th, 1899. The country looks like a vast sea with islands dotted about. The composition of these islands became a matter of interest. They were all of the same pattern, covered with a reddish soil, dotted

as the column of troops wended along they crossed the spurs of boulder-covered high ground, passing along red soil down the talus slope; this thinned down till the white limestone was exposed. The lime would be succeeded by a modern "pan," *i.e.* a water-bed, at this season (beginning of February) generally dry, flat, sandy. In those districts where there is an outlet from the hills a dam is built across, and the dam end of the pan contains some water; and where there is water, there are one or two trees and a farm. On the other side of the pan comes the lime, then the red clay soil, then the hill or kop.

The foregoing is the structure of the country to Ramdam, about nine miles from Enslin. The travertine here is extensive and contains many grains of garnet. A repetition of this sort of country occurs as far as Waterval Drift. At this spot came an interesting, but disappointing, series of strata. The Riet river flows past, the "Drift" being the ford across that stream. The river here has cut its way through thin strata of

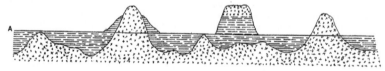

FIG. 1. SECTION OF PLAIN NEAR ENSLIN.

A. General surface level; × Dolerite; ☰ Stratified rock.

with rounded boulders, the tops occasionally presenting a cracked and fissured rock-bed from which the boulders had come, while the slopes were so covered with débris that the rock beneath was inaccessible, except that now and then the débris showed a shaly admixture. The red soil, formed by the destruction of the dolerite, has washed down across the plain, the boulders of the hills seldom travelling far, and being always noticeably smaller as the hills are descended. When the boulders in the plain are larger than those at the foot of the nearest hill, experience teaches that they indicate the site of another doleritic prominence now disguised by débris. (See fig. 1.)

In the centre of the plain the ground is white, and slight cuttings for the railway show that this white soil—which, by the way, has given the homely name of "Chalk Farm" to one of the stations a little further south—is a travertine, varying in thickness from an inch to five or six feet, lying on denuded strata of shales. This white formation is a striking feature of the country. Day after day

sandstones, and the disappointment consisted in not being able to obtain a specimen owing to a hurried move. While halting temporarily at a deserted farm-house on the bank, the paving of the verandah was seen to contain on some of the stones, and on a flag-stone, the remains of plants. However, had a specimen been obtained it could not have been carried away, as a 50-lb. kit does not allow of geological additions. As usual, the strata in the river bed and bank were horizontal, or nearly so, the usual characteristic of strata in this country, even though outcrops or dykes of dolerite occurred in the vicinity, as seen when morning broke on a march commenced at 1 A.M. to the next halt at Weydrei, near Jacobsdal. Circumstances did not allow of any close examination of the few exposures of stratified rocks at Klip Drift, the next halt; they were observed to consist of flags with intervening sandy shales. Further up the river, at Klip Kraal, sections of these flags and shales were exposed; but beyond noticing that the shales were very dark-coloured, no close note was taken, as other urgent affairs demanded attention.

From Klip Drift the next move was to Bank's Vlei, and thence to Paardekraal; the hills or koppies seen on the road all show by their boulder-strewn sides that their tops at least are formed of dolerite, and, as usual, the soil below them is red, while in the lower ground the white lime is exposed. In the bed of the Modder, just above Paardekraal Drift, the shales and flags are seen to have a slight southern dip—the first deviation from the horizontal, or apparently horizontal, seen in this line of country. About a mile further up the river there is a considerable exposure of these rocks, which are "muddy" thin-bedded sandstones, having a shaly fracture, with intervening sandy shales; the whole varying in colour from a light blue-grey to almost black. In some positions were septaria, often of considerable size, while occasional fossilised sand-cracks were seen. On the north bank, just above the Drift, is the Paardeberg, which being translated is "Horse Mountain." This is a mass of dolerite; all the koppies on the south are of the same material, the stratified rocks as far as yet seen being confined to the plain-level, where they have been protected from final denudation by their covering of travertine and the red detrital sandy clay. The only exception to this is a mound opposite the Paardeberg and south of the river, which on one side shows a protrusion of dolerite, and on the other a layer of what appears like chert, lying horizontally on sandstones and shales, whose edges are occasionally seen, and covered by similar strata. At the Drift opposite Cronje's Laager a section was observed of bluish-slaty shale, the top of which was denuded, and is now covered by travertine, and this again by light-reddish sandy soil; below the bluish was a dark shale. To the south-east of Cronje's Laager is a hill called in the map Stinkfontein, but by the British troops re-named Kitchener's Kop; this is a mass of dolerite, on the flanks of which are patches of the débris of chert and sandy shales.

The appended section is suggested as representing diagrammatically the structure of the country (fig. 2). The dolerite, being a hard rock, has acted as a protective cap to the softer sandy strata, while at the same time it has disguised those strata by its débris. The source of the travertine is not apparent as yet. The age of the strata is not disclosed by internal evidence, though from the description of the lithological characters of the Karroo given by Professor Green ([1]) they are probably of that period.

Marching beyond Kitchener's Kop, we come to Osfontein Farm; then, after a flank march, to Poplar Grove; and thence on to Roodepoort, noting nothing new on the way beyond the large variety of bulbous plants that may be seen on the alluvial

([1]) "Quart. Journal Geol. Society," No. 174, pp. 246 and 248.

flats of the Modder River near these places. In the river bed at Roodepoort is a small waterfall over rocks of dolerite, where it was observed that the spheroidal weathering was more marked in this position than on the hills—a fact since observed in every position where this rock has been exposed to water-action. The soil during the march to Roodepoort changed to a light buff sand, 25 to 40 feet of which are exposed in the river banks. About six feet above the level of the waterfall mentioned, a narrow layer contained shells of a species of *Helix* like *H. nemoralis* and a minute bivalve. In another spot further up the river, in a layer of fine gravel about six feet above the present river-level, was a small ungulate bone, a specimen of a bivalve, with broken fragments of a similar shell. In a layer about two feet above this was found a species of *Unio*. A little further up stream these gravelly layers (sections of small streams?) had disappeared, but three strings of kunkur were seen, evidently deposited by another stream. In the grounds of a farm-house on the river bank were found fragments of rock lying about, evidently brought to this spot for building purposes. Among these were pieces of a dark limestone, showing that somewhere not very far away are limestone rocks. This limestone contained minute concretions which at first sight gave the impression of foraminifera. A fossiliferous rock would add to the interest of a region apparently barren of evidence of former life.

The next march was past Abraham's Kraal, halting at Dreifontein, the soil being again red, with occasional exposures of travertine, the koppies boulder-strewn with dolerite as before. The next day saw the column near Kaal Spruit, from which spot the distant hills occasionally showed alternating layers of hard and soft rock, this becoming more evident the following day from Venter's Vallei, though it was not possible to approach near enough to identify the rocks. The stones of which a farmer's hut at this locality was built showed several varieties of sandstones, exhibiting bedding streaks, but evidently hewn from thicker strata than any seen previously on this line of march. Continuing onwards, the next day the column bivouacked on a mass of dolerite called Brand's Kop, the following morning marching into Bloemfontein, the country on that side of the Kop being a grass-covered slope towards the town.

The source of the travertine ([2]) forms to the newcomer a subject for speculation, as no rock was

([2]) In the vicinity of Modder River Station, a spot not included in these notes, this travertine was observed interbedded thinly and unevenly with a fine gravel, the main bed of which lay beneath the lime; in another locality near this was a sandstone, the upper part of which was interbedded with the travertine. At Osfontein, where the travertine lay on a denuded surface of shales, fragments of the shales were imbedded in its lower part, forming a breccia; the same condition was noticed elsewhere, but at this place there was a good section visible in the side of a sluit (a narrow water-channel) cut by man to conduct water from a spring.

observed which could have supplied such vast quantities of lime along the 120 miles or so traversed during the march recorded above. The type of strata, leaving out the igneous, was, first, shales; then thin, flaggy sandstones and shales at Waterval Drift; followed by a preponderance of shales along the Modder River up to Dreifontein; and, finally, an indication in the hills seen from Venter's Vallei of thick bands of sandstone, compared with the previous flaggy rocks, though the shaly strata preponderated. Taking Professor Green's lithological description ([3]), these strata may be set down as "Karroo." It is to be regretted that no fossils could be found to verify the diagnosis. The manner in which the sandstones become more pronounced at the eastern end of the line would point to an ascent in the geological scale, judged by the same standard, and remembering that the country rises slightly from west to east, the strata at the same time being, to all intents and purposes, horizontal, and only disturbed very locally, such disturbance being but once observed, as noted above.

This fact points to denudation having occurred from east to west, though the cessation of the travertine towards the end of the march, soon after leaving the Modder River, may indicate that the denudation of the lime rock occurred from the north-east. This travertine was deposited after the denudation of the country had progressed to almost recent conditions, for it always lies on a denuded surface of the rock of the lower land of the locality—we may call it the valley-rock—generally a shale in the district under review; and as this valley-rock is the lower level of denudation which has progressed downwards from a height that was once considerably above the altitude now reached by the Drakensberg, the deposition of the travertine dates to a period long subsequent to the formation of the highest strata of the country—sufficiently long to allow of the denudation of some 8,000 feet of rock. Professor Green says of the hills in Cape Colony: "They are all of them purely hills carved out by denudation, and they stand as speaking witnesses of what denudation can do, and of the enormous lapse of time during which it must have been at work in this country" ([4]).

The highest beds, excepting the volcanic, of South Africa appear to correspond to higher Triassio beds in Europe. Therefore their period of elevation may be inferred to be fairly remote, and to afford ample time for the denudation which has taken place, but which is now probably proceeding at a comparatively slow pace.

The red soil, very sandy at the points more remote from the modern centres of its formation, more clayey on the hill slopes, is derived from the weathering of the dolerite; and as the dykes and

(3) ' Quart. Journal Geol. Society," No. 174, pp. 246 and 248.
(4) *Loc. cit.* p. 261.

sheets of this material are practically omnipresent, the prevailing colouring of the country is red during the dry season. The fall of the rains rapidly causes the growth of herbage, and converts a barren-looking land into green pastures in a few days.

It was mentioned above that in the centre of the plain-lands was usually to be found a pan, which, as far as can be judged from the character of the narrow strip of country passed through, was seldom in the present day covered with water. These pans represent the denudation of the country, and as seen superficially are composed of whitish sand.

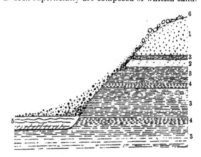

FIG. 2. SECTION OF COUNTRY NEAR PAARDEBERG.

1. Dolerite; 2. Chert; 3. Sandy shales; 4. Thin sandstones; 5. Travertine; 6. Débris of dolerite, red sand, and boulders.

The section mentioned above in the banks of the Modder River at Roodepoort may perhaps represent what would be seen were a section possible through one of these sandy pans, and would point to their formation in recent times. It would show sand, sandy mud, fine gravel, irregularly bedded, with recent and modern shells; it would be interesting to get a section through a large dried-up pan like that between Enslin and Ramdam, as the lowest strata might afford some further clue to the geology of the country. In the case of this pan, its site had evidently been cut out of the travertine now on each side of it; how much deeper it had gone it was impossible to ascertain.

It is with diffidence these rough notes are submitted; they may possibly prove of some interest as referring to a not very frequented part of the Orange River Colony. The difficulties of the occasion and want of transport, necessitating even the omission of a camera from a light scale of equipment, must be the apology offered for paucity of details and of illustrations.

Bloemfontein, May, 1900.

STONYHURST COLLEGE OBSERVATORY. — The " Results of Meteorological and Magnetical Observations, with Report and Notes of the Director, 1899," has been sent us by the Rev. W. Sidgreaves, S.J. It contains the results of an immense amount of careful work, not only at Stonyhurst, but also at St. Ignatius College, Malta.

NOTES ON SPINNING ANIMALS.

By H. WALLIS KEW.

I. A CENTIPEDE'S WEB.

AMONG spinning Myriapods the most notorious, perhaps, is the little slender, elongated centipede. *Geophilus*, whose web, much enlarged, is shown in the accompanying illustration. The drawing is copied from a figure published by Professor Fabre in 1855, and represents the web of *Geophilus convolvens*, of which centipede Fabre kept a number of individuals of both sexes for some time in captivity.

At the end of September this naturalist noticed in passages in the mould in which the creatures lived very small nets, formed of cobwebby filaments, irregularly crossed, and extending from one wall to the other of the passages. Similar nets were seen also above the mould, between

tremity of the body. Numerous attempts were made to ascertain the normal fate of the spermatophore and the manner in which fertilisation was effected, but without success, for the females, being possibly immature, took no notice of the webs. The observer was satisfied, however, that with this animal there is, in the ordinary way, no individual connection between the sexes; and he suggests that the female may deposit her eggs round the spermatophore, causing it to burst; or, more probably, he says, she may come to the web before laying her eggs and take possession of the spermatophore.

Species of *Geophilus* are common in this country, and possibly some reader of SCIENCE-GOSSIP has observed their spinning habits.

Fabre's memoir is in the "Annales des Sciences

WEB OF CENTIPEDE (¹).

sprigs of moss. Near the centre of each net was suspended, free from contact with foreign bodies, a white globule of the size of a small pin's-head, and this the observer at first believed to be an egg. Under the microscope, however, he saw the globule burst, and recognised in it, to his great surprise, a drop of sperm, with spermatozoa in full activity. Several globules were examined with the same result, and it was clear that they were the spermatophores of the *Geophilus*. After two or three days the globules first observed disappeared, having been dried up, or devoured perhaps by mites; but new nets were formed, each with a spermatophore in the centre, and so on, during a month and a half. Therefore, in spite of rapid destruction, five or six were constantly to be seen. These webs, regarded as a sort of nest on which the male lays his product, were believed by Fabre to be spun by the male, from accessory genital glands, which open at the posterior ex-

Naturelles," Zool. (4), III. (1855), pp. 257–316; and one may refer, also, to Ryder, "Proc. Acad. Nat. Sci. Philadelphia," 1881, pp. 79–86; and Zograf, "Zoologischer Jahresbericht," 1883, Abth. 2, pp. 88–92.

II. THE SLIME-JETS OF PERIPATUS.

Peripatus, one learns, possesses paired slime-glands with reservoirs, the contents of which it can suddenly eject through openings at the tips of a pair of oral papillae, in the form of fine jets of tenacious fluid, which, hardening quickly, may form "networks of fine threads, looking like a spider's web" (²).

Originally described as a mollusc, and for years a sort of zoological shuttlecock, *Peripatus*, as most readers will remember, is one of the most remarkable animals in the world. Though now definitely recognised as belonging to the Arthropoda, among which a separate class has been created for it, it has undeniable affinity with higher worms, and is even regarded as a sort of half-way animal between

(1) Web, with spermatophore, of *Geophilus convolvens*, much enlarged. After Fabre, "Annales des Sciences Naturelles": Zool. (4), iii. (1855), pl. IX. fig. 23.

(2) Moseley, "Encyclopaedia Britannica," ed. 9, xvii. (1884), pp. 115–117.

worms and Arthropods. Requiring a moist atmosphere, and living in shady places beneath bark, decaying logs, &c., the various species have soft, somewhat vermiform bodies, two inches to three inches long, and possess, in addition to oral papillae, a pair of antennae, and numerous pairs of little, soft, conical legs, placed laterally, along the whole length of the body. In manner of progression, and in general form, the creatures have been compared to caterpillars: and naturalists who have seen them alive speak with admiration of their appearance. According to Sedgwick [3], "the exquisite sensitiveness and constantly changing form of the antennae, the well-rounded plump body, the eyes set like small diamonds on the side of the head, the delicate feet, and, above all, the rich colouring and velvety texture of the skin, all combine to give these animals an aspect of quite exceptional beauty."

The slime-glands lie on each side of the digestive tract, stretching down nearly its whole length, and having numerous ramified tubes, which twist round the stomach and entangle themselves about the generative organs. The ducts into which the tubes open are enlarged along the greater part of their course into sacs or reservoirs, which serve to store up the secretion of the tubes, and to eject it from the oral papillae in the form of jets or threads.[4]

According to Moseley, these glands are probably homologous with the silk-glands of caterpillars. The oral papillae, he says, are modifications of the second pair of body-members of the embryo.

The ejection of the tenacious fluid has been remarked upon, I believe, by most naturalists who have collected *Peripatus*. Guilding, who was the discoverer of the genus [5], noticed the habit, in St. Vincent, in *Peripatus juliformis* ; Gosse saw it in a *Peripatus* found in Jamaica [6] ; Belt in one found in Nicaragua [7]—Belt refers to the creature as a Myriapod, but dried specimens subsequently shown to Moseley proved to be *Peripatus* [8]—Hutton in *Peripatus novae-zealandiae* [9] ; Oakley in *Peripatus capensis* [10] ; Dendy in a *Peripatus* collected in Victoria [11] ; Steel in *Peripatus leuckarti* from New South Wales [12] ; and Ward in *Peripatus moseleyi*

from Natal [13]. Moritz, Moseley, and Sedgwick, moreover, have recorded observations on this subject. One gathers that it is when the animals are alarmed by sudden exposure to light, or when they are irritated or menaced, that they are most commonly seen to make their discharges. The ejected matter, all the authors agree, is intensely viscid. Gosse compares it 'to birdlime, and Hutton says that it dries so quickly and is so tenacious that the finger is with difficulty removed if stuck with it to the table. The jets stiffen on exposure to the air, according to Belt, "to the consistency of a spider's web, but stronger."

Moritz's notes, written in 1839, are based on a *Peripatus* discovered in St. Thomas, and on numerous specimens observed in Venezuela. He says that the slime is shot out into threads from both sides of the animal, and that the creatures are in the habit of making the discharge at the moment the stones or pieces of wood under which they are found are lifted, the white slime-threads being usually seen before the animals themselves are noticed. Moritz succeeded on one occasion in observing a specimen before it made the discharge ; but even in this case he could not see the actual expulsion, which was as quick as lightning. The slime, moreover, was colourless when it first issued forth, the white colour of the threads and their toughness being acquired upon contact with the air [14]. Moseley writes that *Peripatus capensis* also shoots out the fine jets with such remarkable suddenness that it is almost impossible to observe their passage from the animal's head. The threads thus formed, he says, cross one another in various directions, and form a sort of meshwork, often of considerable complexity, which suddenly appears as if by magic suspended from objects in front of the animal, and has then the appearance of a bit of spider's web dotted with the dew. When examined under the microscope, the threads are seen to be fine and hyaline, with variously sized, highly refractile, spindle-shaped globules situate at intervals upon them ; and they thus much resemble the beaded spiral lines of the snares of Epeirids. Small specimens, Moseley adds, soon exhaust their immediate supply, and cannot be induced, even when squeezed hard, to make more than two or three discharges ; large specimens, however, can make at least a dozen discharges one after another [15].

According to Sedgwick [16], who also observed *Peripatus capensis*, the ejection results from a sudden contraction of the muscular body-wall, by which means the contents of the slime-reservoirs are driven out with considerable force. Belt gives

(3) Sedgwick, "Quart. Journ. Micr. Sci." (N.S.), xxviii. (1888), pp. 431–493 ; "Camb. Nat. Hist.," v. (1895), pp. 3–26.

(4) Moseley, on *Peripatus capensis*, "Phil. Trans.," clxiv. (1874), pp. 757–782 ; Hutton, on *Peripatus novae-zealandiae*, "Ann. and Mag. of Nat. Hist." (4), xviii. (1876), pp. 361–368, and (5), i. (1878), pp. 204–206.

(5) Guilding, "Zool. Journ.," ii. (1826), pp. 443, 444.

(6) Gosse, "Naturalist's Sojourn in Jamaica," 1851, p. 66.

(7) Belt, "Naturalist in Nicaragua," 1874, pp. 140, 141.

(8) Moseley, "Ann. and Mag. of Nat. Hist." (5), iii. (1879), pp. 265, 266.

(9) Hutton, 1876, *l.c.*

(10) Oakley, "Trans. S. African Philos. Soc.," iii. (1884), pp. 35–37.

(11) Dendy, "Nature," xxxix. (1889), p. 366.

(12) Steel, "Proc. Linn. Soc. N.S.W.," xxi. (1896), pp. 94–103.

(13) Ward, "Journ. Quekett Micr. Club" (2), vi. (1897), pp. 424–428.

(14) Moritz, "Archiv für Naturgeschichte," Jahrg. v., Bd. i. (1839), pp. 175, 176.

(15) Moseley, 1874, *l.c.*; "Notes of a Naturalist on the 'Challenger,'" 1879, pp. 160, 161 ; and "Challenger" Reports, i. (1885), p. 285.

(16) Sedgwick, *l.c.*

" about three inches " as the distance to which his Nicaraguan *Peripatus* throws its slime; Ward mentions, however, that his fingers have been struck by the slime of *Peripatus moseleyi* at a distance of six to eight inches, and *Peripatus capensis*, according to Sedgwick, can squirt its slime to the distance of almost a foot; it is reported. moreover, that *Peripatus moseleyi* can thus hit an object even at a distance of two feet ([17]). The slime of *Peripatus capensis*, Sedgwick adds, though extremely sticky, readily comes away from the skin of the animal itself; a fact of interest for comparison with that recorded of the Cuvierian threads of Holothurians, which, while they do not adhere to the slimy body of their possessors, stick to almost everything else with which they come in contact.

Nearly all the species of *Peripatus* are viviparous, the young, at birth, resembling the adults, except in size and colour. The just-born young of *Peripatus capensis* are usually about half an inch long, and of these Oakley has recorded the interesting fact that they are capable, almost from the first, of ejecting "the web of viscid filaments precisely in the manner of their parents." This the infants do, he says, when irritated, and when placed in spirit. The young of small mothers, presumably of the same species, were very tiny, only $\frac{1}{4}$ to $\frac{1}{3}$ inch long; yet these, when irritated, made discharges like the rest. Similarly, according to Ward, the young of *Peripatus moseleyi* are capable of shooting out slime at the earliest age at which it is possible to test them.

From what has been said as to the rapidity of the discharge, it is not surprising that there was at first some uncertainty as to its source. Guilding and Belt supposed it to come from the mouth, and Gosse from the antennae. It is almost impossible, as above noted, to observe the emission of the jets. According to Moseley, however, on close examination with a lens, especially in the case of large specimens, it can be seen that they are projected from the oral papillae; and there is, of course, no doubt on this point when reference is made to the anatomy of the animal.

With regard to the purpose of the discharge, it can hardly be doubted that it is mainly defensive. Moseley observed that when the creatures were pricked about the side or middle, they turned their heads round and aimed the discharge at the place at which the injury was being received; and he further adds, the tenacity of the threads formed by the fluid is so great, and their viscidity so remarkable, that the meshwork of them thrown over an insect or other such enemy would entangle it, and render it powerless for some time, even if it were of considerable size. All the observers above quoted, as we have stated, have remarked upon the viscosity of the ejected matter. Moseley, like Gosse. compares it to birdlime; "whilst I am writing," he continues, "several flies have walked

(17) Gordon, in Ward, *l.c.*

into some of the fluid which I caused a large *Peripatus* to discharge . . . . the flies are helplessly stuck fast; and I believe that the fluid is quite sticky enough to hold small birds, though it dries too rapidly to be used for that purpose." According to Pocock ([18]), a *Peripatus* has been seen, by this means, to overcome a small scorpion. Besides their use for defensive purposes, however, the discharges have also an offensive use. There are statements implying that the creatures habitually strike down small insects for food by shooting slime at them; but the evidence to this effect is not full, for little is known of the animals' feeding habits in a state of nature, and it is not evident that the creatures are constantly in need of capturing active living prey. Belt, a most conscientious naturalist, it is true, has remarked of his Nicaraguan *Peripatus* that it "had a singular method of securing its prey," *i.e.* by discharging viscid slime, with which "it can envelop and capture its prey, just as a fowler throws his net over a bird;" unfortunately, however, Belt does not mention the grounds on which he bases this statement, and one is not sure that he personally saw prey thus captured. Sedgwick, who kept *Peripatus capensis* in captivity, did not see prey taken in this way. Hutton, however, relates that a *Peripatus novae-zealandiae*, which he kept in a jar, shot out its viscid fluid at a fly which had been introduced into the vessel; by this means the fly was stuck down, and the *Peripatus* then went up and sucked out its juices. Steel, finally, who kept numerous specimens of *Peripatus leuckarti* in captivity for more than a year, has recently given attention to this point, the result of his observations being that while the animal does not always use its slime in securing its prey, it certainly does resort thereto when the insect it is endeavouring to secure appears likely to escape, or when it struggles violently, or again when the animal is hungry and wants to make certain of the capture. Under these circumstances, the *Peripatus* "becomes animated, raises the front part of its body, and ejects the viscid fluid from both papillae simultaneously."

(*To be continued.*)

AT THE INVITATION of the President of the Royal Meteorological Society, a meeting was held at the rooms of the Society on the afternoon of May 31st to consider the question of a memorial to the late George J. Symons, F.R.S., the distinguished meteorologist and founder of the British Rainfall Organisation. It was resolved unanimously that the memorial should take the form of a gold medal, to be awarded from time to time by the Council of the Royal Meteorological Society for distinguished work in connection with meteorological science. Contributions will be received by the assistant secretary, Mr. W. Marriott, 70 Victoria Street, Westminster.

(18) Pocock, "Royal Nat. Hist.," vi. (1896), p. 238.

BUTTERFLIES OF THE PALAEARCTIC REGION.

By Henry Charles Lang, M.D., M.R.C.S., L.R.C.P. Lond.

(Continued from page 11[1].)

13. **P. melete** Mén. Cat. Mus. Petr. Lep. ii. p. 113. 1857.

42—58 mm.

Wings white, no dusky shading at base.

F.w. with six black spots extending from apex along ou. marg., often coalescent. A triangular spot between second and third nervules, h.w. with a faint costal spot; extremities of nervules slightly marked black on ou. marg., ♀ much more strongly marked, similar to *P. napi* ♀. Neuration marked

Pieris kreilneri.

with black. U.s. f.w. as above, but with fainter markings. H.w. very pale yellowish-white without markings, the neuration only faintly dusky.

HAB. Various localities in the Amur, Corea, Japan. VII—VIII.

a. var. *veris* Stgr. The spring form. Has the neuration of the u.s. h.w. broadly marked with black (R. H.).

HAB. Amur (Ask. Wlad.) V.

14. **P. krueperi** Stgr. Wien. Ento. Mon. iv. 19. Lg. B.E. p. 29, pl. VI. 3.

44—47 mm.

Wings white, very slightly shaded at base. F.w. tipped with black and with a row of four or five lozenge-shaped spots along ou. marg.; internal to these is a triangular spot with its base on the costa; and below this, between veins 3 and 4, a somewhat reniform black spot. H.w. with basal and central shading as the result of the markings of the u.s. showing through; costa with a triangular black spot. U.s. f.w. white, slightly greenish-yellow at base and apex, costal spot of the same colour, the black spot as above. H.w. greenish-yellow, the outer third lighter in colour. The markings of ♀ are more intense than in ♂.

HAB. Mountains in Greece, Asia Minor, Trans-caucasia. VI.

(1) This series of articles on Butterflies of the Palaearctic Region commenced in SCIENCE-GOSSIP, No. 61, June 1899.

a. var. *vernalis* Stgr. Hor. 1870, p. 34. Larger and lighter than the type. U.s. h.w. marked with green toward base. Spring seasonal dimorphic form. IIIe—IV.

b. var. *prisca* Stgr. This is the eastern form of the species, occurring in Turkestan, Alai, N. Persia. It is larger than the type. U.s. h.w. much whiter and with greenish markings.

c. var. *verna* Grum. The spring form of var. *prisca*, corresponding to var. *vernalis*.

15. **P. canidia** Sparrm. Amoen. Acad. vii. p. 504. 1768.

45—53 mm.

Wings white. F.w. apex and upper half of ou. marg. marked with black. With two black spots. H.w. with a black spot on costa, and with a marginal row of four or five black spots. ♀ marked as in ♂, but all the spots are much more intense, and the bases are dusky.

HAB. Turkestan, Cent. Asia, North India, and China. V—VI.

a. var. *aestiva* Stgr. in litt. The summer form. Whiter, the spots not so intensely marked.

b. var. *palaearctica* Stgr. 35—45 mm. A small light form, in which all the markings are much

Pieris cheiranthi.

fainter than in type. H.w. without marginal black spots. Hardly at first sight to be distinguished from *P. rapae*. The shape of the apical markings, the less pointed outline of f.w., and the larger and rounder spots of the u.s. will be found to be sufficient points of separation. HAB. Turkestan. (? a smaller form of var. *aestiva*.)

16. **P. tadjika**, Gr.-Gr.

43—46 mm.

This species very much resembles the var.

palaearctica of *P. canidia*. It is also very closely allied to *P. rapae*. The ground colour is somewhat purer white than in *P. rapae*, and the spots blacker and more defined. The black shading at the apices of the f.w. is more extended than in *P. rapae* in a downward direction. The yellow tint on u.s. has a much greener tinge than in *P. rapae*. The ♂ has the f.w. marked as in *P. rapae* ♀. It may be merely a local race of *P. rapae*, but Grumgrshomail considers it distinct.

HAB. It was first taken by Grumgrshomail in the mountains of Darwaz and Karategin in South-Eastern Bokhara at 9,000 feet in June.

17. **P. ergane** H.G. 904–7 (1827). Lg. B.E., p. 30, pl. VI. fig. 5; *narcaea* Fen.

33–37 mm.

Wings white in the ♂, apex f.w. with a patch of greyish black much more square in shape than that seen in *P. rapae*. Sometimes one small spot below this, but generally not any. ♀ with two spots on f.w. and one on costa of h.w. U.s. in both sexes without spots, apices of f.w. and the h.w. pale yellowish dusted with grey towards base. It is very close to *P. rapae*, of which it may be a local race.

HAB. Dalmatia and Greece, IV. and VII. Bithynia, VI. and VII. Transcaucasia.

18. **P. rapae** Lin. Syst. Nat. x. 468 (1758). Lg. B.E. p. 30, pl. VI. fig. 4; XV. fig. 3 (larva). "The Small White."

39–50 mm.

Wings white. F.w. dusky at the tip, but not so dark as in *P. brassicae*, nor square as in *P. ergane*. ♂ with a black spot near centre of wing, ♀ with another spot in addition nearer to in. marg. Base generally shaded with black, more especially in ♀. H.w. rounded, white with a small dusky blackish costal spot. U.s. f.w. yellowish at apex and with two black spots in both sexes. H.w. pale ochre, more tinged with yellow than in *P. brassicae*.

HAB. The Palaearctic Region, excepting the Polar portion. Seasonal dimorphism exists in this, as well as in the other species of the genus. There is probably a succession of broods throughout the season, varying in number according to latitude and also altitude, so that in many parts of the region the dimorphism is not so apparent as in the more northern portion, as, for instance, in Britain.

Mr. H. Williams, of Southend, in Essex, England, who has paid much attention to this and the next species, has kindly provided me with the following note :—" Both sexes of the spring brood of *P. rapae* are very lightly marked in Essex, the central spot on the male upper forewing being practically obsolete, whereas in the second emergence both males and females are well marked on the upper sides, and the females are of a more pronounced ochreous tint, and the undersides of both sexes are of a deeper yellow than in the first brood." Mr.

Elwes, in Tr. Ent. Soc., 1899, iii. 316, says :—" In Germany, France, and Spain the difference (between the first and second brood) is rather greater (than in England), and in Germany, Poland, and occasionally in England, we have an aberration of the female which is distinctly yellow both on the upper and under sides. In Algeria, where I found it common in the Province of Constantine, and as far south as Biskra in April and May, and also at Gibraltar in April, some of the males are without a trace of the black spot in cell 4 of the forewing on the upper side. This is also the case in the one taken at Biisk (Altai)."

LARVA. Green, pubescent, with one dorsal and two lateral lines of yellow. On Cruciferae. Generally very common, and often destructive. PUPA. —Ashy speckled with black, often tinged with reddish.

a. var. *orientalis* Oberth., *mandschurica* Speyer. "The forewings in the ♀ speckled throughout their whole area with black. This variety seems to be an accidental form. HAB. Mongolia, Hakodadi, Japan." (R. & H., p. 124.) I have received a pair of *P. rapae* from Vladimar Bay (Amur), taken in August, 1897, by Mr. Lambert. The ♂ resembles that of typical *P. rapae* summer form, but is larger and more strongly marked. The ♀ answers somewhat to R. & H.'s description as above, but the black does not extend further than to the middle of f.w. The apical blotch and spots are very large. I take these specimens to represent the form *orientalis*, which appears to be a distinct local variety. R. & H., in their appendix, p. 713, re-describe *orientalis* so as to exactly correspond to my specimen. The only point of difference is in the size, which is given as smaller than typical *rapae*; this, however, may be an error. HAB.—Different localities in the Amur and Corea are given, and the date from mid July to end of August.

b. var. *similis* Kroulikowsky. "1st generation considerably larger than type. Forewings greyish at the apex, the dis. spots small, grey, often wanting in the male. The ♀ frequently pale yellowish, HAB. Kasan. April—May." R. & H. 124.

c. "var. *messanensis* Zell. Summer brood with very large black spots. The grey colouring of the apices of the wings shaded almost to black. Underside h.w. scantily powdered with black. HAB. Sicily." R. & H. 124.

d. var. *mannii* Mayer. Stett. Ent. Zeit. 1851. p. 151. (? Sp. prop. an *P. rapae* et *erganes* hibr. Staud. Cat., p. 3). The apical spot of the forewings deeply indented internally. H.w. beneath with the nervures shaded along their course as in *P. napi* var. *napaeae*—from which indeed it requires some care to separate. HAB. Tuscany, Parnassus, Turkey, Dalmatia, Podolia. IV., V.

e. var. *minor* Costa. "A smaller form. 37–39 mm. HAB. Tuscany." R. & H. 125.

f. var. *leucotera* Stefanelli. Bull. Ent. Soc. Ital.

I. 147. 1869. Apical spot of f.w. obsolete, or at most represented by a slight grey shading. Tunis. IV., V. (R. & H.) HAB. Italy (Kirby, Cat· 454).

g. var. *debilis* Alph. . Body more slender. The apex of forewings lighter grey, often disappearing. In ♀ the spot of the upper side small, in ♂ scarcely or not at all visible. HAB. Lobnoor (Mongolia), Amdo, N. Thibet. R. & H.

h. var. *kenteana* Stgr. The ♂ specimen I have was received from Dr. Staudinger, and does not differ very much from the type, except that it is somewhat larger, and the grey marking at the apex more extended. U.s. lighter and not speckled with black on h.w. HAB. Siberia, Kentei.

i. var. *immaculata* Fologne. The dark shading at the apex of the f.w. and of the base absent HAB. Belgium. R. & H., p. 125.

j. var. *novangliae* Scudd. Bull. Soc. Ent. Fr. (5). III., p. 57. 1873. A form of the species in which the ground colour of the wings is yellow instead of white. This is the N. American form of *P. rapae.* The species was introduced from Europe in 1856 or 1857 into· North America, where it has spread rapidly and has developed this yellow variety which is very rarely found as an aberrant form in other parts of its range. British specimens of the yellow form are occasionally to be found in collections, but are always considered great rarities.

NOTE.—*Delete* the var. *veris* placed under *P. deota, ante* p. 10.

(*To be continued.*)

BIRDS AT LYNMOUTH.

BY THOMAS H. MEAD-BRIGGS, M.A., F.E.S.

(*Continued from page 4.*)

Corvus monedula Lin. Jackdaw. Resident and common. In June 1898 a pair of these birds built their nest in a fir-tree on the hill above the Esplanade. With field-glasses they could be seen feeding their young. I have observed flocks of young birds in the autumn in the woods near Watersmeet, in East Lyn Valley. The facts of these birds building in a tree, and of the young assembling in autumn, are not, I think, common.

Corvus corax Lin. Raven. Resident; a few seen every year on the coast here, and occasionally they fly down on to the beach below our house. They nest in the neighbourhood, one pair having done so last year close to a high road.

Corvus corone Lin. Carrion-crow. Resident, but not very common.

Corvus frugilegus Lin. Rook. Resident and abundant.

Alauda arvensis Lin. Skylark. Resident and common.

Cypselus apus Lin. Swift. Common, but not plentiful, on arrival and just previous to departure during migration. I was recently told by Archdeacon Bree that he had seen a bevy of these birds at Ilfracombe on September 9th, 1899 ; and on August 3rd last year my brother and I saw about a hundred of these birds flying high in the air and circling over our garden. We did not see any after that date.

Caprimulgus europaeus Lin. Nightjar. One seen flying by my son (H. Mead-Briggs) round our house on May 25th, 1898, and its note heard. I am surprised that I have not seen this bird here more frequently.

Iynx torquilla Lin. Wryneck. I heard and saw this bird once, near Lynmouth Vicarage, on April 26th, 1897. I have never heard or seen it since, and it seems very rare all over the country.

Gecinus viridis Lin. Green Woodpecker. Resident and common.

Alcedo ispida Lin. Kingfisher. Resident, but rare. One seen (November 10th, 1898) flying about, and perched on the edge of the fountain pond in our garden. Another has been seen there since that date, and we were told that one had been shot there before we came here. I have never seen any of these birds up the valley of the East Lyn, where, considering the shallowness of the river, except when in flood, and the multitude of small trout, it ought to be common, especially as all wild birds are strictly preserved in the neighbourhood of Lynmouth.

Cuculus canorus Lin. Cuckoo. Common in the summer. It stays quite as late as in other parts of the country. Mr. E. B. Jeune told me on September 11th, 1898, that he had seen one about ten days before. It was not a young bird.

Buteo vulgaris Leach. Common Buzzard. Resident and fairly common. I have seen young birds in the autumn. The range of country in which I have observed the bird is from Lee Abbey on the west, to two miles beyond Countisbury on the east, and about a mile to the north of Barnstaple inland. Sometimes they may be seen flying in circles, and at others hawking along the top of the hills and cliffs. It is most interesting to watch this bird, especially as it is getting very scarce in most parts of Britain.

Accipiter nisus Lin. Sparrow-hawk. Resident and common. The great disparity in the sizes of the sexes, the male being little more than half the size of the female, has often led the former to have been mistaken for the merlin (*Falco aesalon*). I have on several occasions seen this bird strike at its quarry. It does not hover nearly so long as the kestrel (*Falco tinnunculus*).

Falco peregrinus Tunstall. Peregrine Falcon. Rare but resident. It occurs all along this part of the coast. On June 7, 1899, one near Lee Abbey, which is about a mile and a half from here, hovered over and then struck a rabbit on the ground, where it remained—I suppose, to feed upon its prey.

Falco aesalon Tunstall. Merlin. I saw a pair of these birds on March 25th, 1897, flying about the dwarf woods which are on each side of Lee Bay. They occasionally settled on the ground. I have not seen any since.

Falco tinnunculus Lin. Kestrel. Common and resident, the most frequent hawk of the district.

NOTE.—On March 22nd, 1898, I saw a large bird in the Valley of Rocks which I at first thought was a buzzard, but subsequently it appeared to be a species of *Circus*, one of the harriers. It settled on a low wall and a bush. Its head was pale buff, mantle mottled brown like that of a barn owl (*Strix flammea*), and pale whitish buff over the rump. When settled, it was about the size of a large hen pheasant. The beak was not that of a buzzard. I could not get nearer to it than about forty yards, when it flew slowly a short distance and then settled again. I saw this bird about 10 A.M., and it had not moved far from the same place on my return about 2.40 P.M. If it was a harrier, it could only have been *Circus aeruginosus*, the marsh harrier, a bird now extremely scarce not only in these parts, but in the whole of England.

Phalacrocorax carbo Lin. Cormorant. Common and resident, but not abundant. This bird has a habit of perching on a post, then extending its wings and keeping them in that position for some time. I have not seen here the crested or green cormorant (*P. graculus* Lin.). The latter bird is sometimes called the "scart," but I fancy this name is applied to either species in different localities.

Ardea cinerea Lin. Heron. I have only seen this bird on two or three occasions. There is no heronry, to my knowledge, within twenty miles from here.

Querquedula crecca Lin. Teal. One killed at Brendon, about four miles distant, in the winter of 1899-1900.

Columba palumbus Lin. Wood-pigeon. Very common and resident.

Columba livia Gmelin. Rock-pigeon. I have only seen this bird, as a certainty, upon two or three occasions. Their close resemblance to some varieties of our tame pigeons, which are, of course,

all descended from this species, makes a decision rather difficult; but the wild species never settles on a tree or hedge, and the tame ones very rarely.

Phasianus colchicus Lin. Pheasant. Common and resident.

Perdix cinerea Latham. Partridge. Common and resident.

Gallinula chloropus Lin. Moorhen. I saw one on November 11th, 1899, running about in the meadow just outside our garden.

Vanellus vulgaris Bechstein. Lapwing. Resident and common on the moors a little inland, but on February 11th, 1900, we saw several of this species in the meadow just outside our garden, and two actually came into it. They were evidently driven from the moorlands by the severity of the weather.

Haematopus ostralegus Lin. Oyster-catcher or Sea-pie. Not rare on the coast and resident. During last winter, 1899-1900, there has been a flock of these birds flying along the shore, and they are still here at the end of May.

Scolopax rusticula Lin. Woodcock. Fairly common, and I should think resident.

Gallinago coelestis Frenzel. Common Snipe. Resident. I have seen this bird on Exmoor in the summer and autumn, and on February 8th, 1900, one was flushed from under a fir-tree in our garden.

Tringa alpina Lin. Dunlin. On February 20th, 1900, we saw a pair of these birds flying over and settling on the rocks below the Esplanade here. I had not seen it previously.

Tringa subarquata Güldenstädt. Curlew-Sandpiper. On September 7th, 1898, we saw one of these birds on our lawn, and had a good view of it before it flew away. This bird, or another of the same species, was seen on the following day in another place, about a quarter of a mile distant.

Totanus hypoleucus Lin. Common Sandpiper. Occasionally seen on the beach or up the valleys of the rivers. It probably breeds here.

Numenius arquata Lin. Common Curlew. My brother told me that he heard the "whaup" of this bird on March 13th, 1898, whilst he was walking up the road that leads to the village of Countisbury. It is reported to breed on the moors inland.

Larus ridibundus Lin. Blackheaded Gull. On February 11th, 1900, there were two of these birds walking about in Mr. Jeune's meadow adjoining our garden, and they, or a few others, lingered on the coast for several days later. I only saw two, or the same bird twice, with the rudiments of the brown hood. These are common in South Devon; but I had not seen them here before, and they were probably driven to this place by storms or the severity of the weather.

Larus argentatus Gmelin. Herring-gull. Resident and abundant—"the gull" of Lynmouth. It breeds in the immediate neighbourhood.

Larus fuscus Lin. Lesser Black-backed Gull. I have not seen this bird often here, but a few every summer and winter. It is a resident, and breeds in the neighbourhood.

Larus marinus Lin. Great Black-backed Gull. I have only seen this bird here three or four times, both in summer and winter. It is stated to breed in the neighbourhood.

Rissa tridactyla Lin. Kittiwake. Resident and breeds here; but I only observed it this winter—1899-1900—when it has been in considerable numbers.

Uria troile Lin. Common Guillemot. I have only seen this bird once or twice.

Fratercula arctica Lin. Puffin. This bird ought to be common near Lynmouth, as the coast is most suitable for its breeding haunts, but we have only observed a single dead specimen washed up on the beach.

Colymbus septentrionalis Lin. Redthroated Diver. I have seen young birds of this species on several occasions.

Rock House, Lynmouth,
May, 1900.

GEOLOGY AROUND BARMOUTH.

By JOHN H. COOKE, F.G.S., F.L.S.

(*Concluded from page* 8.)

A CLOSE examination of the Glacial evidences around the Mawddach, point unmistakably to the conclusion that they are not due to the action of one great ice sheet, but that Cader, the Arrans,

the trend of the valleys—east, west, or south, as the case might be.

Among the many striae and groovings in Barmouth itself which will repay a visit are those

FIG 2.—ICE-GROOVINGS NEAR BARMOUTH CHURCH.

and each of the more prominent peaks of the district served as a centre from which were radiated in all directions a number of ice sheets whose directions were determined, to a great extent, by the valleys which they filled. As a rule the striations indicate a westerly movement of the ice; but local circumstances have in many cases modified this, and the direction of the glaciers conformed to

situated near the new church. They are huge semi-circular troughs, lying on a plane inclined at an angle of 30°, and measuring 31 feet by 15 inches and 38 feet by 8 inches. Their direction is north-west by west. One of them is terminated by a large pot-hole, and the other by the merging of two smaller grooves into the main one (fig. 2). At the top of the hill above Belle Vue is another fine

c 4

groove bearing north-west, and measuring 14 feet by 4 inches; and on the left-hand side, just below Sea View Terrace, is a series running parallel to one another and also bearing north-west. At the bend of the road opposite Tyn-y-ffyon Gardens is a fine striated surface, the striae of which culminate in a large groove which extends in a north-westerly direction for 15 feet and finally disappears beneath the soil. A boss of Cambrian grit, mammillated and striated on its eastern face and superficially contorted, lies at the western extremity of Porkington Terrace. Its striae bear north-north-east. The preceding are a few examples, but hundreds of others are to be found within a couple of miles of the town.

The morfa or marshland which borders the coast-line to the north of Barmouth, also affords an abundance of material bearing on the Glacial problems of the district. At Dyffryn, Pensarn, and Harlech it constitutes a broad tract of swampy country fringing the coast-line. It is made up of alternating beds of gravel and peat, through which dykes and drains have been cut to carry the storm waters of the mountain streams to the sea. At Barmouth the drainage is not so effective, so that the waters collect between the lateral moraine of the old Mawddach glacier and the foothills of Cell Fechan.

Between Barmouth and Llanaber the impounding moraine has been completely covered with sand, and an effective barrier is thus offered to the further advance of the waters of Cardigan Bay; but at Pensarn, a short distance to the northwards, the barrier has been breached and the low-lying land has been converted into a typical tidal marsh.

The sand-dunes at Barmouth are very much in evidence on a windy day, and the visitor, whether he will or no, usually becomes pretty intimately acquainted with them before he leaves the district. One need not go farther than the breakfast or dinner table to see that the dunes are made of clean, fine-grained, white quartz, sometimes stained considerably with iron. Around the mouth of the Mawddach estuary, and at several points along the shore-line to the north of Barmouth, it has been drifted in mounds and dunes of considerable size, which are often covered with a thick growth of coast barley (*Hordeum maritimum*), wild wheat (*Triticum junceum*), and marram-grass (*Arundo arenaria*). These accumulations are comparatively recent, and generally overlie the moraine and the old Forest Bed.

In March 1898 a violent storm raged along the coast, and off Llanaber the under-tow of the tide swept the sands away to a depth of 14 feet. A portion of the Forest Bed was laid bare, and the skeleton of a fine specimen of *Cervus elaphus* was exposed. Unfortunately, however, the rapid rise of the tide only allowed of the removal of the skull, antlers, a femur, and a few rib-bones. The remainder was washed away. The beams of the antlers were in part thickly encrusted with calcareous matter, and the tips of the tines were pyritized. The width of the antlers from tip to tip is 3 feet 1½ inches. The right ramus is 13 inches long, and it contains three molars; the teeth of the left ramus are missing.

The Forest Bed in which these interesting remains were found is frequently exposed along the shore-line at low tide, and numerous relics of the flora and mammalia of the Forest period have been found by the fisher-folk. Numerous pieces of bog-oak are drifted ashore every winter. Of the former extent of this forest-land little is known. The numerous exposures that have been made from time to time along the coast-line show that it once extended far out into the area now known as Cardigan Bay, and throughout the length of what is now the Welsh coast. It does not come within the scope of this paper to describe the many remains of early stone monuments with which this delightful locality abounds. Mystic circles, logan-stones, cromlechs and temples connected with the old Druidical rites are surprisingly numerous. The student of prehistoric Britain, whether he be geologist or antiquarian, will find no disappointment at Barmouth.

19 *Ravenswood Road,*
    Redland, Bristol.

SOUTH-EASTERN UNION OF SCIENTIFIC SOCIETIES.

THE Brighton meeting of the Union was held in the first week in June, and constituted its fifth annual Congress. It was successful and well attended. The President-Elect for the Congress was Mr. G. B. Howes, LL.D., F.R.S. At a reception given by the Corporation to the members, Professor Howes delivered his presidential address. He took for his subject the progress of the study of evolution since the time of Charles Darwin, and spoke on the chemical theory of life. He also referred to the "popularisation" of science, and the whole duty of man when scientific. He chided the south for being behind the north in local organisation of science teaching and higher education. The president also reminded his audience that their boasted civilisation was due to the application of science to daily life and domestic customs. Dr. Howes spoke seriously upon the present easy life followed by many youths and maidens, who preferred "sport" and luxury to self-education. He foretold a sad future for these young people if their course of conduct was not arrested in time. The address was well received by a large and influential gathering of members and visitors. One of the most interesting papers was by Mr. F. Chapman, A.L.S. on "The Brighton Raised Beaches and their Microscopical Contents." Mr. Tutt's motion for holding sectional meetings at these Congresses was not carried.

AN INTRODUCTION TO BRITISH SPIDERS.

By Frank Percy Smith.

(Continued from page 16.¹)

(Continued from page 16.¹)

Philodromus aureolus Clk.

Length. Male 5 mm., female 6 mm.

The cephalo-thorax is of a dull yellowish-brown colour, with a tinge of red. Along its central part is a more or less distinct bright yellow band, within which is a reddish-brown mark. The legs are long and of a pale yellow tint. The abdomen is pale yellowish-brown, with a series of reddish-brown bars towards the spinners. In some specimens the abdomen is very pale coloured. ·This species is common, and may be obtained in abundance by beating furze bushes over a piece of white paper.

Philodromus cespiticolens Wlk. (*P. cespiticolis* Bl.) This spider is so like *P. aureolus* Clk. that its specific distinctness has been questioned. The only observable difference between the two is the form of the radial apophysis of the male palpus. It is very common in similar situations to the last-mentioned species.

Philodromus praedatus Cambr.

This spider is very closely allied to both *P. aureolus* Clk. and *P. cespiticolens* Wlk., and can only be satisfactorily distinguished by the form of the radial apophysis. It is rare.

Philodromus constellatus Sim.

Length. Female 7.5 mm.

Cephalo-thorax dark yellowish-brown, with a pale narrow marginal line. A very rare species.

Philodromus emarginatus Schrank.

Length. Male 4.2 mm., female 4.5 mm.

Cephalo-thorax dark chocolate-brown, paler in front, with pale radiating lines on its sides. Very rare.

Philodromus lineatipes Cambr.

Length. Male 4.5 mm., female 4.8 mm.

Similar to *P. emarginatus* Schrank, but a little larger ; the radial apophysis is also different in form.

Philodromus elegans Bl.

Length. Female 6.3 mm.

This spider bears a strong resemblance to *P. aureolus* Clk., but it is much more distinctly and brilliantly marked. It is rare.

Philodromus fallax Sund.

Length. Female 6 mm.

This species may be easily distinguished by its pale, faded appearance. It is rare, and should be looked for on sandy ground.

(1) This series of articles on British Spiders commenced in Science-Gossip, No. 67, December 1899.

Philodromus clarkii Bl.

Length. Male 3.6 mm.

Cephalo-thorax and legs reddish-brown, with darker spots. Abdomen similar in colour, paler at the sides. A rare and very distinct species.

Philodromus variatus Bl.

Length. Female 6.3 mm.

Cephalo-thorax yellowish-brown, with darker lateral bands. Legs pale reddish-brown. Abdomen yellowish-brown, marked with darker brown, and tinged with dull green on the underside. A rare spider hitherto found in North Wales.

Philodromus mistus Bl.

Length. Male 5.5 mm., female 6 mm.

The hairs on this spider reflect beautiful iridescent tints in a strong light.

GENUS *OXYPTILA* SIM.

Eyes in two curved rows. The distance between the hind centrals is less than that between one of them and the lateral next to it. The four central eyes form a quadrangle whose length is greater than its breadth. The spiders included in this genus are usually found among the roots of low plants or under stones.

Oxyptila praticola C. L. Koch. (*Thomisus incertus* Bl.)

Length. Male 3.5 mm., female 4 mm.

Cephalo-thorax pale reddish-brown, with a pale central band. Legs brownish-yellow, mottled with reddish-brown and white. Abdomen yellowish-brown, marked with a large number of reddish-brown and very dark brown spots. This spider is found among low plants and is rather rare, but well distributed.

Oxyptila blackwallii Sim. (*Thomisus claveatus* Bl.)

Length. Male and female 3.5 mm.

This spider may be distinguished from its allies by the possession of a number of short bristles, clubbed at their extremities, on the front part of the caput and on the abdomen. It is very local.

Oxyptila sanctuaria Cambr.

Length. Male 2.5 mm.

This species is allied to *O. blackwallii* Sim., but may be distinguished by its smaller size and by the clubbed bristles being much less clavate. It is rare.

Oxyptila trux Bl.　(*Thomisus trux* Bl.)
Length.　Male 3.5 mm., female 4 mm.
Cephalo-thorax yellowish-red with four very dark brown bands.　Legs yellowish-brown without spots.. This is a rare spider.

Oxyptila atomaria Panz.　(*Thomisus pallidus + T. versutus* Bl.)
Length.　Male 4 mm., female 5 mm.
In this species the colours are not so brilliant as in *O. trux* Bl.　The legs are of a reddish tint.　The colours are very subject to variation, var. *pallidus* Bl. being a very pale variety.　It is not a common spider, but is widely distributed.

Oxyptila flexa Cambr.
A very rare spider, first discovered in 1894.

Oxyptila simplex Cambr.
Length.　Male 3.5 mm. female 3.8 mm.
This spider is nearly allied to *O. trux* Bl., but may be distinguished by its plainer colours and also (in the male) by the form of the radial apophysis.　It is a rare and local species.　I took a single specimen at Hastings in April 1900..

GENUS *XYSTICUS* C. L. KOCH.

In this genus the hinder row of eyes is less curved than in *Oxyptila*.　The hind centrals are never nearer together than each is to the lateral eye next to it. The four central eyes form a quadrilateral figure, which is never longer than broad.　The spiders included in this genus are found in varied situations, both on the ground and upon trees and shrubs.

Xysticus cristatus Clk.　(*Thomisus cristatus* Bl.)
Length.　Male 5 mm., female 7.5 mm.
This species is by far the commonest of the genus, and may be found during the greater part of the year among grass, heather, and low plants of almost any kind.

Xysticus viaticus C. L. Koch.
Length.　Male 5 mm., female 6.5 mm.
This spider is very similar to *X. cristatus* Clk., but among other differences it has a number of bristles on the cephalo-thorax and abdomen which are far stronger and more prominent than in that species. It has a wide range in Britain, but is rare.

Xysticus pini Hahn.　(*Thomisus audax* Bl.)
Length.　Male 4 mm., female 5.5 mm.
This species may be distinguished from *X. cristatus* Clk., both by its smaller size and also by the intensity of its colours, the ground colour often being almost white and the darker parts, especially of the male, being nearly black.　It is not very common, but is well distributed, being usually found on furze-bushes. The colours are subject to a good deal of variation, but the varieties are fairly distinct.

Xysticus lanio C. L. Koch.
Length.　Male 5 mm., female 6.5 mm.
This spider may be easily distinguished from the foregoing by its reddish colouring.　It is not rare, and is usually found near oak-trees.

Xysticus sabulosus Hahn.　(*Thomisus rufo-pictus* Cambr.)
Length.　Male 5 mm., female 6 mm.
The colours of this spider are subject to great variation and are almost indescribable.　It may often be found on the bare ground, but its varied and mottled colours afford it great protection, as it is almost impossible to detect it until it moves.　A very striking variety, with the fore part of the abdomen of a bright reddish tint, is known as var. *rufopictus* Cambr.

Xysticus cambridgii Bl.
Length.　Female 8 mm.
Cephalo-thorax yellowish-brown with pale markings and a central and marginal bands of a yellowish-white.　Legs short, strong.　An extremely rare spider.

Xysticus luctuosus Bl.　(*Thomisus luctuosus* Bl.)　Similar in size and form to *X. lanio* C. L. Koch, but the red markings are replaced by deep brown.　The palpal organs are very simple.　A rather rare spider.

Xysticus lynceus Latr.　(*Thomisus atomarius* Bl.)
Length.　Female 5 mm.
Cephalo-thorax brownish-yellow, with two dark brown longitudinal lines on each side.　Legs, palpi, and abdomen freckled with numerous dark brown spots.　A rare spider.

Xysticus robustus Hahn.　(*Xysticus morio* C. L. Koch.)
Length.　Male 6 mm.
This spider is of a short robust form, and of a deep blackish-brown colour.　It is extremely rare.

Xysticus luctator L. Koch.
Length.　Male 6 mm.
An extremely rare species, allied to *X. bifasciatus* C. L. Koch, but more richly and brilliantly coloured.

Xysticus bifasciatus C. L. Koch.
Length.　Male 6.5 mm., female 8.5 mm.
This spider resembles in form *X. robustus* Hahn., but its colouring is not so dark.　The palpal organs and radial joint also differ greatly from that species.

Xysticus ulmi Hahn.
Length.　Male 4 mm., female 5 mm.
Closely allied to *X. erraticus* Bl., but the abdomen of the male is of a more elongated form, and the digital joint of the palpus is smaller.

Xysticus erraticus Bl.
Length.　Male, 4.5 mm., female 6.5 mm.
Cephalo-thorax pale reddish-brown with a pale

band bordered with a deep red-brown stripe. The female is much paler than the male. This species is rare.

GENUS *DIAEA* THOR.

In this genus the two rows of eyes are nearly concentric, the centrals forming a quadrilateral figure

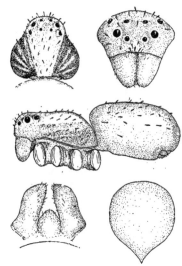

FIG. XIII. CHARACTERISTICS OF GENUS OXYPTILA.

Cephalo-thorax viewed from above ; eyes and falces seen from in front ; side view of spider, legs and palpi truncated ; maxillae and labium ; and sternum.

whose length is greater than its|breadth. The first and second pairs of legs are very long, especially in the male.

Diaea dorsata Fabr. (*Thomisus floricolens* Bl.)
Length. Male 4.5 mm., female 6 mm.
Cephalo-thorax of the female yellowish-green or occasionally bright green. The legs of the female are yellowish-green, but in the male are reddish-brown. The cephalo-thorax of the male is reddish-brown with two dark stripes. This species is not uncommon.

Diaea devoniensis Cambr.
Cephalo-thorax and legs of a brownish-yellow colour. An extremely rare spider.

GENUS *MISUMENA* LATR.

This genus may be distinguished from *Diaea* by the smaller size of the fore-lateral eyes, and the stronger curve of the anterior row. The eyes of the hinder row are equidistant.

Misumena vatia Clk. (*Thomisus citreus* Bl.)
Length. Male 4 mm., female 9 mm.
The central band on the cephalo-thorax is yellowish-white, edged with dull green, in the female, and greenish-white, tinged with reddish-brown, in the male. This species is rather common.

Misumena truncata Pall.
Length of female 7.5 mm.
Similar in many respects to *M. vatia* Clk., but darker. The tubercles on the caput have also a somewhat different form. This rare species has been found in the neighbourhood of London.

GENUS *THOMISUS* WLK.

In this genus the eyes are all very small. The caput has a conical prominence on each side. The distance between the hind-central eyes is greater than that between one of them and the adjacent lateral eye.

Thomisus onustus Wlk. (*T. abbreviatus* Bl.)
Length. Male 4 mm., female 8.5 mm.
The general colour of this spider is yellow, or yellowish-brown, but immature specimens are often found of a pink tint. This pink coloration is an excellent protection for the spider, which usually lurks

Xysticus cristatus (natural size).

in the flowers of the heather. The species is not common, but seems generally distributed over the Southern counties.

It will be noticed that in many cases, as in the last and the next, the dissimilarity between two adjacent families is very great. It should be remembered, however, that exotic groups often exist which help to bridge over the apparent gap. The systematic position of the Thomisidae has been a matter of some discussion, as this family appears to possess affinities in common with several widely separated groups.

(*To be continued.*)

BRITISH FRESHWATER MITES.

By CHARLES D. SOAR, F.R.M.S.

(Continued from page 19.)

GENUS *PIONA*.

THE characteristics of this genus are : body soft-skinned ; claws to all feet ; fourth pair of legs much modified in the male ; three discs let into each plate on each side of the genital fissure, and a small peg on the fourth segment of the palpi partly projecting over the fifth. This peg is on the inner edge of each palpus. The females of some species of this genus appear to be fairly common, but the males are very rare. I have myself taken a number of females whilst collecting, but have never met with any males. Piersig describes five species for Germany, but at present I shall only be able to give two for Britain, I believe I have the females of two others, but until the males are found they must go unrecorded.

1. *Piona ornata* Koch.

BODY.—Female oval in shape, length about 1.40 mm., breadth about 1.14 mm. Colour a bright red, with dark brown markings ; a yellow or sometimes pale-red patch occurs in the centre of the dorsal surface.

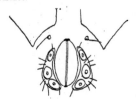

FIG. 1. *Piona ornata.* Genital area of female.

LEGS.—First pair about 1.40 mm. Fourth pair about 2.40 mm., with claws to all feet. The claws on the fourth pair are smaller than the others. The legs are like those we find on the females in the genus *Curvipes*. In colour a pale blue, becoming yellow towards the feet.

EPIMERA.—In four groups, the last pair being much pointed posteriorly (fig. 1). Colour a pale

FIG. 2. *P. ornata.* Peg on palpi.

slaty-blue, as are all the chitinous parts of this mite.

PALPI.—About 0.48 mm. in length, with a peg on the inner edge on the fourth joint (fig. 2).

GENITAL AREA.—Composed of two plates, with three discs on each plate (fig. 1).

MALE.—Smaller than female, being about 0.88 mm. long and 0.66 mm. in breadth. The epimera in three groups (fig. 3), not in four like the female,

FIG. 3. *P. ornata.* Epimera of male.

covering a great part of the body. The third pair of legs are the shortest, and have modified tarsi. The best point of difference between the sexes, and the best point for the recognition of species, is the fourth leg (fig. 4), which is very peculiar with its fourth thick segment.

FIG. 4. *P. ornata.* Fourth leg of male.

LOCALITIES.—Not very common. The first female found in Britain was in Lincolnshire, and sent to me by Dr. George in May 1895, since which I have myself found about six females in different places, at Sunningdale and in Suffolk. The male of this species was only found in the spring of 1900 in this country, a single specimen having been sent to me in May of this year by Dr. George, through whose courtesy I now record it for the first time in Britain.

2. *Piona latipes* Müller.

BODY.—Female very like the female of the preceding mite. About 1 mm. in length and about 0.80 in breadth. It is of a bright red colour, with dark markings, slightly lighter in tone on the front and posterior margins. There is a very light T-shaped patch in the centre. The whole colour is red, including legs and all chitinous parts as well as the softer parts. Male.—This is also of a bright red colour, and as usual is rather smaller in size. The great point for identification is the fourth pair of legs, which are constructed in a peculiar manner (fig. 5).

LOCALITIES.—Females of this mite I have found

FIG. 5. *Piona latipes.* Fourth leg of male.

in two places—Peplow in Shropshire and in Epping Forest, Essex. Only two males have yet been found in Britain, both by Dr. George—one in May 1884, at Ditchington, Norfolk; and one in April 1900, in Lincolnshire. It is from this last specimen the figure here shown of the leg has been drawn.

(*To be continued.*)

PROFESSOR PETRIE'S SCHEME.

LAST month we shortly referred to Professor Flinders Petrie's plan for a National Repository for Science and Art. This he propounded at a meeting of the Society of Arts on May 16th. To say that it has the impress of magnitude is the least attribute, for as a whole it is conceived with boldness and ingenuity, however much we may differ from some of the details. That such a plan is wanted no one doubts, and in creating a basis for discussion Dr. Flinders Petrie has been most successful.

The general idea is that the present plan and accommodation provided by our museums are utterly insufficient and narrow in conception. Even the national collections will in a few years quite outgrow their present development at such a pace that the Treasury must apply a break on expenditure on buildings, and then as educational institutions they will fall into arrear. Again, with a number of competing small local establishments, the "specimens" become so scattered that it is beyond the means, in time and money, of the ordinary student to wander all over the kingdom to hunt for examples of his especial study, so as to obtain a series for comparison; even then they cannot be brought into juxtaposition, and the scholar must largely depend on memory. As Professor Petrie said in his lecture, with the expansion of Western civilisation whole races will disappear before its march, or will adapt themselves to their new conditions, and so we shall lose their characteristics for ever. The same influences cause the annihilation of numbers of species of both the wild fauna and flora of many

districts. Examples from these sources, though only a very small part of all that should be preserved for ready examination by posterity, are being crowded out, and so lost for ever by the limited space at the disposal of directors and curators of our present museums. It is useless to dilate upon this fact, for it is every day in some form or other apparent. This is the beginning of an age of serious exploration of the buried evidences of the past, which will extend immensely as the whole population gets more liberally educated. Take those countries under the influence of Britain, and they could supply more material from that source alone than our present arrangements could house, to the exclusion of all other subjects.

Professor Petrie's proposal is to meet the future in a manner that would provide ample room for all subjects, whether geological, applying to living inhabitants of the earth, to man's ethnographic history, his art progress, his present civilisation, and, indeed, everything of educational value of the past, the present, or the future; whether scientific, artistic, or utilitarian.

For this purpose Professor Petrie proposes that the nation should acquire at least a square mile of land in a dry, healthy situation, within an hour's travel of London, and there create the new town of Sloane, so-called in honour of the founder of the British Museum. The fringe of this estate he would let for building on ground-rent leases for residential purposes, so as to create an income towards the fund for maintenance of the central buildings, especially with regard to renewals and repairs. Thus he would surround the scientific centre with houses that might be occupied by people of culture, who would naturally centralise around the National Collection.

Dr. Petrie proposes that the plan of the museum buildings should be simple as possible, with toplights only, and about 54 feet wide. That they should be inexpensively constructed of iron, brick, and concrete, without any wood-work, as lessening the chances of destruction by fire, at a cost of 200*l.* for each 16 feet. He proposes that these buildings shall be arranged in "gridiron" pattern, with broad spaces between them. When the whole site has been occupied by the first set of buildings, there would be some eight miles of galleries, to which extent they would grow by annual additions of about 500 feet.

When the time came for further expansion a new set of buildings would be commenced in the centre of the spaces, and again between these; so that his square mile would provide comparatively inexpensive expansion for three or more centuries to come.

While unoccupied by buildings the space is to be used for the cultivation of timber-trees of suitable species to the land covering the site. These, of course, would become a source of revenue, and form an ornamental park in the meantime.   J. T. C.

NEW PHYSICAL APPARATUS.

BY JAMES QUICK.

TO comply with the continually increasing demand, chiefly from Secondary and Organised Science Schools, for simple pieces of apparatus to illustrate the fundamental laws of physics, various useful forms have been constructed.

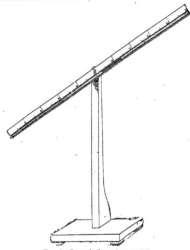

FIG. 1.—BOYLE'S LAW APPARATUS.

The following, designed by Mr. W. Rheam, B.Sc., Senior Physics Master, Liverpool Institute, have been found very serviceable in elementary practical work in physics. Although they are of simple construction, they nevertheless are capable of giving good quantitative results, which are now so essential in nearly all physical laboratory work.

sponding decrease in the volume. This relation is set forth by Boyle's law, which states that " the pressure of a given mass of gas at a constant temperature is inversely proportional to its volume." Fig. 1 shows a simple arrangement for exhibiting the law. It consists of a long glass tube of small bore, closed at one end and fixed to a graduated scale of boxwood, which is supported upon a firm wooden base, and which can be rotated about a central horizontal axis. A thread of mercury is introduced into the tube by means of a pipette. When the tube, and therefore the mercury thread, is in a horizontal position, the air contained will only be subjected to atmospheric pressure. As the tube is rotated towards a vertical position the mercury will exert an increasing pressure upon the air, and the value will be obtained by measuring the vertical distance between the two ends of the thread. This pressure added to that due to the atmosphere will be the total pressure upon the air. The volumes of the contained air corresponding with these different pressures are also noted during an experiment. The product of the two corresponding values will be found to be a constant quantity, or $P V = $ constant, thus proving Boyle's law.

RELATION BETWEEN TEMPERATURE AND PRESSURE OF A GAS MAINTAINED AT CONSTANT VOLUME.

Another law closely related to that of Boyle is the one known as Charles' law, which states that " the volume of a given mass of any gas, at constant pressure, increases for each rise of temperature of $1°$ C. by a constant fraction (nearly $\frac{1}{273}$) of its volume at $0°$ C." By combining both laws we arrive at the statement that if the volume of a gas remains constant, but the temperature and pressure vary, then the pressure is proportional to the temperature when the latter is reckoned from the absolute zero, *i.e.* $-273°$ C. In other words, "the

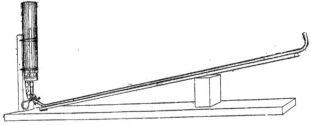

FIG. 2.—APPARATUS TO SHOW THE RELATION BETWEEN P AND T ; V BEING CONSTANT.

SIMPLE FORM OF APPARATUS FOR PROVING BOYLE'S LAW.

If any gas is kept at a constant temperature and the pressure upon that gas is changed, its volume will also be changed in such a manner that any increase in the pressure will result in a corre-

pressure of a gas at constant volume increases by a constant fraction ($\frac{1}{273}$) of the pressure at the freezing point for each rise of temperature of $1°$ C. Fig. 2 depicts an arrangement for showing this relation.

The gas to be experimented on is confined in a vertical glass tube about 10 centimetres long, and is completely surrounded by a water-jacket, by means of which its temperature can be varied as desired. A tube containing mercury is connected to the above tube by means of an india-rubber joint, and is supported on a narrow board hinged to the stand. The pressure upon the gas is regulated by varying the inclination of the mercury tube, which is done by means of an adjustable wedge. A convenient mark of reference on the tube containing the gas is afforded by the top surface of the cork fixing the water-jacket, one level of the mercury being always brought to that position in an experiment, in order to keep the gas at constant volume.

FIG. 3.—SIMULTANEOUS ROTATION APPARATUS.

The different temperatures to which the gas is brought are ascertained by means of a thermometer placed in the water-jacket, while the pressures upon the gas due to the mercury are measured by the vertical differences in height of the two ends of the column. To obtain the total pressure on the gas the atmospheric pressure must be added in each case. A most useful adjunct to both this apparatus and that shown in fig. 1 is a point cathetometer for measuring differences in height. This instrument has already been described and illustrated in the Physics column of SCIENCE-GOSSIP for February 1900.

of silk thread, having tied to its other end a magnetised steel needle. Mercury is poured into the funnel until the free pole of the needle just floats above the surface. To the base of the instrument is fixed a vertical brass rod bent at right angles, and having suspended from the end of its horizontal limb a vertical wire almost overhanging the free pole of the needle, and making contact with the mercury. Rotation of the two wires round one another is set up when a battery current is sent in at one terminal upon the base, through the suspended wire, then through the mercury, and out at the second terminal.

ABSOLUTE EXPANSION OF SOLIDS.

Until comparatively recently there was a dearth of simple pieces of apparatus for the determination in absolute measure of the expansion of solids. I have given a full description in SCIENCE-GOSSIP (December, 1898, pp. 197–8) of a most useful and efficient instrument for this purpose, designed by Mr. F. C. Weedon. Simpler arrangements, but upon the same principle, have been devised, some depending upon the use of micrometer screw gauges for the determination of the differences in length, others upon the readings of spherometers. The form illustrated in fig. 5 utilises a wedge for obtaining the required lengths.

The rod, the coefficient of expansion of which is to be determined, is supported horizontally, and is enclosed, with a thermometer tied to it, in a glass jacket through which steam can be passed from a boiler. One end of the rod presses against the surface of a small fixed glass disc embedded in a rigid vertical support. The other end of the rod is free, and the expansion of the rod for different ranges of temperature is measured, at this free end, by means of a properly graduated right-angled wedge faced with glass, and which slides up and down against a firm vertical support

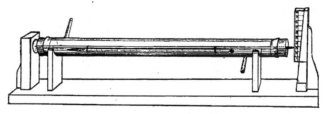

FIG. 4.—CO-EFFICIENT OF EXPANSION APPARATUS.

THE SIMULTANEOUS ROTATION OF A CURRENT-CARRYING WIRE ROUND A MAGNETIC POLE, AND OF THE MAGNETIC POLE ROUND THE CURRENT-CARRYING WIRE.

In the narrow end of an inverted glass funnel (fig. 3) is fixed an india-rubber plug carrying upwards a small hook to which is attached a short bit

attached to the base. The initial length and temperature of the rod are first ascertained. The temperature is then gradually raised, and at several intermediate steps the lengths are observed corresponding to definite temperatures. The measurements with this apparatus are most simple, only two readings being simultaneously required, viz. the thermometer and the wedge readings.

Among the Birds in Northern Shires. By Charles Dixon. x + 303 pp., 9 in. × 6 in., with coloured frontispiece and 43 other illustrations. (London, Glasgow and Dublin : Blackie & Son.) 7s. 6d.

The author is a popular writer on bird-lore, several of his books having been noticed in these columns, The present volume is one that is sure to attain favour among the lovers of wild birds visiting or living in the north of England and Scotland. For twenty years Mr. Dixon has lived among the different kinds of birds depicted in the book before us. He makes no attempt at scientific description, but has pleasantly written half a score chapters on the habits and customs of the feathered denizens of the wild country he loves so well. One finds many curious anecdotes and incidents connected with this subject, and there is not any chapter without interest. The illustrations are luxurious, and many most artistic, as they are drawn by Mr. Charles Whymper. By permission of the publishers we reproduce the one showing gannets and young. We can strongly recommend the book to our readers, as it is beautifully printed on excellent paper, and fully maintains the high reputation of the publishers.

Nature in Downland. By W. H. HUDSON. xii + 307 pp., 9 in. × 6 in., with 12 plates and 14 other illustrations. (London, New York, and Bombay : Longmans, Green & Co.) 10s. 6d. net.

We are glad to find a new writer upon the comparatively little understood beauties of the South Downs of England. This range of undulating chalk hills extends from near Eastbourne westward along the coast away to Portsmouth. They possess at all times of the year a peculiar beauty of fleeting lights and shades that cannot be excelled elsewhere in Britain. We have often wondered that they should be so little known by the general rambler in search of rural beauty. In treating this subject Mr. Hudson has not attempted scientific description of any kind, but devotes his attention to country lore. In this department of literature the author is well known as a successful writer. His references in the book before us to the flora and fauna are of quite a popular character ; in fact, he states on p. 63 his reason for not giving scientific names. We agree with him that in such a work as this to besprinkle its pages with italicised Greek and Latin names in parentheses would be quite unnecessary, and certainly a disfigurement to an otherwise elegant production. Still, we get a good deal of natural history scattered through the work. As a whole, we gladly recommend "Nature in Downland" to both naturalist and general reader. It is liberally illustrated, and many of the drawings are successful, though others are too far leaning towards the impressionist school to be acceptable to those who favour greater exactitude in pictures of places and scenes.

The Bacteriology of Everyday Practice. By J. O. SYMES, M.D., D.P.H. viii + 88 pp., 7⅓ in. × 5 in. (London: Baillière, Tindall & Cox, 1900.) 2s. 6d.

The science of bacteriology is advancing by leaps and bounds, and we may now expect a series of "elementary" text-books on the subject, such as we are already familiar with in the cases of chemistry, physics, and other sciences. The present work is one of the first in the field, and we have no doubt that it will prove useful to the busy medical practitioner who has little time to become acquainted with the later developments of the science, although desirous of keeping himself abreast of the times. The chapters are arranged under the headings of Materials and Instruments, Preparation and Staining of Films, General Infections of the Blood, Suppurative Processes, Diseases of the Respiratory System, Enteric Fever, &c., Serum Therapeutics, and the Use of the Microscope in Bacteriology.

First Stage of Hygiene. By A. LYSTER, B.Sc. viii + 199 pp., 7 in. × 5 in., with 100 illustrations. (London: W. B. Clive. 1900.) 2s.

This is a volume in "The Organised Science Series," and is prepared for those who intend presenting themselves for examination for the elementary stage of the Science and Art Department. The general reader who has not a very perfect knowledge of the human body and its care will find in its pages many useful facts for guidance.

The Sun-Children's Budget. Edited by PHŒBE ALLEN and Dr. HENRY W. GODFREY. Vol. ii., 200 pp., with coloured plate and other illustrations. (London: Wells Gardner, Darton & Co. 1900.) 3s.

The bound parts of this children's periodical make a pleasing volume as a gift-book for young people. It is brightly conducted and pleasantly written in a healthy style. The elementary papers on scientific subjects ought to make their readers think for themselves, and found a taste in them for such things.

Field Columbian Museum. Annual Report 1898-99. 74 pp., 10 in. × 6½ in., with illustrations. (Chicago, 1899.)

When we turn the pages of such a report as this, one sighs and feels ashamed of many that are issued in this country. Its production is simply superb, and bears evidence of the care that has been expended in every department, whether literary, photographic, or typographical. It would be well if those who have to issue reports in England were to examine one of these American publications, and pause. The reproductions of photographs of some of the Museum cases are beautifully printed.

A Guide to Zermatt and *A Guide to Chamonix.* By EDWARD WHYMPER. 224 pp. and 206 pp. respectively, 7⅓ in. × 5 in., with many illustrations. (London: John Murray. 1900.) Each 3s. net.

We again receive these guides, respectively the fourth and fifth editions, with the information brought up to the present year. As models of local guide-books for Alpine visits they are admirable. We can also recommend them for general reading, as their perusal will give anyone a sufficient knowledge of mountain climbing to take an intelligent interest in the subject. For the reader who likes adventure, Mr. Whymper provides the most exciting incidents and stories, enough to turn one's hair gray. They have the further merit of being all true.

White Cattle. By R. HEDGER WALLACE. 237 pp., 8½ in. × 5½ in. With 4 plates and 32 other illustrations. (Glasgow: Natural History Society, 1900.)

This is a paper read before the Natural History Society of Glasgow and reprinted from the Transactions. It is an exhaustive review of what is known of the supposed original bovine inhabitants of Britain and their descendants to the present time. As the author says, there are many points of view from which the subject can be studied : the physiological, osteological, and archaeological. It

those who have not yet met with this very handy little text-book we recommend it, as it is concise, but the contents are sufficiently full and trustworthy.

North Staffordshire Field Club. Report for 1899–1900. 165 pp., 8½ in. × 5½ in., with 5 plates. (Stafford : J. & C. Mort, 1900.)

The Council of this Society is to be congratulated upon the very successful report which they have just issued. It is businesslike, and of more than local interest. Among the papers is one upon

GANNETS AND YOUNG.
From Dixon's "Birds in Northern Shires."

is an interesting topic, and one that well repays research. The author has done well as far as he has gone, and his book will be useful for reference. It is fully illustrated.

Mineralogy. By FRANK RUTLEY, F.G.S. Twelfth Edition, x + 240 pp., 7 in. × 4½ in., with 117 figures. (London : Thomas Murby, 1900.) 2s.

We are pleased to again meet with this, perhaps the best of the less expensive handbooks on elementary mineralogy. The new edition has been revised and corrected, thus bringing it in line with present knowledge of the science. To

the Labyrinthodonts from the North Staffordshire coalfield, by Mr. John Ward, F.G.S., which is illustrated by two effective plates. There is also a paper by Mr. John R. Masefield, M.A., on Staffordshire Helices, indigenous or introduced.

Electric Batteries. By PERCIVAL MARSHALL, A.I.Mech.E. 63 pp., 7 in. × 5 in., with 34 illustrations. (London : Dawbarn & Ward, Ltd., 1900.) 6d. net.

The author puts into a few pages much information on electric batteries, and how to make and use them. It is intended as a practical handbook for amateur electricians.

Elementary Practical Physics. By Professor HENRY STROUD, M.A., D.Sc., xi + 281 pp., 7½ in. × 5 in., with 115 diagrams. (London: Methuen & Co. 1899.) 3s. 6d.

This book forms one of the series of text-books on Technology edited by Dr. W. Garnett and Professor Wertheimer, of which some seven or eight have now been published. Professor Stroud introduces it as a book "intended to form an introduction to practical work in a Physical Laboratory." In reading through its pages one finds it somewhat difficult to decide for what class of students the work is intended. In some instances too much knowledge is assumed on the part of the reader, if the book is to be adapted to the requirements of boys in an organised science school. The general matter is well arranged and in good sequence ; the chapters on specific and latent heats, and on Ohm's law, being very concisely written. In a few cases more care might perhaps have been bestowed on details. For example, on p. 43, in the experiment for obtaining the density of a solid lighter than water, allowance should be made for the fastening thread; on p. 61 we are not told for what particular surfaces is obtained the mean value of ·260 for the coefficient of friction. On p. 228 a simpler form than the Post Office pattern resistance-box should be illustrated in describing a simple one for ordinary purposes. If the beginner is not told of a simple one he naturally would always ask for a P.O. pattern, even in experiments on resistance by substitution. Special mention is made on p. 239 of the Deprez mirror galvanometer ; but the principal feature of such an instrument, viz. that its readings are quite unaffected by external fields of force, is entirely ignored. Nevertheless the book will prove a helpful one in higher schools and in colleges, especially if the teacher can give the student continued help.— *J. Q.*

Photography in Colours. By R. CHILD BAYLEY, F.R.P.S. 74 pp. 7 in. × 5 in., with diagrams. (London: Iliffe, Sons & Sturmey, Limited. 1900.) 1s.

This little book forms over seventy pages of good spare-time reading. Written in a refreshing manner, the subject is put forward without any technicalities, and moreover without necessitating the reader being a photographer. The way is paved with a good introduction upon the nature of colour, and upon the wave theory of light. The various colour processes are then described—viz. Lippmann's interference method, the two three-colour processes of Ives and Joly, and last, but not the least interesting, the diffraction method of Professor R. W. Wood, recently described by him before the Society of Arts and the Physical Society. The author reminds the reader at intervals that the booklet treats of photography in colours, and that photography in natural colours has not only never been, but perhaps never will be, accomplished.— *J. Q.*

Elementary Lessons in Electricity and Magnetism. By SILVANUS P. THOMPSON, D.Sc., B.A., F.R.S. ix + 626 pp., 6½ in. × 4½ in., with 291 illustrations. (London and New York: Macmillans. 1900.) 4s. 6d.

As this book is only another reprint of the 1895 edition, we are unable to call attention to any new matter. Considering that it has been printed upwards of twenty times since its first appearance in 1881, and that it is still held as a standard elementary text-book, too much cannot be said upon its general excellence.

FROG-ORCHIS IN SUSSEX.—I send you herewith a specimen of *Habenaria viridis*, which I gathered at the foot of Beachy Head, near Eastbourne, on the 16th of June this year.—*(Miss) E. Bray, Fairlight House, Hailsham, Sussex.*

PALE-COLOURED BEE-ORCHIS. — I found the flowers of the bee-orchis abundant on the slopes of Beachy Head, in Sussex, on 19th June, but they were all very pale in tint, some almost colourless. They are quite different from the beautiful flowers I have found in other places. I do not know whether the blooms of this orchis are always pale in tint on Beachy Head, or if it is a peculiarity of this season.— *Thomas Hilton, 16 Kensington Place, Brighton.*

KING AND BOTANIST.—A writer in the "Echo de Paris," in mentioning the simple kind-heartedness of the King of Norway and Sweden, a recent royal visitor to France, tells a story good enough to repeat, even if not quite true. It is stated that an eminent French botanist was collecting plants outside Stockholm some time since, when he found another gentleman similarly employed. They struck an acquaintance and collected together for a time, when the Frenchman proposed lunch. The Swede replied, as his new friend was a foreigner he should lunch with him. They sauntered along until they arrived at the palace, which the King entered, remarking : "I happen to be King of this country, so cannot ask you elsewhere." Following lunch, the afternoon was spent in discussing plants.

THE FADING OF BIRDS' EGGS.—It is a curious fact with most birds' eggs that in course of time, certainly in a year or two, they begin to fade. Of these some good examples are those of missel-thrush, chaffinch, and lesser whitethroat. The missel-thrush's eggs lose the clearness of their spots after a little time and become blurred, while in the case of the latter two species the background fades. The first theory, which will explain a good many cases, but most of all the blue eggs, is that the sunlight fades them, and this is especially so with light blue eggs. There is no doubt about this, and the collector is advised to keep his eggs away from the light. There is, however, another theory to put forward. If an egg is not thoroughly blown clean, it will soon fade ; therefore there must be some explanation other than that of the light, and the probable one is that the contents of the egg contain sulphur, and if any is left behind it will decompose, forming hydrogen sulphide. If the egg is exposed to the air, the oxygen will decompose the hydrogen sulphide, forming sulphur dioxide. which has, amongst others, bleaching properties, and in time partially bleaches the egg. This, of course, only refers to eggs not blown quite clean. As some collectors are not very particular on this point, this note may be useful to them. If the eggs are thoroughly blown, immediately washed out, and kept from the light, I do not think they will suffer much from fading.—*F. R. Ward, 11 Cranmer Road, Grange Road, Cambridge.*

DR. J. W. GREGORY'S appointment as head of the scientific staff of the National Antarctic Expedition is good. He will return from Australia in October next, to prepare the details of his side of the expedition.

ONE satisfactory result of Her Majesty the Queen's recent visit to Ireland has been the recognition of the merit as a photographer of our correspondent, Mr. R. Welch, of Belfast. At the request of the Congested District Board, Mr. Welch submitted to the Queen a unique collection of views of Western Ireland, which has resulted in his receiving the appointment of Photographer to the Queen by Royal Warrant.

THE views submitted by Mr. Welch to the Queen were not of the usual commercial character, but in every case illustrated some scientific fact relating either to natural history or ethnography, many being geological sections and other views obtained during the excursions of the Belfast Naturalists' Field Club.

IN connection with the Fauna and Flora Committee of the Royal Irish Academy, Mr. W. West, F.L.S., of Bradford, has, with the assistance of members of the Belfast Naturalists' Field Club, been carrying out investigations with tow-nets on Lough Neagh. The local microscopic and other algae have received especial attention. The headquarters of the party have been at Toome, on the northern end of the lake, by the outfall into the Bann.

MR. ARTHUR PEARSON'S expedition in search of recent specimens of giant sloths in Patagonia will be under the personal direction of Mr. Hesketh Pritchard, who will be accompanied by Mr. J. B. Scribnoer, B.A., as geologist. Dr. Ray Lankester gave an interesting address on the subject of these sloths, in the afternoon of June 21, at the Zoological Society's Rooms. If these gentlemen meet with a specimen, there will not be much difficulty in identification, on account of the animal's enormous size.

TWO eggs of the extinct great auk were sold at Mr. J. C. Stevens's Auction Rooms, Covent Garden, on June 20th last. One was an unrecorded egg from France, and attained a record price of £330 15s. It is said to be the finest example yet sold in the Stevens Rooms. Its length is $5\frac{1}{10}$ inches, though the usual length for these eggs is from $3\frac{3}{4}$ inches to $4\frac{1}{2}$ inches. It is a perfect specimen and of good colour. The second egg passed through the same rooms in 1894, as recorded in SCIENCE-GOSSIP (vol. i., N.S., p. 75), when it sold for 175 guineas. At the sale of June 20th it reached a price of 180 guineas.

THE somewhat notorious William Watkins, who was apt to describe himself as the "Butterfly King," and an advertising dealer in insects, died last month at the age of fifty-one years.

BY the death of Miss Mary H. Kingsley, niece of the genial naturalist of that name, an eminent though unassuming traveller with the power of acute observation is lost to science and sociology.

MESSRS. ROSS, Limited, of London, have sent us their catalogue of this season's optical instruments and photographic apparatus. It contains several items of novelty and much of general interest.

MR. JOHN MURRAY will issue in the autumn the late Charles Darwin's classic, "The Origin of Species," in a cheap edition. We also understand that Dr. Dallinger is engaged upon a new edition of Carpenter.

MR. JAMES SHIPMAN, F.G.S., some time ago compiled a broad-sheet, illustrated in colours, of a "Vertical Section of the Stratified Rocks of the British Isles" for the use of the Nottingham Natural Science Rambling Club. The compiler has now published the sheet at half a crown for the use of the general public.

AT the last meeting of the Royal Meteorological Society, Mr. J. Baxendale gave a description of a new self-registering rain-gauge, designed by Mr. F. L. Halliwell, of the Fernley Observatory at Southport. This instrument is believed to closely approach an ideal standard, and has the merit of being constructed at a moderate price.

WE understand that the second volume of Mr. Tutt's "Natural History of the British Lepidoptera" is completed, and will shortly be published. The book is divided into two sections; the first hundred or so pages are devoted to "Metamorphosis in Lepidoptera, External and Internal Morphology, and Phylogeny of Lepidopterous Pupae"; the second part of the book dealing with species. Following on the first volume, this next portion cannot fail to attract considerable attention among students of Evolution as well as of Classification.

THE remarkable proposal, to which we have referred on page 49, for a national storehouse of objects of educational value, will, we hope, be eventually adopted. Professor Petrie's plan is printed in detail in the "Journal of the Society of Arts" for May 18th last, and is well worthy of examination on account of the care with which he has prepared his subject. Mr. William Matthew Flinders Petrie, D.C.L., LL.D,. Ph.D., is the Edwards Professor of Egyptology in University College, London, and well known for his successful excavations of historic and prehistoric remains in Egypt.

IT is satisfactory to find that the Editor of SCIENCE-GOSSIP has received a number of replies to his invitation to form a London Field-Botany Club. Those who have written will shortly receive an invitation to a preliminary meeting to discuss the plan. It is evident from the replies, and their limited number out of the seven millions of people inhabiting the metropolitan district, that field-botany requires revivification. It is to be hoped that this club will aid in the encouragement of a greater interest in our native plants. A suggestion is made that country members should also join. It seems to be a good one, and will probably be adopted. The Editor will be glad to hear from any who would care to communicate with him.

CONDUCTED BY JAMES QUICK.

ELECTRIC TRAMWAYS IN PARIS.—Visitors to Paris at the present time are finding the electrical tramways and railway a great boon. There are five stations on the latter system, which fringes the Exhibition. For 25 centimes a good view of the exterior of the buildings and the environs of the extensive area covered by the Exhibition can be obtained with ease and comfort. The means of getting about the city by the tramways is very complete. The only forethought necessary is to be provided with a ticket before attempting to enter a car, as without it the chance of securing a seat is practically impossible.

THE STRUCTURE OF METALS.—Some very important work has recently been done by Professor J. A. Ewing and Mr. W. Rosenhain upon the microscopical examination of the surfaces of metals, subjected to various physical changes. Professor Ewing gave a very lucid account of some of these researches at the Royal Institution on May 18th. Every metal or alloy under various physical conditions reveals, when examined under a powerful microscope, a definite granular or even crystalline structure, which cannot be permanently effaced. Even in a substance such as lead, which might from first thoughts be assumed to have no definite structure at all, geometrical form and uniform design are found. Professor Ewing explains the plasticity of metals by the slipping of the granular crystals more or less readily over each other. A further interesting point is the "growth" of these structures. If a metal, say lead, is greatly compressed, the surface reveals a somewhat irregular granular structure, the size of the granules being small. After certain periods, depending upon heat and other conditions, these small granules gradually coalesce and increase considerably in size. They are then found to have again assumed the original structure of the untreated lead.

A NEW X-RAY LOCALISER.—A new method of localisation in X-ray work has recently been devised by Professor F. R. Barrell. In all methods of localisation of foreign bodies it is essential that two distinct radiographs should be taken. This is done by changing the position of the X-ray tube, the distance through which it has been moved being ascertained. The points upon the photographic plate where it is met by lines drawn perpendicular to it from the focus of the tube are located, in hitherto existing methods, by the use of a plumb-line, necessitating, among other things, levelling the plate. In Professor Barrell's method no plumb-line, no threads, and no levelling are required. Two metal cylinders are used that have their ends carefully turned perpendicular to their axes. The size of these cylinders is about 4 inches long and 1 inch diameter. They are placed upright on the plate during the exposure. The shadows thrown by the cylinders are utilised to locate the position of the focus of the tube. After the first excitation the tube is shifted about 8 inches, the cylinders are also shifted towards the opposite end of the plate, and then the second exposure is made, giving rise to the second set of shadows from the foreign body and the cylinders. Lines are now ruled along the edges of the shadows of the cylinders, when these were in the first positions, and produced till they meet. The point of crossing gives the point on the plate which was perpendicularly beneath the focus of the tube during the first exposure. The same process is gone through with the second set of shadows to find the perpendicular point below the focus in the second position. All the necessary data are thus obtained to accurately locate the position in space of the foreign body. The beauty of this method is that it is so very simple. The necessary apparatus costs practically nothing, for two metal bars of square section will serve equally as well as the cylinders, provided they have well-defined edges and flat ends, so that they will stand upright.

KITES AND BALLOONS IN METEOROLOGY.—Some useful work is now being carried on in the study of the upper atmosphere by means of kites and balloons. In September last, observations were made at an altitude of 4,300 mètres, the highest hitherto attained by any form of kite. It is now also possible to send up small balloons during strong gales, even in a wind with a velocity of 14 mètres per second. These captive balloons have attained much greater altitudes than the kites, but a difficulty is encountered by the heating action of the sun. By working at night, by moonlight, over 120 balloons have been launched, by means of which the curves of temperature and pressure have been automatically registered at a height of 13,000 mètres on twenty-four occasions, at 14,000 mètres eight times, and at 15,000 mètres three times.

LICHTENBERG FIGURES.—Every student of electricity knows the beautiful figures that may be produced upon a plate of resin or vulcanite by tracing a design upon it first with the charged knob of a Leyden jar, and secondly with the outer coating, and then dusting a mixture of red lead and sulphur upon it. The sulphur will attach itself to the positive lines, and the red lead to the negative lines. Instead of using the above mixture, one can be made up with 1 part by volume of carmine, 3 of lycopodium, and 5 of flowers of sulphur. The carmine and sulphur are first thoroughly mixed, and the lycopodium is added last. The figures are reversed in colour with respect to the ordinary Lichtenberg figures, as the polarity of the sulphur is reversed in the combination. The figures are of striking beauty; the circles have centres, and outer rings of the opposite colour.

PARIS CONGRESS.—The Paris International Congress of Electricity will be opened on Saturday, August 18th, at 10 a.m., in the Palais des Congrès, Paris, and will remain open up to the 25th. It will consist of the following five sections, under separate presidents: I. Scientific Methods and Measuring Apparatus; II. Generation, Transformation, Transmission and Distribution of Electric Energy; Electric Lighting and Traction; III. Electro-chemistry; IV. Telegraphy, Telephony, &c.; V. Electro-physiology. During the Congress visits will be made to the various installations in Paris, in which only members will be allowed to take part.

CONDUCTED BY HAROLD M. READ, F.C.S.

DEPARTMENT OF CHEMISTRY.—Our readers will notice at the top of this column a new heading. Those who have studied the chemistry of hydrocarbons will recognise in the design their history, from the primeval carboniferous forest, through the coal-mine and the chemist's laboratory, to the benzene hexagons of Kekulé. It is suggested by Mr. Harold M. Read, and our contributor Mr. Frank Percy Smith has kindly designed and drawn the heading.

PYRAXE, A NEW FORM OF PYROGALLOL.—Chemists and photographers have been so long accustomed to the bulky appearance of Pyrogallol, or "Pyro," as it is more generally called, that its appearance in a condition in which a pound occupies no more space than an ounce of the ordinary variety, is a striking innovation. This new modification issues from the well-known house of T. Hauff and Co., and may be obtained in London from Messrs. Fuerst Brothers.

QUARTZ THERMOMETERS.—In a recent issue of the "Comptes Rendus" (130, 775 and 876) M. Dufour describes a thermometer having a quartz bulb and stem, in which the liquid employed is molten tin, since that metal may readily be obtained pure, is not very volatile below a red heat, and has a comparatively low melting-point. The bulbs are made very thick, in order to avoid any fracture which might be caused by the too rapid contraction of the tin. The meniscus is always quite bright and similar to that of mercury.

THE PREPARATION OF CARAMEL.—The many uses of caramel are, perhaps fortunately, unknown to the general public, which does not realise that the bright golden ales, mineral waters, and sometimes vinegars, owe their attractive tints to the addition of the colouring matter produced by the partial burning of sugar and glucose. To the chemist and manufacturer, on the other hand, its uses are familiar, although its satisfactory preparation has long been a source of worry and disappointment. A paper read at a recent meeting of the Society of Chemical Industry, will serve to clear many of the points which previously were doubtful. The requirements of good caramel are: (i) a maximum colour-intensity; (ii) a minimum percentage of residue (ash); (iii) a perfect miscibility with proof-spirit. The importance of the last point is evident when it is remembered that the caramel is used largely in the colouring of ales. It appears that the "caramels" manufactured from cane sugar and from glucose are practically identical; but the cost of the former constitutes a serious drawback to its commercial use, hence glucose is now used almost entirely. The caramel is prepared by first boiling the glucose, then adding to the molten substance a mixture of ammonium chloride and ammonium carbonate, and continuing the heating until the whole mass swells up and gives off pungent fumes. The temperature is then lowered, but the heating is continued until the now black mass becomes quite viscous. It may then be poured off on to iron plates, and, when cool, broken up with a hammer; or it may be dissolved in water and the solution strained to make "liquid caramel." The authors of the paper tried various "charring-agents" in place of the ammonium salts mentioned, but the results with the latter were most satisfactory.

THE VITALITY OF DRIED SEEDS.—The results of a recent investigation carried out by Professor T. Bretland Farmer, undertaken with the view of studying the influence of high temperatures on seeds, are highly interesting and of extreme importance from a chemical as well as from a biological point of view. It is found that, when albumin is dried in an incubator at a temperature of 52°–55°, its solubility and coagulability remain unaltered, and that it resembles in those respects the unheated substance. Further than this, Professor Farmer finds that after albumin has been thoroughly dried it may be heated to 102°–110° for at least thirteen hours without causing any apparent change in its molecular structure. Dr. Morris has already shown that the germinating power of the seeds is not destroyed either by boiling in water or by dry-heating to a higher temperature. In the former instance, however, if the seed-coat be broken or softened, the vitality is destroyed, and it may be that this destruction is caused primarily by the admission of the hot water to the living cell.

THE CONVERSION OF PHOSPHORUS INTO ARSENIO.—A few years ago the scientific world was amused, and the general public interested, by the announcement in the United States of the conversion of silver into gold. A Dr. Emmens, starting with Professor Crookes' conception of the evolution of the elements from a primary "subatom" known as protyle, stated that by the application of repeated blows silver could be converted into gold. Unfortunately from a scientific point of view, though perhaps fortunately from an economic standpoint, we have heard but little of this so-called discovery during the months succeeding its announcement. The statement is vividly recalled by a paper which appeared in the "Leopoldina," Halle, last March, on "The Conversion of Phosphorus into Arsenic," by T. Fittica. Unlike Dr. Emmens, whose gold-producing process was brought out with the air of mystery which one associates with mediaeval alchemy, the author of the present paper gives working details of his method. Ordinary amorphous phosphorus is heated with a concentrated solution of ammonia in the presence of air, or, better still, with the addition of a solution of hydrogen peroxide. This oxidising action furnishes, the author states, ordinary arsenic. By treating yellow phosphorus with strong nitric acid, as much as 2.5 per cent. of arsenic was obtained. Without venturing on publishing the analyses (if any were made) of this substance, the author does not hesitate to publish both a formula (PN_2O) and an equation for its production ($2P + 5NH_4NO_2 = (PN_2O)_2O_3 + 10H_2O + 3N_2$). We leave to our readers the opportunity of forming an opinion on such strange results, but must remark that it would be interesting to have an accurate analysis made of this "black modification of phosphorus."

ASTRONOMY,

CONDUCTED BY F. C. DENNETT.

	1900	Rises.	Sets.	R.A.	Dec.
	July	h.m.	h.m.	h.m.	° '
Sun	.. 9 .. 3.55 a.m.	.. 8.15 p.m.	.. 7.13	.. 22.24 N.
	19 .. 4.7	.. 8.5	.. 7.53	.. 20.54
	29 .. 4.20	.. 7.52	.. 8.33	.. 18.49

		Rises.	Souths.	Sets.	Age at Noon.
	July	h.m.	h.m.	h.m.	d. h.m.
Moon	.. 9 .. 8.34 p.m.	.. 9.38 p.m.	.. 0.52 a.m.	.. 12 10.33
	19 .. 10.58	.. 5.46 a.m.	.. 1.17 p.m.	.. 22 10.33
	29 .. 7.39 a.m.	.. 2.11 p.m.	.. 8.30 p.m.	.. 2 22.17

		Souths.	Semi-	R.A.	Dec.
			Position at Noon.		
	July	h.m.	Diameter.	h.m.	° '
Mercury	.. 9 .. 1.49 p.m.	.. 4.4"	.. 8.57	.. 16.6 N.
	19 .. 1.13	.. 5.2"	.. 9.7	.. 12.59
	29 .. 0.20	.. 5.7"	.. 8.47	.. 12.49
Venus	.. 9 .. 11.55 a.m.	.. 23.9"	.. 7.4	.. 18.9 N.
	19 .. 10.58	.. 27.2"	.. 6.40	.. 17.12
	29 .. 10.5	.. 23.9"	.. 6.31	.. 16.57
Mars	.. 19 .. 9.19 a.m.	.. 2.2"	.. 4.57	.. 22.39 N.
Jupiter	.. 19 .. 8.8 p.m.	.. 12.5"	.. 15.57	.. 19.41 S.
Saturn	.. 19 .. 10.11 p.m.	.. 8.4"	.. 18.0	.. 22.30 S.
Uranus	.. 19 .. 8.49 p.m.	.. 1.9"	.. 16.28	.. 21.45 S.
Neptune	.. 19 .. 10.4 a.m.	.. 1.2"	.. 5.51	.. 22.14 N.

MOON'S PHASES.

	h.m.		h.m.
1st Qr. .. July 5 .. 0.14 a.m.		Full .. July 12 .. 1.22 p.m.
3rd Qr. .. ,, 19 .. 5.31 a.m.		New .. ,, 26 .. 1.43 p.m.

In apogee July 3rd at 3 p.m.; in perigee on
15th at 2 p.m.; and again in apogee on 31st at
9 a.m.

METEORS.

				h.m.	
June 13–July 7 .. Vulpeculids.. Radiant R.A. 20.3	Dec. 24° N.
July 11–19	..a Cygnids ..	,,	,,	20.56	,, 48 N.
,, 12–17	..π Herculids	,,	,,	17.0	,, 37 N.
,, 19–27	..γ Draconids	,,	,,	18.4	,, 47 N.
,, 23–Aug. 4	..a-β Perseids	,,	,,	3.12	,, 43 N.
,, 27–29	..δ Aquarids	,,	,,	22.36	,, 12 S.
,, 30–Aug. 11	..λ Andromedes	,,	,,	23.20	,, 51 N.

CONJUNCTIONS OF PLANETS WITH THE MOON.

					° '
July 9	..	Jupiter	.. 1 a.m.	.. Planet 1.35 N.
,, 11	..	Saturn*†	.. 4 a.m.	.. ,, 0.48 S.
,, 23	..	Mars	.. 1 a.m.	.. ,, 0.44 N.
,, 24	..	Venus†	.. 2 p.m.	.. ,, 3.50 S.
,, 27	..	Mercury*	.. 7 a.m.	.. ,, 0.16 S.

* Daylight.   † Below English horizon.

OCCULTATIONS AND NEAR APPROACH.

		Magni-	Dis-	Angle from	Re-	Angle from
July	Star.	tude.	appears.	Vertex.	appears.	Vertex.
			h.m.	°	h.m.	°
8.. δ Scorpii	.. 2.5	..11.24 p.m.	.. 136	.. 11.54 p.m.	.. 180
12.. ξ¹ Sagittarii.. 3.5	.. 0.19 a.m.	.. 103	.. 1.18 a.m.	.. 202
12.. ξ² ,,	.. 5.0	.. 0.53 a.m.	.. 332	.. Near Approach.
14.. c Capricorni.. 5.2	.. 9.43 p.m.	.. 48	.. 10.17 p.m.	.. 339
17.. 19 Piscium	.. 5.2	.. 5.1 a.m.	.. 347	.. 5.53 a.m.	.. 284
22..53 Tauri	.. 5.5	.. 1.2 a.m.	.. 109	.. 1.52 a.m.	.. 316

THE SUN should be kept under observation, as
interesting groups are at times visible. It is in
apogee at 1 p.m. on July 2nd, when its distance is
about 94,560,000 miles.

MERCURY is an evening star all the month,
reaching its greatest eastern elongation, 26° 2', at
1 p.m. on July 4th, at which time it sets about
1h. 20m. after the sun. The best time for observation will be early in the afternoon, when it is
near the meridian.

VENUS at the commencement of the month is
very near the sun, being in inferior conjunction
with it at 11 a.m. on July 8th, afterwards becoming a morning star, rising nearly two hours
before the sun at the end of the month, when it is
near γ Geminorum.

MARS is a morning star, rising at 1.30 a.m. at
the commencement of the month. and 0.40 a.m. at
the end. Travelling through Taurus, its tiny disc
is too small for useful observation, except with
large apertures.

JUPITER is still not far from β Scorpii. It is
due south on July 1st at 9.23 p.m. and at 7.20 on
31st, and should be looked for as soon as it is
sufficiently dark.

SATURN, southing at 11.27 p.m. on 1st and at
9.21 on 31st, is as well placed as its great south
declination will permit. Its very widely open ring
makes it a most beautiful object.

URANUS is on the meridian at 9.53 p.m. on
July 1st, and at 7.51 on 31st, and near to the 5th
magnitude star ω Ophiuchi.

NEPTUNE is too near the sun for observation.

THREE NEW MINOR PLANETS.—Two were photographically discovered on March 6th at Tokio
Observatory, Japan, by S. Hirayama. A third was
found on a photograph taken by Professor Max
Wolf on May 22nd, at Heidelberg.

ASTRONOMICAL SOCIETY OF WALES.—We have
received the quarterly journal "The Cambrian
Natural Observer," which contains some interesting observations on the great meteor of January
4th. as well as on other phenomena. The notes on
the celebrated meteorologists G. J. Symons, F.R.S.
and E. J. Lowe, F.R.S., who so recently died, are
useful reading.

ALTERATION OF ADDRESS.— Mr. Dennett's address
is now, as previously, 60 Lenthall Road, Dalston,
N.E.

TOTAL ECLIPSE OF THE SUN.—Everywhere there
appears to have been good fortune with the observers along the line of totality. The party of the
Astronomer Royal at Ovar seem to have fared
worst, but not seriously, from the presence of a slight
haze. The form of the corona, as seen by the naked
eye, was very like the observations made in 1867,
1878, and 1898, one ray being visible to a distance of from four to six lunar radii from the sun's
centre. The light at totality seems to have been considerable; more than that from the full moon. The
partial eclipse seen from England was very interesting. The frequent passage of clouds caused
trouble to the scientific observer, but gave very
nice views of the phenomenon to the general public
without the use of troublesome smoked glasses.
The writer was observing with the 3-inch Wray
SCIENCE-GOSSIP Telescope, a solar diagonal, and
power of 75 diameters. The limb of the moon was
noted to be abnormally smooth, and other observers
make the same note. The occultation and re-
appearance of two groups of dark spots proved very
interesting. The mottled surface of the disc
was itself a beautiful object for study. Not
any phenomena traceable to a lunar atmosphere
seem to have been observed. At the time of

greatest eclipse, as seen from Dalston, the sky was very cloudy, but away to the south-east a patch of blue sky was visible, looking dark and dull, more like that of winter than at the end of May. Some observers had an interesting sight where the sun shone between the leaves of a tree and formed on the ground a multitude of little crescents of light. No trace of the moon appears to have been visible projected on the corona. During the eclipse, Professor Howe, of Denver Observatory, Colorado, succeeded in obtaining a photograph of the planet Eros.

CHAPTERS FOR YOUNG ASTRONOMERS.

BY FRANK C. DENNETT.

VENUS.

(*Continued from page 27.*)

TWICE J. D. Cassini thought he observed a satellite to Venus: on January 25th, 1672, and on August 28th, 1686. James Short, the celebrated optician, made a similar observation on October 23rd, 1740. Montaigne on three nights in May, 1761, also thought he observed the satellite; and Baudouin calculated its orbit. Mayer in 1759, also Rödkier, Horrebow, and three others at Copenhagen, and Montbarron at Auxerre, in March, 1764, are all said to have made similar observations. A Belgian astronomer, Stroobant, has been at great pains to show that most of the observers had really seen stars which he identifies; his work, however, is fruitless, as the object observed always had a diameter from a quarter to one-third of that of Venus, and moreover had a phase similar to that of the planet. The true explanation is doubtless a "ghost" in the telescope, one cannot say eyepiece, for Short used three or four eyepieces. Seeing that the majority of these observers were well experienced, their mistakes should act as a caution to those who are as yet only beginners in the science. If any abnormal appearance is observed, it is well to give the eyepiece a quarter turn, or even to change the eyepiece. This will frequently expose or exorcise the "ghosts." If possible, it is also well to appeal to another telescope.

One of the most ancient recorded observations of Venus is that by Timocharis, who witnessed the apparent occultation of a star at the extremity of Virgo's wing, on a date corresponding, according to Hind, with October 12th, 271 B.C. Similar occultations of α Leonis, Regulus, are mentioned by Maestlin on September 16th, 1574, and Kepler on September 25th, 1598. Mars also suffered occultation in 1590 on October 3rd, and that of Mercury occurred on May 17th, 1737.

One striking fact concerning Venus is that the limb, or edge of the disc, is always the brightest part of the planet.

From time to time wonderful stories are circulated about "The star of Bethlehem" appearing in the morning sky. Marvellous descriptions of its apparent form are also given; but invariably the "star" has proved really to be the planet Venus, although some members of the general public are very difficult to convince on the point. Under favourable circumstances, and when at its greatest brilliancy, this planet will readily throw a shadow of earthly objects.

(*To be continued.*)

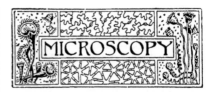

CONDUCTED BY F. SHILLINGTON SCALES, F.R.M.S.

METHOD OF MOUNTING FUNGI IN GLYCERINE.— Mr. A. Lundie, in the Transactions and Proceedings of the Botanical Society of Edinburgh, recommends the following method, by means of which the air can be expelled and the filaments teased out when mounting fungi such as *Eurotium*, &c. A piece of *Eurotium* is placed on a slide, wetted with chloroform, strong glycerine added, and a cover-glass imposed. By heating the preparation over a Bunsen flame the chloroform is made to boil, and so drive off the last traces of air. The bubbles of chloroform vapour, in passing out, scatter the hyphae and tease out the preparation, without breaking it up, as is done when needles are used.

MEETING OF THE ROYAL MICROSCOPICAL SOCIETY.—At the meeting on May 16th the President, Mr. Carruthers, in the chair, Mr. Charles Baker exhibited two microscopes, one of which, a high-class microscope designed for critical work, was fitted with eyepieces of the new Society standard No. 3, *i.e.* 1·27 inch. The other instrument, called the "Plantation" microscope, was very cheap, all complicated parts and movements having been done away with. It is designed for use in the tropics for examining and discovering the ova of internal parasites. Both these instruments we hope to be able to notice at an early date. Dr. Hebb announced that part 8 of Mr. Millett's report of the Foraminifera of the Malay Archipelago had been received, and would in due course appear in the Society's journal. Mr. E. M. Nelson read a paper on "The Lag in Microscopic Vision," illustrated by diagrams and a series of tables showing proportionate values of the performance of various objectives with eyepieces of various powers. As might be expected, achromatic objectives showed the most marked difference; but even in the case of an apochromatic objective the difference between its performance with a low eyepiece and a high one was shown to vary from 14·7 to 7·7. This should finally settle a claim that we have more than once seen advanced on behalf of apochromatics, as compared with achromatics, *i.e.* that the former will bear the highest eyepiecing without *any* detriment. As regards tube-length, Mr. Nelson's experiments showed that microscopes with short tubes had some advantages with regard to lag over those with long ones. Mr. Nelson also read a paper by Mr. E. B. Stringer on "A New Form of Fine Adjustment," and exhibited a microscope made by Messrs. W. Watson & Sons, fitted with the same. There was an excellent exhibition of slides illustrating pond life, due to the courtesy of the Quekett Microscopical Club.

THE QUEKETT MICROSCOPICAL CLUB JOURNAL. The April issue of this journal has duly reached us and is of much interest. The address of the

President, Dr. Tatham, is printed in full and deals mainly with comparatively new mounting media of high refractive indices, such as a solution of biniodide of mercury in potassium iodide, phosphorus, quinidine, realgar, and a combination of piperine with bromide of antimony. All these media have unfortunately their drawbacks, and two at least are really dangerous in unskilled hands. The Proceedings of the Club are reported in full, and notice is given of the various excursions to be held during the summer. There are also several interesting papers, the most noticeable of which is one on Radiolaria, by Mr. Arthur Earland, which might very well serve as an introduction to the study of these exceptionally beautiful Rhizopods. This paper is illustrated by two beautiful plates reproduced from the Report of the Radiolaria of the "Challenger" Expedition, and is followed by a list of fossil Radiolaria from Barbados, compiled by the same author. Mr. J. G. Waller writes on a hitherto undescribed British sponge of the genus *Raphiodesma*, the spicules of which he illustrates; and other members contribute shorter notes on various subjects, amongst which we notice one on the genus *Lacinularia* of the Rotifera, by Mr. C. F. Rousselet.

PREPARING WOOD SECTIONS.—Sections of wood for microscopical observation are of two kinds— viz., those intended to display the anatomical structure of the tissues only, and those intended to teach something about the physiology thereof, as demonstrated by chemical tests. Wood selected for the former may be subjected to pretty violent treatment beforehand. For instance, a piece one inch diameter and about two inches long may be boiled in water, then in alcohol, and extracted in succession by benzene, dilute soda, and dilute acid; then a slit is cut transversely half-way through it at about $\frac{3}{4}$ inch from one of the ends, and the semi-cylindrical plate of wood of $\frac{1}{2}$ inch radius and $\frac{1}{4}$ inch thick removed. The smoothed semicircular surface hereby formed is the place to make the transverse section, which is effected, not by any ordinary knife, razor, or scalpel, but by a specially-constructed instrument. A suitable one is a knife $3\frac{1}{2}$ inches long, $\frac{7}{8}$ths of an inch broad, and over $\frac{1}{16}$th inch thick at the back; it is slightly hollowed on the lower side, very sharp, and set in a solid handle, the whole being about 8 inches in length. It should always be carefully wiped, cleaned dry after use, and kept in a well-closed case. If the wood has been previously treated in the manner aforesaid, no staining or bleaching of the section is required; on the contrary, they must be studiously avoided, except in the case of coniferous wood, when gum styrax is not the mounting medium. Sections of wood required for chemical physiology are prepared either from the fresh material, or from small fragments that have been steeped or boiled beforehand in various solutions. The same kind of knife is as requisite here as in the former case, whilst even greater care, dexterity, and "science" must be employed. For the detection of the more conspicuous bodies, such as starch or oil, the sections may be very thin; but for the study of protoplasm and its contents they must not be too thin. Such, at least, seems to be the opinion of that most expert section-cutter, the late Professor De Bary, whose slides, to the number of some thousands, have lately been acquired by one of our institutions. My own opinion is that a section cut by a properly con-structed and fully-sharpened knife will stand, so to speak, being made considerably thicker than one bunglingly hacked out by a bad or indifferent tool. Moreover, there is no doubt whatever that the thinner the section, provided it hangs well together, the better will the effect of the various test reagents be visible, and the proper interpretation thereof be more readily attained.—(*Dr.*) *P. Q. Keegan, Patterdale, Westmorland.*

NEW COVER-GLASS GAUGE.—It is, of course, well understood that the thickness of the cover-glass slightly alters the correction of an objective, though this becomes serious only with high powers. This necessitates readjustment either by means of a correction collar or by the simpler alteration of the tube-length of the microscope itself. Some day, perhaps, the Royal Microscopical Society will succeed in standardizing the thickness of cover-glass for which objectives are corrected by makers; but at present, unfortunately, there is no uniformity in this respect. Messrs. R. & J. Beck, who themselves correct their objectives for a cover-glass of ·006 inch, have just brought out a new cover-glass gauge with which to determine the thickness of the cover-glass before use. We illustrate it herewith, and it

COVER-GLASS GAUGE ($\frac{1}{2}$ *full size*).

will be observed that its construction is very simple. The long lever is raised and the cover-glass is then inserted against the steel plate or bracket and the lever lowered so that the hardened steel point close to the fulcrum of the lever (shown in the figure) must rest on some portion of the cover-glass. The thickness can then be read off on the scale in ·001 inch; thus the illustration shows a cover-glass ·006 inch thick. We think this gauge should give measurements of quite sufficient accuracy. Its price is 21s.

STEARINE AND NAPHTHALINE IMBEDDING.—I see you ask in SCIENCE-GOSSIP, vol. vi. page 374, for information respecting stearine and naphthaline for imbedding. Perhaps the following extract from Wright's "Popular Handbook to the Microscope," page 132, may be of service. I have not seen the mixture mentioned elsewhere, and know nothing of its value in practice. "Mr. Cole and others recommend for imbedding four or five parts solid paraffin and one part lard, melted together; but this often slips in the microtome tube, and it is better to use a mixture of stearine and naphthaline, using the less stearine the warmer the weather. This is found to hold tightly."—*J. Robinson, 66 Fairhazel Gardens, Hampstead, N.W.*

ANSWERS TO CORRESPONDENTS.

F. B. (York).—Owing to pressure on our space, the answer to your query appeared at the end of the May number (SCIENCE-GOSSIP, vol. vi. p. 380) instead of in the usual place. In the present number you will see also a note on preparing wood sections which may be of service to you.

MICROSCOPY FOR BEGINNERS.

By F. Shillington Scales, F.R.M.S.

(Continued from page 29.)

WHEN the student is making accurate micrometric measurements, it should be remembered that the divisions on the stage-micrometer are not uniform, and that the mean value should be accordingly taken. Further, the stage-micrometer should be used in every set of measurements; the usually recommended plan of making a record with different objectives and eyepieces, once for all, for comparison, being manifestly untrustworthy. Errors due to diffraction in the real edges of objects mainly affect high powers, and we may content ourselves with simply mentioning their existence.

The measurement of the actual magnification due to an objective alone is best made by projecting the image of the stage-micrometer, without eyepiece, on a cardboard screen five feet away, by marking and then measuring the distances between the lines, then dividing by 6, which will give the actual magnification of the objective at the normal visual distance of ten inches. At the five-feet distance the error in the measurement will be inappreciable. The additional magnification due to the eyepiece can be measured by projecting the image of the stage-micrometer through both objective and eyepiece *with the 10-inch tube*, and dividing by the known magnification of the objective. This tube-adjustment must be most carefully made. The distance of the projected image in this latter case, measured from the eye-lens, must also be the normal one, *i.e.* 10 inches, either direct or with the camera lucida. If the objective should be an inch of a proved magnification of ten, and the combined magnification at ten inches from the eyepiece under the above conditions be 60, the magnification due to the eyepiece will be 6. What we have so frequently said about variations in tube-length must be strictly borne in mind in making such measurements or calculations, as every variation in tube-length leads to a corresponding variation in the magnification due to the objective alone. The theory of microscopic vision should be remembered—*i.e.*, that the image of the object, as magnified by the objective, is magnified again by the eyepiece, but that the magnification of the latter is constant, whilst that of the former varies according to whether it is focussed nearer to or farther from the object, thus altering the position of the image referred to above; in other words, and speaking practically, it varies with the tube-length. It must also be borne in mind that the foregoing data are applicable, strictly speaking, only to those of normal sight, and who in consequence form their image at the calculated and ordinary distance of ten inches from the eye. All others having abnormal vision must use lenses or spectacles to correct the difference.

The eyepieces, like the objectives, should be kept clean and free from dust. The dust can be readily localised by reducing the light by means of the iris-diaphragm or otherwise, and then rotating the eyepiece. The objectives should not give much trouble in this way, and it will often be found that what was apparently dust on the back lens of the objective was actually dust on the cover-glass of the object or on the top lens of the condenser. This can be localised in like manner by moving the slide or rotating the condenser.

The use of the polariser and analyser would take more space than we can give to it here. We may simply say that with "crossed nicols" any object showing crystalline structure or possessing double refraction will show more or less brilliant colours on a black ground, and crystals, mineral sections, fibres, odontophores, &c., are very advantageously exhibited by this means. With parallel nicols the background will be bright, but the colours will still be in evidence, though the tints will vary. By rotating the analyser or polariser, most beautiful effects can be produced, and with weakly refracting objects as gorgeous effects can be produced by inserting a slip of selenite. For crystallographic or petrological work an analyser and polariser are necessities, and elaborate microscopes fitted with many other adjuncts are sold for this purpose. Unfortunately the various books on the microscope deal very lightly with the subject, and those who require information on this and other matters connected with the polarising microscope and its uses are referred to a paper by the present writer in the current issue of the "Annual of Microscopy" for 1900.

In concluding these elementary hints on the actual management of the microscope, we would strongly advise our readers to obtain a few "test slides," and diligently to practise themselves in their examination. For low powers there is nothing more suitable than the fine hairs on the tip of the proboscis of the blow-fly. These should come out quite sharp, black, and finely pointed. For medium powers nothing is more suitable than the old-fashioned "podura scale" (*Lepidocyrtis curvicollis*) mounted dry, and if the beginner can get the "hairs" on these sharp and well marked, having the appearance of minute exclamation stops, and not running in the least into each other—we say nothing of interior "marks"—he has learnt much in the management of his microscope, and in obtaining a "critical image." Unfortunately the scales themselves vary very much, and it is not always easy to pick out a coarse enough scale that will be within the beginner's or even the objective's powers. In obtaining such a critical image the worker will bear in mind not only what we have already said in our earlier papers on adjustment of the light by means of the condenser and otherwise, but our warning against an excessive shutting down of the cone of illumination by means of the iris-diaphragm. This latter, whilst apparently at first sight increasing the contrast, really breaks down the definition of the objective, and leads to woolliness of outline in each individual mark on the scale, or even to a ring of refracted light round each. For high powers, especially immersion lenses, we prefer a slide of bacteria, say a slide of *Bacillus tuberculosis*, showing the characteristic beaded appearance. The diatom *Pleurosigma angulatum* is also a useful test object. We mention these as helpful aids in mastering the use of the microscope only. The testing of objectives requires long practice, skill, and experience, though we propose, at a not too distant date, to give the readers of this journal some notes on the matter, and to illustrate them with suitable photo-micrographs.

(To be continued.)

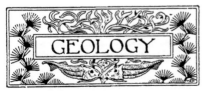

GEOLOGY

CONDUCTED BY EDWARD A. MARTIN, F.G.S.

SOME DEEP LONDON BORINGS.—In addition to the borings, particulars of which have already appeared (SCIENCE-GOSSIP, vol. v. p. 118), the following have been made by Messrs. C. Isler & Co., of Bear Lane, Southwark, who have again kindly placed at my disposal the various sections in detail:

Place of Boring	Dug Well, Made ground, or Ballast	London Clay	Woolwich Beds	Oldhaven Beds	Thanet Sands	Chalk	Total Bored	Water Supply, Gallons per hour	Water Level below Surface
Albert Hall Mansions (Hussey)	—	214	70	13		129½	426½	5000	116
Aldersgate Street (Manchester Hotel)	6	101	33	23	29	158	350	plentiful	127
Bagshot Street, S.E. (White)	20½	—	Nil	13½	37	289	360	6000	23
Battersea, Pure Water Co....	26	142	28½	9½	35	159	400	4000	43
Bermondsey - Hornsey Lane (Barrow) ...	28	27	38	21	40½	95½	250	3000	58½
Bermondsey—Park Street (Bowron)....	29	60½	40	29½	28½	163	350½	2500	70
Bermondsey—Staple Street (Pink) . ..	27	45	32	32	24	140	300	6780	78
Brewery Road, N. (Crosse & Blackwell)....	32	116	42½	32	6½	219	450	3500	210
Bromley-by-Bow (Three Mills Distillery) ..	82	—	—	17	46	105	250	2700	44
Camberwell—Cunard Street (White) ..	18½	—	2½	10	35	296	362	12000	27
Camberwell – Neate Street (White)......	23	—	—	4½	36½	302	366	large spply	20½
Camden Town N.W., Pratt St. (Idris & Co.)	11½	73	31½	19	16	259¾	410¾	4320	160

The figures placed in my hands show the thicknesses of the various layers quoted from a lithological point of view, such as "sand and pebbles" or "blue clay and stones." These I have grouped together under the names of the usually accepted tertiary formations, according as their composition seemed to agree most closely with those formations as seen elsewhere. It will, however, at once be understood that the grouping of the thicknesses assigned to such beds as the Oldhavens and Thanets may be subject to correction. The rise of the Chalk in the south-east of London is clearly brought out by this table, the Camberwell borings showing small tertiary thicknesses over the Chalk. Any material found which may represent the basement-bed of the London Clay has been included with that formation. At Messrs. Barrow's, Hornsey Lane, Bermondsey, 1 foot of "pebbles" was met with; at Messrs. Bowron's, in Park Street, likewise in Bermondsey, there were 6 feet of "dead sand and pebbles." In the Woolwich Beds the word "rock" appears occasionally; there were 6 feet at Albert Hall Mansions. At the Pure Water Company's well in Battersea there was ½ foot of "stone"; at Hornsey Lane, Bermondsey, there were 2½ feet of "very hard rock." Included in the Chalk in the Hornsey Lane section is a layer of ½ foot of "green-coated flints" at the top of the Chalk.—*Edward A. Martin.*

CORRELATION OF RIVER TERRACES.—One of the most puzzling branches of "drift" geology is that relating to the correlation of the various groups of rocks which make up the Pleistocene system. In dealing with widely separated areas we are satisfied if we can say that two deposits are homotaxial, and as a general rule we make no attempt at arriving at a chronological classification of the deposits. One of the exceptions to this rule I propose to discuss is the chronological classification of the river-drifts of the Thames Valley. In this case, as every geologist knows, the factor usually taken is the surface level of the terraces or groups, to which the deposits in question are referable; those occupying the higher elevations being the older chronologically, and those occupying the lower elevations being the newer. So far as any individual district of narrow dimensions is concerned, this plan works admirably; but when we come to compare the strata of one district of the valley with those developed in another portion, we find that the surface levels which have proved such trustworthy allies in the former case begin to break down when applied to a large area; hence we become exceedingly confused in our comparisons. The reasons for this failure are numerous, and the variations spring from a variety of causes, some of which we will briefly consider. In the first place, it is obvious that the surface level of a patch is constantly changing through the agency of meteorological phenomena, inasmuch as the surface level coincides with the contour. Further it is obvious that such changes are not carried on to an equal extent over even restricted areas, for in dealing with such deposits as the gravels and brick-earths, we are dealing with deposits which exhibit ever-changing lithological characters. At one point we may find an extremely coarse gravel, and a few yards further the same beds are seen to pass into a comparatively fine sand. Again, in one place the beds may be firmly compacted by ferric hydrate, in another immediately at hand they may be quite loose and incoherent in texture. Another agent of variation may be found in the erosive power of percolating water acting on the strata upon which the Pleistocene deposits repose. It is sufficient to merely refer to this, at this point, as a fuller consideration of it and an interesting sidelight which it throws will be given in the sequel. If we admit, as I think we must, that the surface levels are essentially unreliable when taken over large areas, it follows that some other character has to be discovered to serve as our principal factor in the case. This cause, I venture to state, will be found in the mean level of the surface of each of the platforms upon which the Pleistocene strata of the different terraces repose. It will be seen that, with certain variations which are hereinafter described, this "basal level" is remarkably constant for each terrace. It may be objected that this line is one that is too difficult to determine, to be of much practical utility in the field. In the Thames Valley, at any rate, such a difficulty will be found to be purely illusionary, for the following reason. The numerous valleys of the side streams which cut transversely through the terraces of the Thames afford magnificent opportunities for the examination of the basal boundaries or levels of the various terraces of drift. All that is wanted is careful and detailed mapping on either side of the valleys to give us the data required. The information thus obtained may be augmented and made of much greater value to us by the details given by junction sections and borings in the districts between the side valleys. A further objection which may be raised is that on the principle of " what is sauce for the goose is sauce for the gander "—namely, that the subterranean erosion of percolating water, which, as I have here pointed out, affects the "surface levels," must in doing that primarily affect the "basal levels." This is true enough, but in being true it opens up an interesting side development. It is clear that such erosion will affect calcareous strata to a far greater extent than it will argillaceous rocks. As the greater portion of the drifts of the Thames Valley are situated upon the London Clay, the effect is but infinitesimal upon the utility of our basal levels as a datum line. Where the chalk crops out and the drifts repose upon it, the effects are far from infinitesimal, for we may find patches of older terraces brought down to approximately the same level as those of newer ones, simply through the dissolution of the calcareous platform, that has gone on since Pleistocene times. If we can determine the amount of lowering that has gone on in this manner in the different districts, then perhaps we shall be able to get some idea as to the rapidity of the process. At any rate, many interesting results are likely to accrue from a more general recognition of the importance of the "basal level" of drift, and from a more general employment of it in our classification. Personally I have of late been determining such lines with good results in several districts in the Lower Thames Valley, which results I hope to publish before long.—*Martin A. C. Hinton, c/o J. C. Graham, Esq., 2 Garden Court, Temple, E.C., June 2nd,* 1900.

NEOLITHIC DEPOSIT NEAR BRIGHTON.—During a visit to Brighton in the early part of this year I came across a small section to the east of the town showing three feet of chalky rain-wash overlying the Palaeolithic rubble-drift. As the exposure was half-way up the slope, the deposit was probably much thicker lower down. The chief interest connected with this rain-wash lies in the fact that it is full of Neolithic flakes ; while it also yielded a couple of burnt stones. The surface above it is strewn with flint flakes, cores, scrapers, &c., which only differ from the underground specimens in being discoloured by oxide of iron and other matter. The occurrence of Neolithic débris in a deposit of this kind is of course no proof that the deposit is of that age ; but in this instance the great abundance of flakes throughout certainly indicates that the rain-wash is, in all probability, a Neolithic one. The surface specimens do not supply any evidence, as they may have been derived from the upper part of the deposit. Near the base I obtained several examples of the following land shells, viz. :—*Pupa muscorum* Lin., *Helix aspersa* Müll., *H. nemoralis* Lin., *Helicella itala* Lin.,

H. caperata Mont., *H. virgata* Da Costa. and *H. carthusiana* Müll. These mollusca are all living in the immediate neighbourhood at the present day. *Helicella virgata* is noteworthy, as. although it is known from the Palaeolithic drift of Barnwell and Ilford, it has not hitherto been recorded from any later deposit in Britain ; that is. on the mainland, for it is known from a hill-wash in the Isle of Wight. *Helix nemoralis* ranges from the Forest Bed upwards ; *Helicella itala* and *H. caperata* from the Pleistocene ; but *Helix aspersa* is probably a Neolithic introduction into this country, its earliest known appearance as a fossil being in the Neolithic alluvium of the Lea and Kennet valleys. *Helicella carthusiana* also seems to be a late arrival in Britain, though it formerly had a much wider distribution than at present. It is now limited to the chalk downs of the south-east coast, and does not range north of the River Stour, though it once inhabited East Anglia, as shown by its occurrence in alluvial beds at Butley and Felstead. —*J. P. Johnson;* 150 *High Street, Sutton.*

A NEW TYPE OF ROCK.—In a paper read before the Geological Society on May 23, Messrs. J. B. Hill and H. Kynaston dealt with "a new type of rock from Kentallen and elsewhere." It was originally described by Mr. J. J. H. Teall under the name of olivine-monzonite, and they regarded it as a type, round which they group a peculiar series of basic rocks discovered in several localities. The rocks consist essentially of olivine and augite. with smaller amounts of orthoclase, plagioclase, and biotite, while apatite and magnetite are accessory. The peculiar feature of the rocks is the association of alkali-felspar with olivine and augite. and the group is related to the shonkinite of Montana and the olivine-monzonite of Scandinavia. The occurrence of the rocks is connected with four neighbouring, but distinct, areas of intrusion. In these areas the new rock is the most basic type, and it occurs in the marginal portions of the areas. Close relationships exist between the different intrusive rocks in each area, so that it may be concluded that these constitute a "rock series" ranging from granite through augite-diorite towards the olivine-bearing rocks, in the plutonic phase, and from orthoclase-porphyry and porphyrite to augite-lamprophyre, in the dyke and sill phase. The whole assemblage appears to have been derived by a process of differentiation from one parent magma ; and the order of intrusion has been. in the main one of increasing acidity. An interesting discussion followed the reading of the paper, but it was felt that the new name of "Kentallenite" which had been applied to it was of too local an origin. It was suggested that it should in future be known as "Teallite," after the well-known petrologist, the President of the Geological Society. A series of slides were exhibited upon the screen, showing well-developed crystals of olivine and augite, with biotite in irregular patches, and orthoclase, &c., fitting in the interstices.

GEOLOGY OF NORTH STAFFORDSHIRE.—The Report of the North Staffordshire Field Club for 1899–1900 contains several geological notes and articles of more than local interest. In the report of the Chairman of Section E, Geology, Mr. Barke. F.G.S., makes reference to the re-survey of the North Staffordshire Coalfield and surrounding tracts, now drawing to a close. It appears that several results will be of great interest to geologists.

CORRESPONDENCE.

WE have pleasure in inviting any readers who desire to raise discussions on scientific subjects, to address their letters to the Editor, at 110 Strand, London, W.C. Our only restriction will be, in case the correspondence exceeds the bounds of courtesy; which we trust is a matter of great improbability. These letters may be anonymous. In that case they must be accompanied by the full name and address of the writer, not for publication, but as an earnest of good faith. The Editor does not hold himself responsible for the opinions of the correspondents.—*Ed. S.-G.*

FIELD-BOTANY CLUB.

To the Editor of SCIENCE-GOSSIP.

SIR,—It was with great pleasure that I read on page 23 of the June number of your journal that you propose to form in London a society for the encouragement of field-botany, and although I observe you intend to give it the name of "The London Field-Botany Club," I trust its committee will permit country members to join.

I am sure that others having an interest in field-botany besides myself would be glad to get in closer touch with each other, and I believe that such a club would be the means of regenerating the interest that was formerly so generally taken in our native plants. We must all of us feel thankful for the higher system of education brought about by the recognised science classes, which have taught us a better knowledge of the structure of plants. Yet we must not allow the morphological teaching to crush our acquaintance with the individuality of our old plant friends, who, I regret to say, now greet most of us each spring-time, with their beautiful blossoms, only to be passed with lordly indifference—because they have not been introduced to us. ARCTIUM LAPPA. *Birmingham.*

NATURE PICTURES OF PLANTS.

To the Editor of SCIENCE-GOSSIP.

SIR,—In the June issue of SCIENCE-GOSSIP Mr. Roger Verity asks if counterfeit representations of the scales of butterflies can be obtained on paper. I have not obtained nature prints of lepidoptera, but the following method of taking off impressions of plants answers admirably, and may be useful to some of your readers. I follow this method with plants: Oil a sheet of fine-woven paper with sweet olive oil; let it stand for about three minutes to soak through, then remove the superfluous oil with another piece of paper, and hang the first up to dry. When the oil is fairly well dried in, take a lighted lamp or candle and slowly move the paper in a horizontal direction over it, so as to touch the tip of the flame, till the paper is perfectly black. When you wish to take off impressions of plants, lay the specimen carefully on the black paper, face downwards, and a piece of clean paper over it, and rub the plant with your finger equally in all parts for about half a minute; then take up the plant, being careful not to disturb the order of the parts, and place it on the paper on which you wish to have the impression. Cover it with a piece of blotting-paper, and rub it with your finger for a short time, and you will obtain an admirable impression. The principal excellence of the above method is that the paper receives the impression of the most minute veins and fibres.

The impressions may afterwards be coloured according to nature. S. ALBERT WEBB. *41 Rothesay Road, Luton, Beds.*

NOTICES OF SOCIETIES.

Ordinary meetings are marked †, *excursions* •; *names of persons following excursions are of Conductors.* § *Lantern Illustrations.*

NOTTINGHAM NATURAL SCIENCE RAMBLING CLUB.
July 14.—" Sandiacre and Stony Cloud. W. Stafford.
  „ 28.—• Attenborough to Stapleford. J. Shipman, F.G.S.
GEOLOGISTS' ASSOCIATION OF LONDON.
July 7.—• Thrapston. Rev. Prof. J. F. Blake, M.A., F.G.S.
  „ 14.—° Croydon to Whyteleafe. President and G. E. Debley, F.G.S.
  „ 21.—• Winchfield and Hook. Dr. P. L. Sclater, F.R.S.

NOTICES TO CORRESPONDENTS.

TO CORRESPONDENTS AND EXCHANGERS.—SCIENCE-GOSSIP is published on the 25th of each month. All notes or other communications should reach us not later than the 18th of the month for insertion in the following number. No communications can be inserted or noticed without full name and address of writer. Notices of changes of address admitted free.

BUSINESS COMMUNICATIONS.—All Business communications relating to SCIENCE-GOSSIP must be addressed to the Proprietor of SCIENCE-GOSSIP, 110 Strand, London.

SUBSCRIPTIONS.—The Volumes of SCIENCE-GOSSIP begin with the June numbers, but Subscriptions may commence with any number, at the rate of 6s. 6d. for twelve months (including postage), and should be remitted to the Office, 110 Strand, London, W.C.

EDITORIAL COMMUNICATIONS, articles, books for review, instruments for notice, specimens for identification, &c., to be addressed to JOHN T. CARRINGTON, 110 Strand, London, W.C.

NOTICE.—Contributors are requested to strictly observe the following rules. All contributions must be *clearly* written on one side of the paper only. Words intended to be printed in *italics* should be marked under with a single line. Generic names must be given in full, excepting where used immediately before. Capitals may only be used for generic, and not specific names. Scientific names and names of places to be written in round hand.

THE Editor will be pleased to answer questions and name specimens through the Correspondence column of the magazine. Specimens, in good condition, of not more than three species to be sent at one time, *carriage paid.* Duplicates only to be sent, which will not be returned. The specimens must have identifying numbers attached, together with locality, date and particulars of capture.

THE Editor is not responsible for unused MSS., neither can be undertake to return them unless accompanied with stamps for return postage.

NOTICE.

SUBSCRIPTIONS (6s. 6d. per annum) may be paid at any time. The postage of SCIENCE-GOSSIP is really one penny, but only half that rate is charged to subscribers.

ANSWERS TO CORRESPONDENTS.

T. (Grimsby).—Some of your plants are aliens—1, 3, 6 we are not able to name; 2 is *Aspenys procumbens*; 4 appears to be *Echinospermum lappa*; 5, *Erysimum orientale*; 7, *Thlaspi arvense.*

EXCHANGES.

NOTICE.—Exchanges extending to thirty words (including name and address) admitted free, but additional words must be prepaid at the rate of threepence for every seven words or less.

OFFERED.—A few first-class lantern slides of Carboniferous fossils. for other l. slides of geological structures ; also a few good Palaeozoic fossils for exchange.—P. J. Roberts, 11 Back Ash Street, Bacup.
WANTED. Carpenter's " Microscope and its Revelations," edited by Dallinger, or other works on microscopy. Exchange Cassell's " Our Earth and its Story," complete, 3 bound vols., original edition, coloured plates, almost new.—John J. Ward, Lincoln Street, Coventry.
WANTED, to exchange Beck's small erect image microscope, in case, without stand, cost 18s. 6d., type slide of foraminifera, cost 12s., and brass mounting table with spirit lamp, cost 4s. 6d., for a telescope.—H. Ebbage, 11 Hall Quav Great Yarmouth.

THE PHOTOGRAPHY OF COLOUR.

By E. Sanger Shepherd.

THE term "photography of colour" is a wide one, and in this article I propose to deal with only two of its many branches, (*a*) the photography of a coloured object in its natural colours, and (*b*) the representation of a coloured object in a monochromatic print—work somewhat unhappily sent the lights and shades of the object photographed as seen by the normal human eye. For instance, blues apparently dark to the eye photograph as white, and bright yellows are represented in the print as black. The object of orthochromatic photography is to obtain in a mono-

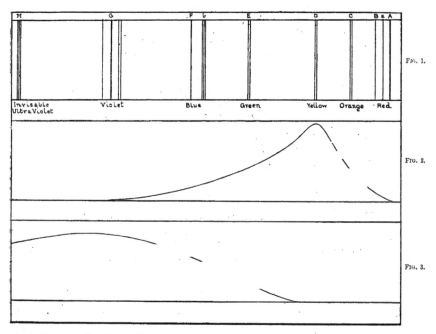

Fig. 1. Spectrum of White Light (Solar Spectrum). Fig. 2. Visual Luminosity of Spectrum (expressed by a curve). Fig. 3. Spectrum Sensitiveness of an Ordinary Dryplate.

known as orthochromatic or isochromatic photography.

The second of these divisions, the representation of a coloured object in a monochromatic print, is perhaps the more simple problem of the two, and may be dealt with first with advantage.

It is well known that when we photograph brightly coloured objects upon the ordinary photographic dry-plate or film, our print does not repre-

chrome print a rendering of all colours more in accordance with the visual impression.

Let us think for one moment of what we are trying to do. It is obvious that in photographing a coloured object we have two scales of contrast—firstly, " contrasts of colour," and, secondly, " contrasts of light and shade," but obviously we cannot represent contrasts of colour in black and white, so that our inquiry is really narrowed down to obtaining a correct representation of the lights and shades of our object.

In questions of colour it is always as well to deal first of all with colours in their greatest purity—that is, white light decomposed into its constituents. In fig. 1 we have a drawing of the prismatic spectrum, and in fig. 2 I have represented by a line the apparent brightness or luminosity to the human eye of the various parts of the spectrum, this line indicating by its height above the base-line the varying depths of light and shade. A perfect photograph of the spectrum should then give us a result something like this, but fig. 3, which represents a photograph of the spectrum taken upon an ordinary plate, shows us how far away from truth is really the ordinary photograph. We here see that nearly all the photographic action has taken place in the blue and invisible violet, whilst the highest visual luminosity of the spectrum—i.e. in the yellow—is represented as black. The difference shown is so great that at first we wonder how ordinary photography can give any representation worthy of the name. Were it not for the considerable amount of white light reflected by nearly all coloured objects in nature, the use of photography for pictorial representation would be practically impossible. Nevertheless, this want of orthochromatism or luminosity sensitiveness is serious in many ways. For instance, the ordinary plate does not distinguish the luminous value of blue sky from that of white cloud, but it is more especially when photographing close objects, notably where the colours are brilliant, that the deficiency of the ordinary plate is most felt.

It is clear, then, from these facts, that what we have to do is to secure such photographic action upon our plates as will give us densities in our negatives directly proportional to the visual intensities of the light reflected from the object, ignoring totally, no matter what they may be, the colours of the object.

Now, nearly all photographic emulsions are sensitive to some extent to all the visible spectrum, but such sensitiveness differs very widely from the eye-luminosity of the curve shown above.

There are two methods by which we might hope to improve our results. First, by so altering the constitution of the emulsion as to render it more sensitive to green and red; and, secondly, by filtering or screening down the action of the blue and violet so as to allow the red and green sensitiveness of the emulsion to impress the plate in its proper proportion.

With regard to the first method, a very great deal has been done by various workers to improve the sensitive film. In 1873, Professor Vogel discovered that certain dyes incorporated with the emulsion had the property of sensitising the resulting film to the portions of the spectrum absorbed by the dye used. Not all dyes, however, possess this property of increasing sensitiveness, and much laborious experiment was necessary before plates of commercial value were obtainable. In fig. 4 we have the sensitiveness of several commercial brands of plates to the white light of the spectrum represented by curves. The best of these curves—that is, the one nearest approaching the curve of eye-luminosities of the spectrum—is the last shown, namely, that of the Cadett Lightning Spectrum plate. On comparing this, however, with the ideal curve, we find it is still far from perfect, the blue and violet still being responsible for by far the greater portion of the photographic action; but one will notice that now we have considerable sensitiveness in the green and the red, and by cutting down the blue and violet by a filter, it is possible to secure, with a reasonable length of time of exposure, an accurate representation of the spectrum. Obviously the construction of a light filter which will cut down the action of the green, blue, and violet in exactly the right proportion is a very difficult matter. The few early experimenters who succeeded in doing this found it a very long and wearisome task; for each trial of a filtering medium negatives of the spectrum had to be taken upon the sensitive plate, and then these negatives had to be measured by a photometer, for the amount of light passed. Only a very narrow band of the spectrum could be measured at one time, and twenty or thirty measurements of the different parts of the spectrum were necessary, before we could tell what amount of success had been obtained. Such a method, therefore, was of little use from the commercial point of view, on account of the cost of the necessary skilled labour to conduct these delicate measurements.

Sir William Abney has, however, by the invention of his Colour-Sensitometer, enabled us to produce light filters of great accuracy at so reasonable a cost that every photographer can avail himself of their use. It is a well-known fact in photometry that, although the eye is a very bad judge of the difference in intensity between two unequal lights, or of the amount of light transmitted by two unequal densities in a negative, it is an excellent judge of the equality of two lights, or of the amount of light transmitted by two equal or nearly equal densities on a photographic plate. Sir William Abney availed himself of this fact in the construction of his sensitometer. The principle of the sensitometer is as follows:—Suppose we take a white, a red, a yellow, a green, a blue-green and a blue glass, and mount these in a row in such a manner that we can photograph them through the light filter we are adjusting, we shall have, provided the glasses are suitably chosen, a very fair representation of the spectrum; with the additional advantage that they will transmit colours which are to some extent mixed colours, and thereby the excess of one spectrum colour may be balanced by the defect of another. If we now photograph such an object, the different opacities of a print, from our negative, compared with the test object will show us very approximately the amount of success obtained. In order to tell with any degree of exactness, however, we have first to measure accurately the luminosity of each of the coloured glasses as compared with the white glass, and then to measure the amount of light passed by each of the patches of deposit in

may be opened or closed whilst revolving at a high rate of speed, and as the sector is divided in degrees, we can measure accurately the difference in the intensity of the two beams. To use the instrument we first balance the two beams until the brightness of the two adjacent patches upon the screen are equal. Into the path of the direct beam we then place the coloured glass, the luminosity of which we wish to measure. The diverted beam

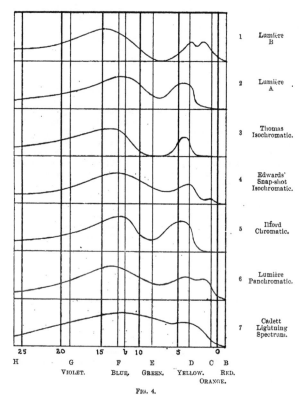

1	Lumière B
2	Lumière A
3	Thomas Isochromatic.
4	Edwards' Snap-shot Isochromatic.
5	Ilford Chromatic.
6	Lumière Panchromatic.
7	Cadett Lightning Spectrum.

25   20   15   10   5   0

H   G   F   E   D   C B

VIOLET.   BLUE. GREEN. YELLOW. RED.
ORANGE.

FIG. 4.

our negative. In order to measure the luminosity of a coloured glass as against white light, we use the whitest light available, that proceeding from the crater of the electric arc. A beam of this white light is allowed to fall as a rectangular patch upon a screen; a portion of this beam is reflected to one side and caused to form another patch close beside the first, a rod being placed between the two beams so that each beam shines into the shadow of the rod cast by the other (fig. 5). In this diverted beam is placed a rotating sector; by a mechanical device the apertures of this sector

will now appear very much brighter than the coloured beam; but by reducing the aperture of the sectors, we shall soon reach a point where the reduced light of the diverted beam appears less bright than the coloured patch. By alternately opening and closing the sectors, we shall find a point where the coloured and the white patch appear of the same brightness. We fix the sectors on this point, and, on stopping their revolution, we can read the aperture at which they are set. That such measurements of colour against white light are readable is proved by the consistency between repeated measurements of the same glass. The sectors are revolving at about 4,000 revolutions per minute, and it is only when the motor driving them has been stopped that we can read their aperture , so that it is impossible for the observer to be influenced in his estimation by his knowledge of previous readings ; yet twenty or thirty successive readings of the same glass will not vary more than 1 per cent. By this instrument we measure once for all the brightness of the light transmitted by each of our test glasses as against white light, and from these data it is easy to calculate what proportion of light the opacities of our negatives of the test object should transmit.

Sir William Abney has also given us the means whereby we may avoid the measurement of each negative. To do this we mount in front of the row of glasses a revolving sector with apertures so arranged that the light coming through each glass of the series may be reduced to equal luminosity— the luminosity of the darkest glass in the series. When once we have this instrument set up, the testing of a light filter is a very easy matter. A photograph taken through the filter under test should give us a row of patches of equal density, and as the eye is a very excellent judge of the quality of the two adjacent opacities, we can tell at once by simple observation the accuracy of our light filter.

With regard to the light filter itself, in addition to its power of correcting accurately the imperfections of the photographic plate, there are two conditions which it must fulfil. First, it must have a fair amount of permanency and durability, and, secondly, it must not affect the definition of the lens. Filters might be made by filling a glass cell with a coloured liquid, but such an arrangement is not very satisfactory in practice, especially if used out of doors, on account of the risk of accidental damage or leakage, and the bad effect on definition caused by the circulation of the liquid with every change of temperature. Therefore, for commercial purposes, we are restricted to a coloured film, sealed in optical contact between two pieces of optically ground and polished plate glass. Fortunately, the aniline dyes give us an almost endless variety of absorptions, and a large number of these which have been carefully selected and tested prove to be quite permanent, when protected by being sealed between two glass plates. By using combinations of these dyes in varying depths of tint, it is possible to adjust a filter to almost any curve of the spectrum. In the early days of orthochromatic photography a piece of ordinary yellow glass, ground and polished, was used as a substitute for a correct light filter ; such glasses, however, cut out the action of the blue entirely, a defect almost as bad as the evil it was supposed to avoid.

Some years ago I showed at the exhibition of the Royal Photographic Society a frame of diaphragm light filters. in which the colour film was

pierced in the centre, so that a small portion of white light would be mixed with a far larger proportion of the filtered light. This, however, was a clumsy device and in the later filters has been done away with, by using two or more dyed films, so that one dye would reduce the action of the green, and another dye the action of the blue. By respectively varying the depth of tint of these colour films, we can correct for definite regions of the spectrum, and it is only by this device that the production of filters of great accuracy has been made commercially possible. For instance, the Cadett light filter gives practically perfect rendering of all colours when used with the Cadett rapid spectrum plate, and yet these filters are but little more expensive than the old pot-yellow glass.

An examination of a number of the Cadett light filters and comparison with the photographic test plates reveals several interesting points. For one thing it proves how utterly unreliable is the eye when used as a judge of a light filter. In reality the light filter is cutting out what the eye does not see, the photographic plate being sensitive to luminosities of blue and violet so low as to be quite invisible to the human eye. Perhaps half-a-dozen filters will appear exactly the same to the eye, although differing very widely indeed in their filtering power, as shown by the deposits in the respective patches of the test negatives. On the other hand two filters, the differences in the colour of which are distinctly visible to the eye, may exactly correspond in their filtering power as shown by the test negative.

When we compare the very wide difference between the curve given by the spectrum on the Cadett lightning spectrum plate without a filter (No. 7, fig. 4) with the ideal curve (fig. 2), we see that the effect of a filter which will sufficiently reduce the action of the blue, so as to allow the reds and greens to impress the plate correctly, must considerably increase the time of exposure necessary for securing a negative. With the Cadett "Absolutus" filter, exposures are increased about forty times, but owing to the extreme speed of the plate, far in advance of any other sensitive surface known, the exposures in the studio are comparatively short, and negatives of brightly-lighted landscapes fully exposed may be obtained in a fraction of a second.

Since the "Absolutus" filter was placed upon the market Messrs. Cadett have introduced another filter, which they term "Gilvus." This filter is made in precisely the same accurate manner as the "Absolutus" filter, with regard to its rendering of blue and green, but is not corrected for red. By ignoring the red the filter can be made much more transparent, so that the increased time of exposure is only about four times that required without the filter. This filter proves very valuable in landscape work, and saves a great deal of retouching in portraiture ; and where the object photographed contains but little bright red, the results are almost as good as those obtained with the "Absolutus." filter.

A very convenient method of using a light filter is to have it mounted in a little brass cap, which will just slip on to the hood of the lens. When mounted in this form it is only a moment's work to put the filter into position, and being outside the camera there is little fear of the filter being left unnoticed when using the ordinary plates. If the ordinary leather cap of the lens is fairly free, the thin brass cap in which the filter is mounted will not interfere with its use in exposure. The working operations of the spectrum plate differ very little from ordinary photography, except that owing to their extreme colour sensitiveness they will not bear exposure to so bright a light in the dark room as the ordinary plate; but, by taking care to shield the dish during the early part of development, the dark room may still. have ample illumination.

Messrs. Cadett have introduced a safe light for use with their plates, which will enable a much brighter light to be used than would be safe with the ordinary ruby light, but such a light, although very convenient, is not an absolute necessity.

In development it is always advisable to use a developer which will give negatives of a neutral grey colour. Most of the modern developers are suitable, more especially metol and rodinal. Hydrokinone, however, should be avoided, as the image takes some considerable time to appear, and should development be stopped in the earlier stages, there is a liability of the blues being proportionally too dense in the negative. With this developer it has been found that the blues of the object appear first, the reds and greens only gaining their proper density after the developer has acted for some considerable time.

(*To be continued.*)

PALAEOLITHIC MAN IN VALLEY OF THE WANDLE.

By J. P. Johnson.

THE fluviatile and other deposits of the valley of the Wandle were described by Mr. W. Whitaker, B.A., F.R.S., in the "Geology of London," and more recently by Dr. G. J. Hinde in his "Notes on the Gravels of Croydon" ([1]); but no mention is made of the occurrence in them of Palaeolithic implements, and, as far as I am aware, there is only one subsequent record. It is therefore desirable to draw attention to some further evidence I have obtained of the former presence of Palaeolithic man in this district. In order to illustrate my finds it will be necessary to give a brief account of these deposits, and in so doing to go over some of the ground already dealt with; but I have much to add that is new.

The drainage area of the Wandle is bounded on the west by the high ground which divides it from that of the Hog's Mill and Beverley Brooks· on the east by the ridge of hills that separate it from the valley of the Ravensbourne; and on the south by a portion of the great chalk plateau known as the North Downs.

Starting from its source at Croydon, the river first runs westwards, skirting the chalk outcrop, to Carshalton, and then turns abruptly north, continuing along the left margin of the valley, until it enters the Thames at Wandsworth. It therefore flows entirely through a tract of soft Eocene strata.

That part of the chalk plateau which constitutes the southern boundary of the above-described district is furrowed by three narrow dry valleys that converge on the river valley at Croydon, and may be considered a southern extension.

(1) "Trans. Croydon Micro. and Nat. Hist. Club," 1896–7.

The Palaeolithic drift occurs in two terraces that are separated from one another by a belt of London Clay, except along the course of the Norbury Brook, where they are joined together by a thin strip of gravel, which is probably due to subsequent slippage—a process that would be facilitated by the flow of water in that direction.

THE UPPER TERRACE.—The upper and older terrace, which is wholly on the eastern side of the valley, extends in a straight line from Croydon to Wandsworth Common, where it joins on to the Thames gravels of the same horizon. It is clearly a river-drift, and consists in the main of ochreous subangular flint gravel; but one meets with flints in every condition of wear, from the well-rolled pebble that has travelled a long distance, to the unworn nodule which has scarcely been detached from the parent rock. They are mixed with a varying proportion of sand and loam, all being the débris resulting from the destruction of the Chalk, the Kentish Tertiaries, the London Clay, and the other strata of the district. There are no stones that have not been derived from local deposits, though one or two of them are far removed from their original source. The chert, for instance, which is said to have been detected in this deposit has come by way of the High Level Drifts of the chalk plateau from the outcrop of the Lower Greensand in the Wealden area. No contemporaneous organic remains have been found in this gravel, for it has been thoroughly decalcified by the action of percolating water. Palaeolithic man has, however, left traces of himself in the shape of "a fine implement nearly eight inches long" found by Mr. A. E. Salter, B.Sc., at Thornton Heath [2]. The drift of Wandsworth Common has yielded a large number of flint implements, but it belongs to the Thames rather than the Wandle.

THE LOWER TERRACE.—After the formation of what is now known as the upper terrace, the deposition of gravel for some reason or other temporarily stopped, and thus failed to counteract, as it had hitherto, the erosive action of the running water on the channel in which it flowed, so that the subsequent accumulation of gravel took place at a considerably lower level. The lower sheet of gravel not only occupies the river valley, but also extends into the dry valleys to the south. It may be divided into two contemporaneous sections, viz. (a) the angular detritus of the chalk area which comprises the dry valleys and coombes which penetrate the chalk plateau, and (b) the subangular gravel of the Eocene tract.

There is one deposit which cannot be definitely referred to either of the above sections. Clothing the eastern end of the chalk slope which extends from Sutton to Carshalton, between the Upper (or Carshalton) and Lower Roads, and resting on the London Clay at its foot, is a mass of rearranged Kentish Tertiaries. This is evidently the deposit which was exposed during the excavation for a new sewer described by Mr. W. W. Watts at a meeting of the Geological Society in 1898. "These excavations are situated at a spot which on the Geological Survey map is coloured as London Clay, and the features of the ground fully justified this conclusion. The excavations, however, have shown that there are loamy and sandy beds of a light yellow colour, some 14 or 15 feet in thickness. At the base these sandy beds become dark and clayey in places, and include flints and pebbles, while below this is London Clay. In the dark pebbly layer were found" various mammalian remains, which Mr. E. T. Newton, F.R.S., determined to be the bones of two or three horses (*Equus caballus*), the skull and part of the skeleton of a woolly rhinoceros (*Rhinoceros antiquitatis*), and a piece of an elephant's tusk. This deposit, some small sections of which are exposed at the present time, was probably found in the same way as the angular detritus about to be described.

THE ANGULAR DETRITUS OF THE DRY VALLEYS.—South of the railway station at Carshalton one of those small coombes that form so picturesque a feature of the North Downs opens on to the river valley. Lining this and projecting on to the lower ground is a mass of angular detritus, which is well exposed in the two pits sunk in the triangular piece of land situated between the Shorts, Alma, and Carshalton roads. The larger section is nearly 120 feet long and 15 feet in depth. It shows a confused mass

of flints and chalky sand, from 5 to 7 feet in thickness, overlaid by alternations of the same materials charged with chalk shots. The majority of the flints, though broken, are but little worn or weathered, and are almost as fresh as those that still remain embedded in the parent limestone. The fine buff-coloured sand, the abundance of green-coated flints, and of small flint pebbles, being all characteristic constituents of the Kentish Tertiaries, point to the destruction and subsequent incorporation of large masses of these very much older strata, such as still exist not far away. The only other foreign materials present are pieces of a compact ironstone. I have obtained from here a trimmed flake of Palaeolithic type.

Similar angular detritus, in which the sand is replaced by loam—evidently decomposed chalk—covers the whole of the bottom of the Chipstead, Hooley, and Caterham valleys. These are the valleys I have previously referred to as converging south of Croydon. The Caterham Valley sheet contains in addition a great number of boulders of pebble-conglomerate. In a now partially filled-in pit at Whyteleafe rounded masses of chalk were also frequently exhumed, and I once noticed one of those blocks of hard sandstone known as greywethers or sarsens. At the bottom of the pit I obtained teeth of the wild ox (*Bos primigenius*) and of the woolly rhinoceros. Dr. Hinde mentions getting a portion of the antler of a reindeer (*Rangifer tarandus*) together with remains of the horse and ox, of *Rhinoceros leptorhinus*, and of the mammoth (*Elephas primigenius*). Mr. A. E. Salter also records being shown a tooth of the last by the workmen. In the Horniman Museum there is, from Coulsdon, a portion of the jaw of the same species of animal retaining one molar. This deposit is identical with the typical rubble-drift of the Hants Basin, and was undoubtedly formed in the same way. It has, however, been suggested that this detritus is of fluviatile origin, that the Wandle during the lower terrace period extended up the Hooley Valley, and that tributary streams occupied the bottom of the Chipstead and Caterham Valleys. The appearance at intervals of a small stream, the Bourne, in the Caterham Valley would at first seem to support this view. It is indeed quite feasible that a small stream might in time remove much of the soluble limestone from the valley bottom, leaving the imperishable flints little, if at all, worn ; but in this instance the occurrence of mammalian bones at the base of the deposit precludes the possibility. On the other hand, a river strong enough to drag stones along its bed and bury animal remains beneath twelve feet of gravel, would certainly impart a considerable amount of wear to the flints. Hence one is forced to fall back on the subaërial theory in order to explain the phenomenon [3].

(2) " Pebbly and other Gravels in Southern England."—" Proc. Geologists' Assoc.," vol. xv. (1898).

(3) See O. Reid, F.R.S., " Origin of Dry Chalk Valleys and of Coombe-rock [=Rubble-drift]."—" Quart. Jour. Geol. Soc."

. In Smitham Bottom, where the three dry valleys converge, the angular detritus presents another facies. This is well shown in the pits at. Purley, where there is hardly any matrix. The flints are weathered, discoloured, and begin to show signs of wear. The only difference between this and the typical rubble-drift is that it has a washed appearance. Though the constituents are the same, there is in addition a quantity of very small, subangular gravel, derived from some older drift such as that filling the "pipes" in the chalk quarry at Purley station. Further north, at Croydon, the angular detritus passes laterally into the river gravel.

THE SUBANGULAR GRAVEL OF THE EOCENE TRACT.—Beyond Croydon the lower terrace drift takes the form of a rather coarse river gravel. It is similar in composition to the angular detritus we have just been considering. It differs from it, in that the great bulk of the flints are of an ochreous colour and have every corner rounded off through long-continued rolling. There can be no doubt that the same force which produced the rubble-drift in the dry valley supplied the greater part of the flints to the ancient Wandle during the lower terrace epoch, for at that time the river flowed, as it still flows, entirely through soft Eocene strata, which only contain flint in the form of small pebbles. This tract starts with a broad sheet extending east and west between Croydon and Carshalton, and narrows irregularly northwards until it is but an eighth of a mile wide at its junction with the Thames gravels of the same horizon at Wandsworth. This sheet is nearly all on the right bank, there being a very little on the left side between Croydon and Carshalton, and north of this only a solitary patch around Merton and Morden.

As is the case with all porous gravels of this type, mammalian remains are very rare and represent only some of the larger animals which existed at the time of their deposition. I have a molar of mammoth from Mitcham, and Dr. Hinde, in the paper already quoted, records in addition the finding of bones of rhinoceros, horse, reindeer, and roebuck (*Capreolus caprea*). It is noteworthy that the reindeer remains both from here and the Caterham Valley are referred to the large form, which is so common in the Thames Valley, and that they therefore belong to the "woodland" group of this species. More important is a Palaeolithic flint implement of the pointed type, which I obtained from a section exposed in Miles Lane. It is much rolled, and the edge is very battered through use. It is not perfect, a portion of the base having been broken off before it found. its way into the old river bed. I have also a trimmed flake, that I found on the common.

In looking over some gravel which had been laid down in a road near Tooting Junction I came across an undoubted Palaeolithic flake. This is of special interest, as although it is stained a deep ochreous colour it is as sharp as when first struck off the original piece. Unfortunately I could not find out where this gravel came from, though it was doubtless of local origin.

NEWER DEPOSITS.—Newer than the Palaeolithic drift is the comparatively modern alluvium which fills the old bends in the river bed. It consists of sand, mud, and peat; but little is known of it, and nothing of its contents. It may date from Neolithic to recent times. While relics of Palaeolithic man are thus so scarce in the valley of the Wandle, Neolithic tools are extremely abundant in the surface soil, especially on the chalk slopes, which are literally strewn with flint flakes, cores, scrapers, etc.

Sutton, Surrey.

YORKSHIRE NATURALISTS' UNION.

UPON the invitation of Mr. George T. Porritt, F.L.S., F.E.S., of Crossland Hall, near Huddersfield, the members of this Union, to the number of about a hundred, gathered from all parts of the county, for a meeting at his residence. Favoured by exceptionally fine weather they had the pleasure of exploring the Colne, Holme, and Meltham Valleys. The visit was preceded at about ten o'clock in the morning by an examination of the considerable Natural History Museum connected with the Huddersfield Technical College. This was explained by Mr. S. L. Mosley. It includes the late Joseph Whitwam's Conchological collection and the late Samuel Learoyd's collection of minerals. Dr. Rawson, F.I.C., the Principal, conducted the party over the rest of the Institute. Subsequently the members divided into three sections, the first party under the guidance of Messrs. T. W. Woodhead, F.L.S., and A. W. Sykes, the second party was conducted by Messrs. Mosley and Bulmer, and the third by Messrs. W. Tunstall, F.E.S., and Harry Mellor. These sections visited the districts above named.

At about five o'clock in the evening the whole of the members assembled at Crossland Hall, which is charmingly situated in a wealth of woodland, where they were entertained by the President and Mrs. Porritt. After the repast addresses were given on various subjects by the President, Messrs. E. Hawksworth (Leeds), Dr. Corbett (Doncaster), Lawton (Skelmanthorpe), Law (Hipperholme), J. H. Rowntree (Scarborough), K. McLean (Harrogate), Crowther (Elland), Tunstall (Meltham Mills), Bayford (Barnsley), Woodhead (Huddersfield), Crosland and Cash (Halifax), the Rev. W. Fowler (Liversedge), and several others. The success of this excursion shows the importance of County Societies being organised in the admirable manner attained by the Yorkshire Union.

BUTTERFLIES OF THE PALAEARCTIC REGION.

BY HENRY CHARLES LANG, M.D., M.R.C.S., L.R.C.P. LOND., F.E.S.

(*Continued from page* 41[1].)

(PIERIS continued.)

19. **P. napi** Lin. Syst. Nat. X, 468. Lg. B. E.
p. 31. "The green-veined white." Pl. VII. fig. 1,
pl. XV. larva.
40—48 mm.

Wings white, with the bases dusky. The nervures
are distinct and black. F.w. with the apices and
sometimes the ends of the nervures dusky, and
sometimes with, but often without, a small black
spot midway between centre and ou. marg. ♀
with two black spots as in *P. rapae*. H.w. with a
black costal spot. ♀ darker than ♂, having blackish
shading along the course of the nervures. U.s.
f.w. white, tipped with greenish yellow, nervures
conspicuous, two black spots as in the allied species.
H.w. pale yellow, with dark scales placed thickly
along the course of the nervures, giving the appearance of green veins.

Pieris deota, ♀.

This species exhibits a more strongly marked
seasonal dimorphism than *P. rapae.* "In the
spring brood the apical blotch is greyer and not so
pronounced as in the summer brood, and the dark
markings are more suffused, whilst the u.s. h.w. is
yellower and veined much more distinctly, the
dusting of the nervures being of almost equal
density to the margin of the wing, but in the
summer emergence it becomes very faint towards
the margin" (H. Williams).

HAB. Europe, Asia Minor, Persia, Siberia, Altai.
Throughout the season.

LARVA. Green with a dark dorsal shading,
spiracles marked with red and yellow, on Cruciferae.
VI.—IX.

a. var. *napaeae* Esp. The summer brood as above
(gen. II. al. post. subt. pallid. Stgr. Cat. 3.)

b. ab. *bryoniae* O. The Alpine form of the species,
and probably the survival of a normal primaeval

condition. ♂ larger than type, and more strongly
veined. ♀ with ground colour yellowish, and the
dark markings very suffused and with dusky powder-
ing more or less distributed between the nervures
over the wing area. In a selected series from
Davos Plaz in my collection there is much variation
in the ground colour, and in the intensity of the
dark coloration. HABITAT. The Alps of France,
Switzerland and Austria, also Scandinavia, at an
elevation of between 4,000 and 5,000 feet. Aberrations seem to occur in Central Asia and Siberia
which seem to be identical with *bryoniae*; they are,
however, said not to be constant. Some very dusky
forms of the ♀ occur occasionally in England and
elsewhere, which very closely approach *bryoniae* in
appearance; the ground colour, however, in these
is white. They belong to the 1st generation, and are
probably the var. *intermedia* of Kroulikowsky.

c. ab. *flavescens* Stgr. Gen. II. ♂ creamy
yellow in the ground colour. ♀ f.w. darker than
bryoniae, h.w. light yellow with black marginal
spots. This is a well-marked local aberration.
The only specimens I have seen are those taken by
Miss M. Fountaine at Modling, near Vienna, in
the summer of 1897.

Pieris mesentina, ♀.

20. **P. ochsenheimeri** Stgr.
30—39 mm.

The apex of f.w. of ♂ rather broadly black,
and this apical patch is continued in a row of
marginal triangular spots at the extremities of the
nervules. Between the second and third median
nervules is a well-defined black spot. H.w. with a
well-marked costal spot, and a marginal row of
triangular black spots. ♀ very like some small
light specimens of *P. napi* var. *bryoniae*, only the
spots and the lines along the neuration are more
defined and the ground colour is white. U.s.
ground colour of h.w. and apices of f.w. much
browner in tint and less yellowish green than in
either *P. napi* or ab. *bryoniae.*

HAB. Namangan, Turkestan, Alai mountains, Taldyk. E. Alai, end of May (R. & H.). I have two specimens from the Alai from (Mr. Elwes) dated June. Two from Staudinger with the locality Pamir. Elwes says that this species seems to represent *bryoniae* in the high mountains of the Pamir.

On the whole, I think *P. ochsenheimeri* may be admitted as a good species, though it comes very near to ab. *bryoniae*. The markings of the ♂ and the white colour of ♀, together with the coloration of the u.s., seem distinctive.

[The following are probably only local forms of *Pieris napi* and *P. oleraceae* :—*P. venosa* Scudd. Proc. Bost. Nat. Hist. Soc. VIII., p. 182. 30—35 mm. Resembles *P. napi*, but is smaller, and in ♂ without markings above, very faintly marked in ♀ u.s. as in *P. napi*, but markings very light and faint. HAB. Alaska. *P. frigida* Scudd. Proc. Bost. VIII. (1861), p. 4. Stgr. Cat., 1871, p. 3. (*P. oleraceae* B. var. ?).]

21. **P. callidice** Esp. 115, 2, 3. Lg. B. E. p. 32. pl. VII. fig. 3.

42—46 mm.

Wings white, ♂ with an elongated black spot on f.w. at the end of the disc. cell. Midway between this and apex a row of small black spots, of three near costa and one below the second nervule, a marginal row of five small triangular black spots reaching from apex. H.w. without pattern, with the exception of what appears through of the underside markings; base dusky. ♀ somewhat larger than ♂ and with the markings greatly expanded, so as to have the following characters. F.w. disc. spot large and sometimes nearly square, marg. and ante-marg. spots forming a broad blackish band enclosing a row of white angulated spots. H.w. with a dusky border enclosing a row of white angulated spots. U.s. disc. spot and apices with greenish yellow. H.w. neuration broadly marked with brownish green, enclosing light yellow spaces mostly of an arrowhead shape.

HAB. Alpine slopes from 5,000 to 8,000 feet in the mountains of Switzerland, France, Austria and the Caucasus. VI-VIII. according to altitude. The flight of this species is powerful.

LARVA. Dark greyish blue spotted with black. On each segment four longitudinal stripes, marked with light yellow spots. IX. on various Alpine Cruciferae. PUPA: Grey, finely powdered with black, and with a yellow dorsal line (Bd.).

a. var. *chrysidice* H.S., 200, 3, p. 97. Somewhat larger than type, ♂ f.w. with the black spots more strongly defined. Disc. spot with a light centre in both sexes. ♀ duskier and more broadly marked. U.s. ground colour nearly white, markings greener. An Asiatic form of the species. HAB. Transcaucasia, Asia Minor, Turkestan, Tarbagtai, Altai (Elwes). VI., VII.

b. var. *kalora* Moore. Larger than type. No white centre to disc. spot f.w. ♀ more dusky and more broadly marked. U.s. h.w. in both sexes with the darker markings predominating. HAB. Lahoul, Himalayas, 13,000 ft. VI.

" I do not see how the line can be drawn between *chrysidice*, *callidice*, and *kalora* of Moore, when a large number are compared, though those from the European Alps are usually more yellowish on the hind wing below." (Elwes., T. E. S. L. 1899. III., p. 318.)

22. **P. daplidice** Lin. Syst. Nat. x. 468. Lg. B. E. p. 33. Pl. VII. fig. 4. " The green chequered-white."

39–47 mm.

♂ f.w. with a black patch at apex extending half-way along ou. marg. and enclosing four white spots alternately from above large and small; at end of disc. cell is a large square black spot with a white line in the centre, where it is intersected by the nervule. H.w. without markings, except sometimes a few obscure black marg. spots. The green pattern of the u.s. shows through as a grey shading. ♀ generally rather larger than ♂, markings more extended. F.w. with an additional black spot near in. marg. H.w. with marginal and ante-marginal rows of greyish-black spots. U.s. f.w. green spotted with white at apices, disc. spot powdered with green. H.w. green with a central wavy white band, a marginal row of white spots, and three more white spots in the anterior part of wing area towards base.

HAB. The Palaearctic Region, excepting the Polar portion. IV–IX. Double brooded. It is occasionally taken in England, principally on the South Coast near Dover, etc. These specimens are probably immigrant from the Continent, and are always those of the summer brood.

LARVA. Greyish blue, covered with small black granulations, with four longitudinal white stripes and a yellow spot on each segment. Legs and ventral surface white. On Cruciferae and Resedaceae. Ve. VIIe. VIII. PUPA. Grey, speckled with black and with reddish stripes.

a. var. *bellidice*. O. i. 2, 154. Lg. B. E. p. 34, pl. VIII. fig. 2. Smaller than type, markings more grey than black, at least in ♂. U.s. h.w. with smaller white spots, and with the green colour much darker. A small form of the first generation. HAB. The more southern parts of Europe, South Russia, and Turkestan. IV., V.

b. var. *raphani* Esp. ? *persica* Biern. U.s. with light yellow markings in place of green. HAB. Turkestan and Persia. VIII.

c. var. *albidice* Oberth. A name given by M Oberthur to a local variety found in Spain and Algiers. VII—VIII. Smaller, more slender, and whiter than type, the markings of the underside not appearing above.

(*To be continued.*)

D 3

ON THE NATURE OF LIFE.

By F. J. ALLEN, M.A.. M.D.

(Continued from page 13.)

I WAS travelling when Mr. Geoffrey Martin's article continuing the discussion on this subject appeared; I was therefore unable to reply to it immediately; but will now try to answer certain of his objections to my views.

Although neglect of the functions of nitrogen is a prominent fault of the older school of biological chemists, there is no ground for attributing to me the opposite fault of neglecting the carbon. In the publication (¹) referred to in my previous paper I have carefully described its great function as the storer of energy, its deoxidation and subsequent oxidation being the principal means of the "energy traffic." I have also given my own views, as well as those of other theorists, concerning the mode in which the carbon is united to the nitrogen and other elements in living and dead material respectively. I must repeat, however, that lability is a rare quality of carbon compounds, unless nitrogen also be present; and that the highest lability is achieved when the nitrogen is combined with oxygen as well as with carbon.

The theory that nitrogen is the central or linking element in living substance is not disproved, but rather supported, by the facts which Mr. Martin adduces against it. I have pointed out (²) that "when nitrogen is the central or connecting element between complex radicles, as in an amine or an alkaloid, the properties are usually pronounced; whereas, when nitrogen is peripherally situated, as in an amide, the properties are usually indifferent." Considering how weak the chemical attractions of nitrogen are, it is only to be expected that the greater the number of heavy groups connected with it, the greater will be the strain and consequent tendency to rupture. In other words, the central situation of nitrogen tends to produce chemical instability or lability. In such a compound as we are describing, oxygen behaves as a heavy radicle if connected with nitrogen; for the retention of oxygen by nitrogen is rendered difficult by the counter-attraction which the carbon, hydrogen, sulphur, &c., of the same molecule exert on the oxygen.

The nitrogenous products whose formulae Mr. Martin quotes are dead substances with peripheral nitrogen. Before the breakdown of the parent living molecule, those nitrogen atoms formed the connecting links by which the radicles held on to the mass, and then the nitrogen was the

"central or linking element." Central nitrogen is characterised by lability and life, peripheral nitrogen by indifference and death.

I have summed up my views with regard to the structure of the active molecule of living substance as follows:—"It is a molecule of enormous size, and, so far as the dynamic (³) elements are concerned, its various groups are linked together by many nitrogen atoms, which are placed *internally but not in chain*. It is not a proteid, a cyan-compound, an amide, an amine, nor an alkaloid; but something that can yield some of these during life, and others at its death. Death consists in the relaxation of the strained relationship of the nitrogen to the rest of the molecule. When thus the silver cord is loosed, the released groups fall into a state of repose. Most of these groups are proteids, in which the nitrogen is peripheral, triad, and unoxidised, having yielded its oxygen to some other element; if, however, such a proteid molecule be applied to a living cell, it can be linked on again by its nitrogen, which thus once more becomes central."

My remarks on silicon do not admit of all the interpretation which Mr. Martin gives to them. As to silicon life, I raised no objection on the ground of strong chemical reagents or high temperature, all of which may be present in some parts of the universe but I objected to the suggestion that the chemistry of the silicates (as known to *us*) was akin to life. Let it never be forgotten that the great phenomenon of life is *not chemical complexity*, but *the energy traffic.* The deoxidation and subsequent oxidation of carbon involves a very great accumulation and dispersion of energy, whereas the changes in silicates involve a comparatively small transfer of energy. There may be some element which under certain circumstances can rob silicon of its oxygen as nitrogen robs carbon. Such an action would involve a great accumulation of energy in the new silicon compound, and by an opposite chemical change the energy could be expended in useful work. This is the kind of silicon chemistry that we have to find before we can attribute to silicon a vital *rôle* comparable with that of the dynamic elements. For many years I have taught the possibility of such a function of silicon, but have held it probable that at high temperatures the energy traffic would be carried on more readily by certain other elements,

(1) "What is Life?" Proc. Birmingham Nat. Hist. and Philos. Soc, vol. xi. part i., 1899.
(2) *Ibid.*

(3) The elements nitrogen, oxygen, carbon, and hydrogen may be called the "dynamic elements," because they are the chief agents in the energy traffic.

such as phosphorus, sulphur, iron, and iodine, whose known chemistry indicates considerable dynamic possibilities.

Silicon is still a fairly abundant constituent of the living body, both of animals and vegetables. Its chief function is to give rigidity to the framework. There is no reason to think that its chemical reactions contribute sensibly to the energy traffic, or that it ever combines directly with any other element than oxygen.

The meteoric theory is the modern version of the nebular theory. It is not customary nowadays to regard nebulae as masses of hot gas. Evidence is in favour of their being collections of matter of all kinds, such as meteoric fragments, dust, and gases, at a low average temperature. Light and heat are produced by collisions of the fragments, but the heat is lost by radiation so long as the material is sparsely distributed. When, however, the material condenses into compact masses, the heat of collision and compression accumulates in the interior of the masses, but escapes more or less from their surfaces. Thus a sphere as large as the sun produces heat enough internally to keep its surface white-hot in spite of loss by radiation. Jupiter can keep his surface perhaps barely red-hot; but a mass so small as the earth can produce so little heat (in ratio to the loss by radiation) that the surface is only "temperate." Since the mass of a sphere varies with the *cube* of the diameter, while the surface varies with the *square* of the same, it is evident that the ratio of radiation to heat-production must be greater in a small than in a large world. Geological evidence indicates that the earth's surface has been at about the same temperature for many millions of years. It may have been cooler at some period, and it may have been warmer; but we can hardly believe that the heat-production of our little world could ever have been sufficient to melt the rocks at its surface.

A few miles below the surface, however, the temperature may be high enough to melt any known rock: and where great lateral pressure exists, with consequent over-riding movements, fusion may occur quite near the surface. This took place not long ago, when the Alps, a comparatively recent mountain group, were thrown up. Under the stress the silica rose to the melting-point, reacted on the adjacent minerals, and produced the granite which forms the core of the mountains. For further information on the movements of the earth's crust and the production of igneous rocks, I must refer the reader to the works of the chief exponent of the subject, Professor Lapworth. The nebular or meteoric theory is lucidly explained in recent publications by Sir Norman Lockyer.

It is quite probable that in some parts of the universe life originates "in a sea of white-hot fluid"; but geologists are less inclined than formerly to believe in the existence of such a state of things at any time on the earth's surface. Mean-while there is abundant evidence of the maintenance of an equable surface-temperature for a vast period; and it behoves the biologist to try whether the known conditions would suffice to produce life as we know it. To begin with white-hot matter is to investigate from the wrong end. We must commence with known conditions: if these be found insufficient, we may go step by step into the hypothetical, and may ultimately arrive at the stage of white heat. For my own part, I am simple enough to believe that the circumstances which support life would also favour its origin.

Beech Lawn, Link Common, Malvern,
July 12, 1900.

NOTES ON SPINNING ANIMALS.

By H. WALLIS KEW.

(*Continued from page* 38.)

III. SPINNING CRUSTACEANS.

IT will probably come as a surprise to many readers to know that a faculty of spinning obtains among Crustaceans. Such, at least, was the feeling of the present writer on first hearing of the facts now re-stated.

SPINNING-LEG OF AMPHIPOD-SHRIMP ([1]).

The Crustaceans to which we refer are those of the order Amphipoda (Amphipod-shrimps), animals of small or moderate size; mostly marine, but more generally known, perhaps, from their fresh-water or semi-terrestrial representatives, such as *Gammarus*, the fresh-water shrimp, and *Talitrus*, the sand-hopper. A very large number of these animals walk, hop, or swim from place to place without making, as far as is known, any definitely constructed home. There is, however, on the

(1) One of the third pair of legs of *Xenoclea megachir*, with the epimeron and gill, seen from the outside, and showing the glandular organ within. (Enlarged 20 diameters.) *a*, The tip of the dactylus, showing the perforation. (Enlarged 100 diameters.) After S. I. Smith, "Transactions of the Connecticut Academy of Arts and Sciences," iii. (1874), pl. III. fig. 3.

other hand, a somewhat well-defined group, the members of which construct abodes in which they take shelter and nourish their young. Such abodes are of various forms; often tubular, but sometimes bird's-nest like; others are built with weed and various objects cemented or fastened together, or they may be of mud, while in some cases the creatures dig passages in clay, or burrow into timber or even harder substances. It is in the construction of certain of these abodes that the spinning powers of Amphipods are chiefly employed; and indeed I do not know that their threads have any other use. The kinds with which we are concerned belong chiefly to the family Corophiidae, at least in the wide sense in which that title was formerly employed; but to conform to the views of recent authorities it is necessary to say that they belong also to Podoceridae, Photidae, and other families.

Among the few of these Amphipod-homes to which naturalists have given attention are the fixed more or less tube-like abodes of *Amphithoë rubricata*. This animal is about half an inch long, and sometimes of a bright crimson colour; and, according to Bate and Westwood, it generally lives in a nest of its own construction, at the roots of Laminariae and other plants, or on the undersides of stones, at a few fathoms' depth of water. In making the nest the animal scratches together various available materials; but the point of interest for us is that a minute examination of the collected matter shows it to be united by a quantity of fine threads, which, crossing each other in a confused manner, are closely woven and knitted together. Some individuals in captivity were observed to build nests against the glass of the vessel in which they were kept; and under the microscope these nests were found by the authors just named to be composed of many bits of weed, matted together by means of exquisitely delicate threads, which had the appearance of having been spun or twisted, numerous small loops being formed by threads intertwined. Some nests of the same Amphipod received by these authors from Banff consisted chiefly of this fine thread, with which only a very little foreign material was built in with the structure. Of *Amphithoë littorina*, formerly considered distinct, but now merged with *Amphithoë rubricata*, the authors describe a nest constructed of bits of weed, sand, etc., bound together by fine threads; but they believed that the creature frequently rolled and cemented together the edges of a leaf of growing seaweed, forming a tube open at each end (²).

More curious are the abodes of *Janassa capillata* (*Podocerus capillatus*), a little Amphipod about a quarter-inch long, and beautifully varie-

gated in colour. It is shown by Bate and Westwood to build, in diminutive submarine forests, little nests, recalling in an unmistakable manner those of birds. These nests, built and firmly established in the branches of zoophytes and algae, consist chiefly of fine thread-like material, woven and interlaced. Some small extraneous fragments are often bound into the structure; but these, it is thought, are more the result of accident than of intention. In form the nest is somewhat oval, the entrance being at the top. It is evidently used as a place of refuge; but it serves, at the same time, as a true nest, in which the mother protects her brood of young until they are old enough to be independent of her care. Bate and Westwood figure a group of these structures from the Cornish coast, built in the slender branches of a *Plumularia*, and certainly appearing surprisingly like diminutive birds'-nests, though the opening at the top is smaller in proportion to the rest of the structure than in the ordinary form of bird's-nest. One of the nests from this group, on being opened, was found to be occupied by a mother and a swarm of young, the latter evidently of two ages, and therefore of two broods. The authors received nests of the same Amphipod from rock-pools near Banff, built in this case in *Corallina officinalis* (³). Further recalling birds'-nests, not in shape, but from the miscellaneous objects of which they are composed, are the tubes of an Amphipod allied to *Ericthonius difformis*, in which McIntosh found, besides mud and cement, "grains of sand, bristles and spines of annelids, hairs of sea-mice, and many fine horny fibres, apparently derived from the byssi of horse-mussels" (⁴).

The small tube-like abodes of the little *Microdeutopus gryllotalpa* have been studied by S. I. Smith, of New Haven. Those examined were built amongst small branching seaweeds, and were found to be composed, largely, of a network of fine threads of cement (⁵). Smith originally described the animal as *Microdeutopus minax*, but subsequently referred it to *M. grandimanus*, which name, according to Boeck and Sars, is a synonym of *M. gryllotalpa* (⁶).

In *Cerapus* the tube, instead of being fixed, is free, and is carried about by the animal, like the case of a caddis-larva (⁷). Some doubt has been entertained as to the origin of this tube, Say having supposed it to be that of some other

(3) Bate and Westwood, *tom. cit.* pp. 442–444.
(4) McIntosh, " Ann. and Mag. of Nat. Hist." (5) xvi. (1885), pp. 484.
(5) S. I. Smith, "Trans. Connecticut Academy," iv. (1882), pp. 274, 275.
(6) Stebbing, "'Challenger' Reports," xxix. (1888), pp. 435–437; Sars, " Crustacea of Norway," i. (1895), pp. 543, 544.
(7) Say, on *Cerapus tubularis* (New Jersey), "Journ. Acad. Nat. Sci. Philadelphia," i. (1817), pp. 49–52; Templeton, on *Cerapus abditus* (Mauritius ?), "Trans. Ent. Soc. London," i. (1836), pp. 185–190; Sars, on *Cerapus crassicornis* (Norway), *tom. cit.* p. 609; Giles, on *Cerapus calamicola* (Bay of Bengal), " Journ. Asiatic Soc. of Bengal," liv. (1887), pt. 2, pp. 54–59.

(2) Bate and Westwood, " British Sessile-eyed Crustacea," i. (1863), pp. 418–425; Bate, " Nidification of Crustacea," "Ann. and Mag. of Nat. Hist." (3), i. (1858), pp, 161–169.

creature appropriated by *Cerapus*. Smith, however, who examined the tube of Say's species, found it to be lined with cement, and covered externally with minute pellets, apparently of the animal's excrement, together with fragments of algae, etc. ([8]); and it cannot be doubted that the structure is, either wholly or in part, the work of the *Cerapus*. In the case of *Cerapus calamicola* (*Cyrtophium calamicola*) Giles has ascertained, curiously enough, that the tube has a vegetable foundation, being, in fact, a short piece of hollow reed, probably trimmed by the animal, and certainly coated, both inside and out, with secreted matter, doubtless laid down in the form of fine threads. In some few of these tubes there was no trace of a vegetable foundation, and it thus appears that the creature is capable of constructing a tube wholly of its secretion.

The spinning-organs of these animals do not appear to have been known until comparatively recent years. Say—who, as we have just seen, did not believe his *Cerapus* to be the maker of its tube—remarked that it had no organ adapted to this task. Bate and Westwood stated, of *Amphithoë*, that they had not been able to discover whether the threads were excreted by the mouth, or whether there was a special organ for their production. The honour of discovering the spinning glands belongs, I believe, to Smith; and one learns with surprise that he found them in certain of the creatures' legs. They are not in the first two pairs, the arm-like hand-bearing gnathopods, but in the two pairs which follow—namely, the first and second peraeopods, which are the third and fourth pairs of thoracic limbs; and it is at or near the tip of the toe of these limbs that the orifice is from which the thread issues. While examining spirit specimens of *Xenoclea* Smith noticed the opaque glandular structure of the spinning apparatus, filling a large portion of the two pairs of legs named—which legs, he says, in most, if not all, of the non-tube-building Amphipods are wholly occupied by muscles. A further examination of the spinning legs of *Xenoclea* showed that the terminal segment (dactylus) was not acute and claw-like, but truncated at the tip, and apparently tubular. Large cylindrical portions of the gland were found to lie along each side of the basal segment, and these two portion suniting at the distal end of that segment, the gland passes through the ischial and along the posterior side of the meral and carpal segments, and doubtless connects with the tubular dactylus. Similar structures were found in the corresponding limbs in *Amphithoë maculata*, etc. ([9]); and also in Say's *Cerapus*, in which the basal segments of the spinning legs are very large and almost wholly occupied by the glands ([10]).

The subject has also received attention from Nebeski, who detected the glands in the first and second peraeopods; as a rule in the basos, the ischium, the meros, and the carpus, from whence ducts let out the secretion at the tip of the toe, with results happily identical with those of Smith. The apparatus occurred in all the Corophiidae examined by Nebeski—namely, species of *Microdeutopus*, *Microprotopus*, *Amphithoë*, *Podocerus*, *Cerapus*, and *Corophium*, the secretion being used, without doubt, for cementing and plastering as well as for spinning ([11]). Giles, writing of *Cerapus calamicola*, doubtfully supposed the spinning glands to be in the huge propodal segment of the second gnathopods; but this suggestion, probably erroneous, was not made in opposition to the findings of Smith and Nebeski, with whose writings Giles was doubtless unacquainted.

Smith, fortunately, has seen the creatures in the act of spinning their threads. In 1874 he watched the construction of the tube in several Amphipods, including a species of *Amphithoë*; and has given an interesting description of the proceedings of *Microdeutopus gryllotalpa* (*M. grandimanus*)—a particularly favourable subject for observation. When placed in a zoophyte-trough with small branching algae, this Amphipod generally commenced at once to construct a tube, and it could readily be observed under the microscope. A few slender branches of the alga were pulled towards each other by the antennae and gnathopods, and fastened by threads of cement spun from branch to branch by the spinning limbs above mentioned. The branches were not usually at once brought near enough together to serve as the framework of the tube, but were gradually brought together by being pulled in and fastened a little at a time, until at last they were brought into the proper position, where they were then firmly held by means of a thick network of fine threads of cement spun from branch to branch. After the tube had assumed very nearly its complete form, it was still usually nothing more than a transparent network of cement threads woven among the branches of the alga, though occasionally a branch of the alga was bitten off and added to the framework. Very soon, however, the animal began to work particles of excrement and bits of alga into the net; the pellets of excreta, as passed, were taken in the gnathopods, maxillipeds, &c., broken into minute fragments, and worked through the web, upon the outside of which they seemed to adhere partly by the viscosity of the cement threads and partly by the tangle of threads over them. Excreta and bits of alga were thus worked into the wall of the tube until the animal within was protected from view, and, during the whole process, the spinning of

(8) Smith, *l.c.* pp. 271–277.

(9) Smith, "Trans. Connecticut Academy," iii. (1874), pp. 32-35 ; "Silliman's Journal" (3), vii. (1874), p. 601 ; and "Ann. and Mag. of Nat. Hist." (4), xiv. (1874), p. 240.

(10) Smith, 1882, *l.c.* p. 271.

(11) Nebeski, "Arb. Zool. Inst. Wien." iii. (1880), pp. 111–163, as abstracted by Stebbing, *tom. cit.* pp. 518–521, 1155 ; "Zool. Record," xviii. (1881), Crust., p. 6 ; "Journ. R. Micr. Soc." (2), i. (1881), pp. 453–455.

cement over the inside of the tube was continued. It was clear that the spinning was done wholly with the first and second peraeopods, the tips of which were touched from point to point over the inside of the skeleton tube in a way that recalled strongly the movements of the hands in playing upon a piano. The cement adhered at once at the points touched and spun out between them in uniform delicate threads, which appeared to harden quickly, and did not seem, even at first, to adhere to the animal itself. In a case in which the entire construction of the tube was watched, the *Microdeutopus* very nearly or quite completed the work in a little more than half an hour. In the *Amphithoë*, Smith adds, the process of constructing the tube was very similar, though less cement and more foreign material entered into the structure ([12]). Giles, who watched *Cerapus calamicola* under the

microscope, one day surprised an individual in the act of repairing the fibrous lining of its reed-tube; the animal had completely withdrawn into the tube, and was keeping the latter slowly but continuously revolving. The tube was transparent enough for the observer to see that the *Cerapus* remained stationary while the tube revolved; but he was unable, unfortunately, to make out the exact manner in which the fibre was being deposited. It may be noted, finally, that this naturalist supposes certain teeth in the animal's second gnathopods to be well suited for cutting the secreted thread, or for trimming the piece of reed used for the foundation of the tube; and he further suggests that the distal segment of the third peraeopods is admirably adapted for guiding the thread.

(To be continued.)

DESMIDS.

By Dr. G. H. BRYAN, F.R.S.

(Concluded from Vol. VI. page 360.)

ACCORDING to the classification in Dr. M. C. Cooke's "British Desmids" there are twenty-one genera represented in Great Britain. In the following columns it is proposed to give a superficial account of the differences between the genera, rather than a detailed description of their botanical features, as the above-named treatise supplies all that is wanted for a fuller study of the desmids.

The genera are divided as follows:—

Section A. LEIOSPORAE. Zygospores usually smooth.

Sub-section a. Individuals more or less closely united in threads or filaments. Genera: *Gonatozygon, Sphaerozosma, Onychonema, Hyalotheca, Bambusina, Desmidium.*

Sub-section b. Cells free, not united in a filament. Genera: *Docidium, Closterium, Penium, Cylindrocystis, Mesotaenium, Tetmemorus, Spirotaenia.*

Section B. COSMOSPORAE. Zygospores normally warted, spinulose or ornate. Genera: *Micrasterias, Euastrum, Cosmocladium, Cosmarium, Calocylindrus, Xanthidium, Arthrodesmus, Staurastrum.*

Commencing with the filamentous forms, the genus *Gonatozygon* contains three species, all of them apparently local, and considerably resembling confervoid algae with their cylindrical filaments formed of elongated cells. Some of the species of *Sphaerozosma* and the two British species of *Hyalotheca* are enclosed in a thick gelatinous sheath which is somewhat difficult to see in un-

stained preparations, and is practically invisible in glycerine. I have been recommended to use Bismarck brown as a stain, but have not yet had an opportunity of trying that pigment. The envelope may be twice or thrice or more times the diameter of the central filament of cells. While the cells of *Hyalotheca* are cylindrical, those of *Sphaerozosma* are aptly described by the late Rev. J. G. Wood, in his "Common Objects," as "looking much like a row of stomata set chain-wise together." In the single species of *Onychonema* recorded from Strensall Common, near York, the alternating and overlapping horns are characteristic. Both *Hyalotheca* and *Sphaerozosma* are represented in this neighbourhood, the latter genus occurring in the *Sphagnum* washings from the Nant Ffrancon valley. The widely distributed sole representative of *Bambusina* has curiously barrel-shaped cells. In *Desmidium* we have triangular cells placed one above the other, forming a filament in the shape of a twisted prism. To get an idea of this arrangement it is only necessary to take a pile of books and twist them slightly from the top downwards. This twist gives the appearance of a spiral band formed by the angles of the cells running diagonally across the filament.

Passing on to the "free" desmids, we find in *Docidium* an elongated straight cell divided into two segments by a constriction in the centre, and having the same peculiar bodies at the tips of the cells that are found in the horns of *Closterium*. The genus *Closterium* contains about forty British species and varieties, two of which are figured in SCIENCE-GOSSIP for April last (page 325). The crescent-shaped outline is sufficient to distinguish at once most of the species of *Closterium.* Specimens of the genus are common in ponds and ditches, and are by no means confined to mountain peat bogs. Ponds in the neighbourhood of London are known in some seasons to yield quantities of them ; and the specimens figured were taken from roadside runnels, the gathering of *C. striolatum* forming dense green patches. The large *C. lunula* or the *C. costatum* figured in the April issue, or one of the other larger species form an interesting and fascinating study in the living state. At the tip of either " horn " is a spherical cavity containing a number of tiny granules in active movement. These exist in all the species of the genus, though they can best be seen with a moderate power in the larger forms.' I have, however, slides containing two small species, *Closterium moniliforme,* and the slender *C. rostratum,* whose fronds terminate in sharp beaks. These desmids were collected from a ditch on the way to Aber waterfall, and in spite of their having been mounted in glycerine for about a year it is easy to find specimens in the mounts in which the granules have retained their position, and are to be distinctly seen with a half-inch, or even with an inch objective. Another feature is the circulation of the cell-contents. The process of reproduction by subdivision has already been described in a paper in the New Series of SCIENCE-GOSSIP, and in gatherings where the desmids are abundant specimens in different stages of subdivision are readily found. Mr. Noad Clarke in his photograph of *C. costatum* on page 325 has succeeded in including one specimen shortly after subdivision, in which the newly formed half of the frond has not nearly reached its full size. Reproduction by conjugation is much less frequently met with, and specimens in this state should therefore, if found, be carefully preserved.

Next to *Closterium* comes *Penium,* a genus containing, according to Dr. Cooke, nineteen British species, independently of varieties. These all have a simple oval or oblong outline without any irregularities (fig. 1); in a few examples the frond is slightly narrowed at the middle, but there is no constriction. The length of the frond is from twice to eight times the breadth ; and the elongated shape, resembling that of diatoms of the genus *Pinnularia,* should suffice to identify some of the commoner kinds, which have, in fact, a very ordinary appearance. The specimens from which the photograph was taken are from a *Sphagnum* bog between Port Dinorwic and Pentir in North Wales, where at suitable seasons they appear to be extremely abundant. Some of the individuals have undergone slight contraction, but this was owing to the material being left standing for some days in a bottle of water before being dealt with, and it did not occur during the process of preparing and mounting. In these, the cell-contents are seen to separate into two parts, the nucleus being

in the centre of the bridge connecting them. Many of the desmids on this slide contain one or more large brown globular bodies, covered with fine setae ; these cannot, of course, be spores, and it is difficult to imagine what they can be, unless they are parasitic. Careful focussing shows them to be inside the desmids.

The genus *Tetmemorus* contains four British species, one of which is figured at BB in the group on page 257 of SCIENCE-GOSSIP for February. They are all of much the same shape, and the genus is distinguished from *Penium* by the constrictions at the middle of the fronds, the ends also having an acute incision. The species is common in *Sphagnum* washings up the Ogwen Valley.

Two other genera, *Cylindrocystis* containing two species, and *Mesotaenium* containing three British species, have shortly oval or cylindrical cells with rounded ends, and without constriction, about twice as long as broad.

The last of the smooth-spored genera is *Spirotaenia,* which can be at once distinguished from all

FIG. 1. *Penium digitus,* with one *Micrasterias crenata.*

other desmids by the spiral arrangement of the chlorophyll ; the cells are oval and enclosed in a thick gelatinous envelope. One of these, *S. condensata,* occurs in some gatherings from the mountain bogs near here, but as yet I have found it sparingly.

Passing on to the rough-spored genera, the members of the genus *Micrasterias* claim attention as including by far the most attractive and beautiful of all the British desmids. *M. oscitans* (fig. 2) is a representative form of the sub-genus *Tetrachastrum,* containing three British forms, while the typical genus, or *Eumicrasterias,* is represented by *M. rotata* in vol. vi. p. 257, figs. 1 and 2, *M. jenneri* (fig. 3 below) and *M. crenata* (in fig. 1). Fig. 9 shows a group of desmids obtained last summer from a ditch containing *Sphagnum* on the old road from Bethesda to Ogwen, and well illustrates what clean gatherings can be obtained by the method of rocking in a dish or soup-plate, as described in previous papers of this series. In this particular gathering *Micrasterias jenneri* was extremely abundant, but a few specimens of *M. oscitans* as well as of the *Xanthidium* occur on most of the spread slides. Some of the larger species of *Micrasterias* are still more graceful, notably *M. radiosa*, *M. furcata*, and *M. denticulata*, of which the two first occur on an old slide of Joshua's from Capel Curig, mounted twenty years ago, that lately I succeeded in restoring. Seventeen species of this genus are figured in "Cooke."

Of the genus *Euastrum*, the species represented in fig. 4, namely, *E. verrucosum*, is a pretty example with its surface covered with dots, but it can hardly be called a typical species, as the majority have the frond about twice as long as broad. Figs. 5 and 6 show the outlines, traced with a camera lucida, of the two species, *E. didelta* and *E. cuneatum*, collected in *Sphagnum* near Llyn

FIG. 2.   FIG. 3.   FIG. 4.

FIG. 2. *Micrasterias oscitans.*   FIG. 3. *M. jenneri.*
FIG. 4. *Euastrum verrucosum.*

Idwal, which appears to be a favourite habitat of the genus. Excluding varieties, Dr. Cooke enumerates twenty-eight species, mostly characterised by the pyramidal shape of their segments with sinuous outlines. A pretty form is *E. oblongum*, with its front of oval shape and either segment deeply 5-lobed, the central lobe being notched at the apex, as in other species of the genus.

Cosmocladium is a small genus containing two very local species with the individuals united by dichotomously branched filaments.

The next genus *Cosmarium* contains at least ninety-seven British species, some of them among the commonest of desmids. A pretty form, *C. ralfsii*, with smooth cells is represented by a camera lucida outline in fig. 7; it is a large species, frequent in *Sphagnum* washings from Llyn Idwal, and of a beautiful green tint when fresh. Another form is seen in vol. vi. p. 257, fig. 1, E. Most of the species of *Cosmarium* have the cells granulated or dotted, and the segments reniform and broader than long. Some species occur commonly in almost any ditch or pool, but often associated with so many other forms of pond life that it would be difficult to clean and mount them. In the Easter vacation of 1891 I found in some of the pools on Reigate Heath, Surrey, green masses which proved to be made up entirely of a small species of *Cosmarium*.

A camera lucida outline of a desmid which I refer to, the genus *Calocylindrus*, is shown in fig. 8, from which it will be seen that the species of this genus have their segments generally more elongated than those of *Cosmarium*, and with little constriction between them; in some species there is not any such narrowing. The specimens from which the outline was sketched formed green masses in

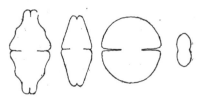

FIG. 5.   FIG. 6.   FIG. 7.   FIG. 8.

FIG. 5. *Euastrum didelta.*   FIG. 7. *Cosmarium ralfsii.*
FIG. 6. *E. cuneatum.*   FIG. 8. *Calocylindrus.*

some roadside pools just after a shower of rain, and looked as if there was some gelatinous connection holding the individuals together as the masses floated about in the water.

Fig. 9, *c*, shows a characteristic form of the genus *Xanthidium* occurring frequently in the *Sphagnum* washings from the Nant Ffrancon valley. They are characterised by the spines upon them, the segments are obtuse, with a deep constriction between them, and ten species are described in "Cooke." The name *Xanthidia* is also given to the curious fossil bodies met with in thin chippings of flint,

FIG. 9. *a, a, a, M. jenneri ; b, b, M. oscitans ; c, Xanthidium.*

from their resemblance to the spores of desmids. A closely allied genus *Arthrodesmus*, likewise with spiny segments, is characterised by the fact that while the species of *Xanthidium* have a circular projection on both faces near the centre, this is absent from *Arthrodesmus*.

The last of the desmid genera, *Staurastrum*, with over ninety species, exclusive of varieties, is no doubt a puzzling genus and most of the species are too small to make good drawings or photographs. One small species may be seen in end view in fig. 1 D of the February article. The end views of many species reproduce the vast variety of triangular and sometimes quadrangular and polygonal outlines that are met with in the diatom genus *Tricoratium*. In some the vertices of the two triangles formed by the two segments are superposed, in others they alternate with each other, and in others again the angles terminate in elongated processes, giving the desmids a star-shaped appearance, with almost any number of rays from three upwards. In an old slide by Joshua from Capel Curig, in which the cells have lost all their chlorophyll, two specimens are eight-rayed, while a third, evidently of the same species, has nine rays.

In conclusion I have to thank Mr. F. Noad Clarke for the excellent series of photographs with which he has illustrated these notes.

Plas Gwyn, Bangor, North Wales.
July 1900.

IRISH PLANT NAMES.

By JOHN H. BARBOUR, M.B.

IN response to various suggestions made to me after my article appeared on this subject in SCIENCE-GOSSIP for last October, that I should enlarge upon it and give the meanings of the Irish words, I now give further details. Before proceeding systematically, it is necessary to say something about the article in order to avoid misunderstanding, and I think the best way to begin is, as "Inisfail" suggested in his letter to this magazine last November (vol. vi. p. 191), with a short list of the trees after which the Gaelic alphabetical letters are called, although I have introduced some of them later, with a word of explanation on each.

A is represented by *Ailm*, strictly the palm-tree, Lat. (¹) palma, but some authorities suggest it is more closely allied to the fir-tree, Lat. abies. The elm is also Ailm, which possibly is a misapplication originally of the Latin.

B is from *Beit*, the birch. Gr. barshon, but it is also by some considered more closely related to Heb. beth.

C is *Coll*, the hazel. Lat. corylus.

D is *Duir*, the oak. Lat. Deus. Heb. Derech.

E is represented by *Eada*, aspen-tree. Lat. tremula. Greek, eta. Heb. heth.

F is *Fearn*, the alder. Lat. alnus. Heb. vau.

G is *Gort*, the ivy.

H is used only as a mere aspirate in Ir., not a letter. It is *Uat*, the whitethorn.

I, called *Iodha*, from *Iodha*, vulgò *iubhar*, yew-tree. Lat. taxus, and is not dissimilar to Heb. jod, and Gr. iota.

L is *Luis*, vulgò *carthan*, the quicken-tree. Lat. ornus.

M is *Muin*, the vine. Lat. vitis, and like the Heb. mem.

N is from *Nuin*, the ash-tree. Lat. fraxinus. Heb. nun.

O diphthong *Oir*, from *Oir*, the spindle-tree. Vulgò *feorus*. Lat. euonymus.

P is not from any tree, but is known in Ir. as *Peat-bog*.

R represented by *Ruis*, old Ir. In vulgar Ir. *Trom*, the elder. Lat. sambucus, and Gr. acte.

S is *Suil*. *Sail*, willow. Lat. salix.

T is *Teine*, from Ir. *teine*, furze.

U is *Ur*, *ubhur*, *iubar*, yew.

In the next place, it will be noticed that one Irish name often refers to several kinds of plants, and of course the reverse is common. This frequently happens, and the following are a few instances of generalised names, but some of them have also been applied to distinct species at times.

Seirg, any clover, trefoil. *Cluain*, a thistle or spurge. *Seisg*, a sedge. *Colubairt*, cabbage. *Abal* an apple-tree. *Ditein* is applied to any tare, as well as specially to the corn marigold. *Copog*, any dock. *Codlan* a poppy. *Mongeac mear* is both henbane and hemlock. *Sailcuac*, the violet or pansy. *Billeog*, any water-lily. *Cona*, the Scotch fir and cotton-grass. Why, I cannot say; in this lies one of the difficulties of such a paper as this. A few Irish names of not indigenous plants are also mentioned, as I met with them.

I might point out that in giving the English names I have often used the most uncommon ones, because some of these are heard only in Ireland, and I might really designate them as Irish-English names—names which have been given by those country folk who habitually speak English, of the past and present, and I daresay it will be found more interesting to others to know about them than to see the usual English ones given; of the latter I hope from time to time to publish supplemental lists as they come to hand.

My endeavour has been to introduce as much material into as small a space as possible, therefore in some cases I have had to resort to merely giving the English, more or less literally, of the Irish in brackets; but in other cases I have ex-

(1) Heb., Hebrew; Gr., Greek; Lat., Latin; Eng., English; Ir., Irish; W., Welsh; Tipp., Tipperary; Linn., Linnaeus; gen., genitive; plur., plural; dim., diminutive; syn., synonymous; var., variety. A query means absolute doubt only.

plained myself more fully. Several changes have been made in this article from my remarks in SCIENCE-GOSSIP of last October as more correct; for instance, I have given restharrow as *Ononis spinosa* (Lin.) rather than *Arvensis* following as I have done this time the ninth edition of the London Catalogue of British Plants. These are most of the points I need refer to. I have followed the London Catalogue so far as I could in my classification, and my material has been gleaned from old books (excluding the Cybele Hibernica) and from the Irish people themselves.

PHANEROGAMS.

RANUNCULACEAE.

NEAD CAILLEAC. nead means "a nest"; cailleac, "an old woman" or "nun." Therefore "nun's nest." *Anemone nemorosa.* wood anemone.

FEARBAN, "sparkling grass." BAIRGIN, "a begotten son." *Ranunculus repens.* creeping crowfoot.

FLEAN UISCE. NEAL UISGE, "river ecstasy." SNAITE BAITE, "drowned threads." LEANANAC, "favourite"(?). LIONAN ABAN, "a very river snare." *Ranunculus heterophyllus.* water crowfoot.

FOLOSCAIN. word also means "tadpole." GRUAG MUIRE. gruag, "hair of the head." "Mary's locks." *Ranunculus auricomus.* goldilocks.

TORACAS BIADAIN. *Ranunculus sceleratus.* celery-leaved crowfoot.

TUILE TALMUIN. tuile, "a flood, rain "; talman, "earth." FEARBAN. *Ranunculus acris* and *R. bulbosus.* butter-flower, buttercup, or gold cup.

GLAISLEUN, "green grief." LAISAIR LEANA. lasair, "a flame "; leana, "a meadow." "meadow flame" or "meadow gold." LONAIG, "jester." *Ranunculus flammula.* lesser spear-wort.

GRAIN AIGEIN. grain, "loathing." Aigein, related to adan, also "a cauldron." "loathing cauldron." SEARRAIG. GRANARCAIN, modification of Grain aigein. LACA, "a duck." *Ranunculus ficaria.* pilewort.

PLUBAIRSIN. BEARNAN BEILTINE. bearna, "a gap "; BEALTAINE, "a little sun," "May flower." LUS BUIDE BEALTAINE. "yellow May flower." LUS MAIRI, "Mary's plant." *Caltha palustris.* meadow bouts.

CRUBA LUSIN, "the little bird's-claw flower." LUSAN COLAM, "the little dove herb." *Aquilegia vulgaris.* Columbine.

EUAT A MADAID. fuat, "hatred "; madad, "a dog "; with allaid added, "a wolf." Hence "wolf's bane." *Aconitum napellus.* wolf's bane.

NYMPHAEACEAE.

DUILLEOG BAITE BUIDE. Duilleog is "a little leaf "; baite, "drowned." "drowned yellow leaflet." CABAN ABAN, "river cup." LIAC LOGAR, "hollow spoon." *Nymphaea lutea.* yellow water-lily.

DUILLEOG BAITE BAN, "drowned white leaf."

LIAGLOGAR. liag, "a blade of an oar"; loga, "splendid." "splendid blade." *Nymphaea alba.* white water-lily.

PAPAVERACEAE.

BEILBAG. "prince" or "virtue of the month." BLAT NA MBODAIG. CANLEAC DEARG, "red moth's cheek." COCCIFOIDE. ? possibly derived from kokkos, coccum, Lat. for scarlet. PAIPIN RUAD. ruad, "red." "red poppy." *Papaver rhoeas.* corn rose, headwark.

COLLAIDIN. collaim, "I sleep." Heb. cholom, "a dream." PAIPIN. *Papaver somniferum.* white poppy.

CEANRUAD or LACA CEANRUAD. ceanruad, "red chief." *Chelidonium majus.* celandine.

FUMARIACEAE.

DEARAG TALMAN. "earth sadness," suggestive of "all ends in smoke." FUAIN A TSORRAIG. CAMAN SCARRAIG. scaraim, "I unfold "; caman, "common." *Fumaria officinalis.* fumitory.

CRUCIFERAE.

AMARAIC. BIOLAR TRAGA or TRAIGBIOLAR. "shore cresses." *Cochlearia officinalis.* scrooby grass.

BIORAR or BIOLAR. "cresses." BIOR-FIER. Bior, "edge of water"; fear, "grass." "brink grasses." *Nasturtium officinale.* water cresses.

GLEORAN. "jollity." BILLAR or BIOLAR GRIAGAN. grian, "lake." "lake-side cresses." *Cardamine pratensis.* lady's smock.

MAEL ISA GARB RAITEAC. *Sisymbrium officinale* hedge mustard.

FINEAL MUIRE. "virgin's fennel." *Sisymbrium sophia.* flixweed.

GARABOG. GAS NA CONACTA. gas, "stalk, stem." Heb. geza. "the Connaught herb." PRAISEAC GARB, "rough pottage." *Brassica sinapistrum.* chadlock, charlock.

CAL, "kail." CADAL. COLUBAIRT. "colewort." PRAISEAC BUIDE, "yellow pottage." *Brassica oleracea.* wild cabbage.

LUS NA FOLA. fola, "garment "; lus, "a plant" or "herb." LUS A SPARAIN. sparain, "a purse, pouch." SRAIDIN, "a lane." *Bursa bursa-pastoris.* shepherd's pouch.

PRAISEAC FIAD. fiad, "food, meat." Hence "pottage meat." PRAISEAC NA CCAORAC. ccaorac, "sheepy," from caor, "sheep." "sheep's food." *Thlaspi arvense.* penny cress. treacle-mustard.

MEACAN RAGUM. rag, "wrinkled, stiff"; meacan, "a tap root." *Raphanus raphanistrum.* wild radish.

MEACAN RAGUM UISCE. uisce, "water, river." *Raphanus maritimus.* water radish.

PRAISEAC or RAITEAC TRAGA. raiteac, "pride "; but it may be derived from rait, "fern" or "brake." Hence "shore pride" or "shore brake," and "shore meat." *Crambe maritima.* Seakale.

RESEDACEAE.

BUIDE-MOR. mor, "noble, great"; related to W. mawr. "noble yellow." *Reseda luteola.* wild woad. dyer's weed.

VIOLARIEAE.

BIOD A LEITID. biod, "a world"; leitid, "a peer." "a world's peer." FANAISGE. aisge, "a present"; fan, "slope." "a present from a slope or bank." SAILCUAC. a narrow or curled guard. *Viola silvestris.* dog's violet.

SAILCUAC. *Viola odorata.* sweet violet.

GORMAN SEARRAIG or GOIRMIN SEARRAD. ?possibly from scarad, "a separation," or sgarad, "a fissure." Gorman and Goirmin, I think, refer to gorm, "blue"; hence "blue fissures" or "blue streaks." SAILCUAC. *Viola tricolor.* herb trinity. kiss-at-the-garden-gate.

POLYGALEAE.

LUS BAINE or LUS AN CAINE, "the little milk herb." *Polygala vulgaris.* milkwort.

CARYOPHYLLEAE.

COGAL or CAGAL. cogal, "ears of barley," from place where found. *Lychnis githago.* corn cockle.

COIREAN COILLEAC. ? coire, "a ring," "girdle," or "cavern"; and coille "sylvan"; while coilleac is "a cock." *Lychnis dioica.* red campion. bladder flower.

CAOROG LEANA. caorog, "a spark"; leana, "meadow." Hence "a little meadow spark." *Lychnis flos-cuculi.* meadow pink. wild-williams.

FLIG. FLIOD. FLIOC or FLIUC. fliod, "a wen, excrescence." *Stellaria media.* chickweed.

TURSARRAIN or TURSACAIN. "dry bags." *Stellaria holostea.* greater stitchwort. fairy flax (Tipp.).

TURSARRAININ. TURSACAININ. *Stellaria graminea.* lesser stitchwort. fairy flax (Tipp.).

CLUAIN LIN. cluain, "a plain, lawn"; lin, "flax." CABROIS. CORRAN-LIN. corran, "sicklehook." *Spergula arvensis.* spurrey.

HYPERICINEAE.

BEACNUAD BEININ or B. COLUIMCILLE or B. FIRION. beacnuad, "new bee," "St. Columcille's bee," "the bees' fair circle" (firion). CAOD-COLUIMCILLE. caod, "tear." "tear of St. Columcille." ALLAS MIURE. "Mary, the most high." EALA BUIDE. "yellow swan." *Hypericum perforatum.* St. John's wort.

MEASTORC ALTA or CAOIL. *Hypericum androsaemum.* tutsaw.

MALVACEAE.

LAEMAD. *Althaea officinalis.* marsh-mallow.

LUS NA MIOL MOR. miol, "any animal"; mor, "great, big." MILMEACAN, "honey root." OCUS, "itch." UCAS FIADDIN or FITRAIN, "wild nap." ucaire, "napper of freize." *Malva sylvestris* mallow.

UCAS FRANCAC, "French nap plant." *Malva rotundifolia.* dwarf mallow.

(To be continued.)

OWENS COLLEGE NEW PHYSICAL LABORATORY.

TWO of the constituent Colleges of the Victoria University—the Owens College, Manchester, and Liverpool University College—are indeed fortunate as regards bequests and donations. They apparently vie one with the other in opening new wings and equipping new departments through the generosity of private individuals. On June 29th last an imposing ceremony took place at the Owens College, when Lord Rayleigh formally opened the new Physics Laboratory. This new building is replete with every arrangement one could wish for carrying on experimental research. Space will only permit attention being drawn to a few of these.

Extreme ranges of temperature are easily obtained, as steam is always available in most of the rooms. An electric furnace is provided capable of producing a temperature of about 6,000° Fahrenheit, while for the other extreme the necessary machinery is installed for making liquid air, which, compressed by means of a pump, produces a temperature of 300° below zero.

The knowledge of accurate time intervals, which is important in many investigations, renders the possession of accurate clocks, expensive as they are, absolutely imperative. The clock to be placed in the basement of the laboratory will probably be the most perfect clock in the country. It is made by Dr. Riefler, of Munich, and has in its original form been tested in many observatories and found to keep more perfect time than the older types of clock. So valuable a timepiece must of course be kept locked up in a room, and will not be generally accessible, but time is supplied to the different parts of the laboratory by electric transmission from a second clock. The lecture-rooms and the larger rooms of the laboratory will all contain dials showing time correct to within a few seconds.

Special attention has been paid to the optical outfit of the laboratory. One of its features will be a room on the top floor which contains Rowland's diffraction grating. An idea of the difficulty experienced in constructing these gratings may be obtained from the fact that it occasionally takes three or four years before a satisfactory grating is manufactured. The one supplied to the College has been certified by Professor Rowland as exceptionally good.

The Photographic laboratory is in the basement. The Observatory, placed on the top of the building, contains the 10-in. telescope presented by Sir Thomas S. Bazley. The Electro-technical laboratory, built and equipped in memory of Dr. John Hopkinson, has been provided, as far as the means at the disposal of the College have allowed, with the most modern form of machinery. Considerable space has been assigned to an electro-chemical laboratory, and as a special lecturer has been appointed in that subject it is hoped that Owens College will take a leading part in its development.

An electric lift connects the different floors, and will serve principally to convey delicate apparatus from one part of the building to another. There will be a small but well-equipped workshop in which instruments can be repaired or even made.

The large lecture-room holds, under ordinary circumstances, 210 students. It has a skylight as well as windows looking to the east, and rapidly moving shutters allow the room to be quickly darkened. The small lecture-room serves for advanced lectures, and rooms are provided for storing the ordinary apparatus and also some historical instruments, as, for instance, Joule's valuable apparatus, some of which have been presented to the College. Electric current is supplied from sixty storage cells, which are connected through a large switchboard over fifteen feet long. Thirty circuits

diverge from this switchboard into the various rooms, the circuits consisting of aluminium or copper wire, stretched principally under the ceiling of each room and passing through specially perforated bricks in the walls. Uncovered wires have been chosen for the purpose, as the expense is reduced and much stronger currents can be sent through the same thickness of wire.

The research-rooms are placed principally in the basement and on the ground floor of the building, where the greatest steadiness and freedom from disturbance can be obtained, and it is hoped that the unusual facilities which in future will be given at Owens College for original research in physics will attract many students. There is also every reason to hope that the John Hopkinson Electro-technical laboratory, with its adjunct the Electro-chemical laboratory, will soon make itself felt in the education of electrical engineers, who will be trained to carry out the highest technical work they may be called upon to perform. JAMES QUICK.

BRITISH FRESHWATER MITES.

By CHARLES D. SOAR, F.R.M.S.

(Continued from page 49.)

GENUS *PIONOPSIS.*

THIS genus is very closely allied to *Piona.* It really holds an intermediate position between *Curvipes* and *Piona.* In both male and female its characteristics are the same as *Piona,* except in the males not having the enlarged fourth segment on the fourth pair of legs, which we found in that genus.

Pionopsis lutescens Hermann.

FEMALE.—Oval in shape. Length about 1.44 mm., breadth about 1.02 mm. It is very like the females of *Piona.* Colour a pale yellow with very dark markings on the dorsal surface, with a bright yellow T-shaped patch in the centre.

LEGS.—First pair about 1.36 mm. Fourth pair about 1.72 mm. Of a pale blue colour, claws to all feet, but the claws on the fourth pair are much smaller than the others.

EPIMERA.—In four groups, similar to the females of *Curvipes* and *Piona.*

PALPI.—About 0.40 mm. long. Pale blue in colour. The last two segments are like fig. 2, page 48.

GENITAL AREA.—Is composed of three discs on each side of the genital fissure, let into special plates like we found in *Piona* and *Limnesia,* but the two posterior discs are side by side, the first one being just above them, not one above the other like those of *Piona ornata* (fig. 1, page 48).

MALE.—Length about 0.72 mm., breadth about 0.60 mm. In form it is very much like the male of *Piona,* but it can be easily recognised by the peculiar structure of the fifth segment of the hind leg (fig. 1), which has six stiff bristles in a row; slightly bent backwards towards the body, in opposition to those we generally find on the legs of water-mites. The length of the fourth leg of the male is about 1.08 mm. PALPI are about 0.40 mm., which is the same length as that of the palpi of the female, but the body of the male being

FIG. 1. *P. lutescens.* Fourth leg of male.

so much smaller than that of the female, gives this mite the appearance of having very large palpi.

LOCALITIES.—Totteridge, Epping Forest, and Norfolk Broads. Not a very common mite.

GENUS *LIMNOCHARES* LATREILLE.

This genus is known by the following characteristics:—Soft-skinned. Legs without swimming hairs. Eyes close together. Small palpi. At present I believe there are only two species known in this genus, the one *L. crinita* Koenike having been recorded from Madagascar, and *L. holosericea*

Latreille which has been recorded from many parts of Europe. The latter is a very slow-crawling mite, and lives a long time in confinement.

Limnochares holosericea Latreille.

FEMALE.—Body about 3 mm. in length and about 2 mm. broad. They vary much in size, even in the adults, but the measures given are mean. I have taken females extended with ova, as much as 4 mm. long. Colour scarlet, legs inclined to yellow at the joints, but the same colour as body in

FIG. 2. *L. holosericea.* Dorsal surface.

other parts. I can compare the shape of the body to nothing better than a miniature sack, or bag of the softest material, which is more striking still when the mouth organs are well thrust forward. The skin is full of folds which are constantly changing their position, as the little creature moves. The cuticle of the body is very soft, and covered with small round papillae.

LEGS.—First pair about 1.20 mm. Fourth leg about 1.76 mm. They are covered with a great number of simple hairs. A few hairs on the joints are plumose, but the limbs are quite without the long swimming hairs we find on some other mites. This species does not swim, but contents itself

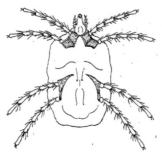

FIG. 3. *L. holosericea.* Ventral surface of female.

with sluggishly crawling among the débris at the bottom of the tank.

EPIMERA.—Chitinous with a border of thicker skin round each epimeral plate, which is fringed with a quantity of fine hairs.

EYES.—Close together in two pairs, on each side of small chitinous dorsal plates (fig. 3) which are situated (fig. 4) near the anterior dorsal margin.

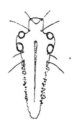 

FIG. 4. *L. holosericea.*   FIG. 5. *L. holosericea.*
Eyes and dorsal plate.   Mouth organs and palpi.

MOUTH ORGANS.—(Fig. 5) Are suctorial, with a short palpus which reaches no further forward than the sucker-like mouth.

LOCALITIES.—Not very common. I have taken it at Woking, Sunningdale, Folkestone, Red Hill, and in N. Wales. Dr. George has taken them in Lincolnshire.

GENUS *MIDEA* BRUZELIUS.

BODY hard-skinned, with a finely granulated surface; a depressed line on the dorsal surface close to margin. Swimming hairs. Epimera in one group. At present I only know of one species in this genus.

Midea elliptica Müller, 1781.

BODY.—Female nearly round, being about 0.64 mm. long and 0.58 mm. in breadth. In colour this is, I think, the most brightly tinted mite we have in this country, when found at its best in

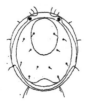

FIG. 6. *M. elliptica.* Dorsal surface of female.

the adult stage. The colouring is due to two causes : first, to the pigment in the skin ; secondly, the contents of the body. On the dorsal surface is a depressed line, which runs a little way in, round the whole surface. Outside of this depressed line the colouring is green and yellow. The eyes, which come in the yellow portion, are a deep red. Inside the depressed line, towards the anterior portion, is a small oval patch, which is white. The remaining part on which are situated most of the dermal glands is blue, which varies much in tint. The dermal glands are yellow. Sometimes on the white patch just mentioned may be a bright patch of crimson-red, which is found in different positions in different specimens. This

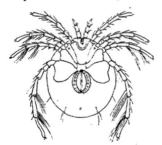

FIG. 7. *M. elliptica.* Ventral surface of female.

latter effect is no doubt owing to the partly digested contents of the inside moving their position.

LEGS.—First pair about 0.38 mm. Fourth pair about 0.68 mm. Colour a reddish-yellow.

EPIMERA.—In one group; colour yellow, running into green on the margins.

FIG. 8. *M. elliptica.*
Genital area of female.

FIG. 9. *M. elliptica.*
Three last segments of third leg of male.

PALPI.—Second and third segments are rather thick. The fourth are long and thin.

GENITAL AREA.—Composed of two plates, with a number of discs on each (fig. 8).

MALE.—Very little difference in structure from that of female, except in the third pair of legs, which have the three last segments like fig. 9. The genital area is also different, having the inner plates of a horseshoe form; but the colouring and arrangement of the dorsal surface are the same as in the female.

LOCALITIES.—Rare. Dr. George has found this species in Kirton-in-Lindsey, Lincolnshire; Mr. Scourfield in Epping Forest; myself on Norfolk Broads.

(To be continued.)

GUINEA-PIGS AND BATS.—In answer to a question in the June number (*ante*, p. 24) with regard to rats and guinea-pigs, my sons have a small pheasant aviary, and as it was visited by rats they put some guinea-pigs into it with the idea that they would drive the rats away. Before many weeks the guinea-pigs had been killed, presumably by rats, but possibly by a stoat.—*K. Deakin, Cofton Parsonage, Alvechurch.*

INSTINCT IN BEES.

BY LIEUT.-COLONEL H. J. O. WALKER.

IN the June number of the last volume of SCIENCE-GOSSIP (p. 11) Mr. Dickson-Bryson deals with bees under the heading of "Instinct." Having myself made the hive-bee a particular study, I should like to point out one or two inaccuracies, although the paper itself as yet remains unfinished. It is stated that the "entire population" of the hive "numbers no less than from eight to ten thousand individuals." This is far below the mark. The summer population of a thriving hive should be about 40,000 bees. A good swarm even should be full 20,000 of these insects.

A little further on, p. 12, the writer states that 15,000 eggs are laid by the queen in one day, and it is hard to see how he reconciles these two statements, the more so that a vigorous queen lays almost continuously throughout the greater part of the year, whenever suitable cells are ready for her. The second statement is quite erroneous. The paragraph begins :—"In the finished hive nearly 50,000 cells await the eggs of the female." As a matter of fact, the greater portion of the combs would be filled with brood and stores; and if a tenth part of the given number of cells, which would be equivalent to 200 square inches of comb, were ready for her it would be quite unusual. Again, no matter how many cells were available, she could lay nothing approaching to the number of eggs stated. As many besides myself have observed, it takes a queen about thirty seconds to deposit her egg and get clear of the cell; so that if any number of cells were ready for her and contiguous, the eggs laid in twenty-four hours could not amount to more than 2,880, even if the queen were to lay all the time without stopping. She, however, rests for about six hours out of the twenty-four, and has to walk about to look for empty cells, so that it is improbable that she ever lays more than 2,000 eggs, indeed seldom so many.

Since writing the above paragraph, I may mention that I have at present under observation an exceptionally quick laying queen, whose average is eighteen seconds; but the time it takes to deposit eggs is small compared with that spent in finding and examining cells. A freshly mated queen lays more slowly.

As regards the method of the hive-bee in comb-building, it would seem probable that it came to be adopted by the working, through natural selection, of the principle of economy in wax; for its secretion is not only a constitutional strain on the worker-bee, but involves a large consumption of the honey which it works so hard to win, and of which from 10 to 21 lbs. may be taken to produce 1 lb. of wax. If instead of the cell-base being angled out so as to form part of those of three cells on the opposite side of the mid-rib it had been left flat, it would have been necessary to lengthen the walls of the cells, and one-fiftieth more wax would thus have been expended, whilst the base itself could not have been left so thin for fear of stretching.

It is well known to observant bee-keepers that Mr. Dickson-Bryson's "precise angle" is by no means always maintained in the case either of cell-walls or base-rhombs, even when no reason can be detected for want of exactness. "The sizes of the rhombs may be so changed that two of them occupy nearly the whole space, while the third nearly disappears, and a fourth makes its appearance." So says Mr. Cowan in his concise and well-illustrated little book the "Honey Bee," which I would recommend to anyone interested.

Finally, although it is not easy to see how natural selection can have influenced bees to produce undeveloped females, it is worth noting that here, too, the principle of economy is involved, seeing that imperfect development is produced by the worker-nurses ceasing after the first three days to feed the larvae with the concentrated and partly glandular food, which they continue to supply in full generosity to those selected for future queens.

Lee Ford, Budleigh Salterton.

THE GREENWICH VISITATION.

THE day for the official visitation is the first Saturday in June; but this year, by an Order in Council, it was postponed until Tuesday, June 26th, owing to the absence of the Astronomer Royal with the eclipse expedition to Ovar, in Portugal.

The Observatory is to have better provision against fire, and an open iron railing is to replace the present wooden fence, so that, with an extension of the boundary of the Observatory grounds, the architectural features of the new buildings will be seen more effectively. During the year from May 11th, 1899, to May 10th, 1900, the transit circle was employed in making 10,712 transit observations and 10,001 circle observations. Amongst the latter are 674 determinations of the nadir point and 637 reflection observations of stars. All these observations are completely reduced up to May 1st.

"The New Ten-Year Catalogue of Stars, 1887–1896," containing 6,892 stars, is printed, with the exception of the Introduction, and that is in the printer's hands. The re-observation of the stars in Groombridge's "Catalogue" will give material for determining the proper motions of more than 4,000 stars from observations about eighty years apart.

The new Altazimuth is now in good working order. One of its pivots has, however, "fired" badly twice, and had to be re-ground. On the last occasion Y's of bell-metal were substituted for the Y's of cast iron. On September 3rd, the object-

glass was found loose, and has now been firmly fixed. A large chronograph, for use with this instrument, has been supplied by Sir Howard Grubb and is found quite satisfactory.

The 28-inch refractor has been used in the micrometric measurement of 492 stars, 268 very close doubles, and the rest chiefly those having very minute companions. Amongst the former ζ Herculis 3, and ·5 magnitude was distant 0''·6; γ² Andromedae 3, 5; 0''·4; and ε Hydrae A.B. 3.5, 6; 0''·2. Mr. Newall, having discovered spectroscopically that Capella is a binary with a period of 104 days, suggested in "Monthly Notices" that possibly it might be within the reach of large telescopes. On fifteen nights between April 4th and May 10th it has been noticed by several observers, who all agree that it is elongated, and, moreover, during the period of observation the position angle of the elongation has changed in accordance with the period mentioned. We believe that the compounds are nearly equal in magnitude.

With the Thompson 26-inch Refractor several photographs of double stars, Swift's comet, and Neptune with its satellite, have been taken for micrometric measurements, as well as plates of the moon, and of Jupiter, Saturn, and Uranus, with their satellites. With the 30-inch Cassegrain several photographs of nebulae have been obtained. On May 11th the large mirror was re-silvered for the first time.

Good progress is being made in the photographic chart of the heavens with the 13-inch Astrographic telescope. Each chart plate is exposed for 40 minutes, and out of the total number allotted to Greenwich 1,076 have already been successfully taken, only 73 remaining to complete it. The plates for the catalogue are each exposed three times—6 and 3 minutes, and 20 seconds. Of these, 1,103 have been taken, and 46 more are required. During the year 88,000 measures of pairs of images on these latter plates have been made.

The photographic spectroscope has been fitted to the 30-inch reflector, and seems to be very satisfactory.

Photographs have been taken of the sun on 180 days either with the Thompson or Dallmeyer photo heliographs.

Photos from India and Mauritius fill up the gaps so effectually that during 1899 there are photos for 364 days.

The magnetic observations give the mean declination for 1899 16°. 34'·2 West, and the dip 67° 10' 13''. No great magnetic disturbances have occurred, and lesser ones were only recorded on 16 days.

During the past 58 years August has only once been so warm—in 1857.

The photographs of the corona and spectrum of the solar eclipses of 1900 and 1898 were exhibited in the same room, and evoked a great amount of interest. The smaller photos taken at Ovar showed considerable over-exposure from the brightness of the sky. The detail of the corona was not so pronounced as in the Indian photos.—FRANK C. DENNETT.

MICROSCOPY

CONDUCTED BY F. SHILLINGTON SCALES, F.R.M.S.

POSTAL MICROSCOPICAL SOCIETY.—During the summer of 1873 a letter appeared in SCIENCE-GOSSIP suggesting that if twelve gentlemen could be found willing to co-operate in forming a little club for the circulation of microscopic slides, and notes thereupon, it might lead to a very pleasant and profitable interchange of thought and study. This letter was replied to by the late Mr. Alfred Allen, of Bath. The scheme met from the very first with much more support than had been anticipated, a code of rules was quickly drawn up, and in September of that year the Society came into existence with a roll of 36 members. Mr. A. Atkinson, of Brigg, the writer of the original letter in SCIENCE-GOSSIP, was fittingly made the first President, and was succeeded in due course by the late Mr. Tuffen West, with whose name every microscopist is familiar. By that time the Society numbered, we believe, over 100 members, and the membership subsequently increased in an even greater degree. The leading spirit of the Society was, however, Mr. Allen himself, and in 1882 he added largely to the usefulness and status of the Society by publishing at monthly and quarterly intervals the well-known "International Journal of Microscopy and Natural Science," which, besides acting as the Society's medium, contained many valuable scientific papers. It is understood that the journal was not self-supporting, but Mr. Allen himself willingly undertook its publication until failing health obliged him to discontinue its issue in 1897, after fifteen years of labour thereon. Mr. Allen's death in the following year (March 24, 1898) was a severe blow to the Society, and it was for a time, we believe, practically in abeyance, until the appointment of a new Hon. Secretary, Miss Florence Phillips, commenced what we hope will prove to be a new lease of life. Unfortunately, since Mr. Allen's death and the cessation of his journal, the Society has had no recognised medium for publishing the many interesting notes that are entered in MS. memorandum-books that have circulated with the slides sent round amongst the members. In consequence, the Editor of SCIENCE-GOSSIP communicated with the Secretary and President of the Society, and offered to place at its disposal a portion of the space in this journal for the publication of such notes. This suggestion has met with approval, and it is intended to occupy at least one page monthly, in the section set apart for microscopy, for notes extracted from the Postal Microscopical Society's memoranda. It is hoped that these will contain information as interesting to our readers as to the members of the Society, and will lead to profitable discussion in our columns. Before closing this announcement we desire to draw the attention of our readers, who are workers in the field of microscopy, to the many advantages accruing to membership of the Postal Microscopical Society. Full particulars may be obtained from the Honorary Secretary, Miss Florence Phillips, "Hafod Euryn," Colwyn Bay, North Wales.

METHOD OF PRESERVING AND MOUNTING ROTIFERA.—The following is Mr. Rousselet's method, communicated to the Manchester Microscopical Society by Mr. Mark L. Sykes, F.R.M.S.: "Rotifera cannot be killed suddenly by any known process without contracting violently and losing all their natural appearance. To kill and preserve them with their cilia fully expanded and in their natural condition, the animals should first be narcotised with a solution consisting of 2 per cent. solution of hydrochlorate of cocain, 3 parts; methylated spirit, 1 part; water, 6 parts. The rotifers should first be isolated in a watch-glass and clean water, and a drop, or two drops, of the solution added at first. After five or ten minutes another drop should be added, and afterwards drop by drop and very slowly, until the animals are completely narcotised. They may then be killed and fixed by adding one drop of a ½ per cent. to ¼ per cent. solution of osmic acid. To clear from the solution they must be washed several times in clean water, until all the acid is completely removed. The rotifers must then be transferred to a 2½ per cent. solution of formaldehyde (2½ per cent. of commercial, 60 per cent. formalin, and 37½ per cent. of distilled water), and should be mounted in this fluid in hollow-ground glass slips. The cells must be well secured after mounting by several coats of cement. The process requires a little practice, and great care should be taken that the animals are always in fluid, and not allowed to become dry in the process of mounting; but the results are excellent, the objects having all the appearance of living animals, the colours, internal structure, and outward form being beautifully preserved in situ."

BAKER'S PLANTATION MICROSCOPE.—This is a cheap microscope designed for use by planters, missionaries, and others who have no practical acquaintance with the microscope, for the detection of the ova of intestinal parasites so common in men and animals in the tropics. It is accordingly simplified to the last degree; there is one objective and eyepiece, giving a total magnification of 150 diameters, and the focussing is done by rotating

BAKER'S PLANTATION MICROSCOPE IN CASE.

the optical tube, which gives a vertical movement by means of a spiral slot and pin. There is a mirror, but no draw-tube, fine adjustment, or condenser, and the stand is a plain non-inclinable one. It fits into a tin case 9 × 2⅝ × 2½ inches, which contains also a supply of glass slips and covers, together with a sheet of printed instructions illustrating the eggs of Ankylostome, Round, and Whip Worms, *Bilharzia* and *Distomo ringeri*, also of *Amoeba coli* and *Trypanosomes*. We do not know what demand there may be in the tropics for an instrument of this sort, but it is certainly designed to stand the maximum of bad usage without

BAKER'S PLANTATION MICROSCOPE.

· ill-effect, and should prove sufficient for its purpose. We would scarcely recommend it, however, for any other than the purpose for which it is designed. The price complete is only £2·5s. This and the foregoing microscope are those recently exhibited before the Royal Microscopical Society.

BAKER'S R.M.S. 1.27 MICROSCOPE.—Mr. Chas. Baker has recently brought out a new microscope, specially designed for advanced workers, which both in design and workmanship deserves notice in these columns. The stand is of the solid tripod type, which, whilst giving nearly as firm a base, even in the horizontal position, as the true tripod, is in some respects preferable to this latter form in the greater facility afforded for getting at the sub-stage adjustments when the microscope is used vertically. The limb is of the "Jackson" form with lever fine adjustment, than which we have found none more sensitive or serviceable. Each revolution of the milled head gives a movement of ·11 millimetre ($\frac{1}{225}$ inch). The body has two draw-tubes, giving a variation of tube-length from 120 to 250 millimetres, thus enabling objectives corrected for both the short and the long tube to be used at will. Both draw-tubes are graduated in millimetres, and the lower one is actuated by rack and pinion ; a very useful addition when adjusting objectives so as to correct them for different thicknesses of cover-glass, especially in view of the growing tendency to make such corrections by this means instead of by the provision of a correction collar to the objective itself. The body is of a large diameter that should lend itself to photography, and the eyepieces are of the new R.M.S. No. 3 standard size. There is a mechanical stage giving a movement of 25 millimetres in either direction, and graduated to half millimetres, and the stage is capable of rotation for about 280°. The top plate is provided with three adjustable stops for 3 inch × 1 inch and 3 inch × 1½ inch slides with a view to greater facility in recording positions, and if required a large flat plate is available. The sub-stage is of the usual form with centering screws, coarse and fine adjustments, the latter being exceptionally neat and so conveniently placed that both adjustments can be controlled without shifting the hand. There are the usual mirrors. All

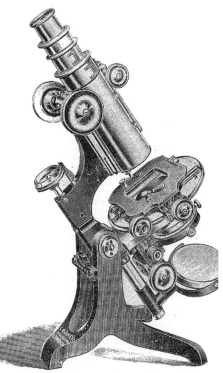

BAKER'S R.M.S. MICROSCOPE.

the fittings are sprung, and have adjusting screws to compensate for wear. The price of the stand alone, without case, is £16 16s.

OVA OF LEPIDOPTERA.—Recently we had the opportunity of carefully examining some hundreds of water-colour drawings of British Lepidoptera. They were the work of our correspondent Mr. E. Wheeler, of Queen's Road, Clifton, near Bristol, who had faithfully delineated under the microscope the external structure and markings. As in most cases he had made drawings at various periods of the development of the embryo within the egg, this study proves to be one of much interest, as is also the ease with which butterflies and moths may be classified by the external structure of their eggs.

MICROSCOPY FOR BEGINNERS.

BY F. SHILLINGTON SCALES, F.R.M.S.

(Continued from p. 61.)

If the microscope requires adjustment, these adjustments should be made with the utmost care. Most microscopes by our best English makers have the wearing parts sprung so that the adjustments may be readily effected, but even then a little attention to the tools with which the work is done may be recommended. The screwdriver, for instance, should be in good condition.

It is well also to bear in mind that the lacquer on the brass-work of the microscope, placed there not so much for appearance as for the prevention of oxidisation, is destroyed by alcohol.

Finally, our advice to the beginner who may wish to oblige a friend by lending him his microscope is—don't!

It now only remains for us to add a few hints on mounting, and these we shall endeavour to make as simple and practical as possible. The beginner must bear in mind that mounting for the microscope has become quite an art, if not a science, and the list of reagents, stains, and media used for special purposes would be quite a formidable one. Fortunately the requirements of beginners and amateurs, especially those for whom we are now writing, are much more easily dealt with, and we shall confine ourselves to the simplest and most commonly used methods, trusting that as knowledge grows and experience comes with it, the beginner will learn more of such advanced methods from works dealing with the subject.

It is of course only with very low powers, and when the nature of the investigation admits of it, that an absolutely unprepared and unarranged object can be examined. For this purpose a pocket-lens is infinitely preferable to the compound microscope with all its complications and refinements. For examination with the latter instrument even opaque objects require to be properly displayed, whilst objects to be examined with transmitted or direct light—that is, by means of light that passes through the object—require very careful preparation beforehand.

Wooden slips and paper-covered slips are now very rarely used, 3 inch × 1 inch glass slips being now almost universal. These can be obtained from any optician. They should, preferably, have ground edges, and for general purposes should be of medium thickness. They will cost from twopence to fivepence per dozen, according to quality, or less for a larger quantity. If any of them should be found to have scratches or specks in the centre, they should be put aside for making opaque mounts. For exceptionally large mounts slips 3 inches × 1½ inch can be obtained. The cover-glasses should be circular, in thickness from ·006 inch to ·008 inch, and might vary in size from ⅜ inch to ¾ inch diameter. It would be well to provide oneself with a stock of ⅜-inch, ¾-inch, and ⅞-inch cover-glasses, and to note their thickness at the time of purchase, and, generally speaking, to adhere afterwards to the same standard for ordinary work. High-power work with objectives of very short focus may require thinner cover-glasses to be used. We would also recommend the purchase of a dozen or so slips with excavated cells of various sizes, *i.e.* with concavities ground in their centres.

Before use, all slips and covers must be scrupulously cleaned. It is generally sufficient to wash them with hot water and soap or soda; but for special work more drastic measures may be necessary. The writer generally uses a fairly strong and hot solution of Hudson's Soap, with subsequent careful rinsing and polishing with an old cambric handkerchief. The great thing to be avoided is any suspicion of grease, even from the fingers themselves. Cover-glasses must be finally polished with chamois leather, and as they are very thin and of course easily broken, various contrivances such as buff blocks are obtainable for the purpose. With a little practice, however, it is quite easy to hold half the cover-glass in a piece of chamois leather between the finger and thumb, but not edgeways, and to polish the other half, turning the glass round meanwhile.

We will first deal with the mounting of opaque objects and of objects that can be mounted dry, this process being comparatively simple. The various apparatus, reagents, media, stains, etc., will be mentioned as we proceed, and their uses will then become apparent at the same time. Accordingly we shall here require a turntable. This is a circular brass plate about 3½ inches in diameter, mounted so as to rotate upon a centre, the upper surface of this plate having concentric rings engraved upon its surface. These latter serve as a

TURNTABLE.

guide in centering the slide upon the rotating plate. There is also a pair of clips to hold the slide in place. The turntable is mounted on a wooden block or iron stand which serves as a support for the hand. The cost will be about six shillings. We do not recommend the "self-centering" turntables. We shall also need two or more good sable brushes, which are best and cheapest in the long run. These should be about ¼ inch and ¾ inch in diameter, costing ninepence or one shilling each. Also a pair of steel or brass forceps, not too narrow, costing one shilling and sixpence, a bottle of gold-size, a bottle of Brunswick Black, and a bottle of gum arabic. All of these are obtainable from the opticians.

The usual plan with opaque objects is to place a slide on the turntable, centre by means of the concentric rings, and then run a disk of Brunswick

Black of the requisite size in the centre, rotating the stage meanwhile by means of the forefinger of the left hand and the milled head beneath. As soon as this black disk is dry, a piece of black paper of the same size is cut out and gummed upon it. The black paper should not have a glazed surface. Then upon the disk is built up a cell of the requisite depth to contain the object. As we have before said, however (SCIENCE-GOSSIP, vol. vi. page 375), this method of mounting opaque objects upon a black background is not only unnecessary, but often inconvenient, as it renders the use of transmitted light impossible, if it should be wanted; neither can such slides be examined by means of a Lieberkühn. We recommend therefore that the black background be omitted, and that instead a similar disk, or two or three disks of various sizes, be put upon thin slips, and one of these can then be placed beneath the slide carrying the object, when it is being examined by reflected light.

The cells are made by running a ring of gold-size of the same diameter as the cover-glass that will be used. This is done by means of the turntable, and is not difficult. It is not advisable to use too full a brush, and the gold-size should be of the right consistency—neither too thick to leave the brush, nor so thin as to run away from position. The tip of the brush is used, and the table rotated not too quickly. For very thin objects one ring will suffice; but thicker objects will need two or three rings, added one on the top of another, each ring being added, however, only when the other is dry. If a few such rings do not give sufficient depth, it is advisable to build up the cell by other means. Rings may be cut out of stout paper or thin and good cardboard, then steeped in paraffin and dried. Stout rings of ebonite, glass, tin, etc., can be obtained from the opticians. It is only necessary to attach these to the slide by means of a ring of gold-size, pressing down the ring firmly, and even giving a very slight twisting motion to make sure of there being no air-bubbles to prevent perfect contact. If the cells of gold-size when dry should not be quite level, they can easily be rubbed down on a piece of very fine emery laid on a flat surface. The object itself must be fastened in place by means of a drop of gum placed upon the slide. Care must be taken that this drop of gum is hidden by the object, unless that is impossible. Thin objects, such as wings, petals, leaves, etc., may generally be kept in place merely by the pressure of the cover-glass. Very minute objects, such as pollen grains for instance, are made to adhere by means of a thin film of very weak gum, which is placed on the slide and allowed to dry. Breathing upon the slide will then moisten the film of gum sufficiently to cause the pollen to adhere when placed thereon. In every case, however, it is of the utmost importance that the gum and gold-size should be allowed to dry thoroughly before the cover-glass is put on, or the remaining moisture will settle on the under side of the cover-glass, and utterly spoil the slide. A final ring of gold-size is then run on, and this last should be allowed to dry until it is just sticky only, when the cover-glass may be gently lowered into place by means of a pair of forceps, and the edges pressed gently down, care being taken that the cover-glass adheres all round its edges. Finally, the slide is finished by a coat of Brunswick Black over all, and just covering the edge of the cover-glass.

(*To be continued.*)

A NEW MINOR PLANET was discovered photographically at Heidelberg on June 4th by Professor Max Wolf and Herr Schwassman.

THE transactions of the "British Mycological Society" for last year's season are of exceptional interest, and graced by a coloured plate. There are several papers, and the Presidential address by Dr. C. B. Plowright, also a list of fungi new to Britain by Annie Lorrain Smith.

PROFESSOR CHANDLER, of New York, was formally admitted to the Chemical Society on July 5th. When he was welcomed by Professor Thorpe, he caused much laughter by remarking that he could only now consider himself a real member, although he had been a "life" member for over thirty years. On July 7th the University of Oxford conferred upon him the honorary degree of Doctor of Science.

WE would draw the attention of photographers to Messrs. Cadett & Neall's plates and printing paper. They will be found admirable, especially the paper for X-ray work. We are reminded of this by the reference to the Cadett light filters in Mr. Sanger Shepherd's article on the "Photography of Colour" in this number. The firm will supply particulars on application to their works at Ashtead, Surrey.

AN interesting paper on abnormalities in the shell of *Helix nemoralis* appears in the July number of "The Irish Naturalist." It is accompanied by a beautiful group, illustrating nineteen examples, from a photograph by the author, Mr. R. Welch, of Belfast. On the same page is also a fine photograph by the same gentleman of a cluster of ova of the Kerry slug, *Geomalacus maculosus.* This latter illustration is accompanied by some interesting notes from the pen of Mr. Thomas Rogers.

IT is satisfactory to find practical applications of the various systems of wireless telegraphy are still being energetically made. It is reported that, in consequence of the successful experiments made with M. Popoff's system, the Russian Naval Minister has decided to introduce it into the navy. The whole of the Black Sea fleet will, it is said, be fitted up this summer with the necessary apparatus. Both in our own navy and in that of Germany the question of signalling in this manner is receiving considerable attention. Experiments were successfully carried out a week or two since between the battleships *Jupiter* and *Hannibal.* Messages were distinctly read when the ships were twenty miles apart, and occasionally when at thirty miles' distance. In Germany an apparatus with a wire 205 feet high, has been put up for the North German Lloyd at Kaiserhafen to effect communication between lighthouses and fireships. The large Lloyd steamers are gradually all to be equipped with apparatus worked upon what is termed the Schäfer system.

CONDUCTED BY F. C. DENNETT.

	Position at Noon.			
1900	*Rises.*	*Sets.*	*R.A.*	*Dec.*
Aug.	*h.m.*	*h.m.*	*h.m.*	° ′
Sun ..	8 .. 4.35 a.m.	.. 7.36 p.m.	.. 9.12	.. 16.18 N.
	18 .. 4.52	.. 7.16	.. 9.49	.. 13.11
	28 .. 5.7	.. 6.55	.. 10.26	.. 9.48

	Rises.	*Souths.*	*Sets.*	*Age at Noon.*
Aug.	*h.m.*	*h.m.*	*h.m.*	*d. h.m.*
Moon ..	8 .. 5.50 p.m.	.. 10.13 p.m.	.. 1.29 a.m.	.. 12 22.17
	18 .. 11.3	.. 6.26 a.m.	.. 2.42 p.m.	.. 22 22.17
	28 .. 8.45 a.m.	.. 2.14 p.m.	.. 7.33 p.m.	.. 3 ·8.7

			Position at Noon.	
	Souths.	*Semi-*	*R.A.*	*Dec.*
Aug.	*h.m.*	*Diameter.*	*h.m.*	° ′
Mercury ..	8 .. 11.17 a.m.	.. 5.0″	.. 8.23	.. 15.26 N.
	18 .. 10.49	.. 3.8″	.. 8.35	.. 17.33
	28 .. 11.5	.. 2.9″	.. 9.31	.. 15.54
Venus ..	8 .. 9.32 a.m.	.. 20.5″	.. 6.38	.. 17.14 N.
	18 .. 9.12	.. 17.5″	.. 6.57	.. 17.38
	28 .. 9.1	.. 15.2″	.. 7.26	.. 17.46
Mars ..	18 .. 8.39 a.m.	.. 2.4″	.. 6.24	.. 23.44 N.
Jupiter ..	18 .. 6.12 p.m.	.. 17.8″	.. 15.59	.. 19.54 S.
Saturn ..	18 .. 8.7 p.m.	.. 8.1″	.. 17.54	.. 22.34 S.
Uranus ..	18 .. 6.40 p.m.	.. 1.8″	.. 16.27	.. 21.42 S.
Neptune ..	18 .. 8.9 a.m.	.. 1.2″	.. 5.54	.. 22.14 N.

MOON'S PHASES.

	h.m.		*h.m.*
1st Qr. .. Aug.	3 .. 4.46 p.m.	Full .. Aug.	10 .. 9.30 p.m.
3rd Qr. .. „	17 ..11.46 a.m.	New .. „	25 .. 3.53 a.m.

In perigee August 12th at 11 a.m.; in apogee on
27th at 10.30 p.m.

METEORS.

				h.m.	
July 23–Aug. 4	.. a-β Perseids Radiant R.A.	3.12	Dec. 43°	N.
„ 30–Aug. 11	..λ Andromedes	„	. 23.20	„ 51	N.
Aug. 5–16	..κ Cygnids ..	„	. 19.28	„ 53	N.
„ 9–11	.. Perseids ..	„	3.0	„ 57	N.
„ 21–25	.. ο Draconids	„	19.24	„ 60	N.
„ 21–Sept. 21	..ε Perseids ..	„	4.8	„ 37	N.
„ 25 „	22 ..γ Pegasus ..	„	0.20	„ 10	N.

The Perseids may be seen not only on the days
mentioned, but, though less frequently, for some
days before and after. The radiant point also is
not stationary, but travels eastwards. On July 25th
it is in R.A. 1h. 45m. Dec. N. 53°, whilst according
to Mr. W. F. Denning by August 18th it has
reached 3h. 41m., N. 59°.

CONJUNCTIONS OF PLANETS WITH THE MOON.

						° ′
Aug.	5	..	Jupiter°†	.. 9 a.m.	.. Planet 1.22 N.
„	7	..	Saturn°†	.. 11 a.m.	.. „ 0.50 S.
„	20	..	Mars°†	.. 6 p.m.	.. „ 2.55 N.
„	21	..	Venus°	.. 10 a.m.	.. „ 1.49 S.
„	23	..	Mercury°	.. 5 p.m.	.. „ 4.59 N.

° Daylight.   † Below English horizon.

OCCULTATIONS.

				Angle		*Angle*
		Magni-	*Dis-*	*from*	*Re-*	*from*
Aug.	*Star.*	*tude.*	*appears.*	*Vertex.*	*appears.*	*Vertex.*
			h.m.	°	*h.m.*	°
13 ..	κ Piscium ..	5.0	.. 4.10 a.m.	.. 15	.. 5.10 a.m.	.. 236
19 ..	ι Tauri ..	4.7	.. 0.42 a.m.	.. 81	.. 1.24 a.m.	.. 346

THE SUN still has interesting groups upon its
surface at frequent intervals.

MERCURY is in inferior conjunction with the
sun at 8 a.m. on August 1st, after which it is a
morning star all the month. This planet reaches
its greatest elongation, 18° 32′ west, at 2 p.m. on
the 19th, when it rises an hour and three-quarters
before the sun, and is in good position for
observation.

VENUS is a morning star all the month, reaching
its greatest brilliancy at 8 a.m. on the 14th. It
rises at 2.20 a.m. on August 1st, and at 1.21 a.m.
on the 31st.

MARS is also a morning star, rising about forty
minutes after midnight at the beginning of the
month, and about eight minutes after at the end.
Its appearance, with a 2-inch telescope, is little
more than that of a large star.

JUPITER is an evening star all the month, but is
in poor position for observation. It comes to the
meridian half an hour before sunset on the 1st, and
sets at 11.35 p.m., whilst on the 31st it sets at 9.41.
It is almost stationary near the beautiful and easy
double star β Scorpii.

SATURN is likewise an evening star southing
about two hours later than Jupiter, and so is in
better position for the observer. Its widely open
rings make it a splendid object.

URANUS is nearly 2° farther south than Jupiter,
and comes to the meridian about twenty-eight
minutes later.

THE ECLIPSE OF THE SUN.—Further details of
the observations are now appearing. On Thursday,
June 28th, there was a joint meeting of the Royal
Society and the Royal Astronomical Society to
receive the preliminary reports of the Expeditions.
On the previous evening a number of accounts were
given at the meeting of the British Astronomical
Association. The Astronomer Royal, at Ovar,
described the corona as very distinctly inferior in
brightness, structure, and rays to the one seen in
the Indian eclipse—appearing, indeed, quite a
different object. Sir Norman Lockyer, at Santa
Pola, considered the corona a repetition of that of
1878. Mr. Geoghegan, with Dr. Downing's party
at Plasencia, 140 miles south-west of Madrid,
observed the shadow-bands for about two minutes
before totality, and again for a similar time after.
Mr. Maunder's party were at Algiers, and the
photographs were very successful. A remarkable
feature, not previously noticed, was the presence
of dark rays in the corona, quite distinct from
mere rifts. Mercury was visible on many of the
photographs. Although some of the plates had a
longer exposure than those employed in India,
none of them exhibited so great an extension of
bright rays as were then portrayed. Mr. Crom-
melin, of the same party, observed the contacts on
a projected image formed by a 3-inch Dollond
achromatic, and for 20 seconds before totality
witnessed the phenomena known as Baily's Beads.
Mr. C. L. Brook, with the same party, observed
the shadow-bands, or, as he suggests, ripples,
three and a quarter minutes before totality. Mr.
Evershed went some 25 miles farther south, to
Mazafram, to get to the south edge of the limit of
totality to have a better opportunity of observing
the flash spectrum. He succeeded in obtaining
some good photographs notwithstanding that he
got just a little too far south. At Elche, Mr. E.
W. Johnson saw the shadow-bands for two minutes
before totality. Some of the observers note that
the longest ray reached nearly to Mercury, about
2° distant from the sun's centre.

CONDUCTED BY HAROLD M. READ, F.C.S.

MANUFACTURE OF ARTIFICIAL PEARLS.—In view of the enormous trade now carried on in artificial pearls, the method of preparing the mother-substance for their manufacture is of interest. This mother-substance is worked up in the Lauscha district, and other glass-making centres of Thuringia, from fish-scales imported from the Baltic. The scales are first thoroughly washed with fresh water to remove impurities, then shaken up with a further quantity of water for about two hours, and finally the whole is subjected to pressure in a linen bag. The silvery lustrous runnings are collected and set aside. This treatment is repeated until the scales have lost their silvery appearance, become transparent, and hard to the touch. The runnings are put aside to clarify, while, to prevent putrefaction, ammonia is added and the mixture kept at as low a temperature as possible. The sediment is now washed repeatedly with water, until the washings are quite clear. Thereupon the lustrous residue is bottled off, and the water gradually removed by successive washings in alcohol. During this treatment the extract assumes the consistency of butter, but it still retains its pearly lustre. For use, the butter-like mass is mixed in small quantities with a hot aqueous solution of gelatine. If the "pearls" are to be coloured, a spirituous solution of an aniline dye is incorporated with the gelatine.

SUGAR AN AID TO THE GROWTH OF PLANTS.—During the last two years some most interesting work has been carried out at Nottingham College, by Mr. J. Golding, on the influence of saccharose or cane-sugar on the growth of plants. The aim of the experiments—the results of which have recently been published—was to compare the increased yield produced by sugar on plants drawing all their nitrogen from the air, with that obtained in the case of plants drawing their nitrogen in a combined form from the soil. If the energy for the nitrogen-fixation comes from the breaking-up of the sugar, those plants which have to fix all their nitrogen might naturally be expected to benefit more by sugar applied to their roots than those which have their nitrogen presented in a combined, and hence more readily assimilable form. Briefly stated, the results of the investigation show that the leguminous plants with healthy root-nodules or nitrogen-fixers benefit by the application of small quantities of sugar to their roots. Further, even in the case of those plants which are devoid of root-nodules, but are supplied with an abundance of combined nitrogen, an increased yield is noticed after the application of sugar to their roots. At the same time, it is found that where the plants are starving for want of nitrogen, the addition of sugar is actually injurious; in fact it is possible to kill plants by the use of too much sugar. These results confirm those obtained some

months ago by Winogradsky and Oméliansky, who found that one part of grape-sugar in 500 entirely prevented nitrification.

PROFESSOR NILSON.—On July 5th the Chemical Society paid fitting tribute to one of their distinguished members, the late Professor Nilson, in a memorial address delivered by one of his co-workers, Professor Otto Petterson. The lecture was delivered in English, and in a style which might well be emulated by many Englishmen. Nilson's life was sketched from his leaving his father's farm in Gottland, a detailed account being given of his training with the great Swedish chemist Berzelius, his researches in pure chemistry at Upsala University, and his work as Director of the Agricultural Department of the Swedish Government. It was in the latter office that Nilson showed not only his exceptional capacity as a thinker, but also his skill as a chemist, and his patriotic enthusiasm for his motherland.

CONVERSION OF PHOSPHORUS INTO ARSENIC.—Last month I referred to a paper by Fittica dealing with the alleged preparation of metallic arsenic from phosphorus. A scathing criticism of Fittica's work by Dr. Clemens Winkler has appeared in the "Berichte," and an excellent translation of this paper is published in the "Chemical News." After pointing out the utter fallacy of Fittica's assertions, and the fact that the percentage of arsenic obtained corresponds almost exactly with that ordinarily found in commercial phosphorus, Dr. Winkler concludes with a paragraph which will be heartily welcomed by all students whose efforts are directed towards the discovery of truth. The remarks may so well be applied, not only to chemistry but to every branch of science, that I reproduce them :—
"It would appear as if in inorganic chemistry a dangerous tendency showed itself of late to enter into speculations without those careful investigations which have hitherto distinguished German chemists. Cases are multiplying which tend to show that a theory is first formed, and that one seeks to find what one wishes to find; or that, in the words of the physiologist Czermak, one starts from 'erroneously observed facts,' and thus falls into mistakes. The reason must be sought, to a great extent, in the fact that the art of analysis is being regrettably neglected. I intentionally say the 'art,' for between analysing and analysing there is the same difference as between artists' and stonemasons' work. One cannot expect analytical aptitude from the physicist, whose field of investigation begins more and more to extend to inorganic chemistry as electrolysis develops ; and he, within his field of investigation, is able to discover useful, even great, facts. But physical chemistry is in no way synonymous with inorganic chemistry ; for the latter, far from being a finished department of science, embraces problems in infinite number which must be solved upon an entirely different road from that indicated by the ion theory. The really successful carrying out of inorganic chemical investigations is possible only to him who is not only a theoretical chemist, but also an accomplished analyst ; not only a practical, mechanical workman, but a thinking and forming artist, who sees clearly the theories of the operations, to whom the knowledge of proportion is a part of himself, and who in all his doings is led by a sense of order and neatness, but especially by a desire for truth."

CONDUCTED BY JAMES QUICK.

THE STRUCTURE OF METALS.—In a paper read before the Royal Society on May 31st last, further results have been given of the investigations of Professor Ewing and Mr. Rosenhain upon the crystalline structure of metals. The first part of the work was briefly described in these columns last month. The authors have examined the changes of crystalline structure, which take place in various metals at comparatively low temperatures. It was noticed that a piece of sheet lead when etched with dilute nitric acid, exhibits a strong crystalline structure, with large crystals. This was afterwards found to be due to a slow process of annealing or recrystallisation at ordinary atmospheric temperatures. The phenomenon was investigated by taking, at low magnifications, a series of micro-photographs of certain marked areas in the surface of a specimen, in order to watch the change which went on through lapse of time, or after application of some thermal treatment. If a piece of cast lead is greatly strained by compression, the original large crystals, after being considerably flattened, are driven into and through one another. A piece of lead strained in this way and kept for nearly six months in an ordinary room, without any special thermal treatment, was found to be undergoing continuous change during that time. A series of photographs of this specimen, taken at intervals during the six months, showed that a great number of the small crystals grew larger at the expense of their neighbours. In similar specimens which were kept at 200° C. the growth was much more pronounced. Experiments were also made at 100° C. and 150° C., which led to the general result that crystalline growth will occur at any temperature from 15° C. or 20° C. up to the melting-point of lead, and that in general the higher the temperature the more rapid is the initial rate of change. A striking feature observed in several specimens was the large and rapid growth of one or two individual crystals. In many instances such individuals grew until they were some hundreds of times larger than their neighbours. Generally the most aggressive crystals were found near the edges of the specimen. It was also noticed that a crystal which had already grown considerably was at times swallowed by a more powerful neighbour. Mr. Rosenhain puts forward an hypothesis to explain the phenomenon by suggesting that the metallic impurities present in a metal play an important part in the action. When a metal solidifies from the fluid state, the metallic impurities finally crystallise as a film of what is known as eutectic alloy between the metal crystals. It is thought then that the observed changes of crystalline structure which go on, whilst the metal is in the solid state, are accomplished by the agency of these eutectic films between the crystals, in dissolving metal from the surfaces of some crystals and depositing

it on others. Several observations and experiments confirm this view.

A REVOLVING MAGNETIC FLAG.—A very pretty lecture-room experiment has recently been described, to demonstrate the existence of a magnetic whirl in the interior of a conductor carrying a current. The conductor itself consists of a beaker of mercury, the "flag" being a small magnet attached with one end at right angles to a glass rod pivoted along the axis of the beaker. The magnet is therefore capable of rotation in a plane at right angles to the direction of the current. A hole is bored through the bottom of the beaker to admit an electrode making contact with the mercury. The return circuit is afforded by a copper vessel into which the beaker tightly fits, the external field due to the current also being thus eliminated. The earth's magnetism is also neutralised by permanent magnets. The glass rod is weighted to obtain neutral equilibrium. Upon passing the current through, the flag rotates with uniform velocity, the north-seeking pole following the lines of force.

PHYSICAL SOCIETY OF LONDON.—The final meeting for the present session of this Society was held on June 22, Mr. T. H. Blakesley, vice-president, occupying the chair. Dr. Harker read a paper entitled "Notes on Gas Thermometry," by Dr. Chappuis. The author discards hydrogen as an absolutely reliable thermometric substance at high temperatures, due to its attacking the walls of glass receivers, and uses a constant volume nitrogen thermometer. He obtains a value of 445.2° C. as boiling-point of sulphur, and criticises Callendar and Griffith's value of 444.53° C. In the discussion upon the paper Professor Callendar, Mr. Glazebrook, and others took part. Professor Callendar remarked that he was unable to agree with the correction to his observations suggested by Dr. Chappuis. Two other papers were read, one by Professor Callendar on behalf of Mr. H. M. Tory on "A Comparison of Impure Platinum Thermometers," the other by Professor S. Young, D.Sc., F.R.S., on "The Law of Cailletet and Mathias, and the Critical Density." The Society then adjourned until October next.

INDUCTION COIL CONDENSERS.—The discovery and applications of the Röntgen rays have given rise to a great increase in the manufacture and sale of induction coils. The article in the "Philosophical Magazine" for February last by Mr. K. R. Johnson upon condensers in induction coils will therefore be of interest. As is well known, if a condenser is connected across the "break" in the primary circuit of an induction coil, the extra induced current flows through it. If the maximum E.M.F. in the condenser is very great, a spark occurs at the break and the oscillations in the circuit are diminished. If the capacity of the condenser is very large, the maximum condenser E.M.F. is small, as is also that at the break. No spark then takes place. The efficiency of the coil is therefore increased by increasing the capacity of the condenser, so long as the discharge by spark at the break is diminished. When this spark is entirely suppressed, the efficiency of the coil is greatest, and the secondary spark-length a maximum. If the condenser capacity is still further increased, the spark-length diminishes, owing to the maximum E.M.F. in the condenser being decreased, as well as the current in the circuit.

UNUSUAL SITE FOR SWALLOW'S NEST.—A pair of swallows have built their nest on the frame of a picture in my bedroom. Early in June I noticed they were constantly flying in and out of the room, and laughingly said to my little daughter, who was ill in bed, that they came to amuse her. Soon the swallows appeared to have great discussions, evidently about a site for their home, and they examined one picture after another. Finally they decided on an oil painting, which had a broad frame, hanging over the fireplace, and one afternoon commenced bringing in mud. I at once covered the whole picture with sacking, thinking, probably, that would cause them to forsake the site; but the birds did not object, and building operations went on very rapidly. The swallows came into the room on an average eight times in five minutes, bringing hay and mud alternately. When the nest was finished, the male bird evidently thought the furnishing was his lady's department, as she brought in all the feathers to line it, though he was always in close attendance on her, flying in behind, and watching the arrangement. Three young ones are now hatched and are growing fast. I have put little silver rings on their legs in the manner pigeons are ringed, so as to recognise them if they come next year. The old birds are wonderfully tame and do not take any notice of us. When the hen-bird was sitting she used to peep over her nest when I went into the room, apparently to see if I were alone; if not, perhaps she would fly off the nest, but now that she has young she does not mind strangers, and has many visitors and human admirers. I have had to take the blind down, put a nail into the window frame to prevent the window being shut by mistake, scatter insect powder over the nest, and take other precautions; but the cleanliness of the parent birds is one of the most interesting points. The old ones take everything away from the nest and drop it outside the window; and any inconvenience is amply repaid by the immense interest and pleasure the little birds are to all of us. I do not know that it is unique that swallows should build in an occupied room, but should be glad to hear if any of your readers have known a similar case.—(*Lady*) *Agnes F. Farren, Bealings House, Woodbridge, Suffolk, 7th July,* 1900.

MOLLUSCA OF SOUTH SURREY.—We are quite four miles from the chalk escarpment, and this clay country is not a very good district for land shells. So far I have only found *Helix nemoralis, Helicella caperata,* and *Cochlicopa lubrica.* In a small brook running past Mason's Bridge, a tributary of the River Mole, *Anodonta cygnea, Unio pictorum, Valvata piscinalis, Limnaea peregra, Sphaerium rivicola* occur, though not in large numbers. As every little "shelf" on the banks of the hook abounds in the spoor of rats, perhaps these little rodents may account for the scarcity of molluscs.—*R Ashington Bullen, F.L S, Axeland.*

HUMMING BIRD HAWK-MOTH.—A fine specimen of this moth was busy on July 1st of this year in my conservatory, dividing his attentions between the roses and heliotrope, at 7 P.M. The weather was rather cold and dull.—*R Ashington Bullen, F.L.S, Axeland, near Horley, Sussex.*

PECULIAR GROWTH OF BEECH-TREE.—The accompanying is a photograph of a peculiar beech-tree in the fir woods on Esher Common. Two main trunks start together from the ground, and at some feet from the ground two large branches effectually reunite, producing a kind of "Siamese twins." The photograph was taken by a friend, Mr. R. T. Morrison, in the winter while the tree was bare of leaves, so that the junction might show the better. It seemed to me of sufficient interest to publish; possibly you may think it

ABNORMAL BEECH AT ESHER.

worth your while, as you have illustrated a good number of vegetable freaks in SCIENCE-GOSSIP.—*W. J. Lucas,* 12 *Caversham Road, Kingston-on-Thames.*

BROWN-TAIL MOTH IN ESSEX.—I think *Porthesia chrysorrhoea* does not now appear to be frequently met with, although it formerly occurred in profusion in various parts of the South-East of England. It may therefore be interesting to record that I found a large brood of the larvae at Clacton-on-Sea in the beginning of June, feeding on a low elm hedge. The remains of the web in which they had hybernated was seen near to the larvae. A few I brought home have produced moths during the last day or two.—*W. Paskell,* 96 *Studley Road, Forest Gate. 16th July,* 1900.

HELIX ROTUNDATA, MONSTROSITY SCALARI-FORME.—Whilst searching under a large tree trunk at North Reston, near Louth, in Lincolnshire, I found a fine specimen of scalarid form of *Helix rotundata*. It was alive, and fairly active ; the shell measures 6mm. in height.—*C. S. Carter, Louth.*

ABNORMAL LIME-TREE FOLIAGE.—We have received a spray from a lime-tree gathered at Colwyn Bay, North Wales. The peculiarity lies in its having, in addition to its ordinary leaves, a large number of small and more or less malformed leaves clothing the various sprays. Our correspondent, Miss Florence Phillips, states that last year a similar sport was exhibited on the same tree, but to a less extent It appears to be a case of proliferation ; but it is odd that it should be found to so large an extent on one particular tree,

HIGH TEMPERATURE IN JULY.—At this address, which is in the London suburb of Hampstead, I have a registering thermometer by recognised makers, enclosed in a Stevenson's screen, raised four feet above the ground, and about twenty yards from the house. On July 16, at noon, the mercury was observed at 93° Fah., and at 1 P.M. it had registered 95°, but had fallen half a degree. As I have not heard of any other observation in these islands, under similar circumstances, exceeding 93°, I feel doubtful of my own, although I have had my thermometer correctly registering in the same position for several years.—*F. W. Watts, 49 Goldhurst Terrace, N.W.*

ABNORMAL EQUISETI.—On April 20th this year a very rare thing happened in Manchester. We had a fine bright day, so I set out with my camera in search of Spring flowers. I met with a group of fertile spikes of *Equisetum telmatia* standing a few inches above the ground, These I photographed in their natural habitat, which was a damp patch on the railway slope at Reddish Vale. Amongst the group were several abnormal forms, having their spikes divided into three, four, and six spikelets. I selected three of these variations and photographed them along with a normal form for contrast. I am not sure if it is a common thing for the fertile heads of *Equisetum* to diverge from the normal type in this way. Perhaps some of your readers could enlighten me ?—*Abraham Flatters, 16 and 18 Church Road, Longsight, Manchester.*

WE hear from Messrs. A. & C. Black that Professor Chrystal's " Introduction to Algebra," which is now in a second edition, is being translated into Japanese.

NOTICES OF SOCIETIES.

*Ordinary meetings are marked †, excursions * ; names of persons following excursions are of Conductors. § Lantern Illustrations.*

NORTH LONDON NATURAL HISTORY SOCIETY.

August 2.—†Selborne Revisited, J. A. Simes.
  „   6.—*Chalfont St. Giles. A. U. Battley.
  „   16 —†Notes on Clearwings. E. W. Lane.
  „   27.—*Warley. R. W. Robbins.

GEOLOGISTS' ASSOCIATION OF LONDON.

August 11.—*Netley Heath and Gomshall. W. P. D. Stebbing, F.G.S.
  „   20-25—§Lake District Excursion. J. B. Marr, F.R.S.

NOTTINGHAM NATURAL SCIENCE RAMBLING CLUB,

August 11.—East Leake. Botanical Ramble.
  „   25.—Charnwood Forest. Geological Ramble.

NOTICES TO CORRESPONDENTS.

TO CORRESPONDENTS AND EXCHANGERS.—SCIENCE-GOSSIP is published on the 25th of each month. All notes or other communications should reach us not later than the 18th of the month for insertion in the following number. No communications can be inserted or noticed without full name and address of writer. Notices of changes of address admitted free.

EDITORIAL COMMUNICATIONS, articles, books for review, instruments for notice, specimens for identification, &c., to be addressed to JOHN T. CARRINGTON, 110 Strand, London, W.C.

SUBSCRIPTIONS.—The volumes of SCIENCE-GOSSIP begin with the June numbers, but Subscriptions may commence with any number, at the rate of 6s. 6d. for twelve months (including postage), and should be remitted to the Office, 110 Strand, London, W.C.

NOTICE.—Contributors are requested to strictly observe the following rules. All contributions must be *clearly* written on one side of the paper only. Words intended to be printed in *italics* should be marked under with a single line. Generic names must be given in full, excepting where used immediately before. Capitals may only be used for generic, and not specific names. Scientific names and names of places to be written in round hand.

THE Editor will be pleased to answer questions and name specimens through the Correspondence column of the magazine. Specimens, in good condition, of not more than three species to be sent at one time, *carriage paid*. Duplicates only to be sent, which will not be returned. The specimens must have identifying numbers attached, together with locality, date and particulars of capture.

THE Editor is not responsible for unused MSS., neither can he undertake to return them unless accompanied with stamps for return postage.

EXCHANGES.

NOTICE.—Exchanges extending to thirty words (including name and address) admitted free, but additional words must be prepaid at the rate of threepence for every seven words or less.

WANTED.—Tooth of mastodon, or tooth of elephant, and a footprint of the Labyrinthodon, for a liberal exchange of fossils or minerals.—P. J. Roberts, 11 Back Ash Street, Bacup.

CAN offer North and South American diurnal Lepidoptera in papers, fine condition, for diurnes from the Malayan Archipelago.—Levi W. Mengel, Boys High School, Reading, Pa., U.S.A.

WANTED.—British and foreign marine and land shells, echinoderms, crustacea, and other marine objects. Offered duplicates of above, also fossils, micro-slides, photo-micrographs, etc.—H. W. Parritt, 8 Whitehall Park, N.

WANTED.—Eggs of Dartford warbler, woodlark, lesser spotted woodpecker, hawfinch, Ray's wagtail, etc. Good exchange offered.—W. Gyugell, 13 Gladstone Road, Scarborough.

DESMIDS.—A few duplicate slides of desmids mounted last year, in exchange for anything interesting, including stamps new to collection, shells, set insects, or plants.—G. H. Bryan, Plas Gwyn, Bangor.

CONTENTS.

ON COLOURING OF MOLLUSCS' SHELLS.

By REGINALD J. HUGHES.

A N inquiry into the colouring of molluscs' shells may be arranged under three heads in order to consider (1) the chemical composition of the various pigments; (2) the object for which they are deposited; and (3) the source from whence they are derived.

1. The most common colouring matter in British

FIG. 1. TYPICAL RECENT SHELLS.

1. *Venerupis enis* (Med. Sea); 2. *Helix pomatia* (Italy); 3. *Cyclostoma plicata* (Philippines); 4. *Pythia leopardus* (New Caledonia); 5. *Tapes decussata* (Ajaccio); 6. *Helicina major* (Cuba); 7. *Cardium edule* (Med. Sea); 8. *Donax reticulata* (Caribbean Sea); 9. *Trochus biaroletti* (Adriatic); 10. *Lucina striatus* (Italy); 11. *Tellina nitida* (Italy); 12. Same as 10, but a white example.

Nos. 5 and 10 coloured by iron, Nos. 2, 3, 7, and 8 by the first, and remainder by second, form of organic pigments.

and North European shells is sesquioxide of iron, the tests for which I have already described (SCIENCE-GOSSIP, N.S., Vol. VI., p. 241). This is found in all British and many foreign bivalve genera, such as *Pecten, Ostrea, Mytilus*, and *Tapes*, also in English Chitonidae, Naticidae, and Buccinidae. Most of the above, it should be noticed, have a uniform tint with little variegation. There is another very common pigment, found in the shells of almost all foreign gasteropods, such as Muricidae, Volutidae, Conidae, Cypraeidae, etc., and some bivalves including the genera *Tridacna, Cardium*, and *Donax* (fig. 1),

SEPT. 1900.—No. 76, VOL. VII.

which is of an entirely different nature and shows no trace of iron, but whose characteristic property is that of being turned violet-blue by pure nitric acid. The colour, however, disappears when the drop of acid applied has dried. It is very soluble, without change of tint, in hydrochloric acid, and to a less extent in caustic potash or dilute nitric acid. A number of tests have convinced me that it contains no metal, but is of organic derivation, and probably composed of the usual elements of such pigments—viz.: carbon, hydrogen, and oxygen. All the brilliantly coloured tropical shells, with ornamental markings, have this pigment, except those of the genera *Turbo, Trochus, Solarium, Nerita*, and the bivalve genus *Tellina* (fig. 1). These are coloured by a distinct form, also organic, that only differs from the last in being unaffected in tint by nitric acid, and is also less soluble. The two colouring

FIG. 2. EOCENE SHELLS.

1. *Voluta luctatrix*; 2. *Lucina concentrica*; 3. *Voluta ambigua*; 4. *Cerithium concavum*; 5. *Natica patula*; 6. *Cerithium serratum*; 7. *Natica parisiensis*; 8. *Cardita imbricata*; 9. *Chama calcarata*. Nos. 7 and 9 (Paris basin) white; pigment of Nos. 1, 3, 5 (Barton), and 2 (Paris basin) iron; remainder (Paris basin) organic.

matters are most likely very closely allied; the last-mentioned may even be the same as the first, with the addition of a mordant. The colour, on

K

the underpart surrounding the opening of some shells belonging to the genera *Aporrhais* and *Conus*, is insoluble, although the upper portions yield the usual blue with nitric acid. Sometimes nitration has already taken place naturally, and then we get violet shells, such as that of *Ianthina communis*, whilst the shell just inside the opening is often of this colour, *Reginula arachnoides* being a good example. The purple dye produced by many Muricidae and Buccinidae may be of the same nature. Even this form of colouring is probably protective. Thus *I. communis* is pelagic in habits, and its bluish colour must cause it to be very inconspicuous when floating on the surface. Most species of *Donax* are covered by a coloured periostracum, which skin is insoluble in acids, but nitric acid will soon break through it and stain blue the true colouring matter underneath. On a few species, such as *D. reticulata* (fig. 1, No. 8), the covering is absent.

No terrestrial or fluviatile shells are coloured by iron, but most, including those of the genera *Helix* and *Bulimus*, from all parts of the world, except Australia, by the second, blue-yielding compound. That in *Planorbis*, *Cyclostoma*. and *Helicina* (fig. 1) does not turn blue, thus being analogous to the third form in marine shells; whilst the colour of those of the genus *Unio* is caused by an insoluble brown or green periostracum, which is stained yellow by nitric acid. The pigment of all the Australian *Helices* I have as yet examined does not turn blue —*H. fringilla* is a typical example—although that in South American species does so. This is interesting if, as I believe to be the case, the invariable form is the elder of the two organic pigments. Specimens from South America are usually found covered with a dark brown periostracum, which must of course be removed by softening with an acid and then peeling off before testing the colour. The absence of iron in land and freshwater shells is very natural, for they obviously have not the same opportunity to secrete it as marine species. It will be seen that the power of depositing an organic pigment must have been acquired independently by different genera, for though some bivalves are coloured in this manner, many univalves are still marked by iron, and a number of allied forms differ in the nature of their colouring matter. It must also be of great antiquity, as land shells have been found in carboniferous strata, and these themselves were the descendants of amphibian and fresh-water species.

2. I think it will not be disputed that the object of coloration is protection, as the habit of producing an organic pigment would hardly have been formed without a purpose, and its independent acquirement by different types of shells points to its being of great utility to them. That the colour is protective is the most reasonable explanation of this, and indeed it is not difficult to imagine that

tints usually ranging from dark brown to yellow harmonise with the sea-floor on which the shells rest. Even the tropical shells, that seem to us so conspicuous, must, when lying on a variegated pebbly or sandy bottom among similarly coloured algae, be distinguished with difficulty by the creatures from which they desire protection. That there is a specific purpose in thus adorning their shells—among, at any rate, those molluscs which form an organic pigment—is shown by the fact that none of the colouring matter is placed where it would not be seen. All is deposited as a thin coating on the outer surface of the shell; none penetrates the lime, and none is found on the inside of the valves. On the other hand, with species coloured by ferric oxide the iron not only impregnates the shell, but is often more thickly deposited inside than out. To illustrate the above compare the interior of an oyster or mussel with that of a tropical bivalve.

3. The obvious answer to the question "From whence is the colouring matter derived?" is "From the food of the mollusc"; and this, no doubt, is in most cases correct. I think, however, that a more rational explanation of the colours of those shells stained with iron is as follows:—During the growth of the shell a fresh layer of cells is being continually exposed to the action of the sea-water, and any particles of iron which were suspended in the latter would very likely be deposited in them. Also, if the mollusc had some control over this deposition, it would be advantageous to have any surplus on the interior, and thus prevent the shell from becoming conspicuous by getting too dark. There is an abundance of iron on the sea-bottom, especially near the coast, where most shells live; every pebble and most sand are stained with it; all mud contains some, and it is found colouring many fossils in an irregular manner, which shows that it was derived after the death of the animal, from the sediment that gradually enveloped the shell.

Supposing, as seems probable, that iron was the original colouring matter on primitive shells, there are three reasons which might cause it to be exchanged for another:—(*a*) In the case of land and fresh-water shells, absence of the means of obtaining sufficient iron to furnish the depth of colour required. (*b*) The desirability of acquiring varied markings, which could hardly be formed by the mechanical deposit of grains of iron oxide; an organic fatty matter is evidently much more suitable. Even then many variegated shells have spikes or other projections to which the colour is often confined, and which undoubtedly assist the accuracy of the patterns and prevent the colour from spreading. It would, however, be wrong to consider the chief use of spines on shells was for this purpose, as many that possess them are of a uniform tint—*e.g. Murex tenuispina*—the left-hand shell on the block heading the "Notes and

Queries "column of SCIENCE-GOSSIP. The principal object of these projections is doubtless to render the inhabitant of the shell more secure from attack. (*c*) It might happen that shells of species originally coloured dark by iron, lived on a white, or nearly white, sea-bottom, such as one formed of limestone or light sand. In time, by the process of evolution, the descendants of these would become white, having gained the power of preventing the iron from impregnating their shells. If after a time the posterity of the latter came to live on a muddy or other dark ground, it would be only in accordance with the principles of evolution that they should not only keep the power of resisting the iron, but also gain that of producing organic pigment.

I think we have evidence that the change has

FIG. 3. *Cardita planicosta.*
Showing position of coloured bands.

been brought about in different cases in each of these three ways, as I shall presently show. It would be incorrect to imagine that some shells were coloured by iron and some by an organic pigment because of the greater abundance of iron in some parts of the sea-bed than in another, for in the same place shells of different genera can be found coloured by both pigments, whilst, on the other hand, shells of the same genus are coloured in the same manner wherever found. Thus, species of *Tapes* from the Philippines are coloured by iron, as are also those from Africa (fig. 1, No. 5) or any other part of the world. There is probably more than enough iron in every part of the sea to supply the needs of all shells that require it. It certainly appears to be true that most northern species are coloured by iron, whilst the organic pigment is characteristic of tropical shells.

(*To be concluded.*)

MOSSES OF LYNMOUTH DISTRICT.

BY CHARLES A. BRIGGS, F.E.S.

IN April last my friend, Mr. John T. Carrington came down here to visit us, for a period of convalescence after a long and serious illness. Being unable, particularly at first, to take any active exercise, and knowing that the Moss Flora of this portion of the country had never been properly recorded, it occurred to him that it would be a good opportunity to make a commencement in that direction, and at the same time to relieve the monotony of his enforced inactivity.

We were both absolutely ignorant of mosses, which no doubt is the cause of the paucity of the present preliminary list. There are doubtless many species which must of necessity be common here, though we have not been able to detect them. Mr. J. A. Wheldon, of Liverpool, kindly volunteered to name our specimens, and gave us many most practical hints as to collecting, distinguishing, and preserving our examples. I cannot sufficiently express our obligation to him for the great trouble he has taken in the matter, and may mention that he has checked the identity of all the species herein recorded.

It appears strange that a district so rich in mosses as this must obviously be, and so frequented by visitors, should have no record ; but such seems to be the case. The only literature that I can find touching on the subject is comprised in (i) "Flora Devoniensis," by the Rev. J. P. Jones and J. F. Kingston (London, 1829). The notes in this work are apparently to a large extent supplied by the celebrated bryologist the Rev. J. S. Tozer, and are almost entirely confined to the Dart, Dartmoor, Plymouth, and other stations in the South of Devon. (ii) "The Mosses of Devon and Cornwall," by E. M. Holmes and Frances Brent, published in the third volume of the Annual Reports and Transactions of the Plymouth Institution and Devon and Cornwall Natural History Society. In this, as in the "Flora Devoniensis," the north of the county seems to have been but little worked by its compilers and their correspondents. A few names of mosses are inserted in it, on the authority of a North Devon guide, but I have not been able to ascertain the particular guide to which the authors referred. Not any I have seen give a list of mosses. Parfitt's "Moss Flora of Devon" I have not yet been able to procure ; but as his other papers on the fauna of Devon more particularly relate to South Devon, I doubt if it would throw much light on the Lynmouth flora.

It may, perhaps, be useful if I shortly describe the character of some of the localities mentioned in the following list :

"Desolation" or "Desolate" is the name of a farm some three miles to the east of Lynmouth, between Countisbury and Glenthorne, the lovely seat of Miss Halliday. The collecting-ground consists of a narrow wooded gorge running from the heathland, here about 1,000 feet above the sea-level, right down to the beach. Throughout the length of this combe a tiny stream rushes and falls in a series of miniature cascades.

" The Tors " consist of a series of peaks rising above the vicarage of Lynmouth, and including the beautiful Alpine garden of Mr. A. L. Ford, of Gwynallt. In these peaks the rocks crop out in a most picturesque manner.

" Lee Bay " is a small bay or cove about two miles to the west of Lynmouth. It is approached by a road made by the side of a little stream in a wood, and contains a fine dripping bank and waterfall.

" Parracombe " will be remembered, by those who have travelled here by coach from Barnstaple, for its steep and dangerous hills; it is a village about six miles west from Lynmouth, on the confines of Exmoor, in which it was, no doubt, formerly included. There appear to be springs and running water in all parts of the village, and it would doubtless repay further research.

" Valleys of the East and West Lyns." These we have not explored beyond Rockford in the former and Lynbridge in the latter. Further exploration is desirable.

"Exmoor" in this list refers solely to that portion of the moor situate near Saddle Gate and Pinkery (more properly Pinkworthy) Pond. Saddle Gate is a gate in the boundary wall of Devon and Somerset, and as this portion is in Somerset, and to a great extent drains to the south into the Barle, it perhaps should not be included in the same sub-district as Lynmouth; but as one of my favourite collecting-grounds is the boundary wall itself and the immediately adjacent moor on both sides which drains northward into the Lyn, it may be convenient to group the locality under Lynmouth, from which it is only some five miles distant. This moor is very different from the ordinary conception of a moor. It consists here of a bog some 1,500 feet above the sea-level, and is chiefly composed of coarse sedgy herbage interspersed with sphagnums and other mosses, with a little heather and whortleberry in the drier portions. The whole moor is dreary and wild, but very interesting from an antiquarian point of view, on account of the number of tumuli and barrows scattered over the district.

A few words are due to the fences in North Devon, which form such splendid collecting-grounds for mosses. They are made in a way that I have not noticed elsewhere. A wall of stones is erected— often some five feet or more thick at the base. The interstices are filled in with earth, and a hedge or frequently trees are grown at the top. The result is that stone mosses, bank mosses, and tree-trunk mosses are all found on the same fence.

In the scientific arrangement and nomenclature I have followed Messrs. Dixon and Jameson's "Student's Handbook of British Mosses." The sign * indicates a sub-species.

The SPHAGNACEAE, the first of the eighteen families of Acrocarpous mosses, are fairly well represented by six species and one sub-species; but as some of them are impossible to recognise in the field, and it is equally impossible to gather a tuft from every patch of sphagnum in such a locality as Exmoor, there are doubtless others yet to be found. *Sphagnum cymbifolium* Ehrh. is common on the moor, while its variety *congestum* Schp. is occasionally to be seen. *S. cuspidatum* Ehrh., *S. intermedium* Hoff., *S. subsecundum* Nees, and *S. acutifolium* Ehrh. are all common, the last also occurring at Desolation. *S. rigidum* Sehp. I have found round Pinkery Pond, and *S. papillosum* grows with *S. cymbifolium* on the highest portion, 1,500 feet above the sea.

Of the ANDREAEACEAE I have not found any representatives; but of the TETRAPHIDACEAE, *Tetraphis pellucida* Hedw. was found at Desolation.

The POLYTRICHACEAE are, as might be expected, one of the great features of the district; one or other of the genus *Polytrichum* being on every bank and wall. The species noticed are *Catharinea undulata* W. and M. in the East Lyn Valley and at Desolation.; *Polytrichum nanum* Neck. on the North Walk, Lynton.; *P. aloides* Hedw. generally distributed, those from Exmoor being very luxuriant.; *P. urnigerum* L. on the Countisbury road; *P. juniperinum* Willd. on the North Walk, The Tors, East Lyn Valley, etc.; *P. strictum* Banks, Exmoor commonly; *P. alpinum* L. Exmoor fairly common; *P. formosum* Hedw. in East Lyn Valley; *P. commune* L. West Lyn Valley, Desolation, Exmoor, etc.

The little family of the BUXBAUMIACEAE is represented by *Diphyscium foliosum* Mohr., found sparingly on spray-covered stones at Watersmeet, and at Saddle Gate.

The large and somewhat cumbrous family DICRANACEAE gives us *Ditrichum homomallum* Hpe., a single tuft of which was found on a dripping bank at Pinkery Pond; *Ceratodon purpureus* Brid. on The Tors, also at Ilkerton, Saddle Gate, Watersmeet, etc.; *Dioranella heteromalla* Sehp. on the North Walk, Ilkerton, Parracombe, and Exmoor; *D. varia* Sehp. on the Esplanade cliffs at Lynmouth, and at Brendon; and *D. cerviculata* Schp. from Exmoor, where it revels in the cuttings in the peat. *Dicranoweisia cirrata* Lab. was found on Summer House Hill, and also on a wall above Saddle Gate; *Campylopsis pyriformis* Brid. sparingly on the Esplanade at Lynmouth, but was abundant at Exmoor. *C. flexuosus*, a single tuft of which I found at Exmoor, was, I fear, overlooked among the ever-varying forms of *Dicranum scoparium* Hedw. that is so abundant throughout the district; while its larger neighbour, *D. majus* Turn., was only noticed in the East Lyn Valley and at Desolation, where the somewhat sphagnoid-looking *Leucobryum glaucum* Sehp. was common.

The next family, FISSIDENTACEAE, is poorly represented by *Fissidens viridulus* Wahl. and *F. taxifolius* Hedw. from The Tors, a somewhat doubtful specimen of *F. bryoides* Hedw. from Ilkertŏn Lane, and *F. decipiens* de Not, found while assisting at the taking of the photograph of "The Nest and Young of the Water Ouzel," illustrated in the June number of SCIENCE-GOSSIP (vol. vii., N.S., p. 3), in the East Lyn Valley.

Among the GRIMMIACEAE, *Grimmia apocarpa* Hedw. is generally common, and *G. pulvinata* Sm. though widely distributed, is not so common as its congener. *Rhacomitrium aciculare* Brid. is to be found sparingly in the valleys of the East and West Lyns, on stones. *R. fasciculare* Brid. is at Watersmeet, and also at Exmoor, on stones. *R. heterostichum* Brid. was found at Saddle Gate, and in the East and West Lyn Valleys ; *R. canescens* Brid. on the heath near Saddle Gate ; and *R. lanuginosum* Brid. in a similar situation at Pinkery Pond and Desolation. *Ptychomitrium polyphyllum* Furnr. is generally common on rocks and walls throughout the district.

The TORTULACEAE are·one of the most difficult families to beginners. This, perhaps, explains why it appears to be so poorly represented ; but no doubt further experience will largely increase the list. At present all that have been noticed are :— *Tortula muralis* Hedw., generally common. *T. subulata* Hedw. at Countisbury. *Barbula fallax* Hedw. commonly in the district, and its variety *brevifolia* Schultz, on the Esplanade at Lynmouth. *B. cylindrica* Schp. in the West Lyn Valley, and its sub-species *vinelais* Brid. freely in the Watersmeet Road. *B. revoluta* Brid. occurred once at Watersmeet, and *B. convoluta* Hedw. at Countisbury. *B. unguiculata* Hedw. is generally common. *Weisia microstoma* C.M. occurred on The Tors, with one specimen of the rare *W. crispata*, and the varieties *amblyodon* B. and S. and *densifolia* B. and S. of *W. viridula* Hedw., the type of which was common everywhere in our district. *Trichostomum crispulum* Bruch. was found on Capstone Hill, Ilfracombe, which is perhaps somewhat out of the Lynmouth district. *T. mutabile* Bruch. and its variety *litorale* Dixon are to be found in the valley of the West Lyn. *Cinclidotus fontinaloides* P.B. occurred on stones near the river bed, in the East Lyn Valley, finishes this scanty list of Tortulaceae.

Encalypta streptocarpa Hedw., found very sparingly in the East Lyn Valley, is the sole representative of the ENCALYPTACEAE.

Of the next family, the ORTHOTRICHACEAE, *Zygodon viridissimus* R. Br. occurs on walls in the East Lyn Valley ; whilst its congener, or sub-species, *stirtoni* Sehp., I have only found at Lee Bay, where also was the curious little gemma-bearing *Ulota phyllantha* Brid. on a tree-trunk.

The SCHISTOSTEGACEAE and SPLACHNACEAE are, so far, unrepresented.

Funaria hygrometrica Sibth., common everywhere, is the only species yet noticed of the FUNARIACEAE.

The MEESIACEAE has only given us *Aulacomnium palustre* Schwgr., which is abundant on Exmoor, where it is found mixed in dense patches of *Polytrichum strictum* and *Sphagnum*.

The one genus that constitutes the TIMMIACEAE we have not found here.

The BARTRAMIACEAE have only given us two representatives, viz. *Bartramia pomiformis* Hedw., the lovely "apple moss," from Desolation, Summer House Hill, Ilkerton Lane, and Lee Bay ; and *Philonotis fontana* Brid., found commonly in one locality near Pinkery Pond, where it was fruiting freely.

Of the fifty-one species and sub-species that constitute the large and important family BRYACEAE, the last of the eighteen families of the Acrocarpous mosses, only fourteen have here, as yet, come under my notice. These are *Webera nutans* Hedw., from the North Walk, Ilkerton, and Exmoor, where it is abundant in the peat cuttings ; *W. annotina* Schwgr. from the wall at Saddle Gate ; *W. albicans* Schp. from a spring in the East Lyn Valley, Rock Lodge, Lynton, and Lee Bay ; and *W. carnea* Sehp., a doubtful specimen, not in fruit, from Lee Bay. *Bryum inclinatum* Bland was frequent at Countisbury, *B. caespiticium* L. at Countisbury, Watersmeet Road, etc., *B. capillare* L. generally common, and its variety *flaccidum* B. and E. on Countisbury Hill. *B. obconicum* Hornsch. and *B. murale* Wils. on walls, especially in the East Lyn Valley. *B. atropurpureum* W. and M. on The Tors, and a solitary specimen, apparently of *B. pallens* S. W., occurred at Lee Bay. *Mnium hornum* L. generally common throughout the district in every kind of locality, but it is especially well grown in the West Lyn Valley. *M. punctatum* is found in the same places.

The group of families that constitute the Pleurocarpous mosses contains so many of the larger and more noticeable mosses that it is naturally better represented in a preliminary list like this than those species in the other division. There are seven of these families, mostly small in number. the first six only containing forty-one species.

Of the first two,· the FONTINALACEAE and CRYPHAEACEAE, no representative has as yet been noticed.

Of the NECKERACEAE, *Neckera crispa* Hedw. is found at Watersmeet, *N. complanata* Hubn. at Brendon on tree-trunks, and *Homalia trichomanoides* Brid. on a wet bank at Lee Bay.

None of the three species comprised in the HOOKERIACEAE have been noticed, and *Porotrichum alopecurum* Mitt., near Rockford and in the West Lyn Valley, is the only one that has occurred of the small family of the LEUCODONTACEAE.

Of the LESKEACEAE, *Anomodon viticulosus* H. and T. occurs on the bridle-path, Lynmouth. and along the river in the East Lyn Valley, two widely differing localities. *Thuidium tamariscinum* B. and S. occurs over the whole district.

The HYPNACEAE, the concluding family of the mosses, is very fairly represented by the following species :—*Pylaisia polyantha* B. and S. at Watersmeet and Lee Bay, on tree-trunks ; *Isothecium myurum* Brid. in the West Lyn Valley ; *Pleuropus sericeus* Dixon is generally common on rocks and walls ; *Brachythecium rutabulum* B. and S. generally common, as its variety *robustum* is at Countisbury. *B. rivulare* B. and S. in the East Lyn Valley and at Lee Bay. *B. velutinum* B. and S. at Countisbury, Parracombe, Lee Bay, etc. *B. populeum* B. and S. in the East Lyn Valley, Parracombe, etc. *B. plumosum* B. and S., West Lyn Valley, and *B. purum* Dixon on the North Walk. *Eurhynchium crassinervium* B. and S. on the old Barnstaple Road ; *E. praelongum* B. and S. on the Esplanade, Lynmouth, Countisbury, Exmoor, etc. ; and its variety *stokesii* L. Cat. ed. 2, in the West Lyn Valley. *E. abbreviatum* Schp. is on the Countisbury Road ; *E. tenellum* Milde in the Watersmeet Road ; *E. striatum* B. and S. at Brendon and Parracombe. *E rusciforme* Milde is abundant in the East and West Lyns and Lee Bay, and its variety *atlanticum* Brid. at Desolation ; *E. confertum* Milde, in the East Lyn Valley, The Tors, The Zig-Zag, and Lee Bay. *E. myosuroides* Sehp. is found in the West Lyn Valley ; *Plagiothecium borrerianum* Spr. in the West Lyn Valley and at Pinkery Pond ; *P. denticulatum* B. and S. West Lyn Valley, Saddle Gate, etc. ; *P. sylvaticum* B. and S. in the old Barnstaple Road and at Lee Bay. *P. undulatum* B. and S. is in the West Lyn Valley ; *Amblystegium serpens* B. and S. West Lyn Valley and Parracombe ; *A. irriguum* B. and S. at Desolation, and *A. filicinum* de Not at a spring at Parracombe ; *Hypnum commutatum* Hedw. at a spring in the East Lyn Valley and on the Watersmeet Road. *H. cupressiforme* Hedw. is generally common ; of its varieties *resupinatum* Sehp. is found in the West Lyn Valley, *filiforme* Brid. at Desolation, and a pretty intermediate form at Lee Bay *ericetorum* B. and S. Countisbury and Saddle Gate, also *elatum* B. and S. at Saddle Gate ; *H. molluscum* Hedw. West Lyn Valley ; *H. palustre* L. in the same place, semi-submerged ; *H. stramineum* Dicks at Pinkery Pond ; *H. cuspidatum* L. at Desolation, East Lyn Valley, and fruiting freely at Exmoor ; *H. schreberi* Willd. at The Tors, Countisbury, and Exmoor ; *H. cordifolium* Hedw. Exmoor, where a single specimen was gathered unwittingly. *Hylocomium splendens* B. and S. occurs very sparingly at Desolation and Exmoor. We have found *H. loreum* B. and S., *H. squarrosum* B. and S., and *H. triquetrum* B. and S. commonly ; the latter once in fruit in July.

To the above list should be added six species recorded in "Mosses of Devon and Cornwall," viz. :—

Weisia verticillata. "On dripping limestone rocks, the fruit rare. Lynton. F. B.". [There appears to be some mistake in this, as I do not think any limestone is here. C. A. B.]

Grimmia maritima. Ilfracombe. Fl. Dev.

Webera tozeri. Combe Martin. N. Devon Guide.

Pterogonium gracile. Lynmouth. N. Devon Guide.

Habrodon notarisii. "On elm-trees. Lynton. Mr. J. Nowell."

Hypnum eugyrium. "Lynton. Rare. Mr. J. Nowell."

It will be seen that we have as yet only obtained about 110 species out of the nearly 600 species recognised as British. Poor as this list is, yet, considering the circumstances, I think one may fairly expect to very largely increase it by another year. The whole neighbourhood is literally carpeted with mosses, which greater experience will, I trust, enable me to recognise and record in the future.

Rock House, Lynmouth,
August 6, 1900.

THE BRITISH ASSOCIATION.

THE last meeting of this century will be held at Bradford this month, after an interval of twenty-seven years since the Association visited that town. Professor Sir William Turner, the eminent anatomist of Edinburgh University, is to preside, and will, we understand, base his presidential address upon the progress of the science of Biology, especially with regard to our advanced knowledge of the function and structure of cells

The presidents of the sections are—A, Mathematics and Physics, Dr. Joseph Larmor, of Cambridge ; B, Chemistry, Professor Perkin, of Owens College, Manchester ; C, Geology, Professor W. J. Sollas, of Oxford ; D, Zoology, Dr. R. H. Traquair, Keeper of the National Museum, Science and Art Department, Edinburgh ; E, Geography, Sir George Robertson ; F, Economic Science and Statistics, Major P. G. Craige ; G, Mechanical Science, Sir Alexander Binnie ; H, Anthropology, Professor Rhys ; and J, Botany, Professor S. H. Vines.

The position of Bradford and its surrounding country is such that an important and successful meeting is anticipated.

ABUNDANCE OF "CLOUDED YELLOWS."—We hear from various parts of the country that there has been in August of this year a sporadic eruption of *Colias edusa* and *C. hyale.* Both these species have appeared in some localities in greater abundance than has previously been seen in this country. The variety *helice* of *C. edusa* has occurred in some numbers in clover fields in East Essex. This season has also been remarkable for the numbers of " holly-blues " (*Lycaena argiolus*) in both spring and summer broods,

GEOLOGICAL NOTES IN THE ORANGE RIVER COLONY.

By Major B. M. Skinner, R.A.M.C.

(*Continued from page 35.*)

2. BLOEMFONTEIN.

THE town of Bloemfontein is built over the northern slope of a valley bounded on the north and south by dolerite hills, open on the east,

Fig. 1. QUARRY BELOW NEW FORT KOP.

and rising gently upwards on the west. A water-course, the Bloemspruit, rising among the hills to the north-west of the town, winds eastwards at the bottom of the valley, passing through the town, and only containing water after heavy rain. The town climbs upwards at its northern end wherever the gentler slope at the foot of the hill allows of standing-space for dwellings; in the gardens of many of the houses dolerite boulders stand out of the red sandy clay, the detrital soil of the dolerite. The houses being for the most part built of red brick, this colour combines with that of the soil to give the impression of extreme warmth when viewed from within the town, this being lost from a distance, whence the corrugated iron roofs form the most striking feature.

Geologically the country is mainly dolerite, intrusive masses of this rock forming the hills and the main portions of the undulations and irregularities of the ground; but in the dongas (stormwater channels) and on the slopes of the hills stratified rocks may often be seen.

North of Bloemfontein is a plateau of dolerite, now called Naval Hill, as the naval guns at present have taken up a position there; to the west of this is a hill of the same material, having originally been part of the same mass, but separated by subsequent denudation; from the northern and western faces of this, spurs are given off. The western hill, which may be called New Fort Kop, as it has at its summit the rudiments of a fort, is of some interest on its southern and eastern faces, as it there shows stratified rocks—non-fossiliferous, but interesting in showing that the dolerite, when bursting through, simply split without disturbing the horizontality of the sandstones, and subsequently overflowed the strata on one of the northern spurs. These strata, measuring 80 feet in height, taken from a buff sandstone at their base, consist first of shales, then thin-bedded sandstone and shales; above these two courses of fine white sandstone separated by a thin layer of slaty shale are seen; then come two compact siliceous layers separated by a clay shale. Of these siliceous layers the lower is grey with irregular white marks; it is of flinty consistency; the upper is black, with conchoidal fracture and white streak, showing lines of bedding differing alternately in consistency. this being markedly shown on the weathered edge. Above these again come shales, buff, grey, ribbed black and grey alternately, and very sandy; while the highest layer of this exposure consists of 6 or 7 feet of sandstone, white and rather soft below,

Fig. 2. ANOTHER VIEW OF NEW FORT KOP.

Fig. 3. QUARRY ON WEST OF SPUR.

harder above. There is only a small portion of
this sandstone remaining. On the eastern face the
lowest shales are well exposed; these rocks have
been quarried to make grey paths in the gardens

result that the junction of the igneous and stratified rocks can be seen, the former having passed
up behind the latter, and baked their constituents
for the depth of 3 or 4 inches.

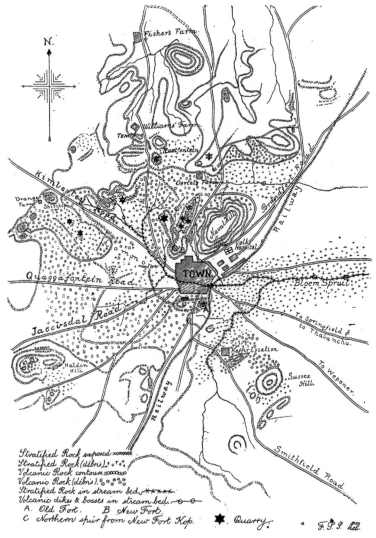

Stratified Rock exposed ×××××
Stratified Rock (débris) + + + +
Volcanic Rock contours
Volcanic Rock (débris)
Stratified Rock in stream bed ★★★★★
Volcanic dikes & bosses in stream bed ○—○—○
  A. Old Fort.   B New Fort.
  C Northern spur from New Fort Kop.   ★ Quarry.   ✦ F.P.S. Del.

MAP OF THE BLOEMFONTEIN DISTRICT.

of the town dwellings, to contrast with the red
soil of the flower beds; while the dolerite behind
has been quarried to make the New Fort with the

On a spur to the north of New Fort Kop, sandstone
has been quarried for building purposes. Of this
sandstone 25 feet are visible on the southern end

of the spur, and 45 feet on the western side. These rocks are more or less visible all round the spur, while the top consists of dolerite. These sandstones consist of grits above, firm sandstone in the centre, and below fine white sandstone, with occasional intervening layers of shale. The grit at the top has the appearance of having been denuded, then irregularly covered by a narrow band of shale, while overlying the shale, and following an uneven surface, comes the dolerite sheet which evidently overflowed from the extrusion at New Fort Kop. The denudation is again shown on a continuation of this spur, where there is another small quarry. Here the top shale and sandstone have been furrowed out to the depth of about 15 inches, the

the centre of a basin, the circumference of which is composed of dolerite. Altogether about 25 feet of sandstone are more or less exposed here, there being a slight outcrop of yellowish sandstone above the quarry, and the same of a buff sandstone below, while the straight-cut wall of the quarry, which now forms a reservoir for water, shows some 20 feet of grey freestone. At two levels in this freestone are large and sometimes curiously shaped boulders, which on section show concentric rings formed round a central rolled stone, sometimes sandstone, sometimes quartzite. The level of the top of this quarry is 90 feet below that of the bottom of the spur.

To the east of Rustfontein Quarry Hill is a valley, the opposite side of which is formed by a hilly

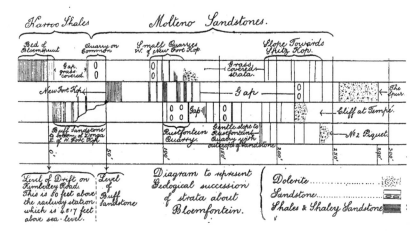

Diagram to represent Geological succession of strata about Bloemfontein.

Level of Drift on Kimberley Road. This is 50 feet above the railway station which is 4517 feet above sea-level.

Level of Buff Sandstone

Dolerite ......
Sandstone ......
Shales & Shaley Sandstone

channel thus formed being filled in by dolerite, which also covered the adjacent surface. The height of the spur is about 80 feet above the top of New Fort Kop, which again is 165 feet higher than the buff sandstone before referred to, as that from which the measurement of the stratified rocks is taken. The top of New Fort Kop is the same level as the lowest sandstone on the spur; so that it may be assumed that there are 210 feet of stratified rock to be accounted for, the upper 35 feet of the above 80 being dolerite. Of these the top 45 feet are seen on the spur, and the lower 80 feet on the south and east of New Fort Kop, leaving a gap of 85 feet. On the west of the spur, as well as on the east, is a gentle slope of diluvial soil, which conceals the missing strata, but they are represented in other localities. The eastern side of the spur has also been quarried, showing at a lower level than the one in fig. 3 sandstones with boulders.

About 1½ mile north of these quarries on the spur is another and more extensive quarry at Rustfontein in layers of sandstone which now form

ridge running about N. and S., and presenting four conical points; the most southerly and the third from the south show on the eastern and southern faces sandstones, the second and fourth and the western side of the others are dolerite, while the undulating country to the north is entirely dolerite. These sandstones on the southern end of the ridge are almost horizontal, and are interstratified with thin layers occasionally of shale; the highest stratum is a fine grit, and is below the level of the Rustfontein Quarry; the series extends downwards 67 feet. The sandstones on the third hill from the south are crushed and distorted in several directions which are difficult to unravel, owing to the very partial exposure of the rocks.

About a mile north-west of the spur is a kop which may be named No. 2 Picquet; its dolerite cap covers sandstones, the topmost being soft and white; at its base is a sandstone, underlain by a thin layer of quartz gravel, and this again by shale. The level of this quartz gravel is 25 feet below that of the base of the spur, and 50 feet below the top sandstone on No. 2 Picquet Hill. From this point
E 3

the country slopes gently east towards Rustfontein Quarry, grazing land covering subjacent sandstone: which occasionally outcrop. To the north-west of No. 2 Picquet the ground is rocky and descends to Tempé Farm. Just south of the farm, about 600 or 700 yards distant, is a cliff of sandstones, lying on shale; of these sandstones there are 75 feet exposed; they are generally soft, pale buff in colour. One layer contains large boulders; most of the remainder show small rolled embedded sandstones. in one of these layers the small rolled stones being much iron-stained and concentric in structure. The topmost layer, which is only slightly exposed, has been baked into a very hard rock. The bottom sandstone of this series is 75 feet below the bottom of the spur.

A little over a mile due south of this cliff is a small quarry, corresponding in level with another small quarry about one mile east, and, like it, containing boulders along one course of the exposed sandstone, of which only some three to four feet have been dug out. These quarries each lie at the foot of a bluff of dolerite, the western bluff having the appearance of being continuous with the dolerite cap to the sandstones at Tempé. In the vicinity of this quarry are some other partial exposures of sandstone, shale in thin layers, and the grey siliceous rock noted on New Fort Kop. The exposures here are very scanty, but it would appear as if the dolerite sheet over these rocks was continuous with that at Tempé, and also extended westward, covering some rocks which are exposed near Spitz Kop.

(*To be concluded.*)

SPIDERS FROM HASTINGS.

By Frank Percy Smith.

THE following list of spiders from Hastings and the surrounding district is compiled from a collection made during the early part of this year. My stay in that locality extended only from March 26th to April 14th. During the first part of the period I was unable to make much progress, owing to weakness following a serious illness. In the latter portion of the time the wind was so violent as to make collecting a matter of great difficulty, and I am myself surprised at the number of species taken.

If straight lines be drawn on a map connecting Battle with the eastern side of Fairlight Glen and a point about a mile to the west of Bexhill, the triangle formed by these lines and the coast will include all the country under consideration. It must, however, be understood that the whole of this tract was not worked, but simply a few tempting localities in various parts of it. Perhaps a few notes as to these localities may be of use to visitors, as they are all within walking distance of Hastings. Space will not allow of my giving minute details for natural history rambles, but the stranger will find all the information required in a shilling guide published by Ward, Lock & Co. Hastings Old Town lies in a hollow, and to the east of this we find an immense cliff, East Hill. Following a path along the cliffs we come to a valley followed by another formidable cliff, and an even more beautiful valley. The first of these depressions is Ecclesbourne; the second, Fairlight Glen, the eastern limit to our district. Both these glens are prolific with regard to spiders, and would well repay working during the autumn months. To the north of Hastings we may find some good collecting grounds; namely, the small woods with which this district abounds. The plantation in which Hollington Old Church is built is a good example. Although I have defined the district worked as reaching to Battle, the only collecting done north of Hollington was in a ditch near Battle station. To the west we find the coast wastes, extending from Bexhill to Pevensey; in which locality I spent one rather unprofitable day.

I have to acknowledge the valuable help rendered by Mr. Connold, the Hon. Sec. of the Hastings Natural History Society, who personally conducted me to a most important locality, which I had previously missed.

Although this list does not include any species of extreme rarity, it contains several which are far from common, and some that are of special interest. Of the first we may notice *Porrhomma inerrans* Cb., *Cercidia prominens* Westr., and *Oxyptila simplex* Cb. In the second may be included, among others, *Atypus piceus* Sultz., *Micaria pulicaria* Sund., and *Walckenaëra acuminata* Bl.

At the head of our list, both in systematic position and general interest, is *Atypus piceus* Sultz., commonly known as the "trap-door" spider, although it makes no such structure. This creature's retreat, which has been so thoroughly explained by Mr. Fred Enock, consists of a tube, sometimes nearly a foot in length, the greater portion of which is buried in the earth, and its detection is a matter of great difficulty. Several adult females were taken, also a nest of young. The family Dictynidae was represented by the three species of *Amaurobius*; namely, *A. fenestralis* Stroem., *A. similis* Bl., and *A. ferox* Wlk., and also by *Dictyna pusilla* Westr. and *D. uncinata* Westr. In the family Dysderidae two species were found, and of these only one specimen of each. These were *Dysdera crocota* Koch. and *Harpactes hombergii* Scop. The Drassidae were not plentiful, the species taken being *Prosthesima pedestris* Koch. *Drassus sylvestris* Bl., and *D. lapidicolens* Wlk. The family Theridiidae was well represented, and many more species would no doubt have been found had time permitted. The genus *Theridion* was not much in evidence, although a fair number of immature specimens were taken, including *T. sisyphium* Clk. and *T. bimaculatum* Linn. A single male of that curious spider *Pholcomma gibbum* Westr. was found on the railway bank near St. Leonards. Of the group Asageneae, characterised by the possession of a stridulating apparatus, two species were taken ; namely, *Crustulina guttata* Wid. and *Pedanostethus lividus* Bl. The important group Erigoneae, which includes Blackwall's genus *Walckenaëra* and a large part of his genus *Neriene*, was well represented by the following species :—*Lophocarenum parallelum* Bl., *Tiso vagans* Bl., *Diplocephalus cristatus* Bl., *D. fuscipes* Bl., *Walckenaëra acuminata* Bl., *Wideria antica* Wid., *Cornicularia unicornis* Cb. (uncommon), *Neriene rubens* Bl., *Dicyphus bituberculatus* Bl., *Gongylidium dentatum* Wid., *G agreste* Bl., *G. fuscum* Bl., and *Erigone dentipalpis* Wid. The Linyphieae were also well represented, including the following :—*Porrhomma inerrans* Cb. (a rare and local species), *Tmeticus concolor* Wid., *T rufus* Wid., *Bathyphantes variegatus* Bl., *B. dorsalis* Wid., *B. concolor* Wid., *Leptyphantes tenuis* Bl., *L. ericaea* Bl., *Labulla thoracica* Wid., and *Linyphia clathrata* Sund. Two species belonging to the family Tetragnathidae occurred, and both in great abundance; namely, *Pachygnatha degeerii* Sund. and *Tetragnatha solandrii* Scop., the former amongst grass and low herbage, and the latter on bushes in Fairlight Glen. Although so early in the season, a good number of the Epeiridae had put in an appearance, including *Meta segmentata* Clk., *M. merianae* Scop., *Cercidia prominens* Westr. (a rare species), *Zilla atrica* Koch, *Z. x-notata* Clk. (common on stone walls), *Epeira cucurbitina* Clk., *E. umbratica* Bl. and *E diademata* Clk. (just hatched). The family Thomisidae was well represented. *Xysticus cristatus* Clk. was common everywhere, and single specimens of *X. pini* Hahn. and *Oxyptila simplex* Cb., the latter a rare and local species, were found in the wood around Hollington Old Church. A specimen of *O. praticola* Koch was also taken. *Philodromus dispar* Wlk. was plentiful; the specimens, however, were all young. As might be expected, the Clubionidae were well represented, especially the genus *Clubiona*, of which six species were taken ; namely, *C. reclusa* Cb., *C. compta* Koch, *C. pallidula* Clk., *C. holcsericea* Degeer, *C. terrestris* Westr., and *C. grisca* L. Koch. *Chiracanthium* was represented by a few immature specimens in bad condition whose identity was uncertain. *Anyphaena accentuata* occurred at Fairlight Glen. Another very interesting species, of which I took

but one specimen, is *Agroeca brunnea* Bl. Its presence in the district was first intimated to me by Mr. Connold, who gave me specimens of its nest. This is a most curious structure, being made in the form of a goblet and coated with yellow mud. This earthy covering is impervious to moisture, and is a great protection against ichneumons, besides making the nest more difficult to detect. *Micaria pulicaria* Sund., which closely resembles the ants among which it lives, occurred in several localities, most abundantly among the coarse grass on the beach at the eastern extremity of Bexhill. *Micariosoma festivum* Koch and *Zora spinimana* Sund. were also found, but not plentifully. The family Agelenidae was represented by *Coelotes atropos* Wlk., *Agelena labyrinthica* Clk., *Tegenaria atrica* Koch, *T. domestica* Clk., *T. campestris* Koch, *Textrix lycosina* Bl., and *Hahnia nava* Bl., this last species belonging to a genus of great interest on account of the extraordinary arrangement of the spinners. The Lycosidae were abundant, including *Pisaura mirabilis* Clk., *Pirata piratica* Clk., *Pardosa palustris* Linn., *P. amentata* Clk., *P. proxima* Koch (not certain), *P. pullata* Clk., *P. annulata* Thor., *Lycosa ruricola* Degeer, and *Lycosa andrenivora* Wlk. Although many of the females were adult, none of them were carrying the characteristic egg-sacs, which was hardly to be expected so early in the season. Two species of Attidae (Salticidae) finish the list ; namely, *Epiblemum scenicum* Clk. and *Heliophanus cupreus* Wlk.

It will be seen from the above list, which contains over eighty species, how much there is to be done with regard to the spider fauna of this district. I feel sure that could I have stayed a few weeks longer, until the multitudes of Linyphias, Theridions, and Epeiras had developed sufficient characteristics to admit of their satisfactory identification, I should have been able to make a considerable addition to the above. This is supported by the fact that from a list of thirty-five species taken by several members of the Hastings Natural History Society, fourteen species do not occur on my present list.

15 *Cloudesley Place, Islington, N.*

SIR WILLIAM STOKES.—The announcement that Sir William Stokes had died suddenly on August 18th at Pietermaritzburg was quite unexpected. His visit to Africa was on behalf of the Government, in connection with the hospital administration of the Army. He was born in Dublin on March 10th, 1839, and was the second son of the eminent Dr. William Stokes of that city, whose life he wrote, a work which was reviewed in these columns a short time since. His first important appointment was that of senior surgeon to Richmond Surgical Hospital. In 1881 he was elected President of the Pathological Society, of which he was a gold medallist, and six years later was placed in the chair of the Royal College of Surgeons, Ireland.

E 4

THE PHOTOGRAPHY OF COLOUR.

By E. Sanger Shepherd.

(*Continued from page* 69.)

DURING the last forty years very many attempts have been made to secure photographs of objects in their natural colours. Glowing accounts have been published announcing a perfect solution of the problem, only to be followed within a few weeks by contradictions and frequently the exposure of some ingenious fraud, until the very mention of a photograph in colours is looked upon with suspicion. To-day, however, photography in colour no longer means a photograph printed in colours, nor yet an ordinary photograph coloured by hand, both of which terms are apt to be associated in the artistic mind with recollections of attempts of a very unsatisfactory nature.

by the second method in accuracy of colouring. An additional disadvantage is that the image can only be seen under particular conditions of lighting and when viewed at a particular angle.

Processes based upon the second method are, however, of a far more practical nature. The Young-Helmholtz theory is based upon the fact that all the colours of the spectrum, and therefore all the colours found in nature, may be counterfeited sufficiently nearly to deceive the human eye by mixtures of three colours of the spectrum itself —a particular red, a particular green, and a particular blue-violet.

First a word about the primary colours—red,

FIGS. 10, 11, 12. (*Vide* p. 112.)

As we speak of it to-day, it means the final outcome and practical result of long series of scientific experiments carried out by various able investigators during the last thirty or forty years. All the methods which have stood the test of time may, however, be classed under two heads:— (1) Those based upon Interference, as the process of Professor Lippmann of Paris; and (2) those based upon the Young-Helmholtz theory of trichromatic vision.

The first process—the Interference method of Professor Lippmann— produces a single positive in the camera, which is incapable of being reproduced; and, as the process is intricate and the requisite exposure long, few results have been shown, and even these have been very inferior to photographs

green, and blue-violet. Even at the present day people find a difficulty in accepting red, green, and blue-violet as the primary colours, this difficulty arising from their ideas of colour mixture being derived from the amalgamation of pigments upon paper.

In order to see a coloured object we have to illuminate it by white light, and white light contains light of all colours. As we have already seen by means of the spectroscope, we can separate white light into its component colours, and in fig. 6 we have a drawing of the spectrum showing curves which, by their height above the base line, represent the power of the three primary colour sensations to affect the human eye.

If we take three coloured glasses, a red, a green,

and a blue-violet, and place them in three optical lanterns, and so arrange the optical lanterns that the projected discs may overlap, as shown in fig. 7, we shall find that where the red disc overlaps the green we get yellow light, where the green disc overlaps the blue we get greenish-blue light, and where the blue disc overlaps the red we get pink light, whilst where all three overlap in the centre

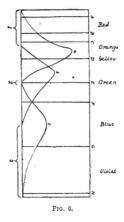

FIG. 5. (*Vide* p. 67.)

we get white light. By no possible combination of coloured lights can we produce a pure red, green, or blue, so we are forced to accept these colours as the primaries. If, however, we examine the reconstituted white light by means of a spectroscope, we shall find that instead of getting a continuous spectrum, as we do, for instance, with the light reflected from a white cloud, we only get three narrow bands of colour—red, green, and blue-violet ; but both the light from the white

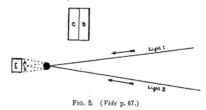

FIG. 6.

cloud and the reconstituted white light appear alike to the eye.

In view of these facts, it would appear very simple to take three photographs of a coloured object, one through a filter admitting red light, one through a filter admitting green light, and one through a filter admitting blue-violet light, and, by projecting prints from these negatives illuminated by red, green, and blue lights upon a screen in

superpositions, secure a reproduction of the object photographed in its natural colours. The first photographs so produced were shown in public by Professor Clerk-Maxwell in 1861. In May of that year Clerk-Maxwell delivered a lecture at the

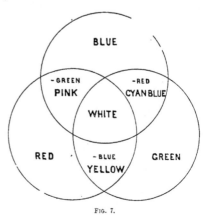

FIG. 7.

Royal Institution, from which the following is extracted :—

"Experiments on the prismatic spectrum show that all the colours of the spectrum and all the colours in nature are equivalent to mixtures of three colours of the spectrum itself, namely, red, green (near the line E), and blue (near the line G).

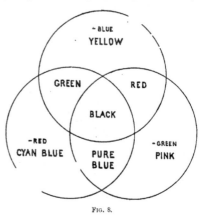

FIG. 8.

"The speaker, assuming red, green, and blue as primary colours, then exhibited them on a screen by means of three magic-lanterns, before which were three glass troughs containing, respectively, sulpho-cyanide of iron, chloride of copper, and ammoniated copper.

"A triangle was thus illuminated, so that the pure colours appeared at its angles, while the rest of the triangle contained the various mixtures of the colours, as in Young's triangle of colour.

"The graduated intensity of the primary colours in different parts of the spectrum was exhibited by the coloured images, which, when superposed on the screen, gave an artificial representation of the spectrum.

"Three photographs of a coloured ribbon taken through the three coloured solutions respectively were introduced into the lantern, giving images representing the red, the green, and the blue parts separately, as they would be seen by Young's three sets of nerves separately. When these were superposed, a coloured image was seen which, if the red and green images had been as fully photographed as the blue, would have been a truly-coloured image of the ribbon. By finding photographic materials more sensitive to the less refrangible rays, the representation of the colours of objects might be greatly improved."

The primary cause of the failure, then, of Clerk-Maxwell's attempt was the insensitiveness of the photographic plates of his day to red and green. Now, however, the case is very different, as plates are commercially obtainable sensitive in some degree to the whole of the visible spectrum, so that all we have to do is to prepare three colour filters admitting light in such proportions that we may secure in our negatives densities corresponding respectively to the curves shown in fig. 6. The red filter must allow light to pass so as to secure action upon the photographic plate from the lines A to E with the maximum of action in the orange between C and D. The green filter must admit light so as to secure an action upon the plate from the line C to midway between D and E, and the blue filter negative must admit light so as to secure action from the line E to H with a maximum of action about midway between the lines F and G. Until quite recently all attempts to adjust the light filters to these curves were made by trial and error, repeatedly photographing the spectrum through various colour filters, until the action secured upon the plate corresponded approximately to the curves shown in fig. 6. By a modification of Sir William Abney's colour sensitometer it is, however, now possible to secure far greater accuracy with considerably less expenditure of time, and it is to the greater accuracy so secured that the great improvement of trichromatic work during the last few months has been due.

Light filters correctly adjusted by means of a measured Abney sensitometer for use with the Cadett Rapid Spectrum plate are now obtainable commercially, and in fig. 9 I have endeavoured to show as clearly as possible the selective action of the light filters when photographing a row of squares of coloured glass.

A (fig. 9) represents the test object, consisting of small squares of white, red, green, blue, and yellow glasses, and also a black or opaque square.

B represents a negative of the test object taken through the red-light filter. The exposure has been made so as to secure a full dense deposit upon the plate by the light coming through the white square; the next square—the red—is also represented by a dense deposit; but the light coming through the green and blue squares has been cut out by the red filter, and they are represented in the negative by transparent spaces. The next square—the yellow—is represented by opacity, because, as we saw in fig. 7, red and green light was required to make yellow; and the black square, allowing no light to pass to the plate, is represented by transparency.

C, the next row of squares, represents the selective action of the green-light filter; here the white square is again represented by opacity, the red square by transparency, the green square by opacity, the blue square by transparency, the yellow square by opacity, and the opaque square by transparency.

D represents the selective action of the blue-light filter; here again the white square is represented by opacity, the light coming through the red, green, and yellow squares is cut out by the filter and they are represented by transparency, the blue square is represented by opacity, and the black by transparency.

Such a set of negatives thus constitutes an accurate *colour record* of the test object, and it only remains to explain how such a record can be used to reproduce the actual colours of the test object.

This object can be achieved by Clerk-Maxwell's device of making transparencies from the negatives and projecting them in superposition upon a screen by means of three optical lanterns, illuminating the red-filter transparency by red light, the green-filter transparency by green light, and the blue-filter transparency by blue light. But very early in the history of trichromatic photography the inconvenient nature of this method, with its consequent enormous loss of light, was felt, and in 1869 the able French experimenter, Du Hauron, came forward with several entirely new methods of synthesis. He invented the first photochromo-scope, or table instrument, for optically combining the three transparencies—a device which led to the Krömskôp of Ives. He suggested making a mosaic screen by ruling alternate very fine lines close together on a glass plate in transparent red, green, and blue ink, afterwards re-invented and commercially worked by Dr. Joly, of Dublin; but, most important of all, he showed us how it was possible to print from the colour record negatives in transparent pigment colours and superpose the prints, thereby securing as a single print a perfect representation of the object, which could be seen in the hand as a complete picture in colours or projected upon the screen by a single lantern.

Now Du Hauron took his negatives through red, green, and blue filters, but in making his triple superimposed prints he did not print from the red-filter negatives in red ink, but in a greenish-blue colour; similarly he printed from the green-filter negative in a pink colour, and from the blue-filter negative in a yellow colour. A little consideration will

nearest the eye from that passed by the other print. Think for one moment what we do when we paint, say, a streak of blue paint on a piece of white card: the white card is reflecting light of all colours to the eye, the streak of blue paint absorbs a part of the white light, the red and green rays, and makes the card appear darker, because the

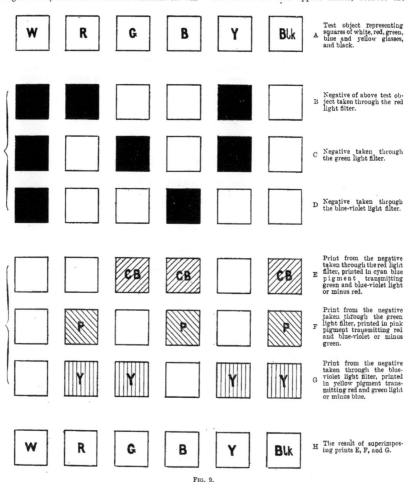

A. Test object representing squares of white, red, green, blue and yellow glasses, and black.

B. Negative of above test object taken through the red light filter.

C. Negative taken through the green light filter.

D. Negative taken through the blue-violet light filter.

E. Print from the negative taken through the red light filter, printed in cyan blue pigment transmitting green and blue-violet light or minus red.

F. Print from the negative taken through the green light filter, printed in pink pigment transmitting red and blue-violet or minus green.

G. Print from the negative taken through the blue-violet light filter, printed in yellow pigment transmitting red and green light or minus blue.

H. The result of superimposing prints E, F, and G.

FIG. 9.

show us the reason of this change. In the experiment of the three overlapping discs of red, green, and blue light shown in fig. 7 we mixed all three lights and produced white light; when, however, we superpose coloured prints we are not adding one coloured light to another, but rather adding darkness, abstracting the light passed by the print

paint only reflects the blue rays to the eye. Therefore, in printing upon paper or in superposing transparent prints for lantern slides, we print, not in the primary colours red, green, and blue, but in their complementaries. The red-filter negative we print in a colour transmitting the complementary of red—*i.e.* minus red, that colour which with red light will form white. This we see from fig. 7 is cyan blue, a light greenish shade of blue. The green-filter negative we print, not in green, but minus green, a pink or light magenta colour ; and the blue-filter negative we print, not in blue, but in minus blue or yellow. Each of these three colours, therefore, reflects two and absorbs two of the three primary colours used by Clérk-Maxwell to form white light.

Fig. 8 shows us the result of printing overlapping discs in these minus colours. The yellow printed over cyan blue forms green, the yellow printed over the pink forms red, the pink printed over cyan blue forms pure blue. and where all three overlap in the centre of the diagram all three colours are absorbed, and we get no light reflected to the eye—black. E, F, and G, fig. 9, represent prints from the test-object negatives printed in the minus colours, and H represents the remainder of the

white light reflected from the white paper after undergoing absorption by the pigments superposed upon it, and is a correct colour copy of the test object A.

We must remember that the diagram fig. 9 represents what happens when we photograph colours in great purity. In photographing from Nature the negatives are not nearly so widely different one from another as they are represented here ; because in Nature objects always reflect a certain amount of white light, and therefore are represented by some amount of deposit in all three negatives. Figs. 10, 11, and 12 are prints from an actual negative of a brightly coloured poster (see p. 108). A is the print from the red-filter negative, B from the green-filter negative, and C from the blue-filter negative.

(*To be continued.*)

BUTTERFLIES OF THE PALAEARCTIC REGION.

BY HENRY CHARLES LANG, M.D., M.R.C.S., L.R.C.P. LOND., F.E.S.

(*Continued from page 73[1].*)

(*PIERIS continued.*)

23. **P. chloridice.**   Hab. 712, 5.   Lg. B. E. 34.   Pl. VII. fig. 2.

27—41 mm.

Wings white. F.w. with ou. marg. slightly concave. The marg. band consists of blackish dashes to the number of about five, running from the margin inwards ; internal to these there is in ♀ a black band running downwards from the costa. The discal spot is white in the centre.   ♀ with the usual black spot near the in. marg., but it is somewhat faintly marked. H.w. white and unspotted in ♂, but with small marg. black spots in ♀. U.s. f.w. arranged as in *P. daplidice*. but the green colour of the ou. marg. is of a much more delicate tint. H.w. of the same tint of light bluish-green, with white spots arranged as in *P. daplidice*, but longer and narrower ; specially so are the marginal spots.

HAB. S.E. Russia, Sarepta, Province of Orenbourg ; Ural ; Turkey ; W. Asia ; Amasia and Tokat (Asia Minor) ; Steppes of Kadomya ; Persia ; Pamir ; Transalai ; Amdo ; Altai, Barabinsky Steppe, Kurai, Tchuja Steppe (Elwes). V., VI. Second brood VIIe. at 6000 ft., Bashkaus down to 2.000 ft. " These differ only in their size from those of the first generation." (Elwes, T.E.S., Sept. 1899.)

24. **P. iranica** Biernert Diss. p. 26. Stgr. Cat. 422.

36 mm.

(1) This series of articles on Butterflies of the Palaearctic Region commenced in SCIENCE-GOSSIP, No. 61, June 1899.

Somewhat like *P. daplidice*, but smaller. Apices of f.w. with black rays, enclosing rounded white spots. H.w. with black streaks and submarginal spots. Neuration of u.s. marked with yellow.

HAB. Pamir—Persia. V., VI. A rare species. occurring at high elevations.

Genus 11.   *ANTHOCHARIS* Boisd.

Moderate-sized or small butterflies, having the antennae short and with a distinct club. Palpi projecting beyond head and hairy. Subcostal nervure with five branches. Wings more or less rounded as in *Pieris*, of a white or yellow colour. Fore wings with a black apical patch, and with a black discoidal spot. Hind wings chequered or striped beneath with green, yellow or orange, and sometimes with silvery patches. Larvae green, narrowed at the extremities. Pupae boat-shaped, with wing-cases proportionately large.

This genus is represented in England by the common "orange-tip" butterfly, which has the wings tipped with bright orange-yellow in the ♂ ; whereas the ♀ is merely white and black above.

The genus may well be divided into two groups (in respect of the Palaearctic species) thus :—

I. Fore wings without orange tips in either sex, ground colour white or yellow. Only the first branch off the subcostal nervure thrown off before the end of the cell. This is the genus *Phyllocharis* of Schatz.

   A. h. wings striped beneath.

       1. *A belemia.*
       2. „ *falloui.*

B. h. wings spotted beneath.
  a. Wings white.
    1. *A. belia.*
    2. ,, *tagis.*
  b. Wings yellow.
    1. *A. charlonia.*
    2. , *levaillantii.*
    3. ,, *mesopotamica.*

II. Fore wings with orange patches at apex, at least in ♂, two branches of subcostal nervure thrown off before the end of the cell.
  A. Wings white in both sexes.
    1. *A. cardamine.*
    2. ,, *bieti.*
  B. Wings white only in ♀.
    1. *A. gruneri.*
    2. ,, *damone.*
    3. ,, *eupheno.*
    4. ,, *euphenoides.*

1. **A. belemia** Esp. 110, 2. Lg. B. E. p. 36. Pl. VIII. fig. 3.
36—41 mm.
Wings white, f.w. rather pointed at tip, costa with small black spots. At the extremity of discoidal cell a black spot. H.w. somewhat more angular than in the other species; with stripes and not spots beneath of white or silvery on a deep bright-green ground easily distinguishing this species from any other Palaearctic *Anthocharis*.
HAB. S. Spain, Portugal, Algeria. I.—IV. according to latitude.
LARVA pubescent, yellow finely speckled with black, with three rosy-red longitudinal bands, on Cruciferae (Boisd).
a. var. *glauce* Hub. 546, 7. Lg. B. E. p. 36. Pl. VIII. fig. 4. A seasonal dimorphic form of second generation. Differs from type merely in its somewhat larger size—the markings of upper side less intensely black. The u.s. h.w. with the white stripes much less defined, and the green markings brownish, with but a slight tinge of green. IV., V. distribution as in type.

2. **A. falloui** Allard. Ann. S. Fr. 1867. Stgr. Cat. p. 3.
35—37 mm.
A single ♂ specimen in my collection, received from Dr. Staudinger from Algeria, resembles *glauce*, but is smaller and more faintly marked. The disc. spot f.w. is much narrower, and not marked with white. It seems to replace the other forms in the hotter and more arid portions of Algeria bordering on the Sahara. III.

3. **A. belia** Cr. Pap. Exot. Pl. 397. (1782.)
36—43 mm.
Wings white. F.w. with a black tip spotted with white, a large quadrate black discoidal spot, and having the costa spotted with black. H.w. white. U.s. F.w. tip greenish-yellow, marked

with pearly or silvery blotches. H.w. dark-green, more or less mixed with yellow and marked with a number of white or silvery spots, mostly rounded in form. ♀ larger than ♂, and with the spots u.s. h.w. less often silvery.
HAB. South Europe, North Africa, Asia Minor. III.—V.
LARVA yellow, spotted with black, with lateral and dorsal pink-red stripes. Feeds on *Biscutella*.
a. var. *ausonia* Hub. 582, 3. Lg. B. E. p. 37. Pl. VIII. fig. 6. Larger and paler in coloration than the type. F.w. without costal row of black dots. U.s., h.w. with the green colouring replaced by yellowish; spots more irregularly arranged and without silver markings. Second generation; in the more southern parts of habitat, the difference from the type is most strongly marked. HAB. as in type. V.e.—VII.e.
b. var. *simplonia* Fir. B. 73, 2 (1829). Lg. B. E. p. 37. Pl. VIII. fig. 7. More heavily marked than type, especially at apices f.w. Basal half of costal f.w. without black dots, which are generally replaced by a narrow black line meeting disc. spot. U.s., h.w. of a dark green, white spots small and not silvery. An Alpine form. HAB. The mountains of Switzerland, Riviera, Austria, and Piedmont. VI., VII.
c. var. *pulverata* Christ. R. H. p. 134. The white spots at the apex f.w. smaller. U.s., h.w. with more numerous and smaller white spots. HAB. Schahrud, Persia; Tekke, Osch. IV.
d. var. *romana* Calb. R. H. p. 134. 38—44 mm. From the description in R. H. this form appears to be very close to *ausonia*, but is larger and lighter. I have not seen a specimen. HAB. The Roman States.

4. **A. tagis** Hub. 565, 6. Lg. B. E. p. 38. Pl. VIII. fig. 8.
33—38 mm.
Very much resembles the allied species last described, but the white spots on the black apical patch f.w. are much smaller. The disc. spot is narrower than in *A. belia*. U.s. f.w. apices broadly marked with bluish-green in ♂, yellowish in ♀ with very faint white spots. H.w. ground colour bluish-green in ♂, yellowish in ♀, having the appearance of being dusted with dark scales. White spots very small and few compared with those seen in *A. belia*.
HAB. Portugal and Andalusia; Elvas, Seville, Cadiz, Granada, Valencia, etc. It does not seem unlikely that small specimens of *A. belia* from Spain and Portugal have often been considered as belonging to this species. The coloured figure in my own former work is by no means satisfactory; and does not give a proper idea of the peculiar cloudy and undefined character of the markings of the u.s. h.w., which are very distinct from those of *A. belia*. IV., V.
LARVA. Pubescent, green, very finely speckled

with black, with a lateral white stripe, above which is a red one. Feeds on *Iberis pinnata.* VI., VII. (Boisd.)

a. var. *bellezina* Boisd. Ind. p. 2. Lg. B. E. p. 38. Smaller than typical *tagis.* F.w. with larger white spots at apex, disc. spot narrower. U.s. h.w. ground colour of a lighter and more yellowish-green, with more numerous and well-defined white spots. HAB. The Basses Alpes and other parts of South-East France. At Digne, at about 2,000 feet altitude, it is common in many places. IV., V.

b. var. *insularis.* Stgr. Cat. 1871, p. 4. Apex of f.w. paler, spots of underside very small, disc. spot f.w. narrower than in *bellezina.* HAB. Corsica and Sardinia. This and the last variety are local races of *A. tagis*, or else two forms of a distinct species. They always seem to me to differ from typical *tagis* almost as much as does *A. belia.*

c. (? var.) *pechi* Stgr. R.H. p. 135. 32—33 mm. Seems to be a small and dark form of *A. tagis.* HAB. Lambessa, Algeria. IV., V.

d. (? var.) *tomyris* Christoph. R.H. p. 135. 35—36 mm. A form described as occurring at Askabad, Caucasus; with greyish-green. coloration in place of dark-green or yellowish. I have not seen specimens of these two latter forms; they may be distinct species.

(*To be continued.*)

TWO NEW VARIETIES OF BUTTERFLIES.

PIERIS RAPAE, VAR. ROSSII.

BEING a member of the Italian Entomological Society, I have received the "Transactions" for April-June, 1900, and have been interested in a catalogue of Professor P. Stefanelli, a zealous collector of Lepidoptera in Tuscany. He describes therein a new variety of the common garden white butterfly, *Pieris rapae.* Thinking that the readers of SCIENCE-GOSSIP who, like myself, study the articles on the "Butterflies of the Palaearctic Region" by Dr. H. C. Lang, would like particulars of this new variety, which must be added to the list of varieties of *P. rapae* given by H. C. Lang in SCIENCE-GOSSIP for July, 1900, I translate Professor P. Stefanelli's description :—

"*Pieris rapae* var. *rossii* Stefanelli. I have here separated from var. *mannii*, giving it a different name, this form, which is really a summer modification of that variety, because of the marked characters of the ♀ sex.

"MALE.—Size of summer brood of typical *P. rapae*, *i.e.* much larger than that of var. *mannii.* External margin of f.w. more rounded than in that variety. Upper side : All the markings much larger than in var. *mannii* and deep black. The medium spot on f.w. nearly always with external margin slightly

concave and with the rest of the margin shading off into the white. Wings white at the base. Under side : First black spot of f.w. very dark and generally square or rectangular. H.w. bright yellow slightly suffused with brown.

"FEMALE.—Corresponds in size to typical ♀ of summer brood. External margin as in ♂. Upper side : markings of f.w. deep black as in ♂, but larger. The triangular apical spot with the internal border very convex. First spot markedly square and often indented ; it generally is joined to the external border by one or two black streaks, sometimes joined by a slight black shading. The second spot is nearly always lunular with its concave side turned towards the base. Under side : first spot of f.w. shaped somewhat the same as on upper side. Tip and a portion of external margin of a fine yolk yellow. H.w. of the same colour slightly suffused with black.

"HABITAT.—Near Florence. Common enough in July and at the beginning of August on the hill of Fiesole, where in 1875 I found it for the first time and where thenceforth I have captured it every year. A fine ♀ which I observed with pleasure in the collection of Mr. Roger Verity was by him found near the Forte dei Marmi, near Pisa, July 27th, 1899."

I may add to Professor Stefanelli's description that the general aspect of *P. rapae* var. *rossii* ♀ is similar to that of *Pieris cheiranthi* ♀. I have this summer made a special study of var. *rossii* in the locality near Pisa, in the pine woods, and have had the fortune of capturing a good number of fine specimens.

I should be very glad, as far as I am able as a private collector, to send specimens to students of Rhopalocera in exchange for other species ; and if any one finds specimens corresponding to my description in other parts of the world, I should be grateful to them if they would let me know, as well as of the variety of the Clouded Yellow (*Colias edusa*).

Professor Stefanelli also describes a new form of *Colias edusa* as follows :—

COLIAS EDUSA AB. ♂ FAILLAE (STEFANELLI).

MALE.—The same as typical *C. edusa* ; but with the antemarginal stripes of both the anterior and posterior wings streaked with yellow along all the nervules which cross them.

HABITAT.—Found sometimes often and sometimes rarely near Florence from April to November, in company with *Colias edusa.*

ROGER VERITY.

1 *Via Leone Decimo, Florence, Italy.*

UNIVERSITY COLLEGE, LONDON.—The Senate of this College has appointed, as Principal, Professor George Carey Foster, B.A., F.R.S., F.C.S. He is a Fellow of the College and Professor of Physics; and has been member of the Senate and Examiner of the London University.

The Text Book of Zoology. By OTTO SCHMEIL. Part I., Mammals. vii + 138 pp., 9½ in. × 6½ in. Part II., Birds, Reptiles, Fishes. vi + 166 pp., with numerous illustrations. (London: Adam & Charles Black. 1900.) 3s. 6d. each part.

This work, which is largely used in the schools and colleges of Germany, is treated from a biological standpoint. It has been translated from the German by Rudolf Rosenstock, M.A., and is edited by J. T. Cunningham, M.A. Considering the range of the subjects included in these parts, and the room occupied by the many illustrations, the author has fairly covered the subjects. In the earlier portion of the work will be found physiological details, and these are followed by classification, commencing with the anthropoid apes. The classification is by no means the most modern, and this is unfortunate, as the general appearance of the work and low price will cause a wide circulation. The nomenclature is in the same category, which is surprising, considering the English editing, where opportunity occurred to bring it even with modern works. The two parts will be followed by one on Insects, the three constituting a complete volume.

Pre-Historic Times. By the Right Hon. Lord AVEBURY. Sixth Edition, revised. xxxii + 616 pp., with xl plates and 243 illustrations. (London: Williams & Norgate. 1900.) 21s.

The fact that this well-known standard work has reached a sixth edition is evidence of its value to general readers, as well as to ethnologists and archaeologists. The present issue has been revised and brought up to recent knowledge, but the general plan of the work is maintained. This is explained by its full title—viz. "Pre-Historic Times as illustrated by Ancient Remains and the Manners and Customs of Modern Savages." The many and beautiful illustrations are most helpful in better understanding the plain but accurate letterpress. Lord Avebury is to be congratulated on this new edition, which is sure to further popularise a fascinating subject for investigation by cultured people.

The Antarctic Regions. By Dr. KARL FRICKER. xii + 292 pp., 9½ in. + 6¼ in., with 12 plates, maps, and 46 other illustrations. (London: Swan Sonnenschein & Co., Limited, 1900.) 7s. 6d.

The recent revival of antarctic exploration has naturally brought forward new literature of the subject. Among the modern books on the Southern ice regions, with accounts of attempts to penetrate the great frozen cap, one of the best we have seen is that now before us. It gives an extensive review of what has been done in times past towards the exploration and investigation of the flora and fauna of those regions. The book is divided into sections—I. Position and Limits ; II. History of Discovery ; III. Conformation of the Surface and Geological Structure ; IV. Climate ; V. The Ice ;

VI. Fauna and Flora ; VII. The Future of Antarctic Discovery ; VIII. Bibliography of the subject. The illustrations are very striking, and doubtless give a good idea to the reader of the fantastically wild character of the ice-shapes and more stable land-contours. This book should have a wide circulation, as within the next few years we shall doubtless hear much from the Antarctic.

The Lepidoptera of the British Islands. By CHARLES G. BARRETT, F.E.S. Vol. VI. 388 pp., 9 in. × 6 in. (London: Lovell Reeve & Co., Limited. 1900.)

We have already on several occasions noticed this extensive work. The present volume includes the remainder of the Trifidae, following with the families Sarrathripidae, Gonopteridae, and Quadrifidae, the Deltoides, the Brephides, concluding with the commencement of the Geometrina, which are carried as far as genus 9, *Halia.* There is an appendix of additions and corrections to the former volume as well as the present.

Experimental Farms. Reports for 1899. 443 pp., 10 in. × 6½ in., with many plates and illustrations in text. (Ottawa : S. E. Dawson. 1900.)

Dr. William Saunders, LL.D., F.R.S.E., F.L.S., the Director of the Canadian Experimental Farms and editor of these reports, is much to be congratulated upon their fulness and usefulness, extending even beyond the Dominion, whose Government publishes them. The year's work which these reports summarise has evidently been successful and voluminous. Among the more interesting articles is one on hybridisation in the genus *Pyrus,* with the object of obtaining a hardy apple-like fruit that will stand the intense cold of winter in the North-West Territories. There are several plates of the trees and the fruit which has resulted from these experiments. The whole report teems with interest in scientific agriculture and horticulture, subjects that have been for many years past successfully investigated under the direction of Professor Saunders.

Year-Book of Photography and Amateur's Guide for 1900. 674 pp., 7 in. × 4 in., with numerous plates and other illustrations. (London: "Photographic News" Office, 1900.) 1s. net.

The editor, Mr. E. J. Wall, F.R.P.S., is to be congratulated on the excellence of this year's issue. It is really a marvel of usefulness and cheapness. The illustrations alone are worth far more than the cost of the whole book. Its literary pages are full of valuable hints for amateurs, and there are a number of most instructive articles. The whole production shows evidence of care and judgment in its preparation.

First Stage Botany. By ALFRED J. EWART, D.Sc., Ph.D., F.L.S. vii + 252 pp., 7 in. × 5 in., illustrated by 236 figures. (London: W. B. Clive, 1900.) 2s.

This is another volume of the "Organised Science Series," and its intention is to be a guide for the elementary stage of the Science and Art Department. It refers only to the flowering plants. Dr. Ewart's reputation, formed while he was Deputy Professor of Botany in the Mason University College, and later as Extension Lecturer of the University of Oxford, is a sufficient guarantee of the correctness of these early lessons in structural botany. We can recommend it to any of our readers who are commencing to take an interest in the organisation of plant life.

The Flora of Bournemouth. By EDWARD F. LINTON, Oxon. viii + 290 pp., 7½ in. × 5 in.. with map. (Bournemouth: H. G. Commin, 1900.) 8s. 6d.

This is a valuable addition to the local lists of plants of this country. It is not confined to Bournemouth proper, but to a twelve-mile radius and the Isle of Purbeck. The district is one of considerable interest to botanists, as it contains no less than 1,137 plants out of 1,218 that occur in the county of Hampshire, including the Isle of Wight, several of which are rare. The introduction to this work occupies a couple of dozen pages, and treats upon topography, climate, geology, and other important features connected with the locality and the plan of the book. The flora contains a phanerogamia and some of the cryptogamia as represented by equiseta, club mosses, and bladderworts. Of course the work is simply a list with localities, and not one descriptive of species; but the stations mentioned are numerous, and the book as a whole will be found a necessity for field botanists visiting the interesting district to which it refers.

Royal Society of Queensland Proceedings. Vol. xv for 1899. 161 + xxv pp., 9¾ in. × 5¾ in. Illustrated by 3 plates. (Brisbane: G. Pole & Co. 1900.)

There are various interesting papers which were read during 1899 before the society. These treat of several subjects, including "Entomology of a Tea-Tree Swamp," "Beginnings of Life." "Nature and Origin of Living Matter," "List of Minerals from North Queensland," "Description of Some Caves near Camaoweal," "Life-History of Mosquito," with illustrations, "New Species of Lepidoptera," and the Presidential address. In this he advocates the greater attention in Australia, on the part of authorities of public libraries, to books on scientific subjects. There is difficulty of obtaining such volumes by scattered students in thinly populated districts.

Manchester Museum. Report for 1899–1900. 280 pp., 10 in. × 6 in. (Manchester: J. E. Cornish, 1900.) 6d.

The Manchester Museum at Owens College is one of the most progressive in the country. Every year we hear of additions and improvements. The period covered by this report is no exception. A new room, measuring 45 ft. × 30 ft., has been made to hold the "Dresser" Collection of Birds, which are arranged in dust-tight cabinets. It is lit by electric light. In other departments additions have come by exchange with the Musée d'Histoire Naturelle in Paris and elsewhere. The geological museum has been improved and the important collections of mollusca made by the late Mr. Layard, his anthropological collection, and the "Schill" collection of Lepidoptera are now features at Owens College. The Section of Botany has also increased. Altogether Mr. Hoyle, the Director of the Museum, and his staff are to be congratulated on the year's work.

American Aleurodidae. By A. L. QUAINTANCE, M.S. 79 pp., 10 in. × 6 in. Eight plates and 16 illustrations. (Washington: Government Printing Office. 1900.)

This is an important addition to the knowledge of a little-known group of animals. A number of new species are described and figured by the author, who has supplied excellent drawings for reference. Anyone working at this group in Europe should not fail to get this work.

The Unknown. By M. CAMILLE FLAMMARION. xiii + 487 pp., 9 in. × 5¾ in. (London and New York: Harper & Brothers, 1900.) 7s. 6d.

This book is a translation of the French work "L'Inconnu," and is an endeavour to reduce the study of what is usually called "Spiritualism" and the "Occult" to a scientific basis. M. Camille Flammarion, who is the popular French astronomer, very justly points out that "the unknown of yesterday may be recognised to-morrow as truth." It is essentially unscientific to condemn any proposition as impossible which has not been thoroughly investigated and disproved. Comparatively a short time since, Auguste Comte limited the study of astronomy to the distances and movements of planets and stars. "We can never," he said, "find out what is their chemical composition." Five years after his death in 1857, spectral analysis had shown the chemical composition of the planets, and stars were classed in the order of their chemical nature. The book before us deals with Telepathic Communications, Hallucinations, the Psychic Action of One Mind upon Another, Dreams, Divination of the Future, and Magnetism. The examples given of the use of hypnotism for medicinal purposes are similar to those described by M. Bernheim in his work on "Suggestive Therapeutics." The instances of psychic force are carefully collected and set before the reader. Whether M. Flammarion can be held to prove, as he claims, that psychic forces exist which "can transmit thoughts and impressions to human beings at a distance, without the intervention of the senses," he at any rate deserves the thanks of all thinking people for the painstaking and scientific manner in which he has treated a subject that charlatanism or actual fraud have done much in the past to drag into disrepute, and prevent earnest thinkers from investigating in a systematic manner. We would, however, suggest that the incidents related would be of greater value if the names and addresses of those testifying were more often given for authenticity. In fact, the letter "X," so commonly used in French newspaper reports, is rather too much in evidence to inspire confidence. The book is ended by a chapter entitled "Conclusion," in which the author sums up his views. Though not denying the possibilities of some of his hypotheses, we fear there is a tendency on his side too readily to accept mere statements on subjects which from their nature render difficult accurate investigation. The whole matter is one of great interest to human beings. To what extent other animals are affected by psychic forces we have no trustworthy means of ascertaining; neither do we know whether they are the result of the recent institution of a "new sense" due to man's civilisation, or a "dying sense" which may be becoming obsolete.—F. W.

Works on Photography in the Library of the Patent Office. 62 pp., 6¼ in. × 4 in. (London: H.M. Stationery Office, 1900.) 6d.

We are glad to observe this, the first of a biographical series of the various subjects included in that useful institution, the library of the Patent Office in London. This number includes not only the works on photography, but also on the allied arts and sciences. The present list comprises 557 works, including seventy-three serials, wholly or in part photographic, and representing 1,300 volumes. To this is added a key to the classification of headings.

In Birdland with Field-glass and Camera. By OLIVER G. PIKE. xvi + 280 pp., 7⅟₂ in. × 5 in., with 83 illustrations. (London : T. Fisher Unwin.) 6s.

As a writer on country lore, the author of this book gives promise of future good work, but needs experience and a closer attention to the style of classical English writers. He evidently possesses a natural aptitude for describing the beautiful side of nature, though at present he is rather colloquial, and his chapters would be improved by a less epistolary vein. The photographic reproductions

Year-Book of Scientific and Learned Societies of Great Britain and Ireland. 296 pp., 9 in. × 5½ in. (London : Charles Griffin & Co., Limited. 1900.) 7s. 6d.

This most useful publication, founded in 1884, should be more widely known, and in the library of every learned society. Year by year its value increases, as it not only contains a directory of scientific and learned societies, but also lists of the papers read before them during the previous session, in fourteen of the leading departments of research. The Year-Book is divided into fifteen

NEST OF SHORT-EARED OWL.
From " In Birdland."

add greatly to the appearance of the book, which, as a whole, is very readable. One point of interest lies in the fact that many of the observations have been made in suburban London. We can recommend " In Birdland," to the general reader. By permission of the publisher we reproduce one of the illustrations representing a nest of short-eared owls. Some of the pictures are decidedly pretty, yet a little more attention is necessary in focussing the backgrounds.

sections, so that any subject may be readily consulted independently of more general societies' work. On referring to some of the earlier annual volumes of this publication, one cannot help being struck with the steady increase in the information gathered by its editor. It is much to be regretted that the information necessary for this Year-Book has not been supplied by so many as about fifty societies. We hope their secretaries will soon amend their ways, and add to its completeness.

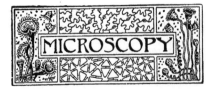

CONDUCTED BY F. SHILLINGTON SCALES, F.R.M.S.

FORMALIN AS A PRESERVATIVE.—A 3 per cent. solution of formalin is preferable to spirits of wine for preserving certain species of insects, as it does not affect their colours. I find, however, that specimens so preserved and afterwards dried deposit an oily dew, or in some cases crystals, on the slide and cover-glass if mounted as dry objects for the microscope. Washing appears to have little or no effect. Can anyone tell me how to obviate this without discolouring the specimens?—*E. G. Wheler, Swansfield House, Alnwick.*

MANCHESTER MICROSCOPICAL SOCIETY.—The Transactions and Annual Report of this Society for 1899 have reached us, and we again congratulate the members on the work they have done, and on their position. The Society is the most enterprising and successful of any microscopical society in the provinces, and well deserves its success. The membership during the year appears to have been well maintained. The Council speak of the attendance and the interest in the meetings as being in every sense satisfactory. The Extension Section appears to have delivered no fewer than thirty-seven lectures in the neighbourhood of Manchester during the winter: by so doing they have benefited many outside their own membership, and doubtless added to the popularity as much as to the usefulness of the Society. We commend this scheme of extension to the notice of other societies, one of which, the well-known Yorkshire Naturalists' Union, has, we believe, a similar scheme in hand. The President of the Manchester Society for the year is Professor Sydney J. Hickson, M.A., D.Sc., F.R.S., of Owens College, and his presidential address on zoophytes is the first paper in the report. Perhaps the most important paper is one by Mr. F. W. Gamble, of Owens College, on "The Power of Colour-change in Animals," a subject in which the author is specially interested, and concerning which he and Mr. F. W. Keeble have been able to make original investigations. The most interesting papers to our readers will be one on "Collecting Lepidoptera," by Mr. H. G. Willis, from which we would have liked to make extracts had space permitted, and another on "Arboreal Aphidae," by Mr. A. T. Gillanders. Amongst other papers we may particularise one on "Termites and Ants of West Africa," by Mr. Mark L. Sykes, and another on "The Pollination of Flowers," by Mr. Charles Turner. There are several excellent plates. Mr. C. F. Rousselet's "Method of Preserving and Mounting Rotifera" is given in full; and as this method has been brought prominently before microscopists, we reprinted it in our last issue. The Report can be obtained, post free for 1s. 9d., from the Hon. Secretary, Mr. E. C. Stump, 16 Herbert Street, Moss Side, Manchester. A list of the Extension Section's lectures can be obtained from Mr. George Wilks, 56. Brookland Street, Eccles New Road, Manchester.

SWIFT'S NEW PORTABLE MICROSCOPE.—Messrs. James Swift & Son have recently brought out a new folding microscope for travelling, for bedside diagnosis, or for field work. It is furnished with both coarse and fine adjustments, the latter being markedly superior to those usually fitted to microscopes of this type. The optical tube carrying the objectives

SWIFT'S PORTABLE MICROSCOPE IN CASE.

is made to slide in its fitting so as to allow very low power objectives to be used. There is a draw-tube permitting of a total extension of tube-length to 7 inches. The stage is larger than usual, and carries a sub-stage ring fitted with Abbé condenser and iris diaphragm. The back leg is divided so as to pass over the fine adjustment screw when folded, whilst the stage is hinged and lies flat against the

SWIFT'S PORTABLE MICROSCOPE.

body of the microscope. The microscope packs thus into a leather case about 9 × 3 × 3 inches, and there is room for two objectives, live-box, small bottle, and sundry minor apparatus, as shown in our illustration. The whole microscope is beautifully finished in bright brass, and is, we think, one of the best travelling microscopes we have seen. The price of the stand, with one eye-piece and the necessary case, but without objectives or other apparatus, is £5.

ROYAL MICROSCOPICAL SOCIETY.—At the meeting on June 20th, the President, Mr. Carruthers, F.R.S., in the chair, Mr. C. H. J. Rogers exhibited a modification of the Rousselet compressor, in which two thin india-rubber bands, sunk into grooves, were employed to keep the cover-glass in position. The advantage of this modification is the facility with which a broken cover-glass can be replaced. Mr. Chas. Baker exhibited an acromatic substage condenser which was a modification of Zeiss's model of the Abbé condenser, the N.A. being 1·0, aplanatic cone 90°, lenses $\frac{7}{10}$-inch diameter, working distance $\frac{4}{10}$ inch. With the front lens removed the condenser is suitable for use with low-power objectives. A short paper by Mr. E. B. Stringer on a new projection eyepiece and an improved polarising eyepiece was taken as read. Miss Loraine Smith contributed a paper on some new microscopic fungi, and Mr. Bennett in commenting thereon referred to the proposed cultivation of fungus parasites on certain insects, especially on the Continent and in Australia and America, with a view of getting rid of insect pests, locusts, and others. The President then read a paper, and gave a lantern demonstration on the structure of some palaeozoic plants.

EXTRACTS FROM POSTAL MICROSCOPICAL SOCIETY'S NOTE-BOOKS.

[In accordance with the announcement in our last issue we commence the publication of extracts from the note-books of the Postal Microscopical Society. Beyond absolutely necessary editorial revision these are printed as written by the various members, without alteration or amendment. Correspondence on these notes will be welcomed.— ED. Microscopy, S.-G.]

SCALE FROM LEAF OF LEMON-TREE.—The scale insects, Coccidae, members of the order Hemiptera, sub-order Homoptera, belong to that division of the sub-order which is distinguished by having only one joint on the tarsus. (^{1}) The females are common on many plants, and are found both on leaves and bark. They appear as small, brown, waxy, convex lumps, more or less elliptical according to species. If the brown case be lifted, one sees a small fleshy mass, with eggs, and usually a cottony substance. The fleshy mass is the female, which dies as soon as the eggs are laid. The eggs hatch into a small, active larval insect with eyes, legs, antennae, and a sucking mouth. After a time these insects fix themselves to a leaf or the bark by means of their sucker. If they are to become females they excrete the waxy shell, throw off all legs and processes, lay their eggs, and die. If they are to develop into males, they draw in their legs and become chrysalids, with a process on each side containing the future wings. The adult males are seldom seen. The male and larva are figured in Miss Ormerod's book. A friend of mine is now trying by an ingenious contrivance to obtain specimens of the males and larvae. He has run the branch of a rose tree, not cut off, into a lamp chimney, and packed the ends with cotton wool. He thinks that when the scale eggs on the branch

(1) Query—apparent joint ?—[Ed. Micros.]

hatch out he will find the larvae inside the glass. The second slide shows a scale in which a parasite, probably an ichneumon fly, has laid its egg, and the grub has hatched and eaten all the scale eggs. The grub is shown lying by the side of the scale. The scales thus attacked are usually of a lighter colour than the rest.

SHEEP-TICK IMAGO.—A tick is a degenerate fly that has lost its wings. (^{2}) The sheep-tick passes the larval state inside the body of the mother. The pupae may be found in little brown shiny cases about three millimetres in diameter. If these are broken open the fully formed tick is seen inside. These are better adapted for mounting than the

SHEEP-TICK.

adult insect which has already sucked blood. The spiracles should be noticed, especially the thoracic ones; also the toothed claws and the tree-like formation between them, answering to the pad on the foot of the house-fly.

APHIS WITH YOUNG.—The peculiarity in this slide is that the winged female in November ought by all rules to be oviparous. The fact of this one being viviparous at that time of year shows how circumstances modify natural habits. The chrysanthemum from which this aphis was taken was in a greenhouse; had it been in the open air the aphis would have laid eggs—at least, if entomologists are to be believed. In the slide showing the pupal aphis the wings are still seen confined in their cases. Both larval and pupal aphides produce young; a phenomenon known as Poedogenesis. Aphides during the summer are viviparous and produce their young parthenogetically. In the autumn the union of the sexes takes place and the result is, not living young, but eggs. The species of aphides are very numerous. The aphis which spins the wool on apple trees has no cornicles. Aphides, like Coccidae, are of the order Hemiptera, and sub-order Homoptera, but belong to that division of the sub-order which has two joints on the tarsus (Dimerae).

(2) The " sheep-tick " here alluded to is *Melophagus ovinus*, belonging to the family Hippoboscidae, order Diptera. The ticks proper belong, of course, to the Acari, and come under the head of Arachnida. *Ixodes reduvius* (*Vide* S. G., Vol. VI., n.s., p. 5) is also known as " the sheep-tick." [Ed. Micros.]

OVIPOSITOR OF TIPULA.—This is a curious organ of the "Daddy-long-legs," but the parts are not arranged on the slide as in nature. There are 250 (²) species of *Tipula* known. They lay their

OVIPOSITOR OF TIPULA.

eggs in the ground or on the surface in batches of 200 or more. The eggs are black. The grub produced from the egg is known as the "leather

BATTLEDORE-LIKE
OBJECT
IN OVIPOSITOR
OF TIPULA.

jacket," and is very destructive. This last remark applies to only two of the 250 species. I should very much like to know what those battledore-like objects lying near may be.

PARASITE FROM HUMBLE BEE.—These are mites and near akin to spiders. They have eight legs and are therefore not insects. The crab-like toothed jaws are curious. I do not know the name

JAW OF PARASITE OF HUMBLE BEE.

this mite, but it cannot be mistaken for *Stylops spencii*, parasitic on bees and other Hymenoptera, which used to have an order all to itself (Strepsiptera), but which is now included amongst Coleoptera.

DISJECTA MEMBRA OF COMMON WEEVIL.—It is no use mounting this very hard and opaque beetle whole, as one cannot see anything in that way. The weevils are an order of Coleoptera, the common weevil being an example, and are furnished with long snouts. Hence they are also called Rhynco-

(²) There are over 1,000.—[Ed. Micros.]

phora = snout bearers. The really distinguishing mark of the order is the elbowed antenna. These common weevils are found in immense numbers in the wheat imported from Calcutta and from Aus-

GIZZARD AND ANTENNAE OF COMMON WEEVIL.

tralia, and infest granaries and mills. They bore a hole in the wheat grain with their long snouts, which are furnished with a row of teeth at the tip, by a sort of turning movement, and in this hole the egg is laid. The hole is then plugged with gum secreted by the female, and the grain looks none the worse. The egg hatches in a few days. The

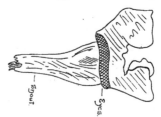

EYES AND SNOUT OF COMMON WEEVIL.

change from larva to pupa, and from pupa to imago, takes place within the grain. The imago eats its way out. They are said to breed only when the temperature is above 65° F. It is a remarkable thing that they never injure the germ of the grain, which therefore grows as well when it has served as a nest and home for this little pest as previously. The eyes at the base of the snout should be noticed, and also the gizzard.

OVIPOSITOR OF WILD BEE.—I do not know the name of this bee. There are, I believe, over 200 species of wild bees in England. This is large, long, and black, one that I have never seen except in the autumn, and which seems especially to frequent the common red fuchsia. The organ is undeniably curious, but lacks the finish generally found in nature.—(*Rev.*) *R. S. Pattrick.*

REMARKS.

APHIS WITH YOUNG.—Is it certain this specimen was in a viviparous (as stated in the notes) condition when mounted? May it not have been in an advanced ovo-viviparous condition at the time, with the outer substance of the ova very thin and delicate, so that each was smashed up in the bursting of the insect by pressure? Do not some of the young still show appearances of having portions of outer envelopes around them? Are not the "battledore-like objects" mentioned in the notes, ova, still having the filaments adhering to them which attached them to the ovary?—*W. Dawson.*

PARASITE FROM HUMBLE BEE.—This belongs, no doubt, to the family Gamasinae. I have a number of species of this family in my possession, but not this particular one. Several species are found parasitic on beetles and bees. I think, on referring to a written description I have by me, that it is *Gamasus coleoptratorum*, Linn., and if so, to bear its name, it ought to have been found on a beetle instead of a bee. Shuckard does not mention it in his "British Bees." I have measured the body of one of these on this slide, and it is $\frac{1}{25}$ of an inch long; but in my description of *G. coleoptratorum* it should be only $\frac{1}{50}$, so the correct name is doubtful.—*Charles D. Soar.*

I have kept this box a few days over time, as I was much interested and pleased with its contents. The scale leaf mounts are very instructive, and it is a subject which has been coming to the front lately. The position of the ovipositor is not shown

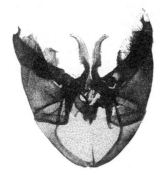

OVIPOSITOR OF WILD BEE.

correctly in the slide, and I have taken a photo of the same object from my own cabinet to show its true position. I have tried to find out what the three little affairs are, but cannot get any information. I see Mr. Dawson suggests that these battledore-like objects are ova, still having the filament attached. I do not think so, for this reason. In Mr. Pattrick's slide there are three out of position. In my photo there are also three, but in position, and it seems hardly likely that there should be exactly three ova in two different objects. In *Knowledge*, of November, 1894, there was a good account of the Crane Fly, or Daddy-long-legs, by Mr. Butler. He says, "The hinder part of the body of the female tapers regularly to a hard and sharp point. This acute tip is the hardest part of the body, and necessarily so, as it has to do the hardest work. It constitutes an egg-laying instrument of superior quality, and is composed of four pieces disposed in pairs. On the upper side are two long and pointed pieces which form the sharp tip, and are used as borers, and underneath these is the other pair, considerably shorter and blunter, their

function being to guide the eggs in their passage into the hole prepared for them by the pair of borers." The whole apparatus, therefore, is something like a combination of an auger and a spoon. The photo will make this clear. When egg-laying the creature balances itself on its two hind legs and ovipositor, whilst the fore front legs are up in the air. I send also photos of Sheep-tick and ovipositor of Wild Bee. Is this the ovipositor? It is a strange one.—*T. G. Jefferys.*

OVIPOSITOR OF TIPULA.—Had there been only two of those battledore forms one might have suggested that they were the "halteres" or abortive wings of the Tipula, but as there are three, that suggestion will not do. The viscera attached show that they are connected with some of the internal arrangements.—*(Rev.) Adam Clarke Smith.*

MICROSCOPY FOR BEGINNERS.

BY F. SHILLINGTON SCALES, F.R.M.S.

(Continued from p. 91.)

Instead of gold-size other cements may be used; but we have found gold-size, especially if old, most satisfactory, save for certain fluid mounts. Bell's Cement is excellent, and so is Ward's Brown Cement, whilst Mr. Cole recommends Watson's Special Club Black Enamel. Marine Glue is to us an abomination, and we have long discontinued its use. Under any circumstance it must be applied hot.

We have dealt at considerable length with mounting opaque or dry objects because it is the easiest, and forms a natural introduction to mounting in preservative media. There are many more or less specialised methods of the latter, but it will be sufficient if we confine ourselves to two—namely, Canada balsam and glycerine jelly. These two methods, and especially the first, are used universally. Objects or sections may need careful preparation beforehand, but we will deal with these methods afterwards, assuming here that, as frequently happens, no such preparation is necessary. Canada balsam is best purchased ready for use, in which case it will be obtained as a solution in benzole or xylol. It should be kept in a wide-mouthed bottle, provided with a glass rod for dropping the contents upon the slide, and with a closely-fitting cap instead of stopper. The bottle must be kept closed as much as possible. Glycerine jelly is practically a mixture of glycerine and gelatine which liquefies when warmed. It can be obtained in shilling bottles fitted with an ordinary cork. The first important distinction to be noticed between them is that whilst objects mounted in Canada balsam must be freed from every trace of water, those mounted in Glycerine jelly must be first soaked in water or some aqueous medium.

In both cases a mounting-table and lamp should be provided. The table in its simplest form is a plate of brass about 4 × 3 inches, standing on four legs about 3 or 4 inches high, and it will cost about 3s. 6d. The lamp is a small glass methylated spirit lamp, with a glass top, such as is used in laboratories to go beneath the table. This can be purchased for 1s.

The actual process of mounting in Canada balsam may be carried out as follows. We will assume that the object has received a final soaking in turpentine. Having carefully cleaned both slip and cover-glass, the latter is taken up in a pair of forceps, and, the slide having been breathed upon to slightly moisten it, the cover-glass is placed on the slide and pressed there to make it stay in position. The slide is then placed on the table and the lamp lighted and put beneath. In less than half a minute the plate will be sufficiently warm—the heat should be no greater than will allow of the finger being placed on the end of the slide. A drop or two of balsam is then placed on the cover-glass which is on the slide, care being taken that it does not overrun the margin of the former. Into this the object is then lowered or slid by means of a section-lifter (a cover-glass held in a pair of forceps may serve) and a needle set in a handle. Care should be taken to get the object right down under the balsam and close to the cover-glass. The object should then be examined with a pocket-lens to make sure that its position is satisfactory, and to see that no air-bubbles are visible in or around it It is then placed under a watch-glass or other cover to protect it from dust, and put aside for twelve hours to harden. It will be found that the balsam skins over very rapidly on exposure to the air, and no time, therefore, must be lost. The warming of the slide partly obviates this. After hardening for twelve hours it is as well to make an examination under the lowest power of the microscope before proceeding further, to make sure that the object itself is properly in position and free from air-bubbles or contained air. The slide is then again placed on the mounting-table, and the insertion of a needle will readily release the cover-glass. Warm · as before, apply a fresh drop to the centre of the hardened balsam, lift with a pair of forceps, reverse quickly, and lower gently down upon the slide, pressing down carefully so as to squeeze out the excess of balsam and to carry any air-bubbles with it. The cover should now lie flat on the slide. It is better to have a slight excess of balsam rather than a deficiency. In case of the latter a drop must be put against the cover-glass, when it will quickly run in by capillary attraction. Small bubbles, other than any embedded in the object itself, may be neglected, one of the advantages of Canada balsam being the readiness with which it will absorb these. There are certain objects, however, which are most difficult to free from larger bubbles than the balsam can absorb. In such cases it is advisable to again heat the mounting-table, whilst holding the cover steadily, but not too heavily, in position by pressing on its centre with the handle of a dissecting-needle, until the balsam is seen to boil. At once remove the lamp, but hold the cover-glass steady until the balsam seems to have set again. By this means, though it needs caution, the bubbles will be driven clear of the cover-glass by the ebullition of the balsam. Wire spring clips can be obtained for a penny each, and it is advisable to slip one of these on before putting the slide on one side to harden. This may take twenty-four hours, or it may take a week, according to the amount of balsam used or exuded. Under any circumstances it is well not to hurry matters. The excess can then be removed with a sharp knife nearly up to the cover-glass, and the remainder cleaned up nicely with a rag dipped in turpentine, methylated spirit, or benzole.

(*To be continued.*)

THE death of Professor Pierce Adolphus Simpson, M.A., M.D., of the Glasgow University, removes an eminent teacher of medicine and a skilled botanist. He died on August 11th at the age of sixty-three.

BY the death of Dr. John Anderson, M.D., LL.D., F.R.S., etc., there passes away, at the age of sixty-six, a well-known zoologist and a former Superintendent of the Indian Museum, Calcutta. Dr. Anderson published some important literature upon science and travel in Asia.

DOUBTLESS some of our readers have kept ants, bees, and wasps for scientific observation. It is well they should know, in regard to bees at least, the County Court judge at Basingstoke has decided that any person suffering through the stings of bees kept by another person has right of action for damages against the owner of the bees, who accordingly in a case brought before him, gave judgment for damages.

THE reports of British Consuls often form interesting reading. A recent one by Mr. Carmichael, in the Consular service at Leghorn, deals with the so-called briar-root industry. The briar-root pipes, as probably most of our readers are aware, are made from the root of *Erica arborea*, which flourishes on the mountain slopes of the Northern Mediteranean and its islands. "Briar" is, of course, a corruption of the French *bruyère*, heath or broom. Although it is not likely this heath will be exterminated by the industry, it is one that will probably exhaust itself through lack of material within another decade.

WRITING to the *Times* on August 13th last, Dr. Mortimer Granville draws attention to a statement which has been going the rounds of the press relating to the alleged recent "discovery of causative relations between filaria and elephantiasis." He points out that as far back as 1878, the late Dr. Spencer Cobbold read a paper summarising what was then known on the subject to the Medical Society of London. Dr. Granville does not consider that the recently published information adds anything to what he and others placed before the Medical Society, the Quekett Club, and other scientific bodies more than twenty years ago.

WITH that aptitude, so usual among our American cousins, of turning to advantage the excitement of a passing event, Nature has been made to contribute to the amusing side of the forthcoming Presidential election. This will turn upon the silver and gold standards of currency question, the former being represented by Mr. Bryan and the latter by Mr. McKinley. We have received a packet of literature from the Jumping Bean Company, who have introduced these well-known seeds into the controversy. Some have been gilded and others silvered. One each of these are sold with a race track for ten cents. The amusement is obtained by watching whether the silver or the gold bean wins the course.

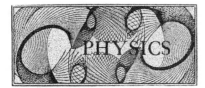

CONDUCTED BY JAMES QUICK.

UNIVERSITY OF BIRMINGHAM.—The principalship of the. recently constituted University of Birmingham has been given to Professor Oliver J. Lodge, D.Sc., F.R.S., Professor of Experimental Physics, University College, Liverpool. Professor J. H. Poynting, D.Sc., F.R.S., has been nominated Dean. In Dr. Lodge, Liverpool loses a man whose remarkable energy and deep knowledge of physics have more than preserved the required status of the college. Professor Poynting's sphere of work will not change, as he has hitherto occupied the Chair of Physics at the Mason College, Birmingham.

ELECTRICAL DISCHARGES UPON PHOTOGRAPHIC PLATES.—Images may be produced upon a sensitive plate by means of an induction coil, using two needles as poles. The positive needle should touch the plate, the negative needle should be fixed about half a millimètre away. A sheet of metal below the plate will facilitate the action. The image produced consists of numerous small black lines, resembling arrows directed from the positive to the negative pole. It is formed without development, and is probably due to the fusion of the silver bromide and of the medium, and after prolonged action to the reduction of the bromide. If iodide of gold is substituted for bromide of silver, the image consists of continuous brown lines connected by many branches.

WAVE-MOTION MODEL.—A striking method of illustrating wave-motion is the following :—An ordinary mercury tray about 70 × 50 × 10 centimètres has its smooth bottom painted white, and is then filled to a depth of about half a centimètre with water slightly darkened with ink. The discoloration of the water must be such that the bottom of the tray is clearly visible through the layer of liquid when at rest. If now a round vessel of about 8 to 10 centimètres diameter is placed in the tray a circular wave is set up spreading outwards. The motion of this wave can be clearly followed, as the thickness of the liquid at the crest of the wave being so much greater than in the trough, the appearance is that of a dark band travelling over a lighter background. Removing the disturbing cause gives a second wave; and if the vessel is kept moving with a period of one or two seconds, a train of waves of the same period is produced. Many other interesting effects can be shown, such as interference, reflection, and refraction. Plane waves can also be produced by substituting a rectangular block for the round vessel.

EXPANSION OF AIR.—A simple apparatus for the determination of the co-efficient of expansion of air can be arranged by taking a glass U-tube 3 or 4 millimètres bore, the limb containing the air being about 30 centimètres long and closed at the end. The second limb is about 50 centimètres long and open at the end. The bend of the tube contains strong sulphuric acid to a depth of 10 centimètres. Another glass-tube is slid into the longer limb of the U-tube and dips into the acid. This tube serves as a plunger, and is used to adjust the level of the acid to the same height in each limb. The maximum temperature during experiments should not exceed 50° C., as above that value the vapour-pressure of the acid would have to be taken into account. The acid, however, has an advantage over mercury, as it keeps dry the air under examination. The U-tube is placed successively in water baths at various temperatures, the plunger being adjusted each time, and the measurement of volume made on withdrawal from the bath. The results obtained with the above arrangement are in close agreement, even if made by elementary students.

WEHNELT INTERRUPTER.—Much work has been done in connection with this recently introduced and valuable interrupter for induction-coil discharges. Investigations have shown that the interruptions are irregular, amounting to 17 per cent. in the case of the original pattern Wehnelt instrument, and 24 per cent. with Simon's pattern. These irregularities have been discovered in an interesting manner by taking a number of mutoscopic impressions of the spark in the liquid when the break was in action. It has thus been shown that if a regular interruption is needed for any particular class of experiments, the electrolytic form must be discarded for a mercury-jet break or a turbine arrangement. It has not been possible until recently to use the Wehnelt break with a low voltage, and various theories have been put forward to account for this. Assuming that it is the oxygen accumulating at the anode which prevents the interrupter being thus worked, any process that counteracts that accumulation must lower the minimum E.M.F. required. This assumption has been put to the test by making a jet of dilute acid impinge against the electrode. With such an arrangement 24 volts suffice to obtain an interrupted current of great steadiness. When the pressure from the jet becomes too high, the current becomes continuous. The impinging jet affords the break another advantage, in that it keeps the whole instrument cool.

WAVY SPARK-DISCHARGE.—Those who have not seen the spark-discharge produced upon an induction coil when worked in conjunction with a Wehnelt break have an astonishing sight to witness. The torrent of sparks produced and the accompanying roar are radically different from the discharge obtained with any form of spring or mercury break. When the distance between the points is reduced, the torrent of distinct sparks changes to a solid flame, curving from one terminal to the other. The frequency of the break can be so increased that the note emitted by this flame-discharge is quite shrill. Intermediate between the above two effects, another peculiar form of spark is obtained when the dischargers are a point and a disc. It consists of a large number of thin spark-lines branching out from the point and extending towards the disc like a brush. Every one of the lines has a sinusoidal wave-shape, due in reality to a spiral form of the spark-path. When the sparks are successively photographed by a mutograph, it is seen that they are very similar in outline, but that they are successively displaced with respect to each other in the direction of the disc.

CONDUCTED BY EDWARD A. MARTIN, F.G.S.

CLASSIFICATION OF FOSSIL CEPHALOPODS.—
The cephalopoda constitute a class of molluscs of
immense interest to geologists, and well repay close
study, since amongst them are classified the numerous ammonites and belemnites which are so
characteristic of Mesozoic times. The class as a
whole is divided into Tetrabranchiata and Di-
branchiata. The former include the ammonites
and like forms, and survive alone in the pearly
nautilus. The latter, the Dibranchiates, include
three sub-orders, two of which are found fossil—
Belemnoidea and Sepioidea—whilst the third,
Octopoda, is very rarely found so in late Tertiary,
and includes the living argonaut, the female of
which has a delicate, single-chambered, spiral
shell. The Belemnoidea are now extinct, except
the genus *Spirula*, whose internal spiral-chambered
shell is found in tropical seas. A complete
belemnite consists of three distinct parts : (1) the
guard (osselet or rostrum), bearing at its anterior
extremity a conical cavity, into which fitted the
(2) chambered phragmacone. This is composed
of a series of chambers (loculi), and the septa are
pierced by the siphuncle. The wall of the phragmacone is prolonged on the dorsal side into a plate
called the (3) proöstracum, and this corresponds
to the "pen" of the cuttlefishes. The Sepioidea
are plentiful, ranging from Jurassic to the present
day, the common cuttlefish (*Sepia officinalis*) being
a well-known species. In this, the proöstracum,
or pen, is so developed as to be of equal length
with the mantle, and is the cuttle-bone of commerce. The two other parts which were found in
the belemnite are here rudimentary, the chambered
phragmacone and the guard being scarcely recognisable. There is no siphuncle. Introduced in
Triassic times, the Dibranchiates culminate in the
Jurassic and Lower Cretaceous, and suddenly de-
cline before the commencement of Eocene times.
The Tetrabranchiata include the ammonites, the
nautilus, orthoceras, ancyloceras, etc., the only
surviving form being the pearly nautilus. It
comprises two orders, the Nautiloidea and the
Ammonoidea. The calcified points of the jaws of
these forms of cephalopod are sometimes found
fossil in Triassic and Neocomian strata, and are
known as *rhyncolites* and *rhyncoteuthis*. *Aptychus*
is a name given to Jurassic and Cretaceous fossils
which, for a long time undefined, have now come
to be regarded as the opercula of ammonoid forms.
They consist of two plates, the name *anaptychus*
being given to the fossil when consisting of a single
plate only.—*E. A. Martin*.

THE RED CRAG.—Mr. F. W. Harmer proposes
to divide the Red Crag into three divisions—viz.
Waltonian, Newbournian, and Butleyan—each
being distinguished alike by the difference of their
faunas and by the position they occupy. The
first, with its southern shells, is confined to the
county of Essex ; the second, containing fewer
southern and extinct, and a larger proportion of
northern and recent species, occupies the district
between the Orwell and Deben and a narrow belt
of land to the east of the latter river ; whilst the
third, in which Arctic forms like *Cardium groenlandicum* are common, is found only further to the
north and to the east. It will thus be seen that
these Crag deposits arrange themselves in horizontal and not in vertical sequence, assuming always
a more boreal and more recent character as they
are traced from south to north. They are the
littoral accumulations of a sea retreating, not continuously, but at intervals, in a northerly direction.
All the beds are believed to have originated in
shallow and land-locked bays, successively occupied
by the Red Crag sea as it retreated northwards,
which were silted up one after the other with
shelly sand. Mr. Harmer suggests that the conditions under which the Red Crag beds originated
seem to exist at the present day in Holland, where
sandy material brought down by rivers, with dead
shells in great abundance from the adjacent sea,
is being thrown against and upon the coast, principally by means of the west winds now prevalent.
From meteorological considerations, it seems probable that strong gales from the east may have
prevailed over the Crag area during the latter part
of the Pliocene epoch. No other explanation of
the accumulation of such vast quantities of dead
shells on the East Anglian margin of the North
Sea at that period can be suggested. At the present day the eastern shores of Norfolk and Suffolk
are almost destitute of such débris.

WESTOW HILL GRAVELS.—In SCIENCE-GOSSIP
of April, 1899, I notice that the plateau or high-
level gravel of Westow Hill, Upper Norwood. in
Surrey, is referred to on the authority of Prestwich
as being but little more than a patch. Recent excavations have shown that the gravel is here of considerable extent, and in a direction north to south
along Church Road it cannot be far short of a mile
in length. The following observations were made
when electric wires were laid from the top of
South Norwood Hill along the Church Road to
Upper Norwood. The excavation in which the
wires were laid was about four feet deep, but in
some places somewhat deeper. From Grange Hill
to All Saints' Church there was little or no "soil,"
the road-material resting upon a sandy loam.
Midway between the two was a peculiar patch of
pinkish-coloured loam with rounded pebbles. On
turning down Church Road orange gravelly loam
commenced to prevail, but opposite to the entrance
to the churchyard a quantity of green earth was
thrown up, containing rounded green-coated flint
pebbles. On passing the exit of Upper Beulah
Hill the orange gravelly loam again set in, and
continued, resembling Croydon gravel. This I have
at other times noticed in the same road when
excavations have been made for drains, etc., as far,
at least, as Westow Street. As the work proceeded
in Church Road, a fine dark yellow clayey sand,
which from "Northwood" to "St. Ives" contained
a large number of rounded flint pebbles. From
"St. Ives" to "Windermere" the trench was in a
pure sand, with scarcely any clayey admixture at
all. The cutting continued along Westow Street to
Westow Hill, being through an impervious clayey
sand, with here and there stones and pebbles.—
E. A. Martin.

ASTRONOMY

CONDUCTED BY F. C. DENNETT.

1900	Rises.	Sets.	Position at Noon.	
			R.A.	Dec.
Sept.	h.m.	h.m.	h.m.	° '
Sun .. 7 ..	5.23 a.m. ..	6.33 p.m. ..	11.2 ..	6.9 N.
17 ..	5.39 ..	6.10 ..	11.38 ..	2.20 N.
27 ..	5.55 ..	5.47 ..	12.14 ..	1.33 S.

	Rises.	Souths.	Sets.	Age at Noon.
Sept.	h.m.	h.m.	h.m.	d. h.m.
Moon .. 7 ..	5.19 p.m. ..	10.43 p.m. ..	2.55 a.m. ..	13 8.7
17 ..	— ..	7.3 a.m. ..	3.2 p.m. ..	23 8.7
27 ..	9.51 a.m. ..	2.25 p.m. ..	6.53 p.m. ..	3 16.3

		Souths.	Semi-Diameter.	Position at Noon.	
				R.A.	Dec.
	Sept.	h.m.		h.m.	° '
Mercury.. .. 7 ..		11.39 a.m. ..	2·5" ..	10.44 ..	10.0 N.
17 ..		0.9 p.m. ..	2·4" ..	11.52 ..	2.16
27 ..		0.31 p.m. ..	2.4" ..	12.54 ..	5.27 S.
Venus .. .. 7 ..		8.56 a.m. ..	13·4" ..	8.1 ..	17.21 N.
17 ..		8.56 a.m. ..	11·9" ..	8.40 ..	16.13
27 ..		8.58 a.m. ..	10·8" ..	9.21 ..	14.18
Mars .. .. 17 ..		8.2 a.m. ..	2·6" ..	7.46 ..	22.4 N.
Jupiter .. .. 17 ..		4.27 p.m. ..	16·4" ..	16.11 ..	20.34 S.
Saturn .. .. 17 ..		6.9 p.m. ..	7·7" ..	17.54 ..	22.39 S.
Uranus .. .. 17 ..		4.44 p.m. ..	1·8" ..	16.28 ..	21.46 S.
Neptune .. .. 17 ..		6.10 a.m. ..	1·3" ..	5.56 ..	22.14 N.

MOON'S PHASES.

	h.m.		h.m.
1st Qr. .. Sept. 2 ..	7.56 a.m.	Full .. Sept. 9 ..	5.6 a.m.
3rd Qr. .. " 15 ..	8.57 p.m.	New .. " 23 ..	7.57 p.m.

In perigee September 9th at 6 p.m. ; in apogee on 24th at 4 a.m.

METEORS.

	h.m.	°
Aug. 21–Sept. 21 .. ε Perseids Radiant R.A.	4.8 Dec. 37 N.
" 25–Sept. 22 .. γ Pegasids "	" 0.20 " 10 N.
Sept. 1–7 .. .. ν Andromedids "	" 23 36 " 38 N.
" 7–24 .. .. ε Taurids.. "	" 4.16 " 22 N.

CONJUNCTIONS OF PLANETS WITH THE MOON.

				° '
Sept. 1 ..	Jupiter ..	8 p.m. ..	Planet 0.51 N.
" 3 ..	Saturn ..	8 p.m. ..	" 1.5 S.
" 18 ..	Mars⊙ ..	Noon ..	" 4.52 N.
" 19 ..	Venus⊙† ..	5 p.m. ..	" 2.50 N.
" 24 ..	Mercury† ..	8 p.m. ..	" 4.59 N.
" 29 ..	Jupiter⊙† ..	9 a.m. ..	" 0.13 N.

⊙ Daylight. † Below English horizon.

OCCULTATIONS.

		Magni-	Dis-	Angle from	Re-	Angle from
Sept.	Object.	tude.	appears.	Vertex.	appears.	Vertex.
			h.m.	°	h.m.	°
3 ..	Saturn ..	— ..	7.16 p.m. ..	126 ..	8.11 p.m. ..	206
4 ..	ξ¹ Sagittarii	5·0 ..	7.35 p.m. ..	66 ..	8.50 p.m. ..	263
13 ..	ρ² Arietis	5·5 ..	4.27 a.m. ..	59 ..	5.40 a.m. ..	226
13 ..	13 Tauri ..	5·4 ..	9.43 p.m. ..	99 ..	10.34 p.m. ..	314

THE SUN should be watched for the spots that sometimes become visible.

MERCURY is a morning star in Leo at the beginning of the month, and an evening star in Virgo at the end, being in superior conjunction with the sun at 5 p.m. on September 13th.

VENUS reaches its greatest elongation, 46° west, at 6 p.m. on 17th, and so is a magnificent object in the morning sky, rising about 1.30 a.m. all the month.

MARS is a morning star rising near midnight all the month, but his tiny disc makes him a poor object for observation.

JUPITER may be looked for as soon as it is sufficiently dark.

SATURN should be observed as soon after sunset as possible. The occultation on September 3rd should be watched. (See "Occultations" on this page.)

URANUS is now too near the sun for satisfactory observation.

NEPTUNE may be found almost in the same straight line with μ and η Geminorum, and produced about as far again westward as the distance between them.

THE LEEDS ASTRONOMICAL SOCIETY. — The "Journal and Transactions" of this useful Society during the year 1899 has been sent to us. Some of the papers are very full of interest and instruction. "The Planet Mercury as a View Point," by the President, Mr. C. T. Whitmell ; "Jeremiah Horrocks and the Transit of Venus," illustrated by a photographic plate, by Mr. A. Dodgson ; and "Astronomical Theories relating to Stonehenge," with two photographic views, by Mr. Washington Teasdale, may be specially mentioned.

METEORS of striking appearance have recently been seen. On July 17th, at 8.47 p.m., in bright twilight, too bright to exactly describe its path, a fireball, thought by Mr. Denning to be a Scorpiid, having a double head, was observed at many places in the North of England. The trail it left was visible for three-quarters of an hour. Another brilliant meteor was observed from Slough, Croydon, &c., at 11.33 on July 18th. and from its extraordinary motion was described by Professor A. S. Herschel as "about the queerest he ever saw."

COMETS.—Giacobini's comet at the beginning of September will be situated in the southern portion of Hercules to the north of α Ophiuchi, but too faint to be seen with common telescopes. A comet was almost simultaneously discovered on the morning of July 24 by M. Borelly, of Marseilles, and Mr. W. R. Brooks, of the Smith Observatory, Geneva, N.Y. At that time it was said to be a beautiful telescopic object, having a stellar nucleus, about 8th magnitude, and a small broad tail. It was discovered closely east of the 5th magnitude star, 38 Arietis, but had travelled rapidly north through Perseus to Camelopardus by August 15th, near the star γ, and is said to be decreasing in brightness. According to the elements calculated by Herr Möller, of Kiel, the comet was nearest the earth about the end of July, and perihelion was passed on August 3rd. It is known as 1900, b. In the middle of August the nucleus was readily visible with a telescope of less than 2 inches aperture. With 10½ inches aperture Mr. Brooks finds it to be duplex, but not clearly separated.

JUPITER IN 1900.—There is a white spot on the northern side of the north tropical belt, bedding deeply into, if not actually breaking through the dark belt. The Rev. T. E. R. Phillips first saw the object in 1899, on January 26th, when it had a longitude of 85·3°, System II. It had, on July 28th, 1900, a longitude of 325·9°, so that in eighteen months it has lost about 120° of longitude, relatively to the zero meridian of System II. This would give a rotation period of 9h. 55m. 50·6s. The object is shown well by Wray's 3 in. SCIENCE-GOSSIP Telescope.

CONDUCTED BY HAROLD M. READ, F.C.S.

THE SOCIETY OF CHEMICAL INDUSTRY.—The recent annual meeting of the Society of Chemical Industry was more than usually interesting, in that, for the first time in its history, the members were addressed by a President from the United States. In the selection of Professor Chandler of New York the unity of men of science throughout the world is again exemplified, and the Society has not only done honour to itself, but has given to English members the opportunity of meeting one who is amongst the leaders of chemical thought across the Atlantic. The subject of Professor Chandler's address was the development and present position of Chemistry in the United States, from both its scientific and technical aspects. It would be unfair to attempt to give an abridgment of an address, the salient points of which will long remain in the memory of those who were so fortunate as to hear it. One of the most striking contrasts to which allusion was made was that while, when Professor Chandler commenced the study of chemistry, there were only four schools of science in the United States, to-day they are so numerous he could not give the precise number. The Society finished its annual meeting with a trip to Paris, one of the most interesting features of the excursion being a visit to the factory of the Société Anonyme Anglo-Français Parfums Parfait at Courbevoie, where the members saw the application of ozone in the synthetic preparation of such perfumes as vanillin, heliotropin, coumarin, and other scents.

GLASS PAVING BLOCKS.—During recent years an industry of considerable importance has sprung up in France in the manufacture of paving blocks, imitation granite, etc., from waste glass. Réaumur pointed out in 1827 that when fragments of glass are softened by heat and then compressed into a mass the glass undergoes considerable change in its physical properties, becoming devitrified and opaque, also showing a marked increase in hardness and capacity for resisting shock or crushing. After seventy years Réaumur's discoveries are being utilised, and two factories in France are now engaged in making this new material. It has already been tried for paving at Nice, and is said to give satisfactory results. The applications of such a substance are so varied that we may, before long, see a huge development of the industry.

OSTWALD'S "FOUNDATIONS."—We are extremely glad to notice the appearance of a second edition of Ostwald's "Foundations of Analytical Chemistry." It is only a few years since the first edition was published, and we may take the present edition as an earnest of the fact that the scientific treatment of analytical chemistry is being more and more recognised. A striking feature of present-day analysis—we refer to that carried out by the ordinary student—is the slipshod and often slovenly manner

in which the work is done, and salts "spotted." There is, no doubt, a tendency in the half-trained mind to endeavour to escape the stern morals which the true spirit of scientific work exacts: any means of repressing this backward tendency is to be heartily welcomed. Ostwald's "Foundations" certainly does not appeal direct to the students fresh to chemistry, but, none the less, its indirect effect through the medium of a teacher is bound to make itself evident. Though we do not consider the ionic hypothesis as the be-all and end-all of chemistry, we cannot but draw attention to the vast fields opened up by this new treatment of analytical chemistry. The book is delightful reading, the equations involved in the theoretical part are extremely simple, and the treatment of what we may call the "practical" part is very thorough. The last chapter of the book is one of the most interesting. In it Ostwald draws attention to the great discrepancy which holds in the stating of analyses of inorganic compounds, as compared with organic ones; while, in the latter case, the analyses are given in absolute percentages of the elements present; in the former the analyst endeavours to group his elements into proximate constituents, in accordance with the old dualistic hypothesis of Berzelius.

EXAMINATION OF FLESH FOODS. — "Flesh Foods," their chemical, microscopical, and bacteriological examination, by C. Ainsworth Mitchell, B.A., F.C.S. (336 pages with illustrations and frontispiece. London: Charles Griffin & Co. Limited. 10s. 6d.), has been recently published. In the preface to this work the author states that "it has been his endeavour to collect and summarise records of investigation scattered throughout English and foreign scientific books and periodicals," and the present work is evidence how well he has succeeded in his task. It is a veritable dictionary, and a work which those who are engaged on the subject will heartily welcome. Not the least pleasant feature is the enormous number of references to original papers. The arrangement of the subject matter and the printing are both excellent.

LIQUID FUEL.—In consequence of the high price of coal, the French Society of Civil Engineers have been again discussing the calorific values of various petroleums. They publish the following table of proportions of carbon, hydrogen, and oxygen in various hydrocarbons, the calories being per kg :

	C	H	O	Calorific power
Light petroleum oil — American	86·894	13·107	—	10,913
Refined	85·491	14·216	0·293	11,047
Petroleum spirit	80·583	15·101	4·316	11,086
Crude petroleum	83·012	13·889	3·099	11,094
Light oil from Baku	86·700	12·944	—	10,843
Petroleum from the Caucasus	84·906	11·636	—	10,328
Ozokerite from Boryslaw	83·510	14·440	—	11,168

Recent investigations show the necessity for exceeding care in their conduct. To obtain the theoretical calorific power and the highest temperature of combustion, the supply of air furnishing the precise amount of oxygen is a necessity, as is, also, the most intimate possible contact between the atmosphere and the fuel.

A HYBRID VIOLET.—At Prestwood, in the vicinity of Great Missenden, Bucks, is a small remnant of common-land, where grows, among the ordinary heath-land plants, the dog violet (*Viola ericetorum*) and the common wood violet (*V. riviniana*). There is nothing unusual in this circumstance, nor in the further fact that here these two violets hybridise with each other. What, however, is worthy of note, is the plentifulness of the hybrid form, and the unusual luxuriance it here attains, forming great patches many feet in diameter. These large patches present a singular appearance, as the apetalous flowers, by means of which violets are chiefly propagated by seed, are invariably sterile, and in their brown shrivelled state contrast greatly with the dark-green foliage. —*C. E. Britton*, 35 *Dugdale Street, Camberwell, S.E.*

VAGARIES OF LIGHTNING.—An oak-tree, about 40 feet high and with a stem of 32 inches diameter, was struck by lightning near here on July 27th last between 3 and 4 p.m. The bark of the stem was almost completely stripped off below the branches, and scattered in fragments round the tree. The largest fragment, 6 feet × 1 foot, lay near the tree, but some pieces were hurled to a distance. I weighed three portions, and paced the distances. No. 1 weighed 6¼ ounces, and was thrown thirty-five steps; No. 2 weighed 1¼ ounce, and went forty paces; and No. 3, which weighed 2¼ ounces, reached twenty-six paces. The trunk shows a split completely through, and the lightning entered the ground between two roots, its passage being evidenced by the split. On August 15th, 1899, at Little Stukeley, Hunts, on the Manor Farm, during a heavy thunderstorm, the electric fluid struck a tree (an elm, I think) and killed a fine young horse which was sheltering underneath. The tree was split down to the point where the lightning entered. Below that point it was uninjured. My mother used to relate two curious facts that happened in the "fifties," in Bermuda. One was the lightning entering her kitchen and passing along the dresser-hooks, leaving only the handles of the jugs hanging on the hooks. The other was more serious, and resulted in the death of the David Island ferryman. The electric fluid passed down one leg and tore the iron nails out of his boot. His little son, or grandson, though also in the boat, was uninjured.—(*Rev.*) *R. Ashington Bullen, F.L.S., F.G.S., Axeland, Surrey.*

SMALL DUCKWEED IN FLOWER.—As the production of flowers by the duckweeds is reputedly rare, I venture to mention that I was out last month in the neighbourhood of Norton Heath, Essex, with a friend, who was so fortunate as to find *Lemna minor* in flower. A small quantity of duckweed, brought from the roadside pond where we found it flowering, and placed on the surface of water in an earthenware basin, developed flowers for several days afterwards. If no flowers were visible, it sufficed to place the duckweed in the sunlight for a few hours, when the minute flowers would appear as whitish specks at the edges of the fronds. I suspect that the flowers of duckweeds are by no means so rare as they are generally considered.—*C. E. Britton, 35 Dugdale Street, Camberwell, London.*

ABNORMAL EQUISETA.—Referring to Mr. Flatters' inquiry last month (*ante* p. 96), I spent some time this spring in collecting fertile heads of *Equisetum maximum* for a friend. Amongst them were a few with divided spikes; but, as far as I remember, not more than two spikelets.—(*Col.*) *H. J. O. Walker, Lee Ford, Budleigh Salterton.*

ABNORMAL GERMINATION OF LEMON.—A few weeks back there was brought under my notice an instance of a lemon, perfectly sound and healthy in appearance, which, on being cut open, was found to contain numerous vigorously-growing seedlings. Germination had evidently taken place some considerable time previous, as the root was well developed, and the same was the case with the young stem. The growing organs had forced their way into the hollow axis of the lemon, and along this the roots were growing. I have never seen before nor read of an occurrence of this kind affecting lemons, and do not know whether this is very unusual or not. These seedlings are now growing in a large pot protected by a bell-glass.—*C. E. Britton, 35 Dugdale Street, Camberwell.*

DEILEPHILA LIVORNICA.—A specimen of this rare hawk-moth was captured at rest upon a garden wall in the village of Offenham, near Evesham, on Sunday, April 29th, 1900, by L. S. Smith, Esq., who brought it to me alive two days afterwards. When first taken it appeared to be quite freshly emerged from the pupa, and although two days' struggling with a pin through it has rubbed off some of the beauty, it is still a very fair cabinet specimen. I made inquiry of Mr Smith as to whether any plant-roots had been imported from abroad amongst which the pupa might have been concealed; but he says nothing of this sort has been introduced into the garden, so that I think this must be a genuine British example.—*T. E. Doeg, Evesham, May 4, 1900.*

UNUSUAL SITE FOR SWALLOWS' NEST.—Replying to the note by your correspondent, Lady Farrent, in SCIENCE-GOSSIP of August (*ante* p. 95), with regard to the building of swallows in occupied rooms, I have pleasure in stating that many years ago my wife, when at school in France at Ste. Foy la Grande, Gironde, occupied, along with quite a number of other young ladies, a large dormitory, the windows of which were always left open, thus giving the swallows fine opportunity of ingress and egress. The roof not being underdrawn, besides making a picturesque display of rafters, yielded many a tempting angle dear to the graceful little migrants; whose return each spring was looked for with greatest interest. The young ones of the previous year that had been decorated with the blue ribbon were recognised among the welcome guests, and who now, with much fussy twittering, at once took their part in building or repairing the habitations for the generation soon to be. There was probably less fear manifested by the swallows than by several of the young ladies when they bethought them of the sad fate of Tobias as recorded in the Apocrypha.—*Samuel Howarth, 26 Grange Crescent, Sheffield.*

TENACITY OF LIFE IN ARGASIDAE.—With reference to the remarkable tenacity of life shown by some of the Argasidae. it may be of interest to note that I have now in my possession a living specimen of *Ornithodorus savignyi* which has certainly survived without visible means of subsistence for a period of at least 19 months. It was sent to me by post from Cape Colony in January last, the sender informing me that it had not, to his knowledge. had any opportunity of feeding during the previous 12 months, and it has since been kept in the corked glass tube in which it travelled. When undisturbed it remains in a quiescent and apparently torpid condition, but resumes its normal activity as soon as it feels the warmth communicated to the tube when this is handled.—*R. T. Lewis, 4 Lyndhurst Villas, Ealing. W.*

TELESCOPIC EXAMINATION OF INSECTS.—Every student of natural history must have experienced the difficulty which arises when attempting to make close observations of the movements and habits of insects under natural conditions. Any approach to them which is sufficiently near to be of much service to the unassisted eye, and certainly the use of a hand-lens, has the instant effect of causing them to suspend their ordinary avocations and to be upon the alert with all their instincts of self-preservation fully armed. It is easy, of course, to examine a captured insect under a lens, and in this way to make out the details of its structure ; but observations of this kind, though useful as far as they go, convey but little information as to the ways of "insects at home." I have, however, myself derived so much instruction and enjoyment from the use of a small telescope for this purpose that I venture to recommend it to others of similar tastes as the best means yet adopted. The instrument I use is an achromatic of ⅜-inch aperture, giving perfect definition of objects at a distance of 3 feet over a field of 6 inches in diameter, with a magnifying power of 6½ times linear, or about 40 times superficial. It measures, when closed, 6½ inches, weighs only 3½ oz.. and can be carried in the pocket without appreciable increase of one's burden. With such a glass it is possible to watch, without disturbance. the busy proceedings of the wood ants, the actions of butterflies and bees upon flowers, the attacks of ichneumons upon aphides, etc., the mode of stridulation in grasshoppers, and the ways of numberless other insects, which could not be observed if once the individuals became aware of the presence of an intruder. At the distance of a yard, an object appears magnified considerably beyond its natural size, and is as distinctly seen as if under a hand-lens of 6 inches focus. If when in the country a higher magnifying power is required for close examination. the draw-tube of the telescope is readily available for the purpose, being practically a compound microscope of about 1 inch focus.—*R. T. Lewis, 4 Lyndhurst Villas, Ealing, W.*

NOTICES OF SOCIETIES.

*Ordinary meetings are marked †, excursions * ; names of persons following excursions are of Conductors. Lantern Illustrations §.*

NORTH LONDON NATURAL HISTORY SOCIETY.
Sept. 6.—†'Fruits and Seeds on their Travels." H. W. S. Worsley-Benison, F.L.S.
,, 15.—*Epping Forest. S. Austin.
,, 20.—†" The Tree in its Relation to Primitive Thought." Mrs. H. M. Halliday.

NOTICES TO CORRESPONDENTS.

TO CORRESPONDENTS AND EXCHANGERS.—SCIENCE-GOSSIP is published on the 25th of each month. All notes or other communications should reach us not later than the 18th of the month for insertion in the following number. No communications can be inserted or noticed without full name and address of writer. Notices of changes of address admitted free.

EDITORIAL COMMUNICATIONS, articles, books for review, instruments for notice, specimens for identification, &c., to be addressed to JOHN T. CARRINGTON, 110 Strand, London, W.C.

SUBSCRIPTIONS.—The volumes of SCIENCE-GOSSIP begin with the June numbers, but Subscriptions may commence with any number, at the rate of 6s. 6d. for twelve months (including postage), and should be remitted to the Office, 110 Strand, London, W.C.

NOTICE.—Contributors are requested to strictly observe the following rules. All contributions must be *clearly* written on one side of the paper only. Words intended to be printed in *italics* should be marked under with a single line. Generic names must be given in full, excepting where used immediately before. Capitals may only be used for generic, and not specific names. Scientific names and names of places to be written in round hand.

THE Editor will be pleased to answer questions and name specimens through the Correspondence column of the magazine. Specimens. in good condition, of not more than three species to be sent at one time, *carriage paid.* Duplicates only to be sent, which will not be returned. The specimens must have identifying numbers attached, together with locality, date and particulars of capture.

THE Editor is not responsible for unused MSS., neither can he undertake to return them unless accompanied with stamps for return postage.

CHANGE OF ADDRESS.

WILFRED MARK WEBB, F.L.S., from Hammersmith to "Oostock," Campbell Road, Hanwell, W.

EXCHANGES.

NOTICE.—Exchanges extending to thirty words (including name and address) admitted free, but additional words must be prepaid at the rate of threepence for every seven words or less.

MONSIEUR ROUSSEAU, La Mazurie, par Aizenay, Vendée, France, offers recent shells. fossils, minerals (including bertrandite), rocks and plants in exchange for similar objects.

FOSSILS AND SHELLS.—The Rev. John Hawell, M.A, F.G.S., Ingleby Greenbow Vicarage, Middlesbrough, offers British and foreign fossils and recent shells for other fossils.

RUSKIN'S "Monera Pulveris," as new, exchanged for "Story of Our Earth," or offers.—W. D. Nelson, 22 Kirk-wynd, Kirkcaldy.

BRITISH DRAGON-FLIES.—Wanted, all species of British dragon-flies ; state name if possible and condition. Name requirements. —H. D Gower, 55 Benson Road, Croydon.

CONTENTS.

SIR JOHN BENNET LAWES, BART.

BY the death, on August 31st, of Sir John Bennet Lawes, F.R.S., LL.D., D.C.L., D.Sc., F.C.S., first baronet, the world at large is poorer by the loss of an unselfish investigator in the science of the chemistry of agriculture, and one with world-wide reputation. The experimental farm at Rothamsted, in Hertfordshire, is known wherever agricultural chemistry is in practice. Sir John Lawes was the eldest son of Mr. John Bennet Lawes, of Rothamsted, where he was born on December 28, 1814, and succeeded to the estate at the early age of eight years. Educated at Eton, in 1832 he went to Brasenose College, Oxford, where he studied for three years. Even then the chemistry of vegetation had attracted his attention, and on his return to Rothamsted he began what afterwards became his life's work. In the early stages he secured the services of a young chemist named Dobson, their experiments being conducted in flower-pots. Mr. Lawes then applied the results of these investigations to fields on his farm and, they proving successful, patented the process, and commenced to manufacture at Deptford, and later at Barking, on premises occupying no less than 100 acres of land, what became known as "super-phosphates" for artificial manure. This industry he fostered until 1872, when he disposed, for a sum of 300,000*l.*, of his interest in the business, and confined his whole attention to the researches in which he was so eminently successful. It is interesting to note that the type of trade he thus founded has grown to be worth upwards of a couple of millions sterling each year in this country alone. In 1843 Dr. Joseph Henry Gilbert joined Lawes as an assistant, the result being that he, like Sir John, has been honoured for the sake of science, and has become the celebrated Sir Henry Gilbert, Director of the Rothamsted Laboratory. It is gratifying to find that that institution will be continued in future under provision made by Sir John Lawes, he having placed at the disposal of trustees the laboratory and certain other premises, with an endowment fund of 100,000*l.* This occurred in 1889, so that the splendid work inaugurated by Sir John will not be

Photo by]   [*Barraud, London*

SIR JOHN BENNET LAWES, BART., F.R.S.

interrupted by his lamentable death. One clause in the trust deed instructs the committee of management to select and despatch, from time to time, a qualified lecturer to the United States of America to spread in that country a knowledge of the successful results attained at Rothamsted.

Sir John's name is so intimately associated with Rothamsted that it is not generally known that he was at one time one of the most important manufacturers of chemicals in this country, he having owned and most successfully conducted a factory for producing tartaric and citric acid at Millwall, on the river Thames.

We believe that Rothamsted was the pioneer station for the systematic investigation of agricultural chemistry and modern experimental farming. It still maintains its position as one of the more important in the world. The fields occupied by the investigations occupy about forty acres of land. The experiments have not been entirely confined to the vegetation alone, but further, on its effect on profitable feeding of farm animals.

As might be expected from a businesslike and systematic mind such as that of Lawes, most careful record was kept of the work done from the early days of his undertaking. The first published report appeared in 1847; the subsequent issues now reaching nine volumes. Sir John was more a man of practice than theory, and consequently his literary remains are chiefly in the form of papers on the science and practice of agriculture than long treatises. These scattered papers number over 120, and from 1847 to a recent date were collected and bound in three quarto and six octavo volumes. These Sir John presented to various institutions throughout the world. He was a frequent contributor to the agricultural newspapers, and his communications always commanded deserved attention on account of their practical value. Such simple but effective advice was given in his articles that they were received with respect by even the most unprogressive country gentleman or farmer.

Such a brain as that of Sir John Lawes was bound in time to receive public recognition. This took the form of honorary degrees given by Universities. and his election as a F.R.S. in 1854. In 1877 Edinburgh conferred the LL.D.. in 1892 Oxford its D.C.L., and in 1894 Cambridge the honorary D.Sc. It was in 1882 he received his baronetcy. and in 1893 the station, on commemorating its jubilee, was honoured by the knighthood of Dr. J. H. Gilbert.

As we have said, the Rothamsted Agricultural

Station still continues at work. It will be managed by a committee nominated by the Royal Society. the Royal Agricultural Society, the Chemical Society. and the Linnean Society.

Sir John's personal character was of the most amiable and enthusiastic ; but his enthusiasm was tempered with shrewd businesslike common sense. In the baronetcy he is succeeded by his son. a well-known sculptor.

NOTES ON SPINNING ANIMALS.

By H. Wallis Kew.

(*Continued from page* 78.)

IV. Snares of Insect-Larvae.

IT has often been said that spiders are the only animals capable of spinning snares for the capture of prey ; and it is certainly true that there is no natural object in the world making even a tolerably near approach to the spider's perfect snare. A few other animals, however, are known to spin structures to help them in catching their prey ; and among these are certain larvae of caddis-flies of the family Hydropsychidae.

The Hydropsychid larvae generally live in running water, and are sometimes found in places where the current is very rapid. Unlike most caddis-worms, they are believed to be mainly predaceous. (¹) The cases in which they live are fixed, not portable like many of the common caddis-cases of our ponds ; and some are remarkable from the fact that they have at the mouth a raised net of silken meshes, small, but of good size relatively to the rest of the abode. This net almost certainly has the function of arresting edible matters, living insect-larvae, etc., chiefly those carried down stream by the current.

Miss C. H. Clarke (whose observations were communicated to the Boston Society of Natural History by Hagen in 1882) has described cases and nets of this kind—the work of larvae of Hydropsyche—found in abundance in swift streams near Boston. Mass. The structures varied considerably, but the typical form was that of a tunnel, without basal wall, loosely attached by its edges to a stone or other object. At the mouth of the tunnel, always facing the current. was a vertical framework with a net stretched across it. The tunnel was usually about half an inch long. and composed of sand or bits of plants. The framework of the little net, formed of vegetable bits, was occasionally stayed or held in position by silken cords stretching from it to suitable points on the stone. Sometimes it had the form of a simple arch. at others of a complete ring, and it was stiff enough to

retain its position when removed from the water. In a certain stream, where the stones were covered with mud, leaves, and rubbish, large communities of these larvae were observed. Looking down upon the stones, numbers of dark holes might be seen, facing the current, often in rows. side by side, stretching obliquely across the stone ; and when removed for examination. the delicate net. supported by its framework, could be observed across each hole. The accompanying illustration is a copy of that given by Miss Clarke. The larvae are strong little creatures, and Miss Clarke concludes that the use of the net is for catching food. It was obvious that without wholly leaving its house the creature could remove from the net anything edible lodged there by the current (²).

Similar structures, also the work of larvae of Hydropsyche, have been described by Howard, who found them in numbers in a swift stream near Washington. They occurred on stones tilted so as to bring a portion of the surface close to the top of the water, and were placed, preferably, at the edge of slight depressions, so that the tubular portion, or case proper. was protected from the full force of the current. The tube was strong, covered with particles of leaves and twigs, and open at either end. It was furnished, anteriorly, with a broad funnel-shaped expansion, woven in wide meshes with strong silk, and supported at the sides and top by bits of twigs and small portions of the stems of water plants. The structure varied in size ; the mouth of the funnel, which in every case was nearly at right angles with the tube, was in some instances not more than 3 mm. in diameter, while

---

(1) M'Lachlan, " Trichoptera of the European Fauna " (1874–1880), p. 349.

(2) Clarke, " Proc. Boston Soc. of Nat. Hist.," xxii. (1882), pp. 67–71 : and " Psyche," vi. (1891), p. 157. A figure, natural size, accompanying the latter paper, gives the impression of a structure somewhat less definite than that shown in the enlarged figure here copied : and the net (compared with the size of the case) hardly appears so large. A figure given by Sharp (" Camb. Nat. Hist.," v., 1895, p. 483)—modified from a re-drawing of Clarke's enlarged figure by Riley (" Report U.S. Department of Agriculture," 1886, pl. ix., fig. 5)—shows a rather large net of wide meshes, and by an oversight it is not stated to be enlarged.

in others it reached 10 mm. On the surface of a stone about 18 inches across, 166 of these cases were counted. In the same stream, larvae of *Simulium* were abundant; and as they were found in plenty on the stones on which the cases of the Hydropsyche occurred, Howard concludes that they furnished the principal food of the latter creature. He thinks it certain that numbers of the larvae of the Dipteron would be washed into the funnel-shaped nets; and the owners of the nets, ensconced within, appeared to be waiting for prey to be thus brought to them [3]. Further, on *Simulium*-covered rocks at Ithaca, N.Y., Howard again found nets and tubes of Hydropsyche in numbers. The nets differed somewhat from those just described, and the observer supposes that they were the work of a different species [4].

Previously to the publication of these facts, Fritz Müller had recorded the occurrence of a net-making Hydropsychid-larva in Brazil, and had established for it the genus *Rhyacophylax*. Its cases, he says, are rather rude canals, covered with irregularly interwoven vegetable fibres and each has, at its mouth, a funnel-shaped verandah, covered with a

principally made on specimens in captivity. When a larva is placed in a vessel of water, he says, it at once begins to explore its new quarters, and eventually selects a site for its dwelling. This is made of silken threads, and when completed the structure consists of a tube considerably longer and broader than its occupant, and open at both ends. It is supported and strengthened by a mesh-work of silken threads, which spread out for a considerable distance, and are attached to surrounding objects. From time to time the larva turns in its case, and even leaves it for a short space. Generally, however, it remains quiet inside, apparently on the alert for prey:

"If a Chironomus or other small aquatic larva approaches, it is almost certain to get entangled in the network of silken threads. At once the Caddis in its retreat perceives the presence of a possible victim. The long hairs which cover the body are possibly tactile, and reveal slight disturbances of the silken network. The Plectrocnemia then proceeds warily to determine the cause of the disturbance. Should the Chironomus be entangled near the middle of the tube, the Caddis-worm does not hesitate to bite its way through the side, and its

AN INSECT-LARVA'S SNARE [5].

beautiful silken net. The creatures live in rapid rivulets, and the entrance of the verandah is always directed up stream, "so as to intercept any eatable things brought down by the water." The cases are placed on stones, and a number are generally built close together, so as to form transverse rows. Müller mentions having once seen, on a large stone, about half a dozen parallel rows, of which one alone was composed of some thirty cases [6].

We have, further, interesting notes by T. H. Taylor, on the larva of *Plectrocnemia*, another Hydropsychid, which occurs in swift streams on a stony bed. When a stone was lifted out of the water its under-side was found to be covered with patches of mud from which the larvae emerged and began to crawl about; the patches were held together by a binding substance, and were evidently the retreats of the larvae. Taylor's observations, however, were

jaws very soon quiet the struggles of the prey. There is some resemblance between the snare of the Plectrocnemia and the web of a Spider, but the Plectrocnemia is effectually concealed by the mud which clings to its retreat. In captivity it forms a web which is free from foreign particles, and allows all its manoeuvres to be observed" [7].

The contrivances about the tubes of Hydropsychids now described make the most satisfactory approach to the unique position of the snare-spinning spiders, of which the writer has yet heard. Their function cannot reasonably be doubted, and it is interesting to note that they are unhesitatingly referred to as "snares" by Sharp as well as by Taylor. The threads are probably not viscid; but even in the case of spiders viscid lines occur only in the snares of one or two groups.

The silk-glands of caddis-worms are often of great development; and the spinneret, like that of larvae of Lepidoptera, Diptera, and Hymenoptera, is oral.

(*To be continued.*)

(3) Howard, in Riley, *l.c.*, p. 510.
(4) Howard, "Insect Life," i. (1888), pp. 100, 101.
(5) Case and net of the larva of a caddis-fly (Hydropsyche), enlarged about four diameters. After Clarke, "Proceedings of the Boston Society of Natural History," xxii. (1882), p. 67.
(6) F. Müller, "Trans. Ent. Soc. London," 1879, pp. 131-144.
(7) Taylor, in Miall, "Nat. Hist. of Aquatic Insects," 1895, pp. 265-267.

POND-LIFE IN THE NEW FOREST.

By G. T. HARRIS, F.R.P.S., AND C. D. SOAR, F.R.M.S.

GENERAL POND-LIFE.

PERHAPS no one appreciates a visit to a fresh district more than the naturalist, especially when the change is quite distinct topographically. In such a case he may hope to meet with forms in his particular branch of natural history hitherto unknown to him; and possibly many that exist in his own district only as rarities are here quite common, owing to a much more favourable environment. Another point in favour of visiting various, and especially widely differing localities, is that the worker obtains a sounder knowledge of the geographical distribution of the group he is working upon, and the especial conditions under which particular species are found most freely.

The Easter vacation of 1899 was spent on the Norfolk Broads by the little party of microscopists who annually brave the inclemencies of that early season. Rich and varied is life in the slow-flowing streams and placid expanses of water, whose sole mission now seems that of mirroring the tall rushes and aquatic plants that fringe their sides. The Easter of the present year was spent in the shelter of the New Forest, where patriarchal trees stand close, guarding one from the cold, searching winds of the early English springtime. Brockenhurst was made our headquarters, and, generally speaking, the district around is prolific in ponds within a radius of five miles. These vary in their character, according to their situation, very advantageously to the student of pond-life. Some are deep forest ponds, others roadside duck ponds, and yet others the remunerative heathland ponds, a perfect Elysium for the algologist, with their varied and beautiful desmids, and general life of freshwater algae. Easter is rather too early for abundant pond collecting, especially when preceded by such a cold, dull spring as that of 1900; hence the results obtained on our excursion are not very striking. As, however, it is desirable to know the forms which exist at each period of the year, Easter collecting is not altogether useless; and, again, an even brief record is always useful to succeeding workers. It is a desire that naturalists visiting the New Forest may pay some attention to its varied pond life that has induced us to give in detail our own limited results.

The Rotifera, so attractive a class to the microscopist, seem well represented in the Forest ponds. *Conochilus volvox*, now apparently growing scarce around London, is here in profusion, principally in the clear ponds of the heath lands. A small duck pond near Balmer Green was especially prolific in Rotifers, and contained some particularly fine

Synchaetae. Anuraea aculeata seemed to be the cosmopolitan of the Forest, for it occurred everywhere in profusion. *Euchlanis triquetra* was another very common species, and several species of *Asplanchna* were not infrequently taken, as were several of the genus *Brachionus*. From a pond by the side of the Lymington Road we took *Daphnia* bearing a commensal *Brachionus* in large quantities, probably *Brachionus rubens*. The pond was little better than a concentrated syrup of *Daphnia*, and as such *Daphnia* was quite laden with this particular rotifer, it may be readily imagined that the environment, from some cause or other, was eminently satisfactory to the Brachionae. Among the sedentary rotifers *Flosculeae* were abundant and generally distributed, as was *Melicerta ringens*. *Melicerta conifera* seemed to favour the heath ponds, though it also occurred sparingly in the forest ponds.

Ophrydium versatile was our best find among the Infusoria; it was floating about in the boggy ditches in masses as large as a walnut. The genus *Stentor* was abundantly represented, including the somewhat sporadic *S. niger*. The form most frequent among the Mastigophora was *Dinobryon sertularia*, unless one includes *Volvox globator* in this order. *Volvox* was certainly one of the most frequent forms of pond life in the Forest; its delicate green spherules were to be met in all kinds of water. The Rhizopoda would undoubtedly well repay careful attention in the Forest district, as the ponds seem specially rich in them. *Actinosphaerium eichornii* is most abundant, as also *Actinophrys sol. Arcella vulgaris* and *Difflugia pyriformis* are, of course, abundant; and in various ponds we found the beautiful Heliozoon *Acanthocystis turfacea*.

Hydra vulgaris and *H. viridis* were the representatives of the Hydrozoa, and in a large pond on Beaulieu Heath the orange variety of *H. vulgaris* was very plentiful. This variety is only occasionally met with, and its colour is due, almost certainly, to symbiotic algae, not, as is often imagined, to any specific difference. In this particular pond all the entomostracans were of the same orange tint. The colour was extremely vivid, often approaching scarlet in its intensity. A tube full of deep orange hydra and entomostracans presented a very beautiful appearance.

The Forest ponds were especially rich in entomostracans; but, as our knowledge of them is particularly slender, their specific enumeration must be left until such time as a specialist in this class visits the district. It was our good fortune, however, to take one specimen of the extremely rare

Chirocephalus diaphanus. Indeed, the capture may be looked upon as a re-discovery of this species in England, for it is so long since an individual was taken that its present existence was regarded as doubtful. In vol. xi. No. 66 of "Natural Science" is an interesting note on "The Apparent Disappearance of the British Phyllopods," in which the writer calls attention to the disappearance of *Chirocephalus* with other Phyllopoda. It is noteworthy that the locality for *Artemia salina* was only three or four miles from the spot where we took *Chirocephalus.*

A brief notice of the principal forms of the freshwater algae must suffice to call attention to the richness of the Forest ponds in this direction. As hinted above, the desmids constitute a prominent feature of the flora of these ponds, especially such as are situated on the open heaths; many species of the genera *Closterium, Micrasterias, Euastrum,* and *Hyalotheca* will be found. *Batrachospermum moniliforme* is not uncommon, while the beautiful *Draparnaldia glomerata* may be seen in profusion. Living diatoms form a not unimportant unit in the pond life of the Forest, and the order Confervoideae is largely represented.

Such, excepting the Hydrachnidae, which Mr. Soar describes, is our modest contribution towards an enumeration of the various forms of "pond-life" occurring in the New Forest, a record, perhaps, hardly worth the making in itself, but published as an incentive to other workers to explore the ponds in this district for material for their own particular groups. It may be some groups are poorly represented in the Forest, but even then it is interesting to know that such is the case, especially if one gleans some idea of why the environment is unfavourable to them. On the other hand, our experience goes to show that eminently successful work could be done on many prominent groups. So far the New Forest has been regarded chiefly as an entomologist's ground, but careful collecting may show that it is equally desirable in other directions.

G. T. HARRIS.

HYDRACHNIDAE IN THE NEW FOREST.

The New Forest is the happy hunting-ground of all entomologists, and as we know that the larval stages of the Hydrachnidae spend a great part of their time on the early and adult forms of aquatic insect life, we concluded it would also be a good field for mite-hunting, but we have been disappointed, the number of species found being very small. This can, however, be accounted for more or less, as the ponds, mostly shallow, had the appearance in some cases of having been completely dried up during the drought of last summer. This, of course, would be fatal to all freshwater mites. Then the weather at Easter was cold and windy, so that all mites that could do so had no

doubt hidden themselves in the mud, in the deepest parts of the ponds, where it was impossible to get at them with the ordinary collecting apparatus. If this last suggestion is correct, another visit during the summer months would give a much better result. Anyway, we think it will be as well to give a list of what was found, as it will form a beginning in the knowledge of their local distribution in that part of Hampshire, and the list can be added to from time to time, as the New Forest gets more systematically worked.

Sub-family HYGROBATINAE.—1. *Curvipes fuscatus* Hermann. A number of them were taken from a pond on the Lyndhurst Road, including both males and females. 2. *Hygrobates longipalpis* Hermann. 3. *H. reticulatus* Kramer. 4. *Limnesia histrionica* Hermann. 5. *L. koenikei* Piersig. 6. *Brachypoda versicolor* Müller. 7. *Arrenurus globator* Müll. 8. *A. albata* Müll. 9. *Cochleophorus vernalis* Müll. 10. *Teutonia primaria* Koenike.

Sub-family HYDRYPHANTINAE. 11. *Hydryphantes ruber* de Geer. 12. *H. ruber* var. *prolongata.* 13. *Thyas venusta* Koch. 14. *Thyas* (?).

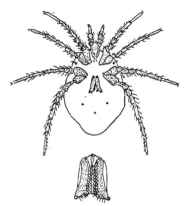

UPPER FIGURE. *Thyas* (?) ♀ Ventral surface.
LOWER   ,,   ,,   ,, ,, Genital plates.

This one we cannot name, so we give a figure. It differs from any *Thyas* we have yet seen. It is of a bright scarlet colour, with straw-yellow legs. Length about 1·50 mm. Mouth organs project beyond anterior margin about 0·25 mm. Only one specimen was found in a small pond at the side of a road near Brockenhurst Park.

Sub-family HYDRACHNINAE. 15. *Hydrachna leegei* Koenike.

Sub-family LIMNOCHARINAE. 16. *Limnochares holosericea* Latreille.

In addition to the above species we took one fine specimen of *Corixa geoffroyi* with larva of an Hydrachna attached to the legs.

C. D. SOAR.

GEOLOGICAL NOTES IN THE ORANGE RIVER COLONY.

By MAJOR B. M. SKINNER, R.A.M.C.

(*Continued from page* 106.)

WORKING downwards from the buff sandstone at southern foot of New Fort Kop, the country slopes gently eastward for some distance, showing at the upper part of the slope sandstones, becoming more muddy at the lower part, and succeeded by shales. These outcrops soon disappear, to be found again in the bed and banks of a donga running between New Fort Kop and Naval Hill, where there is first exposed about 4 feet of grey sandstone, with rounded grey sandstone pebbles; second, a blue-grey shale, very sandy, 2 feet; third, grey thin-bedded sandstone, 1 foot 9 inches, of which two layers were ripple-marked, and others contained worm-tracks; fourth, blue-grey shale varying in shade in different layers, extending downwards some 10 feet, the bottom not being visible. This series gave 45 feet to the top of the buff sandstone.

Proceeding westward from the same level (the buff sandstone) nearly a mile of doleritic débris is crossed, sloping gently west till the "drift" of the Bloemspruit, where the Kimberley road crosses its bed, is reached. This bed consists of soft sandstone with imbedded sandstones at some levels, and is at this spot fifty feet below the above datum; this sandstone extends some distance upstream, being cut across by a dyke, until buried in alluvial sand. It extends a short distance downstream, where it is soon succeeded by shales, which are fairly constant down to the town, occasionally showing a thin sandstone layer, the same condition being visible in the spruit or donga on the Jacobsdal road. Crossing the Kimberley road drift and turning a little south of west, a gentle slope is ascended, grass-covered, without outcrops, almost at the top of which is a small sandstone quarry, 5 or 6 feet deep; at one part large boulders like those at Rustfontein are seen. The top layer of this quarry is sandy shale; the highest point of these strata is 50 feet above the drift.

Returning to the Kimberley road, there is a gentle slope upwards towards the west; at first the road passes across sandy débris, then the red débris of volcanic rock, which is continuous to north of the road, but is mixed with alluvial sand on the south; for the ground takes here an extra upward curve, and, after crossing a dolerite dyke, outcrops of grit and sandstone are met with, the first grit being 140 feet above the buff sandstone, and the highest sandstone in the quarry at the top of the ascent 50 feet higher. Here the grits are very marked. This series to the south-west is cut across by dolerite, continuous through Spitz Kop with the western extension from New Fort Kop. Spitz Kop is evidently the remains of one of the local upbursts which overflowed the neighbouring strata, as may be seen on a hill to its south-west, where sandstone and grit are capped by dolerite. The same condition is seen at Haldon Hill, where a denuded surface of grit, curved at the top, has been covered by dolerite. The grits observed are occasionally very coarse, entirely formed of quartz; above or below some layers is frequently a thin layer of sand with largish fragments of angular quartz. To the north of Spitz Kop are two exposures of stratified rock, both showing a denuded surface of standstone which has been overflowed and baked by the dolerite sheet extending to the north and east, referred to above, and which, when denudation has cut away deep enough, generally shows subjacent stratified rocks which furnish evidence by their denuded condition of having formed a land surface before the upbursts of dolerite, as seen at Spitz Kop, Naval Hill, and elsewhere, occurred. Besides these overflows of dolerite there are, in the lower strata exposed in the dongas, dykes which may or may not have extended upwards to the surface of the then existing land. There are also bosses of the same material causing small local dislocations of the strata from the horizontal, evidently showing that the intrusive material had not sufficient energy to push itself further than the level of some of the lower strata.

On the south of Naval Hill is a succession of alternating shales and sandstones similar to those seen extending upwards from the donga between this hill and New Fort Kop, up to the shale above the white sandstone on the latter hill. The lowest sandstones here show marked current-bedding, with lenticular patches of black shale, sometimes very small, sometimes forming a layer 2 or 3 inches thick and 3 or 4 feet long, but failing to give any evidence of structure. A similar, but less marked, condition is seen in one of the sandstones in the small quarry west of New Fort Kop. This sandstone would be the same as the buff sandstone so often referred to, but the latter shows no sign of current-bedding at its upper part, though it does below; while the corresponding sandstone in the quarry shows in its lower part some boulders.

A short distance north of Deale's Farm is an extensive exposure of sandstones and shales, beginning with shales in a donga, passing up to sandstones, the top one of which shows boulders. One layer shows annelid burrows. These strata extend upwards 160 feet, the lowest corresponding with the shale in the donga east of New Fort Kop.

Piecing together the above fragments, as is shown in fig. 4, the strata appear to arrange themselves into two lithological groups, the lower portion consisting of shales, the upper of sandstones, fine and coarse, with occasional boulder-layers, corresponding with Professor Green's description of the Molteno group of strata ('). The shales pass up conformably, though the transition is marked, into the new era of sandstones; the change, if borne out by fossil evidence, would be extremely interesting. The sandstones at the top give numerous evidences of having formed a land surface before the dolerite overflow took place; this circumstance will prove of interest should opportunity occur of tracing these strata into other localities where Molteno beds are succeeded by others of later date.

Beyond small local disturbances of the strata, their appearance is that of practical horizontality; there may be slight undulations which are not appreciable to the eye in the comparatively small sections seen; there may be faults, but these again are not visible in this locality, and are not indicated by the character of the rocks. Besides the visual impression of horizontality, the rise of country from west to east from 3,873 feet at Enslin to 4,517 feet at Bloemfontein railway station is marked by a gradual change in the character of the stratified rocks from shales to sandstones. The koppies at Enslin showed shales, those in this locality show sandstones; while dolerite once covered the whole expanse. As up to the present no interbedded sheets have been seen, it may perhaps be found that this dolerite cap is the "volcanic" series [2] of the Stormberg Beds, which succeeds the Cave Sandstone, and the Red Beds, these latter strata having disappeared in this. locality; at Enslin the Molteno strata also have disappeared, while the dolerite capping of the hills remains, pointing to denudation, having the same line of action as in the present day, having occurred before the volcanic period. The exact period of this volcanic action remains to be seen.

Bloemfontein, May 31st, 1900.

ON THE NATURE OF LIFE.

By Geoffrey Martin.

(Continued from page 75.)

DR. ALLEN (*ante* page 75) has by no means disproved my assertion that in matter living at ordinary temperatures there is no evidence to show that nitrogen holds a central position. The facts with which he supports this theory may, with a little ingenuity, be twisted in favour of any theory. They support my theory with as much force as they support Dr. Allen's. To vaguely generalise about the activities of nitrogen in different modes of union, and to deduce therefrom the central position of nitrogen in living matter, can hardly be said to be conclusive. Something more definite is required. Indeed, the long strings of linked carbon atoms which are continually being excreted by the living organism tell quite another tale—they indicate a framework of carbon that is continually breaking down. The nitrogen atoms may be—nay, probably are, the centres of molecular break-up in the organism—the points where the linked carbon framework gives way; but this is all that can be said. Indeed, the specific functions he attributes to nitrogen are almost inconceivable from a chemical point of view. For instance, how could an element as feeble as nitrogen—an element which can only with the greatest difficulty be induced to combine with oxygen, or retain its oxygen when combined, ever rob such an element as carbon of its oxygen? An action like this never takes place in ordinary organic chemistry, and very strong evidence must be furnished before we can regard it as occurring in living matter.

In the April number of SCIENCE-GOSSIP (*ante* p. 327) I suggested that living matter was not suddenly created at ordinary temperatures, but has descended in a continuous manner from a time when the world was a white-hot sea. According to this theory, the presence in protoplasm of elements of high atomic weights, such as silicon, phosphorus, arsenic, and so forth, are but the relics of its slow evolution from the molten minerals of the past.

Dr. Allen objects to this view, because he believes "the heat-production of our little world could never have been sufficient to melt rocks at its surface." He quotes the authority of Professor Lapworth. In support of the opposite view— namely, that the world was once a red-hot globe— I refer the reader to one who, at the very least, is an equal authority. I mean, Lord Kelvin. He treats the matter fully in a paper entitled "On the Age of the Earth as an Abode fitted for Life." [3]

Lord Kelvin's first argument is briefly this :— Owing to tidal action the world is continually rotating within a frictional collar of fluid sea. Its speed of rotation is therefore continually diminishing at a known rate. Calculating backwards. we find that 7,200 million years ago the world was rotating twice as fast as present. If the world had solidified at this period, the eccentricity of

(1) Q.J.G.S. vol. xliv. No. 174, p. 246.
(2) *Ibid.* pp. 240, 253.
(3) "Phil. Mag." 1899, vol. i. p. 66.

the earth's shape from that of a perfect sphere would have been far more marked than at present. Even suppose the world to have yielded somewhat as it spun down to its present speed of rotation, yet its shape would not alter to suit the diminished speed—there would always be some "lag." Now, as a matter of fact, the present shape of the world is exactly what it should be were the earth a perfect fluid, rotating at its present speed. Consequently the world must have solidified quite recently—so recently, in fact, that the earth has not had time since to appreciably diminish its speed of rotation.

Lord Kelvin continues: " We may safely conclude that the world was certainly not solid 5,000 million years ago, and was probably not solid 1,000 million years ago."

"The fact that the continents are arranged along meridians rather than in an equatorial belt affords some degree of proof that the solidification took place at a time when the diurnal rotation differed but little from its present value."

Lord Kelvin's second argument is even more conclusive: "The 'Doctrine of Uniformity' in geology, as held by many of the most eminent British geologists, assumes that the earth's surface and upper crust have been nearly as they are now in temperature and other physical qualities during millions of millions of years. *But the heat which we know, by observation, to be now conducted out of the earth yearly* is so great that if the action had been going on with any approach to uniformity for 20,000 million years, the amount of heat lost out of the earth would have been about as much as can heat by 100° C. a quantity of surface rock of 100 times the earth's bulk. This would be more than enough to melt a mass of surface rock equal in bulk to the whole earth. No hypothesis as to chemical action, internal fluidity, effects of pressure at great depth, or possible character of substances in the interior of the earth, possessing the smallest vestige of probability, can justify the supposition that the earth's upper crust has remained nearly as it is, while from the whole, or from any part of the earth, so great a quantity of heat has been lost." "This conclusion suffices to sweep away the whole system of geological and biological speculation demanding an 'inconceivably' great vista of past time, or even a few thousand million years, for the history of life on the earth, and approximate uniformity of plutonic action throughout that time."

Dr. Allen says: " It is not customary nowadays to regard nebulae as masses of hot gas." Is he aware that there is only one theory generally accepted among physicists to account for the origin of nebulae ? namely, that they are produced by the impact of two colliding suns. The vast crash instantly generates a temperature of many hundred million degrees Centigrade, and resolves the colliding bodies into rapidly extending masses

of incandescent vapour, which go to form the nebulae of space. There is an important paper on cosmical evolution in " Phil. Mag." August, 1900, by Prof. Bickerton.

Dr. Allen remarks: " The deoxidation and subsequent oxidation of carbon involve a very great accumulation and dispersion of energy, *whereas the changes in the silicates involve a comparatively small transfer of energy.*" (The italics are mine.) May I ask his authority for this remarkable statement?

The reduction of the silicates involves the expenditure of a very large amount of energy, so large that metallic aluminium will not reduce their silicon, although it reduces most known oxides. Conversely, their formation is attended with a great evolution of energy.

Again, he considers that at high temperatures his so-called " Energy Traffic " would be carried on more readily by other elements than silicon. Is he not aware that carbon and silicon belong to the same family of elements, and most strongly resemble each other? If the lighter element can carry on the function at low temperatures, why not the heavier at high temperatures ?

The universal element at high temperatures is neither phosphorus nor iodine; it is silicon. This element, so passive and inert at ordinary temperatures, awakes to a new life at a white heat. It enters into the most astonishing combinations, and displaces almost all known acids from their combinations with bases. In a word, it becomes the universal element. Bearing in mind this surprising activity, no consideration regarding "the chemistry of the silicates as known to us " need deter us from attributing to silicon " dynamical properties " at a high temperature.

I am doubtful, from Dr. Allen's remarks, whether he believes it to be my opinion that *at ordinary temperatures* such bodies as the silicates are capable of performing vital functions. My theory states that silicon compounds are only capable of acting thus during the " transitional range " of silicon, *i.e.* at a red or white heat.

Again, he considers it sufficient to explain the presence of silica in all living matter by briefly announcing " its chief function is to give rigidity to the framework." This may be so in some of the grasses; but such an explanation will not for an instant hold in the case of living animal protoplasm.

To attribute specific functions to an element is not to explain its presence. Nevertheless, be the function of silica what it may, this in no wise affects the theory that silicon once completely replaced carbon in all living matter, but that at ordinary temperatures it has been completely replaced by carbon and remains merely as an inactive sediment. If this sediment could be put to a useful purpose by imparting rigidity to the frame in some forms of life, the silica would linger longer and be less rapidly eliminated than in other cases where it can serve no such useful purpose ; for example, in animal protoplasm, where only a minute trace remains.

Further, are not bones and shells but the relics of a time when all living matter had an analogous constitution, a lingering-on from generation to generation of conditions which held sway countless million years ago, because the evolution products happen to be useful in imparting rigidity to the frame? Indeed, how else can the calcareous algae and the corals be explained—beings that build up vast reefs in the Indian Archipelago ; or the coccosphaeres, minute spheres of mineral matter, once thought to be mere mineral concretions, but now known to be organisms? Such forms of life flourish in countless millions, yet no theory of life save mine takes the slightest heed of them, or seeks to explain why they are continually depositing mineral matter.

The whole problem of the secretion of mineral matter by living beings is ably explained by supposing the mobile protoplasm of to-day evolved in a continuous manner from the molten minerals of the past.

Nor does this theory depend solely upon the presence of silica in living matter. There are other facts in its favour. For instance, albumen contains a small amount of loosely-bound sulphur, which does not appear to be a very intimate constituent of it. No one knows the function of this sulphur, or can account for its presence. According to my theory, it is simply lingering on, the relics of a time when it almost entirely replaced oxygen in the organism. With the falling temperature of living matter, sulphur was superseded by the lighter and more mobile oxygen, and consequently the sulphur that remains is merely an inert mass in the process of elimination, separating out on account of its heaviness in exactly the same way I have supposed silica to have separated.

Much the same applies to phosphorus in the tissues of the brain and nerves. It has been almost entirely replaced by the lighter and more mobile nitrogen. The small amount of phosphorus present remains merely because it can perform functions of which nitrogen is incapable. Not only is this so, but traces of a still heavier member of the same group of elements—arsenic—have been recently found in certain animals, where it partially replaces the phosphorus in nucleinic matter ([2]).

In these cases, then, we have a whole chain of chemical homologues replacing each other in continually decreasing amounts as they increase in heaviness. Thus:—

Nitrogen   (at. wt. 14), abundant.
Phosphorus (at. wt. 31), less abundant.
Arsenic   (at. wt. 75), minute traces.

[2] Gautier, "Chem. News," March 23rd, 1900.

Oxygen   (at. wt. 16), abundant.
Sulphur   (at. wt. 32), less abundant.
Selenium (at. wt. 79), minute traces, if at all.

Carbon   (at. wt. 12), abundant.
Silicon   (at. wt. 28), traces.

These facts are all strongly in favour of the above theory. and are hard to explain otherwise.

The following words of Gautier are of interest in this connection :—"There is still room for research in every organ [for these rare elements], and, thanks to the most delicate methods, for the different elements that we might consider likely, by being substituted as above for the chemical analogues, to modify the functioning of the organ by reason of certain specific needs. Such would be selenium, in place of sulphur ; negative sulphur, substituted for oxygen ; Cu, Zn, or Mn, replacing iron ; P, As, or even Va itself, playing the part of nitrogen in the more or less complex molecules."

These words, in the light of my theory of evolved protoplasm, are of the utmost significance. Adopting the principle that only those products of evolution were retained which are of use in the animal system, while those which are useless were eliminated with the falling temperature of living matter, we immediately and completely arrive at an explanation of the presence in the organism of all such elements, their minute amounts, and the powerful influence they exercise on the health of the body.

Dr. Allen's theory as it stands is merely of passive interest. It explains no facts, does not involve consequences, and is a self-contained theory. If, however, it is given my slight extension, it immediately becomes pregnant with worldwide consequences, suddenly rolling out the panorama of life from the few million years of Lord Kelvin to countless ages when the world was a surging flood of fire. This at once gives a reason why the vital temperatures of different classes of living beings differ from each other ; for the world and living matter both cooled together until the present temperatures were reached, the living matter continually altering its rate of combustion so as to adjust its temperature to that of the surrounding medium. Nevertheless the temperature of living matter continually lagged slightly behind the temperature of the surrounding medium, in much the same way that a thermometer lags behind the temperature of a cooling liquid.

When modern temperatures were nearly attained, various fragments of living matter began to differentiate themselves from other fragments. Parts of the one primeval organism began to live in great measure in the air, others on land, and a third part kept to the seas. Now it is obvious that life in the air would lose far less heat rapidly than life on land, while those portions of living matter which dwelt in a heat-conducting medium like the sea would quickly take up the continually decreasing surrounding temperatures. By thus assuming that the rates at which varying portions of living matter cooled were different on account of individual surroundings, how easily and naturally we explain all the variations in the vital temperatures of different kinds of beings! Again, the above theory finely explains the steady loss in nervous sensibility in living matter with the falling temperature. A warm-blooded animal has its bodily temperature well within the transitional range of carbon. Its atoms are, therefore, in a finely-balanced condition. The slightest stimulus overthrows this balance and effects a chemical decomposition, which in turn registers itself as a vital sensation. The vital sensations of a warm-blooded animal are therefore very intense. A slight external influence will produce keen pleasure or frightful pain. A cold-blooded animal, however, has a bodily temperature nearer the limit of the transitional range of carbon. Its atoms are, therefore, not so finely balanced as those of a warm-blooded animal. Consequently a greater external influence would be required in the one case than in the other to effect the same amount of chemical decomposition. It follows that the intensity of the sensations of a cold-blooded animal must be less keen than those of a warm-blooded animal. Undoubtedly cold-blooded animals are rapidly approaching in insensibility the lifeless minerals around. The nearer and nearer the bodily temperature of living matter approaches the limit of the transitional range of carbon, the more and more obtuse become their vital sensations. At the transitional limit the eternal breakdown ceases. No stimuli, however great, will awaken in such an animal an animate sensation. It has evolved into a mere mineral or a complex organic compound such as strew the earth in countless millions. Animate nature is therefore only a special case of inanimate nature. From minerals all life evolved; to minerals all life returns.

Animate and inanimate matter are certainly related. Prof. Japp has pointed out that there exists a close connection between the optical properties of compounds and life; while it has been repeatedly emphasised that the synthetic breakdown of such bodies as the sugars is analogous to the process of reproduction. In both cases like begets like.

Indeed, if we study closely any portion of matter, it becomes surprisingly lifelike. Even the inanimate metals are full of crystalline fragments in eternal motion, each growing or diminishing, but never quiescent, full of tumultuous motion like the ever-changing protoplasm. In inanimate matter all this motion runs to waste, but in animate matter there is an agency at work, *directing* it to useful purposes.

13 *Hampton Road, Bristol.*

AN INTRODUCTION TO BRITISH SPIDERS.

By Frank Percy Smith.

(Continued from page 47.)

FAMILY AGELENIDAE.

This is a family of great interest, including several of our largest and best known species. The legs are long and furnished with numerous hairs and spines, the tarsi terminating in three claws. The abdomen is more or less oviform, the spinners being prominent, especially the superior pair, which are usually very long and often consist of two distinct joints. In some genera, each superior spinner is attached to a prominence of the abdomen, and this fact has led to the mistaken idea that they each consist of three joints. Several species included in this family have a general resemblance to the genus *Amaurobius*. They may be readily separated, however, by their possessing no supernumerary spinner, which is always present in the family Dictynidae.

GENUS *ARGYRONETA* LATR.

In this genus the superior spinners are not very long. The falces, especially in the male, are large and powerful.

Argyroneta aquatica Clk.
Length. Male 17 mm., female 10.5 mm.
This spider may be distinguished without difficulty by the fact that it spends the greater part of its life under water. It should be specially noticed that the male is considerably larger than the female.

GENUS *CRYPHOECA* THOR.

The eyes in this genus are in two curved rows, which are almost parallel. The tibiae and metatarsi of the first and second pairs of legs have two longitudinal rows of spines beneath them.

Cryphoeca moerens Cb.
Length of female 2 mm.
The distance between the posterior central eyes is greater than that between a central eye and the lateral adjacent eye. This species is very rare.

Cryphoeca silvicola Bl.
Length. Male 2.5 mm., female 3 mm.
This species may be easily distinguished from *C. moerens* Cb. by the distances between the hind eyes being equal. It is rare and local.

GENUS *COELOTES* Bl.

Eyes in two rows, with the convexity of the curves directed backwards. The anterior row is the shorter. The superior spinners are long and divergent. Maxillae strong, broadest at their extremity. Labium almost oval, truncated at the apex, more than half the length of the maxillae.

(1) This series of articles on British Spiders commenced in SCIENCE-GOSSIP, No. 67, December 1899.

Coelotes atropos Wlk. (*Coelotes saxatilis* Bl.)
Length. Male 10.5 mm., female 12 mm.

The anterior central eyes are distinctly the smallest of the eight. This species, which is not at all common, is liable to be mistaken at first sight for spiders of the genus *Amaurobius ;* but the length of the

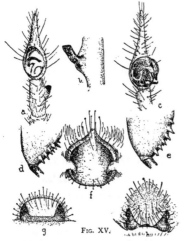

FIG. XV.

Description of fig. 15 : *a.* Palpus of *Tegenaria domestica* ; *b.* Radial apophysis of *T. atrica* ; *c.* Palpus of *T. parietina* : *d.* Denticulation of falces of *T. domestica* ; *e.* Denticulation of falces of *T. parietina* ; *f.* Vulva of *T. atrica* ; *g.* Vulva of *T. parietina* ; *h.* Vulva of *T. domestica*.

superior spinners and the absence of a supernumerary spinner will at once distinguish it.

Coelotes immaculatus Cb.

This extremely rare species may be distinguished from *C. atropos* Wlk. (1) by the reddish-yellow colour of the cephalo-thorax ; (2) by the abdomen being much paler ; (3) by the eyes being almost equal; and (4) by the distance between the hind central eyes being distinctly greater than the distance between one of them and the adjacent lateral eye. The systematic position of this spider is somewhat doubtful.

GENUS *TEGENARIA* LATR.

The cephalo-thorax is of an oval form, the caput being very prominent. The falces are powerful, and armed with teeth opposed to the fang. Superior spinners very long. Maxillae long, and widest at the extremity. Labium short, and somewhat hollowed at the apex. This genus contains the true house-spiders, and also a few which are found in other situations. The snare is a horizontal sheet of web converging to form a tubular retreat. The spiders are very wary, and as they select well-protected situations their dislodgement is usually a matter of some difficulty. If a buzzing fly be held on or near the snare they will usually cautiously emerge, when

their retreat must be skilfully cut off by stopping up the opening with the finger. It is almost impossible to seize the spiders with the forceps, unless some such precaution is taken, as they are very agile and also able to throw off the leg, or even two legs, by which they are held.

Tegenaria atrica Koch.
Length. Male 14 mm., female 18 mm.

The size of this fine spider will usually be sufficient to distinguish it from any others of the genus except *T. parietina* Fourc. and *T. hibernica* Cb. The following characters are valuable :—The legs are not annulated. The sternum is yellowish, with a broad marginal band of a blackish colour. On this band are eight large round yellow spots, one opposite the insertion of each leg. The penultimate joint of each of the superior spinners is of a blackish colour. The form of the radial apophysis and vulva should also be carefully noted (see fig. XV). This species seems to be rather local. I have taken it abundantly in the North London district, but never indoors.

Tegenaria parietina Fourc. (*T. domestica* Bl.)
Length. Male 14 mm., female 18 mm.

In this species the legs are far longer than in *T. atrica* Koch, although the body is seldom much

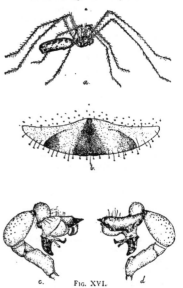

FIG. XVI.

Description of fig. 16 : *a.* Pholcus phalangioides ; *b.* Vulva of *Pholcus phalangioides* ; *c.* and *d.* Palpus of male in two positions.

larger than in that species. The palpus of the male is considerably different in structure, the digital joint being more elongate, and a long curved spine is present upon the front part of the palpal organs. This is a

F 4

true house-spider, is very local, and by no means common. I have received a number of specimens quite recently from Mr. W. H. Terry, of Brentford. In warmer climates this spider lives in the open air.

Tegenaria hibernica Cb.
Length of male from 8 mm. to 16 mm.
I am unable at present to give the characters of this species, not having seen either specimens or description. It seems to be confined to Ireland.

Tegenaria domestica Clk (*T. derhamii* in Spiders of Dorset. *T. civilis* Bl.)
Length. Male 6.5 mm., female 11 mm.
This is the common house-spider, which may be found in disused rooms, cellars, cupboards, and in all such situations. The form of the radial apophysis, the spine on the front of the palpal organs, and the vulva will be quite sufficient to indicate the species with certainty (see fig. XV).

Tegenaria campestris Koch.
Length. Male 5.5 mm., female 7.5 mm.
This spider differs from *T. domestica* Clk. by its smaller size and by its far more distinct markings. The vulva of the female is very typical, and the palpal organs of the male are of an extraordinary form, being highly developed, with several prominent lobes and spines. The digital joint is also very large. This species is usually found out of doors.

Tegenaria pagana Koch.
I had the good fortune to take a female of this species near Birmingham in the autumn of 1899. This is its first, and up to the present time its only, recorded occurrence in Britain. It is a rather small and distinctly marked spider, with the vulva very typical. I am unable to give a detailed description, as the spider is at present in the hands of the Rev. O. P. Cambridge.

Tegenaria cinerea Panz.
Length. Male 6 mm., female 8 mm.
This species is rare. It may be easily distinguished by the pale grey abdomen, which is quite devoid of markings.

GENUS *AGELENA* WALCK.

The spiders in this genus have the legs shorter, in comparison, than *Tegenaria*. The eyes have a different arrangement, the hinder row being strongly curved, the convexity being directed backwards. They are found amongst low bushes and herbage in ditches, spinning a funnel-like snare of a rather substantial character.

Agelena labyrinthica Clk.
Length. Male 10.5 mm., female 12 mm.
Cephalo-thorax reddish-brown, hairy, and with two lateral bands of a somewhat darker colour. Abdomen of a dull brown or blackish-brown colour, sometimes tinged with purple. This purplish tint is more noticeable in young specimens. On the upper side of the abdomen is a series of angular bars of a

paler colour than the general tint. The vulva is distinct and of a rich red-brown colour. The cubital and radial joints of the male palpus have each a prominent apophysis. This is a common, well-known spider, and may be captured as directed for those of the genus *Tegenaria*. Its web is formed amongst grass and low herbage, and should be looked for during the summer and autumn months.

Agelena prompta Bl.
Length. Male 2 mm., female 2.5 mm.
Cephalo-thorax brown, darkest in front. Abdomen dark brown, with a series of pale angular lines. The radial and cubital joints of the male palpus have each an apophysis; and the palpal organs have a long, fine spine attached to and almost encircling them. Very rare.

(To be continued.)

ENIGMAS OF PLANT LIFE.

By DR. P. Q. KEEGAN.

1. EARLY HISTORY OF CHLOROPHYLL.

ACCORDING to a note published in 1773, chlorophyll was first extracted about the year 1750 by alcohol and ether from the tissues by the brothers Guillaume and Hilaire Rouelle. For several years afterwards the green pigment of leaves was considered as analogous to starch, and was called "green fecula." A. Comparetti, in 1791, seems to have been the first to signalise the green granules of chlorophyll. In 1807 Link established the distinction between the starch enclosures and the green pigment, and regarded the latter as a colouring resin; he called it a "resinous colouring matter." In 1817 Pelletier and Caventou named it "chlorophyll"; they extracted it from the marc of herbs with alcohol, and purified the residue of the evaporated solution by washing with hot water. They considered it to be very rich in hydrogen, but free from nitrogen. In 1835 Clamor Marquart obtained it by extracting grass leaves with alcohol, evaporating the alcohol, and separating the chlorophyll from the residue by means of ether. In 1837 Berzelius announced the discovery of three very distinct modifications of chlorophyll—viz. those occurring in fresh, dried, and dark leaves; while about the same time Mohl distinguished between amorphous and granular chlorophyll. He denied its vesicular nature—*i.e.* denied that the granules are cells, strictly so called, or rudiments thereof. It was apparently Mulder, however, who first began to suspect that the pigment was not really homogeneous, for in 1844 he found that by the action of alkalies and acids it could be seemingly resolved into a yellow and a blue substance, also a black matter. Like Berzelius, he likened chlorophyll to indigo, and he essayed for it the formula $C^{16}H^9NO^8$, which, however, he admitted was not conclusive.

In 1850 Morot published the results of the most searching and extensive researches yet undertaken on this subject. He contested the likeness of chlorophyll to indigo, and held that its composition entirely connects it with the vegetable bases, such as alkaloids, etc. He especially emphasised the fact that the development of the green matter is brought about by the intervention of nitrogen, apparently under the form of ammonia. According to him the formula of mallow chlorophyll is $C^{18}H^{19}NO^3$—*i.e.* 3 equiv. starch + 2 equiv. ammonia = 2 eq. chlorophyll + 16 eq. H^2O + 8 eq. O; hence he opined that it is not because *they are green* that plants disengage oxygen, but because *they become green.*

During the latter half of the century the investigators of chlorophyll have been extremely numerous and of various talent, but I need not recite anything further as regards their names or their often clashing and multifarious views. Suffice it to say that up till to-day the solution of the question as to the precise chemical nature, constitution, and origin of this marvellous pigment has not been attained. As Berzelius said : " the entire leaves of a large tree do not contain ten grams of chlorophyll, and the quantity therein is not greater than the colouring matter contained in dyed cotton."

2. THE NATURE OF LIFE.

No narrower or more disastrous mistake could be made than to confine discussions upon the nature and conditions of life to that alone exhibited by animal organisms. The vital phenomena of plants are eminently self-assertive, and leave a trail behind so clear and unmistakable that they cannot possibly be ignored. The fundamental question is, as it always has been, does life depend on organisation, or organisation on life ? What may be called the " journalistic " school of investigators seems to scout the distinction herein involved. Headed by old Treviranus they suggest that action is not an attribute of the organism, but is of its essence— that if, on the one hand, protoplasm is the basis of life, life is the basis of protoplasm ; their relations to each other are reciprocal. We think of the visible structure only in connection with the invisible process. That is to say, the school jumbles up the whole phenomena into a sort of pantheistic unity. The fact that John Hunter expressly taught that organisation depends on life should make us pause and be careful. For instance, as regards the preservation and perpetuation of plants, do cells exist which bear the imprint of the plan of their general organisation and are as the directors of the physico-chemical and mechanical forces which act in living beings ? " At the present time," says Professor Gautier, " science cannot completely answer this question." If it be asserted

with unconquerable assurance that the function depends on the organs, or (more specifically) on the aggregation of cells which compose them, no objection can be offered so far as purely physical observation extends. When, however, it is held that in these cells the transformations which maintain life are derived in great measure from the chemical constitution of the immediate principles which enter into conflict therein, and are of the nature of those which we can produce in our laboratories, all that can be said is that " at the present moment science is not quite certain about it." The causes which maintain life are pretty well known to all of us who have attained a certain age, but what life is in itself is a mystery inscrutable and profound.

It is all very well to define life as " the state of functioning of those aggregates which, borrowing all their energy from the external world, make it, owing to their definite organisation, concur towards a common end." It may be quite true to say that " living organisms function by virtue of an energy which comes to them entirely from the outside " ; but, after all, the functioning is not the life, any more than is the structure of a chemical molecule. In my humble opinion the essential characteristic of life is the primordial, the fundamental, the utterly absolute, constant, and irrefragable quality and attributes of the albuminoid matter, or whatever else it is, wherein it is immanent or implanted, only to be entered therein and be dissolved therefrom under conditions which man can never certainly know.

3. THE CHEMISTRY OF A LEAF.

The wonder of wonders in the scientific world has been to me for many years the crass ignorance and the stupid indifference, of field naturalists and closet " students " alike. as regards the application of chemistry to botanical science. One might imagine, or rather one might unhesitatingly conclude, that anybody possessed of the faintest trace of scientific intelligence, or of one iota of scientific taste, would inevitably be led at an early period of his studies, either in the field or closet, to investigate by chemical methods the constituents that make up one of the most beautiful and fascinating of all creations. " The plant is," as has been said, " and will remain through all time the chemist's ideal."

At this moment I am looking over a copy of the Sessional Report of the Imperial Academy of Sciences of Vienna (Band 57) of the year 1868, wherein I find a paper entitled (*Anglicè*) " On Some Constituents of *Fraxinus Excelsior*," by Dr. Wilh. Gintl, of the University of Prague. It concerns a chemical examination of the leaves of the common ash-tree, and I will endeavour to convey to the reader the methods and results thereof. The leaves were gathered, he says, at the end of Spring, and were extracted with a sufficient quantity of boiling-hot distilled water. The infusion after cooling was fractionally precipitated by neutral, then by basic acetate of lead. Then the residue of the substances in the solution was separated by adding ammonia. The eight precipitates .so obtained were singly treated with sulphuretted hydrogen gas, the filtrates concentrated, purified, and examined. From the first portion of the precipitate by neutral acetate of lead he obtained fat, pectin, a resinous body, and a considerable quantity of a crystallisable acid ; from the last portion of the same, as well as from the precipitate by basic acetate of lead, he obtained a not inconsiderable quantity of a peculiar tannin ; also two substances, one of which, after much trouble in the way of physical and chemical analysis, was determined to be inosite, $C^{12}H^{16}O^{16}$; the other, after a not so searching examination, proved to be quercitrin, $C^{66}H^{34}O^{38}$. The now almost transparent infusion of the leaves, by reason of the complete precipitation with the lead salts, was treated with ammonia, which produced a pretty bulky white precipitate, that was collected and decomposed under water by carbonic acid gas, and the filtrate further purified. There resulted a syrupy, light brown mass that on standing deposited a considerable quantity of small needles, which on testing were shown to be mannite, $C^{12}H^{14}O^{12}$, while in the mother-liquor a gummy substance, also a considerable amount of granular sugar, were found. Dr. Gintl did not find in the infusion any traces of fraxin or fraxetin, nor quinic acid. Such, in brief, is a bare outline of a method of scientific investigation which was much in vogue in certain places about half a century ago and later. The marvel is that the exhibition and publication of it is so rare. Apart from certain researches conducted in the interests of pharmacy, and possibly of the drug trade, it is really excessively difficult to find in the pages of any of the professedly scientific journals or reports of England or Germany anything like a complete account or summary of the chemical constituents of any of our common trees or herbs. One reason of this state of affairs may be that the subject has already been exhausted, some "infallible authority" having spoken ; so that, if we wish to know anything about the subject, we are referred by means of the usual hieroglyphics to some old and musty volume, almost impossible to procure, and if discovered, at least in this country, it will almost certainly be quite " clean " enough to satisfy the most fastidious second-hand bookstaller. It would, no doubt, be voted something quite trite and commonplace, if pen, ink, or breath were wasted in a vain endeavour to show that we really know nothing whatever about a plant if we do not know its physiology through its chemistry. I have only to add that the methods of Dr. Gintl have been modified of late years with great and obvious advantage, but all difficulties in connection with this line of investigation are as yet by no means overcome.

Patterdale, Westmorland.

BRITISH ASSOCIATION.

THE meeting of the British Association for 1900, held at Bradford on the 6th September and following week, was worthy of the expiring century.

Sir William Turner's admirable review of morphological science and its progress during the past hundred years dealt with the knowledge of structural organisation of animals and plants. This, of course, included the cell theory and its development, brought about by recently improved means of observation and more systematic specialism in its investigation. All the sectional presidential addresses were sound and instructive. Among them we may refer to that of Dr. Joseph Larmor, who spoke on the advance of electrical science, the constitution of individual molecules of matter, the functions and dynamics of ether ; all fascinating subjects in themselves. Professor Sollas considered the significance of various physical phenomena in connection with geology with regard to the earth's age. Dr. Ramsay Traquair discussed fossil ichthyology and its bearing on the doctrine of descent. Professor S. H. Vines's address, which had to be read in his unfortunate absence through illness, consisted of a review of the progress of botany during the past century. One of the most remarkable of these addresses was that in the section for Anthropology, by Professor John Rhys, on the Early Ethnology of the British Islands, who approached his subject through folklore and language of the people.

The interesting lecture by Professor E. Gotch, F.R.S., on "Animal Electricity," given in St. George's Hall, commanded much attention from a large gathering of members and their friends. The popular Saturday evening lecture was by Professor Silvanus Thompson on the " Industrial Applications of Electricity"; there was an attendance exceeding 3,500 persons. As might have been expected of the Bradford meeting, the section for Geology was occupied at some length by subjects appertaining to coal. Mr. R. Kidston and Mr. Strahan read papers on the " Flora of the Coal-measures " and the " Origin of Coal " respectively. Both these subjects led to important discussions. A paper, illustrated by a series of photographs of insects in their natural environment, taken by the exhibitor, Mr. N. Annandale, in the Malayan region, created much interest and elicited some valuable opinions from Professor Poulton and others.

The above are only a few of the points of interest at this year's meeting, space not permitting further detail. The Association was most hospitably received by the townspeople of Bradford, and the usual excursions were very successful.

Next year's meeting is to be held at Glasgow, and that of the following year at Belfast.

BUTTERFLIES OF THE PALAEARCTIC REGION.

By HENRY CHARLES LANG, M.D., M.R.C.S., L.R.C.P. LOND., F.E.S.

(Continued from page 114¹.)

ANTHOCHARIS (continued).

5. **A. charlonia** Donz. Ann. Soc. Franc. 1842, p. 197, pl. 8, 1.

25—29 mm.

Ground colour of all the wings yellowish-white, h.w. sometimes more distinctly yellow. General character of pattern much resembles that of *A. tagis*, but the discal spot f.w. is larger in proportion, and there are no costal dots. U.s. f.w. white, yellow at base and along internal half of costa, discal spot black and distinct. Apices dull yellowish-olive. H.w. dull olive, with hardly any markings except an indistinct white central spot.

A. charlonia.

HAB. Algeria, Teneriffe (figured by Mrs. Holt White), Province of Kuliab, S. Turkestan. On the high plateau of Balachan. III.—VI.

a. var. *penia* Frr. 574, 4. 1852. A pale form of the first generation, with discal spot of f.w. larger. Colour of under side greyer. HAB. Amasia, Antioch.

6. **A. levaillantii** Luc. Ann. Soc. Fr. 1847, p. 1. *Charloniae* ? var. Stg. Cat. 1871, p. 4.

25—29 mm.

Resembles the last in pattern of wings, but discal spot f.w. is narrower and less quadrate, marginal fringe pink. The ground colour of all the wings is uniform sulphur-yellow. U.s. h.w. with more light spots than in *A. charlonia*. Wings narrower than in the latter, f.w. more pointed.

HAB. Biskra, etc. (Algeria); Haman (Syria). II.—V.

This form is so constant and so distinct from *charlonia* that I feel bound to admit it as a distinct species.

7. **A. mesopotamica**. Stgr. R. H. p. 136.

29—40 mm.

Larger than the last, and with apices of f.w. less pointed. Colouration of upper side much as in the last. F.w. apices less spotted with yellow, in some specimens hardly at all; discal spot triangular. U.s. apices f.w. and general colour of h.w. light

(1) This series of articles on Butterflies of the Palaearctic Region commenced in SCIENCE-GOSSIP, No. 61, June 1899.

brownish, in place of the dark olive-green seen in the last two species. Costa and cilia f.w. decidedly pink.

HAB. S.E. Asia Minor (Malatia). Ve.—VIe.

A. mesopotamica.

Hadjin and Eibes (Antitaurus). VI.—VIIc. (R.H.)

a. var. *transcaspica* Stgr. R. H. p. 136. A light form of second generation. HAB. N. Persia.

8. **A. cardamines** L. Syst. Nat. x. 468, Lg. B. E. p. 39, pl. xx. fig. 1, pl. xv., fig. 5 (transf.).

34—43 mm.

Ground colour of wings in both sexes white. ♂ with the outer half of f.w. bright orange. A small black spot towards inner edge of the orange patch; apex narrowly black. H.w. white, with the markings of u.s. shining through. Bases of wings black. U.s. f.w. as above, but with green tinting to apex; h.w. marbled with green and white. ♀ as ♂, but entirely without orange tips.

HAB. Throughout Europe, Asia Minor, Syria, Siberia, and the Altai. A well-known and common British insect, "the orange-tip." IV.—VIe., in woods, fields, and hedgerows.

LARVA green, finely speckled with black, with a white lateral stripe less clearly defined at its upper than at its lower edge. On Cruciferae, often on the seed-pods of *Cardamine pratensis*, but more commonly on pods of hedge-mustard (*Sisymbrium officinale*).

PUPA. Boat-shaped, at first green, but changing later to greyish yellow with clearer stripes.

a. var. et ab. *turritis* Och. Schmett. Eur. iv. 156.

A name given to a form of the species in which the orange patch in the ♂ is narrower and the black spot smaller, and on the extreme edge of the orange, not well within it as in type. I do not think that this name is worth preserving, as one frequently meets with specimens, both in England and elsewhere, which answer to this description, which is from a specimen sent to me by Dr. Staudinger as *turritis*.

b. ?var *hesperidis*. Newnham. Ent. Record v., pp. 97, 219 (1894). Kirby, Handbook of Lepi-

dopt. I., vol. 2, p. 189 (1896). This form, which
has been treated by one or two authors as a distinct species, differs from the usual type in its small
size, and in the proportionately large discoidal
spot. I believe this to be merely a dwarf condition
of *A. cardamines.*

c. var. *alberti* Hoffm. U.s., h.w. showing very
little white on account of them being powdered
with dark scales. HAB. Various parts of Germany.

9. **A. bieti** Oberth. R. H. p. 140.
42—44 mm.

This very distinct species seems to be most
properly placed here, though by some it has even
been included in the genus *Midea*, on account of
the shape of the f.w.

The f.w., instead of being rounded, are hooked at
the apex like those of *Rhodocera rhamni*. Otherwise, in size and colouration the insect is very
like *A. cardamines.* The orange patch on the f.w.
of ♂, however, is less extensive and paler in

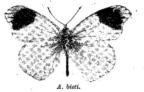

A. bieti.

colour, and with a faint black patch or spot on its
edge below the discoidal spot, which latter is partly
outside the orange patch. Apex greyish, mixed
with white. H.w. above and below much as in *A.
cardamines* but paler. ♀ as ♂, but without
orange patch, and with apex f.w. very indistinctly
marked with grey.

HAB. Amdo, Thibet. On mountains.

10. **A. gruneri** H. S. 551-4, Lg. B. E. p. 40,
pl. ix. 2.
29—33 mm.

Much smaller than *A. cardamines,* which it
greatly resembles in the pattern of the wings, but
in the f.w. ♂ there is a faint grey spot above the
discoidal spot, sometimes joining it, and the
orange patch is often faintly shaded with grey on
its inner edge. The ground colour of the wings in
♂ is pale primrose-yellow, and the colour of the
orange is lighter than in *A. cardamines.* ♀ white,
with broader and paler markings than in *A. cardamines.* Discoidal spot f.w. square. Apices not
so distinctly marked with white. Costa of f.w. in
both sexes shaded with grey.

a. var. *homogena* Christoph. HAB. Mesopo-
tamia.

b. var. *armeniaca* Christoph. HAB. Ordubad
(Russian Armenia).

The two forms *homogena* and *armeniaca* appear
from the description in R. and H. to differ from
the type chiefly in the paler markings, and in the
white ground colour of the wings.

HAB. Turkey, Transcaucasia, Greece, Asia Minor,
Tokat and Amasia, Persia, III-IV.

11. **A. damone** H. S. Feisth. Ann. S. F. 1837.
301. Lg. B. E. p. 40, pl. ix. fig. 3.
36—42 mm.

♂ bright canary-yellow. F.w. with a broad and
very brilliant orange tip, narrowly marked with
greyish black at apex, discoidal spot as in *A. cardamines,* the internal border of the orange patch with
a faint indication of a dark band, more strongly
marked in some examples than in others. H.w.
yellow, with the markings of the u.s. appearing
through as in *A. cardamines.* U.s. h.w. pattern as
in *A. cardamines,* but ground colour deep yellow in
place of white. ♀ very like *A. cardamines* ♀, but
the u.s. h.w. with the ground colour yellow. Apices
of f.w. greenish yellow, without any black shading.
HAB. Sicily (foot of Aetna R.H.), Balkans, Asia
Minor (Tokat, Smyrna), Taurus. IV., V.

12. **A. eupheno** L. Syst. Nat. xii. 762.
29—38 mm.

♂ ground colour of wings yellow, but rather
lighter than in the last. F.w with a bright orange
patch, narrowly black at apex, orange patch bordered internally by blackish band running rather
abruptly outwards from discoidal spot. U.s. h.w.
deep yellow, the spots not arranged as in the last,
but placed in three transverse bands of a reddish
orange colour.

♀ white, h.w. with a reddish-orange tinge at
costa. Apical shading of f.w. more or less distinctly marked with orange-red. U.s. h.w. as
in ♂.

HAB. Granada and Balearic Islands (R.H.);
Algeria. IV., V.

LARVA green, with yellow and black dorsal
stripes. Very similar to that of *A. euphenoides.*
On *Biscutella.*

13. **A. euphenoides** Stgr. Stett. E. Z. 1869,
p. 92. Lg. B. E. p. 41, pl. ix. fig. 4.
30—42 mm.

Upper side greatly resembles the last, but the
internal black border of the orange patch in ♂
does not run abruptly outwards as in that species.
U.s. h.w. much paler in the ground colour; the
pattern more ramified and not forming three
bands; not orange-red, but green. There are one
or two faintly whitish ante-marginal patches.

HAB. S. Portugal and S. Spain, Provence, the
Riviera, Siena, Leghorn, Monte Bré, Tessin, Switzerland. IV. to VI. I have seen this butterfly at
Nice as late as the end of June. Generally found
in mountain districts and in wood clearings.

LARVA. Greenish yellow, with a lateral purplish
stripe above, which is a row of black spots placed
close together. On *Biscutella.*

PUPA. Boat-shaped, but less arched than that
of *A. cardamines.* Grey in colour.

(To be continued.)

ON COLOURING OF MOLLUSCS' SHELLS.

By REGINALD J. HUGHES.

(Continued from page 99.)

LET us now turn to fossil shells, and see how the evidence they afford agrees with the deductions made from present-day forms. Little reliable information can be gathered from shells of pre-Eocene date. The colour is scarcely ever preserved, being destroyed even if the shells are not altogether petrified. I have, however, satisfied myself that some bivalves from the chalk were coloured by iron. There is also one Carboniferous species, *Pleurotomaria carinata*, belonging to the family Trochidae, which deserves special mention. On it the original wavy bands of colour are preserved, of precisely the same type as found in Trochidae of the present day. It is unlikely that a species could alter its pigment without changing the pattern of its coloration, so I may safely say, although I have not myself examined a specimen, that this shell was coloured by the organic and rather insoluble pigment common on recent Trochidae, thus showing this to be an extremely ancient method of coloration.

Among Eocene molluscs, all the terrestrial and fresh-water shells from the London clay and elsewhere which I have examined gave results similar to equivalent shells of the present day. All were coloured by an organic pigment, and that on fluviatile genera, such as *Planorbis*, was unchanged in colour by nitric acid. I need hardly say that in testing fossil shells, especially marine species, great care must be used to discriminate between the original colour of the shell and subsequent iron stains. Several examples of the same species should be compared to ascertain what was the original pattern, as on many specimens the markings are entirely destroyed. They should all, of course, be first thoroughly cleaned, with dilute hydrochloric acid if necessary. I have selected the Barton clay for special examination, as the shells from it are typical of the molluscan life in England during this time, and are in a beautiful state of preservation. I can say with certainty that the following were coloured with iron sesquioxide (see fig. 2): *Crassatella sulcata, Pectunculus deletus, Ostrea flabellula, Chama squamosa, Cardita sulcata, Pleurotoma turbida, P. exarta, Dentalium striatum, Phorus agglutinans, Turritella edita, Rostellaria ampla, Natica ambulacrum, N. patula, N. depressa, Fusus pyrus, F. regularis. F. complicatum, F. longaevus, F. longorus, Seraphe fusiformis, Murex erinaceus, Voluta luctatrix V. ambigua*, and *V. spinosa*. Some seem to have been originally white, as *Buccinum laxatum, Voluta scabricula, V. scalaris, V. sub-ambigua*, etc. Not one was

coloured by an organic pigment, and all the Volutes were marked by iron, although every one of that genus at the present time has an organic colouring matter. Most of these shells were brown and had very little pattern, though some, such as *Natica patula* (fig. 2' No. 5), were beautifully marbled, especially just inside the opening. Other English formations of this period give similar results. From the Bracklesham beds, *Pecten radiatus* is one of the most interesting examples. It was undoubtedly coloured by iron, thus agreeing with recent Pectens.

I would here call your readers' attention to the physical geography of England at the time these molluscs inhabited its waters. A great continent lay to the north and north-west of what is now North-Western Europe, penetrated by a large gulf, which opened into the sea to the south of the Paris basin and extended inland as far as London. Into this gulf one or more large rivers flowed, forming extensive deltas and covering the sea-bed with iron-impregnated mud derived from the erosion of the surrounding country. On this deposit the Volutes and other molluscs lived, in a tropical climate, of which they themselves form part of the evidence. Naturally their colour assimilated to their surroundings, and for this purpose they made use of iron, whose abundance an examination of any piece of Barton clay will demonstrate.

In the Paris basin at the same time we find a very different state of things. The sea-bottom here consisted either of limestone or light sands, and accordingly many of the molluscs were white—*e.g. Cardium obliquum, Cardita arcticostata, Lithocardium avicularis, Fimbria lamellosa, Mesalia sulcata, Cerithium cristatum, C. nudum, C. lapidum, Natica parisiensis, N. sigarotina*, and *Fusus bulbus*. Parts of the sea-bed were, however, covered with sand not perfectly white, and parts in time became again muddy, so often a certain amount of colouring would be useful. And in this formation I have found certain shells coloured by an organic pigment, which is sometimes unaffected in tint by nitric acid and sometimes turned light blue. As the colour, owing to its great age, is generally faint and almost destroyed, it is probable that when living blue would have been produced in the majority of instances. Of these shells *Cardita imbricata* was rose-colour (fig. 2, No. 8); *Cerithium concavum* was white, with a series of orange bands round the shell, the projecting portions remaining white (fig. 2, No. 4); *Cerithium serratum* was salmon (fig. 2, No. 6), and *Rostellaria fissurella* and

Cerithium echinoides also possessed an organic pigment. Washing with a dilute acid often brings out the colour wonderfully, revealing the existence of former markings on shells apparently quite white. In this manner I have found that the following shells were coloured, usually by bands, like many of the same genera at the present day :— *Cytherea laevigata*, *Sunetta semi-sulcata*, *Pectunculus pulvinatus*, and *Corbula gallica.* The latter had violet bands comparable to those on the living shell *Venus fasciata.* But the most interesting example of this kind is *Cardita planicosta.* Shells of this species, from the Paris basin at all events, show, when developed with hydrochloric acid, a system of bands, the positions of which on a medium-sized specimen I have shown in fig. 3. In young individuals the bands marked *a* and *b*, *c* and *d.* and *e, f*, and *g* were joined together so as to form three broader stripes, whilst on larger specimens they are sometimes still further subdivided. It would thus appear that bands marked the positions of most rapid growth of the shell. Many fresh cells were formed along their median lines, and thus they tended to break up with age. It is quite possible that this mollusc—which had a very wide range, from the Mediterranean to North America— may not have been coloured in the same manner in all localities, so that if living to-day it would be divided into separate varieties, if not species. Very few French Eocene fossils were coloured by iron; the only ones I have found to have been so are *Lucina concentrica*, *Lucina sulcata*, *Ostrea mutabilo*, *Chama calcarata*, and *Potamides emarginatus.* All except the last, which probably lived in a muddy estuary, are darkest on the inside; the exteriors of the two *Lucinae* were quite white. As will be seen from the example figured (fig. 2, No. 2), the colouring matter of these latter formed an oval spot covering the whole of the interior of each valve up to near the margin, and was orange in colour, whilst an examination of a few specimens would convince anyone that it was not accidental, but really formed the pattern on the shell. It gives a very light blue with potassium ferrocyanide, so faint as to raise a doubt that it was really iron. A mixture of iron persalts and protosalts would, however, give this result, and that this is what it is can be confirmed by testing a solution in hydrochloric acid with caustic potash, which, as on this hypothesis it should do, will form a green precipitate. Now limestones and shells undoubtedly white, from the Paris basin, are often stained deep yellow by a substance which has exactly the same properties as the above, and must therefore be a mixture of oxides of such a kind that with an acid it always forms both per- and protosalts. The importance of this lies in the evidence it affords that the colouring of shells by iron is entirely mechanical, since a particular mixture of oxides is unaffected by being used on the shell; the mollusc has merely the power of

limiting the places where it is deposited. But the habit, once formed, of being coloured by this combination of oxides has been preserved to the present day, and several recent *Lucinae*, similarly coloured —white outside and yellow inside—give the same result on analysis. As, however, this variety of iron is in many cases not so abundant as it was in the locality where the habit was first acquired, the markings of *Lucinae*, although formerly very constant, are now extremely variable. Individuals can be found of the same species in which the colour of the interior varies from white to orange. A good example is *Lucina striatus* (fig. 1, Nos. 10 and 12). Two Oligocene fossils, *Olivella impressa* and *Auricula douvilloi*, were tinted by the same substance as these *Lucinae*, but on the outside, as they lived on dark estuarine deposits.

Pliocene fossils were coloured with the same pigment as recent shells, except that the rapidly diminishing races of British *Muricidae* and *Volutidae* were still coloured by iron.

Summing up all the evidence here given, the history of the various pigments seems to be something like this :— The most ancient one is iron ; all shells were originally either white or coloured by it. As soon as some species began to exist in fresh water or on land, they learnt perforce to produce an organic pigment. The latter was at first nearly insoluble in all species, after a time, before the Eocene epoch ; but subsequent to the branching off of *Cyclostoma*, and even *Helicina*, from the main body of land shells, a more soluble colouring matter was produced, capable of being stained blue by nitric acid, and which has continued to the present time. Among marine shells the power of producing insoluble markings was soon acquired by the *Trochidae* and their allies, most of which at a very early period found it necessary for their protection to have a waved or dotted pattern. Although these shells were smooth, the pigment was prevented from spreading by a mordant. But all the bivalves and the majority of gasteropods continued to be marked by iron down to Eocene times. It is most improbable that a genus which had acquired the power to produce markings caused by organic matter would lose it and again revert to the iron-stained condition. Accordingly, as we find in British formations typical of the Eocene period—*Volutidae*, *Muricidae*, etc.—coloured by iron, we conclude that shells of those genera had never previously possessed any other pigment. Towards the close of the Eocene epoch, owing to various causes, some of which I have described, most species gained the power of forming an organic colouring matter ; and since, owing to the large size and diverse forms which many shells had acquired, a mordant was no longer necessary, there was produced the pigment characteristic of most univalves, and the more specialised bivalves, of the present day.

Norman Court, Southsea.

IRISH PLANT NAMES.

By John H. Barbour, M.B.

(Continued from page 83.)

TILIACEAE.

TEILE. TEILEAG. *Tilia vulgaris.* lime or linden.

LINEAE.

CAOLAC. caol, "slender, subtle." CAOL MIOS-ACAN. Miosacan means "fairy flax." "slender fairy flax." LION A MBAN SIGE. lion, "flax"; mban, "hand" = "flax fit for fairy hands." MIOSAC. See above. mios, month. CEOLAG. *Linum catharticum.* purging flax.

LIN. *Linum usitatissimum.* common flax.

GERANIACEAE.

BIAD-UR-EUNAIN. biad, "food"; eun, "fowl, bird"; ur, "first, best." "best bird's meat." BILLEOG NA NEUN. Billeog, syn. Duilleog, "leaflet" = "eve-closing leaflet." SAMAD COILLE. SAMA FEARNA. Sama, "sorrel"; coille, "sylvan"; fearn, "pleasant" = "sylvan or pleasant sorrel." SEAM-SOG or SEAMROG. *Oxalis acetosella.* wood sorrel, coocoo sorrel (King's and Queen's Co.).

CREACTAC, "brushwood." RIAN RIG, "road spy." EARBULL RIG. earbull, "tail"; rig, "royal." RIAL CUIL, "road fly." RIGEAN RIG. RIAGAL CUL. RIGEAL CUIL. (These appear to be all corruptions of other forms, and meanings are suggested by those given above.) RIANROIGE. RUIDEL. REILGE. reilge is gen. of reileasg, "a church" or "church-yard." *Geranium robertianum.* Herb Robert. I am inclined to think that more than one species is called by these Irish names, but any plants sent to me from different parts of the country have been this species.

ILICINEAE.

CRAN CUILIN. cuilean cuil, "wicked" = "W. celyn." *Ilex aquifolium.* holly.

CELASTRINEAE.

FEORAS. OIR (letter O). *Euonymus europaeus.* spindletree. Pegtree (Queen's Co.).

RHAMNEAE.

RAMDROIGEAN. droigean, "thorn-tree." RAM-RAGAN, "stiff branch." *Rhamnus catharticus.* buckthorn.

SAPINDACEAE.

CRAN BAN, ban, "white." CRANFION. fion, "pale." FIRCRAN, fir, "white, pale." CRAN SCICE. *Acer pseudo-platanus.* sycamore.

CRAN MAPLAIS. *Acer campestre.* common maple.

GRAN GEANMCU. *Aesculus hippocastanum.* horse-chestnut.

LEGUMINOSEAE.

AITEANN. ait, "sharp." ATTIN. TEINE, "letter T." ON. *Ulex europaeus.* whin.

BALLAN. ballan, "shell, husk." GIOLCAC SLEIBE. giolac, "a place where broom grows" or "a broom." SLEIB. "mountain." *Cystisus scoparius.* broom.

SREANG BOGA, or Bo., or S. TRIAIN. Sreang is "a string," "fibre"; boga, "a bow"; triain, "a third part"; hence "bow strings," or "a third part fibrous"). TRIANTARAN. *Ononis spinosa,* Linn. commock.

LUSNA MEALLA. meala, "honey." SEAMAR CAPUIL. capull, "horse," and seamar, "trefoil." *Trifolium pratense.* red clover.

SEAMAR. SEAMAR BAN. *Trifolium repens.* white clover.

FEANTOG-GREUGAC. feantog, "nettle." GREU-GAIS, "Greek." *Trigonella purpurascens.* fenugreek.

PESSEIR CAPUIL or PIS CAPUIL, PESSEIR DUB. PESSEIR PREACAIN. pesseir and pis, "pease"; capuli, "horse"; dub, "black"; preasac, "pottage"; horse pease, "black pease," etc. *Vicia sativa.* Vetches.

PIS BUIDE. buide, "yellow." *Lathyrus pratensis.* PESSEIR-TUIBE. tuibe, "straw," "thatch"; "thatch pease." *Lathyrus montanus.* heath-pease.

CRUBA EAIN. cruba, "claw"; ean, "bird." *Lotus corniculatus.* bird's-foot trefoil. Crowfoot (Queen's Co.).

(To be continued.)

THE BIRKBECK INSTITUTION.—This Institution, which is in Bream's Buildings, Chancery Lane, and has now completed 77 years of educational work in the metropolis, commences its new session on Monday, October 1st. There are both day and evening courses of study, which comprise the various branches of Natural Science, Mathematics, classical and modern Languages, Economics, Commercial Subjects, Law, Mental Science, Music, and Art. The courses provide for the examinations in the University of London in the faculties of Arts, Science, and Commerce, those of the Conjoint Board, Civil Service, etc. The report for the last session shows that during the year 50 students passed some university examination, while large numbers gained successes at various public examinations. The Institution has had many additions to its appliances in recent years, and the physical, chemical, and metallurgical laboratories are now very thoroughly equipped. The day classes provide courses in Chemistry, Biology, Physics, and Mathematics, for the Science Degrees of London University. There are also day classes in Languages, Music, and Shorthand. The School of Art is open both day and evening. During the recess, considerable additions and improvements have been made by the aid of a gift of 2,000 guineas from Mr. F. Ravenscroft, to commemorate his completion of a membership of 50 years. A new reading room, a new magazine room, and a social room have been provided; a well-appointed metallurgical laboratory has also been added.

Studies in Fossil Botany. By DUKINFIELD
HENRY SCOTT, M.A., Ph.D., F.R.S., F.L.S., F.G.S.
xiii. + 533 pp., 8½ in. × 5½ in., with 151 illustrations.
(London : Adam & Charles Black, 1900.) 7s. 6d.

The author, who is Honorary Keeper of the
Jodrell Laboratory, Royal Gardens, Kew, will be
remembered by his "Introduction to Structural
Botany," issued some little time since. The book
before us is founded on a special course of lectures
given under the same title at University College,
London, in 1896. Dr. Dukinfield Scott has
retained the lecture form, but has largely increased
the matter in these pages. The book does not
pretend to be a manual of fossil botany, but rather
a general introduction to the subject, in view of
coupling up, as it were, in the student's mind, the
plants of the past with those at present existing.
This aspect has the utmost value to the modern
botanical student, who, as a rule, knows far too
little of the evolution of the recent flora. The
book is excellently illustrated with a large number
of new drawings, chosen with much judgment,
from the pencils of well-known palaeontologists,
and easily accessible originals.

The Monthly Review, October, 1900. 198 pp.,
10 in. × 7 in., 14 illustrations. (London : John
Murray. 1900.) 2s. 6d. net.

Mr. John Murray is to be congratulated on the
first number of the long-expected " Monthly Review." As an example of the publishers' art it is
admirable, being beautifully printed in large type
on good paper. The illustrations are worked upon
plate paper, and are artistic examples of tone
reproduction. As becomes a Review of this character, the literary matter is very varied, and we
are glad to see that science is represented in an
article by Professor H. H. Turner on "Recent
Eclipses." The most remarkable article, although
not scientific, is "Details in my Daily Life," by
Abdur Rahman, Amir of Afghanistan. The illustrated article is by Roger E. Fry, on " Art before
Giotto." This constitutes the first of a series of
essays on early Florentine painting. It is superbly
embellished with photographic reproductions. The
" Monthly Review" should have a long and successful career.

The Path of the Sun. By WILLIAM SANDEMAN,
F.C.A. 132 pp., 7½ in. × 5 in., illustrated by 10
diagrams. (Manchester : Sherratt & Hughes.
1900.) 2s. 6d.

The author of this work has evidently given
much thought to his subject ; the pages indicate
careful study and originality. However much we
may disagree with his theories, we must acknowledge his earnestness. Mr. Sandeman does his
best, to quote his own words, " to expose the fallacy
of the 'Precession of the Equinoxes,' though more
properly of Newton's explanation of their cause.
He endeavours to show that the phenomenon is
due to the motion of the sun round an orbit of
rather less than 25,800 years.

Church Stretton. Vol. I. Edited by C. W.
CAMPBELL-HYSLOP. 198 pp., 7½ in. × 5 in., with
illustrations. (Shrewsbury : L. Wilding. 1900.) 5s.

This work appears to be a collection of mono-
graphs on subjects connected with the ancient
town of Church Stretton, in Shropshire ; but, there
being no preface or editorial introduction, we have
not anything to indicate the object of the publication. Vol. I. contains three excellent articles by
local specialists. They are on Geology, by E. S.
Cobbold ; Macro-lepidoptera, by F. B. Newnham,
M.A.; Molluscs, by Robert A. Buddicum, B.A.,
F.G.S. The first of these occupies 115 pages, the
second 63 pages, and the third 22 pages, with an
instructive plate. When complete these volumes
will also contain articles on the Botany, Archaeology, Climatology, and Ornithology. The book
is well produced ; but the title-page would be more
valuable if dated, and the back should bear the
name, as everyone knows the inconvenience of
having untitled books on the shelves of one's
library. More care seems to have been necessary in reading proofs, as on the first page we
happen to open we find that *Vanessa Jo* is very
common, and that : ab. *Joides* differs from *Jo*
only in its much smaller size. Our entomological
readers must understand that the peacock butterfly
is here intended. The author has invented a
number of names for aberrations or sports, which
we hardly think were necessary. For instance, he
has *Anthocharis cardamines* ab. *cinerea*, named
from a single specimen, as also ab. *arsenoides* as
representing a female with partial orange tips,
and some others. We are not quite sure whether
one of them is intended for an aberration, as there
are two dots in front of the word *hesperides*, which
appears to be a small form of orange-tip butterfly.
Mr. Newnham's list of macro-lepidoptera is considerable, also interesting, and is one that will
probably be enlarged. It is a pity he does not also
include the micro-lepidoptera. We do not understand why the editor of this volume, in passing
the scientific names of the lepidoptera, has not
followed the more modern plan used by the
author of the "Monograph on the Molluscs." The
work is a useful addition to county histories.

A Treatise on Zoology. Edited by E. RAY
LANKESTER, M.A., LL.D., F.R.S. PART II.
PORIFERA AND COELENTERA. By E. A. MINCHIN,
M.A.; G. HERBERT FOWLER, B.A., Ph.D. ; and
GILBERT C. BOURNE, M.A. vi. + 405 pp., 9 in. ×
6 in., with 175 illustrations. (London : Adam
& Charles Black, 1900.) 15s. net.

We have already had the pleasure of noticing
the first issued part of this admirable work in our
number for May last (Vol. VII., p. 370). In that
instance we explained the parts were being issued
as ready, The one before us is Part II., which
gives us the most recent information upon a
number of invertebrate animals, including their
latest classification. The sections of the work are
devoted to the Enterocoela and the Coelomocoela,
by Dr. Lankester ; Sponges—Phylum Porifera, by
E. A. Minchin, M.A. ; The Hydromedusae, by
G. Herbert Fowler, B.A., Ph.D. ; The Scypho-
medusae, also by Mr. Fowler ; The Anthozoa, by
G. C. Bourne, M.A. ; and The Ctenophora, by the
same writer. Considering the remarkable advances
which have been made in the knowledge of the
life histories and relationship to each other of
these low forms of aquatic life, and the abilities
of the respective writers, it is only necessary to

point out that before us is a standard work on the subjects, with information up to the most recent date. The arrangement of the letterpress is good, and the pages are not overladen with unnecessary scientific terms. We have the pleasure of reproducing, with permission of the publishers, one of the illustrations as an example of their excellence. We can strongly recommend Part II. of this treatise on zoology.

Handbook of British Rubi. By WILLIAM MOYLE ROGERS, F.L.S. xiv + 111 pp., 9 in. × 5½ in. (London : Duckworth & Co.) 5s.

At length, after a period of thirty-one years, there has been placed before students of our native plants a work devoted entirely to the enumeration and description of the brambles or blackberries indigenous to the British Isles. The only previous work of this nature was the late Professor Babington's "British Rubi," issued in 1869. Since that and it will be found that there are brambles whose characteristics appeal more readily to the eye than the differences between species of buttercups. In this book the section of the genus comprising the brambles proper, excluding the raspberry, cloudberry, etc., is divided by the author into fourteen groups, some of which we think could with advantage be amalgamated. Following the enumeration of the characters limiting these groups is a conspectus of the species, and afterwards comes a full description of each kind, with remarks upon habitat and affinities. An appendix indicates the county distribution of each bramble throughout the British Isles. The author holds the peculiar view that certain brambles have originated by means of hybridisation between other species. This is by no means a new theory, and is shared by prominent workers at this and other highly "critical" genera. Botanists who study plant forms from an evolutionary basis are scarcely

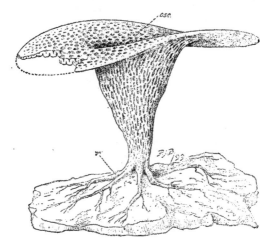

Ventriculites, imagined reconstruction.

From "A Treatise on Zoology."

time immense strides have been made in the knowledge of the forms of *Rubus* inhabiting Great Britain; new species have been described and others identified with Continental brambles. Undoubtedly the plants treated in this Handbook are not popular with even those who take more than a superficial interest in British plants. Brambles, in more senses than one, are a very thorny subject, and it is to be hoped that the Rev. Mr. Rogers's work will remove some of the difficulties attendant upon the study of these very "critical" plants. It may surprise some readers to learn that in the British Isles are found more than one hundred and seventy species, sub-species, and varieties of blackberries, brambles, dewberries, and allies; so that anyone commencing the study of this genus is likely to have plenty of material to work upon for some years. It is, we believe, a rather widely-held opinion that one bramble is much like another. So is one kind of buttercup much like another kind, likely to acquiesce in this solution of the origin of plants which may be inconveniently intermediate between others; and we think the most damaging evidence that can be brought forward in opposition to this view is that bramble hybrids usually fail to perfect seed. Hybrids proper are left untouched by the author as demanding too much space in a handbook; but it would have been well if a list of known bramble hybrids were given, as such hybrids are quite frequent. The preparation and completion of this work has been accomplished in spite of serious physical drawbacks; and now that so necessary a guide has been placed within the reach of field botanists, let us hope a great impetus will be given to the study of what is really a group of very interesting plants. In this way we can express our indebtedness for the years of painstaking work which has resulted in this "Handbook of British Rubi."— C. E. B.

AT the general meeting of the British Association, recently held at Bradford, a report of a committee was presented urging the Indian Government to give more importance to the study of botany in the training of Indian Forestry Department officers. This seems probable, as the Secretary of State for India is in accord on this question.

WE have received a reprint from the "Scottish Geographical Magazine" of a paper by Mr. William S. Bruce, F.R.S.G.S., M.B.O.U., on the proposed Scottish National Antarctic Expedition. It is accompanied by a coloured map of what is known within the South Polar regions. We hope that this expedition will attain the success deserved by its enterprising and energetic projectors.

MR. ABRAHAM FLATTERS asks us to call attention to his "circulating library" of lantern slides, with regard to a convenient extension of the subscription. Hitherto he has supplied three hundred slides for a guinea, providing they were used in a current year. On the suggestion of several artisan societies of Lancashire, the subscription will still remain a guinea, but the time-limit during which the slides must be taken is waived.

STUDENTS of Biology, Chemistry, Geology, and Physics have an excellent opportunity, with almost nominal fees, of attending evening classes in these subjects at the City of London College, White Street, Moorfields, E.C. The laboratory space has been enlarged, enabling one to be devoted solely to biology, as is the case with other subjects. The fifty-third session opens on October 1, 1900. All classes are open to both sexes, with the use of library, reading, and coffee rooms.

WE wonder if "Rectangle" is a Boer; he certainly follows the late President Kruger in trying to flatten out the earth. This is what we gather from "Zetetic Cosmogony," a copy of the second edition having reached us from Durban, Natal, where it is published for half a crown by the author, Mr. Thomas Winship, at 12 Castle Buildings. It will be found amusing reading, even at the price. We learn not only that the earth is not a globe, but that "according to current science the moon was once a piece of molten rock fractured off the earth," also many other interesting "facts." There may well have been war in Natal.

THE following firms connected with the manufacture of scientific and photographic apparatus have received awards for their exhibits in the Paris Exhibition:—James J. Hicks, London, 2 gold and 6 silver; Cambridge Scientific Instrument Company, Limited, Cambridge, Grand Prix. 1 gold and 2 silver; Ross, Limited, London, Grand Prix; James White, Glasgow, Grand Prix; Kodak, Limited, London, Grand Prix; Negretti & Zambra, London, 2 gold; W. Watson & Sons, London, 2 gold; Newman & Guardia, Limited, London, gold; J. H. Dallmeyer, Limited, London, gold; Crompton & Co., Limited, London, gold.

AMONG the important international congresses in connection with the Paris Exhibition, one of the most interesting should be that of the botanists, which will commence on October 1st.

WOMEN workers in science won a well-earned victory at the general meeting of the British Association at Bradford. On the motion of Professor Hartog, seconded by Professor Silvanus Thompson, and ably supported by Sir Henry Roscoe, it was decided that they should be eligible as members of the Sectional Committees.

ARCTIC exploration has been active during this and the last two years. At the front is the successful return of the Duke of Abruzzi's expedition in the *Stella Polare.* He appears to have advanced slightly further than other explorers. We hear that important scientific results will be published later. The return of other expeditions is anxiously awaited, especially that of Lieutenant Peary.

A SERIES of Popular Science Lectures for young people has been organised by the Rev. J. O. Bevan. M.A., F.S.A., F.G.S., Examiner to the College of Preceptors, and Mr. Cecil Carus-Wilson, F.R.S. Edin., F.R.G.S., F.G.S. A course of six, entitled "The World We Live On," will be delivered at the Kensington Town-hall at 4.45 P.M. on Thursdays, commencing October 18. The season tickets for these lectures cost only 5s. for children and 7s. 6d. for adults.

A NEW edition—the tenth—of Skertchly's "Geology," revised by Dr. Monckman, has been issued by Mr. Thomas Murby. A fourth section is added, embracing the more recent requirements of the South Kensington syllabus. This includes many topics which were formerly confined to mineralogy and crystallography. There are also new chapters, some illustrated by material supplied by Mr. W. West, F.L.S., of Bradford, on rock-forming minerals, on crystallography, volcanic and plutonic rocks, and microscopic examination of rocks.

THE Science and Art Museums of the Orange Free State and the Transvaal have both figured unfortunately in the progress of the South African war. It may not be generally known that the unfortunate man who was shot for plotting against the life of Lord Roberts was an assistant in the Pretoria Museum. In an article in the new "Monthly Review" upon "Surgical Experiences in South Africa," Mr. A. A. Bowlby states that so sure were the Boers that the English army was utterly destroyed at Magersfontein that he saw a letter from the Curators of the Bloemfontein Museum asking burghers for relics of the British army for exhibition.

THE University of Cambridge has sustained a severe loss by the death of Professor Henry Sidgwick, M.A., Litt.D., Professor of Moral Philosophy. He was born at Skipton, in Yorkshire, May 31st, 1838. His publications are almost entirely of an ethical character; but his influence on the University of Cambridge was, as a whole, most marked, especially with regard to his support of the higher education of women. He was practically the founder of Newnham College, and on Mrs. Sidgwick becoming Principal, in 1892, it became their regular home. Professor Sidgwick was also President of the Society for Psychical Research.

CONDUCTED BY F. SHILLINGTON SCALES, F.R.M.S.

MOSQUITOES AND MALARIA. — Professor B. Grassi discusses the observations of . Koch in a recent paper published in Italy. He does not consider they have made any contribution to the aetiology of human malaria. It is also indicated that Ross's discoveries are suggested by, and are confirmatory of, Grassi's previous results.

RÔLE OF INSECTS, ETC., AS CARRIERS OF DISEASE.—For the following summaries we are indebted to the Journal of the R.M.S. Dr. G. H. F. Nuttall, in the Johns Hopkins Hospital Report VIII., 1899 (see also *Lancet*, September 16th, 1899), makes a timely and valuable contribution to the literature of animal and vegetable parasites, and their definitive and intermediary hosts. This occurs in a critical and historical study of the part played by insects, arachnids, and myriopods as carriers of bacterial and parasitic diseases of man and other animals. Among the more important and interesting features of the essay may be mentioned the evidence adduced to establish the connection between flies and the spread of cholera, typhoid, and plague, the association of Texas or tick fever with *Ixodes bovis*, of tsetse-fly disease with *Glossinia morsitans* and its recent visit to an infected animal, the subject of filariosis, and the mosquito-malaria theory. The bibliographical appendix is extensive.

REGENERATION IN EARTHWORMS.—A. P. Hazen has made some interesting experiments. It has been shown by Spallanzani, Morgan, and Hescheler that a short piece cut from the anterior end of an earthworm dies without regenerating the posterior end, although such a piece often lives for several weeks, or even months. It was not known, however, whether, if such pieces could be kept alive for a longer time, they would regenerate ; or whether, if regeneration did occur, a head or a tail would develop. By grafting in a reversed direction the small anterior end of one worm upon a large posterior piece of another worm, the small piece can be kept alive for a much longer time. The results showed that a head may regenerate from the posterior end of the seventh segment if it is kept alive for some months by grafting. It seems, comparing this with other experiments, that the part of the body of the normal worm from which the segments are taken determines what will be regenerated, rather than the direction in which regeneration takes place.

STORY OF ARTEMIA RETOLD.—In 1875 W. Schmankewitsch published in the "Zeitschrift für wissenschaftliche Zoologie" a famous paper giving an account of his observations on the brine shrimp, *Artemia salina*, from the Bay of Odessa. He stated that by altering the water he could transform *A. salina* into another species, *A. mühlhausenii* ; and, more than this, that by the addition of fresh water to the habitat in which *A. salina*

lived he could induce a resemblance to the genus *Branchipus* almost amounting to identity. Both results have been repeatedly criticised ; the second has been proved inaccurate, and much doubt has arisen in regard to the first. The most thorough-going criticism, however, has been that of W. P. Ainkin, published in Russian in 1898, but now made available to the unlearned in that language by a summary by N. von Adelung in German. Ainkin points out that the various species of *Artemia* which have been described do not rest on a satisfactory basis—not that they are alone in that—and that some of them are merely cripple-modifications of *A. salina*, induced by sudden alterations in the salinity of the water. His experiments showed that if the degree of concentration was slowly and gradually increased, no structural changes of moment ensued. Some slight changes were, indeed, observed ; but they were only "modifications," not transmissible to the progeny, and disappearing when normal conditions were restored. Moreover, these slightly different individuals were sometimes found together in the same water. It is to be hoped that no one will imagine that the question is closed, but that we shall have more experiments on *Artemia* ; in the meantime, however, Ainkin's four general conclusions will be read with interest. The representatives of the genus *Artemia* show a marked tendency to change, as regards almost all the organs of their body. The form-changes depend mainly on the physico-chemical character of the medium. The changes in individuals which live in salt solutions subject to constant dilution with fresh water do not indicate any transformation of *Artemia* into *Branchipus* ; even those in the least salt solutions retain unchanged the characteristics of their genus, especially in the male sex. The concentration of the salt solution has certainly an influence on the length of the post-abdomen, for in dense solutions those with long post-abdomens predominate, in weak solutions those with short post-abdomens.

MECHANICAL STAGE FOR DIAGNOSTIC MICROSCOPE.—Mr. Charles Baker has added to his "Diagnostic" microscope, noticed by us in this journal, vol. vi. p. 182, a detachable mechanical stage by which the whole of a 1½ inch × ¾ inch cover-glass can be examined. We illustrate this

stage herewith. The lower plate fits on to the stage of the microscope, and has a vertical movement thereon by means of runners, aided by a screw at the top, which, however, gives movement in one direction only. The screw at the side gives horizontal movement for ¾ inch both backwards and forwards, and there is a sliding top plate which can be pushed over so as to increase the travel for a further ¾ of an inch. The stage is thus well designed for systematic examinations over a large field. The price is £2 5s.

EXTRACTS FROM POSTAL MICROSCOPICAL
SOCIETY'S NOTE-BOOKS.

[Beyond absolutely necessary editorial revision
these notes are printed as written by the various
members, without alteration or amendment. Correspondence on these notes will be welcomed.—
ED. Microscopy, S.-G.]

NOTES, BY DR. G. H. BRYAN, F.R.S.

PODURA AND OTHER SCALES.—Although " scales
of Lepisma " and " scales of Podura " find a place in
most cabinets, I do not remember a series of
objects of this class having been circulated round
the P.M.S., and I venture to hope that the present
series may afford some instruction to some of the
members, and perhaps stimulate them to collect a
few of these interesting little objects for themselves. The " Podura scale " seems to have produced a great sensation among microscopists about
1873, when it was brought before our notice by the
late Mr. Beck ; but since then fresh " test objects "
have superseded it to a large extent—viz. *Pleuro-
s'gma angulatum,* and then *Amphiplcura pellucida,*
and these seem to have gradually led to the improvement in lenses required for the study of
bacteriology. At the same time these old test
scales are still worthy of attention. About the
time of their popularity Sir John Lubbock wrote a
" Monograph of the Collembola and Thysanura," published by the Ray Society in 1872, and in the same
year Mr. S. J. McIntyre contributed a paper on
them to SCIENCE-GOSSIP for December 1872 and
January 1873. Sir J. Lubbock's book contains a
valuable appendix on the scales of these insects
by Joseph Beck, illustrated by fine lithographic
plates from drawings by his brother, the late
Richard Beck. *Lepidocyrtus curvicollis.*—Scales
of this insect are often seen in old collections
labelled " *Podura plumbea,*" although the generic
name *Podura* is now given to an insect without
scales, and *plumbea* is the specific name of almost
the only one of the Collembola whose scales do *not*
show the familiar " marks of exclamation." This
is one of the largest species of the genus, and
when alive looks darkish. This slide is best
examined without the cover on, for which purpose
the central part of the ornamental paper is not

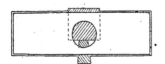

FIG. 1. MOVABLE COVER-GLASS.

gummed down. To remove the cover, carefully
insert a slip of thin card or note-paper under
the edge of the green paper at one side and
push the cover out at the other; replace in the same
way when done with (see fig. 1). If the cover
breaks, a fresh one can be inserted. This is really
an excellent device for mounting dry objects, and
is worth remembering. Note the saltatory appendage from which the name " springtail " is derived ;
also the curious humpback projection of the thorax
characteristic of the genus *Lepidocyrtus,* from which
the specific name *curvicollis* is derived (fig. 9).
L. curvicollis seems to have two varieties, charac-

terised by differences in their scales, which are
called " ordinary " and " test " scales. (¹) Fig. 10
shows scales which I believe to be test scales ;
the markings are bolder, less continuous, and,
when properly focussed, each shows a more distinct
bright line down the middle (see fig. 2). The
notes of exclamation " are thus more easy to
show up separate and distinct than in the ordinary

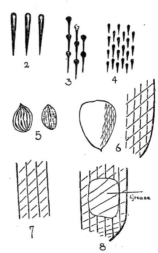

FIGS. 2-8. MARKINGS ON VARIOUS SCALES.

scale. When mounted in balsam the marks become
almost, if not quite, invisible. Dr. J. W. Arnold
succeeded in detaching the " exclamation marks "
by means of an electric spark. [See SCIENCE-
GOSSIP, 1873, p. 40.] *Lepidocyrtus violaceus,* a
smaller species than *curvicollis,* which, like that
insect, is abundant in our cellars.—I refer, with
some doubt, to this species. Both that and the
previous kind seem to congregate under or about
a sheet of paper placed in the cellar, especially if
a little flour has been sprinkled on it. I think
they like the shelter, and dislike light. The scales
are more irregular in shape and the markings
finer, but the lines of marks are wider apart (see
fig. 3); and although the marks are more
continuous, their heads are rather more bulbous
than in *L. curvicollis. Beckia argentea* is a silvery
little insect, far smaller than either of the preceding and much lighter in colour, which I
found in considerable numbers running about
stones on a wall at Colwyn Bay at dusk, when

(1) Students of these scales can purchase them at the
opticians, but they are unfortunately becoming increasingly
difficult to procure. A slide of the " test " Podura only should be
asked for, mounted dry. Under the microscope the coarser
scales and those actually in contact with the cover-glass should
be examined. In this latter case the removal of the cover-glass,
with its manifest dangers, is not necessary. The Podura scale,
on account of its variability, is not a trustworthy guide to the
testing of lenses in unskilled hands, and we put no faith in the
exhibition of the inner markings, on which so much stress is
laid. They are easily shown by an indifferent lens with a little
" stopping down " of the diaphragm, and we believe them to be
caused by an optical effect of diffraction. The true nature of
the markings on a Podura scale has, however, not yet been satisfactorily explained.—ED. Microscopy.

the dew was settling. I could only "bottle" one in a specimen tube, because when I tried to get another the first escaped. The scales are very thin and transparent, the markings very delicate and fine, and the "exclamation marks" in different rows seem as a rule to alternate more than in the preceding species (see fig. 4). It is the exceeding tenuity of the scales which makes me refer the insect to the present species (see Lubbock, p. 253). The scale would probably be a good test object if difficulty of exhibiting the markings were the only qualification. *Seira buskii* (fig. 5) may give some idea of these scales. They are of leaf-like shape, and have a very small number of exceptionally large "exclamation marks"; altogether they look

FIG. 9. *Lepidocyrtus curvicollis.*

as if they might be a primordial or ancestral type of Podura scale. They might be studied with advantage as throwing light on the structure of the scales of other Collembola, the marks, though few in number, being so vastly larger than those of any other species. I found the insect on ivy on a wall by the railway at Shepreth, Cambs, in June 1886. *Tomocerus plumbeus* or *Macrotoma plumbea* is a small, almost black insect, of which I found a solitary specimen under a stone in the woods at Colwyn Bay. This was almost the first stone I examined, and although I looked under many others I could not find any. This is an exceedingly pretty scale, quite different from those of the other Collembola, and approaching more nearly to those of the Thysanura, *Machilis* and *Lepisma*. It has no "note of exclamation," but instead has regular longitudinal striae, and between them faint transverse striae as in *Machilis* (see fig. 11), and

FIG. 10. Test-scale of *L. curvicollis* faintly marked, showing watered-silk appearance under low power.

also several radiating corrugations starting from the pedicle corresponding to those of *Lepisma*. It thus combines two different types of structure. Its dark colour also helps to render it a beautiful object. *Machilis maritima* (cf. fig. 11) is an insect about half an inch long, brown with pretty mottlings and white rings on the three long bristles, from which it receives the name of "bristle-tail." It was simply swarming on rocks just above high-water mark outside the mouth of the Dart, in South Devon, on the Kingswear side, but was very difficult to catch and bottle. I got three after trying for a long time. I have also seen it in abundance on the rocks by the shore at Lynmouth, in North Devon. The scales are to my mind prettier than those of *Lepisma* on account of the

regularity of their transverse striae, which are absent from the latter. On the other hand, this one has not the radiating marks of *Lepisma*, though both structures occur in *Tomocerus*. *Lepisma saccharina* (fig. 12).—This insect was pressed down on the slide instead of the cover ([2]), and the scales have that side uppermost which was nearest the body. In another slide the scales were pressed down on the cover, and so the uppermost side is

FIG. 11. Scale of *Machilis polypoda.*   FIG. 12. Scale of *Lepisma saccharina.*

the outermost, when the scales were on the insect It will be seen that the longitudinal marks are on the outer side, and the radial ones on the side next the body, the appearance presented being roughly sketched in figs. 6 and 7 ([3]). Of course, the markings which are uppermost appear continuous, and these are the longitudinal marks in fig. 7, where the outer surface has been pressed against the cover. The presence of a little grease on the slide represented by fig. 8 also serves to point to the

FIG. 13. Scales of *Polyommatus argus*, showing interference strata.

same conclusion. In a few scales this grease causes the scale to adhere to the cover, thus obliterating the longitudinal marks on the upper side, and the radiating striae on the under side will then be seen to be perfectly continuous and far better shown than in any other part of the scale (see fig. 8). Where only a very little grease is present.

(2) The scales are transferred to a slip of glass, generally the cover-glass, by simply pressing it gently on the body of the insect.—ED, Microscope.

(3) Mr. Joseph Beck, in his appendix to Sir John Lubbock's "Monograph of the Collembola and Thysanura," states that the longitudinal markings are on the under side of the scale.

or at the edges of a grease patch. little air-bubbles will be seen in some places to follow the grooves between the longitudinal striae. showing the thickness of the latter. As *Lepisma* may be found on nearly every kitchen hearth at night time, any reader may verify these facts without any trouble. —*G. H. Bryan, Sc.D., F.R.S.*

MICROSCOPY FOR BEGINNERS.

BY F. SHILLINGTON SCALES, F.R.M.S.

(*Continued from p. 122.*)

The advantage of the method of mounting in Canada balsam dealt with in our last number is not only its comparative facility, but that it results in getting the object close to the cover—a point that may be of importance with high powers. It also ensures the object remaining in position, very minute objects having an irritating tendency otherwise to be carried up to or beyond the margin of the cover-glass as soon as it is lowered upon the slide. At the same time. in many cases it is quite safe to mount the object directly on the slide. The process is the same, except that it is generally carried through in one operation. A drop or two of balsam is placed on the slide, the object worked into it as before, if necessary another drop of balsam added. and then the cover-glass gently lowered and pressed down. If the cover-glass is lowered with one edge first it carries air-bubbles away more readily, but it also has a tendency to displace the object. The beginner will find that at first he uses either too much or too little balsam, but he will soon learn to judge this. Excavated cells, which are used for thick objects, but for those not thick enough to require an actual cell, are rather troublesome at first, as unless there is balsam sufficient to completely fill the cell an air-bubble will be found under the cover-glass, and it is not always easy to get rid of this without displacing everything.

Mounting in glycerine jelly is simpler than mounting in Canada balsam, and the preparation beforehand is also simpler. The object must be well soaked in water. and every trace of alcohol. turpentine, etc. got rid of. Owing to the fact that glycerine jelly does not absorb air-bubbles like Canada balsam. it is well to soak in water that has been recently boiled for about ten minutes and allowed to cool. This steeping is preferably done in a stoppered bottle or jar. Prolonged soaking in water is a great aid in getting rid of air-bubbles embedded or entangled in the object, and will generally prove effectual without the aid of an air-pump. It is advisable to soak finally in a mixture of glycerine and water, say one-third of the

former, before mounting. The process of mounting is carried out as follows: The slide is placed on the brass table, the object is transferred to its centre by means of the section-lifter, and any excess of water removed by the edge of a bit of blotting-paper, care being taken that the latter does not come in contact with the object itself. By means of the point of a knife, a small spatula, or other similar instrument, a small portion of glycerine jelly is then placed on the object. the requisite quantity being easily estimated, the lamp lighted and placed beneath the brass table. In about a minute the glycerine jelly will begin to melt, and the lamp is promptly removed. Any air-bubbles should be skimmed off before the cover-glass is put on; and as the glycerine jelly will only solidify again by cooling, there is no need to hurry the process. After an examination, the cover-glass may be lowered carefully into its place. a clip slipped on, and the whole slide put aside for half a dozen hours or more to set. The excess of jelly around the cover-glass may then be removed by means of a penknife, and the whole slide cleaned by dipping in a saucer of water or holding under a running tap, finally polishing with a bit of rag. Glycerine jelly is often used when mounting in built-up cells, but before doing this it is advisable to run a wetted camel's-hair brush round the cell to make sure that no air-bubbles will cling to the sides or bottom. Pure glycerine is not often used for other than temporary mounts, as it will not set; but a mixture of glycerine and gum arabic, with a little arsenious acid, known as Farrant's solution. is often used, especially in histological preparations, as it dries at the edges. It is best bought. as home-made preparations are not always satisfactory. Glycerine acts as a solvent for carbonate of lime, and should, therefore, not be used for objects of a calcareous nature.

Canada balsam slides do not necessarily need ringing, though our own practice is to ring all our slides, but glycerine slides should be finished off with a couple of rings of gold-size. The process is very similar to that of cell-making. The slide is centred upon the turn-table, taking care to centre by means of the cover-glass and not by the slide, and a ring of gold-size run round the edge of the cover-glass. Care must be taken to just cover the edge of the latter, and not to overlap the glass too widely. Beginners generally take up too much gold-size in the brush. A neat ring is made by attention to this point, by turning the table at a moderate speed and by raising the brush slightly towards the finish. Old gold-size is best, provided it will run easily. The second ring should be added after the first is thoroughly dry. Finally, a ring or two of Brunswick black or white zinc cement may be run over all as a neat finish. We have seen suggestions for various coloured cements to be used for various classes of objects, botanical, zoological, mineral, etc.; but if a distinction is to be made, we think it should refer rather to the mounting medium employed. Our own practice, for instance, is to ring all opaque and dry mounts with Brunswick black, Canada balsam mounts with white zinc cement, glycerine mounts with white zinc cement with a fine black ring in the centre, and other mounts with a white ring on a black one. For use with immersion lenses, *all* mounts must be carefully ringed with shellac cement, or Hollis' liquid glue, or the cedar oil will dissolve the cement.

(*To be continued.*)

whilst the outer side bears the radial markings or corrugations. Mr. R. Beck further pointed out that the crossing of these two. sets of markings produced a curious optical effect. At the extremity of the scale where the markings cross each other very obliquely a series of " exclamation marks " like those of Podura is produced : but where, as at the sides, the crossings are nearly at right angles, the markings appear like rows of beads (see fig. 12). This optical effect is still more strikingly shown in fig. 13, which represents two scales of *Polyommatus argus* lying partly over each other, and producing an appearance very similar to that of a coarse Podura scale. See Carpenter's "Microscope," 7th ed., pp. 899-904, from which figs. 10 to 13 have been copied with a view to illustrating the foregoing notes. Fig. 9 is from the "Micrographic Dictionary."—ED. Microscopy.

CONDUCTED BY JAMES QUICK.

PHYSICS AT THE BRITISH ASSOCIATION.—At the meeting of the British Association at Bradford, just ended, Section A (Physics) was characterised by some brilliant papers, both theoretical and practical. The address of the president of Section A, Dr. J. Larmor, F.R.S., opened the proceedings on Thursday, September 6. It was an excellent *résumé* of the work of some of the foremost physicists and mathematicians of the present century, and also included a lucid explanation of recent researches in the realm of atoms and ionic charges. Among the papers which followed that of Dr. Larmor was one by Messrs. A. Dufton and W. M. Gardner on "The Production of an Artificial Light of the same character as Sunlight," an interesting exhibition of contrast-colours accompanying the reading. It was pointed out by the authors what a common experience it is that many colours alter in appearance when seen by artificial light, and how difficult it is for colour-workers to produce their work satisfactorily when working with anything else but daylight. They use at present the light from an electric arc as being the best approach, but it is not efficient. The peculiar character of daylight is due essentially to the selective absorption exercised by the atmosphere, and the success which the authors of the paper have attained is due partly to their having taken advantage of this fact. They have filtered the light from an arc lamp, by means of transparent solutions of coloured glasses, to as nearly as possible the same extent that does the atmosphere. Friday, September 7, was taken up entirely with mathematical physics papers, although Dr. Larmor, in his communication on "The Dynamical Statistics of Gas Theory, illustrated by Meteor Swarms and Optical Rays," led up to the mathematical considerations by some very clear graphical illustrations. Saturday was a practical day, in that papers were read upon the practical applications of physics. After the reports of one or two Committees upon physical matters had been read, Sir William Preece gave an interesting account of his work on "Wireless Telephony." The first experiments were made in 1894 across Loch Ness. Preece's electromagnetic system for wireless telegraphy was set up, and trials were made to compare telephonic signals with the ordinary telegraphic ones, to ascertain whether speech could be maintained under the same conditions as for Morse signalling. The results were satisfactory, speech being exchanged across the loch at an average distance of 1½ mile between the two parallel wires. Further experiments in 1899 showed that the maximum effects were produced when the parallel wires were terminated by earth-plates in the sea, the conductive effect through the water greatly increasing the efficiency of the apparatus. Ordinary telephonic transmitters and receivers were used. These successful results, as might have been expected, have been quickly followed up by practical and useful applications. It being necessary to establish communications between the islands known as the Skerries and the Anglesey mainland, this system of wireless telephony was determined upon, as the bottom of the channel between the two points is too rough and the currents too violent for a cable to be laid. A wire 750 yards in length was therefore erected along the Skerries, and on the mainland one of 3½ miles from a point opposite the Skerries to Cemlyn. Each line terminates by an earth-plate in the sea. The average distance between the parallel portions of the two wires is 2·8 miles. Telephonic communication is easily maintained, and the service is proving a good one. Mr. C. E. S. Phillips read a paper on "The Apparent Emission of Cathode Rays from an Electrode at Zero Potential," and showed that the green patches which often appear on the surface of the walls of a Röntgen-ray tube, while a discharge is passing, are due to irregularities on the surface of the cathode, as they could be made to alter their position when a movable cathode was employed; and, moreover, that there was strong evidence for regarding an emission of gas in jets as taking place from the electrodes. Probably one of the most important and exciting papers read before Section A was that by Professor J. Chunder Bose, of Calcutta, on "The Similarity of the Effect of Electrical Stimulus on Inorganic and Living Substances." Professor Bose has, indeed, made important strides into the realm of the propagation of electrical waves, and has gone very far towards completely bridging the experimental gap between optical phenomena and Hertzian wave phenomena. He has paid special and untiring attention during the last few years to the subject of the "coherer" principle, and has now put forward theoretical suggestions which, to say the least, are astonishing, not only in the departments of physics and chemistry, but also in biology. We are to think of an electric stimulus as of the nature of a stress, producing in many forms, according to the subject or substance acted upon, what may be termed a strain. Within limits the strain is proportional to stress, but fatigue occurs when the stress is excessive. These actions are to be considered as being upon the atomic, or perhaps sub-atomic, scale. Prof. Bose instanced many physical and physiological phenomena strengthening his views. Indeed, the parallel between the behaviour of a coherer in an electric circuit and the behaviour of living tissue in its ordinary physiological functions is very close; so much so as to be suggestive of a common explanation. The author indicates such an explanation in this electric stress-strain hypothesis, since he regards vital functions as merely the manifestations of an electric stimulus in a substance (tissue), acting molecularly or atomically in the same manner as does a coherer. He, indeed, carries his hypothesis into the domain of every form of differentiated living tissue. On Tuesday, September 11th, a long discussion upon the far-reaching subject of "Ions" was opened by Prof. Fitzgerald, who objects to the term "ionisation." This term, he thought, should only be used to denote the presence of charged atoms. The discussion was taken up by many prominent physicists and also by Prof. H. E. Armstrong, who said that the chemist regarded an atom as a sacred structure.

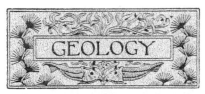

GEOLOGY

CONDUCTED BY EDWARD A. MARTIN, F.G.S.

A LOWER CARBONIFEROUS ISLAND.—There are still some geologists who firmly believe in the permanence of our ocean basins and continental areas, and who refer us for the date of their origin to the earliest years of the separate existence of our globe as a solid body. They contend that a period will only be placed upon them when the globe itself comes to an end, but even they recognise that the edges, at least, of our continents play " see-saw," so to speak; the countries bordering the ocean at one time being submerged beneath the waters themselves, and at others raised far out of reach of the waves. The ocean bed which now closely surrounds the land has in times long past also partaken of this see-saw movement. Once it formed dry land; at other times it approached submergence, forming those raised beaches which are so frequent around our coasts; finally it sank into the trough of the sea, allowing the waters of the ocean to roll where formerly was dry land. The Atlantic Ocean is considered to have at least had its origin prior to the deposition of the Lower Carboniferous, and this view is supported by Prof. Hull and Mr. A. J. Jukes-Browne. All agree that the Atlantic must by that time have been in existence, in order to supply the necessary sediment and other material for the building-up of that formation. According to Jukes-Browne, the sea covered in Lower Carboniferous times the larger part of Great Britain, excepting the regions beyond the Highlands of Scotland. By deduction this may have been an easterly extension of the great Atlantic, and our country was then in one of the downward throes in the great game of continental see-saw. In the midst of the sea which then covered England Mr. Jukes-Browne has ingeniously inferred the existence of a large island of irregular shape, which had its centre in St. George's Channel, touching Ireland, Scotland, and Wales, and pushing an arm out to the east through Wales inland into Shropshire; this land approximately occupying the position of the water area of the present day. The existence of this island is manifested by the series of shore-deposits and conglomerates which are found around its edge. The thinning northward of the Carboniferous Limestone in Monmouthshire and Gloucester is evidence of the shallowing of the sea as approach was made to the island; and a similar thinning occurs throughout the South Wales coalfield. Passing across the water to Ireland, we find Carboniferous rocks in Wexford, which have been judged to have been accumulated in a narrow bay. Northward these rocks occur through Kildare; whilst at Howth, Swords, and Rush there is ample evidence of the close proximity of land. Thus have been traced a southern and a western side to our island. Proceeding, we are reminded of some thick beds of conglomerate with Ordovician pebbles between

Rush and Skerries, in one place there being indeed boulder beds, formed at the base of the early cliffs which bounded the island, and where the fallen blocks had been enveloped in the grey Carboniferous Limestone which was then in the making. Still farther northward Mr. Jukes-Browne finds that around Drogheda the Lower Carboniferous strata appear to have been laid down along a gradually shelving shore, whilst in the north-east of County Down there are limestones associated with red shales and sandstones which are probably shore beds of Upper Limestone age. Passing across into Scotland, we meet the basal conglomerates of Lanarkshire, the so-called Calciferous Sandstone being also contemporary with the lower portion of the Carboniferous Limestone of England. If we now draw our island boundary southward towards Wales again, we find in North Wales a considerable thinning-out of the Carboniferous Limestone towards the south-east, whilst it is entirely absent from the Shrewsbury district. There we are apparently at a point which the Lower Carboniferous sea never covered; and where, so far as actual evidence goes, we have reached the eastern limit of the island. Just within the limits of the sea we find the basement bed of lava and volcanic ash, on the east side of the Wrekin, and this may have been thrown up under water by volcanic agencies. In the land area of South Staffordshire we find the edges of the ancient Silurian rocks, worn, denuded, and lying bare, until the Coal Measures came to be borne in upon them; whilst in Warwickshire the Cambrian are similarly exposed ere the Coal Measures were thrown down upon them, both areas being free from the marine agencies at work during Lower Carboniferous times. Again, the Coal Measures and Millstone Grit rest, in Shropshire, unconformably on Old Red Sandstone and Silurian rocks, these not being submerged till late in the Carboniferous period. The deduction of the former existence of this large island in Lower Carboniferous times, from the nature of certain local deposits and the directions in which they thin out, is interesting as showing what geology can do in the way of building up pictures of ancient land and sea areas by a study of the physical relations of the rocks.—*E. A. Martin.*

THE SALT LAKE OF LARNACA.—In a recent paper read before the Geological Society Mr. C. V. Bellamy described the Salt Lake of Larnaca, Cyprus. The lake occurs in a basin shut off from the sea, its deepest part being about 10 feet below sea-level. The barrier between the salt lake and the sea is made of stiff calcareous clay, associated with masses of conglomerates resting on plastic clay, that on watery mud, and that again on stiff calcareous clay. The sea-water appears to percolate through the highest deposits, meeting with checks in the conglomerates, and thus reaches the basin somewhat slowly, where it is evaporated by the summer heat until it deposits its salt. Artificial channels have been made to carry the flood-water from the land direct to the sea, so that it does not dilute the brine of the lake. The rainfall in the catchment area round the lake is at the most only enough to supply 223,000,000 gallons; and as the lake contains 480,000,000 gallons when full, the balance of 257,000,000 gallons must be derived from the sea. The salt harvest begins in August, at the zenith of summer heat, and it is reported that a single heavy shower at that time of year suffices to cause its ruin.—*E. A. Martin.*

NEOLITHIC HUT-CIRCLES NEAR HAYES AND KESTON.—For the last twelve years Mr. George Clinch, F.G.S., has been engaged in tracing remains of Neolithic man in the neighbourhood of Hayes, West Wickham, and the Shirley Hills, on the borders of North Kent and Surrey. The results obtained have now been printed in the Journal of the Anthropological Institute. In the course of his investigations he caused to be excavated over 150 Neolithic hut-circles, principally near Hayes and Keston, with the result that some hundreds of implements of eighteen different types have been discovered. West Wickham has alone yielded 926 flakes, 141 scrapers, with saws, arrow-heads, spear-heads, etc. At Millfield, Keston, Mr. Clinch found, besides 61 ordinary flakes, 400 flakes from which the pointed ends had been broken, this place having possibly been a factory for arrow-tips or sickle-teeth. The numerous hut-circles examined all show a similarity in method of construction. A ground-space was excavated from 5 m. to 10 m. in diameter, circular in form, and about 1 m. in greatest depth. The removed earth may then have been carefully arranged as a continuous mound around the pit, and in this mound a number of long branches of trees were probably planted, the ends of which met over the middle of the hut. A roof consisting of a thatch of heath or reeds completed the means of protection from the external elements, whilst the encircling mound would help to throw off superfluous rainfall, and afford some degree of warmth and shelter. In some of the larger huts there was a raised mound in the centre, on the sides of which the inhabitants may have reclined when rest was required. Sometimes, however, the mound seems to have been placed in the hut in order to support or steady the lower end of the trunk of a tree upon which the rafters of the roof rested. Owing to the highly inflammable character of the structure it would not have been safe to have a fire within the hut during very dry or windy weather; the cooking fire, therefore, would be made at a short distance from the dwelling. The smaller depressions, from 1·25 m. to 3 m. in diameter, have already been identified as the hearths. Within the Surrey border Mr. Clinch turned his attention to Croham Hurst, and he considers that he has there discovered good evidence of Neolithic times in similar circular mounds, associated flakes and chips of flint, and the form of the interior depressions.

A CARBONIFEROUS CRUSTACEAN.—I have lately found a well-preserved fossil, the abdomen of a crustacean, in the Middle Coal Measures at Poynton, Cheshire. I sent it to the South Kensington Natural History Museum, where Dr. Woodward was kind enough to identify it for me. It was described by him in the "Quart. Journ. Geol. Soc." vol. xxxv., 1879 (p. 551, pl. xxvi.) and was named *Necroscilla wilsoni* H. Woodward.—*J. McDonald, 2 Co-operative Street, Hazel Grove, Stockport; July 6th, 1900.*

EVOLUTION OF GEOLOGY.—In his presidential address before the Geological Section of the British Association at Bradford, Professor W. J. Sollas, D.Sc., LL.D., F.R.S., took Evolutional Geology as his text. Commencing at the very beginnings of this earth, he invited the physicists to more frequently come to the aid of the geologists, and pointed out that when they were studying even the earliest conditions of our planet they were, in fact, geologists.

CONDUCTED BY HAROLD M. READ, F.C.S.

PRACTICAL INORGANIC CHEMISTRY. — "The Modern System of teaching Practical Inorganic Chemistry and its Development" was the subject selected by Professor W. H. Perkin, jun., for the presidential address to the Chemistry Section at the recent British Association meeting at Bradford. The president took as his keynote the question as to whether the modern method of teaching chemistry had kept pace with the enormous strides made in science during recent years. He questioned whether the introduction of chemistry into the syllabus of so many schools had so far resulted in the making of a student capable of grasping the underlying principles, or of showing that originality of thought, the absence of which is fatal to the final success of everyone. So much of the present-day teaching consists of filling a student's mind with bare facts, to be utilised in the examination-room, that the time for laboratory work and the possibilities of sound deductions are reduced to a minimum. The student may, during the later part of his training, acquire a sound knowledge of his subject, but he has been so handicapped previously that he has not the time to make use of the opportunities which come too late. The result is that, though the student may acquit himself to the satisfaction both of his teacher and of himself, yet when he attempts to venture beyond the cut-and-dried facts with which his knowledge has been "bolstered up" he is helpless. No doubt originality of thought is not bestowed with any too lavish a hand; but none the less, a methodical laboratory training, supplemented by the text-book and fostered by the teacher, would help to make up for that which Nature has omitted. In endeavouring to cover the examination field a fearful amount of valuable time is wasted, and therefore, as specialisation is inevitable, owing to the vastness of the general subject, Professor Perkin would have it begun as soon as the broad facts of the science have been assimilated. He suggests that the written examination should be curtailed, while more time is given to practical work, so that the latter may be made profitable, instead of being wasted, as it now is, to a great extent. The advantages of making students take the attitude of discoverers are so manifest, and the results already obtained by the substitution of this new method for the old system of qualitative analysis are so striking, that we look to the address of Professor Perkin to mark the new era, the commencement of which has been so long delayed. We are pleased that our remarks of last month about the disgraceful system of "spotting" salts should have been so borne out by Professor Perkin's comments on present-day teaching. We cannot deny the dangers inherent to any drastic change in a system which has so long held sway, but no one will doubt its advisability.

ASTRONOMY.

CONDUCTED BY F. C. DENNETT.

	1900	Rises.	Sets.	Position at Noon. R.A.	Dec.
	Oct.	h.m.	h.m.	h.m.	° '
Sun	7 ..	6.12 a.m. ..	5.24 p.m. ..	12.51 ..	5.26 S.
	17 ..	6.28 ..	5.2 ..	13.28 ..	9.11
	27 ..	6.46 ..	4.42 ..	14.6 ..	12.44

	Rises.	Souths.	Sets.	Age at Noon.
	Oct. h.m.	h.m.	h.m.	d. h.m.
Moon .. 7 ..	4.33 p.m. ..	11.10 p.m. ..	4.38 a.m. ..	13 16.3
17 ..	0.6 a.m. ..	7.23 a.m. ..	2.27 p.m. ..	23 16.3
27 ..	10.48 a.m ..	2.51 p.m. ..	6.59 p.m. ..	3 22.33

		Souths. Oct. h.m.	Semi-Diameter.	Position at Noon. R.A. h.m.	Dec. ° '
Mercury..	.. 7 ..	0.49 p.m. ..	2·5" ..	13.51 ..	12.22 S.
	17 ..	1.4 p.m. ..	2·7" ..	14.46 ..	18.6
	27 ..	1.15 p.m. ..	3·1" ..	15.36 ..	22.11
Venus	.. 7 ..	9.1 a.m. ..	9·8" ..	10.4 ..	11.39 N.
	17 ..	9.4 a.m. ..	9·0" ..	10.47 ..	8.19
	27 ..	9.8 a.m. ..	8·4" ..	11.30 ..	4.27
Mars	.. 17 ..	7.17 a.m. ..	3·0" ..	8.59 ..	18.32 N.
Jupiter	.. 17 ..	2.49 p.m. ..	15·8" ..	16.32 ..	21.28 S.
Saturn	.. 17 ..	4.17 p.m. ..	7·4" ..	18.0 ..	22.43 S.
Uranus	.. 17 ..	2.50 p.m. ..	1·8" ..	16.33 ..	21.56 S.
Neptune	.. 17 ..	4.16 a.m. ..	1·3" ..	5.57 ..	22.13 N.

MOON'S PHASES.

	h.m.		h.m.
1st Qr. .. Oct.	1 .. 9.11 p.m.	Full .. Oct. 8 .. 1.18 p.m.
3rd Qr. .. „	15 .. 9.51 a.m.	New .. „ 23 .. 1.27 p.m.
1st Qr. .. „	31 .. 8.17 a.m.

In perigee October 8th at 6 a.m. ; in apogee on 21st at 7 a.m.

METEORS.

			h.m.	°
Oct. 11-24.	..	ε Arietids	Radiant R.A. 2.40	Dec. 20 N.
„ 17-20.	..	(ξ) Orionids	„ „ 6.8	„ 15 N.
October	..	δ Geminids	„ „ 7.4	„ 23 N.

CONJUNCTIONS OF PLANETS WITH THE MOON.

					° '
Oct. 1	..	Saturn†	.. 4 a.m. ..	Planet	1.98 S.
„ 17	..	Mars	.. 4 a.m. ..	„	8.29 N.
„ 19	..	Venus†	.. 7 p.m. ..	„	6.11 N.
„ 25	..	Mercury*	.. 8 p.m. ..	„	2.1 S.
„ 26	..	Jupiter†	.. 12 p.m. ..	„	0.27 S.
„ 28	..	Saturn°	.. 1 p.m. ..	„	1.50 S.

* Daylight.   † Below English horizon.

OCCULTATIONS AND NEAR APPROACH.

Oct.	Object.	Magnitude.	Dis-appears. h.m.	Angle from Vertex. °	Re-appears. h.m.	Angle from Vertex. °
7 ..	κPiscium	5·0 ..	1.35 a.m. ..	7 ..	2.29 a.m. ..	236
11 ..	ω²Tauri	4·6 ..	8.47 p.m. ..	74 ..	9.25 p.m. ..	346
13 ..	ζ „	3·0 ..	6.36 a.m. ..	16 ..	7.23 a.m. ..	286
14 ..	νGeminorum	4·0 ..	3.0 a.m. ..	68 ..	3.43 p.m. ..	353
17 ..	αCancri	4·3 ..	0.17 a.m. ..	51 ..	Near approach.
17 ..	κ „	5·0 ..	5.26 a.m. ..	173 ..	6.30 a.m. ..	269
29 ..	dSagittarii	4·9 ..	8.27 p.m. ..	327 ..	8.46 p.m. ..	293

THE SUN.—Small spots and groups may sometimes be seen.

MERCURY is an evening star all the month, reaching its greatest elongation, 23°46' east, at 4 a.m. on October 30th, setting rather more than an hour and a half after the sun at the end of the month. Its path takes it from a little north of Spica Virginis at the beginning of the month, to closely south of δ Scorpii at the end.

VENUS is a morning star all the month. Its path starts from a point about 6° west of Regulus, being closely south of that star on October 7th, and on 31st passes very near to β Virginis. It rises at 1.45 a.m. at the beginning of the month and rather over an hour later at the end of October.

MARS rises at 11.44 p.m. on October 1st, and about 25 minutes earlier at the end of the month. Its path lies almost wholly through the constellation Cancer, about the 24th inst. entering a very barren portion of Leo.

JUPITER is an evening star, but too near the sun for satisfactory observation. It sets at 7.50 at the beginning of the month, and at 6.10 at the end. Jupiter is in conjunction with Uranus at 10 p.m. on the 19th, passing 25' to the north.

SATURN sets at 9.14 p.m. on the 1st, and 7.24 on the 31st, so must be looked for as soon as the dusk will permit.

URANUS is now out of reach of our vision.

NEPTUNE is still on the western borders of Gemini, close to Orion, and Taurus. It rises about 9.12 p.m. at the beginning of the month and two hours later at the end.

THE SOLAR ECLIPSE.—Altogether the finest photographs are said to have been taken by Mr. Charles Burckhalter, with a 4-inch lens of fifteen feet focus. A novel arrangement was some time since invented by him by which it was possible to give different times of exposure to different parts of the same photograph. A plate exposed for eight seconds received :—

$$\text{Exposure at} \quad \begin{array}{ccccc} 0\cdot04 & 0\cdot23 & 1\cdot76 & 3\cdot20 & \text{and} \quad 8\cdot00 \text{ seconds.} \\ 16' & 20' & 32' & 60' & \text{and} \quad 110' \end{array}$$

distance from the sun's limb. The result is that every part of the sun's surroundings, from the prominences to the outer corona, are shown in beautiful detail. Mr. Burckhalter is the director of the Chabot Observatory, in California, and his eclipse expedition was made possible by the munificence of Mr. John Dolbeer, of San Francisco.

DAYLIGHT METEOR.—Another of these remarkable objects was seen on Sunday, September 2nd, at 6 h. 52 m., p.m., just before sunset, at places so far separated as Edinburgh and Wiltshire.

THE OCCULTATION OF SATURN on September 3rd seems to have been well seen by some observers, who were struck by the great difference in the brightness of the planet and of the moon. From one station the moon and planet presented the appearance seen when objects lying at the bottom of a rapidly rushing stream are being examined.

PROFESSOR KEELER, the director of the Lick Observatory, was, we are sorry to hear, snatched away by apoplexy on August 12th, at the early age of forty-three. Previously to succeeding Prof. Holden, in 1898, he had attracted wide attention as director of the Allegheny Observatory. Some of our readers will possibly remember that on p. 100 of our third volume we noted the late professor's demonstration at that observatory of the character of Saturn's rings, photographically showing, by the displacement of the lines of the spectrum, that the inner portions of the ring were revolving faster than the outer sections. The plates had an exposure of two hours. This great work was accomplished in 1895.

CATOCALA FRAXINI IN ENGLAND.—A specimen of this rare moth, known under the English name of " The Clifden Nonpareil," is stated to have been found by a lady on sandhills at Blakeney, Norfolk, about the last week in August. It is probably a migrant from the Continent.

LARVAE OF LYCAENA BAETICA.—I have had the pleasure this summer of finding the larvae of *Lycaena baetica* feeding upon the bladder-senna (*Colutea arborescens*), and I am now breeding some fine specimens of this handsome little tailed-blue butterfly. This plant is not common in our island, so that the butterfly larvae doubtless feed, as elsewhere in Europe, on seed-pods of other Leguminaceae.—*George Baker*, 11 *Saumarez Street, Guernsey*: *September 5th*, 1900.

FASCIATED COTONEASTER.—I am sending you a remarkably fine case of fasciation. It occurred near Abergele, North Wales, on a shrub of *Cotoneaster*, wherein the lower portion of the stem is normal until within about ten inches from the top, when it begins to broaden and flatten to about three-quarters of an inch wide. The small branches on the portion of the stem anterior to the fasciation are changed into leaflets. The broadening divides near its base, forming a short flat branch in the fasciated portion, both being curled at the ends.—*H. Emmett, 33 Selwyn Street, Shelton, Stoke-on-Trent.*

VANESSA ANTIOPA.—The year 1900 appears to have been favourable to the migratory butterflies, as in addition to the invasion of " Clouded Yellows," mentioned last month, numerous specimens of the " Camberwell Beauty " (*Vanessa antiopa*) have been noted. Among the localities are Herne, in North Kent (August 18), Yattendon, Berks (August 18 and September 5), Newlands, North Sussex (August 26), Lindfield, Sussex, Beddington, Surrey (August 31), Watling, Oxfordshire (August 19), Holt, Norfolk (August 30), and in Huntingdonshire (August 19). These are doubtless all migrants from the mainland of the Continent. We do not know of any trustworthy instance of the larva of this species occurring in this country during the last half-century.

NEW ORCHIS.—In the Rev. E. F. Linton's " Flora of Bournemouth," referred to *antè*, page 116, mention is made of the one thing that prevents this work being a mere, however admirably complete, list of plants and localities, and that is the description of a new form of *Orchis*, which is separated from the spotted orchis under the name of *O. ericetorum*, n. sub-sp. It is described as a plant of moist places and bogs on heaths, and is said to range from the extreme north of Britain to the south coast, and is also found in Ireland and the Channel Isles. A number of characters are enumerated as distinct from typical *Orchis maculata*, but whether these are of such value as to

give sub-specific rank to this heathland form is a doubtful matter; though, indeed, the Rev. E. F. Linton seems to be of the opinion that it should take specific rank, if further study shows its distinctive features to be constant. Whatever may be its status, *Orchis ericetorum* Linton is certainly a plant to be looked for next season, and in this connection I may mention that during the past summer my friend, Mr. Jas. Holloway, brought under my notice a very distinct-looking form of *Orchis maculata*, found growing in bogs in Epping Forest, which I now think may prove to be the *Orchis ericetorum*. The variability of *Orchis maculata* has in the past received notice in the pages of SCIENCE-GOSSIP, notably in the article by Dr. G. H. Bryan in 1896 (vol. iii., n.s., p. 175), where are figured examples of variation of the labellum. Some of the figures no doubt illustrate *Orchis ericetorum*, which is, I believe, certainly indicated in the first paragraph of the second column on p. 175.—*C. E. Britton*, 35 *Dugdale Street, Camberwell.*

COLEOPTERA NEAR CARLISLE.—Having during the last few months taken some interesting beetles in this district, I thought a few notes on same might be of interest to some of your readers. *Nebria gyllenhalii* is common under stones in our mountain streams. *Lebia chlorocephala* by pulling grass tufts in winter; not common. *Taphria nivalis* running on roads; two specimens only. *Anisodactylus binotatus* occurs locally under stones and running on roads in spring. *Bembidium rufescens*, *B. aeneum*, *B. minimum*, *B. decorum*, *B. tibiale*, *B. atrocoeruleum*, and *B. saxatile* are all locally common. *B. schüppelii*, *B. monticola*, *B. lunatum*, *B. stomoides*, and *B. paludosum* are rarer, and generally extremely local. *Agabus guttatus* is common, and *A. femoralis* is also common, but local. *Conosoma immaculatus* occurs rarely in putrid fungi. *Tachinus collaris* was very abundant in flood refuse. In about an hour I got some seventy specimens. *Bolitobius atricapillus*, *B. trinotatus*, and *B. pygmaeus* occur in fungi. *Staphylinus erythropterus* is sometimes found running on roads and at tree-roots in winter. *Stenus guttula* is common near water. *Pselaphus heisei* and *Tychus niger* occur among *Sphagnum* moss. *Choleva longula*, one specimen sitting on a blade of grass. *Paramecosoma melanocephalum*, not uncommon in grass tufts and flood refuse. *Anatis ocellata*, sometimes beaten commonly from fir-branches. *Athous niger*, uncommon, generally by sweeping. *Podabrus alpinus* and *Ancistronycha abdominalis* are local and rare. *Telephorus figuratus* is very local, but abundant about the end of June. *Barynotus moerens* is not uncommon on Dogs' Mercury. *Pissodes pini*, fourteen specimens found sitting on the undersides of fir-logs. *Apion carduorum*, *A. striatum*, *A. seniculum*, *A. frumentarium*, *A. nigritarse*, and *A. humile* are common, and *A. affine*, *A. gyllenhalii*, and *A. viciae* are local. *Rhynchites aeneovirens*, not common, were beaten from hawthorn. *Grammoptera tabacicolor* and *G. ruficornis* both occurred, the latter very commonly. *Lema melanopa* occurred occasionally by sweeping. *Chrysomela hyperici* is rare. *Cryptocephalus labiatus*, one specimen beaten from birch in June. Taken all in all, this year up to that month has been favourable for the Coleopterist, some groups—as the Longhorns, for instance—being more in evidence than is usual.—*Jas. Murray*, 11 *Close Street, Carlisle.*

THE PHOTOGRAPHIC SALON.

THE eighth annual exhibition of the Photographic Salon was opened at the Dudley Gallery of the Egyptian Hall, London, on September 21st. and continues until November 3rd. This collection affords an almost unique opportunity of comparing the methods of applying the different styles and processes of artistic photography as used throughout the world. There are upwards of two hundred and fifty pictures shown, and the average merit of this year's collection is markedly in advance of its predecessors. The "Salon" was founded and has been maintained by a private society of leading persons interested in artistic photography. It is known to themselves as "The Linked Ring."

NOTICES OF SOCIETIES.

*Ordinary meetings are marked †, excursions * ; names of persons following excursions are of Conductors. Lantern Illustrations §.*

MANCHESTER MUSEUM, OWENS COLLEGE.

Oct. 13.—"The History of Gold, Tin, and some other Metals." Prof. W. Boyd Dawkins, F.R.S.
„ 14.—"Hill and Valley." Prof. W. Boyd Dawkins. F.R.S.

NORTH LONDON NATURAL HISTORY SOCIETY.

Oct. 4.—† Pocket-book Exhibition and Microscopical Evening.
„ 13.—* Horniman Museum, Forest Hill.
„ 18.—† "The British Flora in Relation to those of Neighbouring Lands." C. S. Nicholson.

NOTTINGHAM NATURAL SCIENCE RAMBLING CLUB.

Oct. 27.—Annual Meeting and Conversazione.

KENSINGTON POPULAR SCIENCE LECTURES.

Oct. 18.—§ "The World when Young." Rev. J. O. Bevan, M.A., F.S.A. F.G.S.
„ 25.—§ "The Chemistry of the Earth's Crust." Cecil Carus-Wilson, F.R.S. Edin., F.R.G.S. F.G.S.
Nov. 1.—§ "How Rocks are made." Cecil Carus-Wilson.
„   8.—§ "The Life of the Past." F. W. Rudler, F.G.S., etc.
„ 15.—§ "Land and Scenery." H. R. Mill, D.Sc., LL.D., F.R.S. Edin.
„ 22.—§ "Some Electrical Discoveries." Prof. Astley Carus-Wilson, M.A., Assoc. M.Inst. C.E., M.Inst. E.E.

Kensington Town-hall, at 4.45 p.m.

LAMBETH FIELD CLUB AND SCIENTIFIC SOCIETY.

Oct. 1.—† "My Friends, the Whistlers and Croakers." Mr. E. W. Harvey-Piper.
„ 13.—* Brockwell Park. Mr. H. J. Stidmersen.
„ 15.—Gossip Meeting. Display of Fungi. Notes on Meteors.
„ 22.—* Annual Soirée and Exhibition.

NOTICE TO READERS.

In consequence of pressure on our space, Mr. E. Sanger Shepherd's continuation of the "Photography of Colour" and other important articles unavoidably stand over.

ANSWERS TO CORRESPONDENTS.

E. R. D. (Salisbury).—We fear the electric lamp you kindly submit is too cumbersome for the use of insect-collectors, as it occupies one band. Have you seen the little electric lamps attached to a coat button-hole, used by omnibus ticket-inspectors in London? They are much more applicable. It is nearly seven years since the address of SCIENCE-GOSSIP was in Piccadilly.

J. H. B. (Chatham).—The spider is *Epeira diademata* Clk., the common "garden spider." The species is subject to great variation in point of colour, so it is only natural that extra large or unusually coloured individuals should be considered as rarities. The most tangible specific distinction of this spider is the vulva (on the underside of the abdomen, towards its fore part). This organ under a hand lens appears to consist of two dark dots, from which springs a long semi-transparent projection which bends down towards the spinners. Formalin is not suitable for spiders. They have to be pulled about for purposes of examination and identification, and if preserved in formalin solution the legs, palpi, and other parts get brittle and break off. Use methylated spirit.—F. P. S.

F. B. (Hailsham).—Ladies' tresses (*Spiranthes autumnalis*).

NOTICES TO CORRESPONDENTS.

TO CORRESPONDENTS AND EXCHANGERS.—SCIENCE-GOSSIP is published on the 25th of each month. All notes or other communications should reach us not later than the 18th of the month for insertion in the following number. No communications can be inserted or noticed without full name and address of writer. Notices of changes of address admitted free.

EDITORIAL COMMUNICATIONS, articles, books for review, instruments for notice, specimens for identification, &c., to be addressed to JOHN T. CARRINGTON, 110 Strand, London, W.C.

SUBSCRIPTIONS.—The volumes of SCIENCE-GOSSIP begin with the June numbers, but Subscriptions may commence with any number, at the rate of 6s. 6d. for twelve months (including postage), and should be remitted to the Office, 110 Strand, London, W.C.

NOTICE.—Contributors are requested to strictly observe the following rules. All contributions must be *clearly* written on one side of the paper only. Words intended to be printed in *italics* should be marked under with a single line. Generic names must be given in full, excepting where used immediately before. Capitals may only be used for generic, and not specific names. Scientific names and names of places to be written in round hand.

THE Editor will be pleased to answer questions and name specimens through the Correspondence column of the magazine. Specimens, in good condition, of not more than three species to be sent at one time, *carriage paid*. Duplicates only to be sent, which will not be returned. The specimens must have identifying numbers attached, together with locality, date, and particulars of capture.

THE Editor is not responsible for unused MSS., neither can he undertake to return them unless accompanied with stamps for return postage.

EXCHANGES.

ENTOMOLOGIST'S ANNUAL, complete set, and many other entomological works; exchange British shells, canaries, fancy pheasants, or glass-topped boxes.—C. S. Coles, Pheasantries, Hambledon, Cosham, Hants.

MICROSCOPY.—Animal hairs, about sixty varieties; what offers? Mounting material not required.—J. T. Holder, 77 Erlanger Road, St. Catherine's Park, London, S.E.

BOTANICAL.—Offered. Lond. Cat. B. Plants 264, 701, 798, 1,063, 1,285, 1,619, 1,786, 1,870, 1,926. Wanted, 219, 324, 342, 350, 384, 387, 615, 800, 847.—A. Hosking, 48 Norwich Street, Cambridge.

DIATOM slides in exchange for other objects, or works on microscopy.—H. Platt, Priory Villa, Victoria Road North, Southsea.

CONTENTS.

SCENERY OF LLANBERIS PASS.

By F. E. Filer.

EVEN those who do not habitually speculate on the causes which have produced the present configuration of the country, generally have their interest awakened when passing through mountainous districts, and nowhere, perhaps, in Great Britain more than when amongst the mountains and valleys of North Wales. Indeed, the varied nature of the rocks, their extreme antiquity, their crumpled and contorted bedding, and the

ice-rounded rocks, perched blocks, and ice-ground furrows, showing that a glacier of great size once descended the pass, fed by tributary glaciers; the signs of the latter being most prominent in the two great hollows of Cwm-glas and Cwm-glas-hach. A short distance along the road from Llanberis the vertically bedded cliffs, on the right, sweep up to the ridge of Llechog, at the foot of which, near the road, is a series of beautifully rounded rocks,

Photo by] [*A. W. Dennis.*

FIG. 1. VALLEY OF CWM-GLAS-BACH, FROM THE LLANBERIS PASS.

very evident traces of later glacial action, cannot fail to attract attention. These features are, perhaps, most marked amidst the grand rock scenery in the neighbourhood of the Llanberis Pass.

There is in the following notes no claim made to anything of an original quality, but they may possibly be of assistance to those who visit the district for the first time, and need a little guidance. The Pass of Llanberis, which lies in a north-westerly direction, reaches at its south-east end an elevation of 1,160 feet, and slopes away north-west down to 340 feet above sea-level at the town of Llanberis. Along the whole of this length—a distance of about five miles—there can be seen

their sides scored with striae, sometimes so deep as to be more like horizontal gullies, with the finer scratchings on their inner surface remaining still fresh and sharp. Here and there perched blocks are left. In many cases their situation leaves no doubt as to the agency which dropped them into position, as it would be impossible for the broken debris of the cliff behind to have reached their present resting-places. Bordering the lake on the left hand, the rocks curve down to the water's edge; their weathered surfaces showing still the mammillated outlines characteristic of glaciation.

Continuing along the pass through the little village of Nant Peris, there presently opens out on the right the tributary valley of Cwm-glas-bach. At

G

its near western end, overlooking the river, is a large bluff having its rounded top liberally scattered with perched blocks. This lies just outside the picture of Cwm-glas-bach, to the right. The mammillated rocks of Cambrian grit show clearly in the foreground, and extending high up the hillsides can be seen the rocks smoothed by the grinding ice, although now much broken up by weathering. The destructive effects of frost will, of course, be most apparent on the lower faces of the rocks which front down the valley, and these broken faces unfortunately are most prominent in the photograph, so the pictorial result is not so striking as the reality. Evidently a tributary

tending down the valley are a series of moraine heaps, some of them of great size, placed more or less concentrically, and scattered with enormous blocks. This can be reached by the easier but less interesting track which ascends Snowdon from Llanberis. The path forks just a little distance before reaching the small lake, and the right-hand branch, which is a road to the abandoned copper-ore workings near the lake, can be followed.

Returning once more to Cwm-glas-bach, and continuing up the Llanberis Pass, the more interesting valley of Cwm-glas is soon reached. This is separated from the former valley by the buttress which is seen descending on the left-hand side of

Photo by]　　　　　　　　　　　　　　　　　　　　　　[A. W. Dennis.

FIG. 2. MORAINE HEAP AT THE FOOT OF CWM-GLAS.
VIEW FROM THE NORTH-WEST.

descended here from the ridge above to join the main glacier of the pass. We found no striae in this Cwm, although I believe they have been seen there. An easy ascent can be made here; and on crossing the shoulder at the top, it is only a short scramble down into Cwm-Brwynog on the north-west of Snowdon.

Perhaps the most impressive evidence of the vast amount of material transported by the old glaciers can be seen at this spot. At the head of the valley, the peace of which is now disturbed by the snort and smoke of the mountain railway, stand the dark and almost inaccessible cliffs of Clogwyn-dur-Arddu. Sheltered in a hollow at their base, at about 2,000 feet elevation, lies a small blue lake. Starting from this lake and ex-

the photograph. The rugged scenery of this grand valley has throughout been modified by the glaciers, which once swept over its craggy slopes.

Situated at the further end of the Cwm and not far from the road is a remarkable mound, partly overgrown with grass and scattered with angular blocks. The most probable explanation of its formation is that it was deposited at the termina-tion of the Cwm-glas glacier, when, owing to the amelioration of the climate, the main glacier of the pass had disappeared. Referring to the photo-graph which was taken when looking up the pass, the glacier descended from the right hand. The house in the foreground shows by comparison the immense size of the mound.

Blackfriars Road, London, October 1900.

THE PHOTOGRAPHY OF COLOUR.

By E. SANGER SHEPHERD.

(Concluded from page 112.)

WE have considered the theoretical conditions necessary for securing a record of the colours of a natural object. Let us see how we may best carry our knowledge into practice.

First, with regard to the most convenient apparatus for taking the negatives. We may mount our three coloured light filters in little brass caps, so that they may be easily slipped over the front

FIG. 13. PHOTOGRAPHY OF COLOUR.

of the hood of the lens, and place our plates in three dark slides in the ordinary way, but changing the dark slides and light filters between the successive exposures takes up time, and there is the liability of shifting the camera during the operation. A very little experience in making trichromatic negatives in this manner will quickly convince the photographer of the inconvenience of so many operations, and the early investigators soon turned their attention to what was at that time a very popular piece of apparatus with the portrait photographer—the "Repeating Back." By mounting a frame at the back of the camera, furnished with a dark slide long enough to take three plates side by side, the plate may be easily changed by sliding the plate-holder along to stops. Such an apparatus is represented in fig. 13 attached to an ordinary quarter-plate camera, and fig. 14 represents the separate parts. The back proper, A, is fitted to the camera by means of a panel cut to match and interchangeable with the dark slide, so that the attachment can be taken on or off in a few seconds; in this back slides a frame B, in which the blue-light filter C, the green-light filter D, and the red-light filter E are mounted. At the top and bottom of the frame are brass plates, which serve to hold the double dark slide F, which takes one long plate 8″ x 3¼″, so that the dark slide and the colour filter can be changed together by one movement. There are three depressions in the side of the colour-filter frame B, into which the

pin of the spring latch H drops when the filters are central with the opening in the back, A.

Such a piece of apparatus is very compact and will enable the photographer to obtain perfect negatives of a very large variety of subjects, landscapes, portraits, still life, etc.; and as it can be attached to any piece of apparatus that will project a sharp image upon the plate, it is perfectly satisfactory for photo-micrography, polariscopic and spectroscopic work. The apparatus occupies only a space of 14½″ x 5″ x 1¾″, and forms but a small addition to the landscape photographer's outfit. For special purposes there are many other forms of camera available. For instance, where a large number of colour photographs are required from flat surfaces, paintings, coloured book illustrations, maps, etc., a camera may be used furnished with three rectilinear lenses, when all three exposures can be made simultaneously, the work being quite as easy as ordinary

FIG. 14. PHOTOGRAPHY OF COLOUR.

photography. Such a form of camera cannot, however, be used for ordinary work from objects in relief, as the points of view of the lenses are separated, and the prints from the three negatives when superposed would not register correctly. In order to take three negatives of such objects simultaneously, a more complicated camera, furnished with one lens and a set of reflectors behind it,

G 2

dividing the beam of light into three portions. is necessary, The Repeating Back is. however. the general favourite of the amateur, as it is so simple in operation and occupies so little space.

In taking the colour negatives. one point only has to be remembered. The negatives through the three filters will require different exposures, for, as we saw (page 111). a white object must be represented by equal density in all three negatives. With each Repeating Back the makers furnish the relative times of exposure for each filter; for instance, the red filter negative may take nine seconds. the green three seconds, and the blue three seconds, and whatever exposures are necessary for our subject they must be given in this proportion. Different batches of emulsion may vary slightly in colour-sensitiveness. and it is very easy for the amateur to test his ratio by photographing such an object as a sheet of white blotting-paper roughly crumpled and pinned upon a black surface. A single negative, exposed and developed, will show at once if any alteration is required.

It only remains then to describe the method of printing the separate sensations, and mounting them in superposition. The minus red or greenish-blue print is made upon a gelatino-bromide of silver emulsion plate, and after development the silver deposit is replaced by ferro-cyanide of iron. the metallic silver acting as a mordant. This substitution product is the particular tint of greenish-blue required, and the image is a very delicate one of great beauty. The prints from the green and blue filter negatives are printed together upon a strip of special film of thin transparent celluloid which has undergone a process of air-drying, in order to ensure freedom from unequal contraction. The celluloid is coated with a soluble gelatine film containing a trace of bromide of silver. The film will keep indefinitely. and it is rendered sensitive for use by immersion in a solution of a chromate salt in the manner of carbon tissue. The gelatine film being, however. very much thinner than ordinary carbon tissue, all the usual difficulties of drying, insensitiveness, etc., are absent, and as the celluloid side of the film is placed in contact with the film side of the negative, so that the exposure takes place through the celluloid film, no transfer with its attendant difficulties is necessary. The image is a visible one and can be examined from time to time. the exposure being complete when all the details are distinctly seen, as a light brownish-grey image, very similar to an undeveloped platinotype print. The printed film is then washed out in warm water, the trace of silver bromide enabling us to see when the development is complete, the image held over a dark surface appearing as a delicate white print in low relief. The silver bromide, having served its purpose. is dissolved out by a solution of hypo-sulphate of soda, and the resulting low relief in clear gelatine washed and dried. The two prints which have so far been treated together are now cut apart, and the print from the green-filter negative stained in a pink dye-bath, so as to get a pink print which allows red and blue light to pass. The print from the blue-filter negative is stained in a yellow dye-bath so as to get a yellow print, which will allow red and green light to pass. When the two prints are dry. they are placed in superposition upon the greenish-blue print and our picture is finished. As the two film prints are reversed prints, in consequence of our printing through the film, and the greenish-blue print is a direct one. the two most important components of the triplet —the greenish-blue and pink prints—are mounted in actual contact, and as the third print, the yellow, is only separated from them by one thickness of very thin celluloid, the finished print behaves in every way as a single picture.

Lantern slides made in this way are as transparent as the best hand-coloured slides, although. of course, infinitely superior in delicacy and accuracy of colouring. They may be shown in an ordinary lantern, without the necessity for any extra attachment ; the mixed jet limelight giving perfectly satisfactory discs up to about 12 feet in diameter. With the electric-arc light pictures of 20 feet diameter or more can be easily shown. For projecting pictures naturally, the whitest light available should be used—preferably the electric arc. but the mixed oxyhydrogen limelight, if a small jet is used, will give a very satisfactory light. although the pictures will be a little warmer in colouring than they would be as seen by daylight. For home exhibition, on a disc up to two feet in diameter, a double acetylene-gas burner will prove quite satisfactory. ,

Before concluding it is desirable to correct an oversight that appeared on line 6 from the top of page 112. The sentence should read : "Each of these colours therefore transmits two, and absorbs one. of the three primary colours used by Clerk-Maxwell to form white light."

London, October 1900.

THE NOBEL BEQUEST.—Just before going to press we have been furnished through the Board of Education, which has received them from the Foreign Office. with copies of the official statutes and regulations of the Nobel Bequest. The large amount of annual interest accruing will be divided equally into five annual prizes. Three of these apply to scientific research. They are: (1) To the most important discovery or invention in the domain of the physical sciences ; (2) to the most important discovery or improvement in chemistry ; (3) to the most important discovery in physiology or medicine. The competition is open to the world, and, it is believed, will be held for the first time in 1901. Next month we shall probably give further details with regard to these valuable prizes.

Two NEW MINOR PLANETS were photographically discovered by Professor Max Wolf and Herr Schwassmann, of Heidelberg, on 15th and 21st of September.

NOTES ON ATYPUS.

BY FRANK PERCY SMITH.

*A*TYPUS *affinis* Eichw. (*A. sultzeri* Bl., *A. piceus* Sultz.), commonly known as the "trap-door spider," is one which, although very local and seldom seen unless carefully searched for, has

FIG. 1. Locality near Hastings for *Atypus*.

attracted a good deal of attention. During the early part of this year I had occasion to visit Hastings, a well-known locality for this species, and was fortunate enough to obtain a number of adult females. Mr. Connold, Hon. Sec. of the Hastings Natural History Society, accompanied

FIG. 2. External portion of tube of *Atypus* at knife point.

me, and took the photographs here reproduced. The greatest difficulty experienced in the photographing was that of bringing the lens within a short distance of the ground, and at the same time preserving the steadiness necessary for a comparatively long exposure. After a considerable amount of manoeuvring, however, the pictures were taken, and with far better results than I anticipated, considering the accuracy required to portray such obscure structures. Two of them represent the tubes as they appeared when first discovered. A third photograph shows a tube with the earth removed so as to expose its entire length. The fourth shows the hillock from which was taken the nest indicated by means of a penknife.

On returning to London I at once visited Hampstead Heath, Mr. F. Enock's famous hunting-ground, but was unable to find a single specimen in this locality. On October 9, however, whilst collecting on Wimbledon Common, I noticed a mound of sandy earth, which from its structure, aspect, and vegetation at once suggested the idea of looking for *Atypus*. During the half-hour of daylight which I had at my disposal, I carefully investigated the mound, and had the satisfaction of finding the remains of an old tube. Living spiders would probably have been found had the light permitted of further search.

A few notes on the creature's structure and habits may be of use. The female is about half an inch in length, exclusive of the falces and spinners; the male is considerably smaller. The falces are extremely prominent and very powerful, and the movable fang is capable of motion in a vertical plane. In all other British spiders the fang moves either in a horizontal or an oblique plane. The spiracular openings are four in number and are very distinct. The eyes are eight in number and are closely grouped on the anterior part of the caput. Two groups of three eyes each are in the form of a triangle, and the two remaining eyes, which are the largest of the eight, are placed transversely between these triangles.

The female spider constructs a tube, partly buried in the earth and partly lying upon it. This tube varies in length according to the age of the spider, and is often 12 inches long or even more, when the creature is adult. The ex-

FIG. 3. External portion of tube of another *Atypus*.

ternal part of the tube is intermixed with fragments of sand and debris, and, after having been exposed to atmospheric influences for some time, is most difficult to detect. There is no trap-door.

FIG. 4. Earth removed to show the entire tube of *Atypus*.

The spider lies in wait in some part of the tube, the external portion being connected with the creature's spinners by means of a number of delicate threads. As soon as any insect ventures to walk upon the tube, the spider, apparently receiving some intimation of the fact by vibration of the silken threads, at once rushes to the spot, seizes the victim through the substance of the tube, and drags it bodily in. The rent thus made is shortly afterwards mended from the inside. It is very interesting to note that when an errant male alights upon the tube he is not seized, although the female has no apparent means of seeing what is going on. He then proceeds to beat with his legs upon the external tube, but still the female remains at the bottom. After waiting some time he apparently becomes impatient, and, tearing up the

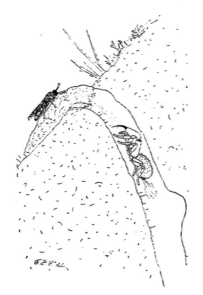

FIG. 5. Section showing *Atypus* in tube.

fabric with his powerful falces, he enters the tube. The pair often live together for some months, at the end of which time the observer is surprised to see the remains of the male brought to the surface and pushed through the tube. It is not a case of "death from natural causes," as careful investigation has proved that the male is devoured by his loving wife. After this tragedy has terminated, the female retires to the widest part of the tube, and lays about a hundred spherical eggs. The female guards the young spiders, which bear a strong resemblance to their parents, until they are large and strong enough to leave the tube and start homes of their own.

15 *Cloudesley Place,*
Islington, N.

AN INTRODUCTION TO BRITISH SPIDERS.

By FRANK PERCY SMITH.

(*Continued from page* 140.)

GENUS *TEXTRIX* SUND.

This genus may be easily distinguished from its allies by the curve of the hinder row of eyes having its convexity directed forward. The spiders included in the genus are usually found amongst loose stones and rocks, more especially on decaying walls, where they spin a web very similar to that of the genus *Tegenaria.*

Textrix denticulata Oliv. (*T. lycosina* BL)
Length. Male 6 mm., female 7.5 mm.
The general colour is brown. The legs are of a pale yellow tint, distinctly annulated with dark brown. The abdomen has a broad central band of a bright reddish colour, edged with yellowish and speckled with dark brown. This is a well-distributed though not common spider, and is easily distinguished

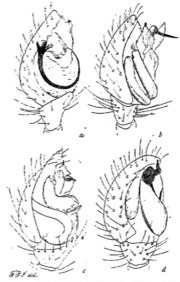

FIG. 1. *a.* Palpal organs of *Pholcomma gibbum*; *b.* Palpal organs of *Steatoda bipunctata*; *c.* Palpal organs of *Pedanostethus lividus*; *d.* Palpal organs of *Theridion lineatum.*

by the bright colouring of the abdomen. On account of the peculiar position of the eyes it may be at first mistaken for one of the *Lycosidae*, but the length of the superior spinners is a sure indication of its true systematic position.

A doubtful species, which may be a young example of *T. denticulata* Oliv., has been described by Rev. O. P. Cambridge as *Textrix boopis.*

GENUS *HAHNIA* KOCH.

The spiders of this genus are easily recognised on account of their short form and comparatively small size, but more so by the curious arrangement of the spinners, which are placed in a nearly straight line upon the under side of the abdomen.

Hahnia elegans Bl. (*Agelena elegans* Bl.)
Length. Male 3 mm., female 3.5 mm.
The general colour of this species is a bright orange or yellowish-brown. The abdomen is dark brown, with a series of broken angular bars of a yellow colour. It is not common.

Hahnia nava Bl. (*Agelena nava* Bl.)
Length. Male 2 mm., female 2.3 mm.
Cephalo-thorax nearly black. Legs brown. Abdomen dull blackish. The spinners have the appearance of being annulated, owing to the pale colour of their joints. . This is a rather rare spider. I have recently taken it at Hastings.

Hahnia montana Bl. (*Agelena montana* Bl.)
Length. Male 2 mm., female 2.5 mm.
This species may be distinguished from *H. nava* Bl. by its paler and more yellowish colour, also by its being less hirsute. It is not common.

Hahnia candida Sim.
Length. Male 1.5 mm., female 1.8 mm.
This species may be distinguished from *H. montana* Bl. by its smaller size and paler colouring. It is very rare.

Hahnia helveola Sim.
Length. Male 3 mm., female 3.5 mm.
This species may be distinguished from *H. montana* Bl. by its larger size, its less distinct markings, and by the greater length of the abdominal pubescence. It is very rare.

FAMILY *PHOLCIDAE.*

The spiders included in this small family are very easy to distinguish. The legs are extremely long and slender, the body is cylindrical, and the eyes are arranged in three groups of 3, 2, 3. The eyes of the lateral groups are large, and the central eyes are small; in some exotic species the centrals are absent.

GENUS *PHOLCUS* WALCK.

The eyes in this genus, which is the only one of this family represented in Britain, are eight in number. The falces are very small and weak.

Pholcus phalangioides Fuessl. Fig. 16.
Length. Male 7 mm., female 9 mm.
This extremely curious spider is by no means common, but when found it cannot well be mistaken. It spins a loose web in the angles of walls in out-buildings, cellars, etc., and when anything touches its snare it rapidly retracts and extends its legs, thus throwing the whole fabric into violent motion. This curious performance, which is indulged in by a number of other spiders such as Epeirids and Theridiidids, may be for the purpose of eluding an enemy, as the creature becomes almost invisible when in motion; or, more probably, for the purpose of entangling its prey. This process seems to reach its maximum with regard to violence of motion in *Pholcus*; but a neater and more scientific method is indulged in by *Hyptiotes paradoxus*. This little creature makes a triangular web with viscid lines crossing from side to side, the base line being produced on each side and fastened to any neighbouring projection. The animal sits at the apex, which is attached by a strong line to some twig, and gathers up a length of this line into a coil, thus drawing the web tight. As soon as an insect touches the web the spider releases its hold, thus allowing the whole structure to be thrown forward against the captive. If necessary the process is repeated several times, and the spider then rushes down one of the lines and seizes its hopelessly entangled victim. This spider has been recently taken, in plenty, in the New Forest (see SCIENCE-GOSSIP, March 1900). I have received *Pholcus phalangioides* from Norwich and Hampshire; and the Rev. E. N. Bloomfield tells me it used to be very common at Guestling Rectory, near Hastings.

(*To be continued.*)

ABNORMAL MUSHROOM.

AN unusually-shaped mushroom (*Agaricus campestris*) with three pileoli on one stem was collected at Corby, near Kettering, on August 28th last, by Mr. Lumby. I enclose a photograph of the example, as it is of such rare occurrence to find monstrosities

ABNORMAL MUSHROOM.

of this character among the fungi. It will be observed that the gills are all turned to the outside. Of course one cap is not visible in the picture, but it was perfect as the rest.   W. W. MIDGLEY.
Chadwick Museum, Bolton,
October 5th, 1900.

NOTE.—Our readers will remember that we figured a monstrosity somewhat like Mr. Lumby's specimen, in SCIENCE-GOSSIP, vol. iv., n.s., page 272. In that case the specimen was double and not a triplet.—[ED. S.-G.]

NOTES ON SPINNING ANIMALS.

BY H. WALLIS KEW.

(Continued from page 131.)

V. SPINNING BEETLES.

AMONG the few insects capable of spinning in the imago state are the water-beetles of the family Hydrophilidae—well known in the person of *Hydrophilus piceus,* the great harmless water-beetle of the aquarium.

Here the females spin cocoons or silken bags, sometimes of complex structure, in which their eggs and newly-hatched larvae are enclosed and protected.

The egg-cocoon of *Hydrophilus piceus* is nearly an inch broad, somewhat roundish, slightly narrowed and truncated anteriorly, where it is furnished above with a spike or "mast," recalling the horn on the anal segment of a hawk-moth caterpillar. The structure is spun from a paired spinneret at the hinder extremity of the beetle. When complete it floats at the surface of the water, usually attached to a leaf or some small floating object, and always with the spike above the surface. The young larvae, which remain in the cocoon for a while after hatching, finally escape near the base of the spike, where the cocoon is only slightly closed. This floating nest, and the manner in which it is spun, were described by the celebrated Lyonnet; and later, in 1809, Miger, who also had witnessed the spinning of the cocoon, published his often quoted memoir and illustrations (¹).

The beetle observed by Lyonnet was kept in a trough, where it was supplied with filamentous algae, floated on the water by means of wooden shavings. Before long Lyonnet saw the beetle setting about the formation of a cocoon. She extended the hinder rings of the body and opened the last of all, exposing a cavity in which was seen a whitish disc, giving off two small prominences side by side, each enclosing a delicate tube about a line in length, stiff towards the base, but flexible and elastic towards the tip. These two tubes formed the spinneret. They always moved together, but each contributed a separate thread. Lying back downwards near the surface of the water, beneath or amongst the algae, the beetle began to weave one side of the cocoon; and as the work proceeded she was careful to press and flatten the growing structure, moulding it with her forefeet against her body, and thus giving it the form of a flattened arch. After the first section, which formed the upper side of the cocoon, was finished, the insect turned over and wove another piece,

(1) Lyonnet and Miger have recently been quoted at length by Miall, *op. cit.,* pp. 61–86; and it is from this source that the present notes of the work of these authors are derived.

exactly the reverse of the first, to form the underside. The two curved pieces were now woven together, and thus the body of the cocoon was constructed. The work so far had occupied about an hour and a quarter. For about two hours the beetle remained still with her body buried in the cocoon, from which, however, it became evident that she was gradually withdrawing. She was in fact laying her eggs. She now withdrew her body completely, and began to spin about the open mouth of the cocoon, gradually narrowing and closing it. Afterwards she proceeded to spin the spike, which gradually rose above the water. The work was complete in five hours, after which the cocoon was left floating.

Miger relates that he kept several specimens of this beetle in a vessel of water with aquatic plants, and that at length he saw a female spin a cocoon. She attached herself to the under-side of a floating leaf, clasping it with her forelegs and applying her abdomen to its under surface. The two tubes of the spinneret could be seen to be pushed in and out with rapidity, while a gummy liquid was passed from them and drawn out into threads ; and these, being attached to the leaf, gradually surrounded the tip of the abdomen, forming at length a semicircular pouch in which that part of the beetle was enclosed. After about ten minutes the beetle turned sharply round, letting go the leaf and bringing her head downwards, but without withdrawing the abdomen from the cocoon. The leaf was now held by the hindlegs only, one being placed on each side of the cocoon. The insect continued to work steadily for nearly an hour and a half; and, through the transparent wall of the cocoon, Miger could see the movements of the spinneret, until at last the gradual addition of threads made the structure opaque. The beetle then laid her eggs, and afterwards closed the cocoon slightly, and began to form the spike. The tips of the wing-cases, which were a little opened, were brought to the surface of the water, and the spinneret was seen to be in continuous and rapid motion. The spinning of the spike took more than half an hour, at the end of which time the completed spike—which, as Miger supposed, served to supply the cocoon with air—rose considerably above the water. The whole work occupied about three hours.

J. Fullagar, among more recent observers, mentions having seen the *Hydrophilus piceus* spin her cocoon in an aquarium. He gives a sketch showing the beetle holding the unfinished cocoon between the hindlegs, and another showing the completed floating structure slightly attached to a bit of weed ([2]).

A. G. Laker also has given particulars of the same beetle's cocoon, of which he found a number of specimens in a pond near London. They varied considerably in size and shape, but averaged about 11½ lines long by 10¾ lines broad, the height to the tip of the spike being about 17 lines. The walls were composed of a substance very like paper; but the part immediately below the spike was of a loose silky material, which, though this part was submerged, was not impervious to water. It was by breaking through this material that the young larvae ultimately escaped. The interior of the cocoon contained similar loose material, as well as the eggs, of which there were fifty to sixty. The cocoons were remarkably buoyant; but they usually floated attached to confervae, to long grass growing in the water, or to the under-side of floating leaves. The spike was of a substance somewhat thicker and stronger than the rest of the cocoon, and was hollow throughout the greater part of its length, except that it was crossed and recrossed inside with a dark thread-like substance. The observer compares it to a horn stuffed with tow ([3]).

There are in various parts of the world a number of other species of *Hydrophilus*, but the writer has not seen observations on their spinning-work, except in the case of the great American *Hydrophilus triangularis*. I am not aware that the spinning process has been watched in this species; but the cocoon itself has been carefully described. Garman, who obtained half-a-dozen cocoons from the surface of a small pool, says that the outside is smooth, of a light brown colour, and much resembles the egg-cocoon of the common black and yellow Epeirid spider of North America. Viewed from above or below, the outline is circular, with a diameter of about 20 mm. The spike, unlike the rest of the cocoon, is of a horny nature. Its base expands into a hatchet-shaped plate, the spike being formed, in fact, by the narrowing and folding back of this plate until its edges almost meet. The process is thus a partial tube, which, projecting above the water, appears to facilitate the entrance of air into the cocoon. Below the expanded base of the spike is a narrow opening through which water passes, and which finally permits the escape of the larvae. The eggs, numbering in one case 107, form a discoid mass surrounded by a loosely-woven silken coat and suspended from the roof of the cocoon. Below, at the sides, and behind the egg-mass, is a space to which water has access; but the silken material above and in front of the eggs is disposed so as to form large cells, and these connect with the air-admitting spike. The chief object of the cocoon,

according to this observer, is to ensure a supply of air for its contents; and this is done, he says, "by excluding water from, and admitting air into, the upper part of the case" ([4]). Riley has written on the same subject, confirming Garman's description. He observes that the cocoon differs from that of *Hydrophilus piceus* "according to the descriptions of this last;" but, when the cocoons themselves are compared, it is possible that they will be found much alike. Riley gives figures and an ideal section of the cocoon of *Hydrophilus triangularis*. It consists, he says, of three distinct parts :—

"There is first what may be called the floater, which itself is composed of two parts, viz. :— (1) a hard spatulate piece of compact brown silk, smooth externally, and with the two edges of the tapering end curled inside and welded at tip, so as to form a stout point [the spike], and (2) a somewhat cuneiform air-chamber. There is, second, the egg-case proper, and, third, the outer bag or covering. The air-chamber has an external slightly bulging covering of the same character as the outer bag, of which it forms a part, but of somewhat darker silk, while the inside consists of loose brown silk, forming large cells and connecting with the spine, the hollow parts of which are in fact filled more or less compactly with these silken fibres. The egg-case proper, which is of a white, rather flimsy or paper-like silk, is partially suspended posteriorly from the roof of the outer bag by white loose silk, but is principally attached to the interior side of the air-chamber."

Between the egg-case proper and the outer covering is the space to which water is admitted; and the larvae, Riley states, breaking through the egg-case proper, remain for a day or two within this space, issuing ultimately through the rent at the foot of the spike. When the cocoon is floating, the spike is directed upward, most of the "floater" being out of the water. The eggs, though bathed in water, are thus freely aerated, the function of the cocoon being, no doubt, to secure a supply of air, and at the same time to protect the eggs and newly-hatched young from numerous enemies ([5]). The contrivance is clearly a complicated and curious one; and it is obvious that the spinning instincts of these beetles are of no mean order.

Cocoon-spinning is not confined to *Hydrophilus*, naturalists having long known it to obtain in the allied *Hydrocharis*, as well as in many of the smaller Hydrophilid-beetles, of which, perhaps, our little *Hydrobius fuscipes* is the best known. Among recent writers, Perkins notes the finding of nests of *Hydrocharis caraboides*. He states that the insect, rolling up a leaf, lines it with a thick cottony web-like substance ([6]). Laker, who writes of the egg-cocoons of *Hydrobius fuscipes*, compares them to the relatively gigantic structures of *Hydrophilus*; but they are flattened on one side,

(2) Fullagar, "Hardwicke's Science-Gossip," xv. (1879), pp. 132, 133.

(3) Laker, "Entomologist," xiv. (1881), pp. 82–84.

(4) Garman, "American Naturalist," xv. (1881), pp. 600, 601.

(5) Riley, "American Naturalist," xv. (1881). pp. 814–817.

(6) Perkins, "Entomologist," xii. (1879), pp. 214–216.

and have no spike. They are about the size of a pea, and are attached to the underside of floating (not detached) blades of grass. As in *Hydrophilus*, the young, on hatching, do not immediately emerge ([7]). The cocoons of this beetle have also been described by W. F. Baker, who states that they are found in abundance at or near the surface of the water in shallow grassy ponds, attached to leaves of water-plants, blades of grass, etc. Being partly filled with air, they float when detached. They are formed of silken threads woven together in various degrees of closeness. One end, that first formed, is firm and smooth; but the other end (that at which *Hydrophilus* makes its spike) is loose, irregular, and furnished with a flap, by which the cocoon is fastened to the objects upon which it rests. The beetles begin to spin a cocoon within three weeks after emerging from pupae; and, at intervals of about a fortnight, each female makes three or four ([8]).

(7) Laker, *l. c.*
(8) Baker, "Naturalist." 1804, pp. 327-333; and in Miall, *op. cit.*, pp. 88-90.

Baker mentions further that the *Hydrobius fuscipes* may sometimes be found with an unfinished cocoon attached to her body; and it is interesting in this connection to note that in *Spercheus emarginatus*, and some other Hydrophilids, it is the rule that the female thus carries her cocoon until the larvae hatch. The *Spercheus* was supposed to be extinct in this country until Billups rediscovered it in 1878-9. Many of the females taken at that time were carrying egg-cocoons. These were of a silky material, pale brown in colour, very closely spun, and slightly inflated like the egg-cocoon of a spider ([9]). Fowler, placing specimens of this beetle in a globe of water, found after a day or two that one had developed a cocoon. It was formed of a tough whitish membrane, and covered the whole of the under-side of the abdomen; after some days it disappeared, about 100 larvae having hatched ([10]).

(9) Perkins, *l c.*
(10) Fowler. "Entomologist's Monthly Magazine," xix. (1882-1883); and "British Coleoptera," i. (1887), p. 233.

(To be continued.)

BUTTERFLIES OF THE PALAEARCTIC REGION.

By Henry Charles Lang, M.D., M.R.C.S., L.R.C.P. Lond., F.E.S.

(*Continued from page* 144.)

Genus 12. *ZEGRIS* Rambur.

Small butterflies having somewhat the aspect of the orange-tipped species of *Anthocharis*. Antennae shorter and with thicker and more rounded clubs. Head more hairy and with shorter palpi. The f.w. are tipped with orange in both sexes, but the orange patch is completely surrounded by dark shading and is proportionally much smaller than in *Anthocharis*. Larva as in *Pieris*. Pupa enclosed in a thin cocoon and attached at both ends with threads.

1. Z. pyrothoe Ev. Nouv. Mém. Mose. 1832, 352, T. 20, 3, 4. Lg. B. E. p. 42, pl. xi. fig. 1 (*Euchloe pyrothoë*).

29—36 mm.

Wings white. Apices of f.w. rather pointed and with a more or less oval orange patch bordered with black; at the costal edge of the patch is a white patch. Black discoidal spot crescent-shaped. H.w. white. showing the markings of u.s. U.s. f.w. as above, but the tips are greenish and the orange patch very indistinct. H.w. bright green with well-defined pearly white spots, few in number, some sub-marginal and of an oblong form, some central of a circular or ovoid outline. ♀ slightly larger than ♂, with the orange apical patch less extensive and more strongly marked with white.

Hab. S. Russia, Orenburg, Kirghis Steppes, Kouldja district, Margellan, Turkestan. Those

from Kouldja are larger than specimens from the other localities. IV.–V.

Z. pyrothoe has long been placed in the genus *Anthocharis* or *Euchloe*. But I have for a considerable time considered that, given the propriety of separating this group from the foregoing genus, this species ought to go with it, and as I now find that in the Brit. Mus. collection it is reckoned as a *Zegris*, I have no hesitation in placing it in that genus.

2. Z. eupheme Esp. 113, 2, 3. Lg. B. E. 43, pl. x. fig. 2.

40—46 mm.

Wings white, sometimes with a very slight yellowish tinge. F.w. with a black crescentic disc. spot. At the apex is a black patch dusted with yellow on its outer side, and enclosing a bright orange oval-shaped blotch, above this a white spot. H.w. white, blackish at base, showing the pattern of the u.s. faintly through. U.s. f.w. white, disc, spot more V-shaped than above, apices greenish yellow, slightly orange about the centre of the patch. H.w. green mixed with yellow, with five or six rather large white spots arranged somewhat as in *Anthocharis euphenoides*. Thorax and abdomen black above and covered with white down. Antennae white above, black beneath, clubs white. ♀ has the apices of f.w. lighter, the orange blotch smaller, and the white spot above it larger.

HAB. Sarepta, S.E. Russia, Crimea, Askabad, Syria, Lepsa, Central Asia, North Tianchan, Byschtun and Tschaptschatschi on the Lower Volga, the Ural River, Amasia. IV., V.

LARVA. Yellow, with a white lateral band and large black points arranged in threes on the sides of each segment. On *Sinapis incana* and other Crucifers.

PUPA. Whitish, not boat-shaped.

a. ab. *tschudica* H. S. 449, 450. Smaller than type. Pure white above. U.s. h.w. w.th large white spots; orange apical patch very small. HAB. Sarepta.

b. var. *menestho* Mén. Cat. Rais. p. 245. A large form of the species. The disc. spot f.w. is narrower and often divided into two ; the greenish-yellow colouring of the u.s. is lighter and more extended. HAB. Amasia, Syria. IV.

c. ? var. *meridionalis* Led. Z. b. v. 1852, p. 30. Lg. B. E. p. 44, pl. x. 3 (*Z menestho*). Larger than *Z. eupheme*. F.w. with the dark shading at the apex more strongly marked ; disc. spot large and doubled, especially in ♀, in which sex the orange is almost absent from apex. U.s. h.w. deeper yellow than in type. HAB. Spain—Leon, Anda-lusia, Castile. Larva, according to R. H., greenish yellow, sometimes entirely suffused with red, finely pilose, with black lateral spots. On *Sinapis, Raphanus*, and *Brassica*. If this form is co-specific with *Z. eupheme*, and not distinct from it, it is remarkable that it should present so striking a solution of continuity in the distribution ; for whilst *Z. eupheme* is confined to the extreme S. E. of Europe and W. Asia, the form *meridionalis* is found nowhere but in Spain, and no other European country has a *Zegris* amongst its fauna. It is quite possible that this is really a distinct species, though in appearance it differs from *Z. eupheme* only by slight characters.

3. **Z. fausti** Christoph. Hor. Ent. Soc. Ross. xii. pl. v., Lg. B. E. p. 69.

35—37 mm.

Wings pure white, ♂ f.w. with apices bright orange, almost scarlet, and very slightly black at

Z. fausti.

outer edge, but with a broader, blacker, and straighter internal band than that seen in *Z. eupheme*. Disc. spot larger and less crescentic. U.s. h.w. green, with large silvery spots and no yellow colouring. ♀ with apex f.w. less brightly red and with more black.

HAB. Turkestan (not Amur), Krasnowodzk, Kisil-arwat, Ferghana, Osch. IV., V.

This is a well-marked and beautiful species, quite distinct from any of the forms of *Z. eupheme*. By an error on p. 334 this species is stated to occur in the Amur. It seems to be confined to Turkestan, in the neighbourhood of the Caspian, and not to occur eastward of that region. I was misled in 1884 when I gave Siberia as a locality for this species.

Genus 13. *LEUCOPHASIA* Steph.

Head moderately large, eyes large and prominent, palpi longer than the head and covered with strong hairs. Antennae of moderate length, furnished with a flattened oval club. Abdomen very slender, reaching beyond h.w. Wings white, rounded, and elongated. Discoidal cells very small. Subcostal nervure five-branched ; all the branches are given off beyond the cell. Larvae pubescent, tapering at the ends. Pupa angular, not boat-shaped.

1. **L. sinapis** L. Syst. Nat. x. 648. Lg. B. E. p. 45, pl. x. fig. 4. " Wood-white."

35—40 mm.

Wings white, f.w. with a circular blackish blotch at apex. U.s. f.w. faintly yellowish-grey along costa and at apex. H.w. yellowish streaked with grey.

HAB. Europe (except the Polar portion). In Britain chiefly in the south and west, but also in the Lake District and other places. On the Continent it is common and widely distributed, in England very local. It extends throughout Western Asia, and occurs in the Amur, Corea, and Japan. IV., V.—VII., VIII.

LARVA. Green with a darker dorsal stripe ; beneath this is a yellow stripe. On *Vicia, Lotus, Lathyrus*, and other Leguminosae.

PUPA. Yellowish-green or grey rusty on the sides and wing-cases VI. and IX.

This species presents marked seasonal variation. There seems to be some difference of opinion respecting the nomenclature of the various forms. I subjoin them as they are given by R. & H.

a. var. *lathyri* Hub. 797-8. Gen. I. The u.s. h.w. shows a darker green colour. Flies with the type in S. Europe and Western Asia. IV., V.

b. ab. ♀ *erysimi*. Bkh. i. 32. Wings entirely white above. HAB. Occurs with type throughout Europe. A dimorphic form of ♀ found with 1. and II. generations.

c. var. *diniensis* Bdv. gen. 6. Apex of f.w. rather less rounded than in type. Wings entirely white beneath. Apex of f.w. above varying in the intensity of the black patch. Occurs as a variety of the second generation. HAB. Principally in S. Europe.

d. var. *sartha* H. R. p. 143. 40—45 mm. F.w. more elongate than in type, with rounded apices.

U.s. f.w. with a yellowish-green apical patch square in outline. H.w. entirely yellow-green with dusky suffusion. HAB. S. Europe, Asia Minor.

2. **L. amurensis.** Mén. Schrk. p. 15, t. i. 45. *L. Sinapis* var. d. Stgr. Cat. 1871, p. 5.
45—50 mm.
A much larger species than *L. sinapis*, from which it is quite distinct. Wings pure white. F.w. slightly angulated at apex and having an apical circular patch of deep black in ♂, lighter in ♀,

L. amurensis.

sometimes in that sex almost absent. U.s. f.w. very faintly yellowish at apex and along costa. H.w. white, with a very indistinct trace of a dusky transverse band.
HAB. Ask., Chab., Vladimar Bay, and other places in the Amur, also Japan. V. and VIII.
a. var. *vernalis* Graeser. R. H. p. 143. 35—40 mm. Smaller than type; ground-colour yellowish-white, with the black apical spot less marked and less defined in outline. Amur. V.

3. **L. duponcheli** Stgr. Cat. 1871, p. 5. Lg. B. E. p. 46, pl. x. fig. 5.
36—38 mm.
Differs from *L. sinapis* as follows : Apex f.w. less rounded ; apical spot less defined at inner edge, extending much further in a downward direction. H.w. greenish-grey, with two white blotches ; bases decidedly tinged with yellow. U.s. h.w. quite different from those of *L. sinapis*, uniform greenish-grey, with two white blotches, one central, the other marginal ; both triangular in shape. Clubs of antennae blacker than in *L. sinapis*.
HAB. S. France, Italy, Asia Minor. IV., V.
a. var. *aestiva* R. H. p. 143. The summer generation. Apical blotch. f.w. paler in colour and less extensive, shaped more like that in *L. sinapis*. HAB. as type. VII.
I have taken both *Leucophasia duponcheli* and var. *aestiva* at Digne, where the species is common. The difference between them appears to be marked and constant. This species is very distinct from *L. sinapis*, but is very local and of limited distribution. The two species, *L. sinapis* and *L. duponcheli*, fly together at Digne ; but the flight of the latter is stronger and more sustained. By this character they may be distinguished when on the wing with some amount of certainty.

Genus 14. *TERRACOLUS* Swainson.
Antennae rather short and thick, with a broad blunt club. First and subcostal nervules of f.w. close together. Colour of wings white, orange, or yellow. This genus contains a number of species of widely differing appearance, some of them having the aspect of *Anthocharis*, and others that of *Colias*. They are mostly African or South-western Asiatic. These species, however, extend into the Palaearctic region.

1. **T. fausta** Olivier, Voyage en Syrie, pl. 33. 4. *Idmais fausta* Stgr. Cat., et al. auct.
35—40 mm.
Ground-colour of wings a peculiar light reddish-yellow, very difficult to describe. Shape of wings resembling that of *Colias hyale*. F.w. with a double row of narrow black marginal spots extending from the costa nearly to the an. anr. There is a small and indistinct black discoidal spot. A black wavy band begins at the outer third of costa and reaches to the first median nervule. Immediately about the submedian nervure in the ♂ is a small patch of thickened scales. H.w. without markings, excepting a row of five very small marginal black spots. ♀ less brilliant in

T. fausta.

colour than ♂, but with all the black markings stronger and much more pronounced. U.s. very light yellowish-orange, especially the h.w., which are nearly white in ♂. There are no proper markings except a very faint trace of a narrow central band of yellow on the h.w. The dark markings of the upper surface show through more or less.
HAB. Syria (Beirut, Jaffa), South of Caspian, Ferghana. IV., V., Osch, Turkestan, Persia, Steppes of Schurab.

2. **T. phisadia** Godt. Enc. Méth. ix. p. 132, n. 40 (1819). (*Idmais.*) var. *palaestinensis.* Stgr. in litt.
30 mm.
F.w. salmon-colour, black at base and along costa, with a broad black marginal band containing three spots of the ground-colour, two near the apex and one in the centre ; discoidal spot black and distinct, but touching the black costal shading. H.w. white with a broad black unspotted marginal band. U.s. light yellow dusted with grey. No marking except on f.w. Disc. spot and three blackish spots near an. ang. I am unable to give

any further account of this beautiful little species, which has the appearance of a diminutive *Colias.*

The type of this species inhabits India and Arabia; the var. *palaestinensis* is found in Palestine. I have received a specimen from Dr. Staudinger, which is here described.

3. **T. nouna** Luc. Expl. Alg. Zool. iii. p. 350. n. 14 (1849). *Anthocharis nouna.* Stgr. Cat. etc. 30—42 mm.

Wings pure white.

F.w. somewhat pointed with bright yellowish orange tips and without spots. H.w. white with some indistinct marginal spots. ♀ larger than ♂, f.w. with some brownish black on outer edge of orange tip h.w. with distinct triangular spots. U.s. h.w. white without markings.

HAB. Algeria (Oran) in the early part of the year and in September.

Probably a variety of the Arabian and East African *T. evagore* Klug.

Genus 15. *CALLIDRYAS* Boisd.

Antennae shorter and thicker than in *Pieris*, but longer than in *Colias*, gradually thickened into

C. pyrene.

a club; generally of a reddish colour. Wings ample. F.w. pointed at apex and h.w. near an. ang. Colour usually white or yellow. Males with a tuft of hair-like scales near the base of the inner margin of f.w. This genus inhabits the tropical portions of both hemispheres, but the species here given occurs in Syria, and therefore comes within the Palaearctic region.

1. **C. pyrene** Swains. Zool. Ill. i. t. 51 (1820).

58—62 mm.

Wings pale yellow or greenish-white. F.w. with a few reddish or dusky marginal spots. Discoidal spot small and black. There are no other markings. U.s. h.w. with discoidal spots sometimes marked with silvery.

HAB. Syria.

(*To be continued.*)

IRISH PLANT NAMES.

BY JOHN H. BARBOUR, M.B.

(*Continued from page 147.*)

ROSACEAE.

CRAN AIRNE or AIRNEAD. airne, "sloes." DRAIGNEAC or DRAIGNEAN. draigean, "a thorn"; draig, "dragon." *Prunus spinosa.* sloe-tree. scrogs. CRAN SIRIS. *Prunus cerasus.* wild cherry. MORGRAIDEAN. MEIRIN NA MAG, "field fingers," or "gladness." *Agrimonia eupatoria* and *A. odorata.* agrimony. AIRGIOD LUACRA. airgiod, "silver"; luacra, "of rushes." LUSCNEAS. cneas, "noble." *Spiraea ulmaria.* meadow-sweet. silver rushes. GREABAN. *Spiraea filipendula.* dropwort. DREISEOG. dreas, "briar," "bramble." DRIOS-GEACTALMAIN, "briar," "thorn of the earth." LUSNA GORM DEARG. gorm, "blue," and dearg, "red"; "make purple." *Rubus fruticosus.* bramble, blackberry. MACALL FIADAIN. fiadain, "wild"; macalla, literally, is "son of the rock." Heb. Traitnin, a little stalk of grass. *Geum urbanum.* herb bennet. MACALL UISGE. "water echo." *Geum rivale.* water avens. SUTH TALMAN. Suth, "juice," "juice of the earth." *Fragaria vesca.* wild strawberry. BAR A BRIGEIN. Bridget's sea-flower. BRIOSG-LAN. briosg, "brittle"; lan, "a blade." "brittle-blade." *Potentilla anserina.* skirret, wild tansy. CUIGEAG. CUIG MEAR MUIRE. cuig, "fine," "lady's or Mary's five fingers. MEANGAC, "crafty," "deceitful." *Potentilla reptans.* common cinque-foil. LEAMNACT. leamnac, "sweet milk." BENEDIN. "septfoil." MEANAIDIN. meanad, "gaping." *Potentilla tormentilla.* tormentil. COINDRIS. coin gen. of *cu* "dog." dris, draseog, "brier," SGEAC MADRA. sgeac, "a bush"; madra, "a dog." *Rosa canina.* dog-rose. COTA PREASAC. cota, "coat"; preasac, from preas, "bush." "bushy mantle." LEAGAD or LEATAC. BIVDE. "yellow band "or "divided yellow." CRUBA LEOMAIN. "lion's claw." DEARNA MUIRE. "lady's palm." FALLAING MUIRE. "lady's cloak." *Alchemilla vulgaris.* lady's mantle. ABALFIADAIN. wild apple. CEIRT-ABAILL ABAL. GORTOG. *Pyrus malus.* crab apple. COILL-CAORTAND. coill, "wood"; caora, "cluster of berries." "woodberry tree." CAORRAN. berries. LUIS, the letter L. *Pyrus aucuparia.* quicken or rowan tree. SCE. SCEAC. SCEAG. "bush." UAT (H aspirate). *Crataegus oxyacantha.* white thorn.

SAXIFRAGEAE.

CLABRUS or CLABRAS. elab, "thick"; rus or ras, "wood." "shrub." LUSNA LAOG. laog, "calf."

GLOIRIS,. " glory," " radiance." *Saxifraga oppositifolia.* opposite-leaved golden saxifrage.

MORAN. " many," " multitude." *Saxifraga granulata.* whisk rushes.

HALORAGEAE.

SNAITEBAITEAD. snait, ": thread," " filament " ; baite, " drowned." *Myriophyllum spicatum.* water milfoil.

DROSERACEAE.

DURACDIN MONA. bowery, " heath plant " or " wile." GEALDRIUG. geal, fan, " white ; " driug, " dew." LUSNA FEARNAIG. *Drosera rotundifolia.* round-leaved sundew. red-hot ; youth-wort.

CRASSULACEAE.

GRAFA NA CLOC. grafa, " a graft " ; cloc, " a stone." *Sedum acre.* wall-pepper.

FEINE GABLA. FEINE TINE. feine, " farmer", tine, " link." " farmer's link." NORP or ORP (the name). SINICIN. TINEAGLAC. "reckoning kindred-'ness." TOIRPIN or TIRPIN. toir, " church " ; pin, " pine." " church pine." *Sempervivum tectorum.* houseleek.

LAMAN-CAT-LEACAIN. laman, " a glove " ; leacain, " a cheek," or " side of a hill," " a kind of cheek glove." CARNAN CAISIL. carnan, "hillock"; carna, " flesh " ; caiseal, " wall," " rock " ; possibly, therefore, name refers to where it grows only, or to its nature also. " stone flesh." LEACAN or LOANCAIT. leao, " a flag," " flat stone "; leaca, " cheek." *Cotyledon umbilicus.* navel-wort or wall-pennywort.

LUS NA LAOG. laog, " calf." W. (lho). *Sedum roseum.* Orpine.

LYTHRARIEAE.

BREALLAN LEANA. breall, " knob," " knot " ; l ana, " meadow," " knotty meadow spike "). *Lythrum salicaria.* purple-spiked willow-herb or loosestrife.

ONAGRARIEAE.

FUINSEAC. FUINSEASGAC. fuin, " west " ; seas, " beyond." *Circaea lutetiana.* enchanter's nightshade.

UMBELLIFERAE.

LUS NA PINGINE. pingin, " penny " ; pingine, a small vessel of water equal to grs. 8 ; corn by weight. Hence an herb not equal to that. *Hydrocotyle vulgaris.* white-rot.

CUILEAN TRAGA. " holly," " shore-holly." *Eryngium maritimum.* sea-holly.

REAGA MAIGE, REAGAM. REAMA. REMA. All these are modifications of same original, I think. reag, "night"; maig, " proud." CAOGMA. MEASCAN. a small dish of butter. BUINE. " ulcer.' *Sanicula europaea.* sanicle or butterwort.

ITEODA. MINMEAR, min, " gently " ; mear,

" lust," " a wanton " ; hence " the flower which gently makes you happy." MUINMEAR. origin may be as minmear or miun, " a bramble." Hence " wanton bush." *Conium maculatum.* hemlock.

LUSARAN GRAN DUB. gran, " grain," " corn "; dub, " black." *Smyrnium olusatrum.* alexanders.

PEIRSIL MOIR. moir, muir, " sea," " sea parsley." *Apium graveolens.* smallage.

FEALLA BOG. feall, " treachery," " deceit"; bog, " soft," " tender." *Cicuta virosa.* water hemlock.

PEARSAIL °or PEIRSILLE. *Carum petroselinum.* parsley.

FOLACTAIN. FUALACTAR. (? origin). *Sium latifolium.* water parsnip.

COIREARAN or TOIREARAN. COIREARAN MUICE. coirearan means a root similar to that of the pignut. Muicin, " a little pig."

CUNLAN. cunla, " modest." *Conopodium denudatum.* jur-nut, earth-kipper.

COSUISGE. cos, " fissure," " foot "; uisge, " water," " river feet or fissure." *Chaerophyllum temulum.* wild cicely, chervil.

DILLE: FENNEUL. FINEAL CUMTRA or FINFAL CUBRA. LUSAN TSAOID. saoid, " hay," " fodder." *Foeniculum vulgare.* finckle, fennel.

TATU. TATU BAN. "white side," " lord," or " ever present." *Oenanthe crocata.* dead tongue, water-dropwort.

SIUNAS. siunan, " a vessel made of straw." *Ligusticum scoticum.* lovage.

AINGEALOG. aingeal, " angel." CUINEOG MIGE. cuineog, " mote," " oats," or " barley " ; MIDE, " Meath " ; migean, " dislike." GALLURAN. GLEORAN. CONTRAN (angelica). *Angelica sylvestris.* wild angelica.

FINEAL SRAIDE. " road " or " lane " fennel. *Peucedanum sativum.* hog's fennel.

MEACAN RIG, " royal taproot," ODARAN. odar, " pale," " dun." *Heracleum sphondylium.* madness. hogweed.

MUGOMAN. MIOIBOGAN. MIODLUCAN. mid, " sight," " aspect " ; luc, " a mouse." Hence a plant with a root like a mouse. A carrot root in shape might easily be suggestive of the shape of a mouse. *Daucus carota.* carrot.

GRIOLOIGIN GRENIUG GEIRGIN GREIMRIC. *Crithmum maritimum.* samphire.

ARALIACEAE.

EADAN EADANAN. ead, " jealousy "; eadan, " face." FAIGLEAD. GORT (letter G). *Hedera helix.* ivy.

CORNACEAE.

COINBILE. coin, plur. of cu, " dog "; bill, " a large tree." " dog's tree." CRAN COIRNEL. CAOR CON. caor, " berry "; cu, " hound." *Cornus sanguinea.* dogberry tree.

(To be continued.)

PARENTAL RELATIONSHIP.

By Gerald Leighton, M.B.

THE question of the relation of parent to off-spring, in its highest development, may be regarded as a problem in ethics; but, like other ethical problems, it can only be understood in its origin when studied from a biological point of view. Once get at the course of evolution of an idea no less than of a thing, and a flood of light is thrown upon all that the idea conveys.

In the highest animals—that is, the mammalia—we see the parental relationship at work every day, the mother nourishing her offspring, the father protecting the family. In the lowest in-vertebrates there is no such thing as the relation of parent to offspring, except the purely physical act of some form of reproduction. At what point, then, in the scale of nature did parental relation-ship first show itself, and what were the conditions necessary for its development?

The question may be put in other words, where do we first see the parents taking care of their young in the scale of nature? We have said it is a familiar picture in the mammals, which indeed take their name from this feature. It is seen well marked too in birds, though in a less degree. What about the class next below them—viz. the reptiles? One finds in text-books on natural history that some authorities deny the existence of parental relations at all in reptiles. This looks as if one must seek hereabouts for the first indica-tion of the phenomenon. But leaving reptiles for the moment, what do we find in the amphibians, which come next on the scale? Here the relation of parent to offspring may be said to be non-existent, the same applying to the fishes, and we look in vain for any sign of it lower in nature. Why is this? No doubt, mainly for two reasons. First, that the last two classes, fishes and am-phibians, reproduce by spawning. In other words, the offspring are not brought forth alive and mature. Secondly, the fecundity of fishes and amphibians is too great, the number of young from each spawning being so enormous as to entirely preclude any idea of parental relationship. No doubt there is care shown in depositing the eggs, but this is a different thing from care of offspring, for the eggs once deposited are no longer an object of solicitude. Now, it is quite true that some fishes do bring forth their young alive and mature; for example, the viviparous blenny. Some of the sharks also are viviparous, the Greenland shark being said to produce only three or four young at a time; the bergylt also is viviparous; but it is doubtful if in any of these cases, which are exceptional in their class, there is a true parental relationship. At any rate here among some fishes we see the beginnings of the first necessity for the development of the parental feeling—viz. that the offspring shall be born in a form recognisable by the parent. This is the first element required to build up the com-plex emotion which is to appear later.

Other factors enter into the process of the development of the parental relations, but at present we are considering the dawn of that phenomenon in nature. We may omit the amphi-bians, as they are spawning or egg-depositing creatures, with few exceptions; and in nearly all the young undergo a metamorphosis—*i.e.* their form at birth is not that of the adult. Next we come to the reptiles, in which group, as before stated, authorities differ, which in itself is an indication that some have seen what others have failed to observe. Theoretically considered, what would one expect to find? Take the case of two British reptiles, the adder and the grass-snake. The former brings forth young adders, the latter deposits eggs and leaves them. The adder family is, I believe, rarely more than thirteen in number; the grass-snake deposits three or four dozen eggs. Therefore, following the reasoning above indicated, one would suppose that one might find indications of parental relations in an adder-mother, which brings forth young adders, not very numerous at a birth. Yet one would not look for the grass-snake to show any such indication. May not this be the very reason that authorities differ about the reptiles, because some are viviparous and some oviparous? May it not be that in the viviparous reptiles parental relationship exists, while it is absent in the egg-layers? If there is anything in this supposition, it is to the adder-mother one must look for signs of some solicitude for her young, for one has to go higher up in nature to find the parental idea developed in the male. One must seek to find out what steps, if any, the adder-mother takes to provide food for her offspring and to protect them from danger, the two direc-tions in which the maternal instinct first shows itself. Those who believe that the female adder swallows her young for protection will at once say there is a marked case of parental relationship. Naturalists hardly admit that phenomenon as yet proved; therefore it is of interest from this point of view that it should be definitely settled. Be that as it may, it is quite evident that it is at the reptile stage in nature that one would naturally expect to find parental relationship first appearing, since here for the first time in the scale are the young of some species habitually developed in the image of the parent, and in sufficiently small numbers for the mother to be capable of being of use to them. When one reaches the birds, one finds a great stride has been made in the develop-ment of the parental idea of duty, instinctive though it still is, and one may watch for hours a father and mother bringing food to a nest of young, and, if necessary, protecting their offspring from intruders.

Grosmont, Pontrilas, near Hereford.

THOMAS HENRY HUXLEY.

TO the student of intellectual progress it must be notable that the two great periods of mental activity should have occurred in this country during the reigns of women. No epoch in English history has shown such literary activity, or writings which have been so enduring as those in the time of Queen Elizabeth, until we reach that of Queen Victoria. We may go so far as to say that in the reign of the former Queen were laid the foundations of modern English Literature. The same may be said with regard to Science during the Victorian era, when this hitherto comparatively neglected, and even publicly scoffed, branch of human knowledge attained a position that has not only commanded respect, but changed the whole conditions of life and customs of a large proportion of the world's human inhabitants. For such benefits as have accrued from the discoveries and teachings of science we are indebted to the perseverance and boldness of preachers like Thomas Henry Huxley.

It is now five years since Huxley passed from among us, and the interval has been well occupied by his son, Mr. Leonard Huxley, in compiling in the two volumes before us his "Life and Letters" ([1]). These volumes are embellished with portraits at different periods of his life, and other subjects of much interest, not the least being one of his study at Hodeslea, the name of his Eastbourne house, where he spent his declining years. Like so much else, he treated this name as a joke, and writing to Sir Michael Foster shortly after settling at Eastbourne he says :—" One is obliged to have names for houses here. Mine will be ' Hodeslea,' which is as near as I can go to ' Hodesleia,' the poetical original shape of my very ugly name. There was a noble scion of the house of Huxley, of Huxley, who, having burgled and done other wrong things (*temp.* Henry IV.), asked for ' benefit of clergy.' I expect they gave it him. not in the way he wanted, but in the way they would like to ' benefit ' a later member of the family. Between this gentleman and my grandfather there is unfortunately a complete blank, but I have none the less faith in him as my ancestor."

What strikes one more than all else in perusal of these two volumes is Huxley's versatility and brilliance as a writer of letters. His correspondence seems to have been most voluminous, and if it were for nothing else the two volumes before us would be amply worth their published price. These letters, extending over the period when science was struggling for its proper place in the public estimation, give us an insight that would otherwise be difficult to appreciate, into the efforts of those who were, after all, but amateurs in comparison with the professionally educated man of science of the present time. We must remember, however, that

it was those of the middle nineteenth century who fought the battle and won the place, for them who now follow. Though so much in earnest and so successful, there was, as we may find in the correspondence of Professor Huxley, a light-heartedness and confidence in ultimate success that should cheer those who, with less difficulties in the present day, are striving to attain a similar, if higher, goal. After all, men of every age are very human, and it is especially when looking over such letters as those before us that we can appreciate how passing disappointments may in fact be only a check leading to greater successes. Such was the case with Huxley in connection with his application for the chair of zoology in the University of Toronto, when he, like Tyndall, who also was a candidate, was rejected.

As a private correspondent Huxley was inimitable. No matter what his subject, gay or grave, he left the recipient of his letter happier for its being written. To take an example, when Charles Darwin sent him a copy of his " Descent of Man and Natural Selection," Huxley wrote :—

" Best thanks for your new book, a copy of which I find awaiting me this morning. But I wish you would not bring your books out when I am so busy with all sorts of things. You know I can't show my face anywhere in society without having read them—and I consider it too bad. No doubt, too, it is full of suggestions just like that I have hit upon by chance at page 212 of vol. i., which connects the periodicity of vital phenomena with antecedent conditions. Fancy lunacy, etc., coming out of the primary fact that one's *n*th ancestors lived between the tide-marks ! I declare it is the grandest suggestion I have heard of for an age. I have been working like a horse for the last fortnight with the fag-end of influenza hanging about me—and I am improving under the process, which shows what a good tonic work is. I shall try if I can't pick out from ' Sexual Selection' some practical hint for the improvement of gutter-babies, and bring in a resolution thereupon at the School Board."

It may be interesting to quote Darwin's note in reply to his receiving an advance copy of Huxley's " Elementary Instruction in Biology" :—

" Many thanks for your Biology, which I have read. It was a real stroke of genius to think of such a plan. Lord, how I wish that I had gone through such a course ! "

It was not always on grave subjects that Huxley wrote. For instance, we quote a note to Matthew Arnold :—"Look at Bishop Wilson on the sin of covetousness and then inspect your umbrella stand. You will there see a beautiful brown smooth-handled umbrella which is *not* your property. Think of what that excellent prelate would have advised, and bring it with you the next time you come to the club."

Anyone who can secure a copy of the " Life and Letters of Huxley " should be congratulated, for it is one of the most entertaining and suggestive books we have read for a long time past .

                                          J. T. C.

(1) " Life and Letters of Thomas Henry Huxley," by his son, Leonard Huxley. xviii+1007 pp., 9 in. x 6 in., in two vols., with thirteen illustrations. (London ; Macmillan & Co., Ltd. 1900.) £1 10s. net.

COCCINELLIDAE IN ARIZONA.

BY PROFESSOR COCKERELL, F.Z.S.

THE following notes relate to some of the native ladybirds of Arizona and New Mexico, which may be useful in destroying scale-insects.

1. *Chilocorus cacti* Linne. A black convex species with two large red spots. This is a species feeding on the native scales, which also now attacks the *Parlatoria* scale on date-palms and is doing very good service. Captain Casey has lately revised the genus *Chilocorus*, and proposed to restrict the name *cacti* to specimens from Central America; while his *Chilocorus confusor*, based on specimens from San Diego, California, would presumably include the Arizona insect. However, after studying the specimens and his descriptions, I believe there is no sufficient reason for the change. I have before me an interesting little pamphlet by one Friderico Friedel, dated 1701, relating to the cochineal insect. In it is the figure of a cactus (*Opuntia*) on which are individuals of the cochineal; and on the ground beneath is a specimen of *Chilocorus cacti*, which must have been observed even at this early date to prey on coccidae.

2. *Thalassa montezumae* Mulsant. This is a large species found by the late Mr. H. G. Hubbard devouring *Toumeyella .mirabilis* near Wilcox, Arizona. Mr. Hubbard suggested that it might feed on the black scale *Lecanium oleae*, which is more or less allied to the *Toumeyella*. For further particulars see Proc. Entom. Soc. of Washington, vol. iv., p. 297.

3. *Chnoodes* n. sp. This preys upon *Icerya rileyi* at Mesilla Park, N.M. It is 3 mm. long, broad oval, convex, black, hoary with a short pale pubescence, moderately shiny, each elytron with an oval dark red spot, and a broad red upper external margin. Under surface, legs, and antennae, dull reddish. Mr. H. S. Gorham at first thought this might be his *C. bipunctatus*, but he now assures me that it is quite distinct, and undescribed so far as he knows.

4. *Hyperaspis lateralis* Mulsant. A small species, black marked with red and yellowish. There is a red spot near the middle of each elytron, one near the end, and a pale bar along the anterior external margin. The head and anterior margin of the thorax are pale yellowish. This is an abundant enemy of *Phenacoccus helianthi* on *Pluchea borealis* in the Mesilla Valley, N.M.; it would probably be useful in destroying other mealy-bugs. In Canad. Entom. 1894, p. 285, I recorded this as *H. undulata*, it having been wrongly identified for me.

5. *H. osculans* Leconte and *H. annexa* Leconte prey upon *Pseudococcus confusus* (*Coccus confusus*, Ckll., Amer. Nat., 1893, p. 1043) at Las Cruces, N.M.

6. *Epismilia misella* (*Pentilia misella* Leconte). A very minute black species found in the Mesilla Valley, N.M., and in many other places, preying on the San José scale. For an account of it, with figures, see Bull. 3, New Series, Div. Ent. U.S. Dept. Agriculture, p. 52. The insect is not a *Pentilia*, and Weise proposed for it the generic name *Smilia*, overlooking the fact that it had been given in 1833 to a well-known genus of *Homoptera*. I have several times directed the attention of coleopterists to this preoccupation, but without result; so, after correspondence with Mr. Gorham, an authority on the *Coccinellidae*, it is agreed to substitute *Epismilia* as above. The *Epismilia* was not observed among the San José scale in the Salt River Valley.

East Las Vegas, New Mexico, U.S.A.

LONG-EARED BAT IN WESTMINSTER.—A specimen of the long-eared bat (*Plecotus auritus*) was captured in the office of the Board of Education on September 29. It was carefully removed, but unfortunately escaped during the night by forcing itself between the bars of its cage. Although a government office is not a probable place to find a bat, I do not imagine these little animals are rare in Westminster, as the parks are within easy flight.—*Frank P. Smith, Islington.*

PALAEOLITHIC MAN IN THE VALLEY OF THE WANDLE.—In your August number you published, under the above title, an interesting article by Mr. J. P. Johnson, dealing with the gravels of the Wandle Valley. These are not of very easy interpretation. He gave one or two records of finds of palaeolithic implements, from which I gather that the author has only been able to ascertain the following evidence of palaeolithic man's existence in the Croydon Valley:—"A fine implement nearly eight inches long from Thornton Heath, a broken palaeolithic implement from Miles Lane, and three flakes." With regard to the implement from Thornton Heath, now in the loan collection of the Croydon Microscopical and Natural History Club at the Town Hall, I think if the author of the paper had seen the implement he would hardly have cited it as evidence. I understand it was represented by the workman who sold it to have come from the base of the gravel, resting upon the London clay. It has not been rolled, and its condition is strongly suggestive of a much later date. The other finds are no doubt valuable as strengthening the evidence afforded by various implements which are reported to have been found from time to time in the Wandle Valley, but the evidence at present is meagre and requires considerable strengthening before we can arrive at any very accurate conclusion as to the palaeolithic inhabitants of the neighbourhood. Great interest attaches to the position of the particular gravels in which any implements may be found, and special care should be taken to notice if they are of plateau type. I have one of the latter in my possession, greatly rolled, but unfortunately am unable to be positive as to the particular gravel from which it was taken, though I think it undoubtedly came from the upper terrace at Thornton Heath.—*N. F. Robarts, F.G.S., 23 Oliver Grove, South Norwood.*

BOOKS TO READ

NOTICES BY JOHN T. CARRINGTON.

The Origin of Species. By CHARLES DARWIN, M.A. xiv + 703 pp., 8 in. × 5½ in., with portrait. (London: John Murray. 1900.) 2s. 6d. net.,

This new impression and cheap edition of the late Charles Darwin's epoch-making work will come as a boon to thousands of young students and their elders. The book is embellished by one of the best portraits of the author that is extant ; in fact it is so good as to be in itself worth the price of the book. This edition is well printed in good readable type, and we have to congratulate Mr. Murray upon enabling the many who have desired to possess this work to find it within their reach.

Thomas Sydenham. By JOSEPH FRANK PAYNE' M.D. xvi + 264 pp., 8 in. × 5½ in., with portrait. (London: T. Fisher Unwin, 1900.) 3s. 6d.

This is another of Mr. Fisher Unwin's "Masters of Medicine," of which series we have already noticed several in these columns. Thomas Sydenham flourished in troubled times, when medicine was little understood as a science. He was born in 1624 and was one of those who laid the foundation of this present highly respectable profession. The author of the book before us has had rich opportunities, denied to previous biographers of Sydenham, in consequence of the recent publication of historical documents relating to Sydenham and his times. The result is an exceedingly readable book of high interest to students of history, medicine, and science generally.

A Year with Nature. By PERCIVAL WESTELL. xvi + 276 pp., with 170 illustrations. (London: Henry J. Drane. N.D.) 10s. 6d.

Though this book is undated, it appears to have been recently issued by the publisher, as the author's preface is dated August last. It is one of the handsome picture-books that have latterly appeared with numerous illustrations depicting country scenes, in this case chiefly dealing with birds. Some of the plates are very pleasing, but others taken, from groups of preserved specimens, are by no means so successful. The book is divided into months as sections, and the letterpress into essays upon country lore. The last section, that devoted to December, includes some useful notes on birds' claws. In another section is a study of birds' beaks, and also one on birds' tails. The book is nicely produced, and will form a handsome present for anyone with a taste for the country.

Flora of Skipton and District. By LISTER ROTHERY. vii + 135 pp., 8½ in. × 5½ in., with diagrams. (Skipton: Edmondson & Co.) 1s. 6d.

At the request of the Craven Naturalists' Association, Mr. Lister Rothery has prepared a list of the flora of Skipton and district. It not only contains the Phanerogamia, but also the Musci and other Cryptogamia. There is a long list of fungi. This local flora will be found useful for comparative purposes.

Birds of Ireland. By RICHARD J. USSHER and ROBERT WARREN. xxxi + 419 pp., 9 in. × 6 in., with a coloured frontispiece, 7 plates, 7 illustrations in text, and 2 maps. (London: Gurney & Jackson. 1900.) £1 10s.

This is one of the most important ornithological works, dealing with a restricted fauna, which has been published for some time past. The authors have for upwards of a decade been preparing the materials from which it is constructed. The design is scientific, and the contents trustworthy. The many original observations are of more than local interest. There is also included every item of ornithological interest which has appeared in various magazines since the days of Thompson, who wrote, half a century ago, so much on the natural history of Ireland. The work is preceded by a table arranged in columns, giving the English and Irish names of the birds, with the pronunciation and their meaning. These are printed in the Celtic characters. There is another valuable table giving the distribution of the species of birds which have bred in Ireland in the nineteenth century. The main body of the book is occupied by notes on each species. These are copious, and deal not only with the distribution, but also with habits and other features connected with each species. The nomenclature followed is that adopted in the second edition of Mr. Howard Saunders' "Illustrated Manual of British Birds." The coloured frontispiece represents half-a-dozen handsome varieties of peregrine falcons' eggs from Ireland. The photographic plates are of subjects well taken and beautifully reproduced. One of the most remarkable of these is by Mrs. Wynne, of Castlebar, representing cormorants' nests in trees. This appears to us to be of exceptional interest, and a not generally known habit in these birds.

Reports of the Moss Exchange Club. Edited by the Rev. C. H. WADDELL and Mr. J. A. WHELDON. 63 pp., 8½ in. × 5½ in. (Stroud : J. Elliott. 1900.)

These reports – the fourth and fifth issued by this society, applying to the years 1899–1900—show the character and extent of the very useful work being done by its members, in the list of whom for 1900 we regret to notice a slight falling off in numbers as compared with that of the previous year. That 5,300 specimens of mosses and hepaticae should in two years have been distributed among the members is pleasing evidence of the increasing interest now taken in this fascinating branch of botany. We are glad to notice that there has been instituted a beginners' section of the club, with the nominal subscription of one shilling per annum. The hon. secretary for this department is Mr. E. C. Horrell, 49 Danby Street, Peckham, London, S.E. This should be a material help to all who, starting on the study of mosses and liverworts, find difficulty in ascertaining whether their specimens are correctly named, or in getting types of the principal genera as a starting-point for individual research. A very important feature of the reports is a statement that an arrangement has been made for every specimen of moss or hepatic to be examined by experts, and the nomenclature checked. This of course entails much labour on the distributor, but is invaluable in eliminating error and its perpetuation. The members ought to be, as they doubtless are, grateful for the admirable manner in which the Moss Exchange Club is conducted by its hon. secretary, the Rev. C. H. Waddell,

B.D., Saintfield, Co. Down, and the hon. distributor, Mr. J. A. Wheldon, of Liverpool. The notes upon species that have passed through the club are valuable, and show that this modestly conducted society is one of the most scientific in Britain.

The Royal Observatory, Greenwich. By E. WALTER MAUNDER, F.R.A.S. 306 pp., 8 in. x 5½ in., with 54 illustrations. (London: The Religious Tract Society. 1900.) 5s.

This is one of the most pleasing popularly written books on a scientific subject that we have met with for some time past. Its object is to give a simple description, understandable by the multitude, of one of the most important institutions in Britain, and of world-wide celebrity. The author's literary work in connection with the science of astronomy is so well known that it would be

Object Lessons in Elementary Science for Standards I., II. and III. By A. H. GARLIC, B.A., and T.F.G., DEXTER B.A., B.Sc. New edition. ix + 254 pp., 7½ in. x 5 in., illustrated. (London: Longmans, Green & Co.) 1s. 6d. each.

These three little books will doubtless be found useful under the present objectionable system of primary education, wherein the most rudimentary knowledge is crammed into children with the hope that at some future time they may take an interest in the knowledge. The information appears to be correct, as far as it goes, in these volumes. Of course we cannot expect detail, as they only deal with object-lessons; but, as the subjects are most diffuse, the chief fear is that the teachers will stop, in most instances, at these suggested lessons and not carry the teaching any further.

DOUBLE STAR OBSERVATION WITH THE SOUTH-EAST EQUATORIAL.

(From Maunder's "Royal Greenwich Observatory.")

supererogation to say more than that he has excelled in the pages before us. They contain not only a history of the Greenwich Observatory, but also a popular description of its interior and the many beautiful instruments which are used by the author, and the other astronomers of the general staff. The illustrations, commencing with a portrait of Flamsteed, the first Astronomer Royal, are chosen with good judgment. They carry the reader from early days to the present time, showing the remarkable progress in optical science since the days of the first Astronomer Royal. We have the pleasure, by permission of the publishers, of reproducing one of the plates, showing a portion of the South-East Equatorial.

Coral Reefs. By CHARLES DARWIN. xx + 549 pp., 7¾ in. x 5 in., with portrait. 7 plates, and 40 illustrations in text. (London: Ward, Lock & Co., Limited.) 2s.

This cheap reissue of Darwin's classic on the structure and distribution of coral reefs, with geological observations on the volcanic islands and parts of South America visited during the voyage of H.M.S. *Beagle*, cannot fail to make the work better known among the general public. The portrait chosen as frontispiece has been copied from the painting in the National Portrait Gallery by the Hon. John Collier. The work is prefaced by a critical introduction by Professor John W. Judd, F.R.S.

Design in Nature's Story. By WALTER KIDD, M.D. F.Z.S. 165 pp., 8 in. × 5 in. (London : James Nisbet & Co., Ltd., 1900.) 3s. 6d. net.

The author has written what appears to have been intended as an essay upon the whole plan of Nature, with the result that he has produced half-a-dozen pleasantly-written chapters that will doubtless find numerous readers among those who do not care to dive too deeply into abstruse subjects.

Guide-Book to Natural Hygiene. By SIDNEY H. BEARD. 103 pp., 7 in. × 5 in., illustrated. (Paignton : Order of the Golden Age. 1900.) 1s. net.

The attractive cover of this little book is suggestive of pleasant reading within. We find, however, that it is chiefly occupied with recipes for cookery of the kind that is apt to leave one hungry, unless associated with more substantial things than vegetables. There are, however, a good many useful hygienic suggestions in the pages, though one is not much encouraged by the heading : " A Bloodless Menu for Christmas."

Dartmoor and its Surroundings. By BEATRIX F. CRESSWELL. 144 pp., 7½ in. × 4¾ in., with illustrations and maps. (London : St. Bride's Press. 1900.) 6d.

This is a second edition of a prettily illustrated guide-book to a beautiful country. It is Number 8 of the Homeland Association's handbooks. It contains much information which is indispensable to those who first visit the district. The edition before us is considerably improved, and has the advantage of having had the supervision of the United Devon Association. There is a chapter on Dartmoor Fishing by Mr. Edgar Shrubsole.

Brain in Relation to Mind. By J. SANDERSON CHRISTISON, M.D. Second edition. 143 pp., 8 in. × 5¼ in., illustrated (Chicago : Meng Publishing Co. 1900.)

The author, who is already known by a book on " Crime and Criminals," was for some time connected with the New York City Asylums for the Insane. His chapters deal with the subject entirely from a physiological point of view. As might be expected, the pages bear evidence of his professional association with defective minds. The book contains a good deal of information which will be found useful.

Elements of Physics and Chemistry. Third stage. By R. A. GREGORY, F.R.A.S., and A. T. SIMMONS, B.Sc. viii + 114 pp., 7 in. × 4¾ in., with 53 illustrations. (London and New York : Macmillans. 1900.) 1s. 6d.

Professor Gregory and Mr. Simmons, both gentlemen with experience as teachers of higher class science, have succeeded in getting together some useful experiments and information, which will lead their students to take an interest in science for its own sake. This primer can be recommended for its simplicity and general correctness. It is quite a book that may be used in the family circle without the aid of a professional teacher.

Contributions to Photographic Optics. By Dr. OTTO LUMMER. Translated and augmented by SILVANUS P. THOMPSON, D.Sc., F.R.S. xi + 135 pp., 9 in. × 6 in., with 55 illustrations. (London and New York : Macmillans. 1900.) 6s. net.

This is an important and highly scientific treatise upon photographic optics, and Dr. Silvanus Thompson deserves thanks for rescuing Professor Otto Lummer's essays from a German magazine. They will be found to give in concise form information not to be found elsewhere. This knowledge is placed before us in so logical and so direct a manner as to be sure to command the attention it should receive. In these pages are contained an exposition of the remarkable theories of Professor von Seidel, of Munich, whose work on geometrical optics in relation to the aberrations of lenses is not sufficiently known. Chapters XII. and XIII., dealing respectively with " Some Recent British Objectives " and " Tele-photographic Lenses," are by Professor Silvanus Thompson and are valuable additions, as are several other portions of the work from his pen.

Bacteria. By GEORGE NEWMAN, M.D., F.R.S.E., D.Ph. Second edition. xvi + 397 pp., 8½ in. × 5¼ in., with 94 illustrations. (London : John Murray. 1900.) 6s.

We have already had the pleasure of noticing the first edition (SCIENCE-GOSSIP, N.S., vol. vi., p. 85), and it is satisfactory to find that a second edition is required within twelve months of the first issue. In this there have been made a number of necessary corrections, and much new matter added. Among the latter information are articles on " The Bacterial Treatment of Sewage," " Industrial Applications of Bacteriology," and " Tropical Diseases." In this edition are fifteen micro-photographs of actual organisms, taken expressly for the book by the celebrated specialist in the photography of bacteria, Mr. E. J. Spitta, M.R.C.S. An extended knowledge of the influence of bacteria on human existence, is so important that this most useful work cannot be too widely known.

One Thousand Objects for the Microscope. By M. C. COOKE, M.A., LL.D., A.L.S. x. + 179 pp., 7½ in. × 4¾ in. 13 plates, containing over 500 figures, and 38 woodcuts. (London : Frederick Warne & Co.) 2s. 6d. net.

Dr. M. C. Cooke's unpretentious little book has long been a favourite with amateurs, and we are glad to find that the publishers have recently reissued it in a new dress. Originally a list of popular microscopic objects, with brief explanatory notes attached, the book has now received an addition in the form of some fifty-six pages dealing with the purchase and management of a suitable microscope, pond and field collecting, and mounting. The treatment of the part dealing with the microscope itself appears to us to be somewhat inadequate to the importance of the subject, though we heartily sympathise with Dr. Cooke's advice in favour of a simple rather than an elaborate instrument. The beginner's difficulties, for instance, with regard to the management of his condenser and its accessories are barely alluded to in his pages. On the other hand, the remarks on pond and field collecting, and on simple preparation and mounting for the microscope, though brief, are eminently practical and to the point. The plates, illustrating about half of the objects described, are apparently the original ones, and the descriptive letterpress thereon, with its practical notes and clear instructions as to where to find and how to deal with these objects, is also unaltered. A new frontispiece has been added, representing typical radiolarians, which is apparently a reduction, unfortunately not equal to the original, of one of the plates in Messrs. Warne's " Royal Natural History."

The book is well printed, and handsomely bound in cloth, and our most serious criticism would be its issue, as a scientific book, without any date on the title-page. A book like this contains a wealth of suggestion that should help to make many a dilettante owner of a microscope an earnest student of nature.—*F. S. S.*

A Monograph of the Land and Freshwater Mollusca of the British Isles. By JOHN W. TAYLOR, F.L.S. Part VI. pp. 321–384. 10½ in. × 6½ in., with 108 illustrations. (Leeds: Taylor Bros. 1900.) 6s.

With the present instalment Mr. Taylor brings to an end the consideration of the shells and soft parts of the molluscs. It appears, however, from an announcement to the subscribers, that yet another part will appear before volume I. is completed. In this will be discussed the important subjects of distribution in time and space, as well as the parasites attacking molluscs and the uses of the latter to mankind. The pages which we have here to consider touch upon some of the most interesting points in connection with mollusca. Under the adductor muscles, but it may be pointed out that the investigations of Mr. S. Pace upon the pearl oyster point to the fact that abnormal conditions in the mollusc itself play a much more important part in the production of pearls. Mr. Pace was able to infect healthy oysters and so obtain pearls from them. The biological remarks upon coloration, though little fault can be found with them, demonstrate how much work there remains to be done upon this fascinating part of the subject. In other places Mr. Taylor discusses molluscs generally; here, however, and perhaps wisely, he confines his attention to the species strictly included under the title of his monograph. Excretory organs, lymphatic glands, and the muscular system are described and illustrated. Finally, the reproductive organs and development bring structural considerations to a close. Among the illustrations kindly lent by Mr. Taylor, we show the love-darts and the annulus which supported one, in a specimen of *Helix aspersa.* Some species spend a considerable time over their courtship, and one of our pictures shows the love-making of a sinistral and a

The courting of a sinistral and a dextral specimen of *Helix aspersa.*
(*From Taylor's "Monograph of Land and Freshwater Shells."*)

heading of glands and secretions the attachment threads (byssus) are described; the embryonic

a. Love-dart of *Helix hortensis.*
b. Love-dart of *Heliciyona lapicida.*

shell is mentioned, while the three pages devoted to pearls are welcome reading. The author alludes to pathological causes giving rise to pearls in the dextral example of the common snail, *H. aspersa.* From one's personal observation one would say that important parts of the drawing illustrating the pairing of *Limax maximus* require to have more detail introduced, and here one misses the

c. The annulus of a dart from *Helix aspersa.*
d. Egg-capsules of *Falcata cristata.*

series of sketches by Mr. Lionel Adams in the "Journal of Conchology" in 1898, leading up to the final position, alone depicted by Mr. Taylor. Numerous figures elucidate the remarks upon development, which, if we except a page and a half of supplementary bibliography, bring to a close a number not descending everywhere perhaps into such detail as in its predecessors, yet well worthy to take its place with them.—*W. M. W.*

PROFESSOR H. J. TODD, having given up the directorship of the United States "Nautical Almanac," has been succeeded for the time being by Professor S. J. Brown, of the Naval Observatory, Washington.

WE observe, in the list of papers to be read before the Royal Geographical Society during the coming session, there will be one by Mr. Vaughan Cornish, giving a further account of his important studies in "wave forms."

WE regret to hear that the eminent lepidopterist, Dr. Staudinger, has died suddenly while visiting Lucerne, in his seventy-second year. Dr. Staudinger had been in feeble health for some time past, which has been the reason for the delay of the long-promised revision of his standard synonymic list of Palaearctic Lepidoptera. We are not aware that this important work has been completed.

THE official journal of the Khedival Government has announced that from 1st September last universal time was adopted in Egypt, and that mean noon of the 30th meridian east of Greenwich is given as noonday signal. There are now eight meteorological stations daily communicating weather reports to the central office, and others are being fitted with automatic registering instruments for early observations. These will probably constitute links of a chain of observatories extending from the North of Europe to Capetown.

WRITING to our contemporary, "Nature," Mr. R. Paulson draws attention to a destructive fungus attack on birch-trees in woodlands around London. The fatal effects have become especially apparent during the past year, when, he states, many healthy trees have died in Epping Forest, on Chislehurst, Hayes, and Keston Commons, where no signs of the disease were evident last spring. Mr. Paulson thinks the trouble is the result of a micro-fungus, *Melanconis stilbostoma* Tul. Have any of our readers noted the same fatality in other parts of the country?

THROUGH the energy of its Committee and Secretary and the help of a public benefactor, the Essex Field Club has arrived within at least one goal of its ambition. On October 18 there was opened the Passmore Edwards Museum at Stratford, which houses the biological and other collections of the Essex Club. The building, which has cost about £4,000, presented to the borough by Mr. Edwards, has been erected in connection with the West Ham Municipal Technical Institute. Thus the students at the Institute have the advantage of the contents of the adjoining museum for reference and study. A feature of importance about these collections is that they are chiefly from the County of Essex, so not only arousing interest among the visitors with regard to their surroundings, but placing on record examples for future comparison.

THE Moss Exchange Club, whose reports we notice under "Books to Read," has distributed a list of the British Sphagna, accompanied by one of those mosses occurring in Europe, after Warnstorf, and corrected up to April of this year.

THE Geological Survey of Western Australia sends us its fourth bulletin. It is a voluminous essay on the mineral wealth of Western Australia by Mr. A. Gibb Maitland, F.G.S., the Government geologist, occupying 150 closely-printed pages, illustrated by maps.

WE have received from the Division of Biological Survey, U.S. American Department of Agriculture, number 18 of the North American Fauna. This is devoted to a revision by Mr. Wilfred H. Osgood of the pocket mice of the genus *Perognathus.* Also a report upon the food of three kinds of birds —namely, the bobolink, blackbirds, and grakles, all more or less destructive of agricultural crops.

SOME of our readers will be glad to have their attention called to the bargains in a number of standard scientific books offered by Mr. H. J. Glaisher, of 57 Wigmore Street, London. Mr. Glaisher purchases the majority of the remaining copies of such works from the publishers, and is thus enabled to dispose of them at an exceptionally cheap rate, although they are quite new copies.

UNDER title of "The South-Eastern Naturalist" are now published the "Transactions of the South-Eastern Union of Scientific Societies for 1900," under the honorary editorial care of Mr. J. W. Tutt, F.E.S. The first thirty-two pages are devoted to the business and official notices of the Union. The remainder consists of the presidential address, and papers read before the last congress. Several of these are of considerable importance.

MR. JOHN NIMMO announces a new and revised edition of a Handbook of British Birds by Mr. J. E. Harting, F.L.S., F.Z.S., with 35 coloured plates from original drawings by the late Professor Schlegel. The price will be two guineas net. It will be remembered that Mr. Harting edited the "Zoologist" for twenty years, and for a still longer period was connected with the natural history columns of the "Field" newspaper.

A new monograph of the Homopterous insects included in the group Membracidæ is in preparation by Mr. George Bowdler Buckton, F.R.S., F.L.S., F.E.S. The author appeals to entomologists and others who have specimens, to confer with him, as there must be many species still unknown to science. The work will be published by Messrs. Lovell Reeve & Co., Limited. Mr. Buckton's address is Weycombe, Haslemere, Surrey.

THE admirable Society for the Protection of Birds, whose offices are 3 Hanover Square, London, is offering two prizes, the particulars of which may be obtained from the Honorary Secretary. The prizes are for £10 and £5 respectively, for the best papers on the protection of British birds. The mode of dealing with the subject is left entirely to competitors, but among the points suggested for treatment are the utilisation and enforcement of the present Acts and County Council Orders; the modification or improvement of the law; educational methods; and the best means of influencing landowners and gamekeepers, agriculturists and gardeners, collectors, birdcatchers and birdnesters. Essays are to be sent in by November 30th next.

CONDUCTED BY F. SHILLINGTON SCALES, F.R.M.S.

MOSQUITOES AND MALARIA.—The evidence in support of the theory that malarial infection is due to the bites of mosquitoes (see SCIENCE-GOSSIP, Vol. vi., p. 182), themselves already infected, seems now to have put the matter beyond a doubt. Drs. Sambon and Low have deliberately taken up their residence in the most unhealthy and fever-stricken spot in the Roman Campagna, a place situated in the heart of the swamp, among the haunts of mosquitoes of the genus *Anopheles*, and of which. the few dwellings near at hand are inhabited by peasants who are constant victims to malaria. These daring investigators have shown that by avoiding mosquitoes they avoid malaria, but a son of Dr. Manson has given an even more striking example of enthusiasm in the cause of science by allowing himself to be bitten by mosquitoes which had been fed on the blood of a sufferer from malaria in Rome. The mosquitoes were sent to London by Professor Bastianelli. and received early in July. The patient, after being bitten, developed well-marked malarial symptoms, though he has never been in a malarial country since he was a child. He has now recovered, but has thus supplied evidence of the positive kind, as Drs. Sambon and Low did of the negative kind. A letter from Mr. H. J. Elwes, F.R.S., to the *British Medical Journal*, calls attention now to the necessity of finding out under what conditions mosquitoes do not produce malaria, and mentions that whilst in certain districts in India he escaped malaria by protecting himself by mosquito curtains, whilst other members of his party who omitted these precautions were attacked, yet in other districts where mosquitoes abound malaria is almost unknown, and these precautions were unnecessary. Amongst recent literature dealing with the subject we may mention a paper on the life-histories of mosquitoes of the United States, published in one of the *Bulletins* of the U.S. Department of Agriculture, and contributed by Dr, L. O. Howard, State entomologist. Descriptions, with illustrations, are given of all the members of the group met with in the United States, with especial reference to members of the genus *Anopheles*, to which suspicion most strongly points. Dr. Howard advocates the use of kerosene for the destruction of the larva, and calls attention also to the agency of fish in this connection. The July number of the *Quarterly Journal of Microscopical Science* contains also a number of plates and diagrams by Major Ross and Mr. R. Fielding Ould, of the Liverpool School of Tropical Medicine, illustrating the life-history of the parasites of malaria. Before leaving the subject we may mention that the second malarial expedition from Liverpool has telegraphed home from Bonny, in Nigeria, news of the discovery of another parasite, found in the proboscis of mosquitoes, which causes elephantiasis. Our readers will be aware of the terrible scourge this disease is to millions of natives

in tropical countries, and that it is due to a small worm which lives in the lymphatic vessels. We understand that the discovery has been simultaneously made in England by Dr. Low, and in India by Captain James.

PARASITE FROM HUMBLE-BEE.—I was much interested in the notes in the microscopical section of the September SCIENCE-GOSSIP, especially those on the parasite of the humble-bee. Last year I mounted some parasites from a humble-bee which

PARASITE FROM HUMBLE-BEE.

may possibly be identical with those referred to. The size, however, is an objection, this one being still larger than the specimen you comment upon. The length, measured to the tip of the extended jaws, is 1·5 mm. The length of the jaws, however,

JAWS OF PARASITE.

would vary, as they seem to be retractile. I enclose a sketch which may be sufficient for identification. —*W. Cran, Mains of Lesmoir, Rhynie, Aberdeenshire.*

MR. C. BAKER'S NEW CATALOGUE.—Mr. Chas. Baker has sent us his new catalogue, which is increased in size by about twenty pages. It contains particulars of his latest instruments, such as the R.M.S. 1·27 microscope, the D.P.H. Nos. 1 and 2, the Diagnostic, and the Plantation microscopes, all of which received notice in SCIENCE-GOSSIP when they first came out for sale. The newest pattern microtomes are also included. The list of stains, mounting media, and other accessories. is exceptionally complete and well arranged. Two or three pages are devoted to pond-life apparatus, but the extra pages are mainly taken up with lists of microscopic slides, arranged mostly in series at a moderate price, though the price of individual specimens seems rather higher than usual. The lists of bacteriological and diatomaceous slides call for special notice. We have already drawn attention to Mr. Chas. Baker's excellent slide-lending department.

R. & J. BECK'S NEW "LONDON" MICROSCOPE.—
In design of microscopes our English makers have
always taken the lead, but the cheaper kind of
Continental stand has had a large sale here, especially amongst our medical and other students.
Recognising this, Messrs. Beck, by laying down
new plant for displacing hand labour by machinery
in accordance with the practice of the day, have
been able to put upon the market a new stand
in which sound workmanship is combined with
cheapness. The stand follows largely the Continental model, but is sold at a considerably lower
figure than any similar Continental stand known to
us. The essential features are as follows :—The
stand is of the horse-shoe form, but the central
pillar is placed farther forward under the stage so
as to give greater facility when the microscope is
in a horizontal position. The base itself actually
rests on three inserted cork pads, which not only give
steadiness, but prevent any possibility of scratching
the table. The coarse adjustment is by spiral rack
and pinion, whilst the fine adjustment is of the

. . NEW "LONDON" MICROSCOPE.

micrometer screw type, in which, however, a
pointed rod impinges upon a hardened steel plate,
which is itself attached to the limb of the microscope and works upon a triangular upright rod.
In the larger model the milled head of the fine
adjustment is graduated and furnished with a folding pointer. The body tube is designed for use
with objectives corrected for the Continental length
of tube, but carries a draw tube capable of variation from 140 to 200 millimetres. The stage is
square and the upper surface is faced with ebonite.
In the larger model it measures 4 × 4 inches. In
the least expensive model a ring, of the Society
size, is fitted beneath the stage to carry the iris
diaphragm and condenser. The latter is specially
arranged to fit above the iris diaphragm, and the
arrangement is both effective and cheap. In the
larger models, or those fitted with more elaborate
sub-stage arrangements, the usual form of Abbé
condenser is provided for in this instrument.

Messrs. Beck make their "London" microscope in
two sizes. The smaller size with sub-stage ring
and iris diaphragm, but without objectives, eyepieces, or condenser, costs, with mahogany case,
only £3 5s. 6d. The addition of a swing-out and
spiral focussing sub-stage increases the price to £4.
The larger model, similar to the last described,
costs £5 5s., or with rack and pinion focussing sub-
stage £6 10s. The necessary eyepieces cost 5s.
each, the condenser in its simplest form 10s.,
whilst the objectives are of Messrs. Beck's well-
known and moderately priced series. This microscope has, like all apparatus mentioned in these
columns, been submitted to our personal inspection,
and we congratulate Messrs. Beck on its production. We illustrate the smaller stand herewith.

EXTRACTS FROM POSTAL MICROSCOPICAL SOCIETY'S NOTE-BOOKS.

[Beyond absolutely necessary editorial revision,
these notes are printed as written by the various
members, without alteration or amendment. Correspondence on these notes will be welcomed. These
extracts were commenced in the September number,
at page 119.—ED. Microscopy, Science-Gossip.]

NOTES BY WM. H. BURBIDGE.

DEVELOPMENT OF BALANUS.—On the rocks of
the southern and western coasts of England, when
the tide is out, we observe that their surface is
roughened up to a certain level with an innumerable multitude of brownish cones. Each appears
as a little castle built of strong plates that lean
towards each other but leave an orifice at the top.
Within this opening we see two or three other
pieces joined together in a particular manner, but
capable of separating. These are Barnacles, class
Crustacea, division Cirripedia (from cirrus, a curl,
and pes, a foot), order Balanidae (from balanus, an
acorn). Fixed and immovable as the barnacles
are in their adult stage, they have passed by metamorphosis through conditions of life in which they
were roving little creatures, swimming freely in
the sea. It is in these conditions that they present the closest resemblance to familiar forms of

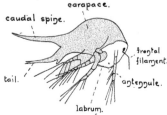

FIG. 1. Nauplius stage of *Balanus.*

crustacea. Fig. 1, represents the Nauplius stage of
the barnacle. It has a broad carapace, a single
eye, two pairs of antennae, three pairs of jointed,
branched and well-bristled legs, and a forked tail.
The skin is cast twice, considerable change of
figure resulting. At the third moult it assumes the
cypris stage as represented by fig. 2, and is enclosed
in a bivalve shell, with the front of the head and
the antennae greatly developed, the single eye

having become two. In this stage the little creature searches for a suitable spot for a permanent residence. The two antennae which project from the shell pour out a glutinous gum which hardens in the water and fixes them. Another moult takes place, the bivalve shell is thrown off, the carapace is composed of several pieces, whilst the legs are modified into cirri and made to execute their grasping movement. Fig. 3 shows the mature barnacle with cirri. Nothing can be more effective or beautiful than the manner in which the cirrus obtains its prey. The cirri are alternately thrown out and retracted with great rapidity, and when fully expanded the plumose

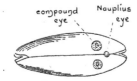

FIG. 2. Cypris stage of *Balanus*. Ventral view.

and flexible stems form an exquisitely beautiful apparatus, admirably adapted to entangle any nutritious atoms or minute living creatures that may happen to be present in the circumscribed space over which this singular casting-net is thrown, and drag them down to the vicinity of the mouth. This action may be easily seen if a small portion of rock be chipped off, having barnacles on it, and placed in a glass with seawater. A hand-glass will show the beautiful little hand with twenty-four long fingers, the net with which this fisher takes his prey, busily at work. Care must be taken that there are living barnacles

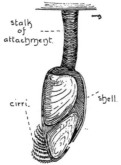

FIG. 3. Mature *Balanus* and cirri.

on the piece of rock, as many are but empty shells. An interesting slide is *Obelia geniculata* (fig. 4). It has double and alternate generations. The polyp bears urn-like reproductive capsules which discharge large numbers of medusiform zooids. Like miniature balloons they float suspended in the water for a while, and then suddenly start into motion with a series of vigorous jerks. They may be considered as swimming polypites with the arms united by a contractile web. They mature and disperse the generative elements, and, having thus fulfilled their function, perish. The ova,

after fertilisation, become ciliated embryos, and when affixed rapidly grow into the plant-like zoophytes we see. *Sertularia pumila*—another hydroid zoophyte—is a very common species, though it makes a beautiful microscopic object. Almost every broad-leaved seaweed has greater or lesser numbers of this zoophyte growing on it.

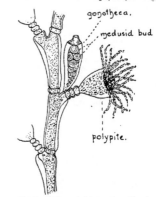

FIG. 4. *Obelia geniculata*, portion of frond.

Coryne vaginata (fig. 5) is one of the Athecata—that is, without any theca or calycle. The capitate tentacles bear on the summit a globular head consisting of a collection of thread cells, a vigorous battery of offensive weapons. They occur in astonishing profusion, and consist of minute sacs

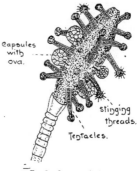

FIG. 5. *Coryne vaginata*.

embedded in the flesh, filled with fluid, which contain a long delicate thread capable of being projected with considerable rapidity. *Corydendrum parasiticus* is a similar creature to the last, but the tentacles are not capitate. There is something singular about the stems that support the polypites; they look as if they acted as capsules and held ova. This is a foreign species, and I cannot find any description of it. I believe Mr. Sinel told me it came from the Mediterranean. *Pennaria carolina* is also a foreign species. Some of the polypites bear gonophores, the buds in some of which the reproductive elements are formed.

(*To be continued.*)

MICROSCOPY FOR BEGINNERS.

BY F. SHILLINGTON SCALES, F.R.M.S.

(Continued from p. 154.)

IN using glycerine mounting media it is well to remember, as pointed out by Dr. Carpenter, that they largely increase the transparency of organic substances; and though this is often advantageous, it may also sometimes result in so great a diminution of their reflecting capacity as to make them indifferent mounts. We have given such instructions in elementary mounting as will, we think, enable a beginner to make rapid progress in the art if he is gifted with only a small amount of perseverance and patience, but it must not be forgotten that the actual mounting is but a part of the work required. Numerous subjects will need very careful preparation beforehand, and on the methods adopted and the skill and judgment with which they are carried out will depend much of the result. Many objects will need dissecting. In one of our previous papers (SCIENCE-GOSSIP, vol. vi., pp. 280–1) we gave a description of a home-made dissecting stand with supports for the hands, and of the necessary apparatus for the purpose. Their uses are self-evident. We may say here, however, that most dissections, and especially delicate dissections, are done under water, with perhaps a little methylated spirit added if the object has previously been soaking for some time in methylated spirit or alcohol. In some cases it will be necessary to fasten the object down, and this may be done with pins on a weighted piece of cork placed inside the dissecting dish, or by running paraffin or some such compound into the bottom as already explained. Watchglasses with flat bottoms make useful dissecting dishes. Two or three needles set in light wooden handles will be required, with both straight and bent points, and these can readily be manufactured at home, or purchased for a few pence. In buying dissecting knives, we strongly recommend that those with ivory handles be chosen; they only cost one shilling and ninepence each, as against eighteenpence for the ebony-handled ones, while the latter are so brittle as to break with very little pressure. There are a good many shapes of blades sold, but perhaps the most generally useful are the usual scalpel forms, the spear, and the spatulate-shaped ones. Forceps may be either steel, brass, or nickel, but we prefer the steel, which should, of course, be carefully kept clear of rust. A few camel-hair brushes are also necessary, and a pair of fine scissors. Insects generally require soaking in a ten per cent. solution of sodium or potassium hydrate (caustic potash), for periods varying from an hour or two up to a week. Too much soaking will destroy the object and also render it too transparent after mounting, whilst too little may leave it hard and difficult to deal with. A little thought and attention will therefore be necessary, and a slight pressure with a blunt needle will tell whether the object is sufficiently soaked. In the case of large insects, like cockroaches, we should soak them for several days until they begin to give off an unpleasant smell. The alkali must then be removed by soaking in several changes of clean water. The inside of the insect can be got rid of by gentle treatment with the camel-hair brushes. Plant subjects are best softened by long soaking in water.

(To be continued.)

GEOLOGY

CONDUCTED BY EDWARD A. MARTIN, F.G.S.

BY the resignation of the chair of geology at University College, London, so long held by the Rev. Thomas George Bonney, D.Sc., LL.D., F.R.S., a scientific prize will fall to someone after Christmas next. Professor Bonney, who is Hon. Canon of Manchester and Fellow of St. John's College, Cambridge, has held the professorship he will so soon vacate for upwards of twenty years. He was born at Rugeley in 1833.

SOME DEEP LONDON BORINGS.—The following particulars of borings in London, which have been made into the Chalk, are in continuation of those given in the July number of SCIENCE-GOSSIP. The same remarks there made will now apply, and the depths there assigned to the ever-changing Tertiaries may be liable to future correction. For these details I am again indebted to Messrs. Isler & Co., the well-known well-sinkers of Southwark:—

Place of Boring	Dug Well, Made Ground, or Ballast	London Clay	Oldhaven Beds	Woolwich Beds	Thanet Sands	Chalk	Total bored, in feet	Water Supply, gallons per hour	Water Level below Surface
	ft.	ft.	ft.	ft.	ft.	ft.	ft.		ft.
Clapham Road, 139 (Causton)	28	77	4½	47½	38	230	425	12,000	54
Croydon (Steam Laundry)	19	47	16	36	57	75	250	1,000	66
Forest Gate (Upton Lane)	28	25	—	23	29½	144½	250	2,340	27
Gray's Inn Road (Perkins)	12	56	—	57	28	147	300	3,000	—
Great Dover St., S.E. (Groves)	49	—	—	62	39	129	279	—	60
Hackney Road (Chandler)....	12½	40½	—	77	17	253	400	10,000	93
London Bridge (Hibernia Chambers)	21	77	—	44½	56	101½	300	1,000	95
Hornsey Road (Baths)	1½	104½	—	28¼	57¾	258	450	10,000	165
Lambeth (Dawn & Valentine)..	31½	90½	—	49½	43½	105	320	4,000	81
Latimer Rd., W. (Phoenix Brewery)	22	126	—	52	22	178	400	4,500	80
Mitcham—(Camwal)	—	—	—	—	—	333	333	1,000	19
Mitcham—(Gas Works)..	148½	22	—	38½	30½	118½	350	3,500	20
Mitcham—(Thunder & Little)	26	67	15	39	39	174	340	2,000	—
Nine Elms—(Thorne)....	22	—	—	151	39	139	351	1,500	54
Peckham—(Rye Lane)....	23	2	17	18½	30½	33	124	1,000	40
Regent's Park—(Zoo)	178	—	—	23	12	239	452	8,000	—
Romford Road, (242), E........	22	—	11	27	44	146	250	4,200	30
Rotherhithe St. (251)	117	—	—	—	7	139	263	2,580	45
Waddon—(Croydon)	4	—	—	—	—	296	300	10,000	42¾

ASTRONOMY,

CONDUCTED BY F. C. DENNETT.

	1900	Rises.	Sets.	R.A.	Dec.
	Nov.	h.m.	h.m.	h.m.	° '
Sun	.. 6	.. 7.4 a.m.	.. 4.24 p.m.	.. 14.45	.. 15.56 S.
	16	.. 7.21	.. 4.9	.. 15.25	.. 18.42
	26	.. 7.37	.. 3.57	.. 16.7	.. 20.35

		Rises.	Souths.	Sets.	Age at Noon.
	Nov.	h.m.	h.m.	h.m.	d. h.m.
Moon	.. 6	.. 4.0 p.m.	.. 11.44 p.m.	.. 6.20 a.m.	.. 13 22.33
	16	.. 1.13 a.m.	.. 7.28 a.m.	.. 1.31 p.m.	.. 23 22.33
	26	.. 10.51 a.m.	.. 3.24 p.m.	.. 8.5 p.m.	.. 4 4.43

Position at Noon.

		Souths.	Semi-	R.A.	Dec.
	Nov.	h.m.	diameter.	h.m.	° '
Mercury	 6	.. 1.11 p.m.	.. 3·8″	.. 16.12	.. 23.53 S.
	16	.. 0.22 p.m.	.. 4·8″	.. 16.2	.. 21.29 S.
	26	.. 10.57 a.m.	.. 4·5″	.. 15.17	.. 15.57 S.
Venus	 6	.. 9.13 a.m.	.. 7·9″	.. 12.14	.. 0.13 N.
	16	.. 9.18 a.m.	.. 7·4″	.. 12.58	.. 4.12 S.
	26	.. 9.23 a.m.	.. 7·0″	.. 13.43	.. 8.36 S.
Mars	.. 16	.. 6.20 a.m.	.. 3·5″	.. 10.0	.. 14.21 N.
Jupiter	.. 16	.. 1.17 p.m.	.. 14·7″	.. 16.58	.. 22.18 S.
Saturn	.. 16	.. 2.30 p.m.	.. 7·1″	.. 18.11	.. 22.46 S.
Uranus	.. 16	.. 1.0 p.m.	.. 1·8″	.. 16.40	.. 22.10 S.
Neptune	.. 16	.. 14.0 a.m.	.. 1·3″	.. 5.54	.. 22.12 N.

MOON'S PHASES.

		h.m.			h.m.
Full	.. Nov.	6 .. 11.0 p.m.	3rd Qr.	Nov. 14	.. 2.38 a.m.
New	.. „	22 .. 7.17 a.m.	1st Qr.	„ 29	.. 5.35 p.m.

In perigee November 5th at 4 p.m.; and in apogee on 17th at 7 a.m.

METEORS.

					h.m.	°
Nov. 2-3	..	ε Taurids	Radiant R.A.	3.40	Dec. 9	N.
„ 10-23.	..	ν Cancrids	„	„	8.52	„ 31	N.
„ 13-16	..	Leonids	„	„	10.0	„ 23	N.
„ 13-28	..	Leo Minorids	„	„	10.20	„ 40	N.
„ 20-28	..	ε Taurids	„	„	4.12	„ 22	N.
„ 23-27	..	Andromedids	„	„	1.40	„ 43	N.

LEONIDS.—In spite of the brilliant moonshine, watch should be kept from 11 p.m. on November 14th until sunrise on 15th, and also from 11 p.m. on 15th till sunrise on 16th. in the hope that the Leonids may make an appearance this year. The Radiant point near the middle of the "sickle" in Leo rises in the north-eastern heavens at about 10.15 p.m. and reaches the meridian at about half-past 6 in the morning. A star map, such as "The People's Atlas of the Stars," published by Gall & Inglis, and costing only one shilling, or some such cheap one, should be on a table at hand, with a pencil and ruler to mark down the paths of any meteors observed, and the exact time of each should also be noted.

CONJUNCTIONS OF PLANETS WITH THE MOON.

							° '
Nov. 14	..	Mars†	..	5 p.m.	..	Planet 7.39	N.
„ 19	..	Venus†	..	1 a.m.	..	„ 5.51	N.
„ 22	..	Mercury†	..	1 a.m.	..	„ 1.30	N.
„ 23	..	Jupiter†	..	5 p.m.	..	„ 1.3	S.
„ 24	..	Saturn†	..	12 p.m.	..	„ 2.8	S.

† Below the English horizon.

OCCULTATIONS AND NEAR APPROACHES.

Nov.	Star.	Magni-tude.	Dis-appears. h.m.	Angle from Vertex. °	Re-appears. h.m.	Angle from Vertex. °
7 ..	ρ³Arietis	5·5	.. 1.47 a.m.	.. 1	.. 2.31 p.m.	.. 273
7 ..	13 Tauri	5·4	.. 6.21 p.m.	.. 27	.. Near approach.	
9 ..	χ¹Orionis	4·7	.. 9.53 p.m.	.. 220	.. Near approach.	
10 ..	χ⁴ „	4·8	.. 2.51 a.m.	.. 184	.. Near approach.	
30 ..	κPiscium	5·0	.. 6.11 p.m.	.. 28	.. 7.7 p.m.	.. 277

THE SUN has during October been showing greater activity. On 22nd the disturbed area was 120,000 miles in length, and covered no less than 2,500,000,000 square miles.

MERCURY is an evening star at the beginning of the month, but rapidly nears the sun, coming into inferior conjunction at noon on 20th, after which it becomes a morning star.

VENUS is a morning star all the month, rising near 3 a.m. at the beginning of the month, and a little over an hour later at the end of November.

MARS rises about 11.18 at the beginning of the month, and nearly three quarters of an hour earlier at the end, but its apparent diameter, though increasing, is only 7·″8 on the 30th, and therefore not much can be seen with ordinary telescopes.

NEPTUNE is now becoming favourably placed for observation, rising a little after 7 p.m. at the beginning, and a little after 5 at the end of the month.

ANNULAR ECLIPSE OF THE SUN.—On the early morning of November 22nd there is an eclipse, unfortunately invisible at Greenwich. The line of central eclipse begins in the South Atlantic Ocean, crosses Southern Africa, the Southern Indian Ocean, and a portion of Australia.

A REMARKABLE SUNSET was noted on September 5th by observers in places so far separated as Barnsbury (London), Bridport, Oxford, the South Coast of England, and Criccieth in Wales. Just after sunset a column of rosy light rose vertically some 20° or 30° above the horizon just ahead of the place where the sun had sunk. The width of the column along its entire length was about 30' or 40', and it remained visible for more than half an hour. The observer on the Welsh mountains saw most of the phenomenon, a second bright ray crossing the first at right angles, doubtless forming part of a circle or halo around the sun.

OPPOSITION OF EROS.—The distance separating the sun and earth is now known with great exactness, but it is hoped that the observations now being made on Eros, the new minor planet whose orbit lies in great part within that of Mars, will give still more accurate knowledge. When in opposition, on November 12th, he will be some 39,000,000 miles from us, but, from the great eccentricity of its orbit, it will not be so near to us then by about 10,000,000 miles as it will be at the beginning of the new century. Measurements are being attempted with the micrometer, the heliometer, and by aid of photography. Observations are being made in Europe and America, as well as in the Southern Hemisphere.

A NEW VARIABLE STAR.—Mrs. Fleming, examining the Draper Memorial photographs, has found a previously unknown variable in the constellation Aquila, R.A. 19 h. 15 m., N. Dec. 9° 41'. It varies from 7th magnitude to 11·5 in about a year. Its spectrum has been found by Professors E. C. Pickering and Wendell to be monochromatic, like those of the gaseous nebulae.

CONDUCTED BY E. FOULKES-WINKS.

[WE have the pleasure to announce that Mr. B. Foulkes-Winks has kindly undertaken the post of honorary departmental editor for Photography in "SCIENCE-GOSSIP." He is a gentleman of great experience in every section of the art, so that our readers may expect to gain much instruction from his columns. Mr. Foulkes-Winks will shortly commence a series of articles on Photography, dealing with the subject from taking the negative, and gradually leading up to the finished print. He will also from time to time give particulars of new apparatus, and any interesting feature that may be of value to our readers working at scientific or artistic photography.—ED. S.-G.]

THE ROYAL PHOTOGRAPHIC SOCIETY.—The most interesting item at present in the photographic world is undoubtedly the annual exhibition of the Royal Photographic Society at the New Gallery in Regent Street. The exhibition was opened to the public on October 1st, and will remain open until November 3rd. The annual soirée was held on September 29th, when there was a large gathering of the members and their friends, who were cordially received by the President of the Society, assisted by the members of the Committee. The exhibition is a distinct advance on all previous shows, and the Society is to be congratulated on the very fine and comprehensive exhibits it has gathered together. In previous years the exhibition was held at the rooms of the Society of Artists in Water Colour in Pall Mall, where every class of work, apparatus, etc., was crowded into one room, and many a good picture was rejected solely for want of space. At the New Gallery there is ample room, and what should be highly appreciated by exhibitors is the fact that there are separate galleries for each class. Thus we have the large centre hall devoted to trade exhibits, and they are very tastefully arranged. Here, on our left, we have an artistically arranged stall by Wellington & Ward, with some fine specimens of bromide work. One picture, on rough drawing paper, and toned a warm sepia, is especially fine. This firm also shows samples of their new films. Opposite the entrance is the stall of Messrs. Goerz, where are exhibits of very good work done with the Anschutz camera, an interesting album of a series of views showing the manufacture of their lenses; here also may be seen some very fine stereoscopic views. The London Stereoscopic Company have a stall on which most of the firm's specialities can be seen. Specimens of work in lenses, cameras, etc., are also shown by the well-known firms of Ross, Ltd., Dallmeyer, Ltd., W. Watson & Sons, and Beck & Co., the last-named firm showing their new "Frena" folding camera, a very neat and complete instrument for hand or stand; and we understand that, in addition to the firm's new Beck-Steinheil lens,

any other lens can be fitted and the camera used for varying foci. Messrs. Griffin & Co. show the working of their Velox papers, and the Platinotype Company exhibit specimens of platinotype printings on their several grades of paper. Among the many beautiful examples of this work, we would draw special attention to a splendidly executed picture in sepia platinotype by Van Mureke, also the practical demonstrations that take place every evening at 7.30 and 8.30 P.M. Kodak, Ltd., show samples of their various cameras, etc., foremost in interest being their a new "Panoram," which, by the way, has been awarded the gold medal of the Society. This camera is certainly very simple and unique in construction, and no doubt will become very popular as an addition to the photographic outfit, as it is most useful for certain subjects where an ordinary camera would be practically useless. We must not pass on to the Pictorial Section without mentioning the most interesting exhibit of colour work by Messrs. Sanger Shepherd & Co.

In the Pictorial Section, there is such a wealth of good things that it is impossible to even mention many of the best. The two pictures that have been awarded medals—"The Orchard,'" by W. T. Greatbatch, and "Venice," by Percy Lewis, are undoubtedly amongst the best at the exhibition, but the Judge must have had a most difficult task to decide which were the two best out of some twenty exhibits, all of which are of equal artistic and technical excellence. Amongst these we would class such work as "The Rivetters' and "Warm Work," by John H. Gash; "The Madonna,' by Rudolf Eckemeyer; "The Water Carrier," by James A. Sinclair; "Clearing the Weeds," by W. Thomas; "Off to the North," by G. H. Faux; "In Adell Woods," by Dr. Llewellin Morgan; and "The Wind Bloweth from the Sea," by James Burn. The predominance of carbon and platinotype this year is a healthy sign, and we sincerely hope these processes will continue to hold their own against all less artistic and less permanent methods. Surely, one would suppose that any negative considered good enough to print from for exhibition purposes would be worth the time and trouble of printing in carbon or platinotype. We would strongly advise all intending exhibitors to select one of these methods for such work. Of course we do not wish to disparage the usefulness of P.O.P. and bromide for home work and quick results. For example, we cannot help regretting that such a subject as "The Wind Bloweth from the Sea" was not finished in a carbon process.

The scientific, technical, and photo-mechanical exhibits will be found in the gallery of the exhibition, and here the student of science will find many items of interest, even though he is not interested in photography: such, for instance, as photographs of the Great Nebulae in Orion and Andromeda, with other astronomical subjects; also several frames of zoological subjects are shown, and photographs of flying bullets and exploding shells. There are thermometric and barometric photographic records and photo-micrographs. Then we have examples of some exceedingly fine photogravures by Pellissier & Allen. T. & R. Annan & Sons are awarded a medal for some of the finest photogravure reproductions of paintings it has ever been our pleasure to examine; whilst Rathby, Lawrence & Co. show some beautiful specimens of their three-colour printing, and the "Joly-McDonough" process is well represented.

CONDUCTED BY HAROLD M. READ, F.C.S.

ARTIFICIAL SILK.—During recent years an industry which shows signs of being considerably developed in the near future has sprung up in the manufacture of artificial silk. Although the idea of making this article dates back to Réaumur in 1734, it took no practical shape until the demand arose for filaments for incandescent lamps. At the present time there are four main classes of artificial silk on the market:—(1) That made by denitrating nitro-cellulose, the denitrating being usually carried out with sulph-hydrates as originally recommended by Béchamp. The product thus obtained is highly lustrous and supple, resembling silk in its affinity for basic colours, and cotton in its composition. (2) "Glanzstoff" or lustre-cotton, made by dissolving cellulose in ammoniacal copper, and precipitating by the addition of an acid. (3) Lustre-cotton, made by treating cellulose with caustic soda, washing, dissolving in zinc chloride solution in the cold, and then precipitating. (4) Viscose Silk, the cellulose xanthate patented by Cross & Bevan, made by treating sodic cellulose with carbon bisulphide. The viscose is forced through capillary holes into ammonium chloride solution, and the thread wound on to reels In this connection it is of interest to note Dr. Liebmann's remarks at the recent Bradford meeting of the British Association. Dealing with the danger likely to ensue from the extremely ready combustion of artificial silk, he pointed out that this drawback has now been overcome, and at the same time the lustre even surpassed that of natural silk. Unfortunately the use of the artificial product is still limited; for example, dress goods made entirely of this article are very brittle while damp, and it cannot be used for warps, but as yet only as weft for silk and cotton fabrics.

SYNTHETIC PERFUMES.—The artificial production of the odorous constituents which characterise both the animal and vegetable worlds goes on apace; and while, from a chemist's point of view, the compositions of both pleasant and disagreeable perfumes are equally important, from the popular and money-making side the preparation of the agreeable odours places the chemist's aspect of the question in the shade. A few years ago Tiemann succeeded in building up chemically ionone, the perfume of orris root, while the recent litigation through the rival claimants to the synthesis of musk is still fresh in our memories. The latest addition is artificial "otto" of rose, which appears on the market simultaneously with a paper in the *Berichte* of the German Chemical Society by Walbaum and Stephan. The research has been carried out in the laboratories of Schimmel & Co., and it appears that German otto of rose consists essentially of a mixture of nonyl-aldehyde and laevo-citronellal with the aldehydes and alcohols normally occurring in such perfumes as bergamot, cassia, and others.

A NEW INDICATOR.—In a recent issue of the *Zeitschrift für Analytische Chemie* a new indicator for both acids and alkalies is described under the name of "Alizarin-green B." It occurs as a greenish-black powder, easily dissolving in water, and belongs to the groups of the thiazines. On the addition of acids to the green solution there is produced a fine red coloration, which reverts to green on the addition of traces of alkali.

THE NATURE OF ALLOYS.—The report of the committee appointed to consider the nature of alloys presented its report to the British Association at its recent meeting at Bradford. Up to within quite recent years the great difficulty in proving the formation of the definite compounds which most chemists considered present has been the abstraction of the particular compound. Ordinary chemical methods are of no avail; fractional solution has been partially successful, such distinct compounds as copper-tin, platinum-tin, zinc-copper having been isolated. Professors Roozeboom and Le Chatelier have now shown that careful determinations of the "freezing-point," supplemented by examination under the microscope, reveal the presence of compounds whose existence was hitherto unknown. A long list of known and supposed alloys was given, and it was stated that the ordinary atomic relations do not hold in the case of alloys.

THE UNIVERSAL EUDIOMETER.—A new eudiometer is described in a recent number of the *Chemical News*. It has been designed by Dr. G. Woollatt, and has been found to be of great advantage in lecture demonstrations, giving accurate results with very little trouble and over a wide range of conditions. There is certainly much to commend it, and it is suitable not only for the demonstration of Boyle's and Charles' laws, for which it was primarily designed, but also for the analysis and synthesis of gases. We recommend it to the notice of lecturers. It is being made by John J. Griffin & Sons, Limited, of London, W.C.

EXPLOSION OF MAGNESIUM.—From time to time we are startled by the want of care on the part of experimenters; one of the most recent instances occurred on September 3rd at the offices of the *Photo-Programme*, 29 Rue du Mail, Paris. It appears that M. Jean Larcher (director of the *Programme*, which is a well-known illustrated review) was engaged with two assistants in developing some negatives in a dark-room attached to his offices. One of them carelessly struck a match to light a cigarette. He happened to be near a big jar of powdered magnesium, and the latter exploded. The house was shaken to its foundations, all the windows were blown out, and a perfect panic seized the inhabitants of the house, as well as the neighbours, for the Rue du Mail is a narrow thoroughfare in a busy quarter, and all the houses are five or six stories high, and let off in flats. When assistance came the men were discovered severely injured, and they were conveyed to the nearest hospital. The other people who suffered personal injuries by the explosion were a M. Pozimonin, who was passing the house and received broken glass on his head. A tenant, Mme. Charlotte, happened to be standing at a window when the accident happened, and she was knocked down and sustained some nasty bruises, but a clerk who was in the next room to that in which the explosion occurred was lucky enough to escape with no injury beyond the shock.

CONDUCTED BY JAMES QUICK.

EXTENSION OF WIRES.—A useful arrangement for the measurement of extension of wires has recently been described by G. F. C. Searle in the Cambridge Philosophical Society's Proceedings. Two similar wires side by side have fastened to their ends small rectangular brass frames 11 cm. long. The stretching weights are placed below these frames. A sensitive level is placed between the latter, resting at one end on a micrometer screw which reads to $\frac{1}{1000}$th mm. This reads off the length by which the loaded wire is stretched.

RÖNTGEN RAYS.—An advance has been made in the production of these rays by Trowbridge in America. He has lately installed an equipment of no less than 20,000 storage cells in the Jefferson Physical Laboratory. By this means he has at his disposal an electromotive force of over 40,000 volts. and moreover a fairly steady current through a large resistance. He has succeeded in obtaining Röntgen rays of exceptional brilliancy with the aid of this E. M. F., which yield negatives of great con-. trast. One great advantage of this new method of generating the rays is the possibility of exactly regulating the current and electromotive force which is necessary to excite the rays. This has not hitherto been possible. When the X-ray tube is first connected to the battery terminals no current flows. The tube must be heated with a Bunsen burner. At a certain critical temperature the tube suddenly lights up with a vivid fluorescence, and the rays are then given off with great intensity.

RADIO-ACTIVE BODIES.—The radio-activity of these bodies has now been tested at low temperatures. and M. Curie has found that at the temperature of liquid air they continue to excite fluorescence in uranyl-potassium sulphate. Radio-active barium chloride becomes more luminous at that temperature. Radium has also been tested at the above temperature by Behrendsen, using the electrometer method. In this case, however, it was found that cooling the preparation reduced its radio-activity by more than one-half. On heating it up again to the normal temperature, a slight increase of radio-activity was discovered.

GALTON'S WHISTLES.—In a modified form of Galton's whistle, devised by Mr. T. Edelmann. the blast of air from a ring-shaped mouthpiece impinges upon the sharp edge of the pipe, which is also circular. and can be brought to within any desired small distance from the mouthpiece. This whistle gives very strong notes. whose pitches can be studied by means of Kundt's dust figures. It has been found that the highest limit of hearing is a little beyond 50,000 complete vibrations per second. By making the diameter of the pipe as small as 2 mm., Edelmann has succeeded in constructing pipes giving the very high pitch of 170.000 vibrations per second.

ATMOSPHERIC ELECTRICITY.—Some work has been done by M. Brillouin upon the nature of positive and negative electrifications in the atmosphere. It is well known, from the researches of J. J. Thomson and others, that ultra-violet rays exert a discharging effect upon negative electrification, and Brillouin has found that cold, dry ice is very sensitive to this action; in fact, about one-tenth as much so as zinc. Water is not so at all. Dry crystals of ice floating in an electric field would become charged positively and negatively, but the negative charges would escape under the influence of sunlight. The air remaining an insulator, the surrounding air becomes negatively charged. When the ice crystals forming the cirrus cloud moves away from this air, it bears a positive charge, and if it evaporates, the air now surrounding it becomes positively charged. When the negatively charged air descends, it charges the earth negatively. At sea the negatively charged air, on expansion, forms negatively charged cumulus clouds. The effects of the travel of these masses of air may explain many of the phenomena of storms; and the blending of the positive and negative charges by night, when there has been no such travel, probably accounts for auroras, luminous clouds, and diffused illumination of the sky on summer nights in our latitudes.

TELEPHONY OVER TELEGRAPH LINES.—Various methods are now being adopted whereby telephonic messages can be sent over single telegraphic lines simultaneously with telegraphic signalling. This possibility is realised by the use of condensers, interposed in the circuit in such a way that the slow telegraphic impulses do not affect the rapid telephony waves. At each station the telephone is connected between line and earth through a condenser. The telegraph instrument is connected between line and earth with no condenser. The capacity of the condenser used need not exceed 0·2 microfarad, and in place of a tin-foil "plate" condenser it is found better to use one formed of parallel insulated wires bound together on a bobbin. For this purpose the wires may be 0·1 mm. diameter. copper, double silk covered, each of the wires having a resistance of approximately 800 ohms. Plate condensers are found to transmit waves more loudly but less distinctly than these wire condensers. It is, however, suggested that loudness and clearness might possibly both be attained by a continuation of a plate condenser and a wire condenser. The Morse-key contacts are made . of carbon, to avoid abrupt changes of current. which would affect the telephone circuit. Another way of bridging the difficulty would be to connect permanently a resistance bobbin between the contacts of the key. When there are intermediate telegraphic stations between the two stations that are to communicate by telephone, each intermediate telegraph instrument must be bridged over with a condenser. So far as the telephone circuit is concerned, this is equivalent to cutting out the resistance of all the intermediate telegraph instruments. In addition to the above, each telegraph instrument, including the terminal instruments, must be provided with an inductive resistance of about 500 ohms, to act as a choking coil for the telephonic circuits. This system is applied to the fire-alarm service in Berlin, and by means of a portable telephone apparatus communication can be made with the central fire station from any of the 800 fire-alarm posts.

BIRD-LOUSE CHANGING HOSTS.—In August last I captured a specimen of the parasitic fly, *Ornithomyia avicularia*, as it was leaving a recently-shot blackbird. On examination I found hanging on to it, like a bulldog, a bird-louse which is figured in Denny's monograph of the Anoplura as *Nirmus merulensis*, one of the species of lice stated by him to be peculiar to that bird. Both species were present on the bird in question, the *Nirmi* being on the quills of the wing-feathers. This is the second instance of a *Nirmus merulensis* attaching itself to an *Ornithomyia* which has come to my notice. In each case the *Nirmus* was hanging on to the posterior portion of the abdomen. Mr. Enock, who has seen the two insects mounted in position for the microscope, believes, I am informed, that the bird-louse takes hold of the *Ornithomyia* with the view of being transported to a fresh bird-host. In that case it might find itself conveyed to a thrush or starling: for, as far as I can see, the Ornithomyiae parasitic on the three kinds of birds are identical. His view is probably correct; the more so that the chitinous integument of an *Ornithomyia* is exceedingly tough, and it is doubtful if the biting organs of the *Nirmus* would be able to penetrate them. It is also well known that Chelifers (pseudo-scorpions) attach themselves to the legs of flies, presumably for transport. One of them was lately found hanging on to the leg of a hive-bee. *Nirmus merulensis* is about three times the size of an ordinary Chelifer.—*H. J. O. Walker (Lt.-Col.), Lee Ford, Budleigh Salterton.*

ABNORMAL CLOVER FLOWERS.—Pupils of mine, Miss Esplin and Miss Meyer, have drawn my attention to the following variations in flowers of the white clover. These variations may be of common occurrence, but I have not seen them alluded to anywhere. The plants were found growing on the chalk, about a mile to the east of Brighton. On certain heads some of the flowers had the ordinary short pedicels, while others had much larger ones, in some cases measuring as much as 18 mm. In some heads all the pedicels were abnormally long. Foliaceous calyx was found to be of common occurrence ; sometimes the calyx was larger than the corolla, and this, combined with the lengthened pedicels, gave the plant somewhat the appearance of certain kinds of rush. The carpel was replaced by leaf in the centre of a number of the flowers where unmodified leaves were observed. These were seldom sessile, sometimes having petioles 8 mm. long. In some flowers the leaf was undivided, in others there were two, three, four, five, or even six leaflets. The flowers exhibited other peculiarities, but these do not show up so clearly in the dried specimens I have before me.—*Florence Itick, Roedean School, near Brighton.*

CATOCALA FRAXINI IN KENT.—A specimen of *C. fraxini* (the Clifden Nonpareil) was taken by Mr. G. Grey and his brother at treacle at Eltham, Kent, on September 3rd last.—*A: J. Poore, 47 Griffin Road, Plumstead.*

THE BRITISH MYCOLOGICAL SOCIETY.—The annual week's fungus foray of this society was held at the Boat of Garten from September 17th to the 22nd, 1900, and by a happy coincidence the Cryptogamic Society of Scotland had also arranged their annual foray in the same locality. The members of the two societies assembled at the Hotel, Boat of Garten, on Monday, September 17th, where they found awaiting their inspection many interesting specimens collected by Dr. C. B Plowright. The hon. secretary also exhibited a specimen of *Strobilomyces strobilaceus* Berk, gathered that day by him in the policies of Murthly Castle, and which he understood was of uncommon occurrence in Scotland. On Tuesday, September 18th, the early train was taken to Aviemore, from whence the members proceeded, under the leadership of the Rev. Dr. Keith, President of the Cryptogamic Society of Scotland, to the Dell of Rothiemurchus Forest. Here the members collected specimens of the excellent edible *Hydnum imbricatum* Linn., the pretty rosy *Armillaria robusta* A. & S., several species of the genus *Tricholomata*, *T. equestre* Linn. being characterised by its bright yellow gills and stem, *T. portentosum* Fr. and *T. virgatum* Fr. by their sombre colour. *Cortinarius (Inoloma) traganus* Fr. was abundant amongst the heather, and its unpleasant smell was only too evident. The rich-coloured *Cortinarius (Inoloma) tophaceus* Fr., *C. (Dermocybe) orellanus* Fr., were secured amongst many others of this interesting genus. *Paxillus atrotomentosus* Fr., the sweet-scented *Hygrophorus agathosmus* Fr, the yellow-milked *Lactarius scrobiculatus* Fr., *Hydnum fragile* Fr., *H. scrobiculatum* Fr., and *H. compactum* Pers. were added to the list. The walk was then continued to the hospitable shelter of the Rev. — McDougal, the members gathering on the way specimens of the destructive parasite *Trametes pini* Fr. On Wednesday, September 19th, the Rev. Dr. Keith conducted the members into the adjacent Forest of Abernethy and round to Loch Garten, which is beautifully wooded and stands out in pleasing contrast to the neighbouring mountains. Dr. C. B. Plowright secured the pretty *Entoloma erophilum* Fr., new to the British Fungus Flora, and some very yellow examples of *Stereum sowerbeii* B. and Br. Professor H. Marshall Ward found a large quantity of the weird-looking parasite *Cordyceps ophioglossoides* Fr. growing on *Elaphomyces rariegatus* Vitt. *Sistotrema confluens* Pers. was fairly common amongst the pine needles. and *Stropharia scobinacea* Fr., *Flammula scamba* Fr., and *Hydnum ferrugineum* Fr. were collected. In the evening Professor H. Marshall Ward, president of the society, delivered a very instructive and suggestive address, entitled "The Nutrition of Fungi." On Thursday the Rev. Dr. Keith led the members to Columbridge and Loch-an-cilan. Many interesting specimens were gathered on the way, including the pretty blue *Entoloma bloxami* Berk., the golden squarrose *Pholiota flammans* Fr., the scaly *Cortinarius (Inoloma) pholideus* Fr., *Lactarius hysginus* Fr.,· *Pleurotus mitis* Pers., and *Naematelia encephala* Fr. In the evening the business meeting of the Society was held. Professor H. Marshall Ward, D. Sc., F.R.S., F.L.S., etc., was unanimously re-elected president, and Mr. C. Rea hon. secretary and treasurer for the ensuing year. Their fellow-member Mrs. Montague's invitation to hold next year's foray in the woods near Crediton, with Exeter as headquarters, was unanimously accepted, and the date was fixed for the last week in September. Professor H. Marshall Ward gave a valuable paper of original research on " Naematelia," which should be read by all mycologists in the Society's Transactions. Mrs. Carleton Rea exhibited drawings of two new species, *Collybia veluticeps* Rea and *Mycena carneosanguinea* Rea, which would be · shortly described in full. On Friday, September 21st, the members proceeded to the woods to the west of the Boat of Garten, but little of any consequence was found beyond a *Boletus* thought to belong to the *Gyrodon* group, so in the afternoon the members again searched Abernethy Forest, obtaining specimens of *Tubaria paludosa* Fr. and *Omphalia umbratilis* Fr. On Saturday, September 22nd, many of the members dispersed, but a few ardent ones remained and took the mid-day train to Aviemore, from whence through birch-clad hills they walked to Lynwilg and on to a pine wood about· a mile further southward, and on the right of the road to Kingussie. This wood was found to be the veritable home of the larger Hydnei, and the collection soon included *H. imbricatum* Linn., *H. fragile* Fr., *H. compactum* Pers., *H. aurantiacum* A. & S.,. *H. zonatum* Batsch, *H. nigrum* Fr., and *H. melaleucum* Fr. So terminated a most enjoyable foray, which was chiefly remarkable for the number of Cortinarii and Hydnei that were found.—*Carleton Rea, Hon. Sec. British Mycological Society, 34 Foregate Street, Worcester.*

NOTICES OF SOCIETIES.

Ordinary meetings are marked †, excursions * ; names of persons following excursions are of Conductors. Lantern Illustrations §.

NOTICES TO CORRESPONDENTS.

To CORRESPONDENTS AND EXCHANGERS.—SCIENCE-GOSSIP is published on the 25th of each month. All notes or other communications should reach us not later than the 18th of the month for insertion in the following number. No communications can be inserted or noticed without full name and address of writer. Notices of changes of address admitted free.

EDITORIAL COMMUNICATIONS, articles, books for review, instruments for notice, specimens for identification, &c., to be addressed to JOHN T. CARRINGTON, 110 Strand, London, W.C.

SUBSCRIPTIONS.—The volumes of SCIENCE-GOSSIP begin with the June numbers, but Subscriptions may commence with any number, at the rate of 6s. 6d. for twelve months (including postage), and should be remitted to·the Office, 110 Strand, London, W.C.

NOTICE.—Contributors are requested to strictly observe the following rules. All contributions must be *clearly* written on one side of the paper only. Words intended to be printed in *italics* should be marked under with a single line. Generic names must be given in full, excepting where used immediately before. Capitals may only be used for generic, and not specific names. Scientific names and names of places to be written in round hand.

THE Editor will be pleased to answer questions and name specimens through the Correspondence column of the magazine. Specimens, in good condition, of not more than three species to be sent at one time, *carriage paid.* Duplicates only to be sent, which will not be returned. The specimens must have identifying numbers attached, together with locality, date, and particulars of capture.

THE Editor is not responsible for unused MSS., neither can he undertake to return them unless accompanied with stamps for return postage.

EXCHANGES.

SCIENCE-GOSSIP (old series), 1865 to 1893, part bound. Wanted, botanical or entomological books, collection of British moths or. plants.—W. R. Hayward, 28 Princess Road, South Norwood, S.E.

OFFERED.—Cypraea fusco-dentata, Calliostoma layardi, Columbella filmerae, and other South African marine shells for other marine, freshwater, or land shells. Send lists.—H. Becher, M.D., Grahamstown, South Africa.

WANTED, good lantern, in exchange for volumes of SCIENCE-GOSSIP, old and new series, and British land and freshwater shells.—A. Alletsee, Claremont, Randall Road, Clifton, Bristol.

FOR EXCHANGE, duplicate specimens and micro-slides of typical British rocks, chiefly Lake District ; own collection Also fossil coal-plants. List.—W. Hemingway, 170 Old Mill Lane, Barnsley.

OFFERED.—Vols. I. and II. of " Insect Life," profusely illustrated, value £2 ; several microtomes and a large collection of current microscopy literature, Geological Survey Memoirs of Lincolnshire, Yorkshire, etc. Wanted, among other desiderata, a good pair of field-glasses by maker of repute.—J. Cooke, 19 Ravenswood Road, Redlands, Bristol.

CONTENTS.

THE STONE CURTAIN AT ROXBY.

BY HENRY PRESTON, F.G.S.

IN a field on Sawcliff Farm, in the parish of Roxby-cum-Risby, North Lincolnshire, there is a deposit of uncommon character and singular beauty. It is particularly interesting to the lover of natural objects. Locally it is known as the "Sunken Church." An ancient tradition informs us that it was a church attached to one of the monasteries, and was buried by a landslip; or, according to Abraham de la Pryme, the Yorkshire antiquary, who visited it in 1696 (Surtees Society, vol. liv.), the tradition is that the church sunk in the ground, with all the people in it, in the times of Popery.

Both the name and the legends do discredit to curtain," a name more in character both with its appearance and manner of growth.

The stone curtain, then, as will already have been gathered, consists of a mass of calcareous tufa deposited by a petrifying spring trickling out of the limestone rocks, as seen in the second illustration. It is a wall-like mass, some ninety feet or more in length, having a varying thickness from fifteen inches to two feet at the top, and a height above ground of nine feet at its highest point. From the higher end where it first leaves the ordinary slope of the hill, there is a gentle fall along the ridge until, about half-way down, a big step of about four feet occurs. Then the ridge

Photo. by]   *[H. Preston, F.G.S.*

THE STONE CURTAIN AT ROXBY, SHOWING GUTTER.

this remarkable structure, inasmuch as the visitor who goes with the idea of finding architectural remains, as some justification of the name, is disappointed to find that no such ancient church exists, and that no human agent has ever been at work in connection with the mass of stone which he has come to see. Such disappointment often fails to yield to the new interest which should be awakened by finding a structure of no mean size, beautifully built and fluted by an artist so diminutive as to be often altogether overlooked. In describing this production of the tiny spring which issues near the foot of the hill, I shall discard the present name and venture to rechristen it the "stone

continues to descend, until at the lower end it almost comes to the level of the ground again. Undoubtedly the most striking feature about it is a groove two inches wide and one and a quarter inches deep, which runs along the ridge from end to end, and also continues down the step above mentioned. This groove is well shown in the first illustration.

At the foot of the hill there is a cattle-trough, into which has been conducted from the top end of the stone curtain a small spring as feeder, and which at once gives a clue to the formation of this interesting piece of Nature's architecture. The foot of the escarpment, which is really part of the

H

Lincolnshire cliff, is formed of the Upper Lias clay, and for ages past this has thrown out carbonated water, which has previously dissolved a considerable amount of limestone from the overlying beds. Upon entering the atmosphere, some of its carbonic-acid gas has escaped, and this liberation of dissolved gas coupled with evaporation has caused the limestone to be released. This has been deposited, and has formed a parasitic hill of considerable size, in which land shells and impressions of leaves are common. The tufa hill extends perhaps seventy or eighty yards on either side of the curtain, and is of considerable thickness—perhaps twenty feet or more, as judged from the natural slope of the escarpment, where the tufa has not accumulated. In process of time the issuing water seems to have gathered itself together and formed

Tufaceous deposits are very common wherever springs issue from limestone rocks; but in this particular case a wonderful balance has been maintained between the size of the spring and the building work it has done. A stronger spring might not have balanced the side deposits so well, and probably would not have produced the remarkable gutter along the ridge. A small stream with a high percentage of evaporation is calculated to produce a larger deposit than a strong spring; and this appears to have been the case in the stone curtain.

The precipitated limestone from water often takes most beautiful and fantastic shapes, as in the stalactitic and stalagmitic deposits in the numerous caverns of limestone districts, and in the wonderful travertine bridge at Clermont, in the

Photo. by]   [H. Preston, F.G.S
                                    STONE CURTAIN AT ROXBY.   LATERAL VIEW.

a single stream, which, taking a fairly direct line over its own bed, has deposited its tufa in this course. During this building period the water has continued to run along the crest of the builded wall, and the tufa growth has been uniform on either side of the stream. Possibly the uniformity of the upward growth is due to spray and to more rapid evaporation along the edges of the stream. In this manner the sides of the groove have been built up, whilst the overflow of water on either hand has by further evaporation formed the curtain. This structure of stone, broadening out in innumerable folds at the base, affords one of the most interesting sights of the county, and, one which every lover of Nature will desire to see preserved in its entirety.

Auvergne, but never before have we known a stream to keep so constant a course while building up its own monument, on and in which for so many ages it must have disported itself.

Grantham, November 1900.

SUN-SPOTS AND RAINFALL.—An important paper by Sir Norman Lockyer and Dr. W. J. S. Lockyer on "Solar Changes of Temperature and Variations in Rainfall in the regions surrounding the Indian Ocean" was presented by Sir Norman at the meeting of the Royal Society of November 23rd. The paper was founded upon the investigation of the abnormal behaviour of the widened lines in the spectra of sunspots since 1894, and the accompanying irregularities in the rainfall of India.

THE NOBEL BEQUEST.

THE Nobel bequest, to which we referred last month (*ante*, p. 164), has for its object the encouragement of research in the departments of physical, chemical, and physiological or medical sciences, the advancement of literature, and the promotion of universal peace. For this purpose Dr. Bernhard Nobel left part of his fortune, to be placed by his executors in safe investments. The interest is to be used for founding prizes for essays or inventions connected with those subjects. It is worthy of note that by the terms of the will, prizes are to be awarded, not necessarily to those who have made the greatest discoveries, but to those who in each branch have rendered the greatest service to humanity. The prizes are to be given at least once in every five years, commencing with the year immediately following the commencement of the Nobel endowment. The sum-total of a prize is in no case to be less than 60 per cent. of the part of the yearly revenues set apart for the distribution of the prizes; neither is it to be divided into more than three awards.

The decision as to the prizes is to be in the hands of the following bodies:—That for physical science and chemistry is to be awarded by the Academy of Sciences of Sweden; for physiology or medicine, by the Carolin Institute of Stockholm; for literature, by the Academy of Stockholm; and for the work of peace, by a commission consisting of five members to be elected by the Norwegian Stortung. The term "literature" is intended to include, not only purely literary works, but all writings having, by their form and style, a literary value. This prize can be divided equally between two works, if each is judged to have merited the prize. If, however, the selected work is the product of two or more persons, the prize shall be given in common.

After the approval of the King of Sweden has been obtained for the statute of endowment, the corporations will nominate the stipulated number of representatives, who will then assemble at Stockholm to elect the board of administration, the members of which will undertake the management of the endowment fund. This will occur early next year. The first distribution of prizes for all sections will, if possible, take place in 1901. From the endowment fund the following sums will be deducted, £16,000 for each section—that is, £80,000 in all. This, with the interest dating from January 1st, 1900, shall serve to furnish the expenses of the organisation of the Nobel institutes. In addition, a further sum, to be fixed by the Council of Administration, shall be utilised for the acquisition of a special site for the administration of the endowment, including a hall for its meetings.

Each candidate for a Nobel prize must be proposed in writing by some one qualified to make such proposal. Those having such rights of presentation are: (1) native and foreign members of the Royal Academy of Sciences; (2) members of the Nobel committees for natural philosophy and chemistry; (3) professors who have received the Nobel prize of the Academy of Science; (4) ordinary and extraordinary professors of natural sciences and chemistry in the universities of Upsala, Lund, Christiania, Copenhagen, and Helsingförs, in the Carolin Institute for Medicine and Surgery, the superior technical Royal School, and also the professors of the same sciences in the superior school at Stockholm; (5) occupants of corresponding chairs in at least six universities or high schools which the Academy of Sciences will select, taking care to divide them suitably between the different countries and their universities; (6) the savants to whom the Academy shall decide to send an invitation to this effect.

The choice of the professors and savants mentioned in numbers 5 and 6 shall be decided in the month of September in each year. Proposals for the prizes must be made before February 1st of the following year. They are to be classified by the Nobel Committee, and submitted to the College of Professors. The College of Professors will decide definitely on the distribution of the prize during October. The vote will be taken in secret, and, if necessary, the question may be decided by drawing lots.

The endowment is to be directed by an Administrative Council which will have its seat at Stockholm, and is to be composed of five Swedish members, of whom one, who shall be the President, is to be nominated by the King, and the others are to be chosen by the representatives of the corporations. The Council shall choose from among itself a Director-General. Corporations having the right to nominate candidates for the prizes shall appoint for two civil years at a time fifteen representatives, six of whom shall be chosen by the Academy of Sciences, and three by each other corporation. Further, the Academy of Sciences shall nominate four, and other corporations two candidates. to take the place of any representative, in case of obstacle to prevent his attendance.

The management and accounts of the Council of Administration are to be examined each year by five revisers, four to be chosen by the corporations before the end of the year, the fifth, who will be the President, to be nominated by the King. The report of the revisers is to be presented to the representatives of the corporations before April 1st in each year, after which it is to be published in the newspapers, together with an account of the work done.

CHAPTERS FOR YOUNG NATURALISTS.

(*Continued from Vol. V., page* 359.)

ROTIFERA.

By Walter Wesché.

ROTIFERA have for the microscopists a fascination which is not difficult to understand. Their high organisation, the uncertainty of their position in the schemes of classification, their curious life-history presenting so many unsolved problems, their charming appearance, and the ease with which many functions of life, such as digestion and mastication, may be watched, all combine to make them one of the most delightful of studies.

A group of naturalists worked at them for years, and brought to that work an artistic freedom of the pencil not often to be found in combination with scientific training. "The Rotifera or Wheel Animalcula," by Dr. Hudson and Mr. P. H. Gosse, is the enduring monument of those labours, incorporating, as it does, the work of many other observers. These authorities have divided the Rotifers into four sub-orders : the (1) Rhizota, those which are fixed or rooted in adult life, build tubes, or excrete various forms of protective covering ; (2) Bdelloida, which swim and creep ; (3) Ploima, the free swimmers, which are further subdivided into "loricated" and "illoricated"; and a small group, (4), the Scirtopoda, "that swim with their ciliary wreath, skip with jointed limbs terminated in fans of setae, and have no foot."

Though the Rotifera derive their name from the appearance of their ciliated heads, the mastax or gizzard is of equal, perhaps of greater importance in classification. This is by itself an interesting study in morphology, as there is great diversity in the forms. The vascular system and the contractile vesicle can in many genera be seen with clearness, as can sometimes the curious vibratile tags, whose use is unknown. The toes have an immense variety as to length and shape, and are furnished with glands which secrete a viscid fluid, that enables the animalcule to attach itself to a smooth surface. In many cases a brain can be seen, with a red eye or pigment spot apparently placed upon it. The effects of light are known to be attractive and exhilarating. The male of a considerable number of species has been identified, and he certainly cannot be said to enjoy life, as he is without digestive organs of any kind. In the Hydatina he is comparatively large and easily studied, appears to show segmentation, the vascular system, and a set of muscles for drawing in the head ; the structure generally can be better seen than in *Brachionus*, the genus in which the male was first observed.

The rotifera shown in the accompanying drawing have all been "dipped" in ponds in the north and west of London, and one or two from Epping Forest. They only represent a percentage of many forms, equally strange and curious in appearance, that have come from the same district. Members of the genus *Floscularia* are the best-known tube-dwellers, and are easily distinguished from others of the *Rhizota* by the large development of the cilia ; these remain motionless in the water, and are spread out like a net. The curious foot forms a suctorial disc, and, like the mastax, are characteristic of the genus. *F. cornuta* is only distinguishable from *F. ornata* by a fleshy process placed on one of the lobes from which the cilia spring. *F. campanulata* is much rarer. I have found it in the Round Pond, Kensington Gardens, in the Upper Wake, Epping Forest, and at Kensal Rise. The tube has always the same gelatinous base, and is quite hyaline when newly made ; but in course of time grains of sand and flocculent matter adhere, and cause it to present the different appearances shown in the drawing. I was only once fortunate enough to take *Stephanoceros eichornii*, and then I had to journey to Epping to the Upper Wake to secure that species. It certainly is the most beautiful and imposing of all rotifers ; it is so well known that description is superfluous, but it may be pointed out that, as in *Floscularia* there are five lobes, so in *Stephanoceros* are five ciliated arms. Two arms in *Stephanoceros eichornii* do not appear in the accompanying drawing, as they are hidden by those on the left and right, and the cilia are depicted as usually seen ; but viewed with " dark ground" illumination, they are found to be very much longer and thicker, and form quite a filamentous net in the water. I am indebted to Mr. C. F. Rousselet for a demonstration of this, by means of one of his admirably mounted preparations of Rotatoria.

Oecistes serpentinus has only an apology for a tube, usually a little flocculent matter. The most curious thing about the animalcule is a pair of minute hooks which show when the cilia are retracted. It is not common ; I have taken it in the Leg of Mutton Pond, Hampstead, and the Round Pond, Kensington. *O. stygis* makes a rather untidy squat tube; this, however, when placed in the centre of a grove of bright green vegetation made a very charming microscopic picture. The ciliated wreath is single, and there are two processes that are possibly used in the construction of the tube. This specimen was taken from a pond near Dollis Hill.

Limnias ceratophylli is also not common, but I have found it in the Round Pond and the Leg of Mutton and Viaduct Ponds at Hampstead. The

small particles into the mouth; there appears to be also a stream of rejected matter, which comes out through a small circular opening and moves away in a curved current in the water. This I have not noticed in any other family.

Melicerta ringens is, if one of the commonest,

A GROUP OF ROTIFERS.

(Drawn from Life by W. Wesché.)

1. *Floscularia cornuta.*	8. *M. solidus* (side view).	15. *Limnias ceratophylli.*	22. *Furcularia gibba.*
2. *F. cornuta*, male.	9. *Cephalosiphon limnias.*	16. *Floscularia campanulata.*	23. *Rotifer vulgaris.*
3. *F. ornata.*	10. *Stephanoceros eichhornii.*	17. *Rotifer tardus.*	24. *Melicerta ringens.*
4. *F. ornata* (cilia retracting).	11. *Colurus deflexus.*	18. *Distyla flexilis.*	25. *Oecistes serpentinus.*
5. *Asplanchna priodonta.*	12. *Stephanops lamellaris.*	19. *Pterodina patina.*	26. *Mastigocerca carinata.*
6. *Melicerta ringens.*	13. *Cephalosiphon limnias*	20. *Oecistes stygis.*	27. *Polyarthra platiptera.*
7. *Metopidia solidus* (dorsal	(cilia retracting).	21. *Actinurus neptunius.*	28. *Furcularia longiseta.*
view).	14. *Diglena catelina.*		

tube is very neat and tidy, and seems made of some chalky substance, appearing light grey in colour. The two lobes of cilia drive a stream of

one of the most interesting species. There is always fascination in watching the animalcule make its pellet, and place it in position on the wall of the tower ; in examining the beautiful arrangements by which different currents of water are formed, some to drive food to the mastax, and others sediment into the lathe-like opening in the centre where the brick is made. The curious little tube-builder *Cephalosiphon limnias* I found at the Upper Wake Pond. The tube is rather dark in colour and unsymmetrical in shape, but the rotifer is readily identified by the single long antenna-like process which rises slowly and cautiously out of the tube, ascertaining if all is safe, before displaying the cilia.

Rotifer vulgaris, said to be the most common of the rotifers, is little understood. The mystery of its sexes is still unpenetrated. I have seen the young animalcule escape from the abdomen of the mother, darting in a fraction of a second of time through a longitudinal slit, of which I could find no trace afterwards, and immediately start life on its own account. I have watched the dust from a dry gutter or from moss gradually swell when placed in water and wake into life as *Philodine* or *Rotifer*; I have had it under repeated observation for years, and so has everyone who has studied Rotifera, and it seems astounding that a male has never been seen or suspected.

Rotifer tardus has a remarkable three-toed foot, and at the base of that foot two more spurs, resembling those of *Furcularia*. It is so slow in its movements that fungoid matter seems to take root to its sides. It is not common. I have taken it at the Leg of Mutton Pond. *Actinurus neptunius* is one of the largest and most astonishing forms; it is quite telescopic, and shuts up into a short angular case, from which the head and tail shoot out with quite startling suddenness. This is constantly happening, as it seldom remains in one position for more than a few seconds. When I have come across it I have found it in water thick with sediment. I have taken it in the culvert of the River Brent at Stonebridge Park, and in the Grand Junction Canal near Acton.

Asplanchna priodonta is a large free-swimming rotifer, whose vascular system and vibratile tags show well. Its method of expelling digested food is considered archaic; it is simply rejected through the mouth. It preys on large-sized infusoria, which are sucked into the stomach. I have noticed the loricated rotifer, *Anurea cochlearis*, in process of digestion. For this species the Grand Junction Canal at Queen's Park, and the Leg of Mutton Pond, have been among my most successful hunting-grounds. *Furcularia gibba* is common round London, and may be found in nearly every pond, but is rarer in the country. It was quite an unfamiliar rotifer to Dr. Hudson, and several observers had named different animalcula by this name.

There is no difficulty in identifying *Furcularia longiseta* with its extraordinary development of toes. I found it in the pond at Dollis Hill on

some duckweed. It remained alive for hours in a live-box, occasionally resting and giving me an opportunity of drawing it, but generally swimming, and sometimes, by suddenly bringing the long toes together, giving a leap after the manner of *Triarthra*. I fancied on these occasions that I could hear a sound produced by the contact of the toes with the cover glass.

Dr. Hudson and Mr. Gosse considered the Notomata family to be the most interesting group of the Rotifera. Many of the genera are predaceous on other rotifers, and show extraordinary signs of fierceness and rage when confined in a live-box or slightly compressed. I have seen *Diglena forcipata* seize hold of a piece of conferva and shake it like a terrier does a rat. The mastax is toothed and capable of being protruded. Similar behaviour may be watched in many of the family. *Furcularia*, though included in the group, shows none of this fierceness, and the small puppy-like *Diglena catelina* is, as far as my observations go, and I have seen it frequently, a quiet, mild-tempered animalcule.

Pterodina patina is very common, but always interesting ; it can be found nearly everywhere. I once dipped hundreds of them, and I remember no more beautiful sight than these seen with a low power and by dark ground illumination. Some were swimming and others had anchored themselves by the foot to weeds and were busily lashing the water with their cilia. Their bodies flashed and glowed with light, their little red eyes twinkled, the conferva was bright vivid green, while the minute particles of sediment spinning round in the currents of the ciliated wheels shone like stars against the dark background.

Polyarthra platiptera is one of the group which have flat blades to aid them in swimming ; the use of these appendages causes them to progress in jerks. The blades, when examined with high powers, are found to be serrated. It is not uncommon. *Colurus deflexus* is one of the smallest of the loricated free-swimmers. It is constantly present everywhere round London ; but farther north it is, I understand, rarer. The little pick with which the head is furnished is best seen from the side. A dorsal view, which from the slight breadth in proportion to the height is rare, shows it to be hyaline, and more like a spade in shape. It is used in breaking up decaying confervae. In the pools left by the tide at Guernsey a rotifer was plentiful that I could not distinguish from *C. deflexus*. In July 1897 I saw two of this genus tied by a ligament which seemed to come from the shoulders of the animalcula ; but they were so restless that it was impossible to say with exactness. I have never repeated the observation.

Metopidia solidus resembles *Colurus* in many respects, but has a flat limpet-shaped body ; it is also common and of the same habits. *Stephanops lamellaris* is easily distinguished by the disc on the head, and from the commoner species, *S. muticus*, by spines on the carapace. It is rare, one only being found in water from Dollis Hill, that yielded many individuals of seventeen different species. The mouth is at the extremity of the head. In Hudson and Gosse no antennae are figured, but I am positive they exist, as later I found a number in a ditch that drained into the Arun at Pulborough, Sussex, and made a very careful examination.

Mastigocerca carinata is an interesting form, as apparently it has only one toe, but rudiments of others can be seen at the base. There are antennae, which are difficult to make out, as they are obscured by the cilia. It is to be found in most of the places I have mentioned.

Distyla flexilis I have only dipped once, when I found some specimens in water from the azalea house of the Botanical Gardens, London. It resembles *Monostyla mollis* in many respects, but is easily distinguished by the two toes. They are, like the *Diaschiza*, on the border line of the loricated and illoricated free-swimmers.

It has never been my good fortune to capture *Pedalion mirum*, whose discovery by Dr. Hudson at Clifton in 1871 led to the formation of the fourth sub-order.

The study of rotifera has one disadvantage: it is only the picture of your capture that is of much practical use. There are methods of mounting them, and these methods are not difficult, but require much patience and care. They must be placed in fluid and consequently in cells, and this prevents examination with high powers. To once fall under the spell of the wheel-bearers is to remain convinced that their study is the most delightful of all work with the microscope.

90 *Belsize Road*,
South Hampstead, London.

BUTTERFLIES OF THE PALAEARCTIC REGION.

By Henry Charles Lang, M.D., M.R.C.S., L.R.C.P. Lond., F.E.S.

(*Continued from page* 178.)

Genus 16. *COLIAS* Fab.

BUTTERFLIES of moderate size. Wings with the margins entire. Sub-costal nervure of f.w. four-branched, the first given off before the end of the cell, the second at the end of the cell or external to it, the third and fourth bifurcating at the apex of the wing. Hind wings always rounded and without any angular projections. Ground colour of wings yellow, varying in different species from pale greenish to deep orange. Hind margins always more or less bordered with black, at least on the f.w.; a black discoidal spot is always present on f.w., except in one or two species. H.w. with a conspicuous light discoidal spot, varying above from orange to white, and of pearly white beneath, generally surrounded by a reddish circle; cilia generally red. Antennae red in colour, short, and rather thick, swelling into a club at the extremity. Head of moderate size, eyes naked and tolerably prominent. Palpi close together and compressed. Thorax rather short. Legs generally red. Abdomen moderately stout, and not reaching to anal angle of h.w.

The females are larger than the males, and generally of a lighter colour; the marginal border is wider but less defined, and is spotted with yellow, with one or two exceptions, even in those species where the border is unspotted in the male. The females of those species whose wings are normally deep yellow or orange are liable to a dimorphism in which the wings have the ground colour nearly white. In this particular they greatly assimilate to what is seen in both sexes of the species of the light-coloured group. This is familiar to British collectors in the case of *C. edusa*, var. *helice*, which greatly resembles *C. hyale*. It has been thought by some that this may be a reversion to a primitive coloration, in fact to the original Pierid type. In several instances, however, the principle is reversed, as in some forms of a remarkable species, *C. wiscotti*, in which the male has the light and the female is of a yellow or orange colour.

It is evident, then, that in the grouping of this genus we must adopt the Horatian saying, "nimium ne crede colori." Because a species happens to have orange-coloured wings, it does not follow that its affinities are necessarily with all orange-coloured species. On the other hand, the genus *Colias*, as commonly understood, forms a very natural group, and to my mind those who would break it up into several parts err in departing from that simplicity which should be aimed at in all zoological classification. If we take some of those species which apparently are widely divergent, such as *C. sagartia*, *C. christophi*, *C. regia*, *C. wiskotti*, and place them side by side, we shall not find one good zoological character by which we are justified in establishing any new genera; although in colouring they differ greatly.

To go further than the character of mere coloration we may take the pattern of the wing-markings, but even here there are not sufficient grounds for the division of the genus, because, as mentioned above, the difference in character is really only confined to one sex. In those species where the marginal borders are unspotted in the ♂, they are invariably more or less maculate in the other sex, and greatly approximate to the species of the group which has the borders spotted in both sexes. Indeed they are in many cases almost indistinguishable from them in the white varieties.

There is a third character which has been made use of by some lepidopterologists for the establishment of a new genus, and that is the presence in certain species of an ovate patch of scales on the costal nervure of the h.w. in the male on the upper surface. This structure is certainly remarkable. It appears to be formed by a group of altered wing scales of a smaller size than those of the rest of the wing and often differing from them in colour. This was described by Boisduval (Gen. et Ind.-Meth.), and called by him a "glandular saccule." I have noticed this sexual structure in "The Butterflies of Europe," p. 60, and fully recognise its importance, but consider it too secondary a character upon which to found a genus; as, for instance, Mr. Watson has done, under the name of *Eriocolias*. It is a useful character, however, in the grouping of the genus, and is so used by Mr. Elwes in his additional notes on the genus *Colias* (Trans. Ent. Soc. Lond. 1884, No. 1).

This genus *Colias* has for the most part somewhat the same geographical distribution as the genus *Parnassius*; but, unlike it, it extends into South America, South Africa, etc.; *C. edusa* being very widely distributed and extending into Syria, the Azores, the Canaries and North Africa. *C. electra* is exclusively South African, replacing *C. edusa* in the Vaal River Colony, Natal, and the Cape. Temperate South America also furnishes several species. Central and Western Asia possess the greatest number of the Palaearctic species, though many are found in Europe.

The Nearctic species are very numerous, and mostly resemble very much those of the Palaearctic region. It is probable, that many of the North American so-called species are nothing else than local forms of species that occur in Europe and Asia. The genus *Colias* extends very much further north than *Parnassius*; indeed it has been found represented in the Polar regions almost as far as any expedition has penetrated. *C. hecla* has been taken between 78° and 83° N. lat.

The larvae of *Colias* are mostly, if not always, green in colour, with lighter lateral stripes. They are cylindrical in shape, but slightly tapering at the extremities, covered with a slight pubescence. They feed on various species of Leguminosae.

The pupae are straight and pointed anteriorly.

I will here endeavour to group the Palaearctic species, and to give a general view of their coloration.

Group I. Female only spotted on the marginal border of f.w.

　a. Having a patch of thick scales at the base of upper side hind wing in male.

C. wiskotti and vars.	Colour varying from greenish to bright orange, marginal border often very wide.
edusa	Orange yellow, sometimes shot with violet.
fieldii	Orange-yellow, but browner.
aurorina	Orange shot with violet.
,, v. *libanotica*.	Orange, powdered with black, shot with violet.
heldreichii	Orange, but darker, strongly shot with violet.
dira	Orange, strongly shot with violet.
aurora	Bright orange, shot with violet.
olga	Brilliant orange, shot with violet.
myrmidone	Bright orange, sometimes shot with violet.

　b. Without basal patch of scales on the h.w. in male.

C. pamiri	Orange.
romanori	Bright orange, shot with violet.
staudingeri	Bright orange, shot with violet.
regia	Very deep orange, almost red, shot with violet.
eogene and vars.	Deep orange, shot with violet, varying to yellowish.
ziluiensis	Orange-yellow.
hecla and vars.	Orange, shot with violet.
chrysotheme	Light yellowish-orange.
marco-polo	♂ greenish-yellow, ♀ light orange.
palaeno and vars.	Greenish-yellow, varying to greenish-white.
anthyale	Greenish-yellow, varying to nearly white.
erate and vars.	Light yellow, varying to light orange.

Group II. Both sexes spotted on border of f.w. :—

C. christophi	Yellowish-brown.
erschofi	Golden yellow.
sagartia	Varying from bluish-green (♂) to nearly white (♀).
sieversii	From light yellow to nearly white.
sifanica	Greenish-white to nearly pure white.
hyale	Sulphur-yellow to nearly white.
aipheraki	Greenish-white.
montium	From pale yellow to greenish-white.
melinos	Greenish-white.
phicomone	Dusky greenish-yellow to nearly white.

C. nastes	. . .	Dusky greenish.
rossii	. . .	Yellow.
werdandi	. .	Greenish-white.
cocandica	. .	Dusky greenish.
maya.	. . .	Greenish-yellow to greenish-white.
tamerlana	. .	Very dusky olive-green.

All the orange species are liable to a dimorphic whitish coloration in ♀.

In Group I., *B.*, *C. palaeno* and *C. anthyale* are somewhat abnormal, insomuch that the females have but a faint trace of spots on the marginal band, and that only occasionally.

In Group II., *C. 'christophi* and *C. erschoffi* are remarkable, the former for the brownish colouring of the wings, the latter for its golden-yellow colour; it, however, approaches *C. sagartia* in general appearance.

(*To be continued.*)

GEOLOGY IN ANTRIM.

By E. Reginald Sawer.

RECENT field-work in Antrim has raised some doubt in my mind as to whether the commonly accepted explanation of the nature of the metamorphosis by which the Chalk of the district has become indurated and crystalline can be accepted as altogether satisfactory. The fine series of sections along the scarps of Antrim show the following Mesozoic rocks in order :—Trias, Lower Lias, Upper Cretaceous. If we take, for the purpose of illustration, the section exposed from Garron Point to Cushendall, near the coast, the eroded surface of the Chalk will be seen to be

of weakness caused by a fault-plane. The chief interest attaching to these dykes lies in the fact that though of considerable dimensions—the latter being fully sixty feet in diameter—the metamorphism induced in the stratified rocks as a result of contact with these igneous masses can only be traced, at the most, within a few inches of the actual point of contact. Contact-metamorphism is, in fact, always distinguished by its extremely local character, and as a rule the extent of metamorphism, and the distance to which it can be traced, bear a marked relation to the volume of the

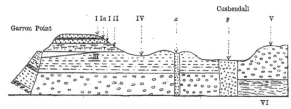

I. Upper and Lower Basalt ; Ia. Pisolitic Iron-ore ; II. Indurated Chalk ; III. Lias and Rhaetic ; IV. Trias ; V. Lower Old Red Sandstone (Dingle) ; VI. Metamorphic Rocks (? Archaean) ; *x*, Basalt Dyke ; *y*, Felstone-porphyry (Orthoclase-porphyry).

overlain by two sheets of basaltic lava, separated by a band of red pisolitic iron-ore ; while a layer of reddish flint-gravel frequently intervenes between basalt and Chalk, representing the product of subaërial decay previous to the eruptive epoch. The Chalk lying beneath the basalt is altered and indurated, and it has very generally been taken for granted that such marmorosis is a result of contact with the lava. It seems, however, to be a question whether this conclusion can be fairly drawn when all the facts bearing on marmorosis are duly taken into account.

Further to the west the section shows the Triassic beds penetrated by two important dykes —the first, lying to the east, consisting of basalt; while the second, on which Cushendall is situated, is composed of Orthoclase-porphyry (Felstone-porphyry), and has been erupted through the line

igneous mass. In this particular case, however, it is the lower bed of basalt alone with which we have to deal, as the band of pisolitic ore affords clear evidence of the subsequent eruption of the upper bed.

Such being the case, it would seem difficult to account for the apparently uniform alteration which has taken place throughout the relatively immense thickness of the Chalk, by mere proximity to the basalt.

In the Crimea, some fifteen miles from Tchatyr Dagh, indurated Chalk occurs similar to that of Antrim, but without relation to any igneous intrusion. Again, in the Caucasus the Chalk is covered by immense sheets of Andesitic lavas, but retains its normal friable character. These two cases sufficiently prove that contact-metamorphism is not essential to the induration of Chalk.

H 3

A microscopic examination of marmorised Chalk shows, that while at the actual point of contact with the basalt, crystals of aragonite have been developed, calcite is the predominant mineral throughout the mass of the rock. Further, while it is but rarely that fossils escape obliteration during the process of contact-igneous-metamorphism, in this case they are remarkably well preserved, and show little or no traces of alteration.

If we substitute regional metamorphism as a primary cause, we are at once met with the difficulty that in such a case the action of heat is greatly modified by, and even subordinated to, that of pressure. The weight of superincumbent basalt is trifling, compared with that to which the Chalk lying beneath the London basin is subjected, and there is no evidence that any induration has taken place in the latter district.

May I propose tentatively that an explanation should be sought in hydrothermal action? The influence of heated water containing dissolved solid matter and also gases, such as hydrochloric acid and carbon dioxide, in solution, is recognised as a potent factor in metamorphism. Such waters retain a remarkable uniformity of temperature throughout long periods. They coincide in distribution with volcanic areas, in which fact we find a satisfactory explanation of the alteration of the Crimean Chalk where this formation lies beyond the area influenced by igneous masses, but yet within the range of thermal waters, which, on the other hand, are characteristically absent from the unaltered Cretaceous beds occurring in immediate contact with beds of Andesitic lava in the Caucasus. A few somewhat parallel cases may be cited in support of our hypothesis. Daubrée has shown that the alkaline waters of Plombières, in the Vosges, conveyed by the Romans to baths through long conduits, have given rise to calcite, aragonite, and fluorspar, together with siliceous minerals, as the result of a rearrangement induced in the bed of concrete made of lime, fragments of brick, and sandstone. M. Foumet, in his description of the metalliferous gneiss near Clermont, in Auvergne, states that all the minute fissures of the rock contain free carbon dioxide, and that the various minerals of the gneiss, with the exception of the quartz, are all softened ; and new combinations of the acid with calcium, iron, and manganese are continually in progress.

In the Lipari Islands the horizontal strata of tuff forming cliffs have been discoloured in places by jets of steam, often above the boiling-point, called " stufas," issuing from the fissures ; similar corrosion of rocks near Corinth, and of trachyte in the Solfatara, near Naples, has been effected by sulphuretted hydrogen and hydrochloric acid. The interest of these instances lies in the fact that the gases must have made their way through vast thicknesses of porous or fissured rocks, and have modified them for thousands of feet.

We are becoming more and more acquainted with the leading part which thermal waters are playing in the distribution of internal heat through the superincumbent strata, and in introducing various chemical compounds into them in a fluid or gaseous condition, which alter, rearrange, and often combine with the component minerals of the surrounding rocks.

20 *Warwick Road,*
Upper Clapton, London, N.E.,
29th October, 1900.

CRUELTY TO WILD ANIMALS ACT.

ON August 6th last there was passed a short Act of Parliament entitled the " Wild Animals in Captivity Protection Act." Although we do not for a moment imagine that any reader of this journal would knowingly come within the range of its penalties, it is perhaps as well that all should be familiar with its chief clauses, which are

1. The word " animal " in this Act means any bird, beast, fish, or reptile which is not included in the Cruelty to Animals Acts, 1849 and 1854.

2. Any person shall be guilty of an offence who, whilst an animal is in captivity or close confinement, or is maimed, pinioned, or subjected to any appliance or contrivance for the purpose of hindering or preventing its escape from such captivity or confinement, shall, by wantonly or unreasonably doing or omitting any act, cause or permit to be caused, any unnecessary suffering to such animal ; or cruelly abuse, infuriate, tease, or terrify it, or permit it to be so treated.

3. Any person committing an offence may be proceeded against under the Summary Jurisdiction Acts, and on conviction shall for every such offence be liable to imprisonment with or without hard labour for not exceeding three months, or a fine not exceeding five pounds, and, in default of payment, to imprisonment with or without hard labour.

4. This Act shall not apply to any act done or any omission in the course of destroying or preparing any animal for destruction as food for mankind, nor to any act permitted by the Cruelty to Animals Act, 1876, nor to the hunting or coursing of any animal which has not been liberated in a mutilated or injured state in order to facilitate its capture or destruction.

5. This Act shall not extend to Scotland.

RADCLIFFE OBSERVATORY, OXFORD, is shortly to have a large double telescope, the work of Sir Howard Grubb. The great dome has a steel framework covered with papier-mâché. The observatory will be furnished with a rising floor worked by an hydraulic lift. The height from the ground to the top of the dome is 53 feet, and the outside diameter of the tower 35 feet. The mass of the great concrete block on which the large brick pier is built weighs some 30 tons, and is placed at a depth of 17 feet below the surface.

BRITISH FRESHWATER MITES..

By Charles D. Soar, F.R.M.S.

(Concluded from page 86.)

GENUS *EYLAIS.*

THIS genus, so named by Latreille in 1796, was supposed to contain only one species, *Eylais extendens* of Müller, until 1896, when Koenike pointed out certain differences in structure, and added to the list of species. This example has since been followed by other writers; so now we have a large number of Eylais named from different parts of the world. This genus seems likely to be yet further enlarged, for I have two or three specimens which I cannot fit in with Piersig's key or figures. Koch, in his great work, 1835–41, described several species; but Piersig has placed these under the one name, *Eylais extendens*. The great point of identification is in the eyes, which are formed in some species very much like a pair of spectacles.

The characteristics of this genus are : Eyes close together, claws to all feet, fourth pair of legs without swimming hairs.

1. *Eylais discreta* Koenike.

Body.—About 3.20 mm. in length and 2.75 mm. in width. Fig. 1, egg-shape in outline, of a bright red colour. The skin of the body is very delicate and easily broken.

Fig. 1. *E. discreta.* Dorsal surface.

Legs.—First pair about 2.40 mm., fourth pair about 3.18 mm., strongly made. The same colour as the body. The first three pairs have swimming hairs, but the fourth pair are quite without them. I do not think this peculiarity is known in any other genus of this family. All the feet have claws.

Epimera.—In four pairs. Of chitinous structure, rather hairy on the anterior edges.

Eyes.—The eyes are set in a chitinous plate, joined together with a ridge of chitin. The great difference in the shape of this plate is the principal point of identification in one species from another in this genus. Nearly the whole of Piersig's key is founded on this eye-capsule and eye-bridge, as

Fig. 2. *E. discreta.* Eye-plate.

he calls it, and he describes twenty-three species known in Germany alone. The eye-plate of this species is shown in fig. 2. In Piersig's key, he says, the fore-rim of the eye-bridge is weakly toothed, the other portion flat and insignificant In the mite I am now describing, I cannot say the anterior edge is as faintly notched as indicated in Piersig's figure of this portion of his mite; but this may be only a local difference. The exact width across the eyes is 0.41 mm. Several specimens were taken and measured, and although the body and leg measurements varied very much in different specimens, the eye measurements came out at about the same dimensions.

Palpi.—About 1.12 mm. in length. Covered with a number of hairs. Piersig figures a number of palpi of different species, but the difference in their structure is so small that I should not like to undertake to identify any species by that character alone.

Locality.—Kew Gardens.

2. *Eylais soari* Piersig.

The shape and general description of all the species of this genus are much the same. This one has all the characteristics of the preceding species, except in size and the shape of the eye-

Fig. 3. *E. soari.* Piersig. Eye-plate.

plate. Length about 2.0 mm., width about 1.70 mm. The eye-plate (fig. 3) is about 0.25 mm. extreme width over all. In all the specimens I have examined the eye-plate is about the same

H 4

measurement, but the actual outline of the plate varies very much. The one I have figured in this paper is as near as possible to Piersig's figure. Both figs. 2 and 3 have been drawn with the camera under a half-inch objective, so that the difference in size and structure of these two eye-plates can be easily seen.

This species received the specific name *soari* from Dr. Piersig in 1899.

LOCALITIES.—Dr. George has taken this mite in Lincolnshire, Mr. Taverner in Scotland, and I have found it at Mill Hill and at Lowestoft.

There is only one more genus to add to those already mentioned in my critical examination of the British freshwater mites, that being the genus *Thyas*. This will be described by my esteemed friend, Dr. George, of Kirton-in-Lindsey. This gentleman has written many interesting papers in this Journal on the Arrenuri, and his kind assistance has been of great help to me in the production of these papers. When the genus *Thyas* is described I think we shall then have recorded all the known British species up to the time of writing. I have several species not yet identified which may have to be added to the list at a later date, and there must also be a great number yet to record when diligent collectors have found them, for we are yet a long way behind the German list. Still, I think, considering the small number of workers in this country on the Hydrachnidae, we have made a very good beginning.

37 *Dryburgh Road, Putney, London, S.W.* .
November 1900.

ARRENURUS ORNATUS N S.

BY C. F. GEORGE, M.R.C.S.

IN the Old Series of SCIENCE-GOSSIP for December 1882, page 273, I figured and described a freshwater mite under the name of *Arrenurus viridis* Duges. Dr. Piersig has pointed out that

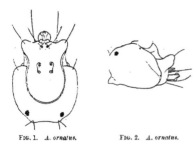

FIG. 1. *A. ornatus.*   FIG. 2. *A. ornatus.*
Male. Dorsal surface.   Male. Lateral surface.

A. viridis of Duges is *A. maculator* of Müller, and that the mite figured by me had not previously been described. This mite must therefore be

called *Arrenurus viridis* George, unless I decided to give it a new name. On consideration, I think it more satisfactory to rename it, especially as by far the greater number of the specimens I have found are not green, but blue.

The petiole of this mite, being very ornamental. also a marked and important feature. I have ventured to name the species *Arrenurus ornatus*. As many of the present readers of SCIENCE-GOSSIP may neither have the old volumes of SCIENCE-GOSSIP, nor ready access thereto, and as I have

FIG. 3. *A. ornatus.*   FIG. 4. *A. ornatus.*
Male. Petiole.   Male. Palpus.

quite recently paid some attention to this mite, I thought I would write another description giving to it the new name, and describing a little more fully this beautiful and interesting creature. I am indebted to Mr. Soar for the accompanying new drawings.

Arrenurus ornatus is one of the tailed mites. so called because the males have a peculiar structure posteriorly, and this differs very curiously in the

FIG. 5. *A. ornatus.* Male.   FIG. 6. *A. ornatus.* Male.
Petiole of, not full developed.   Dorsal surface of, undeveloped.

different species. There are two great divisions—one has the tail cylindriform and more or less narrowed at the base; in the other division the tail is wide, with side corners jutting out, and a petiole projecting freely from the centre of the posterior edge of the tail. Our mite belongs to this second division. The next point is, that there are on the hinder part of the back two elevations like horns, bent so that the points project forwards. These horns are in *A. ornatus* distinctly separated from each other at the base. Below these, nearer to the centre of the hind edge of the tail, are two little knobbed elevations, each carrying a tactile hair; under these in the centre is a sharp point over the petiole, and where it joins the tail is a transparent pellicle called the hyaline membrane. This process, in the mite now being described, is somewhat narrower at the free end than at the base, and the free corners are not drawn out to a point, but rather bluntly rounded. The side corners of the tail are rather thick and project out obliquely. The petiole is fan-shaped and has a central projection, giving it an ornamental character. On each side of the petiole a curved upward-bent bristle projects, about as far as the length of the petiole. The figures 1, 2, and 3 show all these points very distinctly. Fig. 4 shows the inner surface of the male palpus highly magnified.

Another peculiarity of this mite, shared by many other male Arrenuri, is that the fourth internode

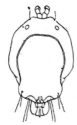

FIG. 7. *A. ornatus.* Male. Dorsal surface more fully developed.

of the last leg is provided with a highly-developed process or spur, as seen in fig. 8. Fig. 9 is a drawing of the ventral surface of the female.

Mr. Soar gives me the following measurements :—

MALE.—Extreme length, 1.12 mm.; length to end of body, 1 mm.; length of petiole, 0.12 mm. width of petiole, 0.10 mm.; width of body, 0.88 mm.; width of posterior part of tail, 0.46 mm.;

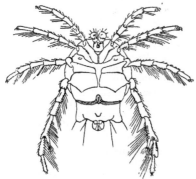

FIG. 8. *A. ornatus.* Male. Ventral surface.

length of first leg, 0.60 mm.; length of second leg, 0.63 mm.; length of third leg, 0.68 mm.; length of fourth leg, 1.08 mm.; length of palpus, 0.21 mm.

FEMALE.—Length of body, 1.12 mm.; width of body, 0.96 mm.; length of palpus, 0.21 mm.; length of first leg, 0.88 mm.; length of second leg, 0.89 mm.; length of fourth leg, 1.04 mm.

During the present year I have been greatly interested in watching the development of this and several other Arrenuri from the nymph or Anurania stage to the full adult form. In the Anurania stage there is no appearance of the petiole, and the creature, which is more or less circular, looks very different from the adult. When this earlier stage is passed the petiole is well developed, but the side corners are not. (See fig. 6.) In a day or two they become evident, as

FIG. 9. Ventral surface of female.

in fig. 7, and soon afterwards they attain their full development, as in fig. 1.

At first the whole creature is of a deep yellow colour, with a shade of green at the edges. This colour is doubtless produced by the yellow contents of the body shining through the light and diaphanous blue chitine. The deepening of the blue colour proceeds, and at one stage I have seen the central part of the petiole of a beautiful yellow colour, whilst the outer parts were of a fine blue, which was very striking and beautiful. Lastly, the extreme ends of the side corners lose their yellow colour, and the whole creature becomes of a beautiful blue, which is rather intensified when the mite is mounted in Canada balsam. The colours during development are pure and beautiful, shading one into another in a manner that must defy the skill of any artist.

Kirton-in-Lindsey,
October 1900.

EDIBLE EARTH OF FIJI.—In a note to the Royal Society of New South Wales Dr. B. G. Corney gives the following result of an analysis by Mr. F. B. Guthrie, F.C.S., of a portion of the edible earth of Fiji which was collected from near the northern coast of Vanua Leva :—Moisture at 120° C., 2.45; combined water, 12.78 silica, 41.53 ; alumina, 35.09 ; Fe_2O_3, 7.66.

AN INTRODUCTION TO BRITISH SPIDERS.

By Frank Percy Smith.

(*Continued from page 169.*)

FAMILY *THERIDIIDAE.*

WE now come to a family of spiders which is at once the largest and most difficult to elucidate. Numbers of the species are extremely minute; and whilst many are very distinct, others are so much alike that they can only be separated with great difficulty. In numerous cases descriptions without accurate drawings are almost useless for purposes of identification. In treating of this family it will be advisable to omit certain species. Some have been described with insufficient accuracy to ascribe to them their true systematic position, the types being now unobtainable; others are of doubtful value as species, and until their distinctness has been thoroughly investigated the insertion of their names in an introduction of this kind would be misleading, as many of them may have to be regarded as varieties. A few species of extreme rarity have been described, of which I am unable to obtain either specimens or good drawings, and these will be simply referred to their discoverers. With regard to the classification, I have adopted many of the numerous genera into which the old Blackwallian groups have been divided, but it will need much time, and an increased number of workers, before anything like finality can be arrived at in the Theridiidae.

In dealing with many of the closely-allied species included in this family, such descriptions as I could insert in the space at my disposal would be useless for purposes of determination. I shall therefore endeavour as far as possible to publish drawings of the palpi and vulvae of many of the best-known forms, as these are in the majority of cases the most trustworthy specific characters.

GENUS *EPISINUS* WLK.

Falces moderately long. Central eyes form a quadrilateral much narrower in front than behind. Maxillae very slightly inclined. Posterior eyes in a strongly curved line, its convexity being directed forwards. Lateral eyes slightly separated.

Episinus truncatus Wlk. (*Theridion angulatum* Bl.)

Length. Male 3 mm., female 3.5 mm.

This not uncommon species may be easily distinguished from its allies by the curious shape of the abdomen, which is narrow in front and widens considerably toward its posterior extremity, where it is steep and truncated.

GENUS *THERIDION* WLK.

Palpus of female provided with a terminal claw. Maxillae inclined, usually more than twice the length of the labium. Central eyes forming a quadrilateral figure, almost or quite as wide in front as behind. Length of falces greater than height of clypeus.

Theridion formosum Clk. (*T. sisyphum* Bl.)

Length. Male 3 mm., female 4 mm.

The abdomen is very convex above, and projects considerably over the cephalo-thorax. The sides are adorned with several oblique markings. It is not common.

Theridion sisyphium Clk. (*T. nervosum* Bl.)

Length. Male 3.5 mm., female 4.5 mm.

The abdomen of this pretty spider has two dark brown longitudinal bands, which are crossed by a number of narrow pale lines. It is very common.

Theridion pictum Hahn.

Length. Male 4 mm., female 4.5 mm.

This very beautiful species, which is not uncommon in gardens and conservatories, may be distinguished at once by the dark red dentated band, bordered with yellow, upon the upper side of the abdomen.

Theridion tepidariorum Koch.

Length. Male 5 mm., female 6 mm.

This species may be recognised by its dull colours and large size, and also by the fact that it is seldom, if ever, found outside greenhouses or conservatories. It is common everywhere.

Theridion denticulatum Wlk.

Length. Male 3 mm., female 3.5 mm.

The upper side of the abdomen is of a grey colour, with a central pale dentated band. It is common in gardens.

Theridion varians Hahn.

Length. Male 2.5 mm., female 3.5 mm.

Similar in general appearance to *T. denticulatum* Wlk., but considerably paler.

Theridion tinctum Wlk.

Length. Male 2 mm., female 2.5 mm.

Similar to *T. varians* Hahn., but considerably smaller, the pale portions being suffused with green. The abdominal pattern also is usually less regular.

Theridion aulicum Koch. (*T. rufolineatum* in "Spiders of Dorset.")

Length. Male 2.5 mm., female 3.5 mm.

Cephalo-thorax dull yellow, with a central and marginal band of reddish-brown. Not common.

Theridion vittatum Koch. (*T. pulchellum* Wlk.)

Length. Male 2.5 mm., female 3 mm.

Cephalo-thorax of a greenish colour, with a broad central and narrow marginal band of black. Abdomen reddish, with a distinct, somewhat narrow, dentated band. Not uncommon.

Theridion simile Koch.
Length. Male 2 mm., female 2.5 mm.
The abdomen is extremely convex above, and has a white dentated band edged with brown. Not uncommon.

Theridion riparium Bl.
Length. Male 3 mm., female 3.5 mm.
Abdomen of a chocolate colour with white markings. Not common.

Theridion familiare Cb.
Length. Male 2 mm., female 2.5 mm.
Cephalo-thorax dull orange-yellow with a dark patch behind the eyes. Found in houses.

Theridion pallens Bl.
Length. Male 1.7 mm., female 2 mm.
This spider is easily recognised by its small size. The upper side of the abdomen is marked with a large pale cross.

Theridion bimaculatum L. (*T. carolinum* Bl.)
Length. Male 2 mm., female 2.5 mm.
The abdomen is of a dark reddish-brown colour, with a distinct yellow band which varies considerably in different individuals. The sternum of the male has a small projection near its centre, best seen when viewed in profile. The palpal organs are large. I have found numerous specimens of this uncommon spider quite recently in Surrey, under the leaves of the burdock. The female carries its egg-sac attached to the spinners, and helps to maintain it in that position by supporting it with one hind leg.

Theridion lineatum Clk. (*Phyllonethis lineata* in "Spiders of Dorset.")
Length. Male 4 mm., female 5 mm.
The whole spider is usually of a pale yellow or greenish-yellow colour, the abdomen being marked with a few black spots and lines. A variety is often found in which the upper side of the abdomen is ornamented with two pale crimson bands. The falces of the male are extremely divergent and armed with a powerful tooth; they vary considerably in different individuals. This spider, which is extremely common, may be found in the autumn enclosed with its pale bluish-green egg-sac in the folded leaf of a bramble or some similar plant.

Theridion lepidum Wlk. (*Phyllonethis instabilis* Cb.)
Length. Male 2 mm., female 2.5 mm.
This species is closely allied to *T. lineatum* Clk., but is much smaller, and the legs are proportionately longer and stronger. The falces of the male are very similar to those of *T. lineatum* Clk., but do not seem to be subject to much variation. I have never seen a variety of this species with crimson markings.

GENUS *PHOLCOMMA* THOR.

Maxillae inclined towards the labium. Falces weak. Clypeus high. Eyes arranged very much as in the genus *Pholcus*. Three large eyes are grouped on each side, and between these groups are two very small eyes.

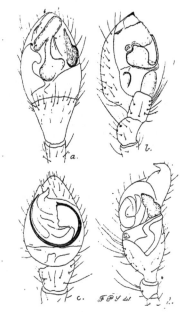

FIG. 1. *a.* Palpus of *Theridion sisyphium. b.* Palpus of *T. pictum. c.* Palpus of *T. pallens. d.* Palpus of *T. bimaculatum.*

Pholcomma gibbum Westr.
Length. Male 1.5 mm., female 1.7 mm.
This little spider may be at once distinguished by the curious grouping of the eyes. It is not at all common, but is widely distributed. I have taken it at Hastings and Epping Forest, and have quite recently received two specimens from Dr. J. W. Williams, who obtained it near Hampstead Heath.

GENUS *THEONOE* SIM.

Maxillae inclined towards the labium. Eyes in two rows. Clypeus high. Sternum very convex. Abdomen globular.

Theonoe minutissima Cb.
Length. Male 1 mm.
The general colouring of the legs and thorax is brown; the abdomen dark olive-green. The most tangible characteristic is the extremely convex sternum. This is a very rare species.

(*To be continued.*)

THE RIDDLE OF THE UNIVERSE.

MOST educated people have been for many years conversant with the philosophical and scientific works written by the master-hand of Professor Ernst Haeckel. To such persons the issue of an English translation of "Die Welt-räthsel"[1] cannot fail to be a source of great satisfaction. Before discussing the work itself, we would congratulate Mr. Joseph McCabe on his clever and effective rendering into English of this important book. Although much of its contents are necessarily of an abstruse character, Mr. McCabe's well-chosen language places Dr. Haeckel's meaning and intention so clearly and pleasantly before the reader as to fix attention from first to last.

Those of our readers who know the author's previous writings can easily anticipate the character of the pages before us, which are founded upon the reaction of thought, brought about in consequence of scientific research and its results during the closing century. To quote the translator, Dr. Haeckel herein summarises the evidence for the evolution of the mind in a masterly and profoundly interesting fashion. The author himself considers this work, " The Riddle of the Universe," collates and marks the close of his studies on the monistic conception of the universe[2]. He says that the earlier plan, projected by him many years ago, of constructing a complete system of monistic philosophy on the basis of evolution cannot now be carried into effect, owing to his failing strength. He adds that being "wholly a child of the nineteenth century, with its close I draw the line under my life's work." There is a touch of sadness in these words, which we trust is only caused by temporary indisposition at the time of writing.

The scheme of the book before us is a brilliant review of the labours and discoveries of most of the important investigators in the field of scientific research. These are collated and applied by the author to the construction of a philosophy of which he is the apostle. The subjects touched upon by him are briefly : the nature of the problem of the universe, including the condition of civilisation and thought at the close of the nineteenth century ;

(1) "The Riddle of the Universe at the Close of the Nineteenth Century." By Ernst Haeckel, Ph.D., M.D., LL.D., Sc.D., and Professor at the University of Jena. Translated by Joseph McCabe. XVI+398 pp., 8 in. × 5½ in. (London : Watts & Co., 1900.) 6s. net.

(2) The philosophy of Monism may be said to be a doctrine which considers mind and matter neither as separated nor as derived from each other, but as standing in an essential and inseparable connection. In other words, it is a philosophy which can as little believe in force without matter as in matter without force. It embraces the pure, unequivocal Monism of Spinoza : matter, or infinitely extended substance, and spirit (or energy), or sensitive and thinking substance, are the two fundamental attributes, or principal properties, of the all-embracing, divine essence of the world.—ED. S.-G.

the fundamental importance of anatomy, both human and comparative ; the cellular theory with regard to man's place in Nature ; the development of the study of physiology from the Middle Ages ; embryonic development and the theories of pre-formation and scatulation ; the history of our species, with an account of the fossil Pithecanthropus of Dubois ; the nature of the soul ; psychic gradations ; the embryology of the soul ; the phylogeny of the soul ; consciousness, giving the various theories of Descartes, Darwin, Schopenhauer, Fechner, and Schultze ; the immortality of the soul ; the law of substance, including the chemical law of the constancy of matter and the physical law of the conservation of energy ; the evolution of the world ; the unity of Nature ; God and the world ; knowledge and belief ; science and Christianity ; monistic religion and ethics ; the solution of the world problems ; and a brief conclusion, in which he points out that the number of world-riddles has been continually diminishing in the course of the nineteenth century, through the progress of a truer knowledge of Nature.

Yet, as says the author, one comprehensive riddle of the universe still remains—the problem of substance, otherwise the Cosmos. The author goes on to say we do not know "the thing in itself" that lies behind the knowable phenomena of matter and energy by which the universe is continued.

Dr. Haeckel urges in a very forcible manner the advisability of a more practical application of scientific knowledge to the daily life of humanity. For instance, with regard to the administration of justice, he points out that though judges and counsel are popularly supposed to be men of highest education, and this is doubtless correct with regard to "legal education," which is of most part formal and technical in character, these gentlemen have only a superficial acquaintance with that chief and peculiar object of their activity, the human organism, and its most important function, the mind. That is evident from the curious views as to the liberty of the will, responsibility, &c., which we encounter in their decisions. Most of the students of jurisprudence have little knowledge of anthropology, psychology, and the doctrine of evolution ; important requisites for a judicial estimate of human nature.

In comparing the two great founders of transformism, Dr. Haeckel says, "We find in Lamarck a preponderant inclination to deduction and to forming a complete monistic scheme of Nature ; in Darwin we have a predominant application of induction, and a prudent concern to establish the different parts of the theory of selection as firmly as possible on a basis of observation and experiment." The latter appears to be the course followed by the author himself in the pages before us; consequently much deduction is left to the reader's own intelligence.

If we have read Dr. Haeckel's conclusions correctly, we feel hesitation in accepting some of them without further evidence in their favour. For instance, when quoting Emile du Bois-Raymond's seven world-enigmas, No. 3 being the Origin of Life, he says the question is "decisively answered by our modern theory of evolution." We fail to see how this applies to a solution of the origin of life, though it accounts for its continuance.

The "Riddle of the Universe" is a very powerful work, and should be read by all people of mature thought interested in science and the progress of humanity. Though one sometimes disagrees with Dr. Haeckel's opinions and conclusions, his theories and facts command consideration.

It is probably well known to our readers that the author in his works takes the materialistic side of argument as opposed to theology. Nevertheless we cannot help pointing out that the earnestness, devotion to duty in view of the benefit to others, and general honesty of purpose exhibited by Haeckel and contemporary scientific workers who agree with him, are in themselves the hereditary development of nineteen centuries of the, perhaps more or less imperfect, teachings of Christian ethics. The monistic philosophy advocated by Dr. Haeckel and his followers does not allow for the qualities of self-sacrifice and self-denial, virtues which, though inherently opposed to scientific theories of physical evolution, yet command the instinctive admiration of all civilised peoples.

F. WINSTONE.

ORIGIN OF SCIENTIFIC SOCIETIES.

AT the opening meeting of the 147th Session of the Society of Arts, held November 21st, the address given was by Sir John Evans, K.C.B., D.C.L., LL.D., Sc.D., F.R.S., upon the "Origin, Development, and Aims of our Scientific Societies." Sir John Evans stated that no learned Society had received a Royal Charter before 1662, when the Royal Society was incorporated. The Society of Antiquaries was, however, much older, having been founded about 1572. Among the meeting-places of this staid and respectable body was the "Young Devil" tavern, in Fleet Street. The Society before which the address was given was founded in 1754, and incorporated nearly a century later, in 1847, as the "Society for the Encouragement of Arts, Manufactures, and Commerce." From the trio of Societies—the Royal, Antiquaries, and Arts—Sir John mentioned that, nearly all the numerous leading learned societies in existence in this country had sprung by a natural process of evolution. The first, perhaps,

was the Medical Society, founded in 1773. The Linnean Society for the Cultivation of Natural History followed in 1788. The lecturer pointed out that during the century now drawing to its close the vast advances in science and the innumerable aspects which it assumed had led to the foundation of the numerous scientific societies with more or less limited scope. These were by no means confined to science as represented by the ordinary acceptance of the word, as many were literary and philosophical in their aims; that of Manchester dating back to 1781. The offshoots of the Society of Antiquaries had not been so numerous, nor so important, as those from the Royal Society; the field of archaeological research being more restricted than that of purely "natural knowledge." The Society of Arts was the first in England to devote attention to the important subjects of forestry and agriculture; the Royal Agricultural Society not originating until 1838. It was the Society of Arts also that laid the foundations for the Institute of Civil Engineers and its offshoots. At the Society of Arts in 1841 there was formed the Chemical Society, from which arose the Institute of Chemistry in 1877. The same birthplace may be claimed for the Society of Chemical Industry and the Sanitary Institute. Similarly originated were the City Guilds Institute, and even the Science and Art Department at South Kensington, though this latter was influenced by the Great Exhibition of 1851. The Photographic Society grew from an exhibition of photographs, the first of its kind, held in the Society's rooms. It was also the parent of the Royal College of Music. Sir John Evans pointed to the fact that without our Societies it would have been impossible for knowledge to have progressed as it has during the past century. They bring about that healthy competition which stirs men from rest or torpor; a state once described by a secretary of the Society of Antiquaries, when he said: "Would to God there was nothing in this world older than a new-laid egg"

A UNIVERSAL WHEATSTONE BRIDGE.—An improved form of a Wheatstone bridge, and one specially designed for the Carey-Foster method of resistance measurements, is described in the "Electrician" for October 5th last. Both the construction and theory of the bridge are thoroughly explained by the writer, Mr. C. V. Drysdale, who is, we believe, the designer.—*J. Quick, Cork.*

THE USE OF FLOATS IN BURETTES.—An article of extreme interest to teachers of chemistry, in the "Zeitschrift für angew. Chemie," points out, as the results of an extended series of observations on the use of floats for reading burettes, (1) that floats should never be employed in reading burettes calibrated without them, and (2) that the results obtained by different observers, and by the same observer at different times, rarely agree. The author, Herr Kreibling, concludes that their use is to be avoided.—*Harold M. Read, London.*

BOOKS TO READ

NOTICES BY JOHN T. CARRINGTON.

Riddle of the Universe. By ERNST HAECKEL.
See page 208 for notice of this work, which is one
of the books of the year.

Problems of Evolution. By F. W. HEADLEY.
xvii + 373 pp., 8¾ in. x 5½ in., with fourteen illustrations. (London: Duckworth & Co. 1900.)
8s. net.

This is another contribution to the already large
amount of existing literature upon the problems of
evolution, which have themselves evolved from
Darwinism and Neo-Darwinism. The work is
written in such a style as to be useful to the
scientific student, and also to the public at large.
For this reason the author has commenced with a
valuable introductory chapter, explaining certain
elementary biological facts, without which the un-
initiated reader might find difficulty in following
the arguments contained in the rest of the work.
This introduction explains Darwin's theory and
the early stages of evolution. There is added a
section on the development of the individual. The
whole work is divided into two parts, the first
dealing with evolution at large, and the second
with human evolution in particular. This latter
portion has been treated in its widest sense; and
not the least interesting chapter of the book is the
one on "The Great Unprogressive People." As
one might expect, the Chinese are the races under
discussion. With regard to the evolution of
animals, the author has selected Dr. Haeckel's
genealogical tree as his type of the ascent. This,
by permission of the publishers, we reproduce.
The "Problems of Evolution" is a book to read
with care and satisfaction.

A Glossary of Botanic Terms. By BENJAMIN
BAYDON JACKSON. x + 323 pp., 8 in. x 5 in. (Lon-
don: Duckworth & Co. Philadelphia: Lippincott
Co. 1900.) 6s. net.

Mr. Daydon Jackson, who for a lengthened period
has acted as honorary secretary for the department
of botany in the Linnean Society, has conferred
upon botanists and the public at large a distinct
boon by compiling this modern glossary of botanical terms, with their derivations and accent. Mr.
Jackson appears to have left no source unsearched
for even the most modern words. Each is treated
with much fulness of explanation; in fact the work
is quite encyclopaedic. In dealing with the terms
used in foreign countries, as far as possible English
equivalents are stated, and we understand that
in all cases the derivations have been carefully
checked. It is needless to remark on the value of
the accentuation, and we trust the book will lead
to greater uniformity in pronunciation. The
magnitude of the work may be estimated from the
fact that nearly fifteen thousand words are dealt
with. There are appendices relating to various
matters incident to the book, such as the adopted

pronunciation of Latin and Latinised words; the
use of the terms "right" and "left"; a bibliography
extending to three pages, etc. No English botanist
can afford to dispense with this work, which should
also be in every important library.

Tales told in the Zoo. By F. CARRUTHERS GOULD
and F. H. CARRUTHERS GOULD. 136 pp., 10 in.
x 7½ in., with one coloured and twenty-three other
illustrations. (London: T. Fisher Unwin. 1900.)
6s.

This is an entertaining book for young people,
and even for some who are their elders. It is a
series of stories founded upon natural history folk
lore. For the most part it deals with birds. These
stories are brightly told, and are really interesting.
They are not all inventions of the author's, but
many are founded on legends which have a distinct
ethnological value. The illustrations are by the
first-named author, and are cleverly drawn. It is
a bright book, and one to make a good Christmas
present.

Geological Antiquity of Insects. By HERBERT
GOSSE, F.L.S., F.G.S. 56 pp., 8½ in. x 5½ in. 2nd
edition. (London: Gurney & Jackson. 1900.) 1s.

This is a re-issue of the author's well-known
small work upon fossil entomology. It has been recognised for the past twenty years as indispensable
to both entomologists and geologists. The pages
before us contain some additions, but, considering
the large quantity of material that has been found
since the work was first published, we had looked
forward to its greater enlargement. We hope Mr.
Gosse will see his way to this, notwithstanding his
prefatorial announcement to the contrary.

Workshop Mathematics. Parts I. and II. By
FRANK CASTLE, M.I.M.E. xviii + 331 pp.
(London: Macmillan & Co. 1900.) 1s. 6d. each
part.

These two little books will no doubt prove a
great boon to the class of students for whom they
are written. Some artisan students are either
ignorant of pure mathematics or are impatient, and
yet require sufficient knowledge of them for workshop calculations. The two volumes before us admirably meet this requirement. The matter is so
smoothly graded that the practical student should
have no difficulty in working through with a little
help from the teacher. The reviewer has personal
experience of the assistance given by these book-
lets. He has a mixed class in magnetism and
electricity and a similar one on mechanics, both on
the same evening. By wedging in half an hour
or so for individual help, this "Workshop Mathematics" is fully appreciated by the students.—
J. Q.

The Construction of Large Induction Coils.
A Workshop Handbook. By A. T. HARE, M.A.
i + 155 pp., illustrated. (London: Methuen & Co.
1900.) 6s.

Much has been said and written within the last
twenty years upon the construction of large induction coils. The "English Mechanic" some years
ago printed many letters and illustrated articles
bearing upon the question. These were from men
who had paid the expense of experience in making
both good and bad coils. There was then a prolonged lull in the matter until Röntgen discovered,
in 1895, the rays bearing his name, when the
making of induction coils, large and small, received

such a spur as it had never hitherto. The interest attached thereto has waned but little, and the work now forms a stable industry in the workshops of the instrument maker and the practical electrician. The author of the present work has spared no pains in bringing forward the fullest described of arriving at the same result, the defects or advantages of each being pointed out. The important question of the interruptor or break receives the attention it deserves, three chapters being given over to that subject. The volume finishes with two useful appendices of theoretical

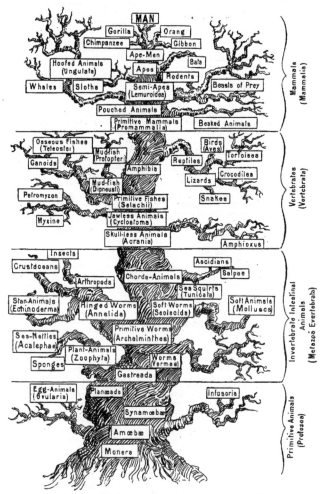

GENEALOGICAL TREE, FROM HAECKEL'S "EVOLUTION OF MAN."

(From "Problems of Evolution," by F. W. Headley.)

details of construction for the benefit of his readers. Every practical point is considered, from the commencement to the finish of the coil, and the book is well illustrated with numerous dimensional drawings. In many cases two methods are considerations. It is to be regretted that the book did not appear three years ago, when coil-making excitement was at its height. Nevertheless it should form a standard reference book for the purpose in view.—*J. Q.*

THE proprietors of "Knowledge" are about to issue an annual for students and workers specially devoted to astronomy. It is to be entitled "Knowledge Diary and Scientific Handbook, 1901," and will contain, amongst other things, useful tables, original articles, calendar of scientific events, etc.

IT is satisfactory to hear that sufficient funds have been collected to assure a memorial to the late G. J. Symons, F.R.S., the founder of the British Rainfall Organisation. The memorial will take the form of a gold medal to be awarded from time to time for distinguished work in connection with meteorological science.

DR. LEIGHTON asks us to insert the following paragraph :—Would readers of SCIENCE-GOSSIP be good enough to send particulars of the relative frequency, average size, etc., of the adder and the ring-snake in the respective districts under observation? His address is—Gerald Leighton, M.B., Grosmont, Pontrilas, near Hereford. The information is required to aid in the preparation of a new work on British snakes.

THE annual course of Christmas lectures, specially adapted to young people, at the Royal Institution, will be delivered by Sir Robert S. Ball, F.R.S., Lowndean Professor of Astronomy in the University of Cambridge, whose subject is "Great Chapters in the Book of Nature." The first lecture will take place on Thursday, December 27th, at three o'clock, and the remaining lectures will be delivered on December 29th, 1900, and on January 1st, 3rd, 5th, and 8th, 1901.

IT is a satisfaction to find from careful statistical figures, prepared by Mr. W. H. Dines, that the English climate must, from the health point of view, in future be considered one of the best in the world. A pleasanter climate may be easily found, but the majority of health resorts to which English people migrate in the winter have a higher death-rate than London at the same season, and a far higher death-rate than that of the country districts of the British Isles. Mr. Dines read a paper on this subject on November 21st before the Royal Meteorological Society.

THE second part of the "List of Private Libraries," compiled by Mr. G. Hedeler, of Leipzig, will soon be ready. It will contain more than six hundred important private collections of the United Kingdom, including supplement to Part I. (United States of America and Canada). Those happy possessors of libraries with whom Mr. Hedeler has been unable to communicate are requested to furnish him with a few details as to the extent of their treasures, and the special direction to which they devote themselves. By doing so they will, of course, not incur any expense or obligation. It is obviously to the interest of bibliographical science that a work of this kind should be as complete as possible. His address is 18 Nürnberger Strasse, Leipzig.

THE death took place on November 22nd, of Somerset Henry Maxwell, J.P., D.L., Lord Farnham. He was a student of natural history and astronomy, and a Representative Peer of Ireland.

THE members of the London branch of the Conchological Society are holding monthly meetings this winter at 11 Queen Victoria Street, E.C. The next is on November 30th at 7 P.M. All conchologists are cordially invited. Further particulars may be had from the hon secretary, J. E. Cooper, 68 North Hill, Highgate, N.

THE Camera Club has lost one of its most useful members by the death of Mr. W. Law Bros, whose illustrations on the ancient monuments of India were given before the club so late as November 5th. In addition to his knowledge and skill as a scientific photographer, Mr. Bros was an eminent antiquary, and an English member of the Association Française. His death took place on November 11th.

THE international meteorological committee has invited the Royal Meteorological Society to co-operate in a series of observations during 1901 of the form, amount, and direction of the clouds, on the first Thursday in each month, as well as the preceding and following days. These observations are to be made in connection with balloon ascents, which will be carried out under the direction of the aërostation committee.

MESSRS. SANDERS & CROWHURST, who have for some time been associated in the employ of Messrs. W. Watson & Sons, of High Holborn, London, desire to say that they have entered into partnership as opticians at 71 Shaftesbury Avenue, London, where, in addition to their general stock of optical and scientific apparatus, they will represent as sole West End agents Messrs. Watson's special instruments.

OUR readers will remember that the Anthropological Institute has established annual lectures in memory of the late Rt. Hon. Thos. Huxley. The first of these was given on November 13th in the theatre of the Museum of Practical Geology in London. Lord Avebury, the first lecturer chosen, gave an admirable review of Huxley's work and his influence, not only upon science in this country, but also on intelligent thought throughout the world.

THE New Mexico Normal University has commenced to issue some leaflets, entitled "Nature Study Bulletins." No. 1 refers to house flies, and is by Wilmatte D. Cockerell. It is illustrated by a magnified example of *Musca domestica*, and a dozen anatomical drawings of that species in the course of development. No. 2 is entitled "Pigments," and is by Professor T. D. A. Cockerell. Both these Bulletins give in simple language most useful information for popular use.

"EQUATORIAL ADJUSTMENTS, popularly explained," is the title of a pamphlet by Mr. William Banks, F.R.A.S., issued by Messrs. Banks & Co., opticians, of Bolton. It is a handy and useful little work of twenty-four pages, illustrated by five diagrams. We would suggest that in another edition signs should be added to the tables on pp. 4, 10, and 11, and initials substituted for signs on pp. 8 and 17. It is a handbook that will be of great use to amateurs desiring to set an equatorial mounted telescope in accurate adjustment. Without such accuracy careful observation is afterwards difficult; and it is a certain means of saving time at critical moments.

CONDUCTED BY B. FOULKES-WINKS, M R.P.S.

ASSISTANCE FOR READERS.—We are pleased to announce that Mr. B. Foulkes-Winks will commence· in the January number a series of instructive papers, which we feel sure will be most valuable to amateur photographers. They will be fully illustrated, and give useful hints, commencing from the initial stages and progressing to the higher technical work. These articles, being from the pen of a thoroughly practical manipulator, and medallist in the art, are certain to be of help to both beginners and to the more proficient —[ED. S.-G.]

PHOTOGRAPHIC QUESTIONS AND ANSWERS.—The Departmental Editor for photography will be glad to answer any questions and to give whatever assistance he can to readers. He will also be pleased to receive examples of work from amateurs for criticism, and occasionally will reproduce those of exceptional interest, with remarks from a technical point of view.

LANTERN SLIDES.—We have lately been experimenting with various makes of lantern plates, and for rich black tones have found the "Mawson" and Thomas's everything that could be desired. The "Gravura" lantern plates give a very wide range of tones, varying from black to almost a red-chalk tint, according to the amount of exposure given and the quantity of A.C. solution added to the developer. Some very beautiful tones can be obtained on "Alpha" plates by developing and fixing, and, after well washing the plate, immersing it in the following solution until the desired tone is attained :—

Toning Solution.—Dissolve 1 dram of ammonsulpho-cyanide in 15 ozs. of water; when dissolved add 5 grs. of chloride of gold, previously dissolved in 1 oz. of water. Use distilled or boiled water. After toning, rinse well and stand plate to dry in usual way. It is always advisable to give a final rinse under the tap, and at the same time to gently wipe the film with a tuft of wet cotton wool. It is sometimes desirable to alter black tone lantern slides to a warmer tint. This may be readily done by toning in the uranium toning bath : but before toning great care must be exercised to render the slide free from every trace of hypo. To ensure this it is as well, after thoroughly washing the plate, to immerse it in the following solution for three minutes :—Peroxide of hydrogen (20 vols.), 1 dram ; water, 5 ozs. ; rinse well and proceed to tone.

Uranium Toning Solution.—I. Ferrocyanide of potassium, 45 grs. ; water, 10 ozs. II. Uranium nitrate, 45 grs. ; glacial acetic acid, 1 oz. ; water, 9 ozs. Mix 1 oz. of No. I. and 2 ozs. of No. II., and immerse plate immediately, keeping the dish well rocked until desired tone is reached. This is for sepia; if a red tone is desired, take twice the quantity of No. I. When the slide is toned, remove

to a dish of water into which a few drops of acetic acid has been added, for two or three minutes. Then wash in running water for at least five minutes, and do not let the water fall directly on plate. In lantern slides, by whatever process, it is always advisable to make them by reduction, as the results are far more satisfactory than when produced by contact. A reduction camera is the most convenient method of procedure ; but a very simple and less costly plan is to use one's own camera, be it quarter-, half-, or whole-plate. First procure a 3¼in. square carrier for the ordinary dark slide to hold the lantern-plate. Then, by any simple means will then, fasten the negative to the window-pane, first seeing that the glass is quite clean, and having previously fixed a piece of white paper on a sheet of card or board, at an angle of about 45°, outside the window. Focus the negative on the ground glass of camera, taking special precaution that the front of the camera is parallel with the negative. When the image on the ground glass is in its right position, and perfectly sharp, insert the slide, and make the necessary exposure. It is obvious that one of the chief advantages of this method is that any particular portion of the negative may be reproduced, and this to any desired size, on the lantern plate. Another advantage is that, should the upright lines in the original negative be converged, or out of the straight, they may be corrected and brought parallel by a careful use of the swing back.

VELOX PAPER AND DEVELOPING CARTOLS.—We have received price list and particulars from Messrs. Griffin & Sons, Ltd., of 20-26 Sardinia Street, W.C., of their Velox papers and Developing Cartols. After considerable experience with this paper we have found the results satisfactory, especially that obtained on the "Special Portrait" paper. Users of this paper must, however, be warned against the tendency of developing and changing the paper in too strong a white light. We are convinced many of the failures are due to this want of care, and strongly recommend that all manipulations should be carried out in a pale yellow light. It is just as easy to work in, and the results are far more satisfactory. Now that the long winter evenings have arrived, this paper should prove of great value. We have just been making some trials with the same firm's new P.O.P. paper "Carbona," and we are very pleased with the resulting prints, especially those on the matt paper, toned with chloro-platinite of potassium. This is the toning bath recommended by Messrs. Griffin, but we have found the ordinary ammonsulpho-cyanide and gold toning bath answer perfectly. For a rich red colour all that is necessary is to fix the print in a weak hyposulphite of soda solution, made by dissolving 1 ounce of soda in 8 ounces of water.

PHOTOGRAPHIC CHRISTMAS CARDS.—We were shown a very pretty selection of Christmas cards the other day at one of the wholesale houses, and these in conjunction with the Carbona paper, form a very pleasing and novel way of sending the Compliments of the Season to our friends. In most designs there is a space left on the card where the photograph is to be slipped in, and if the view has been judiciously selected the whole effect is very pretty. For making these prints we think there is not anything more suitable than the New Double-Weight Carbon Velox, a paper that is very thick, and therefore does not require backing or mounting.

MICROSCOPY

CONDUCTED BY F. SHILLINGTON SCALES, F.R.M.S.

ROYAL MICROSCOPICAL SOCIETY.—The first meeting of the Royal Microscopical Society for the winter session took place on October 17, Mr. Carruthers, F.R.S., in the chair. Dr. Hebb showed samples of stains for microscopic purposes prepared in solid form by Messrs. Burroughs. Wellcome & Co. These stains were noticed in this journal (vol. vi. p. 247) when they were first offered for sale. Messrs. R. & J. Beck exhibited their new "London" microscope, noticed in this journal last month (*ante*, p. 184). Mr. F. W. Watson Baker gave an exhibition of slides and models illustrating the structure and development of skin. Mr. Vesey said the Society was greatly indebted to Mr. Watson Baker for giving this very excellent exhibition at comparatively short notice. Mr. Karop said he had only been able to glance at a few of the specimens exhibited, and he regretted there was no one present to discuss the subject, because several new points had recently been recognised by histologists in the structure of the skin, and it was rather a pity that the opportunity should be lost of having these demonstrated by some one who had made a study of this important and complicated tissue system. Mr. Vesey said that since the last meeting the Society had lost by death a Fellow very well known to many—Mr. Richard Smith. He had devoted his attention to the study of diatoms, and was continually devising new contrivances for use in connection with the microscope. He had likewise undertaken important investigations in the germination of wheat, and had made a large number of observations and experiments in connection with the subject, and published a book relating to it. He would probably be best known as the inventor and patentee of Hovis flour. The President said he regretted to have to announce that the Society had also recently lost by death several other Fellows, one of whom, Mr. Edward George, was personally known to him. He had prepared a short memoir of Mr. George, which he would read to the meeting. The secretary announced that Mr. Millett had forwarded part ix. of his Report on the Foraminifera of the Malay Archipelago, which was printed in the October number of the Journal.

CAUSES OF FRACTURE OF STEEL RAILS.—The value of the microscopical examination of steel will be brought prominently before the public by the recently issued report of the Board of Trade committee appointed to examine into the cause of fracture of a steel rail at St. Neots station on December 10th, 1895, by which a serious accident happened to the down Scotch express. The report itself, dealing as it does with various experimental work undertaken by well-known experts, is somewhat inconclusive, but the microscopic examination by Sir William Roberts-Austen gave results of the utmost interest and value. Briefly stated, it may be said that, according to this eminent authority,

good rail steel consists of "ferrite," or iron free from carbon, and "pearlite," which is a mixture of alternate bands of ferrite and "cementite," the carbide corresponding to the formula Fe_4C. Well-developed pearlite with a conspicuous banded structure is readily shown microscopically, and is characteristic of good rail steel. When, however, steel is hardened by "quenching," pearlite is absent, and "martensite," which consists of interlacing crystalline fibres without banded structure, takes its place. Sir William Roberts-Austen says that "the presence of martensite in a rail should at once cause it to be viewed with extreme suspicion, as showing that the rail is too hard locally to be safe in use." The broken rail at St. Neots showed an outside layer of martensite one hundredth of an inch thick. The report deals further with minute cracks found in this and other rails, and the enormous increase in liability to fracture occasioned thereby, and one conclusion drawn is that patches of martensite can be produced in a rail, when in use, by local treating caused by skidding, followed by the rapid extraction of heat by the cold rail. It is thus evident that the microscope will prove to be an increasingly valuable means of studying the complex structure of steel. For this purpose and for the examination of alloys it is used, and already a quite voluminous literature is growing up around the subject.

NUMBER OF SPECIES OF PLANTS.—Professor S. H. Vines, in his opening address to the Botanical Section of the British Association at Bradford, gave some interesting figures as to the number of species of plants at present known. The figures may be tabulated as follows:—

Phanerogams	Dicotyledons	78,200		
	Monocotyledons	19,600		
	Gymnosperms	2,420		
		100,420		
	Subsequent additions	5,011		105,231
Pteridophyta	Filicinae (including Isoetes) about	3,000		
	Lycopodinae, about	432		
	Equisitinae, about	20		3,452
Bryophyta	Musci	4,609		
	Hepaticae	3,041		7,650
Thallophyta	Fungi (including Bacteria)	39,663		
	Lichens	5,600		
	Algae (including 6,000 Diatoms)	14,000		59,263
Making a grand total of				175,596

which, when compared with the 10,000 species of plants known to Linnaeus in the latter half of the last century, show how vast have been our additions to the knowledge of plants. The amount of work for microscopists, especially in the latter sections, appears to be unlimited.

W. WATSON & SONS' NEW CATALOGUE.—Messrs. W. Watson & Sons, of High Holborn, have sent us a copy of their new catalogue, which seems to contain almost everything necessary to the working microscopist, and will be useful to all who are interested in microscopy. The catalogue is a distinct improvement upon its predecessors. It contains detailed descriptions, and illustrations of the details, working parts, and modes of fitment and adjustment of the various microscopes made by the firm, ranging from the well-known and beautiful "Van Heurck" microscope to the modest but efficient "Fram." The most popular of Messrs. Watson's microscopes appears to be the "Edinburgh" stand, made in many forms of various completeness, and we question if there is a more popular stand upon the market. The newer stands are the "Royal"—an "Edinburgh" microscope in its completest form, with large tube—the "Circuit Stage Van Heurck" (see SCIENCE-GOSSIP, vol. vi. p. 57), and the "Fram" and "School" microscopes. The latter is a perfectly-made stand, with excellent coarse, but no fine adjustment, sold at the extraordinary price of £2 7s. 6d. We are glad to see that one or two older and out-of-date stands no longer find a place in the catalogue. Of most interest, perhaps, are the new Holoscopic series of objectives, eyepieces, and condensers. We spoke in the highest terms of these objectives when they first appeared (see SCIENCE-GOSSIP, vol. vi. pp. 183 and 313), especially of the half-inch of N.A. ·65, which we considered an important advance upon achromatic objectives hitherto obtainable, and we note that the list now includes an inch of N.A. ·30, a quarter-inch of N.A. ·95, an eighth-inch oil-immersion of N.A. 1·35, and a twelfth-inch oil-immersion of N.A. 1·4. Our readers will observe that these apertures equal those of the costly apochromatics. We hope to speak again of these objectives shortly. The holoscopic eyepieces, which can be used with both achromatic and apochromatic objectives, were also duly noticed in this journal (SCIENCE-GOSSIP, vol. vi, p. 183), as was also the holoscopic immersion condenser of N. A. 1·35 (SCIENCE-GOSSIP, vol. vi. p. 57). No less than seven different forms of photo-micrographic cameras are listed, including a new combined vertical and horizontal camera. The whole catalogue is profusely illustrated, and contains no less than 130 pages.

PRESERVATION OF MEDUSAE.— The "Journal of Applied Microscopy" says that medusae may be killed by adding a few drops of concentrated chromic acid to sea-water containing them. Then well wash in sea-water until the chromic acid has disappeared. Gradually add glycerine and alcohol to water, until objects are in pure glycerine and alcohol of same specific gravity as sea-water.

ANSWERS TO CORRESPONDENTS.

G. G. B. (Oldham).—For wood sections we prefer Delafield's Haematoxylin. Squire's formula is as follows :—To 400 cc. of a saturated aqueous solution of ammonia alum add 4 grams of haematoxylin dissolved in 25 cc. of absolute alcohol ; leave the solution exposed to the light and air in a stoppered bottle for three or four days ; filter, and add to the filtrate 100 cc. of glycerine and 100 cc. of methylic alcohol (wood spirit) ; allow the solution to stand in the light until it is a dark colour, re-filter and preserve in a stoppered bottle. If you require a red stain, however, we would suggest a 1 per cent. solution of eosin in alcohol. Immerse in this for ten minutes, wash in methylated spirit, clear in clove oil, and mount in Canada balsam. Borax carmine is generally used as a double stain with acid aniline green. We gave a note on preparing and mounting wood sections in SCIENCE-GOSSIP, vol. vi. p. 214. See a note on staining by Dr. P. Q. Keegan on p. 60 of the present volume ; also our remarks on staining in the present number in " Microscopy for Beginners." F. S. S.

EXTRACTS FROM POSTAL MICROSCOPICAL SOCIETY'S NOTE-BOOKS.

(*Continued from page* 185.[1])

NOTES BY WM. H. BURBIDGE.

Polyp of Alcyonium palmatum (fig. 6). One of the Anthozoa, is of a higher organisation than Hydroida. It is the cream-coloured, fleshy substance commonly called dead men's fingers. The protruded polyp is an elevated tubular column of translucent substance terminating in an expanded flower of eight slender-pointed petals—the tentacles of the polyp. "The spicules in this creature are of interest, being of varying forms ([2]).

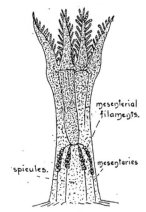

FIG. 6. *Polyp of Alcyonium.*

In *Alcyonium* the sexes are separate, and even the sexes of different colonies are distinct. In any one commonwealth the individuals are either all males or else all females. The ova and sperm masses are borne on stalked capsules upon the free edges of the mesenteries, or straight bands that run down the tube below the curled-up filaments, and development takes place outside the parent. The embryos are free, swimming by cilia. They soon fix themselves, and by continued budding produce colonies " (Hornell). *Stalked larva of Antedon* (fig. 7), better known by the name of *Comatula rosacea,* or "feather-star." Mr. Hornell, in his "Journal of Marine Zoology," describes the delight with which he first pulled up, on a lobster-pot, a colony of this most lovely of star-fishes. I can also recall a red-letter day long ago when I pulled up in a dredge a mass of these beauties in Torbay, one of the greatest prizes, I think, round our English coast. "Its body consists of a disc some half inch across, from which proceed ten long slender arms bearing numerous pinnules on either side. These often reach 3½ inches. so that

(1) These extracts were commenced in the September number, 1900, at p. 119.
(2) The spicules of *Alcyonium* and *Gorgonia* make beautiful objects for polarizer and analyzer.—ED. *Microscopy.*

the creature has a span of 7 inches. The sexes are separate, and the genital organs are situated not in the body disc, but in the tiny pinnules of the arms. The fertilised ova are set free as barrel-shaped embryos which acquire four encircling bands of cilia. Next appear a few minute calcareous plates within this embryo, forming, as it were, a tiny cask set upon a tiny stalk. Free-swimming life being now almost ended, a disc containing a perforated plate appears on the lower extremity of the stalk; and by this, attachment is made to any object that happens to be in the way. The soft, barrel-shaped mass of the swimming larva has now shrunk and adapted itself to the form of the enclosed calcareous skeleton, and the creature is fairly launched upon the stalked and anchored period of its life. In this stage the skeleton is made up of a basal plate, rooting the animal to its host, a considerable number of joints set end to end forming a stalk upon which is seated the cup-shaped framework of the body, consisting of two circles of large perforated plates,

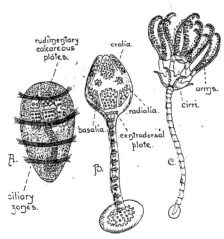

FIG. 7. Stalked larva of Antedon.

respectively the "basals" and the "orals." The former form the base of the cup, and the latter the upper ones. Growth after this is rapid; other circles of plates appear, the ten arms proceed from one of the circles, the top joint enlarges into a plate-like structure and develops claw-like jointed organs, the cirri. The body breaks off from its stalk and becomes free to creep among the rocks at will, or swim gracefully with rhythmic beats of its long feather-like arms. Special interest attaches to this beautiful creature from the great part played by its relations, if not its ancestors, which lived during former periods of the earth's history, for the Encrinites, whose remains contributed so greatly to build up the huge masses of our mountain limestones, were but gigantic Pentacrinoids of structure practically identical with the stalked larva of *Antedon* (Hornell). Dr. Carpenter's "Microscope" has a good plate of the rosy feather-star. My remarks have been largely taken from "Gosse" and from Cassell's "Natural History," also from Hornell's "Journal of Marine Zoology" [3].

MICROSCOPY FOR BEGINNERS.

BY F. SHILLINGTON SCALES, F.R.M.S.

(*Continued from p. 186.*)

Objects to be mounted in glycerine jelly or any similar medium can be transferred directly from water, but those to be mounted in Canada balsam must be first thoroughly dehydrated, and this is done by transference to one or two baths of methylated spirit, alcohol having a strong tendency to absorb water. In some cases—as, for instance, with insects that require arranging—the object should be arranged between two slides, tied together with two slips of visiting, or thicker, cardboard between the ends, and the whole immersed bodily for some hours in the methylated spirit. The spirit has a tendency to harden the structure, so it is necessary to do the arranging in this way beforehand. The object may then be "cleared" of alcohol by transference to clove oil, and thence to a final soaking in turpentine; but in most cases the clove oil may be omitted, and the object simply transferred from the methylated spirit to turpentine. From the turpentine the mounting in Canada balsam may be proceeded with, as already explained in detail. (See SCIENCE-GOSSIP, vol. vii. pp. 122 and 154.) We would again lay stress upon the fact that whilst the turpentine stage always immediately precedes mounting in Canada balsam, and the object must be freed from every trace of water, the reverse is the case when mounting in glycerine jelly or similar media, as the object must then receive its final soak in water, and be free from every trace of turpentine, etc.

The great value of staining is not to make "pretty" objects for the microscopic cabinet—beautiful as the effects often are—but to differentiate the structure. This has now become a high art, and new methods for special purposes are being constantly introduced, especially in histology. With these the beginner has nothing to do, though the time may come, when he is no longer a beginner, when he can refer to larger or special works dealing with the subject, around which so voluminous a literature has already grown. Of all stains, haematoxylin (the active principle of logwood) is the most generally useful, especially for vegetable sections. It is best purchased from the optician's in an alcoholic solution, and improves greatly with keeping. Before staining, many objects, such as vegetable sections, may require bleaching. Steeping in alcohol will generally have this effect; but a solution of chlorinated soda is an excellent bleaching agent, without being too powerful. It is made as follows: One ounce

(3) With regard to the drawings with which we have endeavoured to illustrate the foregoing notes, we may say that figs. 1, 2, 4 and 5 are after figures in Mr. Hornell's most interesting and well-illustrated "Journal of Marine Zoology," now unfortunately in abeyance; fig. 3 is after one in Nicholson's "Manual of Zoology"; fig. 6 is drawn from a slide in our possession, and fig. 7 is after a figure in Claus and Sedgwick's "Text-book of Zoology."—ED. Microscopy.

of dry bleaching powder is dissolved in a half-pint tumblerful of water; two ounces of washing soda are dissolved in another tumblerful of water; after which the two solutions are mixed together, well shaken, and allowed to settle for twenty-four hours or so. The clear fluid is then carefully decanted, filtered, and preserved in a stoppered bottle, away from the light. Sections or objects to be bleached are first soaked in water, and then transferred to a small quantity of the chlorinated soda for a period of time varying from one to a dozen hours or more. They must afterwards be very thoroughly soaked or washed in several changes of clean water until every trace of soda is removed.

The process of staining is as follows: Ten to thirty drops of the stain, according to the requirements of the section or object, are added to an ounce of distilled water, and the section is removed from water to this stain, where it is allowed to stand from ten minutes to half an hour, or even longer. After washing in distilled water it is generally recommended that the section be washed in ordinary *hard* tap-water with a view to deepening the stain. We do not ourselves think this necessary; but if it is done, and if the tap-water should not be sufficiently hard, about ten grains of bicarbonate of soda added to a pint of distilled water will serve the same purpose. The section must then be promptly dehydrated by ten minutes' or more soaking in methylated spirit, cleared in clove oil until it sinks to the bottom, transferred to turpentine, and mounted in Canada balsam.

Overstaining with haematoxylin may be remedied by soaking a few minutes in a ¼ per cent. solution of glacial acetic acid in distilled water, or in a solution made up of one part of 1 per cent. hydrochloric acid in distilled water to two parts of absolute alcohol. This also rectifies overstaining with carmine.

Eosin is likewise a useful stain, and may also be used for double staining together with haematoxylin. For double staining the section should be stained with haematoxylin as above, then transferred to dilute acetic acid in distilled water. After carefully washing away the acid, first with distilled and then with tap water, stain for five minutes or more in a 1 per cent. solution of eosin in alcohol, wash well in methylated spirit, clear in clove oil, and mount in Canada balsam.

Borax carmine and acid aniline green are useful for double staining vegetable sections. Mr. Cole's method is frequently used and quoted, and is as follows: The green stain is made up of two grains of acid aniline green dissolved in a mixture of one ounce of glycerine and three ounces of distilled water. The carmine stain is made by dissolving ten grains of borax in one ounce of distilled water, and adding half an ounce of glycerine and half an ounce of absolute alcohol. Another solution is then made of ten grains of carmine dissolved in twenty minims of ammonia and thirty minims of distilled water. The two solutions are then mixed together and filtered. The process of staining is to place the section in the green stain for five to ten minutes; wash in water; place in the carmine solution for ten to fifteen minutes; wash well in methylated spirit; dehydrate and clear in clove oil; wash in turpentine, and mount in Canada balsam.

The higher branches of section-cutting will be beyond the necessities of the beginner; but plant sections of the simpler sort will be well within his powers. Very fair sections can be cut with a razor by holding the specimen between the finger and thumb. The finger is held horizontally so as to form a rest for the razor, the cut is made towards the operator, and the razor is drawn *through* the object with a diagonal drawing cut. A very simple little hand microtome can be bought, however, for five shillings and upwards. It contains a tube in which the object to be cut is wedged between two pieces of cork, pith, or carrot, or by imbedding in paraffin. At the bottom of the tube is a fine screw which raises the object as required after each cut, and at the top of the tube is a circular flange of brass or glass to serve as a guide for the razor. The Cathcart microtome is, perhaps, the most popular form, and is fitted with two parallel glass runners, whilst the tube is provided with a clamp. This is almost a necessity for paraffin imbedding, as there is otherwise a tendency to slip in the tube. The cheaper form, arranged for imbedding only, would cost fifteen shillings. The short lengths of tubes provided with the instrument as moulds for paraffin blocks are, however, inconvenient. A much better plan is to make little circular boxes of stout paper or thin cardboard, or better still to use a pair of brass L-shaped moulds, which are sold for the purpose. It is as well to buy the paraffin, which should have a melting-point of from 45° to 52° F., according to the temperature of the room in which it is to be used. If the paraffin is too soft the sections will wrinkle, owing to lack of cohesion, whilst if it is too hard they will roll up into tiny rolls. The best way to proceed is to cut a small slice of the specimen about a quarter of an inch thick, dry it, melt the paraffin over a water-bath, dip the section in the paraffin to give it a coat, hold it in the mould in the requisite position for the desired cut, and run in the paraffin around and over it, and allow to cool. As the paraffin will shrink somewhat, especially at the top, the specimen should be well covered. When quite cool the block can be removed from the mould and trimmed to the shape necessary to go into the microtome tube. With microtomes not fitted with a clamp the tube itself must be used as the mould. The razor used for cutting must be kept well sharpened, and constantly wet by being dipped into a saucer full of methylated spirit, the sections being floated off as they are cut. The paraffin is then removed by means of naphtha, benzole, or turpentine, the turpentine washed off with alcohol, and the section placed in absolute alcohol or water, as the case may be. If it requires staining, this may be done either on the cover-glass or the slide, the method having been already explained. The subsequent washing can be done with a small wash-bottle containing water or 75 per cent. alcohol, as required. Dehydrate with methylated spirit or alcohol of increasing strength, clear with clove oil, and mount in Canada balsam. Many sections are preferably stained in bulk before imbedding. More complicated processes, requiring special apparatus and knowledge, such as infiltration or freezing, are better left until skill and experience are gained in the simpler processes.

Spoilt slides, failures, etc., are best immersed in a strong, hot solution of Hudson's soap, after which the slides and covers are well washed with warm water, rinsed with methylated spirit, and polished with an old handkerchief.

(To be concluded.)

LEPIDOPTERA NEAR PORTSMOUTH.—Butterflies and moths have been unusually abundant here during the past season, and the following are a few of the more important species which I have observed this year within the town of Portsmouth and its suburbs of Southsea, Landport, etc. *Lycaena argiolus*, fairly common in 1900. I have not before observed this species here. *Colias hyale*, common. 1900, especially so on the Ports-down hills to the north of the town. Has not been previously observed for many years. *Colias edusa*, a good many seen, but not so common as in 1898 and 1899. *C. edusa* var. *helice*, one specimen caught 1900. *Acherontia atropos*, an imago obtained November 2nd, 1900, in one of the densest parts of the town. *Choerocampa elpenor*, several seen in 1899, but not found this year. *Smerinthus ocellatus*, scarce, but several larvae obtained. *S. populi*, fairly common, larvae frequent on poplar. *Macroglossa stellatarum*, several seen in 1899 and 1900. *Euchelia jacobaeae*, *Spilosoma lubricipeda*, *S. menthastri*, *S. mendica*, and *Arctia caia*, all common, the first three abundant. *Porthesia chrysorrhoea*, scarce, one imago in 1900. *P. similis*, imagos and larvae both common. *Leucoma salicis*, caterpillar and adult very common. *Orgyia antiqua* and *Bombyx neustria* abundant. *Eriogaster lanestris* and *Odonestis potatoria*, fairly common. *Dicranura vinula*, very numerous, especially the larvae. The latter frequently exhibit some curious variations in markings. The extension of the dorsal band on the eighth segment, above the proleg, is sometimes divided, so as to form a separate spot 25 mm. in diameter, purple-brown in colour and surrounded by a white border. This spot is occasionally on both sides of the body, but more often occurs on the left only. One specimen I had this year developed a large number of small dark spots several days before changing to the violet colour which caterpillars of this species assume before pupation. This may be a case of reversion to a primitive spotted type. As regards the exsertible filaments, I am almost convinced they are for the purpose of warding off in some manner the attacks of ichneumon flies; the more so as the easiest way to make the creature protrude them is by touching it on the back. *Notodonta dictaeoides*, rare, one in 1899. *Phalera bucephala*, abundant. *Geometra papilionaria*, scarce, one imago 1900. *Uropteryx sambucaria*, also rare, two in 1899. *Abraxas grossulariata* is not very variable here. The larvae are numerous on *Euonymus europaeus*, but quite two-thirds are destroyed by the grub of a species of *Tipula*. *Cossus ligniperda*, larvae frequent in trunks of the poplars. The above list shows that some interesting species of butterflies and moths can be found within the limits of even a large town. Very many of the commoner kinds are also to be obtained in gardens here, and indeed Portsmouth seems particularly favoured by lepidoptera. This is doubtless owing to its mild climate and the protection afforded on the north by the Portsdowns, and on the south from the storms of the Channel by the Isle of Wight.—*R. J. Hughes, Norman Court, Southsea.*

COLIAS EDUSA IN AUTUMN.—On October 29th this year I saw a freshly-emerged specimen of the clouded-yellow butterfly on a railway bank in Surrey. A few years ago I took a specimen of this interesting butterfly near Brighton as late as November 9th. These dates, though late for England, are by no means unusual in Southern Europe, where *Colias edusa* may be seen until after Christmas, as it there hibernates in the imago stage. We have, however, not any evidence that this species passes the winter in this country, in the perfect state.—*John T. Carrington.*

MONOCHROMATIC VISION.—A test has been made by Sir W. de W. Abney upon a patient possessing only monochromatic vision. All colours were matched with white by him with the same facility as if they were white. The patient's curve of luminosity agrees with and is practically identical with those obtained by the normal eye, when it measures a spectrum of very feeble luminosity.—*James Quick, 3 Hilboro' Place, Cork.*

GOLDEN ROD.—Can any of the readers of SCIENCE-GOSSIP kindly give me information as to the woods in the South of England in which *Solidago virgaurea* is to be met with in great profusion? In the London district one comes across a few plants here and there, and I have seen it in little colonies in various places on banks and roadsides both in England and Scotland; but I believe that in certain woods in the South of England it grows in dense masses and in enormous abundance. I am studying the distribution of our wild flowers, and endeavouring to obtain photographs of the more striking species in their natural surroundings; so I should be very glad to get a photograph of the *Solidago* in its headquarters. If any reader can help me in this matter I should be very much obliged.—*C. H. Jones, Putney, London, S. W.*

ELECTRICAL FIRE-PUMPS ON BOARD SHIP.—The cadet ship "Conway" has been fitted with an electrically controlled fire-extinguishing apparatus. The pumps are driven by electric motors and are capable of producing a very powerful jet of water. The motors are supplied with electricity from a storage battery charged by the ship's dynamos, and the capacity of the battery is sufficient to work the pumps seven or eight hours at full speed.

VALUE OF AMERICAN COAL.—The experiments which have been made with regard to the possible substitution of American for British coal in the English markets are not likely to have any great effect upon the production and winning of coal in Britain. The South Metropolitan Gas Company finds that American coal costs them a third as much again as English coal; and although the former is preferable in some ways, its benefits are not a sufficient set-off against the higher cost. It would not seem at present that there is likely to be any real American competition. The *Engineering Magazine* gives the following as the average cost of coal per ton at the pit-head in the countries mentioned:—United States, 4s. 9¼d.; Great Britain, 5s. 10½d.; Germany, 6s. 11d.; Belgium, 7s. 7d.; France, 8s. 8d.; New South Wales, 5s. 9d.; New Zealand, 10s.; Japan, 5s.—*E. A. Martin.*

TWISTED TREES.—Many trees, as they grow to maturity and enlarge their roots, become more or less ribbed and cracked about the base of the trunk, such ribs and cracks often extending as much as three or even six feet upwards. It is a curious fact, and one which seems to have been hitherto unnoticed, that this ribbing and cracking often shows a spiral twist, the ascending lines sloping sometimes from left to right, and sometimes from right to left; but always in the same direction in the same species of tree. The tree in which this twisting is most clearly marked and most frequently seen is the horse-chestnut. In any half-dozen of these trees the twist is pretty sure to be observable in at least one of them, and, tracing it from the base upwards, the lines will always run from left to right. A twist in the same direction has been observed in the sycamore, the yew, the pear, and the apple; and in the contrary direction in the holly, the hawthorn, the birch, the beech, the plum, and the willow. The oak, the ash, and the elm seem not to exhibit the twist at all, or very rarely. I have not seen, nor have I been able to discover, any explanation of the cause of this curious phenomenon. The law of the "genetic spiral," according to which the growing points of all plants build themselves up in a spiral manner, may have something to do with it; but the fact that the twist in the stem is not shown by any young trees makes it difficult to understand the connection with this law. The uniform direction of the twist, whenever it occurs, in the same species of tree, entirely negatives the suggestion that it is due to wind or any other accidental cause. It appears to be as clearly a specific character as the form of the leaf or the flower; but if this is the case, why is it not exhibited by every individual? I am not sure whether soil or situation has anything to do with it? I have found it more common in some districts than in others. Here is one of Nature's puzzles that will, no doubt, be explained some day. When the attention of botanists has been fixed on any practical question it is never left until a satisfactory solution has been found. All persons who take country walks, and who care to use their eyes, will find much interest in looking out for the twist in the trunks of the trees and in making a record of the species in which they observe it, and of the direction of the spiral. Can any readers of SCIENCE-GOSSIP give records through its columns? —*F. T. Mott, Birstal Hill, near Leicester.*

GREEN VARIETY OF ELDER FRUIT.—I have had sent to me by a lady in North Derbyshire the leaves and berries of a curious variety of the common elder-tree which occurs there. The leaf, and in fact every other part of the tree except the berry, is exactly as in the ordinary form; but the berries are green instead of black, and continue that colour till they fall off or are eaten by the birds. I understand that there are six or seven full-grown trees in a row by a mill near Stoney Middleton, about three miles from Hassop Station on the Midland Railway. There is one tree in the middle of the row which bears the ordinary black fruit, as do all the others in the neighbourhood. I should be glad to know whether this is an unusual occurrence.—*(Rev.) Chas. F. Thornewill, Calverhall Vicarage, Whitchurch, Salop.*

FASCIATED VIOLET.—I enclose a flower of the cultivated variety, "The Czar," of violet, the like of which I have not met with previously. You will observe that the stem is fasciated, and that the two blooms are intermixed.—*(Rev.) W. W. Flemyng, Coolfin, co. Waterford.*

A CUNNING SPIDER.—Whilst recently searching for orthoptera under the bark of old elm-trees at Lonesome, near Mitcham, in Surrey, I was much interested to notice the antics of a spider, *Segestria senoculata* Linn. This creature, with the characteristic cunning of the Araneidea, had taken advantage of the peculiarities of the earwigs with which the tree was infested in order to obtain its sustenance. In the tubular burrows under the bark, the work of wood-boring larvae, the spider had spun narrow silken tubes in the direction of the burrows. As soon as an earwig on its peregrinations entered the tube the spider menaced the creature from the opposite end. According to its usual custom the earwig retreated, at the same time elevating its pincers, with the result that further retreat was summarily stopped by the forceps penetrating the side of the tube, thus rendering the victim helpless.—*John E. S. Dallas, 19 Ulverscroft Road, East Dulwich, S.E.*

ABNORMAL PEARS.—I thought that the enclosed pears might be of interest to some of your readers if described in SCIENCE-GOSSIP, as they are decidedly curious, having almost the exact form of

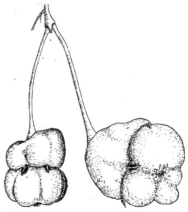

ABNORMAL PEARS.

small "cottage" loaves of bread. They were grown upon a tree trained against the wall in Mr. A. Allcock's garden at Knowle Hill, Evesham. There being two similar ones upon the same bunch looks as if their form was due to a sport rather than to any injury in the early stage.—*T. E. Doeg, Evesham: September 29th, 1900.*

ABNORMAL SPARROW.—Is it not rare to meet with a sparrow possessed of individuality? Here in Lucerne I have seen one for the last two months. He is entirely white, and has only a very moderate appetite for crumbs, seldom coming to the balconies where his brown relations feast, and to all appearance they do not drive him away.—*M. Drummond, Neu Schweizerhaus, Lucerne: October 13th, 1900.*

ASTRONOMY

CONDUCTED BY F. C. DENNETT.

		Rises.	Sets.	Position at Noon.	
				R.A.	Dec.
1900	Dec.	h.m.	h.m.	h.m.	° '
Sun	6	7.51 a.m.	3.51 p.m.	16.50	22.29 S.
	16	8.2 a.m.	3.49 p.m.	17.34	23.19 S.
	26	8,7 a.m.	3.54 p.m.	18.19	23.23 S,

		Rises.	Souths.	Sets.	Age at Noon.	
Dec.		h.m.	h.m.	h.m.	d.	h.m.
Moon	6	4.9 p.m.	—	7.40 a.m.	14	4.43
	16	2.13 a.m.	7.29 a.m.	0.36 p.m.	24	4.43
	26	10.18 a.m	3.53 p.m.	9.40 p.m.	4	11.59

		Souths.	Semi-	Position at Noon.	
			diameter.	R.A.	Dec.
Dec.		h.m.	''	h.m.	° '
Mercury	6	10.27 a.m.	3·4''	15.26	16.14 S.
	16	10.35 a.m.	2·8''	16.14	19.49 S.
	26	10.56 a.m.	2·5''	17.14	22.57 S.
Venus	6	9.31 a.m.	6·7''	14.30	12.45 S.
	16	9.40 a.m.	6·4''	15.18	16.28 S.
	26	9.51 a.m.	6·1''	16.9	19.30 S.
Mars	16	5.6 a.m.	4·4''	10.44	10.59 N.
Jupiter	16	11.48 a.m.	14·5''	17.27	22.54 S.
Saturn	16	0.47 p.m.	7·0''	18.26	22.42 S.
Uranus	16	11.9 a.m.	1·8''	16.48	22.25 S.
Neptune	16	0.14 a.m.	1·2''	5.51	22.11 N.

MOON'S PHASES.

		h.m.			h.m.
Full	Dec. 6	10.38 a.m.	3rd Qr.	Dec. 13	10.42 p.m.
New	„ 22	0.1 a.m.	1st Qr.	„ 29	1.48 a.m.

In perigee December 3rd at 8 p.m.; in apogee on 15th at 1 p.m.; and in perigee again at 4 p.m. on December 30th.

METEORS.

				h.m.		°
Dec. 1–14	Geminids	Radiant R.A.	7.12	Dec.	33 N.
„ 7–10	a Germinids	„	„	7.56	„	32 N.
„ 22–29	Canes Venaticids	„	„	12.56	„	32 N.

CONJUNCTIONS OF PLANETS WITH THE MOON.

					° '
Dec. 13	Mars	1 a.m.	Planet	8.26 N.
„ 19	Venus	7 a.m.	„	2.19 N.
„ 20	Mercury*	0 noon	„	0.2 N.
„ 21	Jupiter*	1 p.m.	„	1.38 S.
„ 22	Saturn*	1 p.m.	„	2.24 S.

* Daylight.

OCCULTATIONS AND NEAR APPROACHES.

		Magni-	Dis-	Angle from	Re-	Angle from
Dec.	Star.	tude.	appears.	Vertex.	appears.	Vertex.
			h.m.	°	h.m.	°
5	ω°Tauri	4·6	6.11 p.m.	67	6.46 p.m.	353
7	ζ „	3·0	3.27 a.m.	330	Near approach.	
10	κCancri	5·0	8.56 p.m.	148	9.52 p.m.	317
31	ρ²Arietis	5·5	7.22 p.m.	355	Near approach.	

THE SUN should be watched for any outbreaks of activity.

MERCURY is a morning star all the month, reaching its greatest elongation west, 20° 50', at 3 a.m. on December 8th. At the beginning of the month it is well placed for observation in Libra. During the first third of the month it rises about two hours before the sun, at 5.50 a.m. on December 6th. Mercury is in conjunction with Uranus

at 3 p.m. on December 22nd, being 34' north; on 30th at 4 p.m. it is in conjunction with and 43' south of Jupiter.

VENUS is also a morning star, decreasing in apparent diameter. rising some time before Mercury. On December 24th and 25th it will be travelling just north of β Scorpii.

MARS is in Leo, rising at 10.34 p.m. on December 1st and at 9.21 on 31st. Its path is from a point just north of ρ Leonis to nearly north of χ Leonis. Its diameter increases to 10'' by the end of the month.

JUPITER, SATURN, and URANUS are all too close to the sun for observation, coming into conjunction with the sun on December 14th at 9 a.m., 29th at 1 p.m., and 5th at 7 a.m. respectively.

NEPTUNE comes into opposition at 4 a.m. on December 20th, and is therefore at its best for observation. It retrogrades through a short path about 54' in length. The nearest bright star is η Geminorum, about 4° 30' east of the planet. Its apparent stellar magnitude is only about 8th magnitude.

BRUSSELS OBSERVATORY.—M. Lagrange, having resigned the directorship of the astronomical work, has been succeeded by Lieutenant C. Lecointe, who was second in command of the Belgian Antarctic Expedition.

LICK OBSERVATORY.—Professor W. W. Campbell has been appointed Acting Director, in place of the late Professor Keeler, until final arrangements can be made.

A BRILLIANT METEOR was seen on Sunday evening, October 21st, at 9.40, at places so far separated as Tewkesbury, Stafford, and London. At the first-mentioned place it was seen to pass through the north-western sky. Its colour was bluish white, but it left a trail of red. About three minutes later a sound like the boom of a heavy gun at a great distance was heard.

NEW NEBULA.—Whilst Mr. R. G. Aitken was watching Comet *b*, 1900, with the Lick telescope he found an object marked as a star of 9·5 magnitude in the "Durchmusterung" to be a planetary nebula with a stellar nucleus. The nebulous envelope has a diameter of some 5'' or 6''. Its position is R.A. 12 h. 29 m. 10 s., N., Declination, 83°21·8', and therefore in the constellation of Ursa Minor.

NEW MINOR PLANETS.—Professor Max Wolf, of Heidelberg, reports the discovery of seven more of these little bodies. One on September 26th, five on October 22nd, and another on the following night, which, however, may prove to be Irina (No. 177), discovered by M. Paul Henry in 1877.

MR. W. F. DENNING has, we are sorry to hear, been obliged to resign the directorship of the Meteoric section of the British Astronomical Association owing to the failure of his health. Mr. W. Besley is his successor.

THE "CAMBRIAN NATURAL OBSERVER" for October has come to hand. It contains a coloured plate of the great meteor of January 4th, 1900, by Mr. Norman Lattey, beside a number of interesting observations. It is the journal of the Astronomical Society of Wales.

DR. FERDINAND ANTON, the director of the astro-meteorological observatory at Trieste, died on October 3rd, in his fifty-seventh year.

GEOLOGY

CONDUCTED BY EDWARD A. MARTIN, F.G.S.

CHALK AND DRIFT AT RÜGEN.—The meetings of the Geological Society of London recommenced on November 7th with a joint paper by Professor T. G. Bonney and the Rev. E. Hill. Their subject was a reconsideration of the "Drifts on the Baltic Coast of Germany." Rügen had been revisited by them in 1838, and the paper was in continuation of their former one, which dealt with the deposits of Drift resting upon the Chalk of that island at Arkona, in Rügen. The Chalk apparently occurs on a sort of island in the Drift. At the well-known locality by the lighthouse it seems to overlie a Drift, but on closer examination the latter appears more probably to have filled a cavity excavated in the Chalk ; this apparent inlier of Drift probably being only a remnant of a much larger mass, so that it is likely this part of the coast nearly corresponds with a pre-Glacial Chalk cliff against which the Drift was deposited. Professor Bonney gave reasons to show that neither solution of the Chalk, nor ice-thrust, nor folding, nor even faulting, can satisfactorily explain the peculiar relations of the Drift and Chalk in Rügen.

FOSSIL BOTANY.—The feeling of melancholy that overspreads the mind on reading Professor Sollas's opening address to the geological section at the recent meeting of the British Association is largely dispelled when that portion of Professor Vines's address is perused which relates to Palaeophytology. "It cannot be said," he states, "that the study of Palaeobotany has as yet made clear the ancestry and the descent of our existing flora." There is no clue as to the origin of the Angiosperms, the pedigree of the Gymnosperms cannot be traced, and no forms of their hypothetical fern ancestry have as yet been recognised. The origin of the ferns is still quite unknown ; in fact he states, "we are not able to trace the ancestry of any one of the larger groups of plants." I, for one, have always felt extremely grateful that such is the case. Indeed, it was hardly necessary for Professor Vines to have reminded his audience that such was the case, for it is negatively confirmed by perusal of any British botanical text-book. That part of the occupation of our text-book compilers which consists in long stories about the ancestry or the evolution of the existing forms of life is certainly gone, so far as botany is concerned—to the eminent advantage of learning and science. Personally I have no more faith in the blood-relationship of plants depending upon common descent than I have in the blood-relationship of lead and gold depending on their pristine unity and amalgamation. How does it happen that those who insist so emphatically on the physical and mechanical nature and origin of the vital forces, should so persistently insist on the blood-relationship and propinquity of existing plants and animals? Therefore I trust that nothing but good can result from specially adverting to and calling attention in this brief note to Professor Vines's most excellent and admirable address.—*(Dr.) P. Q. Keegan, Patterdale, Westmorland.*

IODINE IN CUPRITE.—Mr. Arthur Dieseldorff' M.E., has discovered the existence of iodine in several examples of cuprite from New South Wales. The iodine is present only in non-payable commercial quantities. The total value of the world's production of iodine reached in 1896 the sum of £350,000.

PALAEOLITHIC MAN IN VALLEY OF THE WANDLE.—Referring to Mr. Roberts' note, in the last number of SCIENCE-GOSSIP, on the Thornton Heath implement, I naturally relied on the record in the paper quoted, having no reason to doubt its correctness. I have now seen the specimen, and find that it is a chipped flint celt, which could not, of course, have come from the gravel, as alleged by the labourer who sold it. Mr. Roberts has kindly shown me the implement of plateau type which he mentions, also an implement from the surface at Croham Hurst. I know of no other decisive evidence of the work of Palaeolithic man from the Valley of the Wandle.—*J. P. Johnson, 150 High Street, Sutton, Surrey.*

AN UNDERGROUND TEMPERATURE.—In connection with the deep sinking recently carried on at the Sydney Harbour Colliery, Balmain, opportunity has been taken to note the increase of underground temperature. Taking the mean annual temperature of Sydney to be 63° F., it is found that the temperature increases at the low rate of 1° F. for every 90½ feet. The Narrabeen shales, which crop out on the coast at Narrabeen, eight miles north of Manly, were met at practically the same depth at Sydney as at Cremorne Bore, three and three-quarter miles away.

GEOLOGISTS' ASSOCIATION. — A conversazione took place on November 2nd in the Library of University College, Gower Street, W.C. The exhibits were numerous and interesting. Amongst them were :—A series of minerals from Australia, by F. P. Mennell ; rocks from the Lake District, by A. C. Young ; erratics and "Eoliths" from the East Anglian drifts, etc., by A. E. Salter ; palaeolithic implements from Chard, by E. T. Newton ; geological photographs, by F. H. Teall, and a faulted slate from the Lake District, by J. J. H. Teall ; chalk fossils, by G. E. Dibley ; Miocene fossils from the Faluns of Touraine ; specimens from the Temple of Jupiter Serapis and vicinity, etc., by W. F. Gwinnell ; concretionary types from the magnesian limestone of Durham, by Dr. G. Abbott ; geological photographs, by H. W. Monckton.

THE GOLDSTONE MONOLITH.—An interesting monolith has just been disinterred at Goldstone Bottom, Hove, in the shape of the original and celebrated "Goldstone" or Druidic altar which stood from time immemorial in this well-known valley, but which was in 1883 deliberately buried. The stone is of an irregular, wedge-like shape, and measures about 14 feet by 9 feet, with a thickness of between 5 feet and 6 feet. The stone is described as an ironstone conglomerate, with veins of spar running through it, and when struck responds with a metallic ring. It is proposed to raise the stone on to a suitable base, and place it in the new park at Hove.—*E. A. Martin, F.G.S.*

CONDUCTED BY JAMES QUICK.

BECQUEREL RAYS.—An excellent résumé of our present knowledge on the subject of Becquerel rays is given by Oscar M. Stewart in the September issue of the " Physical Review." The whole history of the subject is well written, and the matter is subdivided so that the different properties of the rays may be treated separately.' The theory that Becquerel rays consist of extremely short ultra-violet light waves is now almost abandoned. In the case of the deflectable rays it is entirely dropped, and it is believed that they, like kathode rays, are negatively-charged particles of matter moving at extremely high velocity. Results from other experiments seem to show that the ratio of the atomic electrical charge to the mass of these small particles always remains the same, and that it is probably the velocity of the particle that varies to produce the different effects observed. The difference in penetration of Becquerel rays and kathode rays is then readily explained by a greater velocity for the Becquerel ray particle.

WIRELESS TELEGRAPHY.—One of the greatest difficulties in the way of the successful and universal working of wireless telegraphy methods is now apparently being surmounted by Marconi. Hitherto it has not been possible to so tune the receivers that they will only respond to waves of definite wave-lengths sent from the transmitter. The method employed by Marconi is the process of syntonising the receivers, each to a special wave-tone. The method apparently does not originate with him, but he appears to be the only experimenter who has given such a practical proof of the idea. He has modified his receiving and transmitting appliances so that they will only respond to each other when properly tuned to sympathy ; the isolation of the lines of communication can therefore be effectively carried out. These latest experiments were made between two stations thirty miles apart : one near Poole, in Dorset, and the other near St. Catherine's, in the Isle of Wight. The apparatus was so adjusted that each receiver at one station responded only to its corresponding transmitter at the other. Two operators at St. Catherine's were instructed to send simultaneously two different wireless messages to Poole, and without delay or mistake the two were correctly recorded and printed down at the same time in Morse signals on the tapes of the two corresponding receivers at Poole. These receivers were then placed one on top of the other and connected both to the same aërial wire hung from the vertical mast. Two messages were then sent from St. Catherine's, one in English and the other in French. Without failure each receiver at Poole rolled out its paper tape, the message in English perfect on one, and that in French on the other. Additional experiments were then made in which the vertical wire attached to the mast was replaced by a hollow

zinc cylinder. No difficulty was experienced in transmitting and receiving isolated messages under these conditions. This great improvement does away with the necessity of the cumbersome tall mast hitherto employed. From the above most successful results it will easily be seen what an immense stride has been made towards the complete practical solution of the problem of independent electric wave telegraphy, in which each wireless circuit can be made as private as one with a wire.

SIMPLE RELAY FOR SPACE TELEGRAPHY.—A simple relay for use in Hertzian wave and similar experiments has recently been devised. It is constructed somewhat similar to an electric bell. The electro-magnets are arranged horizontally, and the vibrator hangs vertically down by their side. The armature is a heavy one, and is also weighted. It is covered with tissue paper next to the magnet, and, to avoid adhesion as far as possible, the platinum contact-piece is so adjusted that it touches the insulated electrode before the armature touches the magnets. The armature is suspended by means of a German-silver spring, and its normal distance from the magnets is only one millimètre. Exemplifying its sensitiveness, it may be mentioned that the waves from a small Hertzian oscillator, consisting of two straight cylinders 5 cm. long and 1 cm. in diameter, can be detected over a length of thirty or forty mètres, and all the Hertzian reflection experiments can be made with it.

KEW OBSERVATORY.—Matters are still proceeding in reference to the apprehended magnetic disturbance at Kew due to the proposed electric tramways in that neighbourhood. Another conference took place recently between the Board of Trade and representatives of the Royal Society and the London United Tramways Company. The proceedings were private, but it is certainly to be desired that an agreement may soon be arrived at as to the means in which the Tramways Company shall arrange its circuits so as not to affect in any way the work conducted at Kew Observatory.

PHYSICAL SOCIETY OF LONDON.—At a meeting of the Physical Society on October 26th Dr. Shelford Bidwell exhibited some interesting experiments illustrating phenomena of vision. "Recurrent vision " was first illustrated. An illuminated vacuum tube rotating about a horizontal axis was seen to be followed, at an angle of about 40°, by a feeble ghost of itself. The same effect was observed in the cases of spots of white, green, and yellow light when these were rotated in a circular path. A spot of red light, however, had no ghost. These phenomena are due principally to the action of violet nerve-fibres. The non-achromatism of the eye was next shown. A six-rayed star, formed by cutting a hole in an opaque screen, was illuminated by a gauze-covered condenser containing an incandescent lamp. The star was somewhat clearly defined, and no coloured fringes were seen. A luminous haze was, however, observed, being brought about chiefly by the three brightest rays—orange, yellow, and green. If, therefore, these rays are obstructed, coloured fringes can be seen, due to chromatic aberration of the eye-lenses. This was done, and the general hue of the star was seen to be purple. Other experiments illustrating further phenomena of vision were also performed.

CONDUCTED BY HAROLD M. READ, F.C.S.

THE BIOLOGICAL DETECTION OF ARSENIC.—A novel method for the detection of arsenic is published in a recent issue of the "Zeitschrift für Hygiene." It has been noticed that when certain moulds are cultivated in the presence of arsenic, arsenical compounds, easily recognised by their characteristic garlic-like odour, are produced. Making use of this fact, Abel and Buttenberg take the suspected material, powder it, and add it to moist breadcrumbs. The mixture is sterilised in a flask and the contents inoculated with *Penicillium brevicaule* or *Aspergillus glaucus*. The flask is then capped and incubated in a hot-temperature incubator for two days. The presence of one-tenth of a milligram of arsenic can be detected by the garlic-like odour which is readily noticed on removing the cap.

ARTIFICIAL COLD.—The first of a course of lectures on "Methods of Producing Artificial Cold" was given at University College on November 9th by Dr. Hampson, the well-known authority on this subject. The struggle towards the absolute zero has been productive of such vast innovations from an economic point of view, and at the same time replete with such epoch-marking discoveries on the purely scientific side, that the chance of making a more exact acquaintance with the subject will, no doubt, be seized by those who have the opportunity. At the lecture referred to the main theoretical question involved in the expansion of a gas under pressure was discussed. The natural deductions that, while a gas expanding without doing work became cooler, the lowering of temperature of a gas doing "power-expansion," such as in working a piston, was infinitely greater, were well exemplified by an experiment in which air under a pressure of 130 atmospheres was allowed to escape into the surrounding air from the bottle in which it had been stored. We hope to again refer to the lectures from time to time.

THE CARBIDES OF NEO- AND PRASEO-DYMIUM. —M. Henri Moissan, in continuing his research on the metallic carbides, describes in a recent issue of the "Comptes Rendus" the crystalline carbides NeC_2 and PrC_2. One of the interesting features of the paper is the comparative result obtained on treating the carbides of the alkaline earths, of aluminium, and of the two metals under notice, with water. While aluminium carbide gives rise to the formation of methane, the carbides of the alkaline earths give acetylene; and M. Moissan now finds the neo- and praseo-dymium give a mixture of hydrocarbons in which, however, methane and acetylene predominate. This fully bears out the position of the two rare metals relatively to aluminium and the alkaline earths, for they had already been placed in the cerium group. The carbides of this group have the general formula RC_2, and give with water similar decomposition products.

CORRESPONDENCE.

WE have pleasure in inviting any readers who desire to raise discussions on scientific subjects, to address their letters to the Editor, at 110 Strand, London, W.C. Our only restriction will be, in case the correspondence exceeds the bounds of courtesy : which we trust is a matter of great improbability. These letters may be anonymous. In that case they must be accompanied by the full name and address of the writer, not for publication, but as an earnest of good faith. The Editor does not hold himself responsible for the opinions of the correspondents.—*Ed. S.-G.*

A PACIFIC OCEAN MYSTERY.

To the Editor of SCIENCE-GOSSIP.

SIR,—Easter Island, the nearest Polynesian island to South America, and in a direct line with Chili, is a small island well known as being crowded with prehistoric remains. Innumerable statues, cyclopean houses, and other relics of some vanished race exist, resembling somewhat in character the Central American and Peruvian monuments of antiquity. Roggewein, a Dutch explorer, visited it in 1721, and found there a semi-civilised race of aborigines, apparently sun-worshippers, as were the Incas of Peru. They worshipped also the gigantic idols which stand on platforms along the coast, and having artificially distended ears, also like the inhabitants of Central America and Peru. They were without weapons, and apparently gentle and most peaceable.

Since then a Polynesian invasion—the most easterly expedition of that maritime race—overwhelmed them, under a leader called Hotu-Matua, so that when Cook reached the island, on his second circumnavigation in 1774, he found the inhabitants to be ordinary Malay-Polynesians, almost savages, knowing nothing of the antiquities of the island or of their origin. Not any of the islanders in Cook's time exceeded six feet tall. This is remarkable, as Roggewein and his companions most solemnly asserted that the aborigines they found there were of gigantic height They said : "All these savages are of more than gigantic size ; for the men being twice as tall and thick as the largest of our people, they measured, one with another, the height of twelve feet. . . . According to their height, so is their thickness, and all are, one with another, very well proportioned, so that each could have passed for a Hercules." It is added that the females do not altogether come up to this, "being commonly not above ten or eleven feet." Cook and all subsequent voyagers have supposed that this was mere romancing ; but I have discovered a slight parallel to this in the account of a recent voyage in Polynesia.

In 1898 H.M. ship "Mohawk" was cruising amongst the Santa Cruz, Swallow, Reef and other islands, about four or five hundred miles east of the Solomon Islands, when there was found an apparently new and overlooked race occupying Tocupia, one of the Reef and Swallow group. They numbered some eight hundred souls, and were gigantic in stature ; one measured 6 feet 10 inches. The women were proportionate. The men had long, straight hair, which they dyed a flaxen colour, and which in thick folds hung over their copper-coloured shoulders. The women, on the contrary, had their hair cut short. Strange to say, these natives had no weapons of defence. They were not Kanakas, ordinary Malay-Polynesians, or woolly-haired or stunted in stature—in fact very different from their neighbours. A remarkable law among them is that they marry only once, the supposition being that if a married man or woman dies the deceased's spirit has gone ahead and is waiting for the other half. It seems possible to me that this is a fragment of some primitive race which once inhabited Polynesia, possibly Easter Island, and by reason of its gentler disposition was exterminated by the Negritos and Malay-Polynesians. It may have been of early Asiatic origin, and perhaps allied to the Caroline and Mortlock Islanders, who are known to enlarge the ear-lobe by heavy pendants, or to have anciently done so. They, too, were rather more civilised than the Polynesians and Melanesians. In any case the discovery is a curious and suggestive fact. Roggewein may not have been so great a romancer as he has been thought to be.

(Rev.) FRANCIS A. ALLEN.

Shrewsbury.

NOTICES OF SOCIETIES.

*Ordinary meetings are marked †, excursions * ; names of persons following excursions are of Conductors. Lantern Illustrations §.*

HULL SCIENTIFIC AND FIELD NATURALISTS' CLUB.
Dec. 5.—† " Local Life of Fungi." R. H. Philip.
  „ 12.—† Microscope Club.
  „ 19.—† " Major-General Overton, Governor of Hull." E. Lamplough.

NORTH LONDON NATURAL HISTORY SOCIETY.
Dec. 6.—† " Trees." L. B, Hall.
  „ 20.—† Annual General Meeting.

SELBORNE SOCIETY.
Dec. 6.—† " The Whitgift Foundation and the Hospital." Dr. J. M. Hobson, B.Sc.

SOUTH LONDON ENTOMOLOGICAL AND NATURAL HISTORY SOCIETY.
Dec. 13.—† " Some Wing-structures in Lepidoptera." Dr. T. A. Chapman, F.Z.S., F.E.S.

MANCHESTER MUSEUM, OWENS COLLEGE.
Dec. 2.—† " The History of Salt." Prof. W. Boyd Dawkins.
  „ 26.—† " Variation." W. E. Hoyle.

LAMBETH FIELD CLUB AND SCIENTIFIC SOCIETY.
Dec. 3.—† " Warning Colours in Insects." Stanley Edwards, F.E.S.
  „ 15.—* " Visit to Museum of Practical Geology." F. W. Rudler.
  „ 17.—† " Tolstoi." Hilda Swan.

PRESTON SCIENTIFIC SOCIETY.
Dec. 5.—†§ " A Journey to Algiers to see the Total Eclipse." E. Dickson, F.G.S.
  „ 13.—† Conversazione in Guild Hall.
  „ 19.—† " An Old Pasture." W. Clitheroe.

NOTICES TO CORRESPONDENTS.

To CORRESPONDENTS AND EXCHANGERS.—SCIENCE-GOSSIP is published on the 25th of each month. All notes or other communications should reach us not later than the 18th of the month for insertion in the following number. No communications can be inserted or noticed without full name and address of writer. Notices of changes of address admitted free.

SUBSCRIPTIONS.—The volumes of SCIENCE-GOSSIP begin with the June numbers, but Subscriptions may commence with any number, at the rate of 6s. 6d. for twelve months (including postage), and should be remitted to the Office, 110 Strand, London, W.C.

BUSINESS COMMUNICATIONS.—All business communications relating to SCIENCE-GOSSIP must be addressed to the Proprietor of SCIENCE-GOSSIP, 110 Strand, London.

NOTICE.—Contributors are requested to strictly observe the following rules. All contributions must be *clearly* written on one side of the paper only. Words intended to be printed in *italics* should be marked under with a single line. Generic names must be given in full, excepting where used immediately before. Capitals may only be used for generic and not specific names. Scientific names and names of places to be written in round hand.

THE Editor is not responsible for unused MSS., neither can be undertake to return them unless accompanied with stamps for return postage.

ANSWERS TO CORRESPONDENTS.

S. G. (Leicester).—The small holes in old furniture, beams, and other wooden structures are caused by the larvae of a beetle of the genus *Anobium,* the perfect insect of one species being known as the " death-watch."

EXCHANGES.

NOTICE.—Exchanges extending to thirty words (including name and address) admitted free, but additional words must be prepaid at the rate of threepence for every seven words or less.

WANTED to exchange guinea Kodak in leather case or ½-plate mahogany bellows camera for a 4-wick oil lantern.—A. Nicholson, 67 Greenbank Road, Darlington.

PLANTS OFFERED.—L. C. 9th ed. 5, 43, 130, 385, 536, 544, 845, 885, 1059, 1088, 1295, 1369, 1499, 1663, 1785, 1788, 1801, 1846, etc. Lists exchanged. - W. Falconer, Slaithwaite, Huddersfield.

MICRO-BOTANY, exclusive of marine algae. I shall be glad to know of young, *i.e.* not advanced, students in Cryptogamic Botany, with a view to mutual assistance.— C. W. Marshall, Yatesbury Rectory, Calne, Wilts.

KIRBY' " European Butterflies and Moths," coloured plates, quite new ; Hooker's " Brit. Flora," 2nd ed., new : " Entomologist's Record," vol. 5 : " Naturalist," vol. 5 : " Economic Naturalist," v. l. 1. Last three in parts. What offers in foreign Lepidoptera, stamps, books, etc. ?— Rev. W. W. Flemyng, Coolfin, Portlaw, co. Waterford.

WANTED, foreign land, marine and freshwater shells, echinoderms, and crustacea in exchange for others. Send lists.—H. W. Parritt, 8 Whitehall Park, London, N.

SLIDES of the eggs of poplar hawk-moth (*Smerinthus populi*), to exchange for anatomical sections or other mounted objects of interest.—L. R. J. Horn, Selborne Villa, Clayton, near Bedford.

BIRDS' EGGS' DUPLICATES. - Curlew, common snipe, whinchat, gold-crested wren, jay, nightjar, sandpiper, dipper, grey wagtail, pheasant, red-legged partridge. Desiderata numerous.—W. S. Lane, 9 Teesdale Street, Hackney Road, London, N.E.

OFFERED, British mosses or lichens, named, in exchange for really good microscopic slides, preferably botanical.—W. Smith, Dalveen, Arbroath, N.B.

A NUMBER of " Amateur Photographers " to dispose of, in exchange for examples of foreign Helicidae : also other photographic papers for shells not in collection.—Charles Pannel, Jun., East Street, Haslemere.

SHELLS OFFERED.—Helix horrida (Pils.), Jourdic, mandarina (Gray), langsonensis, ferruginea (Bavay), elegantissima, latiaxis, Cyclophorus dodranus, courbeti, Hybocystis rochebruni, Cyclotus perezi, Pupina tonkiniana. DESIDERATA, exotic land shells.—Miss Linter, Saville House, Twickenham.

CONTENTS.

PHOTOGRAPHIC WAVE STUDIES.

WE have had submitted to us a number of very beautiful carbon enlargements of photographic studies of the sea under various aspects. Some of these exhibit the ocean greatly agitated, and others with silver moonlight effects. They are the work of Mr. F. H. Worsley-Benison, a gentleman who has made this branch of artistic study a success.

These photographs exhibit remarkable judgment in selecting the exact fraction of a second of exposure, which can have been attained only by much practice. Those who have given attention to this division of photography will remember the natural hesitation which precedes the proper they constitute, especially in the larger sizes, very effective pictures.

We understand that the views were taken from Atlantic rollers on the West Coast of England and Wales, and one can easily imagine in one photograph the boom of the waves on the massive rocks, and the shriek of the accompanying wild wind. In another, where there is only sea tumbling boldly on to the sandy beach, we see the gulls dwarfed by the size of the waves: this also is an effective picture. Another, with a couple of heavy rocks in the foreground, shows the wavelets of a sea in quieter mood. The distance is relieved by a few " white-horses," or, as the French children call them, the

PHOTOGRAPHIC WAVE STUDY, BY F. H. WORSLEY-BENISON.

moment for operating with the instantaneous shutter. The ever-varying waves present constantly changing pictures, each in itself sufficiently good. but one is always hoping for a better. Thus photography of the rough sea is one of the most exciting sections of the art.

Mr. Worsley-Benison's studies are reproduced in colours suited to seascapes. They are in delicate green tints, relieved by grey. Being enlargements printed direct from large negatives, the full effect of detail is retained, as well as the natural crispness of the general effect. In size the specimens before us vary from 14 in. × 10½ in., and 24 in. × 18 in., to 36 in. × 24 in. It will thus be understood that

" little sheep." In this picture one observes the sharpness of the detail in perfection. One of the larger photographs, marked " Westby Series, K," is a really splendid piece of photographic art. It represents the ocean breaking savagely on some rocks at the right, with the water of a receding meeting the incoming wave. It is wonderfully sharp and effective, the foreground and the far distance appearing equally within focus.

Moonlight effects are charming, and several of the couple of dozen or so photographs sent show wild rock scenery, such as the Stack Rocks in Pembrokeshire, with their multitude of bird inhabitants.

SPIRALS IN PLANTS.

BY J. A. WHELDON.

THE subject of the torsion of the trunks of trees raised by Mr. Mott is a most interesting one (*ante*, page 219), and it seems very desirable, as he suggests, that further observations on the subject should be recorded. With a greater amount of evidence before us, it might be possible to propound some theory that would more fully account for all the phenomena than any already offered. It would be advantageous also to know more than we do about other spiral vegetable organs, in addition to those exhibited in the trunks of trees— *e.g.* the twining of the stems of climbers, the coiling of tendrils and petioles, and the twisting of pods and other fruits.

In recording any such observations it is absolutely necessary that everyone should use uniform terms in describing the direction of the spirals, otherwise errors may easily be caused. We must, therefore, always bear in mind that in determining the direction of a spiral the observer should imagine himself to stand within its coil. Then, in a *right* spiral, the ascending lines will be seen to run from left to right across the breast of the observer; if viewed from the outside, the ascending curves appear to run right to left. In a *left* spiral, of course, the reverse of this is seen.

It has been suggested that the twisting of stems is due to the growing apex following the light, but Darwin's observations (*vide* "Climbing Plants") disprove that; nor could this explanation be reconciled with the fact that some stems twist in a direction opposite to that of the sun's apparent course. According to Darwin (*op. cit.* page 7), there is in twining plants a distinction between the revolving motion of the growing apex of a shoot and its axial twisting. By his carefully-conducted experiments and patient observation he discovered that a growing shoot revolved at its apex many times more often than was required to produce the number of twists subsequently developed in its axis ; and, further, that in some instances the apex revolved in a direction opposite to that in which the stem was found to be twisted.

In the "Sagacity and Morality of Plants " Dr. Taylor repeats the theory—originally, I believe, propounded by Sachs—that the twining of stems " is due to the weak and rapidly developing stem growing a trifle faster on one side than on the other, just as a carpenter produces any degree of curl in his shavings according as he presses his plane a little more on one side than the other." Even this demands explanation as to why the growth on the two sides should be unequal in some plants, and not in others ; and why the stems of some plants, such as *Hibbertia*, should twist both ways, and of others, as *Solanum dulcamara*, to the right or left indifferently.

Professor Asa Gray remarked in a letter to Mr. Darwin that " in *Thuja occidentalis* the twisting of the bark is very conspicuous. The twist is generally to the right of the observer ; but in noticing about one hundred trunks four or five were observed to be twisted in the opposite direction. The Spanish chestnut is often much twisted." The same author suggested that the stem gained rigidity by being twisted, on the same principle that a much-twisted rope is stiffer than a slack one.

In my own neighbourhood there are few opportunities for observing trees ; but I have noticed that in some laburnums and elder-trees the stem shows a tendency to twist to the right.

Amongst Cryptogams spirals are of frequent occurrence, as in some of the climbing ferns, the stems of *Chara*, the cells of *Sphagnum*, the capsules, leaves, and setae of mosses, and the elaters of liverworts.

The twisting of the setae of mosses has occasionally been used as a character of diagnostic importance in the separation of species. Mr. Dixon, in his " Handbook of British Mosses," says:—" It is rarely, however, I believe, that the direction is sufficiently constant in any species to afford a good character of distinction, and it is only in a few cases that I have relied upon it."

It would be interesting to compare a large number of observations, and thus ascertain if the twistings so markedly exhibited when dry by the setae of some mosses were constant enough to be of specific value. With regard to the stems of Phaenogams, Darwin says:—" I have seen no instance of two species of the same genus twining in the opposite direction, and such cases must be rare."

In mosses, on the contrary, the converse of this is very commonly observed in the directions assumed by the gyration of the seta, but of course this is hardly comparable, not being homologous with the stems of Phaenogams, nor even with their peduncles. I select a few examples from the Bryales, commencing with the Nematodonteae. In the Polytrichaceae the setae are not conspicuously twisted in the majority of the species, but in some there is occasionally a tendency to gyrate feebly to the right—e.g. *Polytrichum nanum*, *P. urnigerum*, *P. alpinum*, *P. formosum*, *P. gracile*, *P. piliferum*, and *P. aloides*. In a few there is a much more decided twisting in the same direction —viz. *Oligotrichum incurvum*, *Catharinea undulata*, and *Polytrichum sexangulare*. In the remaining orders of this group, Buxbaumiaceae and Tetraphidaceae, I could find no decided twisting in my examples of any of the British species.

In the Aplolepideae the seta is frequently flexuose, and sometimes strongly twisted. The peristome teeth and even the capsule itself are also occasionally contorted. In *Dicranum* the seta is flexed, hardly twisted, to the right in most cases, but in *D. longifolium* occasionally to the left. In *Ditrichum pallidum* and *D. homomallum* it gyrates to the right, but in some specimens of *D. tenuifolium* to the right below and left above. In *Swartzia montana* there is a rather strong twist to the right, whereas in the closely allied *S. inclinata* the spiral turns more feebly to the left. I was unable to satisfy myself of any tendency to twist in *Blindia*; but in some examples of *Seligeria recurvata* and *S. pauciflora* the seta gyrated to the right; in other species, *S. pusilla* and *S. calcarea*, the reverse was the case, and in *S. doniana* it twisted to the right at the base and to the left at the summit. This curious twisting in two directions is very noticeable in the seta of *Brachyodus trichodes*. A decided twist is noticeable in some of the Dicranellae, to the right in *Dicranella varia* and *D. schreberi*, to the left in *D. rufescens*, and occasionally left below and right above in *D. squarrosa*, *D. heteromalla*, *D. cervioulata*, and *D. subulata*. In *Dicranoweisia* and some species of *Campylopus* there is sometimes a feeble twisting to the left, also in *Rhabdoweisia denticulata* and *R. fugax*, in the latter species sometimes very strongly marked. The seta of *Leucobryum glaucum* is often strongly twisted to the right.

The setae of the Fissidentaceae display evidence of only very weak twisting. I found right spirals in some examples of *Fissidens viridulus*, *F. taxifolius*, *F. bryoides*, and *F. decipiens*, and left ones in *F. rivularis*, *F. exilis*, and *F. osmundoides*. In many of these twisted setae are so rare that they might be considered accidental.

In Grimmiaceae, amongst a large number of species of *Grimmia* examined, I did not find any with the seta twisted to the right, though those exhibiting left spirals were not infrequent. On the contrary *Rhacomitrium*, which is so closely allied as to be considered congeneric with *Grimmia* by some distinguished authorities, exhibits a majority of left twisting species, *e.g.* *Rhacomitrium protensum*, *R. aciculare*, *R. heterostichum*, and *R. microcarpon*, only *R. canescens* of those I examined being contorted in the opposite direction. In *Ptychomitrium polyphyllum* also the direction is usually to the left.

Tortulaceae.—In the genera *Pottia* and *Tortula* there is a more or less well developed twist to the left, except in a few instances, *e.g.* *Pottia crinita* and *Tortula angustata*, in which the spiral turns to the right below, but higher up conforms with the direction taken in the other species. In *Barbula*,

on the contrary, the direction appears to be to the right where any twisting at all is displayed, the same being the case with *Trichostomum tortuosum*. The peristome of the Tortulaceae appears to twist to the left much more constantly than does the seta, and the two therefore occasionally form spirals having opposite directions.

In Encalyptaceae, *Encalypta vulgaris*, *E. apophysata*, *E. brevicolla*, *E. rhabdocarpa*, *E. streptocarpa*, and *E. alpina* have the seta twisted to the right with varying degrees of intensity, and only in *E. ciliata* have I seen any tendency of the revolutions to take the other direction.

The Diplolepideae come next, of which I only propose to touch on the acrocarpus section. In the Orthotrichaceae, Schistostegaceae, and Bryaceae, strongly twisted setae are rare. An occasional tendency to twist to the right may be seen in *Zygodon viridissimus*, *Anoectangium compactum*, *Bryum capillare*, *B. warneum*, and *Webera albicans*. In the foreign *Bryum salinum* the rotation is often more strongly evident. It may also be traced in *Mnium cuspidatum*, *M. insigne*, and *M. subglobosum*. Of those twisting to the left with varying intensity, but mostly rather feebly, may be enumerated *Bryum pallens*, *B. calophyllum*, *B. pendulum*, *B. alpinum*, *Webera cruda*, *Oreas martiana*, *Mnium undulatum*, and *M. medium*. I have seen very strongly twisted setae sometimes in *Bryum turbinatum*. *Mnium spinulosum* has the seta twisted to the right below and to the left above, and the same condition is sometimes more feebly exhibited by *M. hornum*, *Webera glacialis*, *W. elongata*, and *Cinclidium stygium*.

Of the Splachnaceae there is sometimes a twist to the left, in *Splachnum sphaericum*, and more strongly and frequently in *Oedipodium griffithianum*. Of the Funariaceae, *Funaria attenuata* and *F. ericetorum* twist to the left, and *Physcomitrium sphaericum* to the right. *Funaria hygrometrica* sometimes twists to the left below, and to the right above. *Meesia uliginosa*, *Amblyodon dealbatus*, and *Timmia bavarica* twist to the left, as do *Conostomum boreale*, *Catoscopium nigritum*, *Aulacomnium heterostichum*, and *A. androgynum*.

All these observations were made from dried specimens, the setae rarely showing any marked tendency to twist in the moist condition. In some of the species quoted the phenomenon is so rare that a large number of specimens were examined before an example could be found, and in such cases the twisting is probably accidental. In other instances, every tuft contains numerous examples, and the twisting appears to be a normal condition of the species, although even in these the degree of twisting varies considerably, being affected by the age of the plants, and perhaps by other circumstances. The spiral may consist of only one or two feeble turns extending the length of the seta, or the torsion may be so strong that the coils resemble the twisted strands of a cable, either in the middle of the seta. at either extremity. or through its entire length. The direction appears to be fairly uniform in many of the species here enumerated. even when collected from widely separated localities. In a few instances, however. I have only been able to examine solitary examples.

Perhaps this brief reference to an interesting subject may attract some of the readers of SCIENCE-GOSSIP to relate their experience on the matters mentioned. By a comparison of notes it should be easy to determine the constancy of some of the conditions described. although, I fear, it will be more difficult to detect their cause.

Walton, Liverpool,
December 1900.

NOTES ON SPINNING ANIMALS.

BY H. WALLIS KEW.

(Continued from page 170.)

VI. PSEUDO-SCORPIONS.

THE Pseudo-scorpions are known to most people, at least by reputation and from book illustrations. Notwithstanding their small size and want of affinity, the creatures are surprisingly like little scorpions without tails. This superficial resemblance is mainly due to the second pair of appendages, the pedipalps, which. compared with the rest of the creature, are enormously developed. They bear, like the pedipalps of the scorpion, large chelae or pincers. The little legs and the flat segmented body give the creatures a heterogeneous form ; and when brandishing the great pedipalps as they move they have a ferocious appearance which. as Dr. Hagen has remarked. is rendered ridiculous by their little bodies and helplessness [1]. Their spinning work does not press itself upon our notice ; and though the creatures are to a large extent predaceous. they are not known to use their silk for spreading a snare after the manner of spiders [2]. They employ it, however, in the fabrication of a kind of nest or cocoon in which they are protected, in some cases at any rate, during hibernation, at the time of moulting, and perhaps also during egg-laying. This is the only use. so far as the writer is aware, which the creatures make of their spinning. It has been stated that they spin an egg-cocoon [3]. but this appears to be a mistake.

Hermann (1804) mentions the finding of a pseudo-scorpion in a silken follicle, covered with dust and attached to a wall [4]. Menge (1855) relates that a *Chernes cimicoides*. which he one day placed in a vessel. had by the morning constructed a web between the wall of the vessel and a fragment of bark. At this web it was still working. and the cocoon thus formed was somewhat

circular, about 4 mm. in diameter by 2 mm. in depth. There was only just room within for the animal, and no opening was observable through which it could come out. According to the same naturalist, the creature constructs its cocoon at the time of moulting. when it remains enclosed about five days, until its new integuments acquire strength. Three months afterwards an individual returned to the same cocoon for hibernation [5]. S. J. M'Intire (1868), who placed a number of pseudo-scorpions in cork cells. found that a chelifer constructed for itself a snug silken cocoon, in texture like the web of the house-spider. The animal was generally seen sitting in the entrance.

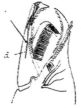

SPINNING-DUCTS AND COMB OF PSEUDO-SCORPION.

Pincers of one of the chelicerae, or mandibles, of a pseudo-scorpion, showing the terminations of some of the spinning-ducts (*d*), and also the comb, etc., believed to be used in manipulating the silk. Much magnified. After Bernard, Journ. Linn. Soc. : Zool. xxiv. (1893), pl. xxxi, fig. 2a.

only withdrawing from observation when disturbed ; sometimes it went abroad and walked about, but it invariably returned to the cocoon. All the healthy chelifers kept by this observer through the winter spun similar cocoons, and hibernated in them. The structures were generally oval, with room for the occupant to turn within, and usually with one or two apertures through which the owner could walk out when the weather was mild. It is supposed that the creatures sometimes made mistakes in returning. for more than once two were seen in the same cocoon. An individual, regarded by M'Intire as

---

(1) Hagen, "American Pseudo-scorpions," Record of American Entomology, 1868, pp. 48–52.

(2) Simon, "Les Arachnides de France," vii. (1879), pp. 12–13.

(3) Cambridge, "On the British Species of False-scorpions.", Proc. Dorset Nat. Hist. & Antiq. Field Club xiii. (1892), pp. 199–231.

(4) Hermann, quoted by Simon. *l.c.*

(5) Menge, quoted by Simon, *l.c.*

an *Obisium*, was also found to spin ([6]). Simon (1879) mentions having frequently observed *Chernes rimicoides* in its cocoon, under the bark of plane-trees at the commencement of winter; the creature being then always found in a circular structure, white or somewhat pearly. In *Chernes cyrneus*, he says, the cocoon is larger, irregular, and very transparent. *Obisium jugorum*, inhabiting the frozen regions of the Alps, according to the same author, constructs, beneath stones, a more solid cocoon, nearly round, without an opening, and covered with earth and vegetable fragments. It recalls the egg-cocoon of the spider *Enyo* ([7]). Zabriskie (1884), writing from Nyack, N.Y., has noted the finding of a pseudo-scorpion which had died in its cocoon, in the act of moulting. The cocoon, made at the edge of a piece of paper on a beam in a garret, was oval, two-tenths of an inch long. The lower surface consisted of a flat white web of extremely thin translucent texture, while the upper surface was slightly convex, firmer and darker than the lower; and "the entire edge was bordered with minute pieces of sawdust, firmly glued to the web" ([8]). Finally, Croneberg (1887) mentions having repeatedly found *Chernes hahnii = Chernes rimicoides* near Moscow during the cold season, in its little watchglass-like cocoons, under the bark of trees ([9]).

With regard to the organs concerned in the production of the silk, a view now believed to be erroneous was long accepted. It originated, apparently, with Menge, who attributed a spinning function to organs now called cement-glands. These structures, of which there are two, open by median papillae beneath the commencement of the abdomen, not far from the genital aperture. The largeness of their openings, it is said, shows that they have nothing to do with spinning; and it is thought that their function is that of sticking the eggs together and fastening them to the animal ([10]). As is well known ([11]), and as the writer has seen in *Chelifer latreillii*, the eggs are attached one to another and to the under side of the abdomen—apparently by a gummy liquid. What are now believed to be the true spinning-glands are found in the cephalothorax; and they have their openings near the tip of the movable finger of the pincers of the first pair of appendages—the chelicerae or mandibles. The apparatus was discovered by Croneberg, whose results have been confirmed by Bernard. The glands when fully developed are of large size, but, like the

cement-glands, they are apparently subject to periodic variations. Croneberg, who studied *Chernes cimicoides*, maintains that the apparatus well fulfils the requirements of a spinning-organ. He found in the cephalothorax, above the brain and the anterior hepatic lobes, two considerable glandular masses, touching each other in the median line, and having much attenuated anterior ends entering the basal joint of the chelicera on each side. Each mass consists of four or five cylindrical closely approximated tubes, the narrow efferent ducts of which pass into the chelicera in a fine bundle, traceable through the basal joint, and into the movable finger, which latter they traverse, so as to reach a soft-skinned process at its tip. This process terminates in short conical points into which the ducts may be traced singly, and in which they doubtless open by fine apertures. Croneberg notes, further, that the chelicerae are provided with a number of processes apparently fitted to pull and arrange threads of silk. Along the inferior surface of the movable finger there is a long comb, consisting, in *Chernes cimicoides*, of eighteen plates; whilst on the immovable part there is a serrated and denticulated process, with a semicircular fold at its base.

Bernard, whose material was derived from bottles labelled *Obisium sylvaticum* or *O. carcinoides*, states that the ducts open on a blunt prominence behind the tip of the movable finger; and that they can be seen, about seven in number, running through each chelicera. The ducts proceed from as many somewhat coiled reservoirs, behind which are the secreting portions of the glands, which latter run backwards, immediately under the dorsal wall of the body, and sometimes extend beyond the cephalothorax into the second or third abdominal segment. Bernard speaks also of the combs of the chelicerae, believed to be used in manipulating the silk, and he regards the spinning-glands as homologous with the poison-glands of spiders.

(*To be continued.*)

PIGEON POST.—In connection with the inauguration of a pigeon post at the Crystal Palace, the first of a number of breeding lofts has been completed, and the remainder are in hand. As a fitting "send-off" a most interesting exhibition of homing birds was organised by Mr. Edgar S. Shrubsole, the Curator. The exhibits, which included many famous war pigeons and the best of the racing birds in the United Kingdom, numbered quite two thousand. Some valuable prizes were offered in connection with speed trials and general excellence. During the meeting discussions were held in the Board Room of the Crystal Palace Company each evening, the subjects being: "General Organisation of the Crystal Palace Pigeon Post. Formation of Council of Advice and Executive Committee;" "Methods of Linking up Lofts throughout the Kingdom, and perfecting the training of the Crystal Palace Birds;" "Appointment of Special Service Sub-Committee to arrange details for the co-operation of the Naval and Military Authorities."

(6) M'Intire, Journ. Quekett Micr. Club. i. (1868), pp. 8–14; and "Hardwicke's Science-Gossip," 1869, pp. 243–7.

(7) Simon, *l.c.*

(8) Zabriskie, "Nest of the Pseudo-scorpion," American Naturalist, xviii. (1884), p. 427.

(9) Croneberg, 1887, as translated in Ann. & Mag. of Nat. Hist. (5), xix. (1887), pp. 316–20.

(10) Bernard, Journ. Linn. Soc.: Zool. xxiv. (1893) pp. 410–430.

(11) Simon, *l.c.*

BRITISH FRESHWATER MITES.

BY C. F. GEORGE, M.R.C.S.

(*Concluded from page* 205.)

Arrenurus tubulator Müller.

I FOUND the male of this mite for the first time on May 18th, 1900. It is strikingly different in appearance from any other *Arrenurus* I have yet observed. Its dark colour and the nearly parallel sides of the tail, best seen when the under side of

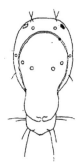

FIG. 1. *A. tubulator.* Dorsal view.

the mite is examined, at once distinguish it from the other members of this family. As far as I know, it had not been noticed since the time of Koch, until Koenike figured it in the "Zoole-gischer Anzeiger." Piersig also gives two illustra-tions on Taf. 40 of his great work. The figures accompanying this article are drawn by Mr. Soar from a living mite sent to him by me. After drawing the dorsal and ventral aspects, he had the misfortune to lose the little creature before he had been able to make a side view. The general colour of the mite, when alive, is a warm reddish-brown, the caeca dark brown. At the outer edge of the body there is a slight shade of green, more evident after mounting in balsam. The impressed curved line on the back is somewhat irregular, but very evident from the sharpness of its edges and its translucency. There are also eight circular, some-what translucent spots, four in front and four behind this line; they appear to be circular plates, and seem to carry a minute and delicate tactile hair. The eyes are bright crimson during life, but become black after mounting in balsam. The tail where it joins the body and for about the anterior two-thirds is much thicker from above downwards than the posterior third, forming a sort of ball, notched at the posterior part, in the centre. The lower third of the tail is somewhat translucent

from its comparative thinness. The legs and palpi are not green, but a semi-transparent reddish-brown. The fourth internode of the last pair of legs is furnished with the spur so often found in male specimens of *Arrenurus*.

I have the satisfaction of knowing that this mite is identical with the creature described and figured by Koenike, this savant having kindly sent his specimen to me for examination. He also informs me he has examined the mite described by "Sig Thor" as *Arrenurus mediorotundata*, and found it to be the same creature. That this mite really is Müller's *A. tubulator* is, perhaps, open to doubt. Müller's figure is not very satisfactory, and his description brief; besides, he says the legs are green. Koch's figure is better than Müller's, but he also says that the legs are green; moreover, he speaks of two large back humps, which I have not been able to discover. He further says that the

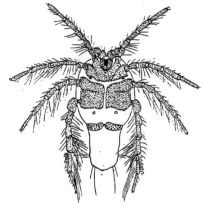

FIG. 2. *A. tubulator.* Ventral view.

mite is common in his neighbourhood. I trust therefore that this matter will very shortly be decided.

Mr. Soar supplies the following measurements: length about 1.10 mm., width about 0.70 mm.; length of first pair of legs about 0.77 mm., length of fourth pair of legs about 0.98 mm.

Arrenurus maximus Piersig 1894.

This large and very beautiful red mite was taken at Enfield in June 1895. It is one of those wide-

tailed Arrenuri having obliquely projecting side
corners, and with a central free petiole; on the
back is a large hump with two points close
together, and on either side at the base is a
secondary smaller hump. The petiole is chisel-
shaped, with almost
parallel sides, slightly
widened at the distal
end; the two curved side
bristles do not extend
beyond the petiole. The

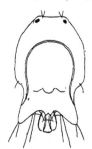

FIG. 3. *A. maximus.*   FIG. 4. *A. maximus.*

hyaline membrane over the petiole is a truncated
cone with pretty sharp corners. These details are
well shown in Mr. Soar's figures. I have not met
with this species in Lincolnshire, and have not
seen it alive. Its length, according to Mr. Soar,
is 1.70 mm.; breadth, 1.04 mm.; first leg, length,
1.20 mm.; fourth leg, length, 1.34 mm.; length
of petiole, 0.12 mm.

Kirton-in-Lindsey, November 30th, 1900.

THE BIRDS OF YORKSHIRE.—Naturalists and
others interested in the subject will be pleased to
learn that arrangements have been made for the
speedy resumption of the publication of Mr. W.
Eagle Clarke's excellent work on the "Birds of
Yorkshire," which has been partly issued in the
Transactions of the Yorkshire Naturalists' Union,
its continuation being interrupted by Mr. Clarke's
removal to Edinburgh. Mr. Clarke and the Union
have secured the services of Mr. Thos. H. Nelson,
M.B.O.U., of Redcar, to continue and complete the
task. Mr. Nelson has now in his possession the
voluminous mass of original and unpublished
observations which Mr. Clarke had at his command when writing the instalments already in
print, including notes, lists, and observations from
most of the naturalists who have studied and
observed Yorkshire birds. In addition to this
is the whole of the information amassed by
the late Mr. John Cordeaux, relating to the
birds of the Humber district, and also the large
amount of notes that Mr. W. Denison Roebuck has
extracted from the very voluminous literature of
the subject, and Mr. Nelson's own accumulated
series of notes on the birds of Cleveland and other
districts, the whole forming an ample mass of
material. Mr. Nelson will be pleased to enlist the
co-operation of those who have it in their power to
assist him with notes on Yorkshire birds, their history, distribution, migration, nidification, variation,
vernacular nomenclature, etc. All assistance will
be duly and gratefully acknowledged. Mr. Nelson
is now actively at work on the families Turdidae
and Sylviidae, which are to be included in the
next instalment sent to press.

ON THE NATURE OF LIFE.

BY F. J. ALLEN, M.A., M.D.

(*Continued from page* 138.)

IF Mr. Geoffrey Martin had acquainted himself
with the work of the chief investigators of
vital chemistry, I think he would have been more
cautious in expressing his opinions in the October
SCIENCE-GOSSIP. These I leave to be estimated
by the learned reader; but his questions I will try
to answer, though one or two of them diverge
somewhat from our original subject.

He asks firstly (page 135): "How could an
element as feeble as nitrogen—an element which
can only with the greatest difficulty be induced to
combine with oxygen, or retain its oxygen when
combined—ever rob such an element as carbon of
its oxygen?" This question contains certain fal-
lacies, which must be corrected before an answer
can be given. Nitrogen, in such forms as ammonia
or amides, combines readily with oxygen, forming
nitrates; and the nitrates retain their oxygen without difficulty. The nitrification of ammonia, etc.,
is performed by the vital action of microbes; and
therefore, on either theory, mine or the older one,
the oxygen is possibly derived from a carbon compound. Certain microbes can oxidise even atmospheric nitrogen, N_2; though that is exceptional,
the usual source of nitrogen in the organic world
being combined nitrogen.

Nitrogen combines with oxygen without difficulty when an electric discharge passes through a
mixture of the gases, a part of the energy from
the electricity being stored in the oxides formed.
Nitric oxide, NO, absorbs atmospheric oxygen with
avidity, forming N_2O_3. The N_2O_3 can then yield
oxygen to other substances, thereby falling back
into its previous state of NO, so that the action
can be repeated indefinitely. This property is
utilised in the manufacture of sulphuric acid.
The reaction is believed to be more complex than
the above description indicates; but whatever it
may be, it is regarded by physiologists as typical
of the "enzyme" actions of nitrogen compounds.
It is also a possible prototype of the action
of nitrogen as oxygen-carrier in the organic
world.

The astonishing thing is, not that nitrogen
should attract oxygen, but rather that carbon
should part with oxygen, seeing that the attraction between carbon and oxygen is of the strongest
kind. It is rational to suppose that oxygen would
never leave carbon (and hydrogen) and escape as
elementary oxygen, unless there were a stepping-
stone in the form of an element with just a
moderate attraction for oxygen. Thus the feeble-
ness of nitrogen, *i.e.* the inconstancy of its combinations, makes it the determinant of this as of
other vital actions.

Though there is little room to doubt that nitrogen does form the stepping-stone by which oxygen leaves carbon and hydrogen, the exact nature of the reactions is still within the realm of speculation, so that I answer Mr. Martin's " how " with caution. The matter may be elucidated some day by laboratory experiment. Many details could be suggested; but the following must suffice in the present circumstances. The action in question is an anabolic[1] one, the kinetic energy being derived from sunshine, heat, or electric disturbance. This energy, acting on the mixture of compounds of N, O, C, H, etc., disturbs their equilibrium, and the lability of the nitrogen compounds leads to the rearrangement of the elements. The first change is probably the assumption of oxidized carbon and hydrogen into a nitrogen compound, and from this compound oxygen is eliminated subsequently. The oxygen is not removed with a wrench, but by easy steps. In the intermediate stages the following groups may be expected to exist:

$$= \overset{|}{\underset{|}{N}} - O - \overset{|}{\underset{|}{C}} \qquad = \overset{|}{\underset{|}{N}} - O - H.$$

If this theory is not satisfactory, no one is more ready than myself to welcome a better.

Whether performed in this way or any other, this deoxidation with absorption of energy, and the converse action of oxidation with the dispersion of energy, are the main physical phenomena of life. The formation of complex compounds may be regarded as a collateral feature which has become more and more pronounced in the course of evolution. In other words, life is not the result of chemical complexity, but complex N, O, C, and H compounds are produced by the accumulation of energy.

I referred to Professor Lapworth as the leading exponent of earth-movements—*i.e.* geological upheavals, depressions, overriding, etc., and the production of igneous rocks. Mr. Martin quotes Lord Kelvin as if he were an opposing authority; but since Professor Lapworth demonstrates what is still occurring, while Lord Kelvin speculates as to what may have occurred in the past, they mostly cover different ground. If their results clash in any way, if we have to choose between geological evidence and a brilliant speculation, the former has a stronger claim. I do not presume to criticise Lord Kelvin's theory, and can only echo the opinions of those who have a right to speak. There are many geologists to whom his calculated results seem irreconcilable with geological evidence; and if that be so, something must have been overlooked in his premises. What, indeed, can be more excusable? Few tasks could be more difficult than to recall correctly all the con-

ditions that have ruled in the moulding of the solar system during unknown millions of years. It has been suggested as a possible correction that the conduction of heat through the earth's substance has not been uniform in place or time. Then it is difficult to estimate the real loss of heat from the earth by radiation, owing to the corrections necessitated by solar heat, presence of water in the atmosphere, etc. Again, while the interior may be very hot, the surface exposed to cold space will be comparatively cool, its temperature depending chiefly upon the sun's heat, and this is the matter that concerns us most.

It has been said that the oblate form of the earth is not sufficient evidence of fluidity; for, if the globe were as rigid as steel, it would still assume the same general shape as a fluid, so great are the pressures concerned. In geological movements hard limestones are folded like paper, solid granite is rolled out like dough, and diamond itself, under the weight of a few miles of rock, would be moulded like wax.

Mr. Martin asks if I am " aware that there is only one theory generally accepted among physicists to account for the origin of nebulae ?—namely, that they are produced by the impact of two colliding suns." No: I have observed that physicists find this theory inadequate. The collision of two suns requires such a remarkable coincidence of orbits and of time that its occurrence may be regarded as exceptional. If, however, they happen to collide, and their substance is converted into incandescent vapour, what then? The vapour, owing to its tenuity, will lose heat very fast, and its constituents will condense into solid particles of low temperature.

In my opinion the collision theory seems not to account for the rotary movement of a nebula. Then there exist groups of stars in the same stage of evolution, *e.g.* the Pleiades, also certain stars in Orion. It seems as if the members of such groups had originated simultaneously in the same event, in which case an explanation is required more comprehensive than the collision of suns. The encounter of broad streams of meteoric matter is more probable; but whence do the streams originate ?

Mr. Martin asks my authority for the " remarkable statement " that the changes in the silicates involve a comparatively small change of energy, No authority is needed for a fact under common observation. The chemistry of the silicates is seen in the manufacture of glass, porcelain, and earthenware, and is comparable, so far as energy is concerned, with the chemistry of such salts as sulphates, the main difference being that the former is carried on in a fused medium at high temperature, whereas the latter occurs in aqueous solution at low temperature.

In the August issue of this magazine (page 74)

(1) *Anabolic*=involving accumulation of energy ; *catabolic* =involving dispersion of energy.

I suggested one kind of silicon chemistry which would involve great changes of energy, the chief factor being the deoxidation and subsequent re-oxidation of silicon. On page 136 Mr. Martin quotes this kind of action as the chemistry of silicates ; but it is really the chemistry of silicon, and differs from that of silicates or silica in the same sense that the chemistry of carbon differs from that of carbonates or CO_2. From Mr. Martin's former articles in this journal I can only gather that he was thinking, when he wrote them, of silicates and SiO_2. This seems quite definite in the passage on page 13 of the June issue, where he says : "The storage of energy by silicon would not be brought about by the addition of hydrogen, as appears to be the case with carbon, in ordinary life. Some other element, perhaps oxygen itself, would perform this function in life at the tempera-ture I am considering."

I agree with Mr. Martin that it is at high tem-peratures that silicon may be expected to play an active part in life. If my references to the subject have not been definite, it may be because I can-not form a definite idea of silicon life with the limited knowledge of silicon chemistry at our disposal.

I recommend Mr. Martin to give further study to the chemistry and physiology of the proteids ; he will then surely speak more respectfully of their sulphur than he does on page 137. As to his arguments based on the presence of heavier elements in living substance, may I suggest that they prove nothing ? The same facts could be equally well used by an advocate for an opposite theory—namely, that the earth has grown hotter, and that the heavier elements have been partially substituted for the lighter.

A theory of life must be wide enough to include the whole universe. The causes which give rise to life are inherent in the Cosmos, as are matter, energy, space, and time. In different parts of the universe the manifestations of life may be widely different, and the elements concerned will differ according to the circumstances. It was formerly taught that the origin of life on this world must be sought in the chemical actions of "a time when the earth was still a glowing fireball." So in 1875 wrote Pflüger, the leader of modern vital chemistry. We admit this to be quite possible ; but since recent investigation tends to show that the earth's surface was never much hotter than it is now, we have to consider the alternative possibility of the origin of life under circumstances not widely different from the present. The matter is not being left to speculation. The phenomena of life, including the exact functions of each vital element, are being subjected to experimental investigation ; and before the end of the twentieth century many vital actions may be imitated in the laboratory and applied to utilitarian purposes.

Malvern, November 1900.

GEOLOGY IN HANTS BASIN AND THAMES VALLEY.

BY J. P. JOHNSON.

IN describing the Pleistocene fossiliferous de-posit of Selsey and West Wittering, Mr. C. Reid, F.R.S., suggested that the overlying rubble-drift was contemporaneous with the Thames Valley gravels ([1]), but this correlation has not been accepted. This was chiefly, I think, because Mr. Reid took the extreme view that not only were the latter contemporaneous with the former, but they were also formed in the same way—that is, by subaërial action, as opposed to fluviatile—whereas they are undoubtedly river-drift. While the one assumption cannot be maintained in the light of recent research, it does not follow that the other is incorrect ; and if it can be proved that the two widespread deposits were built up at the same time, then an important datum line will have been gained. The object of this paper is to draw atten-tion to the very conclusive evidence offered by the valley of the Wandle, when considered in connection with other districts, in support of this correlation.

Rubble-drift or coombe-rock is the name applied to the extensive sheets of chalk detritus which project from the dry valleys of the South Downs on to the Hants Basin. As an example of the ex-tent of this deposit, I may mention that it entirely covers that part of the district which lies between Chichester Channel and Brighton. At its source it is usually of considerable thickness, and is essen-tially a breccia of chalk and flints, with incorpora-tions of newer debris, such as greywether sandstone. It, however, rapidly thins out on leaving the Downs, and at the same time becomes more loamy through decomposition of the chalk, and contains fewer flints. This transition can be studied in the coast sections. East of Brighton, where the sea has breached the chalk hills, sixty feet of typical rubble-drift is exposed ; but a little westward of the town much of the chalk has already been de-composed, and the whole deposit is considerably reduced in thickness. At Bracklesham Bay, where the rubble-drift is about eight miles from the Downs, it has degenerated into a stony loam. Palaeolithic implements have been found in the rubble-drift, also bones of mammoth, rhinoceros, horse, etc., and, more rarely, land-shells ([2]), where it has not been largely decalcified.

The rubble-drift is clearly the result of the sub-aërial denudation of the chalk downs, but of what form of subaërial action is unknown. Mr. Reid has offered an explanation ([3]) which fits in with

(1) C. Reid, F.R.S., "Pleistocene Deposits of Sussex Coast and their Equivalents in other Districts." Quart. Journ. Geo. Soc. vol. xlviii.

(2) *See* [Sir] Joseph Prestwich, "Raised Beaches and Rubble-drift in South of England." Quart. Journ. Geo. Soc. vol. xlviii.

(3) C. Reid, F.R.S., "Origin of Dry Chalk Valleys and of Coombe-rock (=Rubble-drift)." Quart. Journ. Geo. Soc.

the facts at present observed ; but one is naturally shy of theory, especially as applied to such an obscure formation. That the denudation must have been very severe is shown by its enormous extent.

The relation of this to the other Pleistocene deposits of the area at present under consideration may be briefly indicated thus :—

WEST WITTERING.	SELSEY.	BRIGHTON.
Rubble-drift.	Rubble-drift.	Rubble-drift.
Marine shingle (traces).	Marine shingle.	Marine shingle
Estuarine clay.	Estuarine clay.	and sand,
Fluviatile beds, with	Marine saud.	with derived
derived erratics.	Erratic Deposit.	erratics.

As pointed out by Mr. Reid, who discovered it, the " erratic deposit " is clearly the equivalent of the chalky boulder-clay of the Thames Basin, which we know to be older than the Palaeolithic river-terraces. If, then, the rubble-drift can be shown to be contemporaneous with these river-terraces, or · with one of them, then we shall be able to correlate the Pleistocene subdivisions of the Hants Basin with those of the Thames.

It would be strange indeed if a deposit of the nature of rubble-drift, while so extensively developed in the Hants Basin, were yet unrepresented in that of the Thames—if the subaërial agencies that operated on so large a scale on the South Downs had left the North Downs untouched. Rubble-drift, however, does occur in the Thames Basin. In the dry chalk valleys of the North Downs we find a deposit which is identical with the typical rubble-drift of the Hants Basin, and which occupies a similar position as far as it extends. It is true that it is mapped as river-gravel, but there are very serious objections to its being classed as such, and in some instances, as in the case of the Carshalton mass, no river could have existed in the position in which the deposit is found. It may be urged that the insignificant masses of rubble-drift in the North Downs cannot be safely correlated with the extensive sheets which project from the South Downs, but this objection is readily answered when we come to examine the Palaeolithic drift of the valley of the Wandle.

The Palaeolithic deposits of the Wandle Valley were described by me in the August number of SCIENCE-GOSSIP (*ante*, p. 69). They occur in two terraces, the lower one of which not only occupies the river valley but also extends into the dry chalk valleys of the North Downs. This lower mass may be divided into two contemporaneous sections—viz. (*a*) the angular detritus of the chalk area, which comprises the dry valleys and coombes; and (*b*) the subangular gravel of the Eocene tract, or river-valley. The former is typical rubble-drift, and passes sideways into the latter, which is a true river-gravel. I have no doubt that a similar state of affairs will be found

in those other tributary valleys of the Thames that extend into the chalk hills, and I know from personal observation that it is the case with the valley of the Ravensbourne. If then we accept the correlation of the rubble-drift of the North with that of the South Downs, and it would be unreasonable not to do so, then the contemporaneity of the rubble-drift of the Hants Basin with the newer Palaeolithic gravels of the Thames is established. The reason why the rubble-drift has retained the character of a sub-aërial drift throughout the whole of that part of the Hants Basin which we have been considering, and yet has mostly taken the form of a river-gravel in the Thames Basin, is that in the former area the rubble-drift had only a level plain, left by marine denudation, with a few small streams to overwhelm ; but that in the latter tract it had to cope with a well-developed river-system, so that directly it left the dry chalk valleys it was converted into river-gravel.

The relation of the Pleistocene deposits of the two areas may be briefly indicated thus :—

THAMES BASIN.		HANTS BASIN.
Newer Palaeolithic terrace	=	Rubble-drift.
Older Palaeolithic terrace		Fluviatile, estuarine, and marine deposits.
Chalky boulder-clay	=:	Selsey erratics.

150 *High Street, Sutton, Surrey.*

THE completion of the great catalogue of books at the British Museum, after twenty years' labour, is an important event. This index is contained in upwards of six hundred volumes, and enumerates the titles of no less than two million books. The staff is now to be occupied in constructing the subject index, which it is estimated will require fifteen years to complete.

MR. HENRY SPENCE, Hon. Secretary of the Birkbeck Natural History Society, sends us a copy of its fourth annual report, and the Committee desires us to call the attention of those interested in the subject to this society. Its meetings are monthly, and are held at the Birkbeck Institute, Bream's Buildings, London, W.C., where the Hon. Secretary may be addressed. There are also monthly excursions conducted by various members with local knowledge. Visitors are invited to communicate with Mr. Spence.

DISCONTINUOUS DISTRIBUTION.—In the "American Naturalist" for November Professor W. M. Wheeler records a notable instance of "discontinuous distribution." In 1886 a remarkable and aberrant arachnid was reported from Sicily under the name of *Koenenia mirabilis.* Although representing a distinct group it had a superficial resemblance to the whip scorpions. In the spring of 1900 Professor Wheeler obtained in Texas an arachnid specifically identical with the Sicilian form. Professor Wheeler maintains from various evidence that the Koenenia was not introduced, and he regards it as a survival of an ancient fauna. It may be remembered that a similar case occurs in Procapyx stylifer, a primitive thysanurid insect, in Liberia and Argentina.

AN INTRODUCTION TO BRITISH · SPIDERS.

BY FRANK PERCY SMITH.

(*Continued from page* 207.)

GENUS *DIPOENA* THOR.

MAXILLAE inclined towards the labium. Eyes in two rows, the posterior centrals being closer to one another than each to the adjacent

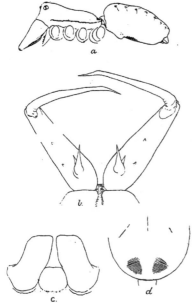

FIG. 1. *Theridion lineatum.* Male. *a.* Profile. *b.* Falces, viewed from underneath. *c.* Maxillae and labium. *d.* Posterior portion of cephalo-thorax, upper side showing stridulating organ.

lateral. The posterior row is somewhat curved, its convexity being directed forwards. Clypeus very high. Abdomen with a strong pubescence.

Dipoena melanogaster Koch. This uncommon spider may be easily recognised by the generic characters.

GENUS *LASEOLA* SIM.

The characteristics of this genus are similar in many respects to those of the genus *Dipoena* Thor. The eyes of the posterior row are, however, nearly or quite equidistant.

Laseola inornata Cb. (*Theridion inornatum* Bl.)
Length. Male 1.7 mm., female 2 mm.

The legs and cephalo-thorax are of a yellowish-red colour, the tibiae of the first and second pairs of legs being darker. The abdomen is convex above, almost globular, and of a blackish-brown hue. This spider is not common.

Laseola tristis Hahn.
This spider, which is extremely rare, is of a uniform sooty black.

Laseola coracina Koch. May be distinguished from *L. tristis* Hahn. by the tarsi and metatarsi being white or yellowish-white. It is a very rare species.

Laseola prona Menge. A rare spider described in "Preussische Spinnen," page 177.

FIG. 2. *Theridion lineatum.* Female. *a.* Profile, legs and palpi truncated. *b.* Extremity of palpus, showing pectinated claw. *c.* Extremity of tarsus of fourth leg, showing row of short spines. *d.* Vulva. *e.* Vulva profile.

GENUS *EURYOPIS* MENGE.

Maxillae inclined towards the labium. Clypeus high. Posterior eyes in a strongly curved line, its convexity being directed forwards. Abdomen strongly acuminate at its hinder part.

Euryopis flavo-maculata Koch. (*Theridion flavo-maculatum* Bl.)
Length. Male 3.5 mm., female 4 mm. Cephalo-thorax brown. Legs paler, with the exception of the tarsi, metatarsi, and tibiae. Abdomen very dark brown, with some distinct reddish and pale markings on its upper side. A rare spider.

Euryopis haematostigma Bl. (*Theridion haematostigma* Bl.) This species may be distinguished from *E. flavo-maculatum* Koch by its smaller size. The cephalo-thorax is also of a paler tint. Found in Ireland.

GENUS *STEATODA* SUND.

Eyes in two curved rows, their convexities being towards one another. The centrals form a quadrilateral figure, almost a square. The anterior centrals are larger than the laterals of the same row. The male is furnished with a very distinct stridulating organ.

Steatoda bipunctata Linn. (*Theridion quadripunctatum* Bl.)
Length. Male 5 mm., female 6 mm.
The comparatively large size of this species and the peculiar colour of the abdomen, which is purplish-brown, make it easily recognisable. It is a fairly common spider. I have recently taken females in a dilapidated barn at Golder's Hill, Hampstead.

GENUS *CRUSTULINA* MENGE.

Similar to *Steatoda* Sund.; but the anterior central eyes are equal, or the centrals smaller than the laterals. The convexity of the anterior row is not so great as in that genus.

Crustulina sticta Cb. (*Theridion stictum* Bl.)
Length. Male 2.2 mm., female 2.7 mm.
Easily recognised by the cephalo-thorax and sternum being studded with minute punctures. Not common.

Crustulina guttata Wid.
Length. Male 1.7 mm., female 2 mm.
Closely allied to *C. sticta* Cb., but smaller. The abdomen is also considerably darker, and is marked with pale spots. Not common.

GENUS *TEUTANA* Sim.

May be distinguished from the genus *Steatoda* by the posterior row of eyes being straight, or slightly curved, with the convexity directed backwards.

Teutana grossa Clk. (*Theridion versutum* Bl.)
Length of male 6 mm. May be distinguished from its allies by its large size. Rare.

Teutana nobilis Thor.
Described in " Spiders of Dorset " under the name of *Steatoda clarkii.*

GENUS *ASAGENA* SUND.

This genus is distinguishable from *Steatoda* by the distance between the posterior central eyes being less than one of them and the adjacent lateral, and also by the lateral eyes being set on distinct oblique prominences.

Asagena phalerata Panz. (*Theridion signatum* Bl.)
Length. Male 4.5 mm., female 5 mm. A rare spider.

GENUS *ENOPLOGNATHA* PAV.

The spiders in this genus are similar in many respects to those preceding, with the exception that the maxillae are only slightly inclined towards the labium, which is hardly half their length. The sternum is rather prolonged posteriorly.

Enoplognatha thoraccia Hahn. Described as *Neriene albipunctata* Cb. in " Spiders of Dorset."

GENUS *PEDANOSTETHUS* SIM.

In this genus, which is closely allied to *Enoplognatha*, the labium is more than half the length of the maxillae. The sternum, also, is truncated at its posterior extremity.

Pedanostethus lividus Bl. (*Neriene livida* Bl.)
Length. Male 3.5 mm., female 4 mm.
Cephalo-thorax and legs reddish-brown, abdomen dark yellowish-brown.

(*To be continued.*)

RAINFALL IN BRITAIN.—Mr. H. Mellish read a paper on "The Seasonal Rainfall of the British Isles" at the last meeting of the Royal Meteorological Society, which was illustrated with a number of lantern slides. He discussed the rainfall returns from 210 stations for the twenty-five years 1866-90, and calculated the percentage of the mean annual rainfall for each season. In winter the largest percentages are found at the wet stations and the smallest at the dry ones. Spring is everywhere the driest quarter, and very uniform over the country. In summer the highest percentages are found in the dry districts and the lowest in the wet ones. As the spring is everywhere dry, so is the autumn everywhere wet, and there is little difference in the proportion of the annual total which falls in the different districts. As regards the relation between the amount of rain which falls in the wettest and driest month at any station, it seems to be generally the case that the range is larger for wet stations than for dry ones. In wet districts rather more than twice as much rain falls on the average in the wettest month as in the driest, and in dry districts rather less than twice. The paper is one of considerable interest, and contains many points which may open the way to further discussion.

BUTTERFLIES OF · THE PALAEARCTIC REGION.

By HENRY CHARLES LANG, M.D., M.R.C.S., L.R.C.P. LOND., F.E.S.

(*Continued from page* 201.)

Genus. *COLIAS.*

C. wiskotti Stgr. B.E.Z. 1882.

45—58 mm.

♂ ground colour of wings greenish, sometimes with a yellowish tinge, marginal border of f. and h.w. very wide, reaching nearly to disc. spot, veined with light yellow. Disc. spot f.w. distinct and black not centred with white beneath, that of h.w. light orange, basal patch lighter than ground colour. ♀ ground colour orange. Border on f.w. narrower than in ♂, with three or four not very distinct orange spots; on h.w. the border is broken up on its internal half by square-shaped light

C. wiskotti ♂.

orange spots; disc. spot bright orange. U.s. ♂, f.w. greenish yellow, yellower toward apex. Near an. ang. a rather black spot, above which is a smaller one. H.w. almost uniform yellowish green, with an indistinct white disc. spot. ♀ as in ♂, but lighter. Base of f.w. orange.

HAB. Alai mountains.

The named varieties and aberrations of this species are numerous, though some of them are founded on very slight differences. There seem to be two separate racial forms, one as in the type, with a very broad border and with light-coloured wings; and the other smaller in size, with a much narrower border and a tendency to orange colouration.

a. var. *soros* Rom. Mem. iv. p. 353 (1890). Size of type, with a broad border. Ground colour orange yellow. HAB. Pamir.

b. var. *separata* Gr.-Gr. Hor. Ent. Ross. xxii. (1888), p. 305. Size of type. Wings greenish-yellow in both sexes. F.w. in ♀ ochre yellow at base; marginal border narrow, intersected by yellow veins. HAB. Altai.

c. var. *chrysoptera* Gr.-Gr. Hor. Ent. Ross. L.c. smaller than type; wings in both sexes ochreous yellow (rarely greenish); marginal band narrower. HAB. Roschan Transalai, Sarykol.

d. var. *draconis* Gr.-Gr. *? aurantiaca* Stgr. Smaller than type. Border narrower, veined with yellow ground; colour deep orange. ♀ with few or no yellow spots on border. HAB. "in montibus ad 'lacum draconis,'" Gr.-Gr. Kara-kul.

e. var. *leucotheme* Gr.-Gr. A light form of both sexes found in the N.W. extremity of C. Pamir.

f. ab. *leuca* Stgr. The white form of ♀ *C. wiskotti.*

g. var. *sagina* Aust. The smallest form of the species. Light ochreous yellow in both sexes. Greenish towards base of h.w. ♂ somewhat resembles var. *chrysoptera*, but smaller, and the black

C. wiskotti ♀.

border very slightly veined. ♀ with more distinct marginal spots. HAB. Kara-Sagin mountains.

2. **C. edusa.** Fab. Mant. p. 23, n. 240 (1787). Lg. B. E. p. 61, pl. xiv, fig. 1, pl. xvi, fig. 3. Larva and pupa. "The Clouded Yellow."

40—50 mm.

Wings orange-yellow, ♂ f.w. with a broad black ou. marg. border, veined with orange-yellow: The disc. spot is large and black. H.w. orange-yellow shaded with dusky towards base, with a large light discoidal spot varying from reddish to pale yellow. Outer marginal black border rather broad and veined with yellow. ♀ larger than ♂. F.w. with black border wider, without veins, but spotted with yellow, the spots varying in size and number. Border of h.w. somewhat similar, but fainter and narrower, the yellow spots usually less defined. Bases shaded with dusky greenish. Disc. spot large rounded and deep orange colour.

This butterfly presents variations in the following particulars: *a.* Size. Normally the ♀ is larger than the ♂, but examples of unusually large males occur as well as of small females. *b.* Shape. The apex of f.w. is usually more pointed in ♀ than in ♂, but occasionally they are abnormally pointed in ♂. *c.* Colour. The male varies in the ground colour from deep orange-yellow to light-yellow ochre. I have in my collection a specimen of this colour taken by myself on Canvey Island in Essex in 1892. Another example, unusually light, taken at Smyrna by Mr. Bliss in 1900. In both these there is a bluish-green shading about the h.w. not usually seen. Some specimens, especially British, are very deep clear orange-yellow, with but little shading on h.w. The female in its normal form is much more constant in colour than the ♂. There is however a dimorphic form in which the wings are greenish-white, the bases of f.w. and the area of h.w. being shaded with bluish-grey, the disc. spot of h.w. still remaining yellow or orange. This is the ab. *helice* of. Hubner. A form of ♀ occurs which may be considered intermediate between *edusa* and *helice* in which the ground colour is very pale yellow ochre. This is ab. *aubuissoni* Car. As stated above, all the orange Coliades are liable to a white dimorphism in the ♀. *d.* Breadth of marginal border. This varies considerably in ♂. In some specimens it is nearly double the width seen in others, though an extremely narrow border is rare. In ♀ the breadth of border is more constant, less so in *helice* than in *edusa*. *e.* Markings of border. The veining of the border in ♂ varies considerably. It is caused by the yellow colouring of the termination of the nervures which cross the black marginal band. In some specimens every nervure is clearly marked on both wings, in others hardly any except one or two near the apex of f.w. The most common condition is a mean between these extremes. This striation of the border seems to be an exclusively male character in this and other members of the genus. The spots on the border on f.w. ♀, as stated above, vary greatly in size and number. There are sometimes as many as seven, most usually three or four; the most stable is that nearest the an. ang. Sometimes the border is almost, and occasionally quite, without spots. The marginal spots on h.w. also vary; sometimes they are large and square or oval, sometimes reduced to mere points. They vary in colour from orange to greenish-yellow, *f.* Costal patch on h.w. ♂ varies in colour from reddish to nearly white. I have never seen a specimen in which it was wanting. *g.* Violet reflection, seen sometimes in ♂, varies in intensity; it is most marked in the darker specimens, but is not a constant character, and does not apparently occur in ♀ in this species. It is seen best in fresh specimens.

I have dwelt thus upon these variations, because it is interesting to notice how far *C. edusa* approaches other species in certain particulars. The proximity to one another of the various species of the orange-coloured group has led several writers to conjecture that many of the types received as specific are only local forms of *C. edusa*.

This species is the most widely distributed of the genus. Its irregular appearance in Britain is well known to every collector, but no satisfactory solution of this problem has yet been given.

HAB. Central and S. Europe, Western Asia, Persia, Daghestan, Algeria, Egypt, Canary Islands. VII.—X., IV.—VI. h.

LARVA. Cylindrical, dark green, with a lateral whitish stripe, spiracles yellowish. On *Trifolium,* *Cytisus, Onobrychis,* and other Leguminosae. VI., VII.

PUPA. Pale yellowish green, with lateral yellow abdominal stripes. The wing cases are streaked with black.

a. ab. *helice* Hub. ♀. The white form alluded to above. It occurs with the typical form, but as a rule is not nearly so common. In the summer of 1900, when we had an abundance of *C. edusa,* in the neighbourhood of Southend, in Essex, ab. *helice* was in many places the commoner form.

b. ab. *aubuissoni* Car. Occurs with type, but is rare. Wings light ochre-yellow in place of orange. A form intermediate between *C. edusa* and its var. *helice.*

3. **C. fieldii** Mén. Cat. Mus. Petr. Lep. I., p. 79 (1855).

47—53 mm.

Very closely resembles *C. edusa,* but the ground colour of the wings is less brilliant than in that species, and has a brownish tint when compared with *C. edusa.* Black border on f.w. less distinctly veined in ♂ and stronger and darker in ♀ with the yellow spot of the same colour as the wings. Disc. spot f.w. large and plain centred with white beneath. Fringes more distinctly rosy red than in *C. edusa.* Disc. spot h.w. of the same colour as ground of f.w.

HAB. Mountains, Amdo, Thibet, Sikkim, Mongolia. VII.

(To be continued.)

THE KAMMATOGRAPH.—We have had our attention directed to an invention connected with the photography and later reproduction of moving objects on a screen. As is now well known, the usual plan for obtaining this effect is by using a long band of small photographs which have been taken rapidly, as transparencies, and then running them in front of a lantern, so reproducing the effects desired as pictures on the screen. Excepting in the hands of a professional operator, the band is said to be liable to give trouble, or at least considerable anxiety. Messrs. L. Kamm & Co., of London, have overcome the necessity of having the band of separate pictures by placing them all on a disc. When exhibiting, this disc is placed in front of the lantern, and by a spiral movement it brings each view exactly opposite the lens, so producing the same effect as if worked by the band. The Kammatograph itself is an arrangement including a camera and projector combined, thus avoiding the use of two apparatus. It appears to be one that will be found useful to amateurs.

A CURIOUS "Weeping Chrysanthemum" was shown at the meeting of the Royal Horticultural Society on December 4th. It was one of eleven all possessing a peculiarity in the geotropic direction of the branches, which were bent like those of a weeping ash, but upturned heliotropically at the ends where the flowers are produced.

THE Board of Agriculture, since the recent retirement of Mr. Charles Whitehead from the position of Technical Adviser, has made some alterations in the means by which the Board obtains advice on questions concerning agricultural botany and economic zoology. It is now decided that the scientific assistance required shall be furnished by the Royal Botanic Gardens, Kew, and the Natural History Department, South Kensington, respectively.

IN the "Journal" of the Franklin Institute Mr. Edward L. Nichols writes on the use of the acetylene flame in physical laboratories. He deals with various interesting points, such as the decrease in illuminating power when acetylene is stored for some time, especially over water. The characteristics of pure acetylene flame, and the uses of acetylene for the lantern, for the production of high temperatures, and for photographic measurements are also treated.

THE Anthropological Institute have decided to issue a monthly record of the progress of anthropological science. It will be entitled "Man," and will include contributions on physical anthropology, ethnography, and psychology. There will also be articles on the study of language and the earlier stages of civilisation, industry, art, and the history of social institutions and of moral and religious ideas. Special attention will be given to investigations which deal with the origins and earlier stages of those forms of civilisation that have become dominant.

THE medals of the Royal Society for this year were presented by the President, Lord Lister, at Burlington House on November 30. They were awarded as follows:—The Copley medal to Professor Marcellin Berthelot, For.Mem.R.S., for his brilliant services to chemical science ; the Rumford medal to Professor Antoine Henri Becquerel, for his discoveries in radiation proceeding from uranium ; a Royal medal to Major Percy Alexander MacMahon, F.R.S., for the number and range of his contributions to mathematical science ; a Royal medal to Professor Alfred Newton, F.R.S., for his eminent contributions to the science of ornithology and the geographical distribution of animals ; the Davy medal to Professor Guglielmo Koerner, for his brilliant investigations on the position theory of the aromatic compounds ; and the Darwin medal to Professor Ernst Haeckel, for his long-continued and highly important work in zoology, all of which has been inspired by the spirit of Darwinism.

A STORY comes from Essex to the effect that a postman was rendered insensible near Great Parndon, near Harlow, on the evening of December 22, by the shock sustained through the fall of a meteorite in his immediate proximity.

THE proprietors of "Knowledge" announce in its issue for December that they have arranged with Mr. M. I. Cross, one of the authors of the "Handbook of Modern Microscopy," to conduct the column on Practical Microscopy in that journal.

THE Royal Society, in selecting for the coming year as its President the eminent astronomer, Sir William Huggins, K.C.B., F.R.S., D.C.L., LL.D., Ph.D., has well commenced the new century. Sir William was born in London in 1824. He built his private observatory at Tulse Hill in 1856, and has long been an authority on spectroscopic astronomy. Among Sir William's recreations is the study of botany.

DEATH, at the age of sixty-three years, removed on December 20 a well-known devotee to natural history in the Croydon district when Philip Crowley, F.L.S., F.Z.S., died at his residence, Waddon Hall, near that town. His connection with the brewery company associated with his family name had produced ample means, enabling him to amass considerable entomological and ornithological collections, also to take some interest in horticulture.

THE lease of Bushey House and the surrounding grounds of thirty acres in extent has been assigned by Her Majesty to the Council of the Royal Society for the National Physical Laboratory. The Government will also add a further sum of £2,000 to the grant for building in order that the extensive alterations and repairs which will be necessary may be effected. This is in consequence of the necessity for removal of the Kew Observatory through electrical disturbance from projected tramways passing from Kew to Richmond.

WRITING to the "Friend," Miss Charlotte Fell Smith discusses "Some Quaker Contributions to Nineteenth Century Literature." In a pleasantly written article she mentions several writers who have added to the knowledge of natural and physical science. Among these are the works of Luke Howard, meteorologist ; "Edward Doubleday on the Nomenclature of Lepidoptera," Joseph Woods in Botany, Edward Newman in Entomology and Ornithology, "Henry Doubleday upon British Birds," and Henry Seebohm on Bird Migration. Miss Fell Smith has, however, transposed the work of the Doubledays, for, although both were eminent entomologists and ornithologists, it was Henry who studied the nomenclature of British butterflies and moths, of which he issued a synonymic list, founded on an arrangement of Guéné, a French savant, no longer adopted by scientific entomologists. Although all these workers did good service for the passing time during which they lived, it is perhaps worth noting that there is not, we believe, one of them who has left any system, or even science, which is in the present day adopted by the more exact investigators in its realm. Neither is there any book written by any of them which is completely relied upon by advanced students of science at this period.

We have received the following books, but in consequence of pressure on our space the notices must be deferred ; "Studies Scientific and Social," Vols. I. and II., by Alfred Russel Wallace ; "The Harlequin Fly," by Miall and Hammond ; "What is Heat and Electricity?" by Frederick Hovenden ; "Lord Lilford : a Memoir," by his Sister ; "By Land and Sky," by Rev. John M. Bacon; "What is Life?" second edition, by Frederick Hovenden ; "Magnetism and Electricity," by P. L. Gray ; "Outlines of Field Geology," by Sir A. Geikie ; "Elementary Mechanics of Solids," by W. T. A. Emtage; "History of Chemistry," by Dr. Ladenburg ; "The School Journey," by Joseph H. Cowham ; "Modern Chemistry," 2 vols., by W. Ramsay ; "Story of Bird Life," by W. P. Pycraft ; "How to Avoid Tubercle," by T. Wise ; "Publications" 46, 47, 48, 49 of Field Columbian Museum ; "Report of Marine Biological Association of the United Kingdom"; "Reports of Director of Experimental Farms, Canada"; "Transactions of Hull Scientific and Field Club," Vol. I., No. III.; "Alga Flora of Yorkshire," by W. and G. S. West ; "Mosquitoes of the United States," by L. O. Howard; "Proceedings of the Society for Psychical Research," Part XXXVII.; "Proceedings of 12th Annual Meeting of (American) Association of Economic Entomologists "; "Millport Marine Biological Station " ; "The Scientific Roll, Bacteria," by Alexander Ramsay ; "Air in Rooms," by Francis Jones; "Annals of Andersonian Naturalists' Society," Vol. II., Part II.; "Godalming" by T. F. W. Hamilton ; "Subject List," No. 3, Patent Office Library ; "Museum Handbooks," Nos. 3 and 4, Essex Field Club, &c.

Our Bird Friends. By RICHARD KEARTON, F.Z.S. xvi + 215 pp., 8 in. × 5½ in. With 100 illustrations by C. KEARTON. (London : Cassell & Co., Ltd. 1900.) 5s.

A book for all boys and girls is the second title for this pretty and exceedingly interesting volume. The pages are full of pleasant reading and good bird stories, told by a true lover of birds in a natural state of freedom. It is much more than a children's book, being one that will readily entertain their elders. The book teems with bird-lore, and, what is of more consequence, the statements are accurate and original, not copied from other sources, as is so often the case with popular bird-books. The hundred illustrations are admirable, as will be readily understood, being from photographs by Mr. Cherry Kearton. By permission of Messrs. Cassell & Co. we reproduce one showing the daily growth in the first week of the life of a young blackbird. This, besides being entertaining, has scientific value, as indeed have most of the pictures. Many of the subjects have been difficult to obtain, and all are most successfully portrayed.

Heredity and Human Progress. By W. DUNCAN MCKIM, M.D., Ph.D. viii + 283 pp., 8in. × 5½ in., with diagrams. (New York and London : G. P. Putnam's Sons. 1900.) 7s. 6d.

The author opens his subject by saying, "Profoundly convinced of the inefficiency of the measures which we bring to bear against the weakness and depravity of our race, I venture to plead for the remedy which alone, as I believe, can hold back the advancing tide of disintegration." That remedy is the painless extinction in lethal chambers, of all human beings of whatever age, who suffer either from moral or physical weakness: the drunkards, the criminals, and other human failures

who live only to reproduce their objectionable qualities through heredity in coming generations. Of course this is simply appalling to the thoughtless person who more or less comfortably passes through life, but who, if Dr. McKim had his way, would as comfortably pass out of life without unnecessary delay. Yet the author and the many who think with him are theoretically right, though who is to decide where to stop in the reduction of the failures is the difficulty. The artificial life under high civilisation of human beings is bound to lead to the disintegration of communities and the return to severer methods of selection of the fittest. The time will come when the form of selection will assert itself. As we are reminded, poverty, disease, and crime are traceable to one fundamental cause, depraved heredity: they are not a necessary human heritage, but result from our toleration of the "weak and vicious." The book before us is carefully thought out and written. Whatever may be one's opinions, it is a book to read.

The Book of Fair Devon. 209 pp., 9 in. × 4¾ in., with numerous illustrations. (Exeter : United Devon Association. 1900.) 2s. net.

This prettily produced handbook to the scenery and places of interest in Devonshire is described as the official invitation of the United Devon Association to visitors and others to become acquainted with that beautiful county. The volume has been compiled by many Devonians under an influential editorial committee. Especial articles have been contributed upon the climate, education, flora and natural history, water sports, sea and freshwater fishing, hunting, and many other subjects. The article on natural history is rather general than specific, and treated from the popular side. We learn that no less than 1,142 species of flowering plants occur in the county ; also that the avifauna embraces upwards of three hundred species of birds. Reference is made to the numerous kinds of fish that occur around the coasts. The illustrations are effective, being from original sketches and photographs. This little work will doubtless attain its purpose of drawing the attention of many persons to the beauties of "Fair Devon."

The Microscopy of the More Commonly Occurring Starches. By HUGH GALT, M.B., C.M., D.P.H. 108 pp., 7½ in. × 5 in., with 22 original photomicrographs. (London : Baillière, Tindall & Cox. 1900.) 3s. 6d. net.

This is an attempt, as stated by the author, to describe and delineate the commoner starch grains as they appear under the microscope. The treatment is slight, but as far as it goes not unsatisfactory, and we agree with the author that the published drawings of starches which appear in ordinary text-books are somewhat misleading, in that the concentric markings are much less evident in starches as shown under the microscope than they appear in the drawings. We also consider that the author rightly lays stress on the respective measurements of the various starches, as well as on their outlines ; but we think his method of illustrating these is, to say the least, unfortunate. The author explains with the utmost frankness that he had had but little experience of photomicrography—a word so generally accepted in England that we must use it in preference to the alternative micro-photography—when he commenced the series of photographs which illustrate his book. As a result, these same photographs in almost every instance show the faults to which the beginner is prone. The prints may be above criticism, in so far as they are reproductions whose outlines at least must be accepted as absolutely accurate; but we fear that this is all that can be the author would have served his aims better had he been content with careful drawings on the same scale of the various starches mentioned. He would thus have been able to give the details of the starch grains as they really appear under the microscope, without necessarily departing from

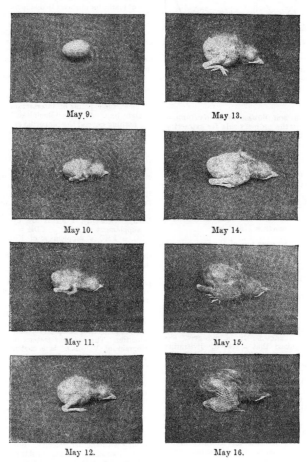

May 9.

May 13.

May 10.

May 14.

May 11.

May 15.

May 12.

May 16.

THE FIRST WEEK IN A BABY BLACKBIRD'S LIFE.

(From Kearton's " Our Bird Friends.")

said of them. It is also a pity the author should mar a praiseworthy attempt to give a practical summary of the leading microscopical characters of the more important starches by the inclusion, as illustrations, of his own earliest attempts in a most difficult branch of photo-micrography. Our own opinion is that under any circumstances photography is not suitable for the purpose, and accuracy with regard to form. It si true these details are rapidly lost when the starch grains are mounted for any length of time; but that only shows the necessity of making the necessary examination on freshly prepared mounts. There is a good comparative table at the end of the book, which we should have liked to see largely extended, and an excellent index.—*F. S. S.*

CONDUCTED BY PROFESSOR G. H. BRYAN, SC.D., F.R.S.

MATHEMATICS.—Dr. G. H. Bryan, F.R.S., having kindly undertaken to contribute occasional columns devoted to popular mathematics, we have pleasure in presenting the first of them. [Ed. SCIENCE-GOSSIP.]

SIMPLE RULE FOR SQUARING A NUMBER.—Most people know the ordinary rule for extracting a square root, but few are aware that by a kind of reverse process the square of any number can be simply found. Thus, supposing we have to find the square of 3,456, the work stands as follows :—

$$3456 \times 3456$$

3000 × 3000	=	9000000
64 × 4	=	256
685 × 5	=	3425
6906 × 6	=	41436

Answer = 11943936

It will be noticed in each line of the first column we double the last figure of the preceding line and bring down one more figure. The numbers in the first column are the divisors which would occur in forming the square root of 11,943,936, and they are written down exactly as in ordinary evolution, when the square root comes out to be 3,456.

TESTS OF DIVISIBILITY.—The ordinary method of "casting out the nines" and "casting out the elevens" is well known. It is, however, less generally recognised that a test of divisibility by 7 and 13 may be found in a somewhat similar form. We have $7 \times 13 = 91 = 100 - 10 + 1$, and $7 \times 13 \times 11 = 1,001$. It follows that numbers greater than 1,000 and less than 1,000,000 will be divisible by 7 or 13 if the difference between the "thousands" and the units be so divisible. Thus 235,683 is divisible by 7 or 13 if $683 - 235 = 448$ is so divisible. This is divisible by 7, therefore 235,683 is divisible by 7. If the number is greater than 1,000,000, we point off the thousands in the ordinary way and subtract the alternate groups from the remaining ones. Thus if the number be 23,749,424,563,215 we have

23	749
424	563
215	

662	1312 difference 650,

which is divisible by 13 ; hence the original number is divisible by 13. The above test of divisibility is made to depend on the divisibility of a number of three figures ; but a further simplification may be made by adding the number in the hundreds place to the tens place and subtracting it from the units

place. Taking 351, we add the 3 to the 5 and subtract 3 from the units place, giving $81 - 3$ or 78, which is divisible by 13, therefore 351 is divisible by 13.

MARKS FOR MISTAKES IN EUCLID.—A correspondent of the "Mathematical Gazette" laments the absence of general agreement as to the deduction of marks for mistakes in mathematical answers. He suggests as an experiment that the readers of the "Gazette" should send the editor postcards stating the deductions they would make in proofs of two propositions in Euclid (I. 37 and IV. 4) for certain specified mistakes. The results, which are promised for a subsequent issue, cannot fail to be interesting.

PROOF OF PASCAL'S THEOREM.—This theorem states that if an irregular hexagon is inscribed in a circle, and pairs of opposite sides are produced to meet, their points of intersection will lie in a straight line. A simple "Euclidean" proof, involving nothing beyond the properties of cyclic quadrilaterals, is given by Mr. R. F. Davis, M.A., in the "Educational Times." In Mr. Davis's figure ABCDEF is a cyclic hexagon, of which BA and DE produced meet in G, and AF, CD in K. CB produced meets KG produced in H, and the circumcircle of DFK meets GK in P, the figure being drawn so that the points on GK occur in the order H, G, P, K. Then, firstly, P D B G are cyclic for DPG = supplement of DPK = supplement of DFK = DFA = supplement of DBA (i.e. of DBG). Secondly, P F B H are cyclic for FPH = supplement of FPK = FDK = supplement of FDC = FBC = supplement of FBH. From the first result BDE or BDG = BPG or BPH, and this by the second result equals BFH. But from the original circle BDE and BFE are supplementary. Therefore BFH and BFE are supplementary. Therefore A, F, E are in one straight line ; that is, EF passes through H; which proves the theorem.

THE PARALLELOGRAM OF VELOCITIES.—There are few subjects so imperfectly treated in most text-books as the composition and resolution of velocities. The ordinary statement of the Parallelogram of Velocities is a contradiction of all common sense. "If a body is moving simultaneously with two different velocities represented by two sides of a parallelogram, it will have a single velocity represented by the diagonal," or words to that effect. The idea of a body moving with two different velocities at the same time is as absurd as that of a person being in two places at the same moment. The usual illustration, which does duty as a so-called proof of the law, is that of a ball moving along a groove with one velocity while the groove is moving with another velocity. But the ball itself does not actually possess either of these velocities, for the first is merely the ball's velocity *relative* to the groove, and the second is the velocity of the groove. What the parallelogram of velocities tells us is that, if the velocity of A relative to B is represented by one side of a parallelogram, and the velocity of B relative to a fixed base—say C—is represented by the other side, then the velocity of A relative to C is represented by the diagonal. Some teachers think that the notion of relative velocity is too difficult for a beginner to understand, but that is no reason for substituting something which nobody can possibly understand.

CONDUCTED BY HAROLD M. READ, F.C.S.

ARSENICAL BEER.—During the past month the sensation-mongers of the daily newspapers have had plenty of opportunity of indulging in a kind of cheap criticism on the so-called "chemicals" in beer. It would appear that some specimens of glucose had become contaminated with arsenic during manufacture from the starch which forms its starting-point. How much or how little is not known, for up to the present there has been no authentic statement. It is so easy to remark that glucose is "loaded with arsenic," and the public would so much rather hear such a remark than that the glucose contained 1 part of arsenic in 25,000, which is really a high estimate, that the newspapers naturally take the course most palatable to their readers. Again, the possibility of the presence of other substances which would produce the symptoms of neuritis has been entirely overlooked. Far be it from us to try to minimise the extreme danger due to the presence of any poison, vegetable or mineral, in any article of food or clothing, but if we approach the question with an open mind we cannot help asking ourselves which patient is more to be pitied, the one who is the victim of arsenic poisoning, or the one who indulges in an excess of the "pure" article? In considering the presence of toxic substances in alimentary articles we may ask how many of the large breweries and glucose manufactories, not to mention hundreds of small ones, employ a capable chemist? A few do so, and a few more employ a "consultant," who in the pressure of other work cannot possibly be in a position to devote himself to the careful study of the one subject. When such an impurity as the one we are discussing is by chance discovered, there is a great outcry against the science in general, while those who are least able to judge, but yet are so placed that they can influence public opinion, endeavour to particularise in a manner which would make the tyro in chemistry blush. It is with extreme interest that we await the publication of authentic analyses and the calculation of the amount of beer daily drunk before the symptoms of neuritis supervene.

ARGON AND ITS COMPANIONS.—In continuing their work on the constituents of atmospheric air, Professor Ramsay and Dr. Travers have found that the gas to which they had provisionally given the name of Metargon is not an element. The spectrum lines supposed to be characteristic were due probably to the carbon monoxide and cyanogen resulting from the presence of carbon in the impure phosphorus they had used to remove oxygen from the original air. Krypton, neon, and xenon have now been obtained in sufficient quantity for the determination of their physical constants. These three elements with argon and helium appear to form a characteristic group by themselves. They exhibit gradations in such properties as refractive index, atomic volume, melting-point, and boiling-point. While their periodicity is so marked, it is curious that there is no place for them in Mendéléeff's Periodic Table. The grouping of the elements which contain them appear as follows:—

Hydrogen 1	Helium 4	Lithium 7	Beryllium 9
Fluorine 18	Neon 20	Sodium 23	Magnesium 24
Chlorine 35·5	Argon 40	Potassium 39	Calcium 40
Bromine 80	Krypton 82	Rubidium 85	Strontium 87
Iodine 127	Xenon 128	Caesium 133	Barium 137

THE PAST CENTURY.—At the close of each year it is a general custom for journals of science to give their readers a brief history of the advances made during the previous twelve months. The practice is so universal that, in view of the many advantages of such a course, it might be thought pessimistic to point out a few of the defects. In the present instance we are confronted, not with the close of a year, but with the end of a century—a century which has practically seen the birth of organised science. If we were to endeavour to give our readers a history, however brief, of the chemistry of the nineteenth century, we should require many pages of SCIENCE-GOSSIP before we had emphasised even the main discoveries, without touching upon the far-reaching effects and economic improvements which have marked each step onwards. It is impossible to speak of a single branch of any industry in which the influences of chemical discoveries have not made themselves felt. In our congratulations on the advances made we are, however, saddened by the thought that our slowly progressive nation still holds aloof from a genuine scientific training for its youth, but prefers the highly interesting, yet practically valueless stories of the sacking of Troy and the loves of Dido.

THE ATOMIC WEIGHT OF NITROGEN.—The classical researches of Stas having so long been regarded as amongst the most exact work ever accomplished in chemistry causes the paper read by Dr. Alexander Scott on December 6th before the Chemical Society to come as a surprise. Dr. Scott finds that during a research having for its chief aim the determination of the hydrogen to oxygen ratio, by comparing the equivalents of hydrazine, ammonia, and hydroxylamine, he could not obtain the same equivalent as Stas for ammonium bromide. The value given by the latter is 98·032, while Dr. Scott finds 97·996, which would lower the atomic weight of nitrogen from 14·046 to 14·010, a number much nearer that deduced from the relative densities of oxygen and nitrogen (16 : 14·003). The silver used was prepared by the reduction of silver nitrate by ammonium formate, and its purity was tested in the severest way. Dr. Scott considers that Stas's bromide must have been contaminated by some impurity, probably platinum, since it turned greyish at 115°, the colour increasing up to 180°, while his own ammonium bromide could be heated up to 180° without change of colour, and could be sublimed in ammonia vapour, the same results being obtained before and after sublimation.

CONDUCTED BY F. SHILLINGTON SCALES, F.R.M.S.

ROYAL MICROSCOPICAL SOCIETY.—On November
21st, Mr. Wm. Carruthers, F.R.S., President, in the
Chair. Mr. Nelson exhibited and described an
erect-image dissecting microscope by Leitz, sent
for exhibition by Mr. C. Baker, figured on page 509
of the Journal, August 1900. The erection of
the microscopic image, effected by means of Porro
prisms, was first described by Behrens in the
Journal of the Society in 1888. This instrument was
valuable as a dissecting microscope; it was provided with hand-rests and three objectives having
a very long working distance. Mr. Disney exhibited a diffraction plate, having the lines ruled
in concentric circles, by which the diffraction
bands were separated with great clearness. The
rulings were about 7,000 to the inch. He also exhibited a steel brooch, the surface of which had
been ruled in the same way. The method by which
the lines were produced was at present a secret.
The articles were of English manufacture, and had
been lent to him by Messrs. Townson and Mercer.
Mr. C. F. Rousselet exhibited an electric lamp for
use with the microscope. After six months' trial
he had found it very satisfactory for work with
low and medium powers. It was manufactured by
Edison Swan Co., and was called the "Focus"
lamp. The President called attention to the exhibition that evening of a number of slides from
the Society's Cabinet, prepared by the late Dr.
Carpenter in connection with his investigations
into the shells of the mollusca. Mr. B. B. Woodward, who has given much attention to this subject,
had also brought down some valuable preparations
for exhibition. Mr. Vezey, at the request of the
President, read a short abstract, copied from the
Report of the British Association for 1846, which
was a *résumé* of the original communication on
shell structure made to that Association by the
late Dr. Carpenter, to illustrate which the slides
exhibited were prepared. The President then called
upon Professor Charles Stewart, who, having referred to the views held upon shell structure at the
present day, and taking the common pinna shell as
an example, proceeded by the aid of drawings on
the blackboard to demonstrate how its structure
was built. Besides studying the sections usually
made, he recommended that the shells should be
broken and the fractured surfaces examined, if a
correct idea of the formation of the shells was to
be obtained.

FRESHWATER ENTOMOSTRACA.—Mr. D. J. Scourfield, in the Proceedings of the South London
Entomological and Natural History Society, calls
attention to the value of Entomostraca in experimental biology. "Their commonness in all parts
of the country, their transparency, the ease with
which they can be isolated and reared under all
sorts of conditions, . . . mark out the Entomostraca
as particularly well fitted for observation in connection with even the most fundamental biological

problems of the day." He adds: "We badly want
detailed studies on local faunas, on the seasonal
distribution and variation of different species, on
the faunas of various types of ponds, on the food
of the most abundant forms, and many similar
subjects."

J. SWIFT & SON'S CONDENSERS.—Messrs. J.
Swift & Son have submitted for our inspection
two excellent condensers of their manufacture.
The first is apo-chromatic, and has a numerical
aperture of ·95. Its aplanatic aperture exceeds,
however, according to our measurements, ·90; and
as the value of a condenser for anything approaching critical work depends on the aplanatic cone of
light that it transmits, it will be seen that this
condenser is eminently fitted for such work. As
an apo-chromatic system it is, of course, distinctly
freer from colour than even the best achromatic
system can be made, and this is very manifest
when using high-angled lenses. The power is
about one-third of an inch, and the price, without
mount, £4 15s. We can recommend this condenser for all high-power work, and it is of
interest to remember that, so far as we are
aware, Messrs. Swift & Son share with Messrs.
Powell & Lealand the distinction of being the
only makers of apo-chromatic condensers throughout the world. The second condenser is achromatic, and is an oil-immersion with a numerical
aperture of 1·4 and an aplanatic cone that we
estimate as exceeding 1·3. The corrections of this
condenser are also excellent, and the working
distance is ample, even with a thick slide. The

FIG. 1. Apo-Condenser.   FIG. 2. Oil Condenser.

power is about ¼-inch, or ⅜-inch with the front
lens removed, and the price without mount is £4.
Both condensers are constructed with the newer
makes of glass manufactured in Jena. It says
much for the enterprise, and perhaps the keen
competition, of our English opticians that we
should have been able to notice in these columns
within a short period three different immersion
condensers of high excellency by three leading
makers.

BAUSCH & LOMB'S NEW CATALOGUE.—Whatever may be the respective merits of English
microscopes and lenses, as compared with those
manufactured abroad, there can be no question as
to the superior nature of the catalogues issued by
our foreign competitors. Messrs. Bausch & Lomb
have just sent us their new catalogue, which is a
large and handsome book of 186 pages, profusely
illustrated, excellently arranged, and strongly
bound in cloth. The copy sent us is numbered
6360! We hope shortly to give a brief notice of
some of Messrs. Bausch & Lomb's instruments and
apparatus, as this firm is now directly represented
in England. In the meantime the American
prices would scarcely serve for the English
market, an apo-chromatic ¹⁄₁₂-inch oil-immersion

objective of N.A. 1.3, for instance, being priced at $120 = £24, and the corresponding objective of N.A. 1.4 costing $160 = £32. Zeiss' prices for similar lenses are respectively £15 and £20. The stands are built entirely upon the Continental model, which is closely adhered to, not only in the horseshoe stand and in the fine adjustment, but also in the later Continental forms of sub-stage arrangements. The list of accessories seems to contain everything that the heart of a microscopist could desire, and sundry apparatus and accessories for mounting are mentioned and illustrated in unusual detail.

ANSWERS TO CORRESPONDENTS.—H. O. W., Budleigh Salterton. The ordinary collodion of the B.P. would serve your purpose. Flexile collodion is used medically and contains castor oil. We would recommend, for the section cutting, Schering's Colloidin, which is largely used by microscopists for section cutting. It can be obtained in chips or solution, A bottle containing 50 grams of the latter can be obtained from C. Baker for 1s. 3d. Beech-tar creosote is to be preferred, especially for clearing colloidin sections. but coal-tar creosote would do equally well, provided it is really white. We shall be glad to give you any further assistance in our power.—F. S. S.

EXTRACTS FROM POSTAL MICROSCOPICAL SOCIETY'S NOTE-BOOKS.

[Beyond necessary editorial revision, these notes are printed as written by the various members, without alteration or amendment. Correspondence on these notes will be welcomed. These extracts were commenced in the September number at page 119.—ED. Microscopy, S.-G.]

NOTES BY W. T. McGHIE.

Bacteria in Water. — A few weeks ago I noticed that a glass ornament on the sideboard in my dining-room contained some flowers in rather cloudy water, and guessing that carelessness had led to the water being left too long I examined a drop of it for infusoria under a moderately high power. I was surprised to find that the water was absolutely thick with every kind of schizomycetes, micrococci, bacteria, bacilli, spirillae, vibriones, and leptothrix forms, besides a few paramoecia, monads, etc. The spirilla forms were unusually plentiful and active, and several zoogloean masses of bacteria were noticeable. I mounted some slides from the liquid, of which the accompanying is one. The microbes are stained with logwood on the cover-glass and mounted in balsam. They will afford a test for the excellence of the objectives, condensers, and fine adjustments of our members' instruments. The vibriones have taken the stain best, the other forms indifferently. By the way, though balsam-mounting is always recommended in the text-books, bacteria show much better mounted dry on the cover-glass.

At least, I find it so. A good quarter-inch objective, which will bear a high amplification, will show well-stained slides surprisingly well. *Scales of Clothes-moth.*—This slide may be considered a trivial one to send round; but though the scales are not rare, they exhibit much beauty of marking in the way of striae and villi when examined under moderately high powers, besides making a charming dark-ground slide under ⅔-inch or 1-inch objectives. I have included the slide, however, principally because I

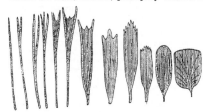

FIG. 1. Development of insect hairs and scales.

think these scales exhibit better than any others the evolution of the insect scale from the simple hair, or rather the probable lines on which it took place. (See fig. 1.) I should much have liked to illustrate this by a picked slide. but could not do so, as I had not the requisite steadiness of hand. On the spread slide, however, will be found plenty of specimens in all the stages of change. The

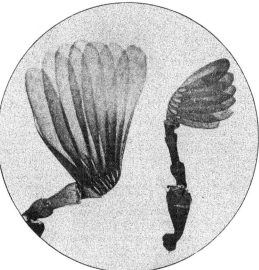

FIG. 2. Lamellae of antennae of cockchafer, *Melalontha vulgaris*, sexes.

piece of wing on the same slide shows well the distribution of the scales: flattest on the centre of the membrane, and shading off into bundles on the edges. In the nervures a crooked system of vessels is perceptible, and these may be traced right through to pedicles of the tufts of bristles at the wing's point, the function being, I believe, to supply the scales with the liquid which, according to Dr. Royston Pigott, is found between the upper and lower membranes of the scales. I am writing without the book, but think this is so · There is certainly, as can be clearly seen with a good objective of wide angle, an intricate system of capillaries feeding every pedicle in the membrane. *Antennae of Cockchafer* (fig. 2).—Of all the remarkable developments which insect antennae attain, I think there are not any more peculiar and interesting than those of the organs of the cockchafer, with their leaf-like expansions, folding out upon one another like the sticks of a fan. This slide contains two of these lamellae mounted in balsam. The accompanying sketches are intended to show the peculiar structure of the leaves. Carpenter (7th edition, 1891, p. 912) says of these markings : "A curious set of organs has recently been discovered in the antennae of many insects, which have been supposed to constitute collectively an apparatus for hearing. Each consists of a cavity hollowed out in the horny integument, sometimes nearly spherical, at others flask-shaped, and again prolonged into numerous extensions formed by the folding of its lining membrane; the mouth of the cavity seems to be normally closed by a continuation of this membrane, though its presence cannot always be satisfactorily determined ; whilst to its deepest part a nerve-like fibre may be traced." Fig. 3 shows the cavities viewed from above under a magnification

Fig. 3. Enlarged view of cavities on lamellae of cockchafer, seen vertically. Fig. 4. Enlarged view of cavities on lamellae of cockchafer, seen sideways.

of 1,000 diameters, and fig. 4 the aspect they present when seen partly sideways at the edges of the lamellae. A memoir of the structure by Dr. Hicks is to be found in the "Transactions of the Linnaean Society," xxii. page 147. [We have copied Mr. McGhie's drawings, but we think our readers will find that these censory organs are really pits containing minute papillae. They are generally assumed to be used for swelling.—. ED. Microscopy, S.-G.] *Paddle-leg of Dytiscus marginalis* (fig. 5) needs little accompanying notice. *Dytiscus marginalis* is one of the best-known of the beetle tribe. A friend of mine, who has a large conservatory containing an artificial pond—the waters of which, by the way, have developed a remarkably rich growth of diatomaceae —found a fine large specimen of the larval form of this formidable insect that was working great havoc among the tadpoles of the aquarium. He kept it alive in a glass jar for several weeks, and we were able to watch its habits. Its fierceness and voracity correspond with its repellent aspect. [This is, of course, the hind leg, which is specially adapted for swimming by the flattening of the

tibiae and tarsi, and by their being furnished with rows of long bristles. The fore-legs of the males are even more interesting, the basal joints being expanded into broad flat plates, furnished with curious sucker-like discs, which secrete an adhesive fluid similar to that in the foot of the house-fly. Figs. 2 and 5 are original photographs from

FIG. 5. Hind leg of *Dytiscus marginalis.*

Nature, for which we are indebted to our friend, Mr. R. C. Nelson. Fig. 5 was taken with a Dallmeyer stereo-portrait lens working at f. 4.— ED. Microscopy, S.-G.] *Acute tubercle in human spleen* (fig. 6).—This slide is self-explanatory. A high power is needed to properly display the stained *Bacillus tuberculosis. Epidermis of Leaf of Auricula.*—This was stripped from the underside of a leaf, treated with dilute nitric acid, and stained and mounted in Canada balsam. Its main interest lies in the glandular hairs, which are best seen with a ½-inch objective, and in the stomata.

Remarks on the Foregoing Notes.—I have been much interested in some of Mr. McGhie's slides, and always like to see members preparing their own. There are two slides upon which I would offer some remarks. *Bacteria in Water.*—I would suggest another stain. Logwood is not a good stain for bacteria. Some of the aniline stains, such as methyl blue or gentian violet, give much better definition. A preliminary fixation by heat or absolute alcohol is also desirable. Mr. McGhie would then have no hesitation in mounting them in Canada balsam, as the staining would be very pronounced. *Tubercle in Spleen.*—I think this slide is not so self-explanatory as Mr. McGhie believes. I have examined the section carefully with a 1/12-inch oil-immersion and fail to detect any tubercle bacilli. The abbreviations at corner of label of slide—which I construe to mean : par. = paraffin, as embedding agent; al. car. = alum carmine, as stain ; Or. = oil of origanum, as clearing agent ; C.B. = Canada balsam, as mounting medium—represent a method of preparation not calculated to demonstrate tubercle bacilli. The nuclei of the tissue are well stained, and the tubercle, which in the spleen is always secondary to tubercle elsewhere, is seen as miliary granulations, but no bacilli are visible. The bacilli are not readily stained in tissue such as this. The Ziehl-Neelsen method is the best. The special advantage of this method is that not only does it demonstrate the tubercle bacilli, but it is at the same time diagnostic, as no other bacilli are stained in this way except the bacilli of leprosy. The method is as follows : The sections are transferred from weak spirit to carbolic fuchsine stain for fifteen minutes, then decolorised in weak sulphuric acid (sulphuric acid, 10 c.c. ; distilled water, 30 c.c.), and afterwards rinsed in 60 per cent. alcohol and washed in large quantity of water to remove the acid. The sections may be then counterstained with methyl blue, then dehydrated in absolute alcohol, cleared in cedar oil, and mounted in Canada balsam. The bacilli will then be stained red and the surrounding tissue blue.—*J. R. L. Dixon, M.R.C.S., L.R.C.P.*

Examining Bacteria.—For the information of members I give a short explanatory note on the examination of bacteria. If the slide is prepared for photo-micrography, gentian violet is the most suitable stain. Take a drop of the bacterial solution on a platinum wire, and touch with it a clean cover-glass that has been washed with water and alcohol. Then take a second glass, rub the two together so as to get a nice clear film on the glasses. Then filter some fuchsine in aniline, and place the cover-glasses in the pigment. Occasionally take one out with the forceps, and if stained wash well in alcohol. Now stain again in the methyl blue, wash in dilute sulphuric or nitric acid, then again in alcohol, and when the cover-glasses are dried with a piece of filter-paper or blotting-paper they can be mounted in balsam. It seems a tedious process, but when understood is very easy. I often examine spleen for tubercle bacilli by this method in fifteen minutes. A splendid double stain can be purchased from Messrs. R. & J. Beck, of Cornhill, London, for one shilling per bottle, which saves great work, but the solution must be warmed before use.—*John Swift Walker, M.D.*

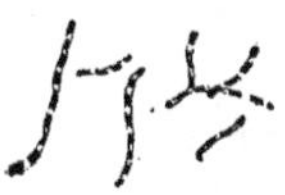

FIG. 6. *Bacillus tuberculosis* in sputum, from photograph.

[We gave a note on staining tubercle bacillus in SCIENCE-GOSSIP, vol. vi. p. 247. Bacillus tuberculosis occurs in long, thin, non-motile rods, straight or curved, rounded at the ends, frequently beaded, as shown in the illustration, 2 microns to 4 microns and occasionally 8 microns long. The micron is the standard of microscopic measurement, and measures one-thousandth of a millimetre. It is generally represented by the Greek letter μ.—ED. Microscopy, S.-G.]

MICROSCOPY FOR BEGINNERS.

BY F. SHILLINGTON SCALES, F.R.M.S.

(*Continued from p. 217.*)

All slides should be carefully labelled and kept in proper boxes. The boxes should contain trays, so that the slides may lie flat, and these can be obtained very cheaply.

This series of papers, begun some eighteen months ago, has now drawn to a close. We have endeavoured throughout to keep in view the wants of the *beginners* for whom we were writing, and to give just that practical information and advice as to the choice of a microscope and its accessories, their uses, and the more elementary methods of preparing objects for the microscope, that experience has taught us the beginner most needs. We have tried also to explain, without technicalities, the principles underlying microscopical practice, methods, and technique, and though we feel we have barely touched upon the subject, we have been encouraged, by numerous letters from readers hitherto unknown to us and by personal expressions of interest, to believe that our—we trust unassuming—efforts have been of real service.

So much is this evident that we may add that it has been decided to shortly republish these papers in handy book form from the offices of this journal. This will enable us to largely extend their scope, and to deal more fully with many matters but lightly touched upon hitherto. We shall welcome gladly any suggestions from readers who may require any point further elucidated or explained.

It may be helpful if we give a brief list of the most useful books dealing with the microscope, and with the animal and plant life for which the microscope is so essential a means of study, to which also we have appended short explanatory notes. This list appears only as a rough draft, and we trust to have the assistance of our readers in making it more complete or in amending it in any way.

In conclusion, may we earnestly urge upon our readers the necessity of taking up some particular subject of study, and of using the microscope as a means to that end, rather than as an interesting optical toy for idly examining heterogeneous slides, which by themselves will soon lose their novelty and interest? Rightly used, the microscope is a means to an end, rather than an end in itself, and is capable of opening out to its owner ever-widening fields of fascinating and absorbing study and occupation. We cannot all be great scientific discoverers, but we may all be builders of the temple of science, if it be only to lay one single brick in that rapidly growing structure, or to supply a little of the clay or straw out of which such a brick may be constructed by others.

BIBLIOGRAPHY OF THE MICROSCOPE AND MICROSCOPY.

TECHNOLOGY.

BAUSCH, E.: Manipulation of the Microscope. Ill., post 8vo. *Bausch & Lomb Co.* (For beginners.) 1$.

BEALE, Dr. L. S.: How to Work with the Microscope. 600 ills., cr. 8vo. *Harrison,* 1886. (Rather out of date.) 21s.

BOUSFIELD, Dr. E. C.: Guide to the Science of Photomicrography. Ill., cr. 8vo. *Churchill,* 1892, (A thoroughly practical book.) 6s.

CLARK, C. H.: Practical Methods in Microscopy. $1.60.

CROSS, M. I., and COLE, M. J.: Modern Microscopy. Ill., demy 8vo. *Baillière,* 1895. (An excellent book. Deals with the use of the microscope and mounting.) 3s. 6d.

DAVIES, T.: Preparation and Mounting of Microscopic Objects, Ill., 12mo. *Gibbings,* 1896. (Fair.) 2s. 6d.

FREY, Prof. H.: Technology of the Microscope. (Trans.) Ill., cr. 8vo. *New York,* 1880. $6.0.

FRIEDLÄNDER, Prof. C.: Microscopical Technology. (Trans.). Ill., 16mo. *New York,* 1886. $1.

GAGE, S. H.: The Microscope. 165 ills. (*New York.* An Introduction to Microscopic Methods.) $1.50.

JAMES, F. L.: Elementary Microscopical Technology. (A Manual for Students.) $0.75.

LEE, A. B.: The Microtomist's Vade-mecum. Demy 8vo. *Churchill,* 1900. (Standard work on advanced histological methods.) 15s.

MALLEY, A. C. : Micro-photography; Wet Collo-
dion, Gelatino-bromide Process. Cr. 8vo.
H. K. Lewis, 1885. 7s. 6d.
MARSH, Dr. S. : Microscopical Section-cutting.
Ill., 12mo. *Churchill*, 1882. (Chiefly animal.)
3s. 6d.
MARTIN, J. H. : Manual of Microscopic Mounting.
Ill., 8vo. *Churchill*, 1878. (Chiefly medical.)
7s. 6d.
MILLS : Photography Applied to Microscope. 2s.
NÄGELI, Prof. C., and SCHWENDENER, Prof. S. :
Microscope, in Theory and Practice. (Trans.)
210 ills., 8vo. *Sonnenschein*, 1888. (Chiefly
theory.) 21s.
PENDLEBURY, C. : Lenses and Systems of Lenses.
Cr. 8vo. *Bell*, 1886. (After Gauss.) 5s.
PHIN, J. : How to Use the Microscope. (For
beginners.) $1.40.
PRINGLE, A. : Practical Photo-Micrography. Ill.,
sm. 4to. *Iliffe*. (Perhaps the most practical
book on the subject.) 5s.
SCHACHT, H. : The Microscope and Its Applica-
tion. (Trans.) Post 8vo. *Churchill*, 1855.
(Chiefly vegetable.) 6s.
SEILER, C. : Compendium of Microscopical Tech-
nology. 12mo. *Philadelphia*, 1880. $1.
SPITTA, E. J. : Photo-Micrography. 6 plates and
63 ills., la. 4to. *Scientific Press*, 1899. (Excellent plates.) 12s. 6d.
SQUIRE, P. W. : Methods and Formulae Used in
the Preparation of Animal and Vegetable
Tissues for Microscopical Examination. Post
8vo. *Churchill*, 1892. (Standard book.) 3s. 6d.
VAN HEURCK, Dr. H. : The Microscope, its Con-
struction, Manipulation, and Technique.
(Trans.) 4to. *Lockwood*, 1893. 18s.
WHITE, T. C. : Elementary Microscopical Manipulation. Ill., fcp. 8vo. *Roper*, 1888. 2s. 6d.

APPLICATIONS.

BEHRENS, Prof. H. : A Manual of Micro-chemi-
cal Analysis. (Trans.) Ill. or. 8vo. *Macmillan*,
1900. (A well-known book.) 6s.
BEHRENS, J. W. : The Microscope in Botany.
(Trans.) Ill., 8vo. *Boston*, 1885. $5.
CARPENTER, Dr. W. B. : The Microscope and its
Revelations. (Ed. Dr. Dallinger.) 21 plates
and 800 ills., dy. 8vo. *Churchill*, 1891. (The
Standard work. New edition in the press.) 26s.
CLARKE, L. LANE : Objects for the Microscope.
8 col. plates, cr. 8vo. *Groombridge*, 1871.
COLE, H. C. (ed.) : Studies in Microscopic Science.
Col. plates, 8vo. *Baillière*, 1883-6. 4 vols.
Vol. I. 27s. 6d.; Vols. II., III., IV., ea. 31s. 6d.
COOKE, M. C. : British Desmids. 66 coloured
plates, 8vo., issued in 10 parts. 1866-87.
Now sells for £2. 10s.
COOKE, M. C. : 1,000 Objects for the Microscope.
Ill., cr. 8vo. *Warne*, 1900. (For amateurs.)
2s. 6d.
DAVIS, G. E. : Practical Microscopy. 310 ills. and
col. front., 8vo. *W. H. Allen*, 1889. (Useful
to beginners.) 7s. 6d.
FURNEAUX, W. : Life in Ponds and Streams. 8 col.
plates and 331 figs.; cr. 8vo. *Longmans*, 1897.
6s. net.
GOSSE, P. H. : Evenings at the Microscope. Ill.,
cr. 8vo. *S.P.C.K.*, 1895. (Elementary. A
delightfully written book.) 4s.
GRIFFITH, J. W., and HENFREY, A. : Micrographic
Dictionary; ed. Rev. M. J. Berkeley and

Prof. T. R. Jones. 15 plates and 818 ills.
representing 2,680 figs., 8vo. *Gurney*, 1883.
(A standard work of reference.) 52s. 6d.
HOGG, JABEZ : Microscope : History, Construction,
and Application. 20 plates and 447 ills.;
cr. 8vo. *Routledge*, 1898. (Ranks next after
Carpenter.) 10s. 6d.
KERR, R. : Hidden Beauties of Nature. Ill., 8vo.
Religious Tract Society, 1895. 3s. 6d.
LANKESTER, E. : Half-hours with the Microscope.
Ill., 12mo. *W. H. Allen*, 1900. 2s. 6d.
MARTIN, J. H. : Microscopic Objects Figured and
Described. 194 ills., 8vo. *Van Voorst*, 1871.
14s.
"QUEKETT CLUB MAN" : Handbook of the
Microscope. 38 ills, cr. 8vo. *Roper*, 1887.
(Selection and management.) 2s. 6d.
"QUEKETT CLUB MAN" : My Microscope and
Some Objects from my Cabinet. 5 ills., fcap.
8vo. *Roper*, 1888. 2s. 6d.
SCHERREN, H. : Ponds and Rock-pools. Ill., cr.
8vo, *R. T. S.*, 1894. (Good.) 2s. 6d.
SCHERREN, H. : Through a Pocket Lens. Ill.,
cr. 8vo. *R. T. S.*, 1897. 2s. 6d.
SHELLEY, H. C. : Chats about the Microscope. Ill.,
post 8vo. *Scientific Press*, 1899. 2s.
SLACK, H. J. : Marvels of Pond-life. Ill., cr.
8vo. *O. Newman & Co.*, 1892. (Well-known
book.)
STOKES, Dr. A. C. : Microscopy for Beginners.
Ill., cr. 8vo. *New York*, 1887. $1.50.
STRASBURGER, E. : Handbook of Practical Botany.
(Trans.) 149 ills., cr. 8vo. *Sonnenschein*,
1900. (Well-known book.) 10s. 6d.
WOOD, Rev. J. G. : Common Objects of the Micro-
scope. Ill., cr. 8vo. *Routledge*, 1900. (Well-
known book.) 2s. 6d.
WRIGHT, LEWIS. : A Popular Handbook to the
Microscope. Ill., or. 8vo. *R.T.S.*, 1898. (An
excellent book for beginners.) 2s. 6d.
ZIMMERMANN, A. : Botanical Microtechnique.
Demy 8vo. *Constable*. 12s.

PERIODICALS.

JOURNAL OF THE ROYAL MICROSCOPICAL SOCIETY
every two months. *Williams & Norgate*. 6s.
JOURNAL OF THE QUEKETT MICROSCOPICAL
CLUB : every six months. *Williams & Norgate*.
3s. 6d.
SCIENCE-GOSSIP : monthly. 110 *Strand, W.C.* 6d.
KNOWLEDGE : monthly. 326 *High Holborn*. 6d.
QUARTERLY JOURNAL OF MICROSCOPICAL SCIENCE :
not strictly quarterly. *Churchill*. 10s.
JOURNAL OF APPLIED MICROSCOPY, monthly.
Dawbarn & Ward. 4d.
AMERICAN MONTHLY MICROSCOPICAL JOURNAL,
monthly. *C. W. Smiley, Washington, U.S.*
per ann. $2.
ANNUAL OF MICROSCOPY. *Lund Humphries*.
1899-1900, each 2s. 6d.

The foregoing list is, as we have before said,
only intended as a draft, and we hope to have the
assistance of our readers in its completion or correction. Some of the books are already out of
print. In every case when the date is given it is
the most recent. We have included only books and
periodicals written in or translated into the English
language.

7 *The Elms, Sunderland,*
December, 1900.

CONDUCTED BY EDWARD A. MARTIN, F.G.S.

THE MORTIMER MUSEUM AT DRIFFIELD.—A descriptive illustrated catalogue of the Mortimer Museum at Driffield has been prepared by Mr. T. Sheppard, F.G.S. He is to be congratulated on the result of his work. Mr. Mortimer, who is now advanced in years, being anxious that his labours and those of his brother should not be lost to the district in which they had worked together for years, offered to the East Riding County Council for half its value the Museum, both building and contents. This liberal proposal was declined last year, and it is now hoped that the Hull Corporation may be induced to purchase the Museum. The specimens were obtained by the brothers Mortimer many years ago. Being first in the field, and situated in a convenient centre on the Yorkshire Wolds, then by far the most prolific collecting grounds for prehistoric relics in England, thousands of spear-heads, arrow-heads, "scrapers," axe-heads, and other flint and stone implements were got together by these enthusiastic collectors. The farm labourers for miles around were induced to spend their spare time in walking over the ploughed fields in search of stone implements, and prizes were given to those who collected the greatest quantity. At that time it was a not uncommon occurrence for a bucketful to be brought to Mr. Mortimer. Now, however, the Wolds having been carefully traversed over and over again, it is a difficult matter to get together half-a-dozen specimens. The Driffield Museum is not exclusively made up of these relics picked up from the ground. It contains numerous skeletons, together with objects of stone, bone, bronze, or other material found in association. therewith, which have been dug from the "barrows" or burial mounds. The opening of these "barrows" has been Mr. Mortimer's chief hobby, and scores have been excavated during the last forty or fifty years, and their contents removed to Driffield. There are now very few "barrows" left unopened, consequently this branch of the collection cannot be replaced. Besides containing many objects of antiquarian interest the geological collection is not less valuable. Students of geology and palaeontology in that part of the country have every reason to complain of the absence of good rock exposures from which to collect chalk fossils. Mr. Mortimer's collection of sponges, echinoderms, inocerami, and other fossils is certainly by far the finest from the Yorkshire Wolds. There can be little doubt that local collections are of far greater educational value when retained in the county from which they have been derived, and we hope that Hull will not be guilty of that lack of enterprise which lost to Brighton many years ago the magnificent collections of the great Dr. Mantell.

A BOULDER MONUMENT.—Shap Fell, in Westmoreland, was an important centre of dispersion in the Glacial Period, and boulders of Shap granite are found, which have been spread over the Yorkshire plain, and carried to the coast between Whitby and Scarborough, and also into Lincolnshire. One of these boulders, weighing about twelve tons, has been taken up out of the bed of the Tees, and reared on a pedestal in the Darlington Public Park to the memory of Dr. R. T. Manson, F.G.S. We cannot conceive of a more appropriate monument to a geologist.

INCREASE OF COAL CONSUMPTION.—The amount of coal raised each year in the United Kingdom grows apace. In 1899 the increase over 1898 was no less than 18,040,265 tons. The increase of exported coal was 6,121,902 tons.

RAISED BEACHES OF THE RED SEA.—Mr. R. B. Newton reports, in the "Geological Magazine," the results of his examination of a collection of over 1,500 specimens of shells from the raised beach deposits of the Red Sea, submitted to him by the Geological Survey of Egypt. Some of these were collected by Dr. Hume from the western shore of the Gulf of Akaba, but the majority of them came from the west side of the Red Sea and the Gulf of Suez. The species are stated to exhibit the true Red Sea or Indo-Pacific facies, with a very slight commingling of Mediterranean forms. A few are extinct, but Mr. Newton classifies the beaches as Pleistocene, on account of the large number of recent species.

THE BRIGHT STREAKS IN COAL.—Mr. W. S. Gresley, F.G.S., is still crying in the wilderness. He has revived the ancient theory that some of the coal-plants were water-living forms, and actually grew where they are now fossilised. With the latter opinion every one agrees, except perhaps Mr. A. Strahan ; but the idea that certain bright streaks of shiny coal which can be seen in almost every lump were aquatic forms of vegetation was scouted by a former Secretary of the Geological Society as being "unsupported by any evidence whatever." Mr. Gresley now returns to the charge, both in the "American Geologist" and in the "Geological Magazine." He cites some striking evidence in support of his theory.

SELBORNIAN ROCKS OF THE SOUTH OF ENGLAND.—A new Geological Survey memoir, of considerable interest to geologists in the home counties, has just been published, entitled Vol. I. of "The Cretaceous Rocks of Great Britain." This volume deals with the Gault clay and the Upper Greensand strata of England, and is especially noticeable from the official acceptation of the term "Selbornian" for the group of rocks described. Mr. Jukes-Browne has been, perhaps, somewhat prolific in the re-naming of rocks, but in this case no re-naming is involved. He merely associates under one name two formations which, although distinct in many places lithologically, are indivisible elsewhere. 'Much of the Folkestone Gault, for instance, and of the Upper Greensand are "correlative deposits formed at the same time in different parts of the same sea." In fact, the Gault and Upper Greensand are not distinct stages, when the former in one locality is compared with the latter in another, the littoral Upper Greensand being formed in one place where the sea had shallowed, whilst Gault was still forming in deeper water. We can therefore give a cordial welcome to the term "Selbornian," the village of Selborne being a locality where the development of the Upper Greensand is almost unique.

ASTRONOMY

CONDUCTED BY F. C. DENNETT.

	1901 Jan.	Rises. h.m.	Sets. h.m.	Position at Noon. R.A. h.m.	Dec. ° '
Sun	.. 5 ..	8.7 a.m. ..	4.3 p.m. ..	19.3 ..	22.40 S.
	15 ..	8.1 a.m. ..	4.17 p.m. ..	19.46 ..	21.12 S.
	25 ..	7.51 a.m. ..	4.33 p.m. ..	20.29 ..	19.4 S.

	Jan.	Rises. h.m.	Souths. h.m.	Sets. h.m.	Age at Noon. d. h.m.
Moon	.. 5 ..	5.9 p.m. ..	0.3 a.m. ..	7.57 a.m...	14 11.59
	15 ..	3.12 a.m. ..	7.38 a.m. ..	11.59 a.m. ..	24 11.59
	25 ..	9.35 a.m. ..	4.22 p.m. ..	11.25 p.m. ..	4 21.24

	Jan.	Souths. h.m.	Semi-diameter. ° '	Position at Noon. R.A. h.m.	Dec. ° '
Mercury..	.. 5 ..	11.23 a.m. ..	2·4'' ..	18.20 ..	22.24 S.
	15 ..	11.53 a.m. ..	2·3'' ..	19.30 ..	23.41 S.
	25 ..	0.24 p.m. ..	2·4'' ..	20.41 ..	20.30 S.
Venus	.. 5 ..	10.4 a.m. ..	5·9'' ..	17.1 ..	21.40 S.
	15 ..	10.18 a.m. ..	5·7'' ..	17.55 ..	22.47 S.
	25 ..	10.32 a.m. ..	5·6'' ..	18.49 ..	22.46 S.
Mars	.. 5 ..	4.3 a.m. ..	5·2'' ..	10.59 ..	10.6 N.
	15 ..	3.26 a.m. ..	5·6'' ..	11.1 ..	10.17 N.
	25 ..	2.44 a.m. ..	6·1'' ..	10.58 ..	10.56 N.
Jupiter	.. 15 ..	10.20 a.m. ..	14·8'' ..	17.56 ..	23.10 S.
Saturn	.. 15 ..	11.4 a.m. ..	7·0'' ..	18.41 ..	22.32 S.
Uranus	.. 15 ..	9.18 a.m. ..	1·8'' ..	16.53 ..	22.37 S.
Neptune	.. 15 ..	10.9 p.m. ..	1·2'' ..	5.48 ..	22.11 N.

MOON'S PHASES.

		h.m.			h.m.
Full	.. Jan. 5 ..	0.14 a.m.	3rd Qr. Jan. 12 ..		8.38 p.m.
New	.. „ 20 ..	2.36 p.m.	1st Qr. „ 27 ..		9.52 a.m.

In apogee January 12th at 11 a.m. ; in perigee on January 24th at 11.30 a.m.

METEORS.

				h.m.		°
Jan. 2–3	..	Quadrantids	Radiant R.A.	15.28	Dec.	49 N.
„ 2	..	„	„	15.20	„	53 N.
„ 5	..	„	„	9.20	„	57 N.
„ 17	..	κ Cygnids	„	19.40	„	53 N.
„ 18–28	..	θ Coronids	„	15.32	„	31 N.

CONJUNCTIONS OF PLANETS WITH THE MOON.

Jan. 9	..	Mars †	..	8 p.m. ..	Planet	9.10 N.
„ 18	..	Jupiter*	..	9 a.m. ..	„	2.13 S.
„ 18	..	Venus*	..	2 p.m. ..	„	2.12 S.
„ 19	..	Saturn†	..	4 a.m. ..	„	2.41 S.
„ 20	..	Mercury*	..	3 p.m. ..	„	6.34 S.

* Daylight.　　† Below English horizon.

OCCULTATIONS AND NEAR APPROACHES.

Jan.	Star.	Magnitude.	Disappears. h.m.	Angle from Vertex. °	Reappears. h.m.	Angle from Vertex. °
2 ..	ω³Tauri	4·6 ..	5.1 a.m. ..	325 ..	Near approach.	
3 ..	χ'Orionis	4·7 ..	5.55 p.m. ..	220 ..	„	„
3 ..	χ⁴ „	4·8 ..	10.47 p.m. ..	191 ..	„	„
23 ..	13 Tauri	5·4 ..	8.3 p.m. ..	354 ..	8.28 p.m. ..	310
31 ..	χ'Orionis	4·7 ..	2.41 a.m. ..	69 ..	3.36 a.m. ..	229

THE SUN needs careful watching for sudden outbreaks of activity. At 9 p.m. on January 2nd the Earth passes that portion of its orbit which is nearest to the Sun.

MERCURY is a morning star until 2 a.m. on January 22nd, when it is in superior conjunction with the Sun, after which it becomes an evening star. At no time is it in good position for observation. At 11 p.m. on the 7th Mercury is in conjunction with Saturn, being 1° 51' south of that planet.

VENUS is also a morning star badly placed for observation. At 9 p.m. on January 3rd she is in conjunction with Uranus, being 1° 10' north of that planet. At 9 p.m. on the 15th she is in conjunction with, and only 22' north of, Jupiter ; whilst at 8 p.m. on the 24th she is in conjunction with, and only 20' south of, Saturn.

MARS is in Leo all the month, rising at 9.17 p.m. on the 1st, and at 7.7 p.m. on January 31st. It is well placed for observation, but its apparent diameter is only 12.8'' at the end of the month.

SATURN, JUPITER, and URANUS are all too near the Sun for successful observation.

NEPTUNE is well placed for observation all the month.

THE LEONIDS.—So far as a great shower was concerned, the Leonids were not seen this year, only a few meteors from this radiant being recorded. It seems certain that their orbit has become changed.

MINOR PLANETS.—Another was discovered by Professor Max Wolf, on October 31st, at Heidelberg. One of the supposed recent discoveries appears to be an observation of No. 244 Sita. The two found in Japan in March last have not been again seen.

THE MOON.—We are very glad to hear that the members of the British Astronomical Association are now going to make the attempt to construct a trustworthy map on a scale of 200 inches to the Moon's diameter. This was the scale adopted by the Lunar Committee of the British Association for the Advancement of Science in 1864–1869. If any of our readers would like to assist in the good work, Mr. W. Goodacre, F.R.A.S., Director of the Lunar Section, 1 Birchington Road, Crouch End, N., would be glad to hear from them. It is not so much the size of the instrument, as persistent observation, which is useful in this work.

THE NINETEENTH CENTURY, with its wonderful discoveries and advances, has passed away, leaving us on the threshold of the new era, the history of which is as yet all unknown. The opening day of the nineteenth century was marked by the discovery of the first of the minor planets by Piazzi, and to-day the number known is over 460. The next year, 1802, was marked by Wollaston's discovery of gaps in the solar spectrum, being the first step towards spectrum analysis, by the aid of which not only the atmospheric constituents of the heavenly bodies can be known with certainty, but movements, otherwise invisible, are measured. Binary stars beyond the power of the greatest telescopes are now not only discovered, but their movements are studied. Photography, too, has come, making records otherwise utterly beyond the power of man, besides revealing the secrets of the heavens untraceable by the most delicate eye, even when aided by the largest telescopes. Saturn and Jupiter are now each known to have an additional satellite to those previously recognised. Mars has had two revealed, and two more have been found accompanying Uranus. Neptune with his moon has also been found, the former by the mathematical skill of Leverrier and Adams. These are but a few of the discoveries of the past. What shall be those of the new century ?

CHAPTERS FOR YOUNG ASTRONOMERS.

BY FRANK C. DENNETT.

(*Continued from p.* 59.)

JUPITER.

NEXT to the Moon, Jupiter is the most easily studied object in the solar system, having an apparent diameter of from 30″·0 to 46″·0. Its appearance in the heavens is so striking that under favourable conditions it will throw evident shadows on the Earth. The real distance separating the Earth from Jupiter when in opposition is about 390,000,000 miles, so that its magnitude must be very great. Its equatorial diameter is 88,000 miles—more than eleven times that of the Earth— whilst 1,389 globes the size of this world would together only equal Jupiter in volume, and have a surface 124 times that of our planet. The density

that time such spots have been frequently seen, some of them being very persistent, and others very evanescent.

After the middle of the nineteenth century, and especially subsequent to the advent of the silver-on-glass reflectors, attention was directed to the colours which were apparent on different parts of the disc. The brighter band surrounding the equatorial regions of the planet was often observed to have a tint like yellow ochre or pink. The dark belts bounding this band were at different times seen to be copper-coloured, slaty-blue, or chocolate hue. The polar regions sometimes showed tints of delicate blue or green. But to see these colours at all well, telescopes of considerable aperture were necessary.

Fig. 1 shows the planet as seen by the writer on April 19th, 1898, at 10.55 P.M., with a 9¼-inch Newtonian, by the late Mr. G. With. It shows the nomenclature of the various parts of the disc

FIG. 1. JUPITER, April 19th, 1898, 10.55 p.m.[1]

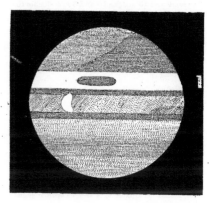

FIG. 2. JUPITER, July 28th, 1878, 1 a.m.

of its contents is, however, less than one-quarter that of the Earth, the result being that 317 globes the weight of ours would just balance the planet Jupiter. The mean rate of the Earth in its orbit is 1,100 miles a minute; but Jupiter only travels 483 miles in the same time. When the first telescopes were turned on Jupiter his large size was noticed; but some years elapsed before it was discovered that, instead of being circular, the polar diameter was about one-sixteenth less than the equatorial—a consequence of its rapid rotation, which as early as 1665 Cassini found to take place in about 9h. 56m.

Long before this, in 1635, it had been noticed by Zuppi and Fontana at Naples that there were dark grey markings on the planet—belts, as they were called—running parallel to each other across the disc. Traces of these may be just seen with a 1-inch achromatic telescope, or even less, bearing a magnifying power of about 40 diameters. Nothing more was specially noticed until 1664, when Hooke observed a dark spot on one of the belts. Since

(1) A, South Polar Region; B, Southern South Temperate Band; C, South Temperate Band; D, South Tropical Belt; E, Equatorial Zone; F, North Tropical Belt; G, North Temperate Band; H, Northern North Temperate Band; I, North Polar Region.

adopted by the Jupiter Committee of the British Astronomical Association.

In the summer of 1878 attention was called to a remarkable red spot which had appeared in the bright band south of the south tropical belt. Fig. 2 represents the first drawing of this object made in England, at 1 A.M. on July 28th, 1878. This object led to a complete revolution in the method of observing Jupiter. The marking, which at first was thought to be new, has been traced back, and appears to be identical, in the light of subsequent study, with objects seen at intervals, so long since as 1869, if indeed it was not the object seen by many observers at still earlier dates. In 1869 it presented the appearance of an oval or elliptical ring; but at the time of its appearance in 1878 it was bright red, a colour which it retained for a few years, but subsequently lost. The spot was still to be seen in 1899, but it had become very faint, and was only visible with a good instrument in clear air. The writer only suspected it once with a 3-inch aperture, but the dim object is shown in Rev. T. E. R. Phillips's beautiful drawing of May 6th, 1899. The length of this notable spot was about 27,000 miles, and its breadth about 8,000 miles.

(*To be continued.*)

CONDUCTED BY B. FOULKES-WINKS, M.R.P.S.

EXPOSURE TABLE FOR JANUARY.

The figures in the following table are worked out for plates of about 100 Hurter & Driffield. For plates of lower speed number give more exposure in proportion. Thus plates of 50 H. & D. would require just double the exposure. In the same way, plates of a higher speed number will require proportionately less exposure.

Time, 10 A.M. to 2 P.M.

Between 9 and 10 A.M. and 2 and 3 P.M. double the required exposure.

SUBJECT	F. 5·6	F. 8	F.11	F.16	F.22	F.32	F.45	F.64
Sea and Sky..	$\frac{1}{120}$	$\frac{1}{75}$	$\frac{1}{50}$	$\frac{1}{30}$	$\frac{1}{8}$	$\frac{1}{4}$	$\frac{1}{2}$	1
Open Landscape and Shipping	$\frac{1}{30}$	$\frac{1}{15}$	$\frac{1}{2}$	$\frac{1}{4}$	1	2	4	
Landscape, with dark foreground, Street Scenes, and Groups ..	$\frac{1}{8}$	$\frac{1}{4}$	$\frac{1}{2}$	1	2	4	8	16
Portraits in Rooms ..	8	16	32	—	—	—	—	—
Light Interiors	16	32	1	2	4	8	16	32
Dark Interiors	1	2	4	8	16	32	64	128

The small figures represent seconds, large figures minutes. The exposures are calculated for sunshine. If the weather is cloudy, increase the exposure by $\frac{1}{2}$ as much again, if gloomy double the exposure.

EXPOSURE TABLES.—I am exceedingly pleased to notice that SCIENCE-GOSSIP is recognising those amongst its readers who have taken to photography, and that the editors are catering for them and placing this section under the care of Mr. Foulkes-Winks. Possibly you will be inundated with suggestions, and perhaps these, for the time being, cannot be considered apart from the plans you have already decided to follow. If one more suggestion is acceptable, might I ask that you provide us with a concise monthly table of exposures? I know that an actinometer is the best to use, but a table which would be approximately correct would be extremely useful. Amateur photographers could then from month to month follow its variations, and become ultimately good judges without such aids. Thanking you in anticipation of giving this suggestion your favourable notice.—*Thomas W. Brown*, 80B *Church Lane, Old Charlton, Kent.*

[We have pleasure in adopting our correspondent's suggestion, and, as will be seen above, have now commenced the monthly table.—ED. Photo., S.-G.]

PRINTING BY GASLIGHT.—In the dull weather and long evenings that we get during the month of January, the value of printing on bromide paper is appreciated by all. There are a great number of bromide papers on the market, such as platino-matt, bromide velox, gravura, etc., but for rich black tones on a matt surface we have never seen anything to surpass if indeed to equal the "New Gaslight Bromide" paper of the Imperial Dry Plate Company. This is a very slow bromide paper, and most suitable for contact printing. We have found an exposure of 16 seconds, eight inches from an incandescent gas-burner, sufficient for a negative of average density. The developer we prefer is made up as follows :—Amidol, 100 grains ; sulphite of soda, 3 ounces ; bromide of potass, 30 grains ; distilled water to make up to 20 ounces. This will give a very rich black tone. If a greyer tone is required, dilute the developer with an equal quantity of water. When the print is sufficiently developed, transfer direct to the fixing solution, which should be made as follows :—Hyposulphite of soda. 2 ounces ; sulphite of soda, 2 drams ; sulphuric. 2 or 3 drops ; water, 20 ounces. The prints should remain in the fixing bath for at least ten minutes. A longer immersion will do no harm. Conduct all operations in a pale-yellow light. After fixation is complete, wash the prints in running water for an hour, when they will be ready to take out of the water and dry.

NATIONAL PHOTOGRAPHIC RECORD ASSOCIATION.—At the last Council meeting of the above Association 366 photographs were presented from all parts of the kingdom, forming a valuable addition to the collection, recording as they do some of the most interesting subjects, both from an antiquarian and historical point of view. The President, Sir J. Benjamin Stone, M.P., sent in 100 prints taken in Warwickshire, including a series of Stratford-on-Avon and an interesting record of collecting the "wroth money" at sunrise at Ryton-on-Dunsmore. Mr. Sulman gave a set of the Old Historical Houses of Hornsey and Highgate, many already removed ; 103 from Geo. Scamell, Hon. Sec., of the Historical Houses of London, and the old Sussex Churches, including Bosham, Sompting, Shoreham, &c. ; some of Old Newgate by Mr. T. Bolas. Canonbury Tower and other contributions were by E. Scamell ; Worcester Cathedral by F. Littledale ; many especially interesting records of Irish Life and antiquities by Mrs. Muriel and A. Hogg, the latter sending a particularly fine series of the Tumulus of New Grange, the interior views being splendid specimens of flash-light work. The Rev. A. C. Hervey contributed an interesting set of photographs of the Old Parish Register showing extract of Act for burying in woollen—affidavits that such had been done—and another page certifying that certain families had paid the penalty of £5 that their friends might be buried in linen. Mr. Clark forwarded a set of old crosses at Llantwit Major ; Mr. Calcott an interesting record of many of the Old Houses of Bristol, several of which have been already removed ; Mr. Felce a fine series of Norman Capitals and Misereres in the Northampton Churches ; Mr. Hodgson a long record from Kingston-on-Thames ; and Mr. F. Parkinson a very complete set of the Easter Sepulchre at Heckington Church. These photographs have now been forwarded to the British Museum, and together with those already deposited make up a collection of nearly 1,600 prints contributed by members of the Association.—*Geo. Scamell, Hon. Sec.,* 21 *Avenue Road, Highgate.*

PHOTOGRAPHY FOR BEGINNERS.

BY B. FOULKES-WINKS, M.R.P.S.

SECTION I. CAMERAS.

THE student commencing to practise photography is always much in doubt what kind of camera he should buy, and for his guidance we propose to treat this section rather fully, as such a variety of cameras are produced. All the types have some distinct advantage, each being suited for a particular line of operation. There are two definite types of cameras, stand cameras and hand cameras, and these are used for classes of work so widely different that the beginner should decide which he intends adopting before purchasing his instrument.

STAND CAMERAS.

Stand cameras are used more especially for architectural studies, portraits, groups, landscape, and still life, and for all these subjects are by far

the most useful. The various stand cameras may be again subdivided into several classes, according to the uses for which they may be required. All, however, should have the following movements :—

I. LONG AND SHORT EXTENSION.—This should be adjusted by means of a rackwork and pinion. This is necessary, as it gives the operator power of using lenses of varying foci. The camera should rack out to at least double the focal length of the shortest foci lens likely to be used. This will enable the student to reproduce any object the same size, but if the camera will rack out to double the focal length of a mid-angle lens, so much the better. The subject will be treated when considering lenses. The camera should be made so as to allow of the front and back being brought together near enough to enable a short focus lens to be used. As a guide we would suggest that this distance should not be greater than the lesser diameter of the plate ; for instance, if a half-plate (6½ inches × 4¾ inches) camera is to be used, it should be capable of an extension of from 4¾ inches to 13 inches.

II. SWING-BACK.—No camera should be accepted without this movement, which will enable the operator to bring the several parts of a picture in focus, when at varying distances from the

camera, such as the foreground and distance. To achieve this, the student will find it necessary to bring the lower part of the focussing screen a little nearer to the lens than the top part, and, after making this alteration on the focussing screen, it is advisable to re-focus for the centre of the plate. The swing-back is also most essential in architectural subjects, as it enables the photographer to get the upright lines perfectly parallel. Many cameras are made with the horizontal swing only, but we strongly recommend one with both horizontal and vertical swing. The vertical swing is used in views and buildings in which one side of the picture is nearer to the camera than the other. In photographing buildings, the swing-back is not used for the purpose of getting all portions into sharp focus, but rather for bringing all vertical lines parallel. The degree of sharpness or definition required is obtained by using a small stop or diaphragm in the lens. It may be stated as a fixed rule in architectural photography that, no matter what the position of the camera may be, the swing-back must be absolutely perpendicular. We therefore recommend that a plumb level should be attached to the swing-back.

III. RISING FRONT.—This is of great importance, and should be in constant use. With it the photographer is able to cut off much of the foreground

FIG. 1. CAMERA WITH CONIC BELLOWS, CLOSED AND OPEN.

of a picture and take in more sky without tilting the camera, which should be avoided under all circumstances. The rising front is also an absolute necessity in photographing houses, churches, public buildings, etc.. In such cases, the front will often require raising to the full extent, and for this reason the beginner will do well to see that the camera he selects has a good rising front, say at least one-third the diameter of the plate. The rising front of a camera should also permit of being lowered beyond the centre of the camera, as there may be occasions when it is desirable to take in more than the normal amount of foreground.

IV. SWING FRONT.—Under certain circumstances it is useful, but we do not attach much importance to this particular movement.

V. REVERSING BACK.—This is a very useful movement, and enables the operator to take his picture either horizontally or vertically at will, without having to remove his camera from the tripod, as is the case in many older forms of instrument.

VI. REMOVABLE FRONT.—It is advisable to always have a camera in which the front panel that carries the lens is removable, as it enables the operator to readily exchange the lenses. The same object can be attained by having adapting rings or flanges to screw one into the other.

VII. TURNTABLE.—This is a rather modern addition to the camera, and is very convenient, as it dispenses with the tripod top and screw. The tripod legs are fitted on to the camera by its means, and it also allows the camera to be turned in any direction.

VIII. DOUBLE DARK SLIDES, OR PLATE-HOLDERS.—There are two forms of slides in the market, one called the book form, and the other solid or shut off. Of these two we much prefer the book form, as it is more convenient in use, and as a rule better made than the shut-off type. It is a matter of absolute necessity that the dark slide should be light-tight, and this is more readily secured in the book form than in the shut-off. We consider it advisable to have at least three double slides to the camera, thus carrying six plates, two in each slide.

IX. TRIPOD.—This is the stand upon which the camera is erected. The legs are made in many forms, but whatever style is selected, we

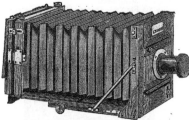

FIG. 2. CAMERA WITH SQUARE BELLOWS.

would strongly impress upon the beginner to choose one that is perfectly rigid. The three-fold stand is, we think, the most convenient, as it folds up fairly small, and can be had in conjunction with great rigidity. In most three-fold stands the bottom portion of the leg is made to slide, which will be found very useful in working on uneven ground, and also for lowering the camera. Fig. 1 shows a drawing of a type of camera containing all the movements we recommend, which are found the most convenient style for the average amateur. It is light, and folds into a small space, being easily carried and readily erected. There is, however, another good form of camera which we advise, where weight is not so much an object as extreme rigidity. This camera is generally referred to as the square-bellows type, and is perhaps the more useful all-round camera of the two. It can also be readily adapted for stereoscopic photography. If it is proposed to take up this most interesting branch of photography, we recommend the student to start with a $7\frac{1}{2}$ in. × 5 in. camera, as it is most convenient size and lends itself readily to adaptation for stereoscopic work, the stereoscopic plate being $3\frac{1}{4}$ in. × $6\frac{3}{4}$ in. Although we recommend this size where possible, yet the half-plate camera can be used for stereoscopic photography, and the view may be taken on an ordinary half-plate. We shall touch upon the various requirements for this purpose under a future section. Fig. 2 illustrates this square type of instrument.

(*To be continued.*)

GREEN ELDER-FRUIT.—The Rev. C. F. Thornewill will be interested to know that there is a bush of the green-fruited elder. in this parish ; also that I remember seeing when a boy bushes of it—one near Wrottesley Park, and another at Tettenhall, both in the neighbourhood of Wolverhampton.—(*Rev.*) *K. A. Deakin, Cofton Hackett, Worcestershire.*

GREEN VARIETY OF ELDER-FRUIT.—The green-fruited elder which the Rev. Mr. Thornewill had sent him by the lady of North Derbyshire is not at all an uncommon plant in this neighbourhood. Those who are thoroughly familiar with the variety find little difficulty in distinguishing it from the dark-fruited and commoner sort, even in winter time. It usually grows more compact, is lighter coloured where the old bark begins to rend, and the buds show a much brighter and more lively green when they begin to expand. Botanically it is regarded only as a variety of the commoner form.—*John Wilson, Leazes Park, Newcastle-on-Tyne,* of *Sambucus nigra.*

THE MILD AUTUMN.—It may interest your readers to know that on October 21st, at Pirton, near Hitchin, I picked several wild roses in full bloom and with buds. On the 25th of the same month I took 111 larvae of *Bombyx rubi,* and could have taken nearly as many again ; they were feeding in the sunshine on grass and plantain. On the same day I picked two fine pieces of honeysuckle in bloom, a very late date for this shrub to flower.—*Fredk. Jas. Bridgman, "Kenmore," Avenue Road, Highgate, N.*

RELATION OF PARENT TO OFFSPRING.—In his interesting paper on "Parental Relationship" in the November number of SCIENCE-GOSSIP, Mr. Leighton remarks that in the amphibians "the relation of parent to offspring may be said to be non-existent, the same applying to the fishes, and we look in vain for any sign of it lower in nature." May I be permitted to remind him that signs of the relationship are to be found well marked in some insects—*e.g.* the common earwig, the female of which sits on her eggs and young, and looks after them as well as does any hen-bird ? Again, amongst spiders the females of many species watch and even carry about with them the sac containing their eggs, and when the young are hatched feed them till they are able to fend for themselves. Amongst fishes, again, the common stickleback is a well-marked example of the existence of parental care. In the case of this fish it is the male who is the nurse. A most interesting account of the labours of papa stickleback may be found in a paper by the late Mr. Grant Allen which appeared in the "Strand Magazine" for May 1899. In the same article he refers to the parental duties as carried out by the tube-mouth fish, the pipe-fish, and some amphibians and reptiles.—*Albert May, Hayling Island.*

BECQUEREL RAYS.—Among a large number of papers on Becquerel rays now before us, two are worthy of especial notice as giving a comprehensive and general account of the phenomena. For those who propose to study the subject more fully, no better guide could be found than Professor Elster's report in Eder's "Jahrbuch für Photographie und Reproductionstechnik" for 1900. The references to original papers in footnotes form a complete bibliography of the subject up to the time when the article appeared, and it is surprising that Professor Elster should have succeeded in summarising so large an amount of matter in eleven very small pages. Dr. B. Walter's article in the "Fortschritte auf dem Gebiete der Röntgenstrahlen," illustrated by a plate of radiographs, is somewhat more popular, and the most important phenomena are therein dealt with at rather greater length.—[Dr.] *G. H. Bryan, Bangor, North Wales.*

ROYAL INSTITUTION.—Among the lecture arrangements at the Royal Institution before Easter are the following: Sir Robert Ball, six lectures (adapted to young people) on "Great Chapters from the Book of Nature"; Professor J. A. Ewing, six lectures on "Practical Mechanics (experimentally treated), First Principles and Modern Illustrations"; Dr. Allan Macfadyen, Fullerian Professor of Physiology, R.I., four lectures on "The Cell as the Unit of Life"; Dr. Arthur Willey, three lectures on "The Origin of Vertebrate Animals"; Mr. F. Corder, three lectures on "Vocal Music, its Growth and Decay" (with musical illustrations); the Right Hon. Lord Rayleigh, six lectures on "Sound and Vibrations." The Friday Evening Meetings will begin on January 18th, when a discourse will be delivered by Professor Dewar on "Gases at the Beginning and End of the Century"; succeeding discourses will probably be given by Dr. A. W. Ward (the Master of Peterhouse), the Right Rev. Monsignor Gerald Molloy, Professor G. H. Bryan, Professor J. J. Thomson, Sir W. Roberts-Austen, Mr. H. Hardinge Cunynghame, Mr. W. A. Shenstone, Dr. Horace Brown, the Right Hon. Lord Rayleigh, and other gentlemen. Sir Robert Ball's lectures commenced on December 27th, being continued on the 29th, January 1st, 3rd, and 5th, at 3 o'clock each day. The admission is by subscription of one guinea and half-a-guinea for adults and children respectively.

HOMOCHRONOUS HEREDITY AND PRONUNCIATION.—It is generally supposed that Frenchmen and Germans have a particular difficulty in the pronunciation of the English Th sound. Now, seeing that this sound must have been used by the Celtic ancestors of the former, and that it is also found in the old High German and Gothic alphabets, ought not the children of these nationalities at least to inherit the power of distinguishing and pronouncing these discarded but ancestral sounds, even though from want of practice they lose the power as they grow older? For any information given through your columns, especially as to authorities or to systematic experiments and investigations on this subject, I should be much obliged.—*Charles G. Stuart-Menteath, 23 Upper Bedford Place, W.C.*

HOMOCHRONOUS HEREDITY AND INHERITANCE OF MUTILATIONS.—If an animal, after sustaining the loss of a limb, gives birth to young exhibiting an analogous deficiency, this is regarded either as a great coincidence or as an irrefragable proof of the doctrine of the inheritance of acquired characters. Is this, however, consistent with the ontogenetic recapitulation of phylogeny, which could scarcely allow the inheritance until the attainment of nearly the age or stage of development of the parent when the latter's accident occurred? There seems, therefore, little coincidence after all.—*Charles G. Stuart-Menteath, London.*

BRERA OBSERVATORY, MILAN.—Professor Schiaparelli, who has been director since 1862, retired in November, and is succeeded by Professor Giovanni Celoria, who has occupied a post there for some years, besides filling the chair of Geodesy at the Technical Institute.

EROSION OF SHELLS.—I have recently been examining specimens of *Linnaea stagnalis, Planorbis complanatus Pl..corneus,* and *Paludina* from a locality in which nearly all specimens show some degree of erosion, often considerable. The erosion seems always to commence in a punctiform manner. Appearances seem to lend much support to the view of K. Semper ("Animal Life," 1890, p. 213), and others, that the destruction of the periostracum is immediately due to the boring of small algae. Once the periostracum is gone, the calcareous matter is probably easily dissolved by the CO_2 in the water. The *L. stagnalis* suffer most, the *Planorbis corneus* least.—*Arthur E. Boycott, Hereford.*

MALARIA AND MOSQUITOES.—On November 28th Major Ronald Ross, D.P.H., M.R.C.S., gave a lecture at the Society of Arts on his discoveries with regard to the relations existing between mosquitoes and malaria. After giving a full and interesting account of the disease itself, its terrible prevalence in hot climates, and the old theories concerning its origin from the miasma of the blue mists, Major Ross proceeded to describe the insect that appears to be inseparably connected with malaria, as it always appears where this disease is prevalent. It belongs to the genus *Anopheles*, the females of which differ from *Culex*, the ordinary British gnat, in having long palpi, and spotted wings instead of plain ones. There is also another important difference which requires special mention in connection with malaria. The larvae of *Culex* live in almost any stagnant water, such as tubs, flowerpots, broken bottles, drains, etc. It is, however, otherwise with *Anopheles*, the larvae of which prefer collections of water on the ground. Hence they abound only in low-lying localities where water suitable for them exists, and also delight in thick rank vegetation. It is the female of the Culicidae that feeds on the blood of animals, including man. The male, as a rule, is not bloodthirsty, but lives chiefly on fruit. As means of prevention, Major Ross suggests that care should always be taken in the prophylactic use of quinine, the careful employment of mosquito-nets and window-screens, and the destruction of larvae round the house. After an examination of various public methods of prevention, the lecturer stated that he personally adhered to the method proposed by himself to the Government of India nearly two years ago, namely, that of careful surface drainage. To this he adds removal of undergrowth, and the use of powdered culicicides in certain cases, also segregation and large airy houses for Europeans. Major Ross also stated that he did not suggest this would exterminate mosquitoes throughout Africa or anywhere else. He referred only to their extinction in large towns.

CORRESPONDENCE.

WE have pleasure in inviting any readers who desire to raise discussions on scientific subjects, to address their letters to the Editor, at 110 Strand, London, W.C. Our only restriction will be, in case the correspondence exceeds the bounds of courtesy; which we trust is a matter of great improbability. These letters may be anonymous. In that case they must be accompanied by the full name and address of the writer, not for publication, but as an earnest of good faith. The Editor does not hold himself responsible for the opinions of the correspondents.—*Ed. S.-G.*

To the Editor of SCIENCE-GOSSIP.

SIR,—You are probably aware that the Royal Society has been engaged for some years past in arranging for the publication of an International Catalogue of Scientific Literature, to begin from January 1st, 1901.

Each science will be represented in an annual volume containing lists, arranged under authors and subjects, of all books and papers published during the year; these will be contributed through official channels of information—abroad, by direct control of the respective Governments—at home, by means of the various Societies which devote themselves to particular sciences; those Societies whose domains overlap having arranged for mutual co-operation.

The collection of title-slips for the United Kingdom of Great Britain and Ireland as regards botany has been undertaken by the Council of the Linnean Society, and they would appeal to all botanic workers for support in their endeavour to compile a complete record, by asking every publishing body to send notices promptly of all botanic issues to the undersigned. Societies are requested to help by sending their issues as soon as possible after publication, either by gift, loan, or exchange, so as to co-operate in producing a yearly record of botanic literature throughout the world.

Communications for this Catalogue should be addressed to me as below.

Trusting that I may rely upon your aid,

B. DAYDON JACKSON,
Secretary.

Linnean Society of London,
Burlington House,
Piccadilly, London, S.W.,
15th December 1900.

NOTICES OF SOCIETIES.

Ordinary meetings are marked †, *excursions* *; *names of persons following excursions are of Conductors. Lantern Illustrations* §.

HULL SCIENTIFIC AND FIELD NATURALISTS' CLUB.

Jan. 9.—† "Vegetable Galls." J. F. Robinson.
„ 23.—† "Notes on an Ornithological Tour in Norway." E. W. Wade.

MANCHESTER MUSEUM, OWENS COLLEGE.

Jan. 5.—† "Defence and Defiance in Nature." W. E. Hoyle.
„ 12.—† "Volcanoes and Earthquakes." Prof. W. Boyd Dawkins.
„ 13.—† "The Fossil Mammals." Prof. W. Boyd Dawkins.
„ 19.—§ "Economic Botany: (1) Sugar and its Sources." Professor Weiss.
„ 26.—§ "Economic Botany: (2) Wheat and Sago." Prof. Weiss.

SOUTH LONDON ENTOMOLOGICAL AND NATURAL HISTORY SOCIETY.

Jan. 10.—§ "Photo-Micros of Lepidopterous Ova." F. Noad Clarke.
„ 24.—Annual Meeting.

SELBORNE SOCIETY.

Jan. 3.—† "The Poetry of Science and Religion." Rev. E. S. Lang Buckland.

BIRKBECK NATURAL HISTORY SOCIETY.

Jan. 12.—* Zoological Gardens, Regent's Park. H. W. Unthank, B.A., B.Sc.
„ 12.—† "The Reasoning Process in Man and the Lower Animals." G. Armitage Smith, M.A.

PRESTON SCIENTIFIC SOCIETY.

Jan. 9.—§ "Some Recent Researches in Sun-spots." Rev. A. L. Cortie, S.J., F.R.A.S.
„ 122.—§ "The Cruise of the *Defiance*: a Photographic Holiday on a Canal Boat." Edgar Bellingham.

NOTICES TO CORRESPONDENTS.

EDITORIAL COMMUNICATIONS, articles, books for review, instruments for notice, specimens for identification, &c., to be addressed to JOHN T. CARRINGTON, 110 Strand, London, W.C.

BUSINESS COMMUNICATIONS.—All business communications relating to SCIENCE-GOSSIP must be addressed to the Proprietors of SCIENCE-GOSSIP, 110 Strand, London.

NOTICE.

SUBSCRIPTIONS (6s. 6d. per annum) may be paid at any time. The postage of SCIENCE-GOSSIP is really one penny, but only half that rate is charged to subscribers.

EXCHANGES.

NOTICE.—Exchanges extending to thirty words (including name and address) admitted free, but additional words must be prepaid at the rate of threepence for every seven words or less.

WANTED, Pleistocene Cave mammalian remains. Offered, Teeth of Goniopholis (Hastings), Odontasis, etc. (Bracklesham), Scales of Lepidotus (Hastings), and Middle Eocene Mollusca.—G. White, 31 North Side, Clapham. S.W.

PHOTOGRAPHIC Appliances wanted, ½-plate lens R.R. or W.A. Camera Body or Oil Lantern. Exchange Best Microscopic Slides:—Thomas Brown, Rosemount, 80B Church Lane, Old Charlton.

OFFERED, Field or first power Opera Glasses, case and shoulder-strap, or books on Geology. Desiderata, Side Silver Reflector Paraboloid, or other illuminating apparatus.—John J. Ward, Lincoln Street, Coventry.

OFFERED.—The new British Physa from Guidebridge, Lancs., Vertigo moulinsiana, etc., and N. American L. and F.W. Shells. Wanted, British Physa and foreign L. and F.W. Shells.—Fred. Taylor, 42 Landseer Street, Oldham.

"LITERATURE" from the commencement (Vols. I. to V. inclusive), in Parts as issued, offered in exchange for works on Natural History, Conchological preferred.—M. V. Lebour, Corbridge-on-Tyne.

QUARTZ CRYSTALS from Ceylon. Small, singly or in clusters, large number on hand, excellent specimens. Will exchange for crystals of other minerals.—Sergeant C. Barter, 2nd K.R.R., Rawal Pindi, Punjab, India.

OFFERED, singly and in clutches, Red-tailed, Red-shouldered and American Rough-legged Buzzards, American Robins, Mocking, King, Cat, and Cedar Birds, Red-winged and Meadow Starlings, Hairy and Golden-winged Woodpeckers, etc., for semi-common English eggs.—Thomas Raine, Chapel Allerton.

CONTENTS.

APHIDES IN ANTS' NESTS.

DESCRIBING A NEW SPECIES.

BY G. B. BUCKTON, F.R.S.

THE obscure subject which rather inappropriately may be styled "Sycophancy amongst Insects" is both curious and invites attention. The recorded number of insects which affect the society of ants is somewhat large, and contains members of several entomological orders, the principal of which are Coleoptera and Hemiptera. Of the aphides there are several species. Only now we are beginning to realise that in many cases these Hemiptera undergo part of their metamorphic stages underground. Some affect ants' nests, others, as *Phylloxera*, attack the roots of vines, etc., simply for the sustenance they obtain from the sap there elaborated.

Some radicle forms of aphides may be here mentioned (¹) : *Siphonophora millefolii, Siphono-*

or six facets or simple lenses. The males, if aërial in habit, have their eyes fully developed.

Some genera, however, possess dimorphic males, as may be seen in *Phylloxera vastatrix*. In these the underground males are either blind, like the females, or have exceedingly small eyes, and also are mouthless. This is a remarkable instance of blindness in the parents which is never transmitted to the offspring until the final stage of development occurs for reproduction.

The example I now describe was turned out of a nest of *Formica rufa* at Oxshott, and it appears to be a male of a species of *Lachnus*. Mr. Donisthorpe could detect no other example of either sex in the heap. Are certain insects treated by ants as pets, just as we treat our lapdogs?

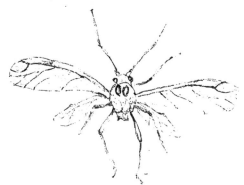

Lachnus formicophilus N.S.

phora rubi, Aphis subterranea, Pemphigus lactucarius, Tetraneura ulmi. Mr. H. Donisthorpe has forwarded to me an aphis that appears to be new. The Lachninae are in many cases gall-inhabiters. The question arises, What becomes of those forms of aphides which occur in swarms on some plants in summer, but cannot be found at all in autumn, either as true sexes or in the condition of ova, the plants themselves being green and vigorous?

Like some of the numerous beetles comprised in Mr. Donisthorpe's list (²) of Myrmicophilous Coleoptera, the females of these aphides are wholly or partially blind. At least, their visual organs, if present, are rudimentary, or often restricted to five

(1) See "British Aphides," notes by Balbiani, vol. iii. p. 74 and Lichtenstein, vol. iii. p. 112.

(2) "Entomologist's Record," vol. xii. p. 172.

The hypothesis is that at the end of the generation the oviparous female descends from the food-plant into the ants' nest, and that subsequently the male seeks her there.

LACHNUS FORMICOPHILUS N.S.

MALE. Body small, globular, black, and covered with white flocculent matter. Pronotum with two shining black bubbles on the back. Head rather large. Antennae very long and six-jointed. Legs brown, slender, and long. Fore wings ample, with fine black nervures and with a hyaline stigma bordered only with brown. Expanse of wings 11·0 millimetres. OVIPAROUS FEMALE unknown.

HABITAT. Oxshott, Surrey, found in an ant-heap of *Formica rufa*.

Weycombe, Haslemere, January 10th, 1901.

K

DENDRITIC SPOTS IN PAPER.

BY F. SHILLINGTON SCALES. F.R.M.S.

IN the first volume of the present series of SCIENCE-GOSSIP, on page 85, there appeared an article by Mr. A. F. Tait, entitled "Crystals bred in Books." In this Mr. Tait dealt with the dendritic spots frequently found in books and papers, one of which he illustrated. Concerning the origin of these spots the author wrote as follows : " The dendritic crystal is formed by chemical action set up by the accidental deposition of a minute fragment of copper upon the surface of paper during the processes of manufacture or of printing : the presence of the minute fragments of copper deposited being probably due to the wear and tear of the paper-making or the printing machinery, so far as the mechanism is built up of copper. The agency of manganese and the action of heat or of moisture in building up the dendritic crystal . . . must be left for a future article. The dendritic crystal requires, as far as I have observed, rather more than twenty years before reaching its fullest development." On page 268 of the same volume Mr. Carrington observed in connection with the same subject : " Mr. Archibald Liversidge, in 'Journal of Chemical Society' (vol. x. 1872, p. 646), mentions that previous to his chemical examination in 1872 dendritic or plumose spots appearing on paper do not seem to have received scientific treatment in view of ascertaining their origin, though botanists and microscopists had frequently examined them without arriving at any satisfactory conclusion. Indeed, Agardh and Lyngbye named the spots *Conferva dendritica* ; and Schumacher placed them amongst fungi under the name *Dematium oliraceum*. Mr. Liversidge's experiments with the blowpipe, which are set forth at length in his paper now referred to, proved the dendritic spots on paper to be caused by crystals of copper in combination with sulphur." Mr. Carrington, in discussing the sources of origin of other dendrites, further observes that they may in some instances be caused "possibly by cryptogamic vegetable patterns being replaced or filled up by the infiltration of some solution containing metal oxide, which crystallises in the cells formed by the vegetation ; for instance in some Mocha stones." Mr. Tait's article called forth several notes by various correspondents, one of whom elaborates the suggestion as to the agency of manganese by saying : " In a small elementary book on minerals, I see that manganese dioxide is employed largely in the manufacture of chloride of lime for the use of linen bleachers ; if so, might not this account for the presence of these beautiful little dendrites on paper, which is so frequently made of old rags ? " Another correspondent says

he always supposed these dendrites to be "some form of lichen," and another again suggests the manganese origin.

The subject was therefore one that would well bear further examination, and the present writer, who, as it happens, has a very practical acquaintance with the details of paper-making, accordingly decided to make further investigations. I have unfortunately not read the referred-to paper by Mr. Liversidge, and consequently cannot judge of his reasons for suggesting the agency of sulphur as well as copper, but the knowledge of paper-making above referred to made me consider the theory as inherently improbable. The agency of manganese can be also dismissed for the same reason, for though bleaching-powder is largely used in paper-making, manganese dioxide is not an adulterant of bleaching-powder, its use being confined to the generation of chlorine gas by a mixture of HCl and MnO_2, the gas itself being, in the process referred to, subsequently passed into contiguous chambers and absorbed by slaked lime.

By the kindness of numerous correspondents of this journal I have been furnished with many specimens of these dendritic spots. For one of these I am indebted to Mr. W. H. Harris, of Cardiff. From this I have made photo-micrographs, and they are here reproduced. Fig. 1 represents the crystal in its entirety, and fig. 2 a portion of the same at a higher magnification. In both photographs I have endeavoured as far as possible to include the fibres of the paper. In fig. 1 the nucleus of the dendrite is very distinct, and this nucleus is characteristic of all such spots without exception. It will further be found that the ramifications of the dendrite correspond exactly to the lie of the fibres, carrying out in a different sense Mr. Carrington's remarks, already alluded to, as to the crystallisation of a metallic oxide in the place occupied by a vegetable pattern. As a matter of fact, the crystallisation, as it slowly formed, has run along the line of least resistance, namely, in the direction of the principal fibres, and preferably those with large, hollow, central canals. This is clearly shown in the original photomicrograph represented by fig. 2, where two branches can be seen joined by the unfilled remainder of the fibre causing them. The ramifications of the dendrite depend, therefore, entirely upon the fibres of the paper in which it occurs.

There are many fibres used in paper-making, but white papers in Europe are made almost exclusively from linen, cotton, esparto grass, and wood-pulp, the latter being both chemical and mechanical. The use of wood-pulp is of comparatively recent date, and both it and esparto are confined mainly to the less expensive classes of printings and writings. Of all these fibres only one has a large central cavity combined with thin side-walls, *i.e.* cotton. All the papers examined by me contain cotton fibres, and I have found no example of a dendrite in any other fibre, so am inclined to question its possibility, on account of the small central canal.

I hope to be able, at a not distant date, to contribute to this journal a paper dealing with the microscopic examination of fibres, more especially those used in paper-making. By means of these notes the readers can themselves verify these statements, though the identification of certain fibres, more especially after being made into paper, is a matter requiring much practice.

With regard to the metallic origin of these spots,

$3 H_2O$, and into cupric ammonia nitrate, $Cu(NO_3)_2$, $4 NH_3$; its precipitation as $Cu(HO)_2$, and subsequent characteristic solution in ammonia; and its precipitation also with potassium ferrocyanide as Cu_2FeCy_6, further verified with acetic acid. The process in this case requires no higher power than the use of an objective of an inch nominal focus. Copper is usually stated as being unoxidisable at ordinary temperatures, but this is manifestly not correct, as it is certainly oxidisable in air containing any moisture, but only with difficulty and after the lapse of considerable time. It is possible that the use of bleaching-powder and other chemicals used in the process of manufacture may aid the oxidisation. I have no hesitation, therefore, in saying that these dendritic spots may now be definitely stated to be formed by the slow oxidisation of a minute spot of

DENDRITIC SPOTS IN PAPER.

(Highly magnified and photographed by F. Shillington Scales, F.R.M.S.)

I was inclined to credit the theory of deposition of a minute particle of copper, because brass enters largely into the machinery used in paper-making, notably, so far as our present interests are concerned, in the broad sheet of endless wire gauze on which machine-made paper is felted, and in the brass "beater plates," by means of which the finer qualities of papers are reduced to pulp instead of by the usual steel plates. Small particles of brass and other metals, such as brass hooks and eyes, are also a frequent trouble in rag-made papers.

It appeared that this was a typical case for micro-chemical analysis, in which the writer has long been interested, and small though the nuclei of these spots are, I was able to obtain unmistakeable reactions proving that the nuclei were indeed nothing but minute particles of copper. For those of our readers interested in such matters I may mention that the analysis included the conversion of the nuclear spot into cupric nitrate, $Cu(NO_3)_2$

metallic copper, deposited in the process of paper manufacture, into the black cupric oxide (CuO). The slowness of the oxidisation is evidenced by the length of time required for the formation of these dendrites, as stated by Mr. Tait. Their absence from modern papers is due partly to this cause, and partly, unless this theory requires modification, to the fact that the comparatively solid fibres of wood-pulp and esparto used in modern paper-making do not lend themselves readily to the branching out of these formations. We must therefore look in rag-made papers of mature age for them, especially in ledgers and account books, to which, in this age of cheapness, rag-papers are now chiefly confined. That such papers are generally made with a more or less pronounced blue tint is the only explanation I can give of the fact that five-sixths of the dendritic spots submitted to me were found in blue-tinted papers.

Sunderland, January 1901.

LIFE OF A PET NEWT.

By M. E. ACKERLEY.

HAVING recently lost a newt, which I kept in captivity for about fifteen years, it appears to me that some little account of its life may be of interest to the readers of SCIENCE-GOSSIP.

During the summer of 1885 or 1886 my brother went fishing for tench in a pond in the neighbourhood. He was not successful in obtaining any tench, but he had one catch, which proved to be a newt. He brought it to me, and together we made a home for it out of an old soap-box. We put in soil and moss, also a dish of water sunk to the level of the surrounding earth, and covered the box with wood and perforated zinc. The newt was placed in the cage and kept in the garden. The animal soon grew accustomed to its new surroundings, and would readily take worms from our hands.

I cannot say with certainty to what species of newt it belonged. Judging from the description in Dr. Cooke's "British Reptiles," I thought it was the male of the great water-newt (*Triton cristatus*); but I have been told that in some parts of England this species is much larger than my newt, or any we find about here, so it is possible I may be mistaken. It was about five inches long, including the tail; head, tail, and back were black, though, if looked at closely, the black proved to be a sort of chocolate brown, with patches of darker brown that appeared black. The underside was brilliant orange, with black patches, and along the sides were white dots exactly like the picture in Dr. Cooke's book. The skin was rough with little warts. In the autumn and winter there was a small ridge along the middle of the back, from head to tail. In spring and summer this ridge developed into a crest.

We thought this newt would be happier with a companion, so my brother went on another fishing expedition, and returned home with a female newt of the same species, without the ridge and crest. This second newt never settled down to captivity, and a short time afterwards it disappeared. Thinking it had probably escaped, we made a better cage, with a piece of glass to cover the top. This was brought into the house and placed in front of a window on the staircase. At different times I gave the newt companions, but none of them became tame, and they generally vanished in a few weeks. I suspected my old newt to have made a meal of some, but as one had evidently escaped by raising the glass, I had a still better cage made. The floor and framework of solid oak fitted into a zinc tray, the four sides and the top being of glass. All the glass was fastened firmly, except the top glass which slid in and out of a groove, but was so tightly pressed down by the top of the groove that it was impossible for a newt or anything else to raise it. In this cage my newt lived happily till last autumn. I used to feed it at intervals, generally on worms, putting a number in at once. It would come out from its hiding-place, under half a small flower-pot, when I whistled, and would take worms from me, usually not caring for more than three at a meal. The remainder, being put in the cage, would burrow the soil and come up in the evening, when the newt was generally prowling round seeking for prey. If he had finished the worms in the cage and I did not bring more, he would try to attract my attention by coming near the glass when I passed. When worms were scarce I would give him slugs, but he was not very fond of them, and in winter I have tried him with underdone beef, but that he refused. He always liked his food to show signs of life. He would take quite large worms, and I remember an occasion when he had begun on one end of a moderately large worm, but meanwhile the other end of the worm was hurriedly making its way into the soil. The newt, finding that no amount of pulling would draw the worm from the earth, tried a very ingenious plan. Holding the one end of the worm tightly in his mouth, he suddenly began to roll over and over with, for a newt, great rapidity, the result being that the worm broke in two, and so, though the newt could not have the whole worm, he obtained half. He always ate his worms from one end in a series of gulps with intervals between, and would keep lifting the worm to prevent its burrowing. When the last bit had disappeared into his mouth he would be very quiet for a few moments, and then, after one or two yawns, he would be ready for worm number two. Sometimes I put aphides in, and, though I never saw him eat them, they always vanished. He would also eat green caterpillars.

Early last November I got some worms and took them to the cage, but no newt was to be seen, and as he did not come when I whistled I was rather alarmed, and decided to empty the cage, taking all the soil out by teaspoonfuls, and searching carefully among the moss and sphagnum and in the water, also under the cage-floor. The newt could not be found, living or dead, but any quantity of worms, showing he must have been missing for some time. It still seems an absolute impossibility that he could have escaped, and yet if he had died his skeleton at least would have remained. I may possibly have overlooked the skeleton, otherwise his end is wrapped in mystery. The last time I saw him he was slowly changing his skin and seemed to feel the process more than usual, although generally not very lively at the skin-changing period, but brightening afterwards. On a former occasion I watched nearly the whole process of skin changing. He always seemed fatter than usual beforehand and refused food. The following description of the process I quote from my notebook, dated December 30th, 1891 :— " The newt has been very quiet for a day or two, and refused to eat slugs. To-day it changed its skin, and I watched nearly the whole process. The skin had rolled down as far as the neck. The newt then crawled about, and by swelling itself out above and contracting below the skin, it gradually worked down. The animal then drew out its front paws, pushing the skin down as far as it could along the back with them. Then the same process of wriggling, swelling, and contracting commenced, with long rests between whiles, until the skin arrived at the hind legs ; these were drawn half out, and then the tail was pulled out with the hind legs. Finally it got the covering off the one hind leg and nearly off the other, and took a very long rest, then walked away, and pulled it off the last hind leg by pressing the ground with the tail. The newt walked right away, and did not eat the skin. The operation lasted over an hour. I have just been looking at the newt again, about four hours later, and find that the skin has gone. I wonder has it eaten it."

It will be seen from the above that my newt did not sleep all the winter. Probably captivity altered its habits, or the warmth of the house was deceptive. It would sleep at times, and then be lively and hungry, just when worms could not be obtained. My pet was able to endure great cold without apparent inconvenience, and several times in the winter has taken a bath, but stayed too long in the water, and I have found it the next morning in the centre of a solid block of ice. I always allowed the ice to thaw naturally, and when free my pet would appear as well as ever. The newt was fondest of the water in the early spring or late winter, when its crest was appearing, but would go in all the year round, especially in the evening.

I never heard him utter a sound but once, an account of which I again quote from my notebook, dated July 5th, 1894 : "To-day I was pulling at some plants in the cage, and suppose I must have hurt the newt, for I distinctly heard him utter a little cry. Never before have I heard a newt make any sound."

If Dr. Cooke is correct in saying that newts take five years to come to their full growth, mine must have been about twenty years old, for he was fully grown when I got him. His cage was always an attraction to our friends, and there are many who, though wondering at my taste for such a strange pet, regret his loss.

Mytton Vicarage, Whalley, near Blackburn.
January, 1901.

BUTTERFLIES OF THE PALAE-ARCTIC REGION.

BY HENRY CHARLES LANG, M.D., M.R.C.S., L.R.C.P. LOND., F.E.S.

(*Continued from page* 238.)

4. **C. aurorina** H. S. 453-6 (1850). *Tamara* Nord. ; *chrysocoma* Eversm.
50—61 mm.

Shape of *C. edusa*. Markings almost similar, but the expanse of wings is much larger. ♂ has the ground colour deep orange with violet reflec-

C. aurorina. Male.

C. aurorina. Female.

C. aurorina var. libanotica.

tions, considerably less free from basal shading than *C. edusa*. Marg. border very distinctly veined on both wings. It seems to occur only in the light form. It greatly resembles a large ab. *helice*, but has the ground colour altogether white and less shaded, the black borders less intense in tone, disc. spot h.w. bright orange and very conspicuous.

HAB. Mountains in Armenia, Transcaucasia, Amasia. Taurus, Persia (Schahku). VI.–VIII.

LARVA on *Astragalus caucasicus* (R.H.).

a. libanotica Led. Wien. Mts. 1858, p. 140. Differs from *C. aurorina* as follows: ♂ ground colour of wings much duller orange, in consequence of an admixture of black scales; marg. borders sprinkled with yellow scales, not striated except very slightly at apices f.w. Disc. spot f.w. smaller. ♀ usually found in the orange form. Ground colour light orange, shot with violet. Border of f.w. very broad at its costal end, and more irregular on its internal edge than in *C. edusa*. HAB. Mountains in Syria (Lebanon), North Persia.

b. var. *transcaspica* Christoph. Differs but little from type. The disc. spot f.w. ♂ is said to be smaller and lighter. ♀ light orange, with seven yellow spots in the black border f.w. HAB. Transcaspian (June).

5. **C. heldreichi** Stgr. Stett. e. Z. 1862, p. 257. (*Aurorinae* var.) Lg. B. E. p. 63, pl. xiv. fig. 3. 45–50 mm.

Size and pattern of wings of ♂ almost identical with that of *C. edusa*. The ground colour is dark orange, thickly mixed with black scales, so as to produce a much more dusky appearance than that seen in var. *libanotica* of *C. aurorina*. Basal patch in h.w. is red. Discoidal spot large and reddish-orange. All the wings have violet or blue reflections, more strongly marked than in any other *Colias*. ♀ much resembles that of var. *libanotica*, but is somewhat stronger in markings and colouring. A white form occurs. U.s. in both sexes resembles that of *C. edusa*, but is much greener in tint, especially in ♀. Disc. spot f.w. has a white centre which is not seen in *C. edusa*.

HAB. Veluchi and other mountains in N. Greece at high elevations.

Miss M. Fountaine, F.E.S., of Bath, is the first British collector to take *C. heldreichi*. She writes to me as follows: "I found it on Mount Chelmos in the last fortnight of June, not lower than 4,000 feet and, then, not higher than 5,500 feet, though the last time I was there, quite at the end of June, I saw one specimen on one of the summits above 6,000 feet, so that no doubt in July it would occur at the higher elevations. I heard from Herr Krüeper, at the Museum in Athens, that *C. heldreichi* occurs much in the same way on the Parnassus and Veluchi, but it seems only to frequent the high mountains. It is doubtful whether it has ever been taken below 4,000 feet. Its flight was not exactly like the other Coliades I have seen. The males flew with great rapidity, but nearly always close to the ground, hovering rapidly over the low scrub on the sides of the mountain, as though hurriedly searching for something. The

females were more sluggish, much rarer, and more difficult to find. I would take in one day from twenty to thirty males easily, but three was the greatest number of females ever together in my collecting-box. Mr. Elwes's courier was collecting with me, and together we must have secured nearly 150 specimens. I alone had seventy-eight. Amongst these were three white females. I also saw one other white female, which I failed to catch. Considering how exceedingly plentiful was this butterfly in this locality, I do not think the proportion of white females as great as that of var. *helice* to *C. edusa*, in the south of Europe. I have never seen *C. edusa* flying in large quantities in one place, as in the case of *C. heldreichi* on Mount Chelmos." (M. Fountaine. January 9, 1901.)

This species is commonly considered a variety of *C. aurorina*. In company with a few others I have always looked upon it as specifically distinct. I am bound to say, however, that the specimens kindly sent to me by Miss Fountaine taken in 1900, as above mentioned, differ somewhat from specimens previously sent to me by Dr. Staudinger, and others, from the Veluchi Mountains. The former have less of the violet gloss on the wings, and have the disc. spots of the h.w. smaller and yellower. The costal patches also are not so decidedly red, but rather yellowish. In fact, the specimens from Mount Chelmos approach more nearly the var. *libanotica* of *C. aurorina*, but of course are smaller.

C. diva. Male.

C. diva. Female.

6. **C. diva** Gr.–Gr. R. & H. p. 190. 30–40 mm. (25–29 mm. H. & R.)

♂ wings deep orange with purple reflections. F.w. marginal band much as in *C. edusa*, though perhaps somewhat straighter and narrower. Yellow veinings very distinct. Disc. spot small, but distinctly black and oval. H.w. of the same colour as f.w., marginal band as in *C. edusa.* Disc. spot small, but of a bright reddish-orange, costal patches yellowish-orange. ♀ marginal band of both wings with very distinct greenish-yellow spots, large and evenly placed. F.w. ground colour orange, dusky at base, disc. spot distinct. H.w. almost entirely greyish-black, disc. spot bright orange, and there is sometimes a very slight trace of orange coloration beneath it; marginal spots large and of a greenish colour. There is a white form analogous to *C. edusa* var. *helice,* and also an

intermediate form in which the white ground colour of f.w. is slightly tinged with orange. U.s. tinged with light green. Disc. spot f.w. white centred, that of h.w. round and silvery.

HAB. Dschachar Mountains, C. Asia (H. & R.). Most, if not all, the specimens in collections in this country are, I think, some of those taken by Grum-Grshmailo in the Amdo Mountains in 1890. The series in my own collection are from that source, obtained through Mr. Elwes. The male is said to be very rare, perhaps on account of the difficulty of its capture owing to rapid flight, as in the case of the last species.

(*To be continued.*)

NOTES ON SPINNING ANIMALS.

BY H. WALLIS KEW.

(*Continued from page* 229.)

VII.—SPINNING FLIES.

AMONG the few insects capable of spinning in the imago-state are certain little pseudo-neuropterous flies of the family Psocidae. Some insects of this family are wingless throughout life, and of these the little book-lice, which are commonly seen running over paper, are known by sight to most people.

Those Psocidae with which we are here concerned have not received any popular name. They are winged in the imago-state, and are prettily coloured little flies, of curious and pleasing appearance, living often gregariously on tree-trunks, palings, etc., as well as among the foliage of trees [1]. Their spinning habits have long been known. Westwood (1840) gives a statement communicated to him by Audouin, that the latter naturalist had observed a female winged *Psocus* weave a web over eggs which she had deposited in the depressions formed by the veins of leaves [2]. Hagen (1861) says generally that the females cover their eggs with a tissue, so that they form flat silvery spots [3]; and M'Lachlan (1867) states that the eggs are laid in patches on leaves, bark, and other objects, the females covering them with a web [4]. Further, in 1884, Packard mentions having observed Psocids lay eggs on leaves of lilac, pear, and horse-chestnut; and he states that they are covered with a flat round web like the "cocoon" of a spider, though only about a line in diameter [5]. The principal observations, however, are those of

(1) Hagen, Synopsis of British Psocidae, Entomologist's Annual, 1861, pp. 18–32 ; M'Lachlan, Monograph of British Psocidae, Entomologist's Monthly Magazine, iii. (1867), pp. 177, etc. ; Sharp, Cambridge Nat. Hist., v. (1895), pp. 390–8.
(2) Westwood, Modern Classification of Insects, ii. (1840), p. 19.
(3) Hagen, *l.c.*
(4) M'Lachlan, *l.c.*
(5) Packard, Standard Nat. Hist., ii. (1884), p. 142.

Pierre Huber, published in 1843, and based on two Psocids, one with wings of uniform colour and the other with those organs spotted with brown. The first individual watched was on the upper side of a *Cytisus* leaf, in the depression formed by the mid-rib, where the creature was making a little shining white spot, oval in shape, and consisting, as was seen by means of a strong lens, of a great number of silken threads, placed closely together and variously crossed. On examining the leaf against the light the little satiny patch of web was found to cover six small eggs. Subsequently, on leaves of trees of various kinds, Huber found many such patches, all covering clusters of from six to fourteen eggs. By placing several of the flies under glass vases, with leaves from the trees on which they were found, he saw them spin repeatedly, observing the whole process from the laying of the eggs to their final concealment by the completed silken covering [6].

Huber observed that individuals believed to be males were capable of spinning ; and M'Lachlan has expressed the opinion that both sexes spin. The web, the latter naturalist remarks, is undistinguishable from that of spiders ; and he adds that if a number of specimens be enclosed in a pill-box it will be found, at the end of several hours, that the interior is traversed in all directions by numerous lines of web. This circumstance appears to have been noted also by S. F. Aaron in the case of *Psocus sexpunctatus,* caught near Philadelphia and taken in paper boxes [7]. It was noticed by Huber that some clusters of eggs had, in addition to the ordinary close-fitting covering, another protection, a little distance above the first,

(6) P. Huber, Mémoire pour servir à l'histoire des Psoques, Mém. Soc. de Phys. et d'Hist. Nat. de Genève, x. (1843), pp. 35–47.
(7) McCook, A Web-spinning Neuropterous Insect, Proc. Acad. Nat. Sci. Philadelphia, 1883, pp. 278–9.

of less close tissue, composed of stronger and longer threads placed parallel. This upper tent was closed on one side, but always open on the other ; and Huber thought it might serve as a place of refuge for the mother. He mentions that the species with spotted wings always made for its own residence a sort of web composed of long parallel threads. Some of the individuals which thus lurked under a trellis of silk were believed to be males. The creatures had space beneath their coverings to move about a little, and they could easily escape when they wished to do so. According to McCook, *Psocus purus* in America makes a tubular tent-like web in crevices of bark, etc. Under this web the insect lives after the manner of a tube-weaving spider, and entomologists who wish to capture it have to push it out by pressing upon the tent [8]. As to the purpose of these webs, Huber thinks that the lines may serve to warn the creature of the approach of enemies ; but not to protect it from attack, for the tents were so frail that a gnat might almost break them. It is certain, he adds, that Psocids do not lie in ambuscade like spiders, for they are not predaceous.

With regard to the position of the spinning-organs, Huber has made it clear that the threads come from the mouth. He was at considerable pains in attempting to establish their exact source, but found observation difficult on account of the smallness of the flies and the rapidity of their movements. After placing some of the insects in glass tubes and following them with a strong lens, he was inclined to believe that the silk proceeded from the upper lip (labrum). Hagen, however, who examined the labrum, found no spinning-organ there, and he suggests that it is probably situated in the tongue (hypopharynx), a structure associated with the lower lip (labium). Burgess has since found within the tongue a pair of peculiar organs—lingual glands, which open by a common duct into the mouth near the throat [9] ; and these are thought to be concerned with the spinning work [10].

VIII.—EMBIIDS.

Other spinning imagines are found in the family Embiidae, also Pseudo-neuroptera, a small group of insects of moderate size, of no popularity, and imperfectly known even to entomologists [11]. The creatures are said to be related in certain respects to Psocids ; but they are dissimilar, having perhaps a slight resemblance to Termitids, or white ants, with which also they are supposed to have some relationship.

The spinning faculty, though occurring in the perfect insect, is better known in the larva ; and it

has been observed, further, in a form regarded by M'Lachlan as the nymph. The precise period at which it is noted, however, is not of great importance, for there is no marked metamorphosis, the insect having much the same form throughout life. Lucas states that the larvae of *Embia mauritanica* are found under stones in small silken tunnels ; and similarly those of *E. solieri* are stated by Lucas and by Girard to be found in similar situations also in tunnels of silk. A box in which some of the larvae of the first-named species had been placed in 1850 was forgotten till 1858, and when opened was found to have its walls clothed with fine white silk, forming circular tunnels, in which dead larvae were lying [12]. Another Embiid, *Oligotoma michaeli*, found chiefly in a wingless, perhaps larval, condition in a hothouse among orchids, was observed by Mr. Michael to have made a large number of webs on the roots and stems of the orchids and of neighbouring plants [13]. These webs had the form of silken galleries. They were not perfect tubes, but more of the nature of coverings, protecting the creatures from above and permitting them to feed upon the surface of the plant on which the web rested [14]. One individual spun a web on the side of a box in which it had been confined for two hours [15], and some time afterwards the web had been so much extended that it attached the lid to the box. When the lid was removed the insect ran about with activity, "ever and anon retreating into a small, loose silken den or tunnel it had woven for itself at the bottom of the box" [16]. A figure of this individual shows small undeveloped wings, so that it had perhaps passed the larval condition. M'Lachlan states that silken tunnels were spun by nymphs as well as larvae of this species ; but it has been suggested that the form regarded as the nymph may have been the adult short-winged female [17].

It is in *Embia latreillii* = *Oligotoma saundersii* that spinning is definitely recorded for the perfect insect. According to Lucas, both winged and wingless individuals of this species were found in silken galleries about the base of leaves of a *Cycas* (from Madagascar), the silk being woven by the winged perfect insects as well as by the wingless larvae.

It has been supposed that the threads come, as in *Psocus*, from the mouth ; and it has even been said that spinning organs open on the labium [18]. Michael saw *Oligotoma michaeli* making its web apparently "with some organ situate in or near

(8) McCook, *l.c.*

(9) E. Burgess, Anatomy of the Head, etc., in the Psocidae, Proc. Boston Soc. of Nat. Hist., xix. (1878), pp. 291-6.

(10) Sharp, *l.c.*

(11) Sharp, *tom. cit.*, pp. 351-5.

(12) H. Lucas, quoted by Hagen, Monograph of the Embidina, Canadian Entomologist. xvii. (1885), pp. 141-206.

(13) W. H. Michael, Gardeners' Chronicle (n.s.), vi. (1876), p. 845.

(14) M'Lachlan, Journ. Lin. Soc., Zool., xiii. (1878), pp. 273-84.

(15) Michael, *l.c.*

(16) W. G. S., The Web of the Embia, Gardeners' Chronicle (n.s.), vii. (1877), p. 50.

(17) Hagen, *l.c.* ; Sharp, *l.c.*

(18) Hagen, *l.c.*

the head " ([19]). Grassi, however, who has studied two wingless Embiids, has maintained that the webs are spun by the first pair of legs, of which the first tarsal segment is enlarged and contains glands whose secretion escapes by orifices at the tips of setae interspersed between short spines that are placed on the sole ([20]).

With regard to the use of the webs, Lucas believed them to serve as traps for insects, on which he supposed the Embiids to feed. Yet he does not appear to have seen insects entangled ; and it is improbable that these feeble creatures are predaceous. There is reason to believe that they

(19) Westwood, Gardeners' Chronicle (n.s.), vii. (1877), pp. 83–4.

(20) Grassi (1893), quoted by Sharp. *l.c.*

live chiefly on vegetable matter; and M'Lachlan remarks that the silken tunnels of *Oligotoma michaeli* are not at all of the nature of spiders' snares, but are similar to the webs of many caterpillars. Another suggestion made by Lucas is that the threads may serve to warn the Embiids of the approach of enemies ; and this may well be true ; but no doubt the webs are mainly valuable as a covering and place of abode. It is interesting to find, moreover, that according to Grassi the silken tissue is intimately associated with the insects' progression ; and that, in conformity with their mode of life within the galleries of web, the two posterior pairs of feet are much modified, possessing papillae and a comb which act upon the silk.

(*To be continued.*)

AN INTRODUCTION TO BRITISH SPIDERS.

By FRANK PERCY SMITH.

(*Continued from page 236.*)

GENUS *CERATINELLA* EMERTON.

TIBIAE of fourth pair of legs without spines or with a very small one. Tarsi very little shorter than metatarsi. Caput of the male not elevated to any extent.

Ceratinella scabrosa Cb. (*Walckenaera scabrosa*, in " Spiders of Dorset.")

Length. Male nearly 2 mm.

The palpal organs are large and complicated, and have connected with their extremity a strong coiled spine.

Ceratinella brevis Wid. (*Walckenaera depressa* Bl.)

Length. Male 1.8 mm., female 2 mm.

This uncommon spider may be distinguished from *C. scabrosa* Cb. by the smaller size of the palpal organs and the spine at their extremity. The cephalothorax, sternum, and abdomen are minutely punctured.

Ceratinella brevipes Westr. (*Walckenaera brevipes* in " Spiders of Dorset.")

Length. Male 1.5 mm., female 1.7 mm.

This spider may be distinguished from *C. brevis* Wid. by its smaller size and by the greater height of the clypeus compared with the width of the ocular area.

GENUS *WIDERIA* SIMON.

The curious little spiders included in this genus have the labium recurved at the extremity, the maxillae short and broad, and the cephalo-thorax strongly impressed in the region of the caput. There is often a very distinct cephalic lobe in the male. The posterior row of eyes is strongly curved, its convexity being directed backwards. The tibiae of the fourth pair of legs are each provided with a very minute spine.

Wideria antica Wid. (*Walckenaera antica* in " Spiders of Dorset.")

Length. Male 1.7 mm., female 2.2 mm.

In this species the cephalic lobe viewed from above presents an oval appearance, two eyes being visible upon it.

The tibiae of the first two pairs of legs are darker than the rest of the joints.

In front of the caput are two processes curving upwards.

This spider is not uncommon, and has been frequently taken in the neighbourhood of London.

Wideria cucullata Koch. (*Walckenaera cucullata* in " Spiders of Dorset.")

Length. Male 1.7 mm., female 1.8 mm.

In this species the radial joint of the male palpus is of a very curious form. When viewed from above it is seen to consist of two branches, one of which is curved inwards to such an extent as to point almost directly backwards. This spider is rare.

GENUS *WALCKENAERA* Bl.

In this genus the posterior row of eyes is straight, or slightly curved, its convexity being directed forwards. The eyes are restricted to the central part of the caput, which is rather broad. The form of the cephalic region varies greatly in different species, and in the male is, in some cases, elevated in an extraordinary manner.

Walckenaera acuminata Bl.

Length. Male 3.5, female 4.2 mm.

This species can be recognised at once by the enormous elevation of the caput of the male, which is produced into a slender pedicle upon which the eyes are placed. It appears to be somewhat uncommon.

Walckenaera obtusa Bl.

Length. Male 3.7 mm., female 4 mm.

In this species the caput is not elevated to any extent. The palpi are long, and the palpal organs have a long coiled spine connected with them.

Walckenara nodosa Cb.

Length. Male 1.7 mm., female 2 mm.

The caput of the male is furnished with a small very distinct lobe upon which the two central posterior eyes are placed. The radial joint is provided with a sharp apophysis, which projects over the digital joint.

have recently taken this uncommon species at Southgate, near London.

GENUS *PROSOPOTHECA* SIM.

The caput is narrower in this genus than in *Walckenaera*, the eyes occupying its entire surface. The spines on the tibiae of the fourth pair of legs are very small and difficult to distinguish. The eyes of the posterior row are large and close together, the intervals between them being less than their diameters.

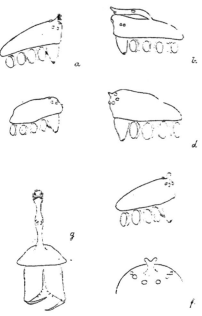

Fig. 1. CEPHALOTHORACES OF MALE SPIDERS.

a. Prosopotheca monoceros. Profile. b. Tigellinus furcillatus. Profile. c. Tigellinus saxicola. Profile. d. Cornicularia cuspidata. Profile. e. Cornicularia unicornis. Profile. f. Cornicularia unicornis. From above. g. Walckenaera acuminata. From in front.

FIG. 2. CEPHALOTHORACES OF MALE SPIDERS IN PROFILE.

a. Ceratinella scabrosa. b. Ceratinella brevis. c. Wideria antica. d. Walckenaera acuminata. e. Walckenaera obtusa. f. Walckenaera nudipalpis. g. Walckenaera nodosa. h. Walckenaera jucundissima.

Walckenaera jucundissima Cb.

Length. Male 1.7, female 2 mm.

In this species the male caput is very similar to that of *W. nodosa* Cb., but the clypeus is higher and the cephalic lobe more detached.

Walckenaera nudipalpis Westr.

Length. Male 3.5 mm, female 3.7 mm.

The male caput is slighly elevated and projects somewhat forwards. The palpi are long, and the radial joint is furnished at right angles to it with a long, slightly curved apophysis. This apophysis is best seen when the palpus is viewed from above. I

Prosopotheca monoceros Wid. (*Walckenaera monoceros* in "Spiders of Dorset.")

Length. Male 2.2 mm., female 2.5 mm.

The caput of the male is furnished with a small pointed projection, which is ornamented with a number of hairs. The radial joint of the palpus is divided into two parts, one of which is bifurcated at its extremity.

GENUS *TIGELLINUS* SIM.

The tibial spines are large and distinct in this genus. The curve of the front row of eyes is very slight, and is directed forwards.

Tigellinus furcillatus Menge. (*Walckenaera furcillata* in "Spiders of Dorset.")

Length. Male 2.5 mm., female 2.5 mm.

The cephalic region of the male is of a most extraordinary form. A long process springs from its hinder portion, and, bending forwards so as to be almost parallel with the surface of the cephalo-thorax, terminates a short distance from the foremost part of the caput. The posterior central eyes are placed some distance back upon this process.

Tigellinus saxicola Bl. (*Walckenaera saxicola* in "Spiders of Dorset.")

Length. Male 2.5 mm., female 2.7 mm.

In this species the caput is not elevated. The tibiae of the first two pairs of legs are darker than the rest of the joints.

GENUS *CORNICULARIA* MENGE.

The spiders included in this genus are very similar to those in the genus *Tigellinus*; but the tibial spines are very small and weak, and the tarsi, compared with the metatarsi, are much longer than in that genus.

Cornicularia cuspidata Bl. (*Walckenaera cuspidata* in "Spiders of Dorset.")

Length. Male 2.2 mm., female 2.5 mm.

In the male of this species a small tooth or "cusp" springs from the centre of the ocular area.

Cornicularia unicornis Cb. (*Walckenaera unicornis* in "Spiders of Dorset.")

Length. Male 2.2 mm., female 2.5 mm.

This may be distinguished from *C. cuspidata* by the form of the cephalic process, which, instead of being a simple cusp, is bifurcate. Its true form is best seen when viewed from in front.

Cornicularia vigilax Bl. (*Neriene vigilax* in "Spiders of Great Britain and Ireland.")

Length. Male 2.2 mm., female 2.5 mm.

This species may be distinguished from *C. cuspidata* Bl. by the absence of the cusp.

Cornicularia clara Cb. This species, described by the Rev. O. P. Cambridge as *Nereine clara*, seems referable to this genus.

GENUS *BARYPHYMA* SIM.

The spiders in this genus have the anterior row of eyes straight, the posterior row being very slightly curved. The spines on the tibiae of the fourth pair of legs are very small.

Baryphyma pratensis Bl.

Length. Male 2.5 mm., female 3 mm.

The caput of the male is slightly elevated and is somewhat massive, with an indentation extending backwards from each pair of lateral eyes.

GENUS *PANAMOMOPS* SIM.

This genus may be easily distinguished from *Baryphyma* by the posterior row of eyes being rather strongly curved, the anterior side of the medians and the posterior side of the laterals being in a straight line. The distance between these medians is considerably less than that between one of them and the adjacent lateral. The spines upon the tibiae of the fourth pair of legs are fairly strong and of considerable length.

Panamomops bicuspis Cb.

This minute spider may be at once recognised by the presence of two small cusps, one on each side of the ocular area.

(*To be continued.*)

CHAPTERS FOR YOUNG NATURALISTS.

(*Continued from page* 199.)

THE LAWS OF HEREDITY.

By HENRY E. M. MORTON.

THE most casual observer of natural phenomena must have frequently been struck with the persistency of various types and impressions which are met with in different families. If he has ever wandered through an ancestral hall and gazed upon the portraits that are hanging upon the walls of the picture gallery, he could not have failed to recognise that the same traits are indelibly impressed upon the physiognomy of the individuals of the family, with a persistency which is very forcible. In the limited circle of one's own acquaintance a superficial observation will suffice to demonstrate how the peculiarities of habits pass from parent to child. The same gestures may be noticed in conversation; the inflections of the voice are so alike as to be confusing; and the mother and daughter, whilst engaged in singing or conversation, have frequently been confounded with each other, even by members of their own family, in consequence of the difference of the timbre of the voice not being easily discernible. Between father and son the resemblance is often very marked. There is the same gait, the same sharp swing of the arm, and the son appears to be an exact repetition of the father. I knew a case where a father, watching his son, a boy of about ten years of age, engaged

K 4

at play with his schoolmates, noticed the same characteristics that the father had displayed when of the same age, and he stated that he saw himself living over again in his child.

Whence comes this heredity? We cannot regard as accidental this transmission of ancestral types; but what are the influences at work to account for this strange persistency? Why should an ancestral peculiarity or weakness be absent from some generations and then make its appearance in a remote descendant? Nature has certainly some law to which this atavism is subservient. Its persistent and well-defined recurrence cannot be regarded as an accidental variation from the immediate parental type, and some influence must be at work to maintain latent for generations the peculiarity of a former ancestor, which subsequently appears in the posterity.

In the very remote ages—long before the legendary halo of history was conceived—let us imagine primordial man, yet but little above some of the other animals that surrounded him. When the reasoning faculties commenced to dawn, he must have taken some cognisance of the animals and plants he met with in his peregrinations, and noted the striking resemblance which the progeny of the one and the offshoots of the other displayed towards the parental type. Nothing, I imagine, would be more apparent to his mind than that the young was an exact image of the parent. It is, therefore, not at all surprising to find the fixity of the idea that "like begets like" is found in the folklore of all nations. Theoretically we know that there are no two beings exactly alike, any more than there exist two plants without some minor modification of their structures; but in all essential particulars there is no apparent difference, and the chief characteristics are transmitted, so that the offspring possess the same appearance as the parental type. The acorn of the oak grows and becomes the oak, whilst the child develops into the man, the same as the former stock, and each breeds true to its kind.

Nearly every naturalist has dealt with this matter, quoting many instances in which persistency of impressions has been transmitted by the female to her offspring: but this form of accidental congenital variation has in most cases failed to be perpetuated unless the variation is the evolution of some characteristic resulting from natural selection, as shown by Darwin in his "Origin of Species," for the better preservation of the type, arising either from the disuse or modification of some organ in consequence of the "struggle for existence." That congenital malformations are capable of transmission I do not for one moment attempt to deny. A well-authenticated case in point is that of a Maltese man with six toes on each foot. His progeny had the malformation of six toes on one foot transmitted to the members of the male line, whilst the feet of the females were normal, but the latter found some of their offspring ushered into the world with six toes on one of their feet.

Now it is a very remarkable fact, and none the less so because it has been noted by nearly every well-known biologist, that not alone peculiar formations, but habits and idiosyncrasies are transmitted. Medical science is not at all backward in asserting that tendencies to specific diseases are also faithfully transferred from the parent to the offspring, and what is known as the "heredity tendency" or diathesis plays a very important part in the transmission of disease.

It is an established fact, long since removed from the debatable ground of theory, that life and growth in either the animal or the vegetable world —whether the lowest form of vegetable life, as represented by the bacilli, or man, the highest animal form—is carried on by the means of the propagation of cells. The microscopic protoplasm, which is regarded as the essential condition of life, is but the aggregation of these cells, in each of which is a nucleus for the creation of a new cell. Such are the elementary conditions under which life is carried on from generation to generation. The lower the scale of life, the more simple is it, each cell producing an individual life acting independently for itself, multiplying into an innumerable number of other cells, each of which rushes about as an individual creation, the animal becoming, as one well-known writer says, practically immortal. With the higher form of animal existence the reproduction is delegated to a special form, the ovum, and the due fertilisation of this is a *sine quâ non* for the proper development of this germ-plasm into that of a highly sensitised creation of the animal kingdom.

To account for the transmissibility of ancestral types, Darwin in his work on "Pangenesis" promulgated a theory which, in his usual exceedingly cautious manner, he stated was only advanced tentatively, that each cell threw off what he designated "gemmules," which formed the nuclei of another series of cells, whose sole destiny in the economy of Nature was the propagation of its species. These "gemmules" formed the blastema, in which was contained an exceedingly microscopical impression of the animal which might ultimately be called into being. If this were the case, some might add we should be able to submit the miniature image to our investigation by means of the microscope. Now, it is fully admitted that the microscope has revealed to us much of the mysteries of the hidden world, and we may readily assume that even with some of the high-power lenses used at the present day by many of our investigators much is still invisible. There is scarcely anything which appeals to our imagination so much as figures, and in order to explain more lucidly this theory it may be advisable to have recourse to them. The red corpuscles of human blood have been measured, and the diameter is found to be about $\frac{1}{4000}$th part of an inch. The number of these red corpuscles which would adhere to the point of a needle would not be less than a million. As a further illustration, we will take one of the hydrozoa as being almost at the bottom of the scale of animal creation. If we carefully watch it through the microscope, we shall find that the ciliary movement is performed with great regularity. Here we have life ; and here we have blood coursing through the veins of these animalcula. How minute must be the blood corpuscles of this creature ! They are there, but the microscope does not reveal them. Take the molecular theory. Everyone is aware this theory teaches that the final division of matter is the atom, and in this matter-of-fact world the atom has been measured. It is calculated that if a cube of water $\frac{1}{50000}$th part of an inch were taken, there would exist 30 billions of atoms. After this we need not express any further astonishment at the microscopical nature of the "repetition" impressed upon the germ-plasm.

To account for the non-transmission of certain characteristics from one generation to another it is assumed that during this period the germ cells which contained them were latent through the preponderating influence of other cells. When the factor became removed or had undergone neutralisation, then the impressions asserted themselves, and we are able to observe the phenomenon of the absent peculiarities making their appearance, and thus seemingly producing a variation from their immediate race. This reversion has been very aptly designated by stock-breeders as "throwing back."

Galton, who has devoted a large amount of his life to anthropological science, and collected an immense store of data dealing with the subject, advances the theory that the germ-plasm is composed of two distinct cells—viz. the body-cell and the germ-cell. The former is destined to build up the body, and the functions of the latter, being confined solely to the future reproduction of the race, are finally deposited in special organs set apart for this purpose. There appears to be no conflict whatever between Galton's double-cell theory and Darwin's gemmule theory ; in fact, the former may be regarded as an evolution of the latter.

Thus far have we advanced along the beaten track of Science, endeavouring to find out the reasons why "like begets like," and the views set out by Galton are the most acceptable as proving rationally the phenomena which are associated with our every-day life. In his recent rectorial address to the Scottish students Lord Rosebery, in dealing with the position we have held in the world during the past century, also with the struggle we should have for existence in the future to maintain our superiority amongst the nations, in pointing out the difficulties which now beset our race, asked what had the coming century in its awful womb for us ? From a scientific point of view we can echo the question. Patiently and unostentatiously are the earnest workers plodding on, seeking to wring from Nature her secrets. From the success which has attended their efforts in the past, we can surely look forward with every confidence to the unravelment of the great mysteries surrounding life, which will be the means of broadening our knowledge of the remarkable gift of inheritance.

Finsbury Park Road, London, N.

GEOMETRICAL SNOW CRYSTALS.

By Dr. G. H. Bryan, F.R.S.

MOST of us are familiar with the figures of feathery six-rayed stars given in many popular text-books under the title of "Crystals of Snow seen under the Microscope." If, however, ordinary snow be magnified, no appearances of the kind figured will be seen. I can well remember my disappointments in bygone days on placing some snow under a microscope.

On the afternoon of January 7th, at Edgbaston, Birmingham, during the frost, I was much delighted to see perfect six-rayed discs about a quarter of an inch in diameter falling on my coat, hat, and umbrella, reproducing all the numerous varieties of form figured in the books. In each the symmetry of the six feathery rays was as complete as that of a pattern seen in a kaleidoscope. This was a little after 3 p.m. ; somewhat later rather larger discs, almost $\frac{3}{8}$ inch in diameter, were falling, with more feathery structure, but many of these were slightly imperfect. The discs on either occasion were very sparsely scattered through the air, and could hardly be said to constitute snow ; indeed, in the course of a short walk there were not sufficient falling to form a coating on one's umbrella. I had never previously seen anything like this phenomenon.

It is evident that such perfect crystalline forms can only exist under peculiar meteorological conditions, and not in ordinary snowflakes in which a quantity of crystals are agglomerated together. The air must be so slightly saturated with vapour that few crystals are formed, and there must be sufficient absence of wind for these crystals to remain suspended in the air and continue to grow for a considerable time without coming into contact with each other.

Bangor, North Wales, January 1901.

THE Belgian Government is evidently determined to adopt wireless telegraphy, not only on their mail steamers plying between Ostend and Dover, but also overland from Antwerp to Brussels. The experiments from the steamer to shore have been very successful, but those overland disappointing. Still it is to be hoped they will ultimately succeed.

THE KAMMATOGRAPH.

LAST month we drew attention to this remarkable invention, by which the use of the long tape celluloid film is avoided during the exhibition of moving pictures. Those who have exhibited rapidity. Fig. 2 shows the extreme simplicity of Mr. Kamm's invention, which has the advantage of a minimum of parts, so that it is almost impossible for the mechanism to become disorganised.

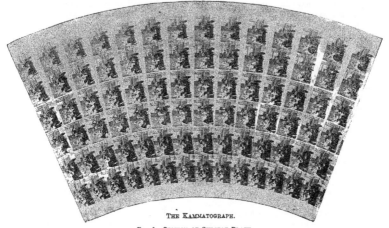

THE KAMMATOGRAPH.

FIG. 1. SECTION OF SUBJECT PLATE.

with that type of film are familiar with the anxieties that arise during its rapid run past the lantern lens. Mr. Kamm some time ago realised this difficulty, and consequently set to work to design an instrument through which the use of the tape is dispensed with, by printing the pictures in spiral form ¡upon a circular· glass plate or disc. Fig. 1 shows a section of this spiral.

Fig. 3 exhibits the instrument closed and ready for operation, its dimensions being 14 inches high, 13½ inches broad, and 3½ inches in width, the total weight being no more than 8 lbs.

Of course it is necessary to have a lantern for exhibiting the pictures, and fig. 4 represents the arrangement of the kammatograph placed ready for work in front of the lantern. These lanterns are

FIG. 2.

FIG. 3.

Having conceived this idea, it became necessary to design certain mechanical movements enabling the operator to revolve this plate in such a manner as to exhibit every picture in sequence and with designed by Mr. Kamm for use with a small electric arc lamp, though other sources of light may be used, accessories for these purposes being available.

A remarkable feature in connection with this invention is, that it serves the purpose of a camera and projector combined, thus dispensing with the use of two apparatus, one for taking the photographs, the other for displaying them, as is the case with the ordinary celluloid film. All that is necessary is to have a circular glass plate ready sensitised, and then placed within the instrument, in a dark room, when the amateur photographer, after arranging the apparatus in position, simply removes the cap and turns the handle at the required speed until he reaches a check which informs the operator

FIG. 4.

LANTERN AND KAMMATOGRAPH.

that about 600 pictures have been taken on the plate. The cap is replaced, the instrument returned to a dark room, the negative plate developed, and, when dry, printed in the ordinary manner by contact. Thus it will be seen that moving pictures of important or entertaining events may be taken with this simple arrangement by persons of ordinary ability, and exhibited to their friends within an hour or two on a screen showing moving pictures about 6 feet square in clear detail. The cost of this apparatus is £6 10s., the negative and positive plates being 2s. 6d. and 3s. for subject plates. On account of its simplicity and comparatively small cost Mr. Kamm's new instrument will make quite a revolution in the exhibition of moving pictures. Especially is this the case, as by this means the duration of exhibition of each picture is lengthened, thus obtaining a prolonged view of the subject under observation.

"JOURNAL OF MALACOLOGY."—The December number of the "Journal of Malacology" completes its seventh volume and its tenth year. In the part under consideration Mr. Walter Collinge describes a new species of *Veronicella* from the Fiji Islands, but gives no figure. Mr. H. H. Bloomer gives notes on some *Anodonta cygnea* with deformed gills, and Mr. F. J. Partridge records a reversed shell of *Helicigona lapicida*.

NOTICES BY JOHN T. CARRINGTON.

By Land and Sky. By the REV. JOHN M. BACON, M.A., F.R.A.S. x + 275 pp., 9 in. × 6 in., with 4 illustrations. (London: Isbister & Co., Ltd., 1900.) 7s. 6d.

As a popular account of many aerial voyages and other expeditions, Mr. Bacon's work will find numerous interested readers. It is rather a popular book for the encouragement of ballooning, than a scientific treatise, consequently its chapters follow each other with a pleasant flow of words that captivate the reader even against his will. Many of the observations with regard to the movements of sound-waves are interesting, as are some of the author's conclusions. Still, we do not always agree with him. One of these is, that in no case could he trace an echo from a cloud. Possibly his sound-waves were not sufficiently strong, and we should like to ask how he explains the roll of thunder. The curious effects of sound-waves rising from earth to the higher altitudes will be read with interest. The author has given considerable attention to the travelling of sound, so has much to say about the subject. One of the pleasures of the book is that one frequently comes on an unexpected and suggestive paragraph giving a new idea of facts. For instance, it is pointed out that what strikes an observer from a balloon while crossing the dense city of London is "the large proportion that open spaces bear to the actual mass of bricks and mortar." This arises from the fact that to passers in the streets the rows of houses always hide the gardens and spaces between, all of which are visible from a balloon. As an example. we may mention that there is an illustration given, taken from a photograph by Mr. Bacon whilst passing at an elevation of 1,000 feet over Clifton, near Bristol. The author's style of writing is picturesque and very readable.

Lord Lilford: a Memoir by his Sister. xxiii + 290 pp., 8¼ in. × 5½ in., with portrait, 17 plates and illustrations in text. (London: Smith, Elder, & Co. 1900.) 10s. 6d.

There can be few readers of this magazine who are unaware of the eminence as an ornithologist of Thomas Littleton, fourth Baron Lilford, F.Z.S., a past President of the British Ornithologists' Union, whose demise was noted in these pages in July 1896. This memoir by his sister will be welcomed, as it contains many extracts from Lord Lilford's correspondence, showing throughout that real love of Nature which was his characteristic. The book is prefaced with an introduction by the late Bishop of London, which is an appreciation of Lord Lilford's personal character, Bishop Creighton having frankly confessed it was beyond his power to estimate him as an ornithologist. The book is handsomely illustrated by Messrs. A. Thorburn, G. E. Lodge, F D. Drewitt, F. Smit, and from photographs.

Studies, Scientific and Social. By ALFRED RUSSEL WALLACE, LL.D., D.C.L., F.R.S. 2 vols. xviii + 1,067 pp., 7¼ in. × 5 in., with 114 illustrations. (London and New York: Macmillan & Co., Ltd. 1900.) 18s.

In these two volumes are reprinted some of the articles by Dr. Russel Wallace, contributed to various periodicals during the thirty-five years from 1865 to 1899. In entitling these reprints "Studies," Dr. Wallace does so because they deal, in a large part, with problems in which he has been particularly interested. Among them is of course the modern theory of evolution, with which the author has been so intimately connected. The others are chiefly educational, political, and social questions. These have been illustrated copiously so as to aid some readers to get a better appreciation of their intention in exposition or criticism. With regard to general divisions, the articles are grouped into "Earth Studies," or geological, "Descriptive Zoology," "Plant Distribution," "Animal Distribution," "Theory of Evolution," "Anthropology," and "Special Problems," which make up the first volume, the second being devoted to political, social, and ethical essays. The whole of the chapters are written in Dr. Russel Wallace's easy, charming style, and several contain novel views or new arguments.

The Harlequin Fly. By L. C. MIALL, F.R.S., and A. R. HAMMOND, F.Z.S. viii + 196 pp., 9 in. × 6 in., with frontispiece and 129 illustrations. (Oxford: at the Clarendon Press. 1900.) 7s. 6d.

This is an important book from the strictly educational point of view, and one that will long serve as a text-book in entomology, and practically as an introduction to the anatomy and morphology of a section of the great Order Diptera. The full title of the work is "The Structure and Life-History of the Harlequin Fly (Chironomus)." In the larval condition *Chironomus dorsalis* is popularly known as "blood-worms," and is to be met with at the bottom of many slow-flowing streams. They are often abundant and easily reared in captivity, thus forming a good subject for study. Professor Miall's name is sufficient guarantee for the accuracy of the work in the book, and Mr. Hammond's figures are admirable as usual in his drawings. The book is a worthy companion to Miall and Denny's similar treatise on "The Cockroach."

Arsenical Poisoning in Beer-drinkers. By T. N. KELYNACK, M.D., M.R.C.P., and WILLIAM KIRKBY, F.L.S. xii + 125 pp., 8½ in. × 5½ in., with 16 illustrations. (London, Paris, and Madrid: Baillière, Tindall, & Cox. 1901.) 3s. 6d. net.

After the wide attention to the epidemic of disease caused, as is supposed, by arsenical poisoning of beer, we might naturally expect some authoritative book upon the subject. The authors are respectively, the former Medical Registrar and late Pathologist at the Manchester Royal Infirmary, and the latter Lecturer on Pharmacognosy at Owens College, of the same city. The book is divided into sections, dealing with "Introductory," "Clinical," "Chemical," and "Biography." In their extensive investigations the authors have had much assistance and ready help from several firms of brewers "implicated in the brewing of the arsenical beer." Peripheral neuritis in alcoholic subjects is not very new to the medical profession in certain districts, and was diagnosed as early as 1884-5. Indeed, for the past twenty years it has been a very common affection in Lancashire amongst alcoholics, and a prolific cause of paralysis, invariably among beer-drinkers. Latterly these evil effects have reached the proportions of an epidemic, chiefly in working-class neighbourhoods. For some time past the authors have been investigating this subject, with every assistance from various authorities interested. Dr. Reynolds was the first who really directed attention to the association of arsenic with the epidemic, and this has abundantly been confirmed by the authors of the work before us. The illustrations are remarkable, and are from photographs taken from individuals actually suffering. The effects are apparently typical in nearly all cases, though varied and certainly most unenviable.

Elementary Mechanics of Solids. By W. T. H. EMTAGE, M.A. viii + 333 pp., 7 in. × 4⅜ in., with 157 illustrations. (London and New York: Macmillan & Co., Ltd. 1900.) 2s. 6d.

The author, who has had much experience in teaching and examining in theoretical mechanics, gives the benefit of his observation, and points out the importance of carefully explained examples. It is a book that may be used without any mathematical attainments beyond an ability to solve easy algebraical equations, and will be found generally useful for reference or more extended education.

Scientific Foundations of Analytical Chemistry. By WILHELM OSTWALD, Ph.D., translated by GEORGE M'GOWAN, Ph.D. Second English Edition. xx + 215 pp., 8 in. × 5 in. (London and New York: Macmillan & Co., Limited. 1900.) 6s. net.

This volume is a translation of the second German edition, to which has been added alterations and additions by the author, who has also examined the translator's pages. The subject is treated in an elementary manner by Dr. Ostwald, who is Professor of Chemistry in the University of Leipzig. In the present edition there is also a section upon electro-chemical analysis, and other additions and revisions have taken place.

British Flies. By G. H. VERRALL. 691 + 121 pp., 10¼ in. × 9¼ in., with 458 figures. (London: Gurney & Jackson, 1901). £1 11s. 6d. net.

The long-expected first volume of the fourteen to be issued by Mr. Verrall upon the diptera inhabiting Britain is now before us. Mr. Verrall has indeed undertaken a stupendous work, if the rest is to be judged by this one, which is admirably produced alike in arrangement, scientific description, and illustration. We are pleased to note that with regard to synonymy, the author treats this debateable subject in a separate paragraph after his descriptions, instead of loading the first line, as is so commonly the case in other works on Natural History. The volume to hand is Vol. VIII., of the proposed fourteen, and we congratulate Mr. Verrall upon his wisdom in making it complete in itself, which is his intention with future sections, so that they may become separate works of reference, in case, which would be greatly to be deplored, the author is not able to finish the whole. They would otherwise remain as an unfinished fragment. In an explanatory page the author sketches the contents of the proposed volumes as divided by families. Volume VIII. contains the Platypezidae, Pipunculidae, and Syrphidae. These

families, containing some of the largest and
most showy of the diptera, will appeal to
the general entomologist, and probably the
book will do more than anything which has
hitherto been published to create a taste
for the study of the diptera. The beginner
will find within its covers ample material to
fully occupy his time for long hence, without troubling himself about other families
of our two-winged flies. This fact may be
gathered on examining the catalogue of
species at the end of the book, which occupies no less than 120 pages. In this catalogue are included references to the description of each species in the chief works on
diptera by other writers. The beginner also
will find a nine-page introduction to the
study of Diptera Cyclorrhapha that will
clear the way of initial difficulties, and by
aid of the score or so of illustrations enable
him at once to commence his study without further assistance. The illustrations
in this work are by Mr. Verrall's scientific
assistant, Mr. J. E. Collin. F.E.S., and are

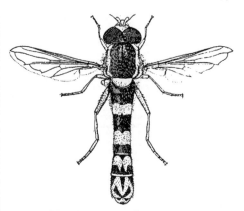

Sphaerophoria scripta. Male. Length 10 mm.

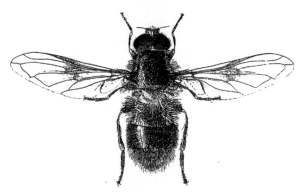

Chilosia illustrata. Length 9 mm.

hunting-ground for diptera.
It is about an acre in extent,
on the borders of Newmarket,
in Cambridgeshire. He esti-
mates that he has there taken
at least 500 species of diptera,
and about a hundred species
on his study window. In
the book is a list of reputed
British Syrphidae, and also
of species occurring in Britain and North America. The
work is embellished with a
frontispiece portrait of Dr.
Joh. Wilh. Meigen, an eminent
dipterist, who lived between
1764 and 1845, being copied
from a quaint print. No
entomologist or scientific
library can afford to be
without Mr. Verrall's new
volume on British flies.

characterised by the extremely careful
manner in which they have been portrayed, accuracy evidently being the
artist's strong point. The figures are
necessarily enlarged, but the natural
size of each species is found with the
description. By permission of the publishers we reproduce some of these as
examples. Without finding fault with
the drawing, perhaps it would be better
if the artist would in future use a fine-
pointed pen, as we think he could
therewith attain greater delicacy in
delineation. A curious feature appears
on page 666, being "Species in horto
meo." This list of forty-seven, equal-
ling twenty-three per cent. of the
known British species, should encourage
would-be dipterists to look to their own
gardens for a commencement. Mr.
Verrall's has, indeed, proved a prolific

Ascia podagrica. Male. Length 5½ mm.

(From Verrall's " British Flies.")

Story of Bird-Life. By W. P. PYCRAFT. 244 pp., 6 in. × 4 in., with seven illustrations. (London: George Newnes, Ltd. 1900.) 1s.

This is another of the Library of Useful Stories published by Messrs. Newnes. It contains a sketchy compilation giving a general account of birds and bird-life. It deals with their form, object of colours, food, flight, courtship, migration, distribution, and other objects of interest.

What is Life? By FREDERICK HOVENDEN, F.L.S., F.G.S., F.R.M.S. Second edition. xiv + 210 pp., 9 in. × 6 in., with twenty-three illustrations. (London: Chapman & Hall, Ltd. 1899.) 6s.

The author must feel considerable satisfaction in re-issuing this book. Opportunity has been taken to make several corrections and additions, including an appendix which is really a subject to be inserted in his chapter dealing with the evidence proving the statement of his case.

What is Heat and What is Electricity? By FREDERICK HOVENDEN, F.L.S., F.G.S., F.R.M.S. xvi + 320 pp., 9 in. × 6 in., with 99 illustrations. (London: Chapman & Hall, Ltd. 1900.) 6s.

The author, who is a Fellow of the Physical Society of London, and Honorary Secretary of the London Institution, has also written the book, " What is Life?" In a similar manner to that work Mr. Hovenden discusses the qualities of heat and electricity. The pages are illustrated to show experiments and demonstrations in connection with the investigations set forth in the book before us. The subjects are divided into fourteen parts, and the book bears evidence of considerable labour and thoughtful digestion of the works of leading physicists. Amateurs will find the work entertaining and suggestive of many experiments.

Edible British Fungi. By E. W. SWANTON. 43 pp., 8¼ in. × 5 in., with six coloured plates. (Huddersfield: Charles Mosley. 1900.) 2s. 6d. net.

The full title of this pamphlet is " An Annotated Catalogue of British Fungi," the author being Conservator to the Haslemere Educational Museum and a member of the British Mycological Society. The pages are occupied by short descriptions of our edible fungi, many of which are figured on the plates. These, however, are somewhat disappointing, being flat and often wanting in coloration. Still, for those who have no more pretentious works, this will be useful.

Whence and Whither. By Dr. PAUL CARUS. vi + 188 pp., 8 in. × 5 in. (Chicago Open Court Publishing Co. 1900.) 3s. 6d.

This psychological work bears the sub-title of " The Nature of the Soul, its Origin and its Destiny." Doubtless some of our readers would like to see this book and consider the views expressed by the author, but it is needless to point out that they are necessarily highly speculative.

Outlines of Field Geology. By Sir ARCHIBALD GEIKIE, F.R.S. Fifth edition. xvi + 260 pp., 7 in. × 5 in., with 88 illustrations. (London and New York: Macmillan & Co., Ltd. 1900.)

In re-issuing this standard popular work on field geology the author has thoroughly revised and brought up the information therein contained to present knowledge. Consequently there are numerous alterations and additions, without altering the simple style of arrangement. This handy little work therefore remains one of the most useful for beginners in geology.

Text-Book of Zoology. By Dr. OTTO SCHMEIL. Part III. Invertebrates. viii + 186 pp., 9¼ in. × 6½ in., with numerous illustrations. (London: Adam & Charles Black. 1900.) 3s. 6d.

We have already noticed former parts of this popular work, which is translated from the German by Rudolph Rosenstock, M.A., and edited by J. T. Cunningham, M.A. It is an entirely elementary work, intended for use in schools and colleges, and it is one that will doubtless lead to a taste for natural history.

The School Journey. By JOSEPH H. COWHAM. 79 pp., 8¾ in. × 5½ in., with 50 illustrations. (London: Westminster School Book Depôt. 1900.) 2s. 6d.

The author, who is Lecturer on Education at Westminster Training College, and also of several books on education, in this sketches his views upon the value of teaching from object views met with during school excursions. These are illustrated from subjects seen during actual excursions with boys under his direction.

England's Neglect of Science. By JOHN PERRY, M.E., D.Sc., F.R.S. 115 pp. 8¾ in. × 5½ in. (London: T. Fisher Unwin. 1900.) 2s. 6d.

We are very pleased to find a sort of awakening in this country to a recognition of the claims of science. It is quite time that the professional man of science should take his place in the social organisation, as does the doctor of medicine or barrister. The modern education of the scientific man is quite as serious an undertaking as that in any other profession. Professor John Perry, the President of the Institute of Electrical Engineers, and of the Royal College of Science, London, reprints in book form seven articles which have appeared elsewhere, including a paper read before the Society of Arts. Several of them are important, and cannot be too widely known and read.

What the World wants. By G. B. MOORE. 102 pp., 7½ in. × 5 in. (Chicago and London: Self-Culture Society. 1901.) 1s.

The Self-Culture Society has for its objects the refinement and self-education of the people, two most important factors in the betterment of society. The days are long past when all that was considered necessary was to pack off a girl or youth to a " finishing school." The fact is now recognised that to be successful every unit of a highly civilised nation must continue the habit of self-education throughout life. This little book is intended to found in young people that habit, in view of leading them to success in after-life, its closing motto being " There is no Darkness but Ignorance."

How to avoid Tubercle. By TUCKER WISE, M.D. Third edition. 24 pp., 7 in. × 4½ in. (London: Baillière, Tindall, & Cox. 1900.) 1s.

Too much cannot be known by the public at large about the circumstances which are favourable for the cultivation and spreading of tubercular consumption. In the pages before us will be found, concisely put before their readers, much valuable information, written in popular language, that should be widely spread. The day will doubtless come, and perhaps not long hence, when consumption will be as rare a disease in this country as is smallpox. There is really no reason why such should not be the case. Our King has for some time past devoted much attention to this scourge, and will doubtless use his influence towards its suppression.

WE DESIRE to ask from among our readers for a volunteer who will kindly undertake the honorary editorship of the department of botany in SCIENCE-GOSSIP. We shall be glad to welcome communications in this matter, and suggestions as to the arrangements for the page.—ED. S.-G.

. IN common with the British people throughout the world, we deplore the nation's loss of Her Majesty Queen Victoria. It may interest some of our readers to know that when she was the Princess Victoria, she was Patron of the Entomological Society of London on its formation in 1833.

A COMPLIMENTARY dinner has been arranged to take place early in March to commemorate the retirement of Sir Archibald Geikie from his post of Director-General of the Geological Survey of the United Kingdom, which appointment he has held since 1892.

THE Hampstead Astronomical and Scientific Society propose to form a natural history section, with the primary object of working at the local fauna and flora. Members and others who are willing to identify themselves with the section are asked to send their names to Mr. J. W. Williams, M.R.C.S., 128 Mansfield Road, Gospel Oak. It has also been decided to alter the name of the society to the Hampstead Scientific Society.

MR. THOMAS SHEPHERD has been appointed curator of the museum attached to the Royal Institution, Hull. The committee, in making their selection, have carefully considered the applicants with the object of obtaining a man of sufficient technical knowledge to enable him to properly arrange and classify the various exhibits. Mr. Shepherd is well known to our readers as the hon. secretary of the Hull Scientific and Field Naturalists' Club, a skilled geologist, and member of the Boulder Committee of the Yorkshire Naturalists' Union.

WE have received an intimation from Mr. E. A. Martin, the hon. secretary of the Croham Hurst Preservation Committee, that the Croydon County Council have decided by unanimous vote to purchase the whole of the beautiful wooded hill to the south of Croydon, known as Croham Hurst. Our readers may remember that thirty-five acres were acquired about two years ago. Public attention having been turned to the matter and interest aroused, several meetings were held, resulting in a committee of seventy members, to whose exertions is due the present arrangement to purchase. The County Council are offering £15,000 for the remaining forty-five acres.

THE annual report for the year 1900 of the Royal Meteorological Society is satisfactory, there being an addition of fifty-five Fellows on the roll. It mentions that the late Mr. G. J. Symons, F.R.S., had bequeathed, among other things, to the Society about 2,200 volumes and 4,000 pamphlets from his library.

WE have before us a handy little catalogue of binocular field and opera glasses, telescopes, and compasses, issued by Messrs. Ross, Limited. As the illustrations are numerous, this list will appeal to many of our readers who need portable instruments of such kind. A useful one for amateurs is their seaside and tourist's telescope, which is supplied with a tripod stand.

THE last effort of Nature in this country in the 19th century was a further destruction of one of the most ancient human monuments. Two of the largest stones of the outer circle of Stonehenge fell during the storm of New Year's Eve. Considering the comparatively small cost that would be required to place these prehistoric remains beyond further immediate decay, it seems a pity that this cannot be at once effected. We trust a movement will be inaugurated having this object in view.

ONCE again a successful Conference of Science Teachers has been held in London, this, the third time, at the South-Western Polytechnic, Chelsea, on January 10th. The addresses were as good as ever, but owing to the lack of time the discussions cannot be considered to be so satisfactory as those of last year. Next year three days will most likely be devoted to the meetings with good effect. On the present occasion many points of interest were touched upon, the more important being the co-ordination of workshop and laboratory practice, and the making of useful and simple apparatus in place of expensive and obscure instruments which are of little educational value. Then the fitting up of chemical, physical, and mechanical laboratories was gone into. The scientific education of girls was discussed, and the value of psychological knowledge in the case of young children, as well as of older students, most properly emphasised. Full reports will doubtless be published in the "Technical Education Gazette" of the London County Council, which should be read by all interested in the teaching of science, who were not present at the conference.

THE Science Masters of our public schools are greatly to be congratulated on publicly expressing their opinions as to the reforms needed in science teaching, which they have not power to make. It is to be hoped that all those directly interested in the carrying on of one of the most important crusades of the day will read the excellent papers read before the Conference of January 19th at the London University. Those also in authority, whether governesses or headmasters, must at all costs be convinced that under the ordinary circumstances of school work science teaching is a mind-destroying farce. What the nation demands is that her sons shall be prepared to uphold her traditions by means of a training in methods of science as well as mechanical culture. We might mention that Sir Henry Roscoe, who is Vice-Chancellor of London University, presided. He is also a governor of Eton College, from which we believe the idea of the Conference emanated, the circular convening it being signed by four of its science masters, with Mr. M. D. Hill as secretary.

CONDUCTED BY B. FOULKES-WINKS, M.R.P.S.

EXPOSURE TABLE FOR FEBRUARY.

The figures in the following table are worked out for plates of about 100 Hurter & Driffield. For plates of lower speed number give more exposure in proportion. Thus plates of 50 H. & D. would require just double the exposure. In the same way, plates of a higher speed number will require proportionately less exposure.

Time, 10 A.M. to 2 P.M.

Between 9 and 10 A.M. and 2 and 3 P.M. double the required exposure.

SUBJECT	F. 5·6	F. 8	F.11	F. 16	F. 22	F. 32	F. 45	F. 64
Sea and Sky..	$\frac{1}{250}$	$\frac{1}{120}$	$\frac{1}{64}$	$\frac{1}{32}$	$\frac{1}{16}$	$\frac{1}{8}$	$\frac{1}{4}$	$\frac{1}{2}$
Open Landscape and Shipping	$\left. \right\} \frac{1}{64}$	$\frac{1}{32}$	$\frac{1}{16}$	$\frac{1}{8}$	$\frac{1}{4}$	$\frac{1}{2}$	1	2
Landscape with dark foreground, Street Scenes, and Groups	$\left. \right\} \frac{1}{16}$	$\frac{1}{8}$	$\frac{1}{4}$	$\frac{1}{2}$	1	2	4	8
Portraits in Rooms ..	$\left. \right\} 4$	8	16	32	—	—	—	—
Light Interiors	8	16	32	1	2	4	8	16
Dark Interiors	$\frac{1}{2}$	1	2	4	8	16	32	64

The small figures represent seconds, large figures minutes. The exposures are calculated for sunshine. If the weather is cloudy, increase the exposure by half as much again; if gloomy, double the exposure.

PHOTOGRAPHIC CONVENTION OF THE UNITED KINGDOM.—We understand that the Convention will be held this year at Oxford under the Presidency of Sir William J. Herschel, Bart. The meetings will take place during the second week in July, at the Municipal Building. We hear that the secretary and local committee are already making extensive arrangements for visiting the most interesting and picturesque places in the locality. In Oxford alone the amateur might easily occupy the time at his disposal; but when we consider that such places as Abingdon, Woodstock, Warwick, etc., are within easy reach, we certainly think that the committee are to be congratulated upon their selection for this year's Convention. We expect to see one of the largest gatherings on record, and, judging from our experience at Oxford and its vicinity, we are sure the amateur and professional will spend a very enjoyable week, and return with plenty of charming views. Readers wishing to attend should write to Mr. F. A. Bridge, East Lodge, Dalston Lane, London, N.E., who will give every information.

SNOW PICTURES.—Amateurs who put away their camera in the late autumn, and do not bring it out again until the spring, often miss some very charming effects when the snow is on the ground. We believe there are a number of workers who do this every year, for we often hear the remark, " Oh, it is no use bringing the camera out in the winter." Quite recently we saw some exceedingly pretty effects taken when the snow had just fallen, and as we shall probably experience severe weather during February, the amateur will be well advised to have his camera ready for emergencies in this direction. The best results are generally obtained on slow plates with a full exposure. For developing these snow pictures we prefer a weak developer of the metol type so as to ensure getting all possible detail. The selection of the process by which the negative is to be printed is worthy of consideration. The most natural and pleasing results are obtained on platinotype paper. If this process is to be used, it is advisable to make the negative fairly strong, or, in other words, to let the negative show greater contrast between light and shade. As platinotype printing, however, is very troublesome at this time of the year, we would suggest that the bromide process may be used to advantage. In this respect we would refer the reader to last month's issue of SCIENCE-GOSSIP, where we describe the working of a slow bromide paper which is most suitable for this purpose. If, however, it is decided to finish pictures of this class in gelatino-chloride printing-out paper (P.O.P.), we are strongly in favour of a matt paper toned with platinum. The following formula, given by the Eastman Kodak Company, we have found work well and have very satisfactory results. STOCK SOLUTION.—Potassium chloroplatinite, 5 grains; citric acid, 40 grains; sodium chloride (common salt), 40 grains; water, 20 ounces. This bath will keep well for a month. Wash the prints for about five minutes in several changes of water and then immerse in the above bath; examine the prints by transmitted light. Tone the prints to a deep chocolate colour, rinse slightly, and then place in the following bath to stop action of the toning bath if a warm sepia tone is desired. BATH.—Sodium carbonate (washing soda), $\frac{1}{2}$ ounce; water, 20 ounces. Rinse, and transfer to the fixing bath. FIXING BATH.—Sodium hyposulphite, 3 ounces; water, 20 ounces. The prints should be kept moving about in the fixing bath for about ten minutes. Wash thoroughly and dry between blotting boards or by suspending them from print clips.

A TONING BATH FOR P.O.P.—Whilst on the subject of printing on P.O.P. we should like to draw attention to a new formula which has, in our hands, given most pleasing tones, and is free from the objectionable double-toning so often met with in ordinary toning baths. We append the formula which we consider to give the best results:—STOCK SOLUTION—Ammon. sulphocyanide, 20 grains; common salt, 200 grains; water, 20 ounces. To tone, add one or two drams of the following solution:—Chloride of gold, 15 grains; water, 2 ounces. The prints may be immersed in the solution without any previous washing, and will tone in a few minutes. Fix for about ten minutes in a bath as follows:—Hyposulphite of soda, 3 ounces; water, 20 ounces. Wash thoroughly and dry.

FADING NEGATIVES.—Writing to "Nature," on January 17th, Mr. William J. S. Lockyer draws attention to the disappearance of images on astronomical photographic plates, and gives instances of faint objects after a time disappearing. He mentions a process accomplished by Sir William Crookes for successfully bringing them back to view, and gives Sir William's formula.

PHOTOGRAPHY FOR BEGINNERS.

BY B. FOULKES-WINKS, M.R.P.S.

(*Continued from p. 264.*)

SECTION I. CAMERAS (*continued*).

HAVING described the two principal types of stand cameras, we must now consider the hand cameras in general use.

HAND CAMERAS.

There are so many hand cameras on the market that it would be impossible to mention all in the space at our disposal. We therefore propose to describe those that are best known and most likely to be met with. We have personally tested and examined all the following types:—

No. 3 FOLDING POCKET KODAK CAMERA.

First of all there is the camera made specially for use with rollable films; then the cut-film camera; and, lastly, the hand camera adapted only for plates. These latter may be divided into three distinct systems of changing—that is to say, the automatic or falling-plate system, the bag changing, and the dark-slide.

DARK-SLIDE CHANGING.—Of these, the dark-slide hand camera is undoubtedly the simplest and least likely to get out of order, but it has the distinct disadvantage for hand-camera work of requiring considerable time for changing the plates. That is to say, there is a long interval involved from making one exposure to the moment when the next plate is ready for use. It has also the drawback of being rather bulky, as six double dark slides, holding twelve plates, will take up much more room than is occupied by the same number of plates in either of the other systems. On the other hand, it is absolutely sure in changing, and is undoubtedly a type that renders the plate most free from dust and scratching. This method of changing plates is used in such cameras as the Ross "Twin-Lens" camera, the "Adams Reflex," Houghton & Son's "Sanderson" hand camera, the "Goerz-Anschutz" folding, Lizars's "Challenge," Shew's "XIT," Watson's "Gambier-Bolton," and most of the American folding cameras.

AUTOMATIC CHANGING.—In general use the automatic, or falling plate system, is probably the most popular form of plate changing. The principle of this system is that twelve unexposed plates are inserted into twelve sheaths and are placed in a chamber facing the lens; the first plate being in the focal plane of the lens and thus

ready to receive an exposure. When this plate has been exposed, the mechanism is so arranged that by touching a lever or button outside the camera the exposed plate will fall into a chamber at the bottom of the camera and automatically bring the next plate into its previous position, ready to receive another exposure, and so on, until the twelve plates have been exposed. Although cameras of this type are sold under a great variety of names, they are, so far as changing is concerned

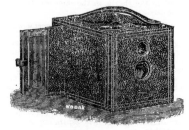

No. 2 BULLET KODAK CAMERA.

more or less alike. To give an instance of this class of camera, we may mention Lancaster's "Stopit," the City Sale and Exchange Company's "Salex," Benetfink's "Lightning," "The Optimus," "Lynx," and Butcher's "Midg." This method has the distinct advantage of occupying very little space, and is exceedingly quick in changing the plate, a point of great value when it is desirable to take several negatives of the same moving object. It has, however, the objection common to all automatic changing systems, of occasionally failing to change the plate. It is also more liable to create dust spots on the negatives, owing to the falling plate disturbing any dust in the camera,

BUTCHER'S "MIDG" HAND CAMERA.

which may afterwards settle on the plate and so cause dust spots. Hence it is very important that this class of camera should be well dusted inside before being loaded with plates.

BAG CHANGING.—The bag system of changing the plate seems at present to be the most perfect all-round method. It is quickly changed, occupies less space than any other, and, if well made, is free from both dust and scratches. The cameras made upon this system are undoubtedly the highest class instruments in the market and the most complete in every respect. The principal makes and some of the best instruments are : Adams & Co.'s "Yale" and " De Luxe," Newman & Guardia's "N & G" camera, Watson's "Vanneck," and the "Rouch." The two firms first mentioned make detachable changing boxes, which are supplied with several other makers' cameras, such as the "Gambier-Bolton," etc. This system of changing is used largely for cut films, in which case the cut film is inserted in the sheath, in the same way that the plate is carried.

CUT FILMS.—There are very few cameras made only for cut films, the most important of these being the "Frena" and the "Tella." The changing system is somewhat similar in each case : that is to say, films are put into packets of twenty or twenty-five, as the case may be ; each camera holding as many as forty or fifty films. When the first film is exposed it is passed away into the receiving chamber, and the next film comes into position ready to receive an impression. The mechanism in each case is exceedingly ingenious, and achieves a very difficult operation in separating one film from another.

ROLLED FILMS.—Films capable of being rolled are used almost exclusively in the Eastman "kodak." They consist of a very thin film of celluloid coated with a sensitive emulsion. This is wound, in conjunction with a strip of black paper, upon a spool which fits into the camera. It is then passed across the focal plane of the lens on to another and empty spool, and is gradually wound off one spool to the other. Thus we get a series of pictures upon a long strip of sensitive film, which, when developed, may be cut up into its several parts, as may be desired. The great charm of this changing system is the fact that a spool can be inserted into the camera in broad daylight, six or twelve exposures made, and the film wound off, taken out of the camera, and a new spool inserted. It occupies little space, and is very light in weight.

Having decided upon the changing system, there are other essential points to consider in a hand camera.

LENSES.—A good quick-acting lens is necessary, one with iris diaphragms working from F. 8 to F. 32 if desirable, and in the best instruments with lenses of the anastigmat type from F. 6·3.

SHUTTERS.—A time and instantaneous shutter is required, giving exposures varying from $\frac{1}{4}$ to $\frac{1}{100}$ part of a second; also for indefinite time. With cameras fitted with the quickest lenses, such as the "Planar," a range of speed from $\frac{1}{4}$ to $\frac{1}{1000}$ part of a second is desirable. The shutter should not open when being set. We think the most satisfactory method of timing a shutter is beyond question the pneumatic regulation principle, as used by Newman & Guardia, Adams & Co., and Bausch & Lomb.

RISING FRONTS.—In addition to the above points, we strongly advise the student to select a hand camera having a rising front. (See Stand Cameras, *ante*, p. 253.)

FOCUSSING.—A rack-work focussing arrangement is most important.

(*To be continued.*)

JOHN HENRY LEECH.—What at one time appeared to promise a brilliant scientific career was closed, we understand by the breaking of a blood-vessel, on December 29th last. Mr. Leech, born thirty-eight years ago, was educated at Eton and Trinity Hall, Cambridge. He was the son of the late John Leech, Gorse Hall, Dukinfield, Cheshire, and on attaining his majority he inherited very ample means. J. H. Leech was a born naturalist and no mere collector. At an early age he cut himself adrift from the narrowing influences of a mere British collector, following the pursuit of entomology and sport as a traveller in little-known districts of Central Brazil, Northern India, Corea, and Japan. During this period Mr. Leech passed through many dangerous adventures and serious risk from climatic diseases ; in fact, it is probable that the seeds were sown which ultimately so sadly cut short a life useful to science. After these extensive wanderings Mr. Leech commenced to scientifically investigate his large collection of lepidoptera, and soon became a specialist in those species constituting the fauna of Eastern Asia. To the collections already in his hands he largely added by employing both European and native collectors in China, Japan, Corea, and portions of the Palaearctic region, thus amassing material to an extent rarely exceeded by any private collector. The result has been the publication of an important illustrated work on the butterflies of China, Japan, and Corea, and the formation of a magnificent type collection of Eastern lepidoptera. For some time past the butterflies in this collection have been offered for sale, but as the price was several thousand pounds we hear that they have not yet found a purchaser. A large supernumerary collection of Heterocera from the same region, in addition to Mr. Leech's own types and series, was some time ago acquired for the national collection at South Kensington. Mr. Leech was a B.A. of Cambridge, a F.L.S., F.Z.S., F.E.S., and author of a small work on British Pyralides, published in 1885. He was an ardent sportsman, which led to the unfortunate loss of his left forearm by the bursting of a sporting gun while pheasant shooting in Cambridgeshire, and the bearing with extraordinary fortitude of two amputations with a considerable interval between. With his usual resolution, he habituated himself to his loss, and became, with one hand only, as expert as most men with both hands. He was also an exceptionally clever fisherman, and contributed piscatorial articles to the "Badminton Magazine," "Baily's Magazine," and elsewhere. For some years Mr. Leech was the proprietor of the "Entomologist," which was edited by his scientific assistant, Mr. Richard South. The funeral of Mr. Leech took place at Baverstock, near Hurdcott House, Salisbury, where he died. We have not heard the destination of the collections ; it is improbable his widow will encounter the responsibility of their care and possible rapid deterioration.

CLIMATE OF NORWAY.—At the Annual Meeting of the Royal Meteorological Society Dr. C. Theodore Williams delivered the Presidential Address, taking for his subject "The Climate of Norway and its Factors." He considered that its meteorology should prove an attractive study for the Society, as having much in common with that of our country. Both the Norwegian and the British shores are influenced by the same Gulf Stream, and have their ,winters and summers tempered by the same equalising agency. The factors which influenced the climate were: (1) The insular character of the country ; (2) the distribution of the mountain ranges, which explains to a large extent the rainfall ; (3) the waters of the ocean, which from a variety of circumstances come into close connection with much of the country and thus temper extremes ·of climate ; and (4) the sun, which in this latitude remains in the summer long above the horizon and in the winter long below it.

THE TELEGRAPHONE.—This instrument, invented by M. Valdemar Poulsen, is described by the inventor in the "Journal de Physique" for December. The principle is simple. An electromagnet, with a soft iron core about 8 mm. long and ·75 mm. connected with a microphone, is moved over a steel wire. On speaking into the microphone, the currents traversing the electromagnet will permanently magnetise the portions of the wire successively in contact with it, and a permanent record or "writing" will thus be impressed on the wire. On substituting a telephone for the microphone, and again passing the electromagnet over the wire, currents will be induced by the variations in the magnetisation, and these will reproduce the words originally spoken. Finally, by connecting the electromagnet with a pile, and again passing it over the record, the latter will be, so to speak, erased, and the wire, now uniformly magnetised, will be ready to receive another inscription.—*(Dr.) G. H. Bryan, Bangor, N. Wales.*

MOUNTING BEETLES ON CELLULOID.—Take a piece of card and spread on it a thin layer of fairly thick gum. While the gum is still wet, place the beetles on the card and arrange legs and antennae in desired position. Allow to set, and leave until the insects are thoroughly dry. Remove them from the card by dissolving the gum in water and dry on blotting paper, taking care not to allow them to stick to the paper. Procure thin celluloid, such as is used to support the sensitive film in the kodak cartridges, and cut a number of triangular pieces, each about three-eighths of an inch long and an eighth of an inch wide at the base. The apices may be more or less pointed, according to the size of the insect. Take one of these triangles and pierce it at the base with a Continental pin, pushing the latter three-quarters of the way through. Apply a little gum to the apex at the upper surface of the triangle and mount the insect thereon. Allow to set. Descriptive labels may then be placed on the pin. The advantage of this method is the facility with which both the upper and under surfaces of the insect can be examined, the celluloid being perfectly transparent.—*G. Granville Buckley, Norwood, Oldham, January* 1901.

ANODONTA CYGNEA.—At a recent meeting of the Linnean Society of London the Rev. John Gerard exhibited a number of specimens of *Anodonta cygnea* of large size. They were sent by Mr. W. Fitzherbert Brockholes from a pond at Claughton, Garstang, Lancashire. When the water was let out of the pond, it is said that one specimen measured nine inches in width, twenty-eight others upwards of eight inches, and a hundred or more upwards of seven inches.

GREEN ELDER FRUIT.—With reference to the Rev. C. F. Thornewill's note on green-fruited elder in Derbyshire, the following may be of interest. Mr. G. Greensmith, of Thorpe, near Ashbourne, writes me :—"I have two or three trees in my garden which produce very fine white elderberries. A few years ago I made some wine from them, and it was very fine. I made it extra strong, and after remaining in the cask for about three years it was almost as good as yellow Chartreuse, and very much like it." When these trees are in fruit again I will examine them, and send a note describing this form more exactly. I understand, though I have never seen it, the colour of the fruit is whitish-green.—*John E. Nowers, Burton-on-Trent.*

SCIENCE APPOINTMENTS.—Instead of dwelling on the achievements of the past masters in Chemistry, and by so doing lull ourselves into a state of self-gratification, we would rather draw the attention of our readers to the boundless horizon opened by the following advertisement, which we take from a recent journal devoted to General Education :—"Wanted, a Senior Master, able to teach Chemistry, Physics, French, and Mathematics. Salary £75 a year, non-resident." Are the achievements of science real when the future freemen of our land are to be trained by one man in four subjects; each of which ought to be the life-work of the teacher—this, too, whilst the tutor is himself trying to work out the bread, butter, and silk-hat factors of £75 a year ?—*H. M. Read, London.*

AUSTRALIAN CERATODUS.—I should be exceedingly pleased if any of your readers can enlighten me through your columns with regard to the Australian Ceratodus or burnet salmon. Is its air-sac a swim-bladder or a lung? Dr. Mivart, in the "Common Frog," says that it is a swim-bladder and not a lung, because blood is not brought to it directly from the heart. Drs. Parker and Haswell, however, in their "Text-Book of Zoology," distinctly refer to it as a lung. There is, however, a pulmonary artery given off from the afferent branchial system carrying blood directly to the air-sac. It certainly has a highly developed apparatus for the respiration of air, because farther on Dr. Mivart says that "at night it leaves the brackish streams it inhabits and wanders among the reeds and rushes of the adjacent flats." The air-sac well performs the functions of a lung. There are now two specimens of this interesting and curious fish in the Reptile House at the London Zoological Gardens. — *Arthur R. Mynott, 78 Hunsdon Road, New Cross, London, S.E.*

ASTRONOMY,

CONDUCTED BY F. C. DENNETT.

1901	Rises.	Sets.	*Position at Noon.*	
			R.A.	Dec.
Feb.	h.m.	h.m.	h.m.	° '
Sun .. 4· ..	7.36 a.m. ..	4.51 p.m. ..	21.10 ..	16.20 S.
14 ..	7.18 a.m. ..	5.10 p.m. ..	21.50 ..	13.9
24 ..	6.58 a.m. ..	5.28 p.m. ..	22.28 ..	9.36

	Rises.	Souths.	Sets.	Age at Noon.
Feb.	h.m.	h.m.	h.m.	d. h.m.
Moon .. 4 ..	6.23 p.m. ..	0.25 a.m. ..	7.24 a.m. ..	14 21.24
14 ..	3.50 a.m. ..	8.0 a.m. ..	0.11 p.m. ..	24 21.24
24 ..	9.10 a.m. ..	5.1 p.m. ..	— ..	5 9.15

		Souths.	Semi-	*Position at Noon.*	
			diameter.	R.A.	Dec.
	Feb.	h.m.		h.m.	° '
Mercury.. .. 4 ..		0.55 p.m. ..	2·6″ ..	21.51 ..	14.47 S.
14 ..		1.18 p.m. ..	3·1″ ..	22.53 ..	7.13 S.
24 ..		1.11 p.m. ..	4·1″ ..	23.26 ..	1.5 S.
Venus .. .. 4 ..		10.47 a.m. ..	5·4″ ..	19.42 ..	21.36 S.
14 ..		11.0 a.m. ..	5·3″ ..	20.35 ..	19.21 S.
24 ..		11.11 a.m. ..	5·2″ ..	21.26 ..	16.11 S.
Mars .. .. 4 ..		1.56 a.m. ..	6·5″ ..	10.50 ..	12.2 N.
14 ..		1.5 a.m. ..	6·8″ ..	10.38 ..	13.25 N.
24 ..		0.11 a.m. ..	6·9″ ..	10.23 ..	13.51 N.
Jupiter .. .. 14 ..		8.48 a.m. ..	15·6″ ..	18.23 ..	23.6 S.
Saturn .. .. 14 ..		9.20 a.m. ..	7·1″ ..	18.55 ..	22.17 S.
Uranus .. .. 14 ..		7.26 a.m. ..	1·8″ ..	17.1 ..	22.45 S.
Neptune .. .. 14 ..		8.9 p.m. ..	1·2″ ..	5.45 ..	22.11 N.

MOON'S PHASES.

		h.m.			h.m.
Full	.. Feb. 3 ..	8.30 p.m.	3rd Qr. Feb. 11 ..		6.12 p.m.
New	.. " 19 ..	2.45 a.m.	1st Qr. " 25 ..		6.38 p.m.

In apogee February 9th at 7 a.m.; in perigee on 21st at 3 a.m.

METEORS.

		h.m.	°
Feb. 5–16 ..	a Aurigids	Radiant R.A. 4.56	Dec. 43 N.
" 15–20 ..	a Serpentids	" " 15.44	" 11 N.
" 20 ..	.. ..	" " 17.32	" 36 N.

CONJUNCTIONS OF PLANETS WITH THE MOON.

					° '
Feb. 5	..	Mars	.. 10 p.m. ..	Planet 9.54 N.
" 15	..	Jupiter	.. 5 a.m. ..	" 2.51 S.
" 15	..	Saturn†	.. 7 p.m. ..	" 3.3 S.
" 17	..	Venus†	.. 9 p.m. ..	" 5.49 S.
" 20	..	Mercury°	.. 1 p.m. ..	" 3.29 S.

° Daylight.   † Below English horizon.

The only Occultations visible are those of stars below 5·5 magnitude.

THE SUN should be watched for occasional outbreaks of activity.

MERCURY as an evening star is fairly placed for observation after the first few days of the month. It reaches its greatest eastern elongation, 18° 6', at 10 p.m. on 19th, on which day it does not set until 1 h. 46 m. after the sun. The greater part of the month it is in the constellation Aquarius, but after the 20th in Pisces. At about 5 o'clock on this evening Mercury may be found about 5° below the thin crescent moon.

VENUS is a morning star, badly situated for observation.

MARS, being in opposition at 6 a.m. on February 22nd, is as well placed for observation as he will be this year. At 8 a.m. on the 25th he is in the part of his orbit most distant from the sun; as a consequence his apparent diameter is very small, being only 13·8″ at the best. Notwithstanding, he is a beautiful little globe to study.

JUPITER, SATURN, AND URANUS are too low down and too near the sun for observation.

NEPTUNE may be looked for as an 8th magnitude star, retrograding along a very short path, about 2° north-west of the 4th magnitude star χ' Orionis.

The minor planet VESTA is in opposition on February 1st, when it appears as a star slightly above 7th magnitude. Its nearness to ξ Cancri, 5th magnitude, on February 4th, and to the pretty double star Σ 1311 on the 7th, should assist in its identification.

COMET c 1900.—On December 20th, M. Giacobini, of Nice, discovered a comet moving in a south-easterly direction in the southern part of the constellation Aquarius.

NEW MINOR PLANETS.—Whilst photographing Eros and its surrounding stars, Professor W. R. Brooks, of Geneva, U.S.A., discovered three new planets within 1° of Eros, one exceeding it in magnitude. On December 20th, Professor Max Wolf, of Heidelberg, also observed two more.

QUITO OBSERVATORY.—M. F. Gonnessia, of the Lyons Observatory, has been appointed to act as director for five years, during the work of the French Commission engaged in re-measuring the arc of the meridian in Peru.

LICK OBSERVATORY.—According to "Science," Mr. D. O. Mills, of New York, has promised about 24,000 dollars to defray the cost of an expedition either to South America or Australia, to study, under good conditions, the motions of stars in the line of sight.

MARS.—During the past few weeks some of the newspapers have given much prominence to an observation by a Mr. Douglass at Lowell Observatory, Flagstaff, Arizona. A bright patch was observed for over an hour on the border of the Icarium Mare, when it disappeared. M. C. Flammarion is doubtless right in ascribing it to the reflection from a cloud. It is by no means the first time such an observation has been made on the planet Mars.

JUPITER.—The most recent measures are those by Dr. T. J. J. See with the 26-in. Washington Refractor. Taking the solar parallax as 8·″796, the equatorial diameter is 89,919 miles, and the polar 84,111 miles. The diameters of the larger moons are said to be I.—1,574 miles; II., 1,461 miles; III., 3.187 miles and IV., 2,884 miles.

A USEFUL BLOTTER.—We have received from the Scottish Provident Institution the fourth issue of their blotting-book, containing a chart of the heavens looking north and south from London at 10 p.m. on the 1st of each month. The stars are in gold on a blue ground, and the positions of the Moon and Mars, Jupiter and Saturn are also marked. A description is given by Mr. W. B. Blaikie, which makes the whole quite a handbook for studying the face of the sky. It is interesting to note, that commercial institutions find a scientific advertisement sufficiently attractive.

CHAPTERS FOR YOUNG ASTRONOMERS.

By FRANK C. DENNETT.

(*Continued from p. 261.*)

JUPITER.

DURING 1900 Mr. Stanley Williams thought the spot was plainer, and that the reddish colour was again appearing in its following or eastern half. Immediately north of this spot there is a sort of

THE PLANET JUPITER, Oct. 17th, 1880, 12 p.m., G.M.T.

"bay," in the south tropical belt, which broadens eastwards, forming a sort of dark elbow, which is well shown in Rev. T. E. R. Phillips's drawing, fig. 4. This is readily shown with a telescope of no more than 2 inches aperture, and is traceable in many drawings so far back even as September 1831, when it was delineated by Schwabe.

Bright spots have been seen on the planet, especially in the equatorial zone, from very early times. Schmidt long since had pointed out that the rotation of the planet was apparently found to be shorter or longer when bright spots or dark ones were observed. This was abundantly con-

firmed when observers were attracted to the planet by the advent of the great red spot. The bright spots were found to overtake the red spot and to pass. Several such spots are beautifully shown in fig. 3, bedding into the south tropical belt.

On October 17th, 1880, at midnight, the writer found two small dark spots on the northern north-temperate band, shown on fig. 5, which is from a drawing made at the time, by the aid of a 9½-inch Calver reflector. These objects proved of great interest. Their colour was a deep Prussian blue; they increased in number quickly, and were found to gain even upon the white spots, so that there were at the same time on the planet three classes of maculation, each yielding a different answer to the question, "What is the period of Jupiter's rotation?" Mr. F. W. Denning determined the three periods to be from

Red spot	9 h. 55 m. 34 s.
Bright spots	9 h. 50 m. 5 s.
Dark spots, slightly exceeding .	9 h. 48 m. 0 s.

This is not all; these rates, as pointed out by Denning, are not constant. For instance, in 1878 the red spot yielded a rotation period of 9 h. 55 m. 33·5 s., but in 1898 it had increased to 9 h. 55 m. 41·5 s., and this increase was not regular. A variation is similarly noticeable in the case of the bright spots. Thus it becomes evident that the surface of the planet seen is by no means solid, otherwise the motions would be regular. To simplify the recording of observations, a zero meridian was chosen, that passing the centre of the disc at midnight on December 31st, 1871. Two rotation periods were reckoned. System I., 9 h. 50 m. 30 s. (877°·90 of Jovian longitude passing the centre in 24 h.), and System II., 9 h. 55 m. 40·63 s. (870°·27 passing in 24 h.): System I. very nearly accords with the period of the bright spots, whilst System II. almost agrees with that of the red spot. In dating drawings it is common to note the longitude of the central meridian according to both systems, as it simplifies the study of the movements on the planets. Even a telescope of

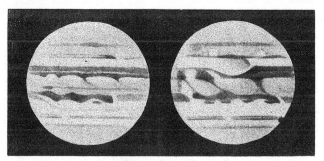

April 15th, 12 h. 10 m., G.M.T.   May 6th, 10 h. 20 m., G.M.T.

THE PLANET JUPITER IN 1899. *Drawn by Rev. Theodore E. R. Phillips*

3 inches aperture will show much of the beauty of the belts if the air is good and a power of from 130 to 180 employed.

(*To be continued.*)

CONDUCTED BY F. SHILLINGTON SCALES, F.R.M.S.

ROYAL MICROSCOPICAL SOCIETY. — December 19th, William Carruthers, Esq., F.R.S., in the chair. Notice was given on behalf of the Council that at the next meeting of the Society the name of Dr. C. T. Hudson would be submitted for election as an Honorary Fellow. Mr. E. M. Nelson exhibited a small pocket microscope made by Dr. Gilbertson, of London, of unknown date. Dr. Hebb read the list of nominations by the Council for election at the Annual Meeting on January 16th, as follows:—As President, Mr. William Carruthers; as Vice-presidents, Dr. Braithwaite, Messrs. Michael, and Nelson, and the Right Hon. Sir Ford North; as Treasurer, Mr. J. J. Vezey; as Secretaries, Dr. Dallinger and Dr. Hebb; as Council, Messrs. Allen, Beck, Bennett, Browne, Rev. E. Carr, Messrs. Dadswell, Disney, Karop, Plimmer, Powell, Professor Urban Pritchard, and Mr. Rousselet; as Curator, Mr. Rousselet. Mr. Barton, on behalf of Messrs. Ross, Limited, exhibited some new forms of lanterns which could be used for ordinary projection purposes either with or without the microscope. The first was constructed so as to exclude all light from the room except what passed through the lenses; the second was larger and more complex, and could be used for all purposes, including enlargements. Both gave excellent definition. Mr. Barton also exhibited and described several new forms of microscope with detachable circular staging and new form of electric arc lamp for lantern use. A new form of limelight was also exhibited which attracted much attention from its great brilliancy and steadiness, and the silence with which it burned. These effects were produced by causing the gases to impinge upon each other previous to their entrance into the mixing chamber, and by the construction of the chamber itself.

JOURNAL OF THE QUEKETT MICROSCOPICAL CLUB.—We regret that space has prevented our noticing earlier the November issue of this interesting journal which contains the index for the volume comprising this and the preceding five numbers. The present number contains an article by Mr. T. B. Rossiter on the anatomy of the tapeworm *Dicranotaenia coronula*, with two excellent plates; a note on four rare British fungi by Mr. Ernest S. Salmon, with one plate; and another note by Mr. E. M. Nelson on *Actinocyclus ralfsii*, with special reference to the cause of the colour exhibited by this diatom when examined with low-and high-angled lenses respectively. Mr. R. T. Lewis adds a most interesting contribution to the life-history of *Ixodes reduvius*, also with a plate, which we recommend to the notice of the many readers of SCIENCE-GOSSIP who were interested in Mr. E. G. Wheler's recent articles (S.-G. vol. v, N.S. pp. 5, 46, 108) on "Ticks and Louping Ill," and which is a further elaboration of the life-history

dealt with therein. These articles were subsequently published in a more extended form in the Journal of the Royal Agricultural Society, and have been since reprinted for private circulation. Mr. A. A. Merlin deals with the structural division of the endoplasm observed by him in the bacilli of bubonic plague and of other microbes; Mr. Chas. D. Soar contributes a list of fresh-water mites found near Oban, N.B.; and Mr. D. J. Scourfield, the editor of the "Journal," adds a note on the swimming peculiarities of *Daphnia* and similar Entomostraca, with some original and very practical hints as to examining living specimens. Several minor notes, reviews of books, and the usual reports of the proceedings of the Club make up a most interesting number.

BAUSCH & LOMB'S NEW CATALOGUE.—We desire to say that since the appearance last month of our notice of Messrs. Bausch & Lomb's new catalogue we have received from their London agents, Messrs. Staley & Co., 35 Aldermanbury, a copy of the same catalogue specially prepared and priced for the English market. It is not for us to enter here into political and fiscal matters, but it is worth noting that the heavy import duties in the United States apparently represent a difference of about 25 per cent. in favour of the English buyer. The prices in the English catalogue, accordingly, compare favourably with those current in America and on the Continent. We take the first opportunity of making the necessary correction. Our notice, of course, was of the American catalogue, which had been sent to us direct from America.

NOTES ON CRUSTACEANS.—Professor G. S. Brady has kindly sent us reprints of two papers read by him and reprinted from the "Natural History Transactions of Northumberland, Durham, and Newcastle-on-Tyne." One of these contains a new description of *Ilyopsyllus coriaceus*, a remarkable little crustacean notable for its brilliant colour, eel-like movements, and peculiar mouth-parts. The almost obsolete mandibles, and the reduction of all the other mouth-apparatus—maxillae and maxillipedes—to a very few minute filaments or setae, preclude its coming into line with any of the three divisions established by Thorell. The author proposes, therefore, to establish a new section for the reception of *Ilyopsyllus* under the name of Leptostomata. The divisions of Copepoda, would then stand as follows: Gnathostomata, Poecilostomata, Leptostomata, and Siphonostomata. In the same paper is given a list of littoral forms of crustacea collected by the author at Alnmouth, in Northumberland, such as *Paratylus uncinatus*, *Apherusa borealis*, *Siriella norvegica*, and *S. armata*, which are considered by Dr. Norman to be new to the local fauna. Some new forms are described in *Cyclops salinus*, *Ectinosoma melaniceps*, *Stenhelia limicola*, *Echinocheres violaceus*, and *Cyclopicera berniciensis*, all of which are illustrated by plates. The author notes that *Acartia clausii* is infested with what is probably an immature fluke, of which the dab (*Limanda limanda*), for instance, may be the final host.

NATURE AND ORIGIN OF FRESH-WATER FAUNAE.—The other paper deals with the nature and origin of fresh-water faunae. Professor Brady points out that, accepting the orthodox zoological theory that life originated in the sea, we may suppose that the first tentative efforts at life occurred in the shallow littoral region where warmth, shelter, and the action of the tides would greatly favour both animal and vegetable growth. He reminds us of the fact that one very important group of marine animals, the Echinoderma, is altogether absent from fresh-water fauna; and that another large group, the Coelentera, is very feebly represented. Only two classes have a preponderance in fresh water, the Amphibians (frogs, newts, and the like), which are altogether terrestrial or fresh-water in habitat, and Insects, which, although not largely represented in fresh water, are even more scarce in the sea. Protozoa, however, as represented by Radiolaria and Foraminifera, are almost exclusively marine, though they are near relatives of Amoeba, which is ubiquitous in fresh water. Foraminifera, though abundant in estuaries and even in pools of marshes where the water is only very slightly saline, do not seem to be able to penetrate further than this, and even in such localities they are invariably depauperated, the tests becoming thin and deficient in lime. Professor Brady adds the natural reflection that it is possible that absence or diminished quantity of lime in the water may be a chief cause of the absence of Foraminifera, just as the sub-acidity and "softness" of the water of peat-mosses seem to render it unfit to support Microzoa with calcareous shells, while those with merely chitinous valves are often abundant. Sponges are almost wholly marine in their distribution, of the forty odd families only one being found in fresh water. Of the Coelentera (jellyfishes, zoophytes, etc.) only three families out of about seventy are represented in fresh water, these being the well-known *Hydra* and *Cordylophora*, and two small and comparatively infrequent Medusae. The heterogeneous group Vermes includes many fresh-water forms, but the larger and more important section Arthropoda is very imperfectly represented. Of this division the microscopic Entomostraca are the only crustaceans found abundantly in fresh water, except the common crayfish; the Spiders, on the other hand, are absent from the sea, except the Mites and Pycnogons. Centipedes are wholly terrestrial, and insects almost wholly so. The Mollusca are chiefly marine, though some families are largely represented in fresh water. Of aquatic vertebrates, fishes are, of course, the most important, and of these, of 137 families, 35 only are found in fresh water, and many of these but sparingly. The Amphibians are, as already mentioned, entirely absent from the sea. This is an interesting epitome, and Professor Brady refers to a paper published fourteen years ago by Professor Sollas for the following principal causes why comparatively few animals have been able to establish themselves in fresh water:—Firstly, the prevalence among marine animals of larval forms so feeble as to be unable to withstand river currents; secondly, prejudicial fluctuations of temperature in fresh water; thirdly, disturbing physical causes, such as floods, droughts, upheavals and depressions of surface, etc.; and to this Professor Brady himself adds the question of the non-salinity of fresh water. In this connection he quotes the well-known example of the transmutation of *Artemia salina* into *A. mülhausenii*, but this result has been much criticised and has recently been discredited. (See a note in this journal, on p. 151 of the present volume.)

ROSS'S NEW "STANDARD" MICROSCOPE.—The popularity of the Continental form of stand in our hospital laboratories and medical schools, alluded to on p. 184 in a recent notice of a new microscope by Messrs. Beck, is further evidenced by Messrs. Ross's latest stand, especially designed for this and similar classes of work. This stand has the usual Continental form of horseshoe foot, but with the pillar brought as far forward as possible with a view to obviating the want of steadiness so characteristic of this type of foot. The coarse adjustment is the usual diagonal rack working at the end of the limb in a cylindrical slide, and the other end of the limb, also in accordance with Continental practice, is borne upon a triangular bar and actuated by a direct-acting micrometer screw. One drawback to this arrangement, whatever may be its advantages, is manifestly that the whole weight of body-tube and limb is borne by

ROSS'S "STANDARD" MICROSCOPE.

the fine adjustment, and that the limb the position of which makes it peculiarly liable to strains and ill-usage, is not rigid. Opinions, however, differ on this as on other points connected with the design of microscopes. The stage in the various models varies in size from 3 inches to $4\frac{1}{2}$ inches square, and is covered with vulcanite. The substage in its simplest form is represented merely by a ring beneath the stage, but in the more expensive stands by a rack and pinion sub-stage. An iris diaphragm is fitted immediately beneath the stage in the cheaper instruments, an arrangement which, whilst effective with low powers, is manifestly not advantageous when a condenser is being used. The length of the tube is 160 millimetres, which in the larger stands extends to 250 millimetres. The cheapest stand is priced at £4 10s., whilst the one illustrated costs with eye-

ASTRONOMY,

CONDUCTED BY F. C. DENNETT.

1901	Rises.	Sets.	Position at Noon. R.A.	Dec.
Feb.	h.m.	h.m.	h.m.	° '
Sun .. 4 ..	7.36 a.m. ..	4.51 p.m. ..	21.10 ..	16.20 S.
14 ..	7.18 a.m. ..	5.10 p.m. ..	21.50 ..	13.9
24 ..	6.58 a.m. ..	5.28 p.m. ..	22.28 ..	9.36

	Rises.	Souths.	Sets.	Age at Noon.
Feb.	h.m.	h.m.	h.m.	d. h.m.
Moon .. 4 ..	6.23 p.m. ..	0.25 a.m. ..	7.24 a.m. ..	14 21.24
14 ..	3.50 a.m. ..	8.0 a.m. ..	0.11 p.m. ..	24 21.24
24 ..	9.10 a.m. ..	5.1 p.m. ..	—— ..	5 9.15

	Souths.	Semi-diameter.	Position at Noon. R.A.	Dec.
Feb.	h.m.		h.m.	° '
Mercury .. .. 4 ..	0.55 p.m. ..	2·6″ ..	21.51 ..	14.47 S.
14 ..	1.18 p.m. ..	3·1″ ..	22.53 ..	7.13 S.
24 ..	1.11 p.m. ..	4·1″ ..	23.26 ..	1.5 S.
Venus .. .. 4 ..	10.47 a.m. ..	5·4″ ..	19.42 ..	21.36 S.
14 ..	11.0 a.m. ..	5·3″ ..	20.35 ..	19.21 S.
24 ..	11.11 a.m. ..	5·2″ ..	21.26 ..	16.11 S.
Mars .. .. 4 ..	1.56 a.m. ..	6·5″ ..	10.50 ..	13.2 N.
14 ..	1.5 a.m. ..	6·8″ ..	10.38 ..	13.25 N.
24 ..	0.11 a.m. ..	6·9″ ..	10.23 ..	14.51 N.
Jupiter .. .. 14 ..	8.48 a.m. ..	15·6″ ..	18.23 ..	23.6 S.
Saturn .. .. 14 ..	9.20 a.m. ..	7·1″ ..	18.55 ..	22.17 S.
Uranus .. .. 14 ..	7.26 a.m. ..	1·8″ ..	17.1 ..	22.45 S.
Neptune .. .. 14 ..	8.9 p.m. ..	1·2″ ..	5.45 ..	22.11 N.

MOON'S PHASES.

		h.m.			h.m.
Full ..	Feb. 3 ..	3.30 p.m.	3rd Qr.	Feb. 11 ..	6.12 p.m.
New ..	" 19 ..	3.45 a.m.	1st Qr.	" 25 ..	6.38 p.m

In apogee February 9th at 7 a.m. ; in perigee on 21st at 3 a.m.

METEORS.

					h.m.	°
Feb. 5–16 ..	a Aurigids	Radiant R.A.	4.56	Dec.	43 N.
" 16–20 ..	a Serpentids	"	"	15.44	"	11 N.
" 20 ..	..	"	"	17.32	"	36 N.

CONJUNCTIONS OF PLANETS WITH THE MOON.

					° '
Feb. 5 ..	Mars	.. 10 p.m. ..	Planet 9.54 N.
" 15 ..	Jupiter	.. 5 a.m. ..	" 2.51 S.
" 15 ..	Saturn†	.. 7 p.m. ..	" 3.3 S.
" 17 ..	Venus†	.. 9 p.m. ..	" 5.49 S.
" 20 ..	Mercury⊙	.. 1 p.m. ..	" 3.29 S.

⊙ Daylight.   † Below English horizon.

The only Occultations visible are those of stars below 5·5 magnitude.

THE SUN should be watched for occasional outbreaks of activity.

MERCURY as an evening star is fairly placed for observation after the first few days of the month. It reaches its greatest eastern elongation, 18° 6', at 10 p.m. on 19th, on which day it does not set until 1 h. 46 m. after the sun. The greater part of the month it is in the constellation Aquarius, but after the 20th in Pisces. At about 5 o'clock on this evening Mercury may be found about 5° below the thin crescent moon.

VENUS is a morning star, badly situated for observation.

MARS, being in opposition at 6 a.m. on February 22nd, is as well placed for observation as he will be this year. At 8 a.m. on the 25th he is in the part of his orbit most distant from the sun ; as a consequence his apparent diameter is very small, being only 13·8″ at the best. Notwithstanding, he is a beautiful little globe to study.

JUPITER, SATURN. AND URANUS are too low down and too near the sun for observation.

NEPTUNE may be looked for as an 8th magnitude star, retrograding along a very short path, about 2° north-west of the 4th magnitude star χ' Orionis.

The minor planet VESTA is in opposition on February 1st, when it appears as a star slightly above 7th magnitude. Its nearness to ξ Cancri, 5th magnitude, on February 4th, and to the pretty double star Σ 1311 on the 7th, should assist in its identification.

COMET c 1900.—On December 20th, M. Giacobini, of Nice, discovered a comet moving in a southeasterly direction in the southern part of the constellation Aquarius.

NEW MINOR PLANETS.—Whilst photographing Eros and its surrounding stars, Professor W. R. Brooks, of Geneva, U.S.A., discovered three new planets within 1° of Eros, one exceeding it in magnitude. On December 20th, Professor Max Wolf, of Heidelberg. also observed two more.

QUITO OBSERVATORY.—M. F. Gonnessia, of the Lyons Observatory, has been appointed to act as director for five years, during the work of the French Commission engaged in re-measuring the arc of the meridian in Peru.

LICK OBSERVATORY.—According to " Science," Mr. D. O. Mills, of New York, has promised about 24,000 dollars to defray the cost of an expedition either to South America or Australia. to study, under good conditions. the motions of stars in the line of sight.

MARS.—During the past few weeks some of the newspapers have given much prominence to an observation by a Mr. Douglass at Lowell Observatory, Flagstaff, Arizona. A bright patch was observed for over an hour on the border of the Icarium Mare. when it disappeared. M. C. Flammarion is doubtless right in ascribing it to the reflection from a cloud. It is by no means the first time such an observation has been made on the planet Mars.

JUPITER.—The most recent measures are those by Dr. T. J. J. See with the 26-in. Washington Refractor. Taking the solar parallax as 8·″796, the equatorial diameter is 89,919 miles, and the polar 84,111 miles. The diameters of the larger moons are said to be I.—1,574 miles ; II., 1,461 miles ; III., 3,187 miles and IV., 2,884 miles.

A USEFUL BLOTTER.—We have received from the Scottish Provident Institution the fourth issue of their blotting-book, containing a chart of the heavens looking north and south from London at 10 p.m. on the 1st of each month. The stars are in gold on a blue ground, and the positions of the Moon and Mars, Jupiter and Saturn are also marked. A description is given by Mr. W. B. Blaikie, which makes the whole quite a handbook for studying the face of the sky. It is interesting to note, that commercial institutions find a scientific advertisement sufficiently attractive.

CHAPTERS FOR YOUNG ASTRONOMERS.

BY FRANK C. DENNETT.

(Continued from p. 251.)

JUPITER.

DURING 1900 Mr. Stanley Williams thought the spot was plainer, and that the reddish colour was again appearing in its following or eastern half. Immediately north of this spot there is a sort of

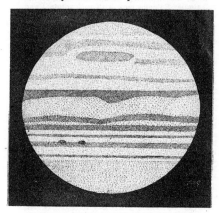

THE PLANET JUPITER, Oct. 17th, 1880, 12 p.m., G.M.T.

"bay," in the south tropical belt, which broadens eastwards, forming a sort of dark elbow, which is well shown in Rev. T. E. R. Phillips's drawing, fig. 4. This is readily shown with a telescope of no more than 2 inches aperture, and is traceable in many drawings so far back even as September 1831, when it was delineated by Schwabe.

Bright spots have been seen on the planet, especially in the equatorial zone, from very early times.

firmed when observers were attracted to the planet by the advent of the great red spot. The bright spots were found to overtake the red spot and to pass. Several such spots are beautifully shown in fig. 3, bedding into the south tropical belt.

On October 17th, 1880, at midnight, the writer found two small dark spots on the northern north-temperate band, shown on fig. 5, which is from a drawing made at the time, by the aid of a 9½-inch Calver reflector. These objects proved of great interest. Their colour was a deep Prussian blue; they increased in number quickly, and were found to gain even upon the white spots, so that there were at the same time on the planet three classes of maculation, each yielding a different answer to the question, "What is the period of Jupiter's rotation?" Mr. F. W. Denning determined the three periods to be from

Red spot	9 h. 55 m. 34 s.
Bright spots . . .	9 h. 50 m. 5 s.
Dark spots, slightly exceeding .	9 h. 48 m. 0 s.

This is not all; these rates, as pointed out by Denning, are not constant. For instance, in 1878 the red spot yielded a rotation period of 9 h. 55 m. 33·5 s., but in 1898 it had increased to 9 h. 55 m. 41·5 s., and this increase was not regular. A variation is similarly noticeable in the case of the bright spots. Thus it becomes evident that the surface of the planet seen is by no means solid, otherwise the motions would be regular. To simplify the recording of observations, a zero meridian was chosen, that passing the centre of the disc at midnight on December 31st, 1871. Two rotation periods were reckoned. System I., 9 h. 50 m. 30 s. (877°·90 of Jovian longitude passing the centre in 24 h.), and System II., 9 h. 55 m. 40·63 s. (870°·27 passing in 24 h.): System I. very nearly accords with the period of the bright spots, whilst System II. almost agrees with that of the red spot. In dating drawings it is common to note the longitude of the central meridian according to both systems, as it simplifies the study of the movements on the planets. Even a telescope of

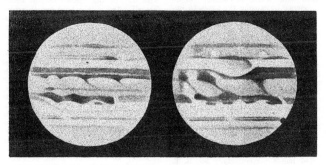

April 15th, 12 h. 10 m., G.M.T.   May 6th, 10 h. 20 m., G.M.T.

THE PLANET JUPITER IN 1899. *Drawn by Rev. Theodore E. R. Phillips*

Schmidt long since had pointed out that the rotation of the planet was apparently found to be shorter or longer when bright spots or dark ones were observed. This was abundantly con-

3 inches aperture will show much of the beauty of the belts if the air is good and a power of from 130 to 180 employed.

(To be continued.)

CONDUCTED BY F. SHILLINGTON SCALES, F.R.M,S.

ROYAL MICROSCOPICAL SOCIETY. — December 19th, William Carruthers, Esq., F.R.S., in the chair. Notice was given on behalf of the Council that at the next meeting of the Society the name of Dr. C. T. Hudson would be submitted for election as an Honorary Fellow. Mr. E. M. Nelson exhibited a small pocket microscope made by Dr. Gilbertson, of London, of unknown date. Dr. Hebb read the list of nominations by the Council for election at the Annual Meeting on January 16th, as follows:—As President, Mr. William Carruthers ; as Vice-presidents, Dr. Braithwaite, Messrs. Michael, and Nelson, and the Right Hon. Sir Ford North ; as Treasurer, Mr. J. J. Vezey ; as Secretaries, Dr. Dallinger and Dr. Hebb ; as Council, Messrs. Allen, Beck, Bennett, Browne, Rev. E. Carr, Messrs. Dadswell, Disney, Karop, Plimmer, Powell, Professor Urban Pritchard, and Mr. Rousselet ; as Curator, Mr. Rousselet. Mr. Barton, on behalf of Messrs. Ross, Limited, exhibited some new forms of lanterns which could be used for ordinary projection purposes either with or without the microscope. The first was constructed so as to exclude all light from the room except what passed through the lenses ; the second was larger and more complex, and could be used for all purposes, including enlargements. Both gave excellent definition. Mr. Barton also exhibited and described several new forms of microscope with detachable circular staging and new form of electric arc lamp for lantern use. A new form of limelight was also exhibited which attracted much attention from its great brilliancy and steadiness, and the silence with which it burned. These effects were produced by causing the gases to impinge upon each other previous to their entrance into the mixing chamber, and by the construction of the chamber itself.

JOURNAL OF THE QUEKETT MICROSCOPICAL CLUB.—We regret that space has prevented our noticing earlier the November issue of this interesting journal, which contains the index for the volume comprising this and the preceding five numbers. The present number contains an article by Mr. T. B. Rossiter on the anatomy of the tapeworm *Dicranotaenia coronula*, with two excellent plates ; a note on four rare British fungi by Mr. Ernest S. Salmon, with one plate ; and another note by Mr. E. M. Nelson on *Actinocyclus ralfsii*, with special reference to the cause of the colour exhibited by this diatom when examined with low- and high-angled lenses respectively. Mr. R. T. Lewis adds a most interesting contribution to the life-history of *Ixodes reduvius*, also with a plate, which we recommend to the notice of the many readers of SCIENCE-GOSSIP who were interested in Mr. E. G. Wheler's recent articles (S.-G. vol. v, N.S. pp. 5, 46, 108) on "Ticks and Louping Ill," and which is a further elaboration of the life-history

dealt with therein. These articles were subsequently published in a more extended form in the Journal of the Royal Agricultural Society, and have been since reprinted for private circulation. Mr. A. A. Merlin deals with the structural division of the endoplasm observed by him in the bacilli of bubonic plague and of other microbes ; Mr. Chas. D. Soar contributes a list of freshwater mites found near Oban, N.B. ; and Mr. D. J. Scourfield, the editor of the "Journal," adds a note on the swimming peculiarities of *Daphnia* and similar Entomostraca, with some original and very practical hints as to examining living specimens. Several minor notes, reviews of books, and the usual reports of the proceedings of the Club make up a most interesting number.

BAUSCH & LOMB'S NEW CATALOGUE.—We desire to say that since the appearance last month of our notice of Messrs. Bausch & Lomb's new catalogue we have received from their London agents, Messrs. Staley & Co., 35 Aldermanbury, a copy of the same catalogue specially prepared and priced for the English market. It is not for us to enter here into political and fiscal matters, but it is worth noting that the heavy import duties in the United States apparently represent a difference of about 25 per cent. in favour of the English buyer. The prices in the English catalogue, accordingly, compare favourably with those current in America and on the Continent. We take the first opportunity of making the necessary correction. Our notice, of course, was of the American catalogue, which had been sent to us direct from America.

NOTES ON CRUSTACEANS.—Professor G. S. Brady has kindly sent us reprints of two papers read by him and reprinted from the "Natural History Transactions of Northumberland, Durham, and Newcastle-on-Tyne." One of these contains a new description of *Ilyopsyllus coriaceus*, a remarkable little crustacean notable for its brilliant colour, eel-like movements, and peculiar mouthparts. The almost obsolete mandibles, and the reduction of all the other mouth-apparatus—maxillae and maxillipedes—to a very few minute filaments or setae, preclude its coming into line with any of the three divisions established by Thorell. The author proposes, therefore, to establish a new section for the reception of *Ilyopsyllus* under the name of Leptostomata. The divisions of Copepoda, based upon the structure of the mouth-organs, would then stand as follows : Gnathostomata, Poecilostomata, Leptostomata, and Siphonostomata. In the same paper is given a list of littoral forms of crustacea collected by the author at Alnmouth, in Northumberland, such as *Paratylus uncinatus*, *Apherusa borealis*, *Siriella norvegica*, and *S. armata*, which are considered by Dr. Norman to be new to the local fauna. Some new forms are described in *Cyclops salinus*, *Ectinosoma melaniceps*, *Stenhelia limicola*, *Echinocheres violaceus*, and *Cyclopicera berniciensis*, all of which are illustrated by plates. The author notes that *Acartia clausii* is infested with what is probably an immature fluke, of which the dab (*Limanda limanda*), for instance, may be the final host.

NATURE AND ORIGIN OF FRESH-WATER FAUNAE.—The other paper deals with the nature and origin of fresh-water faunae. Professor Brady points out that, accepting the orthodox zoological theory that life originated in the sea, we may suppose that the first tentative efforts at life occurred in the shallow littoral region where warmth, shelter, and the action of the tides would greatly favour both animal and vegetable growth. He reminds us of the fact that one very important group of marine animals, the Echinoderma, is altogether absent from fresh-water fauna; and that another large group, the Coelentera, is very feebly represented. Only two classes have a preponderance in fresh water, the Amphibians (frogs, newts, and the like), which are altogether terrestrial or fresh-water in habitat, and Insects, which, although not largely represented in fresh water, are even more scarce in the sea. Protozoa, however, as represented by Radiolaria and Foraminifera, are almost exclusively marine, though they are near relatives of Amoeba, which is ubiquitous in fresh water. Foraminifera, though abundant in estuaries and even in pools of marshes where the water is only very slightly saline, do not seem to be able to penetrate further than this, and even in such localities they are invariably depauperated, the tests becoming thin and deficient in lime. Professor Brady adds the natural reflection that it is possible that absence or diminished quantity of lime in the water may be a chief cause of the absence of Foraminifera, just as the sub-acidity and "softness" of the water of peat-mosses seem to render it unfit to support Microzoa with calcareous shells, while those with merely chitinous valves are often abundant. Sponges are almost wholly marine in their distribution, of the forty odd families only one being found in fresh water. Of the Coelentera (jellyfishes, zoophytes, etc.) only three families out of about seventy are represented in fresh water, these being the well-known *Hydra* and *Cordylophora*, and two small and comparatively infrequent Medusae. The heterogeneous group Vermes includes many fresh-water forms, but the larger and more important section Arthropoda is very imperfectly represented. Of this division the microscopic Entomostraca are the only crustaceans found abundantly in fresh water, except the common crayfish; the Spiders, on the other hand, are absent from the sea, except the Mites and Pycnogons. Centipedes are wholly terrestrial, and insects almost wholly so. The Mollusca are chiefly marine, though some families are largely represented in fresh water. Of aquatic vertebrates, fishes are, of course, the most important, and of these, of 137 families, 35 only are found in fresh water, and many of these but sparingly. The Amphibians are, as already mentioned, entirely absent from the sea. This is an interesting epitome, and Professor Brady refers to a paper published fourteen years ago by Professor Sollas for the following principal causes why comparatively few animals have been able to establish themselves in fresh water:—Firstly, the prevalence among marine animals of larval forms so feeble as to be unable to withstand river currents; secondly, prejudicial fluctuations of temperature in fresh water; thirdly, disturbing physical causes, such as floods, droughts, upheavals and depressions of surface, etc.; and to this Professor Brady himself adds the question of the non-salinity of fresh water. In this connection he quotes the well-known example of the transmutation of *Artemia salina* into *A. mülhausenii*, but this result has been much criticised and has recently been discredited. (See a note in this journal, on p. 151 of the present volume.)

ROSS'S NEW "STANDARD" MICROSCOPE.—The popularity of the Continental form of stand in our hospital laboratories and medical schools, alluded to on p. 184 in a recent notice of a new microscope by Messrs. Beck, is further evidenced by Messrs. Ross's latest stand, especially designed for this and similar classes of work. This stand has the usual Continental form of horseshoe foot, but with the pillar brought as far forward as possible with a view to obviating the want of steadiness so characteristic of this type of foot. The coarse adjustment is the usual diagonal rack working at the end of the limb in a cylindrical slide, and the other end of the limb, also in accordance with Continental practice, is borne upon a triangular bar and actuated by a direct-acting micrometer screw. One drawback to this arrangement, whatever may be its advantages, is manifestly that the whole weight of body-tube and limb is borne by

ROSS'S "STANDARD" MICROSCOPE.

the fine adjustment, and that the limb the position of which makes it peculiarly liable to strains and ill-usage, is not rigid. Opinions, however, differ on this as on other points connected with the design of microscopes. The stage in the various models varies in size from 3 inches to 4½ inches square, and is covered with vulcanite. The substage in its simplest form is represented merely by a ring beneath the stage, but in the more expensive stands by a rack and pinion sub-stage. An iris diaphragm is fitted immediately beneath the stage in the cheaper instruments, an arrangement which, whilst effective with low powers, is manifestly not advantageous when a condenser is being used. The length of the tube is 160 millimetres, which in the larger stands extends to 250 millimetres. The cheapest stand is priced at £4 10s., whilst the one illustrated costs with eyepiece £6, or with racked sub-stage 25 shillings extra. The addition of an Abbé condenser adds a further 25 shillings. The objectives supplied with the stands are a ⅜-inch of N.A. 26, and a ¼-inch of N.A. 65, priced at 25 shillings and £2 respectively. Both of these are excellent lenses, of good definition, and arranged so as to work approximately in the same focal plane when rotated on a nosepiece.

EXTRACTS FROM POSTAL MICROSCOPICAL SOCIETY'S NOTEBOOKS.

[These extracts were commenced in the September number at page 119. Beyond necessary editorial revision, they are printed as written by the various members.—ED. Microscopy, S.-G.]

NOTES BY F. PHILLIPS.

The twelve diatom slides referred to in these notes will probably form a pleasant change. Diatoms are a family of confervoid algae of very peculiar shape. They are found in almost every pool of fresh, brackish, or salt water, sometimes forming a yellow or brown layer at the bottom of the water; at others, attached to various plants, or

few washings, then treat in a similar manner with nitric acid; wash repeatedly with distilled water until perfectly free from acid. Separate the various diatoms according to their different specific gravities by allowing them to pass through water in the following manner:—Take a long glass tube, about two feet in length and half an inch in diameter, fitted at the bottom with a stopcock to facilitate the letting out of some of the diatoms at any stage of the process; place a drop on the slide and examine. If suitable for mounting, proceed as follows:—Fix a needle obliquely through a collar of cork which is itself fitted to a low-power objective, and to the end of the needle cement a bristle, preferably one from a rat's tail. The end of the bristle can now be focussed along with the diatom and brought in contact with the latter. The diatom will adhere to the bristle, and can be raised and transferred to a clean slide. In this manner any single diatom can be selected out of a gathering and mounted separately. Of course great care and patience are requisite, but practice and perseverance in this, as in everything else, will bring good results in course of time. The following brief description of the characteristics of the individual slides may be of further interest:—

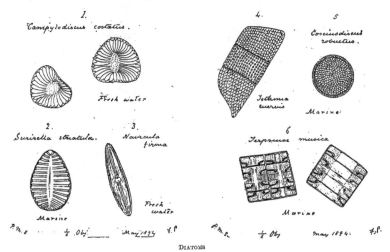

DIATOMS

(From Drawings by Miss F. Phillips.)

on stones and decaying plant-stems—in fact in almost any moist place. The individual cells are called frustules, and are furnished with a coat of silica, noteworthy on account of its beautiful markings, which take the form of bands or lines, either parallel, radiate, or crossing each other; also of dots. These markings however cannot be seen until the diatoms have been properly cleaned. When the diatoms have been collected, one of the most troublesome operations is freeing them from impurities. Several text-books give different methods, but the following is the one I adopt. Place the diatoms in a test-tube with strong hydrochloric acid and boil for five minutes; after allowing the boiling to subside, get rid of the acid by a

Campylodiscus costatus (fig. 1): frustules saddle-shaped and contorted, valves circular with radiating channels, the centre minutely dotted; found principally in boggy pools. *Surirella striatula* (fig. 2): frustules somewhat wedge-shaped, margins produced into a kind of wing with distinct and parallel channels, valves ovate and striate; found in brackish water. *Navicula firma* (fig. 3): valves narrow, gradually tapering to the rounded ends, a longitudinal line near each margin, central nodule conspicuous, striae fine; common in ditches. *Isthmia enervis* (fig. 4): frustules depressed, valves of a reticular or cellular appearance, uniformly covered with depressions; found in salt water only. *Coscinodiscus robustus* (fig. 5): frustules

free, single, disc-shaped, valves circular, flat
or slightly convex, exhibiting a cellular or
areolar appearance, no internal septa nor lateral processes; marine and fossil. *Terpsinoe
musica* (fig. 6): frustules faintly punctate, front
view rectangular oblong, side view equally inflated
in middle and at the ends, nodules separated by
septa; found in salt water only. *Arachnoidiscus
ehrenbergii* (fig. 7): frustules adherent, disc-shaped,
valves plain or slightly convex with radiating and
concentric lines (rows of dots) and a central pseudo-
nodule; marine only. *Trinacria regina* (fig. 8):
frustules with three broad bispined, equal-lengthed
processes, margin pearly, angles naked; marine
only. *Heliopelta leeuwenhoeckii* (fig. 9): frustules
single, valves circular with imperfect radiating
septa, markings absent in centre, but as many
large submarginal apertures present as there are
rays, and numerous erect opposite submarginal
spines on each side; the spines connect the pairs
of young frustules; fossil. *Actinocyclus ehrenbergii*
(fig. 10): frustules free, single, or disc-shaped,
valves circular, exhibiting cellular markings with
rays or bands radiating from centre, which is free
from cellular appearance, no internal septa; fossil.
Actinoptychus senarius (fig. 11): frustules free,
single, or disc-shaped, with six rays and internal
radiating septa, valves apparently cellular except
opposite the rays; fossil. *Aulacodiscus kittonii*
(fig. 12): frustules single, disc-shaped, circular,
without internal septa, valves furnished with
tubular or spiniform processes; marine and fossil.

REMARKS.—This series of slides contains a very
perfect example of *Isthmia enervis*. I never saw
the secondary markings more clearly. On ex-

photograph this appearance, and regret I have not
time to try. I insert a photo of *Isthmia nervosa*
(fig. 13) which I made some time ago with a
$\frac{1}{12}$ oil-immersion lens. This shows distinctly the
secondary areolations surrounding the primary
ones, as may be seen in *Isthmia enervis*; but the
perforations in the centre of the primary areolations were not very visible in this specimen, and
are not reproduced in the photo. *Aulacodiscus
kittonii* should also show interesting secondary
structure.—*J. R. L. Dixon, M.R.C.S., L.R.C.P.*

[Those of our readers who require further information on the preparation of diatoms are referred
to a couple of articles by Mr. Edward H. Robertson

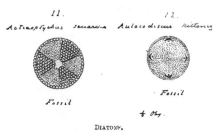

11.

Actinoptychus senarius

Fossil

12.

Aulacodiscus kittonii

Fossil

+ Oliy.

DIATOMS.

in SCIENCE-GOSSIP, vol. v., N.S., pp. 172–174 and
211–212. In vol. iii., pp. 32–34, appeared an article
by the Rev. Adam Clarke Smith on the "Generic
Names of Diatoms," which will be found useful to

7

Arachnoidiscus
Ehrenbergii

8

Trinacria
regina

Marine

9

Heliopelta
Leeuwenhoeckii

10

Fossil

nodule

Fossil

Actinocyclus Ehrenbergii

Ph 3.

½ Oliy.

may 1894   7 P

DIATOMS.

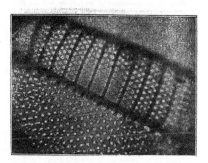

FIG. 13. *Isthmia nervosa.*

amination with an oil-immersion lens, the primary
areolations are seen to be surrounded by small
secondary areolations, and in one specimen, on the
valve surface particularly, the centre of the primary
areolations may be distinctly seen to be perforated.
I do not know whether it would be possible to.

those who are not familiar with the nomenclature.
The figures illustrating these notes, except fig. 13,
are by Miss Florence Phillips, the hon. secretary
of the P.M.S., and are drawn from the original
slides which accompanied the notes. We have
reprinted the latter in the hope that some of our
readers may be induced, by the simplicity of the
methods given, to take up the study of diatoms—
perhaps the most fascinating of all microscopic
studies—and we have for this reason abstained
from further elaborating them. Dr. Carpenter, in
"The Microscope and its Revelations" (seventh
edition), devotes many pages to the subject, and
to these we may refer the reader.—ED. Microscopy,
S.-G.]

CONDUCTED BY WILFRED MARK WEBB, F.L.S.

HELIX NEMORALIS AND H. HORTENSIS IN YORK-SHIRE.—Having recently visited an old chalk-pit in the Yorkshire Wolds for *H. nemoralis*, which occurs there in some variety, I send some notes on the variation of this species within the very limited space of about twenty square yards, together with other notes on that species as found near Scarborough. In the old chalk-pit my attention was first attracted to the varieties *hyalozonata* and *citronozonata*, which together made up about 10 per cent. of the specimens present. In addition to these varieties, *albolabiata* also occurs, having the normal five or a less number of bands. The five bands, when present, are thinner than usual, and either the first or second is absent from about 50 per cent., though bandless forms are rare. In colour they are all of the form *libellula*, with sometimes an inclination to the colour form of *rubella*. All specimens are either of normal size or larger, some exceeding an inch in diameter. The most remarkable variation is in the weight of the shells. Of some seventy specimens, taken at random, the weights in decimals of an ounce are as follows: five specimens weighed ·02 each; thirty-three, ·03; nineteen, ·04; seven, ·05; two, ·06; three, ·07; two, ·08; and one example, the heaviest of all, turned the scale at ·12 oz., nearly two drams and actually six times the weight of the lightest shells. I need scarcely say that before weighing I had been careful to remove every particle of animal matter from the shells, which were all full-grown. All the specimens having been taken from one small chalk-pit, and at one visit, of course the variation in weight cannot be explained by environment, or varying climatic or local conditions. I presume it is entirely due to age, the heaviest specimens, not always the largest, being the most eroded. In contrast to this narrow-banded heavy form are the specimens which we commonly find here on the sloping sea-cliffs. These shells show marked tendency to coalescence in the banding. A hundred specimens taken at random are banded as follows: 33 per cent. (12345), 40 per cent. (123) (45), 26 per cent. have the five bands more or less distinct and separate, whilst 5 per cent. are bandless *libellula*, and all of normal weight. Variety *rubella* is not uncommon in this district, and a pretty form of var. *libellula* shows a dark clouding near the margin. Var. *castanea*, so far as I know, does not occur in the neighbourhood. Of band-varieties here the formula 00300 is decidedly rare. Although the largest specimens which I have found occur in the old chalk-pit, the chalk formation alone is probably not responsible for their size, as the very smallest I possess, having deep black mouths, I took at the Dane's Dyke, Flamborough. *H. nemoralis* though common, is locally distributed here as elsewhere. It occurs from just above highwater mark on the sea-cliffs to the top of one at least of our

higher moors, where a small colony thrives at an elevation of six hundred feet on a bare stony ridge where the stone curlew (*Oedicnemus scolopax*) still breeds. *H. nemoralis* seems to differ in habits in different districts. Here I have never seen these snails climbing trees, which is a general habit of the species at Sledmere in the Wolds. *H. hortensis* is more generally distributed than the former species, and shows a predilection for the hedgerows. If evidence were needed to prove the specific distinctness of these two helices, the occurrence of their varietal forms in this district would prove useful. Here *H. hortensis*, var. *olivacea*, is not uncommon, but the complementary colonr. variety *castanea*, in *H. nemoralis* is, I believe, unknown. *H. hortensis*, var. *incarnata*, I have sought here in vain, but *H. nemoralis*, var. *rubella*, is fairly well distributed. During the last five years a change has occurred in the comparative abundance of some varieties of *H. hortensis*, var. *olivacea* having become scarcer, and var. *arenicola* much more common. In one locality last year I found the latter variety to the number of 70 per cent. in one ditch, and this year 50 per cent. was the proportion in another district.— *W. Gyngell, 13 Gladstone Road, Scarborough.*

PALUDESTRINA JENKINSI.—On December 15th Messrs. A. S. Kennard and B. B. Woodward made an interesting communication to the Essex Field Club regarding *Paludestrina jenkinsi*. This mollusc has given rise to a good deal of controversy. First of all, it had to be recognised that although Mr. Edgar Smith described the species from specimens found by Mr. Jenkins on the south side of the Thames, Mr. Crouch had previously presented examples to the British Museum from the north. Many other localities have since been added, and questions as to the introduction of *P. jenkinsi* with timber from the Baltic, and so on, have been raised. The form has not, however, been discovered in any of the supposed original habitats abroad, though two specimens have existed for a long time under another name in the celebrated collection of British land and freshwater shells made by the late Dr. Gwyn Jeffereys, and now, unfortunately for us, in America. Messrs. Kennard and Woodward have now made out a series of specimens from a surface deposit on the Essex side of the Thames of such depth as to point to their having lived in this county early in, if not before, the historic period. This disposes, once for all, of the theory of recent introduction.

COLONIES OF SNAILS AT HIGH ALTITUDES.—Up the romantic pass of the Winnats, above Castleton, Derbyshire, at least 1,250 feet O.D., I found a flourishing colony of *Helix nemoralis*, with *Helicigona arbustorum* and *H. lapicida*, under and around broken pieces of carboniferous limestone. On August 28, 1900, it was a very dry spot at the time, there being no evidence of a spring or a watercourse. *H. lapicida* had lost its periostracum. I took very few specimens, as such a colony is worthy of encouragement. Above Poole's Cavern, on a mound of fine concreted lime-ash from old lime-works, I found evidences of the existence of *H. arbustorum*, many bird-broken shells occurring. This set me searching for the snail's habitat. Though my first hunt was unsuccessful, I obtained *Vitrea glabra* and *V. alliaria* among mossgrown stones in an ancient watercourse. The *V. glabra* there gave as strong a scent of garlic as does *V. alliaria*. On a second visit I found the colony

I sought. It was about twelve feet above the path, hidden under tussocks of grass. The individuals were of all ages and very vigorous. Here the altitude was about 1,250 feet O.D. I found another flourishing colony on the banks of an intermittent watercourse above Lovers' Leap, Ashwood Dale, not a mile out of Buxton. The snails were crawling on the underside of leaves of meadowsweet, dock, and potentilla. One specimen, young and recently dead, occurred on the cliff to the south of the stream-course. By the banks of the stream I counted quite three dozen in about fifty feet. The location was about 950 feet O.D., and the superior size and better appearance of the colony may be attributed to greater abundance of food and more favourable general conditions. In no instance here had the periostracum been worn or dried off the shell. The animals were well grown and the shells very beautiful and glossy. Two varieties occurred, one dark, the other albino. I found traces also, alive and dead, of *H. arbustorum* on the cliff-sides of Ashwood Dale, south of Lovers' Leap. None of the colonies were scattered over a large area, so far as I was able to judge. In Derbyshire, as elsewhere, the species is limited and local in distribution. – *(Rev.) R. Ashington Bullen, F.L.S., F.G.S.*

SCIENTIFIC INVESTIGATION OF MOLLUSCA.—A Committee, consisting of Messrs. J. R. B. Masefield, F. Taylor, R. J. Welch, and A. E. Boycott, has been appointed by the Council of the Conchological Society of Great Britain and Ireland for the purpose of conducting a collective investigation of phenomena connected with the variation and life-history of British Land and Freshwater Mollusca. The object of the investigation is to inquire, by collecting the results of the individual experience of many naturalists, into points liable to general uncertainty and to local or other variation, and into the diffusion and dispersal of species. A certain small number of subjects for investigation will be published each year, and it is hoped that an abundance of replies will be received, so that the results may be thoroughly representative. The following five subjects have been selected for 1901:—(1) How far is the smell of garlic constantly associated with *Vitrea alliaria*? Under what circumstances and at what seasons of the year is it most noticeable? Does *V. alliaria* seem to escape destruction by other organisms more than the rest of the genus? Is the smell of garlic found in other species, and under what circumstances? (2) Have you in any case found any species or variety of land snail constantly associated with any particular plant? (3) Is any preference shown by (i) *Helix aspersa*, (ii) *H. rufescens* for the neighbourhood of human habitations and buildings; if so, what explanation do you consider the most probable? (4) What localities produce the largest specimens of *Anodonta*? Describe the nature of the water, soil, geological formation, etc., and give the dimensions, and, if possible, weight. (5) In the genus *Helix*, where not indigenous, when and how were any of the species introduced? It is desired to put on record so far as possible the date of introduction of any species into any given locality, both from abroad into the British Isles and from one part of the country to another. The locality for which each answer is recorded should be carefully given, with any details of geological formation, altitude, vegetation, etc., which may seem desirable. Returns should reach me by September 1, 1901.—*A. E. Boycott, The Grange, Hereford.*

GEOLOGY

CONDUCTED BY EDWARD A. MARTIN, F.G.S.

COLUMNAR STRUCTURE IN CLAY-SLATE.—An interesting exhibit at a recent meeting of the Geological Society was a piece of clay-slate, in which columnar structure had been produced through being subjected to the heat arising from the spontaneous combustion of the waste-heap in which it was exposed. The specimen came from Shipley Colliery, Derby, and was exhibited by the Rev. J. M. Mello, M.A.

NEW RECORD FROM THE KELLAWAY SANDS.—Mr. Thomas Sheppard, F.G.S., has unearthed from the western slope of Mill Hill, Brough, Yorks, some interesting remains of *Cryptocleidus* from the Kellaway Sands, a saurian hitherto unrecorded from rocks older than the Oxford Clay. The remains are highly ferruginous, and, being imbedded in a soft sandy deposit, they have been excavated from the matrix with but little damage to the specimens. Twenty-five portions of the animal were submitted to Mr. E. T. Newton, F.R.S., who identified them. Mr. Sheppard records his find in the "Geological Magazine."

THE CORALLIAN ROCKS OF ST. IVES AND ELSWORTH.—The Elsworth Rock, resting immediately upon the Oxford Clay, is identified by Mr. C. B. Wedd, F.G.S., with the St. Ives Rock of Huntingdonshire. In a paper which he read before the Geological Society, he records tracing the rock at various spots, although, owing to an area of fen intervening, he was unable to follow the actual connection between the two named. The Elsworth Rock, where it has been identified, is bounded on the west by the Oxford Clay and on the east by the Ampthill Clay. The two rocks are considered identical in consequence of the similarity of consistency, the absence of any other rock-bed, the dip of the strata, and the presence of the Ampthill Clay above.

CLEAVAGE OF SLATES AND SCHISTS.—We have received a pamphlet, reprinted from the "Proceedings" of the Liverpool Geological Society, dealing with analyses of various slates and schists from North Wales, by T. Mellard Reade, F.G.S., and Philip Holland, F.I.C. Amongst other things, they wished to ascertain upon what composition or causes the perfection of slaty cleavage depends. They conclude: "At all events it seems pretty clear that slaty cleavage is not alone a mechanical effect. That is to say, mere pressure and shearing will not produce of themselves true slaty cleavage, unless accompanied by chemical action within the material itself of which the slate is in process of being formed." As regards the strength and durability of slates they find: "As a matter of experience we may say that, as regards Welsh slates, the finer varieties split thinner and truer, but the granular are the strongest and most lasting." The analyses contained in the pamphlet, together with other useful economic information therein, give it a high technical value.

NOTICES OF SOCIETIES.

*Ordinary meetings are marked †. excursions * ; names of persons following excursions are of Conductors. Lantern Illustrations §.*

SELBORNE SOCIETY.

Feb. 7.—† "A Chat about the Sun." H. Keatley Moore.

MANCHESTER MUSEUM, OWENS COLLEGE.

Feb. 2.—§ " Economic Botany : Nutmeg and other Spices." Professor F. E. Weiss.
 „ 9.—§ " History of Coal." Professor W. Boyd Dawkins, F.R.S.
 „ 10.—§ " Arrival of Man." Professor W. Boyd Dawkins, F.R.S.
 „ 16.—§ " Some Connecting Links; Peripatus." Professor S. J. Hickson.
 „ 23.—§ " Some Connecting Links : Mudfishes." Professor S. J. Hickson.

SOUTH LONDON ENTOMOLOGICAL AND NATURAL HISTORY SOCIETY.

Feb. 14.—§ " Fossil Insects." W. West, L.D.S.
 „ 28.—† " The Breeding of Lepidoptera." A. M. Montgomery.

HULL SCIENTIFIC AND FIELD NATURALISTS' CLUB.

Feb. 6.—† " Borrowed Plumes." T. Audas, L.D.S.
 „ 20.—† " Horticulture and Herb Lore before the Norman Conquest." J. R. Boyle, F.S.A.

BIRKBECK NATURAL HISTORY SOCIETY.

Feb. 9.—* Natural History Museum, South Kensington. J. L. Monk.

PRESTON SCIENTIFIC SOCIETY.

Feb. 5.—† " Thornton Films." W. D. Welford.
 „ 7.—§ " The Metaphysical Aspect of Evolution." Rev. O. Coupe, S J.
 „ 13.—† " Mosses and Moss Hunting." H. Beesley.
 „ 19.—† " Alpine Photography." J. G. Shaw.
 „ 27.—† " Some Biological Questions of the Day." John T. Lingard.

NORTH LONDON NATURAL HISTORY SOCIETY.

Feb. 7.—† " Notes on the Natural History of Arnside and Witherslack." A. J. Rose, F.E.S.
 „ 16.—* Visit to the Natural History Museum, South Kensington.
 „ 21.—§ " A Dragon-fly," W. J. Lucas, B.A., F.E.S.
 „ 23.— Ninth Annual Exhibition.

STREATHAM SCIENCE SOCIETY.

Feb. 2.—† " Animal Architects." H. E. Cocksedge.

LAMBETH FIELD CLUB AND SCIENTIFIC SOCIETY.

Feb. 4.—† " Prehistoric Man in Britain." W. Johnson.
 „ 17.—* Visit to British Museum (Prehistoric Section). W. Johnson.

ROYAL INSTITUTION OF GREAT BRITAIN.

Feb. 1.—† " Electric Waves." Rt. Rev. Monsignor Gerald Molloy, D.D., D.Sc.
 „ 5.—† " Practical Mechanics." Professor J. A. Ewing, M.A., F.R.S.
 „ 8.—† " History and Progress of Aerial Locomotion." Professor G. H. Bryant. Sc.D., F.R.S.
 „ 12.—† " Practical Mechanics." Professor J. A. Ewing, M.A., F.R.S.
 „ 15.—† " The Existence of Bodies smaller than Atoms." Professor J. J. Thomson, M.A., Sc.D., F.R.S.
 „ 19.—† " Practical Mechanics." Professor J. A. Ewing, M.A., F.R.S.
 „ 22.—† " Metals as Fuel." Sir W. Roberts-Austen, K.C.B., D.C.L., F.R.S.
 „ 23.—† " Sound and Vibration." Lord Rayleigh, M.A., D.C.L., F.R.S.
 „ 26.—† " The Cell as Unit of Life." Dr. Allan Macfadyen.

NOTICES TO CORRESPONDENTS.

EDITORIAL COMMUNICATIONS, articles, books for review, instruments for notice, specimens for identification, &c., to be addressed to JOHN T. CARRINGTON, 110 Strand, London, W.C.

BUSINESS COMMUNICATIONS.—All business communications relating to SCIENCE-GOSSIP must be addressed to the Proprietors of SCIENCE-GOSSIP, 110 Strand, London.

NOTICE.—Contributors are requested to strictly observe the following rules. All contributions must be *clearly* written on one side of the paper only. Words intended to be printed in *italics* should be marked under with a single line. Generic names must be given in full, excepting where used immediately before. Capitals may only be used for generic, and not specific names. Scientific names and names of places to be written in round hand.

TO CORRESPONDENTS AND EXCHANGERS.—SCIENCE-GOSSIP is published on the 25th of each month. All notes or other communications should reach us not later than the 18th of the month for insertion in the following number. No communications can be inserted or noticed without full name and address of writer. Notices of changes of address admitted free.

THE Editor will be pleased to answer questions and name specimens through the Correspondence column of the magazine. Specimens, in good condition, of not more than three species to be sent at one time, *carriage paid*. Duplicates only to be sent, which will not be returned. The specimens must have identifying numbers attached, together with locality, date, and particulars of capture.

THE Editor is not responsible for unused MSS., neither can he undertake to return them unless accompanied with stamps for return postage.

SUBSCRIPTIONS.—The volumes of SCIENCE-GOSSIP begin with the June numbers, but Subscriptions may commence with any number, at the rate of 6s. 6d. for twelve months (including postage), and should be remitted to the Office, 110 Strand, London, W.C.

NOTICE.

SUBSCRIPTIONS (6s. 6d. per annum) may be paid at any time. The postage of SCIENCE-GOSSIP is really one penny, but only half that rate is charged to subscribers.

EXCHANGES.

NOTICE.—Exchanges extending to thirty words (including name and address) admitted free, but additional words must be prepaid at the rate of threepence for every seven words or less.

SMALL CAMERA wanted in exchange for the " Erato " chromatic Auto-harp.—W. D. Nelson, 22 Kirk Wynd, Kirkcaldy.

MICROSCOPIC SLIDES (entomo'ogical) exchange others, Lantern Slides, set Diptera, Neuroptera, Hymenoptera, or Coleoptera.— J. H. Bromwich, Tyburn, Erdington, near Birmingham.

NEW ¼-plate Monroe Hand Camera, 3 double dark slides, for exchange Will accept recent edition " Micrographic Dictionary," or Cabinet to hold 500 microscopic objects.— H. Ebbage, Chemist, Great Yarmouth.

OFFERED, Triaulopsis iridentata (American), H. ericetorum and many English species, and Foreign Stamps. Wanted, offers of young greenhouse plants, ferns or cuttings. No shells.— A. Whitworth, Greeta, I. of Man.

ENTOMOLOGICAL and other natural history works for exchange. Wanted, Sowerby's Illustrated Index of British Shells, Canaries, or Fancy Pheasants.—O. S. Coles, The Pheasantries, Hambledon, Cosham, Hants.

OFFERED, Marlstone Fossils for those of same or other formations.—T. Stock, Springfield Place, Clandown, Radstock, Som.

WANTED, oblong window aquarium or ¼ objective. Exchange cock siskin and hand-reared cock bullfinches, chaffinches, and greenfinches, and hen canaries.—Robt. Hall, 9 Russell Street Leek, Staffs.

CONTENTS.

SOME BRITISH DIVING BEETLES.

By E. J. Burgess Sopp, F.R.Met.Soc., F.E.S.

Genus *DYTISCUS* Linn.

IF the Cicindelidae be looked upon as the "tigers" of the terrestrial Adephaga, the Dytiscidae can lay equal claim to be regarded as the of which Dickens humorously said that they died on land and were unable to live in the water. Notwithstanding this assertion by our popular

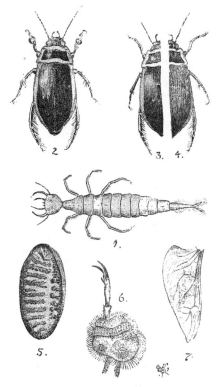

Fig. 1. Dytiscus marginalis L.

Drawn by E. J. Burgess Sopp.

1. Outline sketch of larva (dorsal surface). 2. Male beetle. 3 and 4. Female beetle, showing the two forms assumed. 5. One of the smaller spiracles, enlarged. 6. Under side of fore foot of male beetle, showing the two larger discs (at bottom of figure) and general arrangement of the smaller ones (magnified). 7. Wing of *Dytiscus*.

"sharks" of the aquatic branch of that most important division of the Coleoptera.

Unlike the Carchariidae, however, the Dytisci are amphibious, belonging to that group of animals novelist, however, the various species of the genus manage to exist pretty comfortably under either condition, leading as it were a dual life; for although passing by far the greater portion of their time in water, they not infrequently crawl up and sit for considerable periods on the protruding stems of aquatic plants, or take distant flight in quest of further food supply. Newly constructed ponds on downs and in similar situations, far removed from other water, are often early found to be colonised by these enterprising insects.

The commonest of our larger water-beetles, the ubiquitous *Dytiscus marginalis* (fig. 2), so called from the yellow lines with which the thorax and elytra are bordered, is an insect easily examined, since with a little perseverance it can usually be captured in pools and ditches throughout the kingdom. The young coleopterist should provide himself with a water-net made of mosquito netting or similar stout material, sufficiently strong to withstand rough usage and the possible entanglement amongst hedge-trimmings and other sunken debris ; and thus equipped he may set to work by any roadside with the full knowledge that he need, as Crabbe says, ". . . fear no bailiff's wrath, no baron's blame; His is untax'd and undisputed game."

By many entomologists and authors this is called "the Great Water-beetle" ([1]), but, although to a certain extent applicable, the term does not altogether commend itself as a happy one, since it might easily lead one to the supposition that it was our largest aquatic species, whereas that distinction really belongs to *Hydrophilus piceus*, a beetle which, although formerly included by Linnaeus in the present genus (*Dytiscus*), is now recognised as belonging to an entirely different division of the Coleoptera. It measures fully one-fourth longer and about a fifth broader than *Dytiscus marginalis*. Both Linnaeus, in his "Systema Naturae," and Fabricius threw together the aquatic groups.

All the Dytisci are carnivorous in both their larval and imaginal states, and are extremely fierce and voracious, preying upon young fish, newts, small frogs, tadpoles, various larvae, and other aquatic organisms, not even excepting individuals of their own and kindred genera. They have also been known to attack the giant *Hydrophilus piceus* above alluded to, whose seemingly invulnerable armour they pierce in its only vulnerable place—namely, at the narrow interstice between the plates of the head and thorax. The beetles prefer stagnant to running water; but, owing to the necessity for frequent access to the atmosphere, they do not appear so partial to ponds entirely covered with "weed" as are many other aquatic animals. In common with others of the family, however, they are fond of lying along under the banks, and by quietly approaching the margin of a pool one or more of the beetles may occasionally be seen resting at the surface during aëration.

The eggs of most amphibia are laid in water. Those of the Dytisci, which are elongate and

cylindrical in form, instead of being laid in packets, like those of the Ephemera (Houghton), or dropped at random in the water (Lyonnet), as with *Acilius*, are deposited singly in the stalks of rushes, pond-weed, and other aquatic plants ([2]), the ovipositor of the female being furnished on either side with sharp instruments, for the purpose of making incisions in the submerged stems to receive them (Régimbart ; Miall). Although under natural conditions egg-laying, for the most part, takes place during the spring and summer, it may not perhaps be entirely confined to the warmer months of the year. A female *Dytiscus punctulatus* captured at West Kirby on December 19th, 1900, and kept in a cold unused upper room, commenced to oviposit on the 21st of that month, depositing twelve eggs in sixty hours. These eggs are of an opaque white, with just the faintest tinge of yellow, and, as seen under a microscope, have a slightly roughened surface, formed by low rounded prominences, with scattered brown and black irregular markings. They are slightly curved, and measure about $6\frac{1}{2}$ mm. in length by barely 1 mm. in diameter.

Should all go well, and the ova escape the attention of various enemies, the larvae are hatched in about three weeks, only one brood only being brought forth in a year. The exact period would, however, be liable to be influenced by temperature and other causes, cold weather tending to considerably retard development. The eggs laid towards the end of December, above alluded to, commenced to hatch out on January 31st, and on the following day my wife and self were fortunate in observing the process of liberation of the young. For a short time previous to the birth of a larva the egg increases slightly in size, whilst the dark rings which had gradually appeared to encircle it—being really the segments of the larva within seen through the stretched outer covering—become increasingly marked. Shortly before one o'clock a slight movement of an egg under observation could be noticed occasionally, and at about twenty minutes past the hour the larva was seen to be escaping from one end of the ovum ; a process which, although of a somewhat laborious nature, was soon accomplished. It is then seen that the escaped insect is considerably longer than the egg-shell from which it has emerged. The head and adjoining section, together with the last segment of the body and legs, are quite white at birth, all the middle portion of the larva being from the first slightly tinged with brown and having the rims of the segments darkly and conspicuously marked by comparison. The early life of the larva is not without interest, but space pre-

(1) Rev. W. Houghton, M.A., "The Great Water-beetle (*Dytiscus marginalis*);" Rev. J. G. Wood, etc. etc.

(2) The writer in the "London Natural History of Insects," vol. ii. p. 159, who states that the eggs of *D. marginalis* are laid "in a singularly formed nidus of a silky substance which is allowed by the parent to float on the surface of the water, with its tapering mast, etc.," has really attributed the egg-cocoons of *Hydrophilus piceus* to this beetle.

vents my at present dwelling further on the sub-ject.

The larvae of the Dytisci have not inaptly been called " water-tigers," and were considered by early naturalists (Mouffet, etc.) to be closely allied to the shrimp; thus for a time we find them included amongst the crustaceans, where in the genus *Squilla* they figured as *S. aquatica*. They commence to feed shortly after birth, and, like the young of the generality of Coleopterous carnivora, soon attain their greatest size (fig. 1). During growth the skin is repeatedly cast, but I am unaware of the exact number of ecdyses effected. In from a month to six weeks the larvae become full grown, at the ex-piration of which period they often measure as much as two inches in length. One in my posses-sion taken at Hoylake during the summer of 1899 exceeds 58 mm., or a fraction over two and a quar-ter inches long. They may be captured, in various stages of growth, in similar situations to the perfect insect, and can be kept alive and fed in the same way, so that their movements in the water are easily studied. The general form may be said to be spindle-shaped, something like a torpedo with a head and legs. Like a torpedo on occasion, too, the larva is equally capable of rendering itself eminently disagreeable to the various dwellers in its immediate vicinity; for its appetite equals, if possible, that of the adult insect, which is saying much, whilst its energy is untiring.

The large, flat, circular head is armed with strong sickle-shaped jaws, traversed throughout their length by hollow tubes, so that when the insect has fastened them upon its prey it is enabled to suck up the juices of the unfortunate victim. Many authors state that this is the larva's only mode of feeding (Siebold, etc.), and Nagel was of opinion that the method was rendered easier by the injection into the body of its captive of a kind of " digestive fluid," which rapidly dissolved all the more solid parts of its organisation and thus pre-pared them for assimilation. These theories can scarcely claim the support of modern investigators. Researches by De Geer, Meinert, Burgess, Miall, and others have served to show that in addition to the mandibular perforations the larva is possessed of a small but true mouth, by which it is enabled to consume minute portions of solid food. The head is joined to the body by a distinct neck, and, after gradually increasing in size until near the middle, the posterior portion of the insect tapers to the apex, where the extremity of the eighth abdominal segment terminates in two short cerci, which are prominently supplied with hairs. These appendages, in addition to being used on occasion as auxiliary organs of propulsion, further serve to sustain the larva at the surface of the water during respiration. The legs are long, the rows of stiff hairs running along their inner surfaces well fitting them for swimming purposes, whilst the cruel double tarsal hooks are used in the capture and

tearing of its prey. The general colour of the larva is of a dirty greyish-brown tint, thus usually harmonising with the muddy bottoms of the pools and sluggish streams in which the larva lies in wait for its victims.

The period spent in the larval state probably varies with the season of the year; those hatched out during the early summer may attain their full size in from fifteen (Fowler) to twenty days, whilst others, emerging later, bury themselves in the muddy banks of their pools should cold weather set in, pupating only at the approach of spring (Miall).

As the time draws near for its transformation the larva crawls out of the water and constructs a rude cell in the ground close by, where it under-goes its change to the pupal state. This inter-mediate change is always terrestrial, and lasts about three weeks or so, at the expiration of which period the perfect form is reached.

The newly emerged imago is at first quite soft and of a light testaceous hue, but gradually hardens and darkens, until at the end of a week it presents pretty much the general colour and solidity with which we are familiar. In its perfect state *Dytiscus marginalis* again resumes its aquatic existence, and, once more adopting its piratical life, no insect or other creature capable of being overcome is safe from its ferocious attack. Having enormous appetites, the beetles are often very destructive to young fish; Frank Buckland, amongst others, having called attention to the damage wrought by them to young salmon in Galway during 1865 (" Field," November 26th). In writing of *Cybister roeseli*, a closely allied species, at one time in-cluded in our British list, Burmeister records that in forty hours one of his captive insects ate and digested two frogs.

In their broad structural features the Dytiscidae resemble the Geodephaga, or predaceous ground beetles; but in the same marvellous manner that the general form of the *Cicindela* is especially suited for speed on land, is that of the *Dytiscus* eminently adapted for rapidity of movement in the water. In shape it is oval and somewhat flattened, the broad head fitting deeply into the pro-thorax, which in turn closely adjusts itself to the base of the elytra, so that a continuous outline is im-parted to the body as a whole. Having, moreover, a smooth polished surface, more particularly in the male sex, and being without pubescence of any kind, there is nothing to exert any retarding in-fluence as the beetle passes swiftly through the water. The 11-jointed antennae are smooth and glabrous, devoid of hairs, and inserted imme-diately in front of the large compound eyes. The jaws are curved, as in the terrestrial Adephaga, and are now solid and much stronger than in its larval state, as becoming an insect whose whole muscular system has considerably developed since its pre-pupal days.

L 2

In all our British species the clypeus and a narrow marginal stripe the elytra are yellow, the labrum being broad and distinctly notched. The legs of *Dytiscus* are slender, and appear somewhat disproportionate to its size. The posterior pair, which are supplied with powerful muscles, are longer and stronger than the others, and have the tibiae and tarsi flattened and furnished along their edges with long swimming hairs. By a peculiar keel-like prolongation of the thorax these organs are set far back below the centre of the beetle, where their strongly developed flat coxae are firmly fixed to the metasternum. The arrangement and articulation of the limbs render it easy for the legs to be brought at right angles to the body, thus allowing of a long sweep being made, the flattened portions of the tibiae and feet with their stiff setae acting like the blades of a pair of oars and serving to propel the insect forwards. These paddles, as a rule, strike the water simultaneously, although they are also capable of being used singly and in conjunction with the intermediate leg on the same side for turning rapidly in a small space; on the same principle as that by which an expert sculler manœuvres his rudderless skiff. The anterior and middle legs are shorter; the latter, in addition to affording some assistance in propulsion, being used for steering purposes. Both pairs are employed in the seizure and rending of its victims, and to anchor the insect when desirous of remaining at rest beneath the surface of the water. For these latter uses the last tarsal joints are, as in the larva, furnished with two sharp hooks, weapons which in the oar-like posterior feet are much less strongly developed.

The sexes differ considerably in appearance, so much so that they were formerly described as distinct species. The males are slightly larger than the females, have the swimming legs more powerfully developed, and the elytra smooth and highly polished (fig. 2). The females are dimorphous— i.e. exist under two different forms, of which the commoner has the elytra deeply grooved from near the base to about midway between the disc and apex (fig. 4). Females are, however, occasionally found with elytra resembling those of the males (fig. 3), but these are of rare occurrence in Britain. Simmermacher has stated that the sculptured females belong more particularly to northern latitudes; and the observation of Redtenbacher, that near Vienna the smooth form of female is as common as the rough, would seem to support this assertion. It may also be interesting to note that in the closely allied Australasian genus, *Hyderodes*, the females, although dimorphic, are almost invariably smooth, like the males.

No very satisfactory explanation as to the use of the furrowed elytra has yet been advanced. Earlier entomologists (Kirby and Spence) usually associated this roughness of the female with the perpetuation of the species, regarding it as a

special provision of Providence and accessory to the dilated pads on the anterior tarsi of the males; but Plateau was the first of several naturalists to prove by experiment the fallacy of these ideas, by demonstrating that a roughened surface was opposed to the action of the suckers. Careful observation will disclose the fact that the suckers on both the anterior and middle tarsi of the males are always applied to the thorax and smooth edges of the elytra, and never to the furrowed portion of the wing-cases. These sucker-like discs on the fore and intermediate feet of the males constitute another important distinction between the sexes; they being entirely absent in the females. In the anterior pair the three basal joints are very remarkably dilated so as to form large saucer-like discs, the function of which is for mating purposes, as suggested above. They appear to be a specialised modification of the pulvillus met with in many phytophagous beetles, flies, and other insects, from which, however, they must be carefully distinguished; the latter being for creeping purposes, and therefore on all six feet, and present in both sexes. In the coleoptera they are particularly noticeable in the family Chrysomelidae, and amongst the Rhynchophora or Weevils, although by no means confined to those groups of beetles. Under a microscope the peculiar arrangement of this curious enlargement of the anterior legs of the *Dytiscus* becomes increasingly interesting, and the pad of the foot is seen to possess a series of sucker-like prominences, two of which are considerably larger than the others (fig. 6). The under side of the middle tarsi is also densely clothed with small discs raised on foot-stalks which exactly resemble the smaller of those on the anterior feet. Much careful attention to the subject has convinced Dr. B. T. Lowne that the structure of these discs is identical with that of an insect's hairs; that they are, in fact, only different in form. These hairs are, of course, distinct from those with which we are familiar in the higher animals, in being mere processes of the integument itself. There appear to be nearly 200 disc-bearing hairs on the anterior tarsus of our three commonest species of *Dytiscus*, and Dr. Lowne's observations tend to prove conclusively that Blackwall was right in his contention, early in the last century, that these hairs are hollow and exude a viscid fluid, which enables the insect to adhere to polished surfaces. Derham, White, Banks, Home, and many other naturalists have associated the action of the pulvilli, or "cushions," with atmospheric pressure; and, considering the general interest of the subject and the wide prevalence of the latter popular belief, it may not be out of place to cite an experiment carried out by Dr. Lowne before the Royal Microscopical Society so far back as 1871. Following upon an explanation of the nature and structure of the discs, he proceeded: "I have here a male *Dytiscus* rendered insensible by the action of chloroform. I apply its four anterior tarsi to the interior of an air-pump receiver. By pressing these slightly with the side of a needle, they adhere firmly to the glass, and the insect remains suspended in its interior. I now exhaust the air, and the insect still remains suspended. It is necessary to render the insect insensible, otherwise it voluntarily detaches itself" ("Trans. R. Micr. Soc.,"

vol. v. 267). The sticky marks of the individual hairs in the print of the foot of a *Dytiscus* on a glass slide afford additional proof of the presence of an adhesive fluid, which Plateau found was sufficiently powerful in a newly killed beetle to sustain a weight more than thirteen times that of the insect itself.

(*To be continued.*)

BRITISH FRESHWATER MITES.

BY C. F. GEORGE, M.R.C.S.

(*Continued from page* 231.)

Arrenurus geminus n.s.

THIS pretty little mite belongs to Piersig's first division of *Arrenurus*—that is to say that the back part of the body is lengthened into a cylindriform tail, narrowed at the base as in *A. caudatus* De Geer, which mite it somewhat resembles. Its general colour is blue, with a triangular white mark in the centre, the apex directed backwards ; the eyes are crimson, and the fourth internode of the last pair of legs is furnished with a spur, as in many other male Arrenuri. At about the lower third of the tail

there is a marking somewhat resembling two lancet-shaped church windows placed side by side (see fig. 1). Below this the tail is sloped off in an irregular manner, forming three tiers, as shown in Mr. Soar's drawing (fig. 2). These markings are best seen in the living mite, for shortly after being kept in preservative solution the blue colour

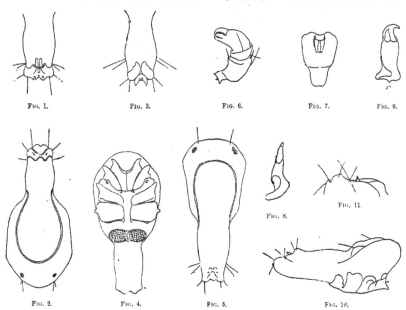

FIG. 1. FIG. 3. FIG. 6. FIG. 7. FIG. 9.

FIG. 2. FIG. 4. FIG. 5. FIG. 8. FIG. 11. FIG. 10.

FIGS. 1 to 11. *Arrenurus geminus*

disappears, and the whole mite becomes a nearly uniform warm brown, the eyes turn black, and the markings are not so easy to differentiate.

When I examined one of these mites I observed two transparent glass-like conical projections at the end of the tail (see fig. 3). These gradually shortened and disappeared in a day or two, leaving only circular marks as in fig. 1. This shows that the mite had at that time not quite reached its full development.

I took this mite on the last day of July, and found two specimens only in the gathering. One of these I sent to Dr. Koenike, who pointed out to me that the mite was not *A. caudatus* De Geer, but a new species. He also sent me for comparison the drawings of the mouth parts of this mite and

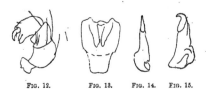

FIG. 12.   FIG. 13.   FIG. 14.   FIG. 15.

FIGS. 12 to 15. *A. caudatus.*

of *A. caudatus* ; also the view from above, the side view, and the highly magnified sketch of the end of the tail. These Dr. Koenike kindly allows me to use for publication. The figs. 1, 3, and 5 are from drawings made by myself from the living mite ; and figs. 2 and 4 are drawings by Mr. Soar from my mounted specimen.

This mite is much smaller than *A. caudatus* De Geer, as will be seen by the following measure. ments supplied to me by Mr. Soar :—

Length of *A. geminus*, 1·27 mm.; of *A. caudatus*, 1·48 mm.
Width of   „   0·66 mm. ; of   „   0·74 mm.
Length of first leg of *A. geminus*, 0·70 mm.
        „   „   *A. caudatus*, 0·92 mm.

Arrenurus soari n.s.

This mite was found by me in July 1898. I forwarded a specimen to Mr. Soar, who considered

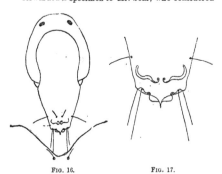

FIG. 16.   FIG. 17.

FIGS. 16 and 17. *A. soari.*

it not to be fully developed. I have on several occasions since visited the place where it was

taken, with the intention of finding others and trying to keep them alive for some time, to watch possible development. I have, however, not suc-ceeded in finding it again, so think it ought to be put on record, as, in my opinion, it is a very distinct as well as beautiful species. If Mr. Soar's figures be compared with any of the *Arrenuri* previously appearing in SCIENCE-GOSSIP, this will be very

FIG. 18. *A. soari.*

evident; but the difference is still more striking if the mites themselves, either alive or mounted, are compared.

During life its chitinous coat is of a transparent blue colour, and the circular discs of which it is composed are thicker at the edges, and therefore of a darker blue, than the central portion, giving the mite the appearance of network. To avoid loss of colour, it is well before mounting not to keep the creature long in preservative solution. The extremity of the last leg, near to the claws, is remarkably bent—more so than is usual in most other male *Arrenuri*.

This mite is smaller than *A. caudatus* De Geer. Mr. Soar gives its measurements as follows :— Length, 1·12 mm.; breadth, 0·52 mm. Palpus: length, 0·17 mm. Fourth leg : length, 1·28 mm.

Arrenurus robustus Koenike.

Described by Koenike in "Zoologischer Anzeiger," No. 453, July 30th, 1894. It is extraordinarily broad for its length. My specimen was taken on October 8th, 1900, and kept alive until Novem-ber 21st, then placed in preservative solution. Its colour was brick-red with green legs. Koenike says that his specimen was yellowish-green (*gelblich-grün*). Below the impressed curved line on the back are two large dimples or hollows. The petiole is short and chisel-shaped, rather wider at the free end and slightly convex. There is a peculiarity in the two curved hairs nearest to the petiole, for these are recurved at the point. I have not seen this feature in any other male *Arrenurus*, so that, at least for the present, this may be considered a characteristic of this mite. The hind legs are furnished with a highly developed spur on the fourth internode.

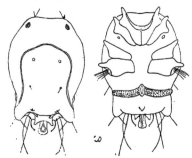

FIG. 19. Dorsal view.　FIG. 20. End of tail, highly magnified.

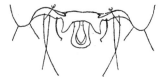

FIG. 21. Ventral view.

FIGS. 19 to 21. *A. robustus.*

I believe it has not been before recorded as British.

Mr. Soar gives me the following measurements, taken from the living mite :—Length, 1·28 mm. ; breadth, 0·84 mm. ; palpus, 0·15 mm. ; petiole length, 0·12 mm. ; breadth, 0·07 mm.

Curvipes aduncopalpis Piersig.

In Mr. Soar's papers on *Curvipes* (SCIENCE-GOSSIP, January to May 1899) he describes eleven

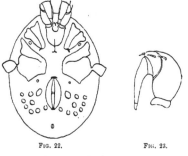

FIG. 22.　FIG. 23.

FIGS. 22 and 23. *C. aduncopalpis.*

species of that genus as British. On April 21st, 1900. I had the pleasure of taking the female of another species, *C. aduncopalpis* Piersig. This is the mite which Piersig takes as his type of this genus. The genital discs are embedded in the skin, and not placed on chitinous plates. These discs vary in number, not only in different specimens, but sometimes, also, on the different sides of the same specimen. The most remarkable feature of this species is that the second internode of the palpus is unusually dilated, and very much thicker than the first pair of legs. This increases the number of described British species of *Curvipes* to twelve.

Kirton in Lindsey, December 1900.

[The description of *A. geminus* was in type for publication in December last, and that of *A. soari* in January, but they have had to stand over for want of space in our pages.—ED. S.-G.]

THE ZODIACAL LIGHT may be seen as soon as it is sufficiently dark, nearly following the course of the ecliptic, in the western sky after sunset when the moon is absent.

NEW LISTS OF CHEMICAL APPARATUS —Messrs. Gallenkamp & Co., Ltd., have recently issued a large list of chemical and physical apparatus. An enormous variety of glass and metal wares are described, and, on the whole, the publication is a decided advance on the majority of English lists. There is one very advantageous feature, a good index. Messrs. Griffin & Sons, Ltd., have issued, as an adjunct to their "Sets of Apparatus," a descriptive price-list of all the apparatus and chemicals required for working through Gregory and Simmonds's "Elementary Chemistry and Physics."

SOUND FROM LONG DISTANCES.—Sound carried far on the mournful event of the first portion of Queen Victoria's funeral. Whilst the body was conveyed from the Isle of Wight to Portsmouth on February 1, minute guns were fired by the battleships during the progress between their two lines. The day being calm, the sound of the guns was heard at long distances. We have extracted some records from various sources, including reports sent to the *Times* and other daily London newspapers, from *Nature*, and direct correspondence. At Axeland, near Horley, they were heard by Mrs. Ashington Bullen ; at Boars Hill, Oxford (67 miles), by Professor E. B. Poulton and others; at Sutton, Surrey (60 miles), by Professor F. J. Allen ; Wimbledon Common and Richmond, Surrey (64 and 62 miles), by many people ; Beachy Head, Sussex (84 miles), Woodchurch (84 miles), Croydon (66 miles), Marcham (64 miles), Tunbridge Wells (66 miles), King's Langley (74 miles). Leighton Buzzard (84 miles), and Great Missenden, where windows were shaken at a distance of 69 miles. At Brightling, 69 miles away, cock pheasants crowed as is their habit during thunderstorms. There can be little doubt about the source of the sounds, as in most of these places they were noted to be at minute intervals.

EXPERIENCES IN FLOATING FORAMINIFERA.

BY DR. G. H. BRYAN, F.R.S.

THE failure of my attempts to obtain satisfactory results in cleaning foraminifera by the method of surface flotation, elicited about two years ago a series of interesting articles in SCIENCE-GOSSIP from the pen of Mr. Earland. With the information obtained from these, and the experience derived from the somewhat similar operation of cleaning desmids, I have succeeded in securing several really excellent "cleanings" of foraminifera. It has seemed to me that an account of the results may be of interest to your readers in explaining the failures of the "flotation" method in dealing with certain gatherings. I propose to consider separately the several classes of material used in my experiments.

1. FINE SPONGE SAND.

The greater portion of sponge sand passes through the finest gauze sieves that I possess. On introducing some of these fine powders into water, it was impossible to prevent a thick film of the grains from floating on the surface in consequence of capillarity, and even when a paper funnel was used, bubbles would be carried down and rise to the top with some of the fine sand adhering to them. The cause of failure of the flotation method was evident on examining the sediment, for it contained practically all the foraminifera, which had sunk to the bottom, their chambers still being filled with air. Even the substitution of salt-water for fresh-water did not give the foraminifera sufficient buoyancy to rise to the surface. By placing the sediment in water at one side of a saucer or soup plate, and shaking it, the foraminifera rose to the surface of the sand; and by carefully rocking the plate they could be drawn round to the other side, leaving the sand high and dry behind. The foraminifera were transferred to another plate, and any shells which got stranded were removed by a second washing; the rejected material then consisted practically of pure sand with only a very occasional foraminifer still left. The number lost in this way must have been insignificant.

After thus dealing with small batches of material and getting rid of the bulk of the sand, the cleaned material, now reduced to a very small quantity, was once more washed in the same way and the residual sand removed. A teaspoonful of the sponge sand yielded a small pinch of cleaned material, barely sufficient for mounting three very pretty slides, thickly spread over ¾-inch covers. In some cases over 50 per cent. of the total number of particles were foraminifera. This method of washing, as applied to sponge sand, eliminated the whole of the fine sand grains constituting the great bulk of the material. It gives washings rich in foraminifera, but these were mixed with debris, such as broken shells, fragments of coral, sponge fibre, etc. I have tried to separate these by repeating the washings, but after a certain stage both the foraminifera and the associated particles began to separate into different sizes without further selection from each other. I then tried placing the washings on an inclined plane of paper, but found that, so far from the foraminifera separating from the debris, the larger and more unsightly particles of the latter were the first to skip off the paper. So I tried tapping the paper from below several times, and it soon appeared that the foreign bodies with comparatively few of the larger foraminifera jumped the greatest distance down the plane, and the material which remained furthest behind, on being spread, yielded slides of far greater purity than could be obtained by spreading the washings in bulk. This action of the inclined plane was in some respects the reverse of what I had previously experienced with sand from Colwyn Bay.

2. COARSER SPONGE SIFTINGS.

On sifting a quarter of a pound of sponge sand through the same fine sieve, about a teaspoonful of white grains remained, and a coarser sieve removed the large fragments of shell, seaweed, and other large foreign bodies mixed with the siftings. Practically none of the foraminifera floated in water; but by rocking in a soup-plate, as before, sufficient foraminifera for about seven or eight closely spread slides were separated from the rest of the mass, mixed, however, with an equal bulk of vegetable debris. To separate the latter I waited till the chambers of the foraminifera had filled with air, and then rocked them in a soup-plate, this time much more gently; and it was thus easy to draw the vegetable matter away, leaving the shells behind. The rocking process rendered it possible, first, to draw the foraminifera from the sand, and then the vegetable matter away from the foraminifera, the material being separated into three different portions, the foraminifera constituting that of intermediate density. The spread slides of these foraminifera contain hardly any foreign particles, though many of the shells themselves are somewhat broken.

3. COARSE SHORE-SAND.

A heap of shore-sand used for building purposes in this neighbourhood had been left for some weeks exposed to the weather, and a number of little white patches collected in the hollows. These were seen to contain foraminifera, chiefly large Miliolinae; so I collected enough to fill a two-ounce tobacco tin nearly. The foraminifera would not float in water, but by rocking successive portions in a soup-plate, and then rocking again, and finally rocking more gently to separate the vegetable debris, a nearly pure gathering was obtained, mixed with a small percentage of fragments of shells and a few "blacks" composed of coal or coke dust, soot, etc. In mounting spread slides I picked out most of the blacks without a hand lens, using a fine brush, and placing the spread slide or cover on a white ground.

I have lately tried floating the cleaned material to separate the blacks, but the foraminifera again sank at once.

4. MUD FROM MENAI STRAITS.

In several places along the Penrhyn foreshore and elsewhere small patches strewn with white particles were seen left by the receding tide. I scraped up the mud containing these particles until there was sufficient to fill two-thirds of a two-ounce tobacco tin, and, after straining through a wire-gauze coffee-pot strainer, rocked batches of the material in water; then repeated the process with the separated portion, finally rocking more slowly to remove vegetable and other debris. The result was a teaspoonful of a mixture of foraminifera and blacks, with a small percentage of mica. The proportion of blacks was considerable—say, one-third to two-thirds of the total material—so that mounting spread slides was out of the question, and I determined to have another attempt at floating. Before doing so the material was dried and sifted with the finest sieve I had, the foraminifera practically all passing through. On introducing into water through a paper funnel, pouring a little water down the funnel to clear the last foraminifera away, and stirring the material, floating forms immediately rose to the surface and mostly collected round the edges of the glass. By lifting out some of the surface floats with a little water in a teaspoon, so as to lower the water in the glass, the floats were chiefly left adhering to the sides, so could be collected with a brush and transferred to water in the teaspoon, poured into a filter and dried. The floats only amounted to a small pinch of material, but they were almost pure foraminifera, including a sparse sprinkling of flask shells or Lagenidae; these last are of several species, hardly two on any slide being alike.

On drying and refloating the sediment I obtained a very little more foraminiferous material, but the majority of the foraminifera which would float had evidently come off in the first operation. The sediment contained some forms that would not float, and with the small quantity of material operated upon it was impossible in any way to

separate these foraminifera from the blacks, especially as the blacks now largely preponderated over the foraminifera. This material having yielded successful surface floats, I tried to dispense with some of the preliminary washings; but the floats were neither so abundant nor so free from foreign matter. In one case the floats had to be dried and refloated to give satisfactory results.

5. BLACK-ROCK, BRIGHTON.

Wishing to show some foraminifera to a friend at Brighton, I gathered somewhat hastily some sand from the rock-pools in that locality. This sand apparently consisted of patches of coarse material containing large Miliolinae, and other patches of finer sand containing smaller shells, the two kinds being mixed in collecting. The gathering was rather roughly cleaned in a soup-plate; but on returning home and floating the small quantity obtained, both the large Miliolinae and the smaller forms floated. By sifting through the finest gauze the large Miliolinae were obtained in a state of purity, and the smaller forms which passed through the sieve were sufficient to yield one interesting slide, in which, however, they were mixed with much foreign matter, especially blacks. To separate these it would have been necessary to operate on a larger quantity of material than I had gathered.

Thus, out of five kinds of material, the foraminifera in two floated, after the bulk of the sand had been washed away; those in the remaining three refused to float. The experiments fully confirmed Mr. Earland's remarks as to the trouble of dealing with "blacks." I tried burning out these, by letting the material fall through a gas flame, but the results were unsuccessful. The blacks are evidently separable from the surface floats, and the only possibility I can suggest of eliminating them from foraminifera which sink at once is that the foreign bodies light enough to be washed forward with the foraminifera when the latter are full of air may perhaps be carried in front of them when the water has filled their chambers. Such a solution of the difficulty is probably easier in theory than in practice.

Foraminifera seem to me to be more difficult to clean than desmids, because when the material has been separated into two or more portions by any process of washing, it is necessary to use a hand lens to see which portions contain the foraminifera, whereas the bright green colour of the desmids sufficiently indicates when these have been separated from the foreign matter associated with them.

Plas Gwyn, Bangor, North Wales.

THE ISAAC NEWTON STUDENTSHIP for encouraging research in astronomy and physical optics, of the value of £200 per annum for three years, has been bestowed, by election, on Mr. S. Bruce M'Laren, B.A., of Trinity College, Cambridge.

ARSENIC AND ARSENIC-EATERS.

By C. A. MITCHELL, B.A., F.I.C.

THE name arsenic is derived from a Greek word signifying "masculine," which is fancifully explained as due to its masculine force in destroying man. It has long been known in the form of one or other of its compounds. It is mentioned by Pliny and Dioscorides as a remedy for certain diseases, and is said to have been used from remote times for similar purposes by the Chinese and natives of India.

When isolated from the numerous bodies with which it combines, arsenic is a brittle metallic substance of a steel-grey colour. On being heated, it volatilises as a colourless vapour, which has a characteristic odour, recalling garlic. It burns with a livid bluish flame, combining with atmospheric oxygen to form the white powder which is the white arsenic of commerce.

The chief source of commercial arsenic and its derivatives is arsenical iron pyrites, which is obtained from mines in Styria. It also occurs occasionally as a compound with sulphur, and with metals other than iron.

There are factories at Reichenstein and Altonberg, in Silesia and Styria, where the arsenic is separated from the ore. The process employed is to roast the mineral in a furnace, and to conduct the vapours of liberated white arsenic into condensing chambers of special construction, on the walls of which they deposit as a white crust. Every few weeks the deposited arsenic is removed through special doors, and purified by heating and re-condensation.

The removal of the arsenic from the chambers is a highly dangerous operation, owing to the deadly nature of the fumes. The workman has a closely fitting leather dress, and his head is covered with a helmet having glass in front. As a further precaution his nostrils and mouth are protected by means of a wet sponge or cloth.

The workmen at the arsenic factories live largely on leguminous food, very little meat being eaten by them. Each man is given a small quantity of olive oil to drink daily, and eats as much butter and other fat as possible. All alcohol is supposed to be avoided, though there is reason to believe that many of the workmen, being arsenic-eaters, can take it with comparative impunity.

The white arsenic thus obtained is a substance closely resembling flour in appearance. It can be made to crystallise from an acid solution with a beautiful phosphorescent flash as each crystal separates. It is the basis from which other commercial preparations of arsenic are manufactured, such as "realgar," a beautiful orange-red colour; "orpiment," a golden-yellow, both of which are compounds of arsenic and sulphur; and "Scheele's green" and "Schweinfurt green," which consist of arsenic in combination with copper compounds. The last two were at one time largely used for colouring wall-paper, but at the present day they are principally employed in the manufacture of oil paints.

White arsenic is the form in which arsenic is eaten by the peasants of Styria and the Tyrol. Professor Schallgrueber, of Graetz, was the first to call attention to this practice, in a report which he made in 1822 to the Austrian Government on the cause of the numerous deaths from arsenic poisoning in those districts. He found that arsenic was kept in most of the houses in Upper Styria under the name of "hydrach," evidently a corruption of "Hüttenrauch," or furnace smoke. His statements were subsequently confirmed from personal observation by a Dr. McClagan, of Edinburgh, but for many years afterwards the arsenic-eaters were generally disbelieved in; and it was not until 1860 that Mr. C. Heisch published convincing evidence.

Arsenic is principally eaten by hunters and woodcutters with the object of warding off fatigue and improving their staying powers. Owing to the fact that the sale of arsenic is illegal in Austria without a doctor's certificate, it is difficult to obtain definite information of a habit which is kept as secret as possible. According to a Dr. Lorenzo, in that district the arsenic is taken fasting, usually in a cup of coffee, the first dose being minute, but increased day by day until it sometimes amounts to the enormous dose of twelve or fifteen grains. He found that the arsenic-eaters were usually long-lived, though liable to sudden death. They have a very fresh, youthful appearance, and are seldom attacked by infectious diseases. After the first dose the usual symptoms of slight arsenic poisoning are evident, but these soon disappear on continuing the treatment.

In the arsenic factories in Salzburg it is stated that workmen who are not arsenic-eaters soon succumb to the fumes. The manager of one of these works informed Mr. Heisch that he had been medically advised to eat arsenic before taking up his position. He considered that no one should commence the practice before twelve years old, nor after thirty, and that in any case after fifty years of age the daily dose should be gradually reduced, since otherwise sudden death would ensue.

If a confirmed arsenic-eater suddenly attempts to do altogether without the drug, he immediately succumbs to the effects of arsenic poisoning. The only way to obviate this is gradually to acclimatise the system by reducing the dose from day to day. As further evidence of the cumulative properties of arsenic, it is interesting to note that when the graveyards in Upper Styria are opened the bodies of the arsenic-eaters can be distinguished by their almost perfect state of preservation, due to the gradually accumulated arsenic.

There have been several criminal cases in Styria in which the prisoner has been accused of poisoning with arsenic, and acquitted by bringing evidence to prove that the deceased was an arsenic-eater. The interesting legal problem of an arsenic-eater being wilfully deprived of his daily arsenic does not, up to the present time, appear to have required solving.

Chancery Lane, W.C.

BUTTERFLIES OF THE PALAEARCTIC REGION.

By Henry Charles Lang, M.D., M.R.C.S., L.R.C.P. Lond., F.E.S.

(Continued from page 263.)

Genus *COLIAS* (*continued*).

7. **C. aurora** Esper. 83, 3. Sibirica Ld. Z. b. V. 1852.

48—60 mm.

♂ ground colour of wings deep clear orange, nervures black except where they traverse the marginal borders, where they are more or less greenish-yellow. The amount of striation of border varies in different specimens, but it is constant near the apex of f.w. Marginal borders themselves narrower in proportion than in *Colias diva*, especially on h.w. Black discoidal spot of f.w. variable in size, but generally smaller than in *Colias edusa*. H.w. greenish-yellow at inner marginal fold and an. ang. Discoidal spot very little brighter than ground colour. Basal patch very distinct and light ochreous. The outline of the h.w. often exhibits faint traces of outer marginal indentation, but this is not a constant character. All the wings are more or less shot with violet. ♀ larger in expanse than ♂. F.w. ground colour as in ♂, bases with greenish dusky shading. Marginal border as in *C. edusa*, with five or six chrome-yellow spots. H.w. greenish- or yellowish-orange, more or less shot with violet. Marginal border broad with large yellow spots; disc. spot light in centre. U.s. canary-yellow in both sexes. F.w. orange towards central area. Disc. spots well marked, and with silvery centres both on f. and h.w. Fringes of all the wings red, together with the head, antennae, and prothorax.

Hab. Transbaical, S. Siberia, Amur, E. Altai, Kuldja. The specimens from the Amur seem to be the largest and brightest.

a. ab. ♀ *chloe* Eversm.—Bull. Mosc. 1847, ii. p. 73. The form of the ♀ corresponding to *Colias edusa* ab. *helice*. Ground colour of wings white or greenish-white; disc. spot of h.w. whitish and never orange. Hab. (?)

b. var. *kentcana* R. H. p. 731. A smaller and less brightly coloured form. Colour of wings yellowish, instead of being brilliant orange. Disc. spot f.w. light-centred, violet reflection slight or absent. Hab. Transbaical. Kentei mountains. VI.–VIII.

8. **C. olga** Romanoff. Mem. Lep. 1883; cos. H. S.

50—60 mm.

♂ bright orange shot with violet, f.w. with the border as in *Colias edusa*, that of the h.w. being narrow as in *Colias myrmidone*. Neuration of wings not black as in *C. aurora*. Marginal border finely dusted with yellow scales, and only very slightly veined with yellow towards apex of f.w. Disc. spot f.w. well defined and round; that of h.w. deep orange. Fringes red, streaked with light yellow. ♀ usually found in the white form (forma

Colias aurora. Male and Female.

alba). It much resembles the ab. *chloe* of *C. aurora*, but differs in the following particulars : the marginal borders are not so deeply black ; the neuration is of the same colour as the wings, which are of greenish-white. The bases are not so deeply shaded, and the discoidal spot of h.w. is very distinctly orange. The orange form (forma *aurantiaca*) greatly resembles *C. aurora* ♀. but differs in having less basal shading and in the colour of the neuration; also in the intensity of the disc. spot h.w. U.s. disc. spot f.w. without any silvery centre. A submarginal row of widely separated black spots on all the wings.

HAB. Daghestan, Transcaucasia. V.–VII.

9. **C. myrmidone** Esp. 65, 1, 2 ; Lg. B. E. p. 59, pl. xiii, fig. 4.

42—50 mm.

Wings orange-yellow, deeper than in *Colias edusa*, but not so intense as in *C. aurora* and *C. olga*. ♂ has the marginal border of f. and h.w. much narrower than in *C. edusa*. It very rarely shows any pale rays, but is finely dusted with light yellow scales. H.w. have a row of indistinct light blotches immediately internal to the narrow marginal border. ♀ greatly resembles *C. edusa* ♀, but the marginal spots are larger, better defined, and more regular ; they contrast more with the ground colour of the wings, which is redder than in *C. edusa*, and are more greenish in tint.

HAB. Eastern, Southern, and Central Germany; Austria, Hungary, North Turkey, South-Western Russia, Asia Minor, Caucasus, Altai. VII.—VIII.

LARVA. Green, with a darker dorsal stripe and lateral green streaks of a lighter colour. The surface is covered with short black hairs growing from minute tubercular elevations. On *Cytisus biflorus* and other Leguminosae. IX. and V.

a. ab. ♀ *alba* Stgr. Cat. 1871, p. 6. The white form of ♀. Very much rarer in proportion to the orange form than ab. *helice* of *Colias edusa*.

b. var. *caucasica* Stgr. Cat. 1871, p. 6. *myrmidone* var. Ld. Ann. S. Belg. xii. 20. "*major*, saturatius flav.*" Stgr. Some entomologists have thought that the species now understood as *C. olga* Romanoff is identical with this form.

It is probable that many specimens of *C. myrmidone* were formerly confounded with it. Until the closing years of the nineteenth century the genus *Colias* was much less perfectly understood than at the present day. Even in Staudinger's Catalogue of 1871 there is much evidence of imperfect knowledge and confusion of forms.

There is no doubt, however, that *C. myrmidone* does present a form in Armenia and also in the Ural Mountains which is larger and more intensely orange-yellow than the type. A specimen in my collection, marked "Ural," obtained through Mr. Elwes, has a very large disc. spot on f.w. ; the marginal band on h.w. is much more sinuous than normal, and so indeed is the outline of wing itself

along its on. marg. The expanse is greater than in the type; but. in spite of these characters, it is easily distinguishable from *C. olga*.

We now come to the end of that group of *Colias*, the species of which possess the character of the peculiar sex mark of a costal patch of thickened scales on the h.w. of the ♂. Setting aside *C. wiskotti*, an altogether abnormal species—some of the forms of which, though now considered as varieties, will probably be eventually accepted as specific—we may take *C. edusa* and *C. myrmidone* as the types of the group, all the other species being more or less closely connected with either one or the other. It will be noticed that the forms increase in size and intensity of colour as we proceed in an easterly and south-easterly geographical direction of the Palaearctic region. In *C. aurorina*, *C. aurora*, and *C. olga* we have probably the highest developments of what is represented by the two well-known European species above mentioned.

(*To be continued.*)

REPTILES IN WINTER.

BY GERALD LEIGHTON, M.B.

THE ardent field naturalist whose affections are specially developed in the direction of reptiles is apt to find the winter months of a climate such as ours rather long and tedious. It seems so tiresome to wait until the spring to settle a point in which one had just become interested when the reptiles inconsiderately retired for the winter. After all, however, the winter months ought not to be wasted by any observant field naturalist, though they may seem out of proportion to the length of time allowed by summer weather for more practical work out of doors. If one has worked hard and been fairly successful. there ought to be a considerable mass of more or less illegible writing in the pocket notebook carried by every observer who wishes to make use of what comes under his own notice; and all this material has to be written, classified, systematised and corrected, either as a matter of accuracy, or for the use of others. Then, if one has a reptilian collection of one's own, there are probably a few specimens to be re-labelled, re-measured, or more carefully preserved, and their detailed descriptions recorded in a club "Transactions" or one's own catalogue. Most important of all perhaps, for no lover of animals cares to spend much time indoors in the summer, is the correspondence with fellow-observers in other localities, for which the winter months leave time. It is also a good plan to form some definite ideas for the next season's investigations; what to look for and where to go, and to make arrangements accordingly. Apart altogether from all this, which applies to our own little sphere of labour, is the consideration that before the season of practical work comes round, there are a number of books and papers on various points to read or be referred to, so as to thoroughly digest the full significance of what has been seen while observing the reptiles.

Now that the serpents have disappeared for the winter hibernation, it is interesting to think of the strange physiological picture they present—alive, yet continually unconscious, without food, but not wasting away. The adaptation of the reptile to the winter environment is one of the most marvellous arrangements in Nature. In hot and tropical climates " aestivation " takes the place of " hibernation," and the reptiles retire in the heat and reappear on the approach of the rainy season. In our region it is a matter of food and temperature. They cannot stand the rigour of our winter above ground, and that same cold drives the animals on which they depend for sustenance to a similar retirement. It is an error to suppose that snakes can endure any amount of cold ; on the contrary, they are most susceptible to its influence, and if roused from their winter sleep usually die. When in that torpid condition, respiration, always slow, is almost stopped, and the heart beats very feebly and at long and irregular intervals. Digestion is absolutely at a standstill ; and of course the digestive fluids must cease to be secreted, or they would dissolve the tissues of the reptile itself, at that time of extremely low vitality.

The result of observations by Dr. Guyon has led him to the conclusion that the venom of the adder is more to be feared after the season of hibernation than at any other time, because it has then had a longer time to accumulate. I think this could hardly be the case in our country, as one can scarcely believe the poison glands go on secreting at a time when all chemical activity in the body is arrested. Indeed, some say an adder bite sustained at this time is practically harmless, and most English observers agree that the results to the person bitten are most grave at the hottest time of the year. Of course, that effect is not so serious if the adder has just previously emptied its poison gland, as when the secretion has not been discharged, say, for a week before the accident. As a rule, our serpents show little sign of life if roused in winter, and it requires a good deal of very patient searching to find them in their hibernating quarters.

Our ring snakes and adders are apt to congregate in large numbers in their winter retreats of holes and crevices, presumably for the warmth they afford each other by close proximity. I remember a correspondent telling me of a case in Kent where some workmen who were digging out a rabbit from a hole came across a mass of adders coiled together —he said, to the number of over a hundred. Similarly in an old quarry near Llanelly some men who were getting stone disturbed immense numbers of ring snakes which had retired there for the winter.

Grosmont, Pontrilas, Hereford.

LAND AND FRESHWATER MOLLUSCA OF HAMPSHIRE.

By Lionel E. Adams, B.A., with kind assistance of B. B. Woodward, F.L.S., F.G.S.

THE assemblage of land and fresh-water mollusca in the county of Hampshire is especially interesting, as it contains exceptional forms. The number of species so far recorded is 106 out of the 139 that have been hitherto found in Britain. Five others have been obtained from alluvial deposits, and it is probable that most of these may yet be found in a living state when a more thorough search has been made for them in the county.

In looking through the following list it will be noted that the county possesses at least two forms of exceptional interest. They are *Helicella barbara*, which originally reached this country from the south-west, over regions now submerged; while others, like the well-marked *Helicodonta obvoluta*, came from the Continent to the south-east.

The initials following localities are those of the following workers in the county: Late Chas. Ashford (C. A.), A. Loydell (A. L.), Miss Fl. Jewell (F. J.), Ch. E. Wright (C. E. W.), late Wm. Jeffrey (W. J.), H. P. Fitzgerald (H. P. F.), Lionel E. Adams (L. E. A.).

The list of non-marine mollusca of Hampshire, including the Isle of Wight, is as follows :—

Testacella maugei Fér. Porchester.

Testacella scutulum Sby. Newport, I. of Wight. J. W. Taylor. Journ. Conch. v. 343. See Journ. Malac. vi. p. 26.

Limax maximus Linn. Winchester (L. E. A.); Christchurch (C. A.); Isle of Wight (A. L.). Var. *maculata* Pic., Christchurch (C. A.).

Limax flavus Linn. Ditcham Wood (C. E. W.) ; Christchurch and Mudeford (C. A.); Isle of Wight (A. L.).

L. arborum Bouch. Chant. Christchurch (C. A.). Var. *nemorosa* Baud. ; var. *bettonii* Sord. One specimen of each of these found under felled timber in the New Forest, in Wootton enclosure (C. A.). Hambledon.

Agriolimax agrestis Linn. Commonly distributed. Var. *sylvatica*. Frequent (C. A.).

Agriolimax laevis Müll. Christchurch (C. A.). Hambledon ; Beckford Green.

Amalia sowerbii Fér. Frequent round Christchurch (C. A.).

Amalia gagates Drap. Var. *plumbea* Moq. Christchurch. The type does not occur here (C. A.). Hoe Moor.

Vitrina pellucida Müll. Commonly distributed.

Vitrea crystallina Müll. Common.

Vitrea alliaria Müll. Winchester (L. E. A.). Not frequent in Christchurch district (C. A.).

Vitrea glabra Brit. auct. Hayling (C. E. W.); Isle of Wight (A. L.).

Vitrea cellaria Müll. Commonly distributed. Var. *albina* Moq. 2 specimens at Christchurch (C. A.).

Vitrea nitidula Drap. Winchester (L. E. A.); district round Christchurch (C. A.); Isle of Wight, fine (A. L.).

Vitrea pura Alder. Isle of Wight (A. L.).

Vitrea radiatula Alder. Crabbe Wood, near Winchester (L. E. A.).

Vitrea excavata Bean. Numerous where it occurs, but very local. Chuton Glen, Hengistbury Head, Boscombe, Hoborne, in all which places the type only has been found in a small copse. At Roeshot Hill, near Christchurch, a variety with a transparent light yellow shell only occurs, to the exclusion of the type (C. A.).

Vitrea nitida Müll. Langstone (C. E. W.); side of ponds, Bonchurch to Ryde (A. L.).

Vitrea fulva Müll. Winchester (L. E. A.); sparsely round Christchurch (C. A.). Var. *mortoni* (Jeff.); a few specimens in woods round Winchester (L. E. A.).

Arion ater Linn. Both black and rufous varieties commonly distributed. Var. *rufa* Linn., frequent near Christchurch (C. A.).

Arion hortensis Fér. Christchurch (C. A.). The variety known as *A. celticus* Poll. has been recorded from Southampton.

Arion circumscriptus Johnst. Christchurch (C. A.); Winchester (L. E. A.).

Arion intermedius Norm. Woods near Winchester (L. E. A.).

Arion subfuscus Drap. Christchurch, Southampton (C. A.).

Punctum pygmaeum Drap. Christchurch (C. A.).

Sphyradium edentulum Drap. Highcliff (J. H. Ashford); Hoborne Common (C. A.). See Journ. Conch. v. 163; also Alverston and Steep Hill, Isle of Wight ? author; possibly Leconte, Ann Mal. Belg. iv. p. lxi.

Pyramidula rupestris Drap. Christchurch (C. A.); Warblington, Southampton (C. E. W.). See Proc. Malac. Soc. i. p. 296.

Pyramidula rotundata Müll. Common. Var. *alba* Moq. Christchurch (C. E. W.); var. *turtoni* Flem. Isle of Wight (A. L.).

Helicella virgata Da Costa. Common in the chalk. A reversed monstrosity has been found at Yarmouth, Isle of Wight. M. *sinistrorsum* Taylor, Yarmouth (C. A.); var. *lutescens* Moq. var. *maculata* Moq., var. *nigrescens* Grat., var.

submaritima Jeff., all Isle of Wight (C. A.); var. *albicans* Grat. Newtown (C. A.), Isle of Wight (A. L.).

Helicella itala Linn. Common on the chalk. Var. *hyalozonalis* Ckl. Isle of Wight (A. L.); var. *alba* Sharp., var. *minor* Moq., both common with type; m. *sinistrorsum* Taylor. A reversed monstrosity has been taken at Havant (H. Beeston).

Helicella caperata Mont. Common on the chalk. Var. *alba* Pic., Yarmouth, Isle of Wight (C. A.); var. *fulva* Moq., Isle of Wight (A. L.); var. *obliterata* Pic., Colony at Freshwater Bay (C. E. W.).

Helix barbara Linn. Freshwater Bay (C. A.). Var. *bizona* Moq., Freshwater Bay (C. A.); var. *elongata* (Cr. & Jan.), Needles (C. A.) and (A. L.); var. *strigata* Menke, Freshwater Bay (A. L.).

Helicella cantiana Mont. Common all over the chalk districts. Var. *rubens* Moq., var. *albida* Taylor, one small colony at Mudeford containing these two varieties (C. A.).

Hygromia fusca Mont. Two specimens have been reported from the New Forest, near Holmsley.

Hygromia granulata Alder. Abundant close to river Avon, Christchurch; a few at Winkton; Ringwood (C. A.); one specimen at Avingdon Park (L. E. A.). Var. *cornea* Jeff., North Hants; rare (H. P. F.).

Hygromia rufescens Penn. Common all over the chalk districts, Isle of Wight, local (C. A.). Var. *rubens* Moq., Winchester (L. E. A.); North Hants, common (H. P. F.); var. *alba* Moq., Isle of Wight (A. L.), Christchurch (C. A.); var. *albocincta* Ckl., Isle of Wight (A. L.).

Hygromia hispida Linn. Abundant at Christchurch (C. A.); common in woods round Winchester (C. E. A.); Isle of Wight (A. L.). Var. *hispidosa* Mouss., with type.

Acanthinula aculeata, Müll. Two specimens from Somerford (C. A.).

Vallonia pulchella Müll. Widely distributed. Var. *costata* Müll., more common than type in Isle of Wight (A. L.).

Helicigona lapicida. Widely distributed. Var. *albina* Menke, near Christchurch (C. E. W.).

Helicodonta obvoluta Müll. Crabbe Wood (L. E. A.), Petworth, Ditcham Wood, Miscombe (C. E. W.).

Helicigona arbustorum Linn. North Hants (H. P. F.) Very local round Christchurch (C. A.); Havant, rare; Hayling, few (C. E. W.); Isle of Wight (A. L.). Var. *alpestris* Zeigh. South Hants (A. L.); var. *flavescens* Moq., North Hants, rare, H. P. F.; var. *albina* Moq., Isle of Wight, one specimen (A. L.).

Helix aspersa Müll. Common. Var. *zonata* Moq., North Hants (H. P. F.); var. *undulata* Moq., North Hants (H. P. F.); var. *exalbida* Menke, Isle of Wight (A. L.).

Helix pomatia Linn. Petersfield, quoted by the late

Wm. Jeffery from "Zoologist," 1878. North Hants, rare (H. P. F.).

Helix nemoralis Linn. Besides the common varieties in colour and markings there are here and there in the south of the county colonies of the crimson-banded form with pink lips. There are also some forms apparently hybrid between this and the next species.

Helix hortensis Müll. This is rather more local in the county than the last species. On Hayling Island a remarkable form occurs, yellow with a black mouth, but with the characteristic dart of the species.

(To be continued.)

AN INTRODUCTION TO BRITISH SPIDERS.

By Frank Percy Smith.

(Continued from page 267.)

GENUS *TAPINOCYBA* SIM.

Anterior eyes in a straight line, close together and almost equidistant. Posterior eyes large, their intervals seldom greater than their diameters. Ocular area considerably wider than clypeus. Tarsi of first pair of legs as long as metatarsi. Metatarsi of fourth pair shorter than tibiae. This genus contains several British species, none of which can be called common.

Tapinocyba pallens Cb.

A rare species, the type of the genus.

In this and in many other cases where obscure and closely-allied spiders come under consideration it will, as before mentioned, be found impossible to describe the species with sufficient accuracy to admit of their differentiation. The form of the processes connected with the male palpus and of the vulva of the female are in many cases the only satisfactory characters. Illustrations of these parts will, I feel sure, be more useful to the student than any description, however elaborate and detailed. Therefore figures will be prepared and given later of such parts. These, in conjunction with the generic characters, will tend to remove the difficulty of identification.

Tapinocyba subaequalis Westr.

Length. Male 1.5 mm.

A rare species described under genus *Walckenaera* in "Spiders of Dorset" and figured as *W. fortuita* in "Trans. Linn. Soc." for 1870. The caput is not furnished with any projections. The palpal organs are rather complex.

GENUS *CNEPHALOCOTES* SIM.

The spiders in this genus may be distinguished from those of the genus *Tapinocyba* by the posterior eyes being rather small, their intervals considerably exceeding their diameters, and by the distance between the anterior centrals being less than that between one of them and the adjacent lateral eye. Tibial spines very short.

Cnephalocotes obscurus Bl. (*Walckenaera obscura* Bl.)

Length. Male 1.75 mm.

The caput has a distinct prominence. The palpal organs are very complicated, and have connected with their structure several conspicuous spines, one of which, springing from the lower part of the organs, is extremely long and slender. This species is the type of the genus.

GENUS *TROXOCHRUS* SIM.

In this genus the curve of the anterior row of eyes

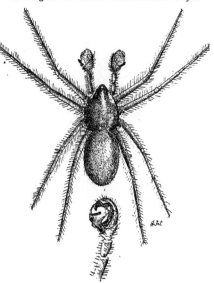

Fig. 1. *Savignia frontata* Bl.
Male and palpus viewed from beneath.

is directed forwards. The integument of the abdomen is strongly coriaceous.

Troxochrus scabriculus Westr. (*Walckenaera aggeris* Bl.)

Length. Male 1.5 mm., female 1.75 mm.

The caput of the male is rounded and considerably raised. The body is glossy black and the legs yellowish-red. It is a very rare species. .

GENUS *STYLOCTETOR* SIM.

The spiders included in this genus are very similar to those of the genus *Cnephalocotes*, differing chiefly in the size and position of the tibial spines. In the present genus they are long, or, at least, longer than the diameter of the joint from which they spring, and, in the case of the tibiae of the fourth pair of legs, are placed nearer the end of the joint than in *Cnephalocotes*.

Styloctetor penicillatus Westr.

This species, which is not very rare, may be easily distinguished from its allies by the possession of a tuft of bristly hairs upon the radial joint of the male palpus.

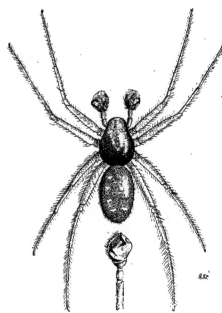

FIG. 2. *Cornicularia unicornis* Cb.
Male, and palpus viewed from beneath. See p. 267.

Styloctetor broccha L. Koch.

A most curious and interesting species. The spiracular plates of the male are strongly corrugated, and are acted upon by projections on the coxae of the fourth pair of legs. This arrangement enables the creature to make a sound which, though imperceptible to human ears, is no doubt of use to the spider, probably as a call-note to his mate.

GENUS *METAPOBACTRUS* SIM.

This genus differs from *Tapinocyba* in the following points. The anterior row of eyes is curved, its convexity being directed backwards. The ocular area and clypeus are about equal in width. The tarsi of the anterior pair of legs are hardly more than three-quarters the length of the metatarsi.

Metapobactrus prominulus Cb. (*Walckenaera prominula* in " Spiders of Dorset.")
Length. Male 2 mm.
Cephalo-thorax blackish-brown. Legs reddish-yellow. Abdomen black. A very rare species.

GENUS *POCADICNEMIS* SIM.

In this genus the hairs and spines upon the legs are long and rather robust. The posterior extremity of the sternum is very broad. The eyes of the posterior row are equidistant.

Pocadicnemis pumilus Bl. (*Walckenaera pumila* Bl.)
Length. Male 1.5 mm., female 1.7 mm.
The palpal organs of the male are complex, and have connected with their structure a long coiled spine. Blackwall states that this species is found under stones in meadows. It is not common.

GENUS *ARAEONCUS* SIM.

The spiders included in this genus bear a strong resemblance to those of the genus *Troxochrus*. The following characteristics of the present genus, however, will separate them without difficulty. The distance between the posterior central eyes is never less than that between one of them and the adjacent lateral eye. The spines on the tibiae of the fourth pair of legs are not near the middle of the joint, as they are in *Troxochrus*. The integument of the abdomen is not coriaceous.

Araeoncus humilis Bl. (*Walckenaera humilis* Bl.)
Length. Male 1.5 mm., female 1.7 mm.
The caput of the male is elevated and rounded, but has no distinct eminence. The anterior central eyes are small and close together.

GENUS *MINYRIOLUS* SIM.

This genus may be distinguished from *Panamomops*, to which it is closely allied, by the eyes of the posterior row being equidistant and close together.

Minyriolus pusillus Wid. (*Walckenaera pusilla* in " Spiders of Dorset.")
This very rare spider is extremely minute, the adult male scarcely exceeding one millimetre in length.

(To be continued.)

TRICENTENARY OF TYCHO BRAHE.—The Royal Academy of Sweden intend holding a special meeting on October 24th to commemorate the three-hundredth anniversary of the death of the founder of modern practical astronomy.

MECHANICS OF CONDUCTION OF SAP.

By HAROLD A. HAIG.

AS we trace upwards through the vegetable kingdom the various modes of life of different plants, from the unicellular organism low down in the scale to those of higher grades of evolution, we find a gradual differentiation of the methods whereby the fundamental life-factors—viz. absorption, respiration, nutrition, and excretion—are performed ; and also of the agents that are, during these processes, called into requisition.

The first of these, absorption—including conduction—is one in which we can find, perhaps, the most complete variation as regards the accessories that are brought into play ; but here again, as in most of the others, we can recognise the same underlying principles. What strikes one most is, in fact, the obvious division of labour that obtains in the higher plants, a division which requires that many tissues shall take part in the same common work, whereas in the lower organisms the whole of it may be thrown on to one tissue, or even on to one special portion of a single cell.

I propose here to study the mechanism of absorption and conduction of that most important element in the higher land-plants—namely, the sap—the word " sap " meaning that portion of the more fluid constituents of a plant whereby all the inorganic and most of the organic food-materials are conducted and distributed.

We may, for convenience, class this fluid into two distinct portions—the first of these being the " raw sap," that part which is taken in and conducted upwards by the root and stem ; the other part being that which forms the " elaborated " sap that is derived from the leaves and other organs of the plant in which assimilation is going on. It is well to bear in mind that this distinction is in no wise meant to trespass on the chemical nature of these two, the intention being only to separate them for greater simplicity of description.

We now proceed to the consideration of the manner in which the raw sap enters a plant ; to do this we must first of all examine with the microscope the delicate roots of some seedling (¹). Under a low power we can easily make out, a short way from the tip of one of these roots, a portion that is covered by a great number of hair-like processes. Under a higher power (600–800 diams.) these processes are seen to have a definite structure, and to be prolongations of the cells of that outermost layer of the root, the " epiblema," which corresponds to the epidermis of the stem. It is not

(1) These roots must be well washed, in order to detach particles of adherent soil. When this has been done they may be examined microscopically in the usual way.

every epiblemal cell in this area that sends out processes, or " root-hairs," like this, as may be evinced from the fact that many of them are quite like ordinary epidermal cells, though perhaps somewhat more elongated. Each root-hair may be thus seen to consist of a delicate cell-wall and a very delicate layer of parietal protoplasm that sends " bridles " across from one side to the other, and a well-marked nucleus somewhere in the hair, or may be in the originating cell. With a careful preparation the characteristic " rotation " of the protoplasm may be seen, but it is slow. This structure thus identified, we at once proceed to the most interesting and essential factors in the phenomenon of absorption.

We know from physical principles that a fluid of a certain degree of concentration enclosed in a tube, one end of which is closed by a membrane permeable to that fluid, will, if placed under these conditions in a beaker of the same fluid, but of less concentration, take in a certain volume of liquid until an equilibrium is set up. Further, if the open end of our tube is connected with a manometer, we can measure the pressure that arises from this infiltration. It is, however, probable that at the same time some of our fluid in the tube will have escaped through the membrane in an opposite direction. Nevertheless, since in this case infiltration is so much more rapid than filtration in the opposite direction, we get a certain pressure, and in the cell constitutes the so-called " turgidity." In fact, there is a powerful force at work which, as it were, draws in the fluid of lesser concentration ; and it depends entirely on the difference in concentration whether fluid shall proceed in the one or the other direction. The process in hand is known as " osmosis," and it is this that determines the inflow into the root-hairs of the dilute watery solution of earthy salts, otherwise the raw sap. As a matter of fact, something does escape into the soil from the sap inside the walls of the root-hair, and so we have not in this case what physicists know as a " semi-permeable membrane "—namely, a membrane that allows a certain substance in solution to pass in one direction only. It is, however, a very near approach to this condition, and, as we shall see, there is something—namely, the " power of selection "—that acts in much the same manner. This selective power depends upon the layer of parietal protoplasm, or, as it has been often called, the " primordial utricle " (Von Mohl), some substances in solution being allowed to pass through the cell-wall, but not through the protoplasmic lining ; some others through this latter as well. In this manner a most efficient apparatus is afforded for the taking-in of those materials which are of the utmost importance in the· nutrition of plants—namely, salts of sodium and potassium, and of some of the other metals.

(To be continued.)

IRISH PLANT ·NAMES.

By John H. Barbour, M.B.

(Continued from page 174.)

Caprifoliaceae.

· Baine-gamnac.· gamnac, "a stripper "; baine, "milk," " stripper's milk." Cas fa cran. cas, "twined"; fa, "to"; cran, ":tree"; Duilleur Feitlean, Featlead. Feleog. Felix. All words having reference to "clinging." Iadslat. Iadad, ":I surround "; slat, "switch," " surrounding rod." Lusan crois. "the crowning ·herb "; Taitfui-lean. "pleasure holly." *Lonicera periclymenum,* woodbine.

Ceiriocan. ceir, "berry"; ioca, "healing." *Viburnum opulus.* water elder, guelder rose.

Cran dromain or Cran tromain.· Trom cran trium. tron and trom,·"elder "; ruis, "letter R." *Sambucus nigra.* elder.

Craob-dromain. Malabar. Malabard. Mu-laburd. latter three names are but modifications and have reference to its being second to the former tree, in that " bar " means a son. *Sambucus ebulus.* dane's-blood. dwarf elder. dane's-blood, because it was supposed to grow from the blood of fallen Danes.

Rubiaceae.

Madar, Madra. "dog." *Rubia peregrina.* madder.

Madra fraoig. fraoid, "skirts," " dog's skirts." Balad Enis. balad, "smell," " scent," Enis, "opening made in the warp by the gears of a loom in weaving." Cucullon. Ru. *Galium verum.* yellow lady's bedstraw.

Lusgarb. "rough herb." Airmeirig. *Galium aparine.* clivers. robin-in-the-hedge. catchweed.

Valerianeae.

Caoirin leana. caoirin, a little " berry " or " sheep "; I think it is "meadow sheep." Cartan-curaig. cartan, "charity"; curad, "champion," "knight," "a knight of charity." Castran arraig. Lus na tri ballan. tri, "three"; ballan, "teat," " shell." *Valeriana mikanii* and *V. dioica.* valerian, all-heal.

Dipsaceae.

Caba desan. a very pretty cap. *Scabiosa arvensis.* field scabious.

Leadan. hair of the head. Leadan an ucaire. liodan, "litany"; ucaire, "a cloth napper," "a fuller," "fuller's litany." Lus an fucadoir.· fucadoir, "a fuller," "fuller's herb." Taga. taca, "pin, peg, nail." Dipsacus syl-vestris. teasel.

Greim a diabail. greim, "hold," "grip." Devil's grip. Uracalac or Uraballac. ura, "contention," "strife"; calla, a "hood," "cowl"; callac, a "boar"; ballac, "spotted," possibly "spotted strife," referring to its appearance. Odarac mullac or Uarac mullaig, both probably same words and meaning, "dun head " or "dullish coloured head." *Scabiosa succisa.* devil's-bit scabious.

Compositae.

Noinin. noinin, "a little noon." *Bellis perennis.* daisy.

Cadltib. Gnablus. cnaub, "hemp." Liatlus roid. *Filago germanica.* chafeweed, cud-weed.

Ellea. Ailgean. "noble offspring." Meacan uillean. *Inula helenium.* elecampane.

Marb droigean. marb, "dead," "heavy," W. marw. droigean, "deep." Sceacog muire. seçao, ": briar," " Mary's briar." *Bidens cernua.* nodding bidens.

Scatog muire. scatog, "blossom." "Mary's blossom." *Bidens tripartita.* water-bur. mari-gold.

Atairtalmiun. atàir, "father"; talmiun, "of the ground.". Eartalmiun. ear, "head." "earth-head." *Achillea millefolium.* yarrow.

Cruaid lus. cruaid, "hard." " severe." "the callous plant." Meacan-ragaim or Meacan roibe. ragaim is the name, also roibe. meacan, "a taproot." *Achillea ptarmica.* sneezewort.

Coman mionla. coman, "a shrine," "a holy shrine." *Anthemis nobilis.* chamomile.

Bilic cuige. bilic, "a tuft"; cuige, "secret." Ditein a tare. Liatan. Buacalan buide possibly means "yellow bleacher," but buacalan may stand for buafanan, "a toad," "yellow toad." *Chrysanthemum segetum.* goulans, goldins.

Easpuig ban or Easpuig speain. white bishop. Noinin mor. mor, "great"; "greater daisy." *Chrysanthemum leucanthemum.* ox-eye daisy.

Mead duac. proper drink for fever. *C. par-thenium.* feverfew.

. Eansead. Lusna ffranc. "French herb." *Tanacetum vulgare.* tansy.

Buacalan or Buafanan ban or Liat. buafan, "a young toad"; ban, "white"; liat, "grey," "hoary," "white or hoary toad." Liatlus. grey plant. Mongan measga. Mugard. mugan, "a mug." *Artemisia vulgaris.* mugwort.

Adan. adan, "a cauldron"; adancium, "to kindle." Billeog an or Duilleur spoinc. spuinc, "claw"; "claw-like leaf." Cluais liat. "grey ears." Fatan or Fotanan. *Carduus (?)* "thistle."

(To be continued.)

NOTICES BY JOHN T. CARRINGTON.

Text-Book of Vertebrate Zoology. By J. S. KINGSLEY. viii + 439 pp., with 378 illustrations, 8¾ in. × 5½ in. (London: George Bell & Sons; New York: Henry Holt & Co. 1900.) 12s. net.

Following the aphorism that observation and uncorrelated facts do not make science, Professor Kingsley has produced an addition to the already numerous text-books on Vertebrate Zoology. The

national body has fixed a uniform system which we may all adopt. The figures forming the illustrations are generally well drawn and instructive, the subjects being chosen with discretion.

A Primer of Astronomy. By Sir ROBERT BALL, LL.D., F.R.S. viii + 183 pp., 7 in. × 4½ in., with 11 plates and 23 diagrams. (Cambridge University Press. 1900.) 1s. 6d. net.

Sir Robert Ball has placed before his readers a primer on the subject of astronomy, which will be found admirable for those who have not yet entered upon the pleasures of its study. Sir Robert's lucid style and easy diction render his pages so simple as to appeal without difficulty to the popular mind. The fifteen chapters are divided into numbered paragraphs, which render reference easy. The plates are from subjects obtained from the best sources, and are well reproduced. By permission of the publishers we give, as an ex-

THE FULL MOON. Age 14 days, 8 hours. *Photo. at Lick Observatory.*

(*From Sir Robert Ball's "Primer of Astronomy."*)

first part of his volume is devoted to an outline of morphology of vertebrates, based upon embryology, and the remainder presents an outline of the classification of vertebrates—a subject, as Professor Kingsley reminds us, that has in recent years been too much ignored in college work. For this purpose reference is made to fossil as well as recent forms, since the existing fauna of the earth should be studied in the light of the past. In the classification the author differs in some points from other students, and in regard to nomenclature he prefers to keep well-known generic names in spite of the law of priority. There is some comfort in this—at any rate, till some recognised inter-

ample of the illustrations, a picture of the moon at full, as photographed at the Lick Observatory. This little work is sure to largely increase the popular knowledge of celestial bodies.

Michael Faraday. By SILVANUS P. THOMPSON, D.Sc., F.R.S. ix + 308 pp., 7½ in. × 5 in., with portrait and 22 illustrations. (London, Paris, New York, and Melbourne: Cassell & Co., Ltd. 1901.) 2s. 6d.

This constitutes a volume of the "Century Science Series." Professor Silvanus Thompson has been successful in giving in his pages a history of the life and work of this eminent scientific worthy.

Of course many biographical works on Faraday have already appeared, and the author has had not only the advantage of them for reference, but also permission of the Royal Institution to take hitherto unpublished extracts from Faraday's note-books. Further, he has examined a number of his private papers, by permission of Miss Jane Barnard, and some extracts from these appear in the pages before us. The whole book is exceedingly entertaining and pleasantly written.

In Nature's Workshop. By GRANT ALLEN. viii + 240 pp., 7½ in. × 5 in., with 100 illustrations by F. ENOCK. (London: Geo. Newnes, Ltd. 1901.) 3s. 6d.

On examining this book we are reminded of the hand that is passed away, and the many pleasant chapters written by the late Grant Allen. His popular writings on natural science subjects always commanded attention, even if in some cases accuracy was sacrificed to the picturesque. Still such works are exceedingly useful in drawing attention to the odd sides of Nature. The addition of Mr. Enock's beautiful drawings to the pleasant pages of this book makes it one to be recommended to the general reader, who prefers to be amused while instructed.

Modern Astronomy. By HERBERT HALL TURNER, F.R.S. xvi + 286 pp., 7½ × 5 in. With 30 illustrations. (London: Archibald Constable & Co., Ltd., 1901.) 6s. net.

This is an account of the revolution of the last quarter-century in instruments, in methods, and in theories. It is an expansion of three lectures given at the Royal Institution in February, 1900, and is written with an accuracy worthy of the Savilian Professor of Astronomy at Oxford. It is given in such a manner that all may understand, so that it should be a welcome addition to the shelves of every student and of every public library.—*F. C. D.*

Inorganic Chemistry. By RAPHAEL MELDOLA, F.C.S. Revised by J. CASTELL EVANS, F.I.C. xvi + 320 pp., 7 in. × 4½ in. Illustrated by 36 figures. (London: Thomas Murby. 1900.) 2s. net.

The fact that Professor Meldola's work has reached a fifth edition speaks well for its use amongst the students of schools. It made its first appearance more than twenty years ago. The present edition has been brought up to date by Mr. J. Castell Evans, a co-worker with Professor Meldola at the City and Guilds of London Institute and Technical College. It is somewhat changed in the sections treating of the metallic elements. These have been arranged in accordance with Mendelejeff's law.—*H. M. R.*

Modern Chemistry. By WILLIAM RAMSAY, LL.D., D.Sc., Ph.D., F.R.S., F.C.S. 327 pp., in 2 vols. Vol. I., Theoretical Chemistry; Vol. II., Systematic Chemistry. (London: J. M. Dent & Co. 1900.) Is. per vol.

If the publication of text-books be any criterion as to the popularity of any subject, we may safely assume that the growth of chemistry is at last becoming real; for not only does the issue of new works on the subject show no signs of abatement, but there is, on the contrary, a steady increase. Moreover, the increase is not altogether devoted to advanced theories on the science, but more to a consolidating and retrospective treatment, which is further enhanced by the introduction of the essentially new ideas which took their inception

from the masterly applications of Van t'Hoff and Ostwald. As one of the foremost of English chemists, Professor Ramsay's work is naturally regarded with interest. It would be invidious to describe it as good. Clear and concise, it bears on every page the impress of a master-hand. The enormous amount of fact and theory condensed into two such small volumes is hardly noticed, so flowing and well organised is the language, until one pauses to review his reading of each chapter. The whole treatment of the subject is refreshing, for Professor Ramsay leaves the beaten track followed by the majority of former writers. From the standpoint of the periodic law and the reactions of ions he impresses on the reader the importance of the acidic as well as of the basic radicles. The ionic hypothesis should now be no stumbling-block to the average student. Its importance to the proper understanding of chemistry is so self-evident that it is hard to understand why it was, till within recent years, considered both impractical and too advanced. The printing is excellent, and the work is much embellished by the two very good portraits of Dalton and Boyle which appear as the frontispieces to the two volumes. One or two manifest errors, such as "chloride" for "chlorate" at the bottom of page 117 of the first volume, and the formula "Mt₃" for nitrogen given on page 32 of vol. ii., will no doubt be corrected in the next edition. It would be of advantage, too, if the "equals" sign were printed in thicker type, as it would not then be confused with the expression of a dyad bond.—*H. M. R.*

Introduction to Modern Scientific Chemistry. By Dr. LASSAR-COHN. Translated by Professor Patterson-Muir. viii + 348 pp., 7½ in. × 5 in., with 58 figures. (London: H. Grevel & Co. 1901.) 6s.

Dr. Lassar-Cohn is already known to the majority of English general readers through his "Chemistry in Daily Life." The present work will not appeal so directly to so large a class as did the former, yet none the less its educational value is sure to be considerable. As the author remarks in his preface, "the book can be followed easily by anyone who takes a serious interest in natural science," while the younger of our teachers of chemistry will undoubtedly be aided by the presentment of the various branches which is here adopted. The only points calling for alteration are the illustrations, which could be vastly improved, and the outer cover rather belies the word "Modern."—*H. M. R.*

History of Chemistry. By Prof. LADENBURG. Translated by Dr. LEONARD DOBBIN. xiv + 373 pp. (Edinburgh: The Alembic Club. 1900.) 6s. 6d. net.

We are reminded by Professor Ladenburg of the earlier days of chemistry, when modern theories were in their infancy, old theories were "dying hard," and the discussions of questions were not carried on in so scientific a spirit as is usually met with in these days. Dr. Dobbin's translation is good, while Professor Ladenburg's remarks which wind up each chapter are valuable, and serve to broaden the interest that the reader has already begun to feel. The author's remark as to the priority in the discovery of oxygen, and his statement in this connection that "Lavoisier repeatedly tried to appropriate to himself the merits of others," are of great interest to Englishmen, whose claims on behalf of Priestley have not always been so openly acknowledged by our Continental fellow-workers.—*H. M. R.*

PHOTOGRAPHY

CONDUCTED BY B. FOULKES-WINKS, M.R.P.S.

EXPOSURE TABLE FOR MARCH.

The figures in the following table are worked out for plates of about 100 Hurter & Driffield. For plates of lower speed number give more exposure in proportion. Thus plates of 50 H. & D. would require just double the exposure. In the same way, plates of a higher speed number will require proportionately less exposure.

Time, 10 A.M. to 2 P.M.

Between 9 and 10 A.M. and 2 and 3 P.M. double the required exposure. Between 8 and 9 A.M. and 3 and 4 P.M. multiply by 4.

SUBJECT	F. 5·6	F. 8	F.11	F. 16	F. 22	F. 32	F. 45	F. 64
Sea and Sky ..	$\frac{1}{305}$	$\frac{1}{150}$	$\frac{1}{120}$	60	$\frac{1}{40}$	$\frac{1}{10}$	$\frac{1}{6}$	$\frac{1}{4}$
Open Landscape and Shipping	$\frac{1}{150}$	$\frac{1}{64}$	$\frac{1}{32}$	$\frac{1}{16}$	$\frac{1}{8}$	$\frac{1}{4}$	$\frac{1}{2}$	1
Landscape, with dark foreground, Street Scenes, and Groups ..	$\frac{1}{32}$	$\frac{1}{16}$	$\frac{1}{8}$	$\frac{1}{4}$	$\frac{1}{2}$	1	2	4
Portraits in Rooms ..	2	4	8	16	32	—	—	—
Light Interiors	4	8	16	32	1	2	4	8
Dark Interiors	$\frac{1}{4}$	$\frac{1}{2}$	1	2	4	8	16	32

The small figures represent seconds, large figures minutes. The exposures are calculated for sunshine. If the weather is cloudy, increase the exposure by half as much again ; if gloomy, double the exposure.

ROYAL PHOTOGRAPHIC SOCIETY.—The annual general meeting was held on Tuesday, February 12th, at 66 Russell Square, London, when the result of the election of officers for the year was made known. Mr. Thomas R. Dallmeyer was elected President. The Vice-Presidents are the Right Hon. the Earl of Crawford, Messrs. Chapman-Jones, J. W. Swan, and General Waterhouse ; Treasurer, G. Scamell ; members of the Council, A. Cowan, T. Bolas, F. A. Bridge, E. J. Wall, J. A. Hodges, W. Thomas, J. Spiller, H. Snowden-Ward, G. B. Wellington, A. Mackie, J. W. Marchant, R. Meldola, J. A. Sinclair, E. S. Shepherd, Rev. F. C. Lambert, H. V. Hyde, W. B. Ferguson, P. H. Emerson, T. Bedding, and C. H. Bothamley. The following gentlemen were also elected to act as judges in the Technical and Scientific Section of the Autumn Exhibition :—General Waterhouse, Messrs. Chapman-Jones and E. J. Wall. The judges for the Pictorial Section will be Colonel Gale ; Messrs. G. A. Storey, A.R.A., P. H. Emerson, J. B. Wellington, and F. M. Sutcliff. The Council is strong, and under its guidance the Society will doubtless make the same satisfactory progress it has done during the past year or two. We must, however, regret that Messrs. W. E. Debenham and J. J. Vezey are no longer on the

Council. The meetings for March are :—5th, Lantern meeting, an illustrated lecture on "Rome," by Mr. W. B. Ferguson, K.C., M.A. ; 12th, Dr. Harting will read a paper on "Recent Rapid Lenses"; 19th, a paper by Mr. Chas. B. Howdill, A.R.I.B.A., entitled "Photographing Stained Glass by the three-colour Process"; 26th, "Some Improvements in Optical Projection," by Mr. J. H. Agar-Baugh.

PHOTOGRAPHIC DIARY.—We have just received a copy of Wellcome's "Photographic Exposure Record and Diary for 1901," published by Messrs. Burroughs, Wellcome & Co. It is produced in very attractive pocket-book form, and contains much useful information. Development with the different "Tabloid" developers is very fully treated in a thoroughly practical way, showing clearly the many advantages claimed for this method of development. We can speak from experience of the great convenience of this system of carrying the developer, &c., when on tour; also of the accuracy in the proportions of the various chemicals used. Amongst other information, there is a most useful series of "Tables of Exposure," dealing with light values, lens aperture, subject-matter, and plate speeds. These are arranged in a manner somewhat similar to that at the beginning of this article. The section devoted to recording the number of

plates exposed is neatly arranged with the following headings : No. of Slide, Plate, Subject, Date, Time of Day, Light, Stop, and Exposure. There are also several pretty little illustrations as samples of subject-matter. We do not know of a more useful note-book and guide to exposure than this, and we would recommend all our photographic friends to obtain a copy.

HAMPSTEAD SCIENTIFIC SOCIETY. — In the spring of last year the Council of this Society authorised the formation of a photographic section. The first Hon. Secretary of this section was Mr. F. Lubbock Jermyn, but as he has recently been obliged to retire, his place has been taken by Mr. Philip Joshua. Several indoor meetings have been held, when demonstrations have been given on photographic subjects. Mr. T. Manly gave a lecture in October on "Ozotype," the new printing process, and Mr. F. L. Jermyn on the History of Photography. Two outdoor excursions also took place—one to Tring and Ashridge, and the other on Hampstead Heath.

PHOTOGRAPHY FOR BEGINNERS.

By B. Foulkes-Winks, M.R.P.S.

(Continued from page 278.)

Section I. Cameras (*continued*).

View-Finders.—The camera should be fitted with properly adjusted view-finders, one to show the picture horizontally and one vertically. There are three distinct makes of view-finders in general use, viz., the ground-glass finder, the "Brilliant" or concave lens finder, and the "Real Image Brilliant" finder. Of the three patterns the ground-glass finder is the most perfect instrument, and can easily be made to give an absolutely correct representation of the picture, as it will be seen on the negative. Were it not for the difficulty of seeing the image on the ground glass in a bright light, it would be far and away the most perfect finder yet invented. It has, unfortunately, the very serious drawback that in certain lights it is almost impossible to see the picture in the finder. This, however, is overcome to some extent by sinking the finder in a kind of well, or by supplementing the finder with a rising hood, so as to prevent the light falling directly on to the ground glass. The concave-lens finder we do not recommend, as it is most difficult to judge the exact amount of picture embraced, or whether the same picture seen in the finder will be on the resulting negative. The cause of this is that the angle of view varies according to the distance of the eye from the finder, also the view included alters according to the position of the eye with regard to right and left. The third type of finder, viz., the "Real Image Brilliant," if correctly made, is one that should embrace the advantages of both the former. It is constructed solely of convex lenses, and therefore gives a real image which is not movable, as is the case with the concave-lens finder. It is also very brilliant, and the picture can be seen in the brightest light. Those with three lenses can be so adjusted as to represent the angle of view of any particular lens. There are, however, many of these finders on the market which do not represent the picture given by the lens with which they are sold, and are therefore very misleading. Under these circumstances we strongly recommend that the ground-glass finders should be selected, except where the triple convex-lens Brilliant finder can be secured, and can be relied upon as having been properly adjusted to represent the picture given by the lens upon the plate in conjunction with which it is to be used.

Folding Pocket Cameras.—The No. 3 Folding Pocket Kodak shown on page 277 is for $3\frac{1}{4} \times 4\frac{1}{4}$ rollable films—that is, the ordinary plate size. It will take spools of 12, 6, or 2 exposures; measures only $1\frac{5}{8} \times 4\frac{3}{8} \times 7\frac{3}{4}$; weight, 22 ounces. It is fitted with a five-inch rapid rectilinear lens, Brilliant finder, and focussing arrangement; has three stops, working at about F. 8, F. 11, and F. 16. The whole camera is covered in black leather, with nickel bright parts. The shutter is an ever-set one, giving both time and instantaneous exposure. A very similar camera, but with single lens, is made in the following sizes: $3\frac{1}{4} \times 2\frac{1}{4}$, $3\frac{1}{4} \times 3\frac{1}{4}$, and $4\frac{1}{4} \times 2\frac{1}{4}$. Any of these cameras, when folded, can easily be carried in the pocket.

The No. 2 "Bullet Kodak" (see p. 277, *ante*) is a fair type of the box form of hand-camera for roll-able films. The one illustrated is for taking pictures $3\frac{1}{2} \times 3\frac{1}{2}$; it can also be adapted for taking plates $3\frac{1}{4} \times 3\frac{1}{4}$, if this is desired. All that is necessary is to procure some double-plate holders, made for this purpose by the Kodak Company. The camera is fitted with time and instantaneous shutter, and single achromatic lens of $4\frac{1}{2}$-in. focus with three stops. This camera is made in two qualities, and also in 5 × 4 size, the best quality of which is fitted with rapid rectilinear lens, iris diaphragms, time and instantaneous shutter, also rackwork focussing arrangement. It has two Brilliant finders, and can be used either for roll-able films or plates. There is also the folding "Cartridge Kodak," of a slightly different type of camera, although built very much upon the same lines, but better finished and a more complete instrument. The No. 3 is for pictures $3\frac{1}{4} \times 4\frac{1}{4}$; the No. 4 is for pictures 4 × 5 ; and the No. 5 for 7 × 5 pictures. These are all fitted with best rapid rectilinear lens, time and instantaneous shutter, iris diaphragms, rising, falling, and sliding fronts, two finders, rack focussing, and socket for tripod screw. They are very compact, and most suitable for carrying on a bicycle. The "Cartridge Kodak" can be fitted with a wide angle lens if required. This will be found a very useful addition, and should always be carried. The other makes of cameras of this firm are the "Panoram":—No. 1, for panoramic pictures 7 × $2\frac{1}{4}$, capacity six exposures; No. 4, for panoramic pictures 12 × $3\frac{1}{2}$, capacity five exposures. The No. 3 and No. 4 Zenith Cameras ; the No. 3 is for $\frac{1}{2}$-plate pictures, and the No. 4 is for pictures 5 × 4.

We show an illustration of the No. 1 "Panoram Kodak" for pictures 7 in. × $2\frac{1}{4}$ in. It will be understood from this measurement that the size of

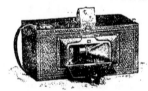

No. 1 "Panoram Kodak."

the picture renders this camera exceedingly useful for such subjects as river scenery and open landscapes ; also for very open street views, such as public buildings, squares, etc. The twelve-exposure spool made for the " No. 1 Folding Pocket Kodak " is used for this camera, which will give six exposures just double the size of the F.P.K. picture. The exposures are made by the lens rapidly moving across the field of view, thus acting as the shutter with which two speeds can be given.

The firm also make a Cartridge Film Roll Holder which can be easily fitted to almost any existing camera without interfering with the camera or plate-holders. It carries a spool of film sufficient for twelve exposures, and forms a very useful addition to any camera to which it can be fitted. They are made in vertical and horizontal forms ; the Vertical Roll Holder is the one that is usually adapted to field cameras and most hand-cameras ; but for some forms of hand-cameras, such as the "Zenith," the horizontal will be found the more convenient. The prices of "Kodaks" range from 5s. to £7 7s.

We have to express our thanks to Mr. Harold M. Read for having undertaken in this volume the honorary departmental editing of the subject of Chemistry. Mr. Read now retires in consequence of his professional work fully occupying his time. We are pleased to announce that Mr. C. Ainsworth Mitchell, B.A. (Oxon.), F.I.C., F.C.S., Member of Council, Soc. of Public Analysts, has kindly undertaken the duties of Honorary Departmental Editor for this section, and will commence his duties with the April number.

CAPTAIN ABNEY'S invention of an apparatus for measuring the luminosity of colours will probably have an important effect on the progress of photography in colours. The result is attained by rotating discs.

THE theory, founded upon researches on the ocean floor during the "Challenger" expedition, that chalk is absent in great depths has been disturbed through the discovery by the Prince of Monaco of calcareous mud at soundings of 600 to 3,000 fathoms, or 18,000 feet below the surface.

ARRANGEMENTS are being made for creating another National Park in North America. The district chosen is of great anthropological interest, being the region of New Mexico so rich in Communal dwellings and their remains. One of these groups is said to contain no less than fifteen hundred rooms.

WE would draw attention to the greatly improved tone of the "Entomologist's Record." The two last numbers are most interesting to those who study entomology. The January number contains a masterly summary by various writers on the entomological work of the past century. It is notable that the science of entomology has been raised from one subject to sneers, to a place among the most scientific of studies.

THE Provisional International Committee appointed at the Conference held in London in June, 1900, to undertake the preliminary work of publishing a complete catalogue of all the scientific literature of the world has recently issued its report. It is proposed that the annual cost of a set of seventeen volumes shall be £17. Several countries have already subscribed for sets on this basis. The United States of America takes sixty-eight copies.

M. ADOLPHE CHATIN, the well-known French botanist, died on January 13th at Essarts-le-Roi, near Rambouillet. His contributions to science have been of great value in helping to revive the study of plant anatomy, which now occupies an important position as a factor in the evolution of a natural system of plant classification. M. Chatin did not, however, confine himself to the study of botany. He published papers on the results of some important investigations, especially on the occurrence of iodine in air and water.

HASLEMERE has been selected for the next annual congress of the South-Eastern Union of Scientific Societies, which will take place in the first week in June.

PROFESSOR P. G. TAIT, author of "The Unseen Universe" and one of the founders of the meteoric hypothesis in astronomy, has been for some time past engaged on a biography of his friend Lord Kelvin. Professor Tait is retiring from the Chair of Natural Philosophy in Edinburgh University, so will have greater leisure for extending his general scientific work and conducting his literary labours.

LEPIDOPTEROLOGY is the poorer by the loss through death, after a long and painful illness, of Herbert Williams of Southend-on-Sea. He was a painstaking investigator, and successful in rearing British butterflies through all their stages. At one period he was honorary secretary to the South London Entomological Society. Mr. Williams was only 32 years of age when he died, on January 5th.

WE have received a copy of the "Natural Science Gazette" or Journal of the Streatham Science Society. It is a remarkable literary production, edited by Mr. Fred. G. Palmer, being all in that gentleman's handwriting. The cover is illustrated also by his pen. The amount of work necessary to produce each number must be indeed great, and shows much enthusiasm on the part of the Society and its editor.

THE last meeting of the Entomological Club was held on January 15th at the Holborn Restaurant, Mr. G. H. Verrall being in the chair. Mr. Verrall, who was then President of the Entomological Society of London, afterwards entertained some sixty or so of the leading entomologists of this country at supper. These charming meetings, when at Mr. Verrall's invitation, are too well known to describe; but it ought to be stated that they are among the most valuable and pleasant factors that at present exist for encouraging entomology.

THE eminent European entomologist, Baron Michel Edmond de Selys-Longchamps, was born May 25th, 1813, in Paris, and died in December last, at his son's house in Liége, where he was visiting. His literary productions were considerable, a list of over 250 articles having been printed. His chief work was among the neuroptera and orthoptera, though other subjects relating to birds, mammals, fishes, and reptiles were also considered, as well as scientific agriculture. He was an honorary member of many of the societies devoted to natural science throughout the continent of Europe.

AMERICAN women are becoming recognised as workers at the biological laboratories of Europe and the United States of America. In the latter country a society of considerable importance exists for maintaining women at the zoological stations of Naples and Woods Hall, also for encouraging ladies to conduct scientific research. The association offers a prize of $1,000 or over £200 for the best thesis presented by a woman on a scientific subject, embodying the results of her independent laboratory research, in any branch of biological, chemical, or physical science. It must be in the hands of the secretary before December 31st, 1902, who is Florence M. Cushing, 8 Walnut Street, Boston, Mass.

PROFESSOR G. H. BRYAN, F.R.S., has been awarded a gold medal for his paper, read last year before the Institution of Naval Architects, upon "Bilge Keels."

THE extensive museum of curiosities and natural science exhibits, with the building containing them and surrounding ground, near Forest Hill, in the southern suburbs of London, formed by Mr. Frederick John Horniman, M.P., has been presented by him to the London County Council, with other property as an endowment.

ON January 21st there died Elisha Gray, the eminent American electrician, who succumbed to heart disease. He was inventor of the telephone, electro-harmonic telegraphy and telautography, also lately renowned for researches on submarine signalling by sound. Elisha Gray was a self-educated man, and originally a working engineer.

AT a meeting of the Royal Meteorological Society, held on February 20th, Mr. E. Mawley presented his report on the phenological observations for 1900. During the greater part of the winter and spring the weather proved cold and sunless, but in the summer and autumn the temperature was as a rule high, and there was an unusually good record of bright sunshine. As affecting vegetation, the two most noteworthy features of the phenological year ending November 1900 were the cold, dry, and gloomy character of the spring months and the great heat and drought in July. Throughout the whole of the flowering season wild plants came into blossom much behind their average dates; indeed, later than in any year since 1891. Spring migrants, such as swallows, cuckoos, and nightingales, were also later than usual in visiting these shores.

THE Fifth International Congress of Zoology will be held in Berlin in August of this year, the opening meeting being on the 12th of that month. Professor Dr. K. Mobius will be President, and Professor Dr. F. E. Schulze representative of the President. The following are the names of some of those who have undertaken to give addresses at the general meetings:—Professor Dr. W. Branco, of Berlin; Professor Dr. O. Butschli, of Heidelberg; and Professor Dr. A. Forel, of Morges; Professor Dr. Yves Delage, of Paris, who will read a paper on "The Theories of Fertilisation"; Professor Dr. G. B. Grassi will contribute one on "The Malaria Problem from the Zoological Standpoint"; and Professor Dr. E. Poulton, of Oxford, on "Observations in Mimicry and Natural Selection." Those interested in zoology may become members of the Congress by paying twenty marks, or one pound sterling, which will also include a report of the Session. Ladies visiting in the company of a member may become Associates on the payment of ten marks, which will entitle them to attend all the meetings and receptions. Applications for tickets should be addressed to Praesidium des V. International Zoologen-Congresses, 43 Invalidenstrasse, Berlin N. 4. The programme as at present arranged includes visits to many places of interest in Berlin and elsewhere, among others to the Zoological Museums and Institute, the Geological-Palaeontological Mining and Metallurgy, Pathological, Botanical and Ethnographical Museums, the two Anatomical Institutes, the Botanical and Zoological Gardens, and the Treptow Observatory. Excursions will also be made to Potsdam, Hamburg, Heligoland etc.

THE astronomical observatory of Seeberg, near Gotha, was destroyed by fire on February 19th. This institution was known for the work done there by the astronomers Encke, Zach, and Lindenau.

THE study of the rotifera appears to be extending. We notice that Mr. Abraham Flatters, of Longsight, Manchester, is issuing, by request of students of the group, a series of photographs of species that have been described since the publication of the great work by Hudson and Goss on these interesting animals. The price appears to be exceedingly low.

ABOUT twenty leading Australian ornithologists dined in Melbourne a short time ago under the appropriate chairmanship of Dr. Charles Ryan, "Consul for Turkey." After disposing of the main business of the evening, which was an appreciation of Mr. D. le Souëf, assistant director of the Zoological Gardens of that city, it was proposed to form an Australian Union of Ornithologists, with the hope of possessing its own organ, to be entitled "The Emu." A committee was formed for carrying out the proposal.

FOR some time past most sensational paragraphs have appeared in American and some European newspapers with regard to the effect of the use in large quantities of common salt by human beings for the sake of health and longevity. These theories were attributed to some extent to Professor Jaques Lobb, of Chicago University Laboratories. It is needless to say he has denied the story, which was hardly necessary; for, as he himself says, no one there takes any notice of "science" in American newspapers. We fear, in some instances, the same may be said of some of those in this country.

WE have received intimation that a proposed scientific expedition will leave Boston, U.S.A., on June 26th and return in September. It is under the auspices of the Harvard University, but partakes of a public character, as any person with slight scientific knowledge may join on payment of $500; half on application before March 15th, and the rest by June 1st next. The steamer will visit Labrador, Greenland, and Iceland, and opportunity will occur for about three weeks' hunting on shore in Labrador and Greenland. Explanatory lectures will be given by the leader of the expedition. Applications for membership are to be made to Mr. R. A. Daly, Department of Geology, Harvard University, U.S.A.

WE may mention that we have received a copy of the "British Weather Chart, 1901," by B. G. Jenkins, F.R.A.S., being the fifteenth year of publication. By this chart we are to understand approximately the kind of weather to be expected during the coming twelve months. The author claims that his forecasts are based on some "important features resulting from a lengthy investigation as to the variation of actual barometric and thermometric readings from those predicted by calculation." Also, that "as the weight of evidence appears to indicate that the barometric movements of the atmosphere are, as a rule, about four days later than the corresponding thermometric." Mr. Jenkins' forecasts have been revised accordingly. We would suggest that, as the chart only costs 6½d., post free, our readers interested in meteorology should obtain a copy and study it for themselves. It may be had from R. Morgan, 65 Westow Street, Upper Norwood, S.E.

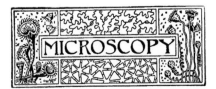

CONDUCTED BY F. SHILLINGTON SCALES, F.R.M.S.

ROYAL MICROSCOPICAL SOCIETY.—Annual meeting, January 16th, 1901, William Carruthers, Esq., F.R.S., President, in the chair. Mr. Hugh M. Leake exhibited a new form of rocking microtome, designed to cut perfectly flat sections. Dr. Hebb said it seemed to remedy certain defects of the ordinary rocking microtomes, it appeared to be easily manipulated, and was very stable and solid in construction. The President said Mr. Leake had taken great pains to bring this instrument to perfection, and it appeared to be very efficient. Dr. Hebb read the report of the Council for the year 1900, and Mr. Vezey, the treasurer, read the annual statement of accounts and balance-sheet. The President announced that the whole of the Fellows nominated for Officers and Council had been duly elected, and expressed his thanks to the Fellows of the Society for again placing him in the position which he had occupied during the past year. He congratulated the Society upon the improved conditions indicated in the report. The library had been gone through carefully, and much that was useless was eliminated. Their thanks were due to Mr. Radley for the great pains he had taken in preparing the card catalogue. Their collection of instruments had also been put into excellent order. He congratulated the Fellows upon the state of their funds. The President then read the Annual Address, which consisted chiefly of an interesting epitome of the life and work of John Ellis—known in his time as " Coralline Ellis." Mr. A. D. Michael in asking the Fellows to give their thanks to the President for his Address, said he had unearthed one of those attractive bye-paths of science which, when brought to light, so often proved to contain lessons that all might learn with advantage. Ellis, originally attracted by the picturesque side of the subject, was gradually drawn on toward the scientific, and then endeavoured to turn that scientific knowledge to the benefit of the human race. There was no field of research more enticing than that border-land which lies between the animal and vegetable kingdoms, and the steps by which the existing knowledge on this subject had been acquired were of the greatest interest. From the growth of knowledge the gap between the lowly hydrozoa and the highly-organised polyzoa seemed a wide one, but the keen insight into Nature shown by the man or men who first appreciated the differences between these very similar groups of creatures went far to show how great an observer Ellis really was.

POSTAL MICROSCOPICAL SOCIETY.—The annual report of this Society has reached us. It contains the report of the Hon. Secretary, Miss Florence Phillips, of Hafod Euryn, Colwyn Bay, the financial report, rules of the Society, and list of members. The Secretary's report contains a reference to the late Mr. Alfred Allen, of Bath, to whom the Society owed so much from its first inception, and alludes to the new step in advance taken by the Society's recent connection with SCIENCE-GOSSIP, which now regularly publishes excerpts from the note-books circulated by the members as accompaniment to their boxes of slides. The President for the year is Sir Thomas Wardle. The balance-sheet shows a satisfactory financial position, but we could wish to see the number of members largely increased. There must be many of the readers of this journal who could both receive help from and give help to a Society such as this, and though we are glad to know that several of our readers have recently joined we cannot help thinking that others are deterred by a too modest consciousness of their own shortcomings. Briefly put, the method of working is to divide the members into circuits of six members and the secretary, and to circulate amongst each circuit boxes of slides accompanied with descriptive notes, elaborate or simple, original or gathered from others, as the owner's knowledge warrants. Each member provides twelve slides per annum, and the notes for them. The slides are not always original. At the end of their travels the books and slides belong to the original owner, and the former are generally enriched by the notes and queries of the other members of the circuit. No attempt whatsoever is made at fine writing or description, or at pedantry. The Society exists solely to help and encourage, and the present writer can say frankly that in all his examination of the books incidental to the work of making the necessary selections likely to interest the readers of this journal he has not come across a single unpleasant remark or discouragement of a beginner. We make this observation thinking that there may be amateur microscopists who are deterred by some such thought from joining a Society which could help them in so many ways. The annual subscription is so small as to be within the reach of all.

FORMALIN AND ALCOHOL AS PRESERVATIVES FOR ZOOLOGICAL SPECIMENS.—The Journal of the R.M.S. briefly alludes to a discussion by Mr. J. Hornell, the well-known preparer of slides of marine zoology, as to the respective values of formalin and alcohol as preservatives for museum specimens. Mr. Hornell expresses the opinion that the best effects of formalin are seen with Medusae and Tunicata, and says that most animals should be mounted in formalin-alcohol after previous fixation. For some animals which contain lime salts, such as echinoderms and crustaceans, formalin is unsuitable, as it slowly decalcifies them and renders them very brittle. In collecting trips formalin is more useful than alcohol, as in its concentrated form it does not occupy so much space, and is, therefore, more easily stowed. In microscopical technique, maceration of the objects, sections, etc., may be obviated by the use of a two per cent. solution, either as an addition to staining solutions, or to replace pure water in washing out fixatives.

FORMALDEHYDE AS A KILLING AND FIXING AGENT.—Formalin, formic aldehyde, or formaldehyde, which are all practically the same thing, are now so generally used for killing, fixing, and preserving that the following formula, as used by Prof. T. P. Carter for killing and fixing, and given by him in the American Monthly Microscopical Journal, may be of service to some of our readers. It is as follows: Formaldehyde, 40 per cent. solution, 50 ccm. ; distilled water, 50 ccm. ; glacial acetic acid, 5 ccm. By this solution tissues are killed and fixed in from six to twelve hours. but the immersion may be continued for twenty-four hours without damage. The pieces are then transferred to 50 per cent. alcohol for one hour, and afterwards to 75 per cent. and 95 per cent. alcohol for half an hour each.

NEW $\frac{1}{12}$-INCH IMMERSION OBJECTIVE.—Messrs. Watson & Sons have sent us for inspection a new $\frac{1}{12}$-inch oil-immersion objective of their ordinary or "Parachromatic" series. The aperture is only moderate, *i.e.* 1·10 N.A., but this is no disadvantage for ordinary work, whilst it enables the cost of the objective to be proportionately reduced. The lens submitted to us was an early one, but we can speak highly of its performance. It was unusually flat in the field, its definition was excellent, and it bore high eye-pieceing well. The corrections were for the short tube-length. The cost of the objective is only £4.

NEW MACERATION MEDIUM FOR VEGETABLE TISSUE.—Herr O. Richter finds that strong ammonia will macerate vegetable tissue without injury to the cell contents, such as starch and aleurone grains, chlorophyll granules, etc. The fluid was used boiling, cold, and at a temperature of about 40° C. The rapidity of the maceration depends on the temperature.

[For further articles on Microscopic subjects see pp. 294, 296, 303 and 305 in this number.—ED. Microscopy, S.-G.]

ANSWERS TO CORRESPONDENTS.

W. D. N. (Kirkcaldy).—The condenser you allude to at 17s. 6d. is an optical part only, and would cost 32s. 6d. with mount complete. The other condenser is, I think, complete, and being made by the maker of your instrument would be preferable. I believe I answered some queries on this matter on a previous occasion. A camera lucida is used at a distance of 10 inches from the table, because that is the normal visual distance: but if the image is too large you can reduce this distance, or preferably you can use a lower power objective. It is not easy for me to advise you as to a course of work, but I sympathise with your desire to devote your time and your microscope to more definite study. As you take an interest in botany I think you could not do better than for a beginning than obtain a book on practical botany and work steadily through it, say Strasburger and Hillhouse's "Practical Botany," published by Swan Sonnenschein at 10s. 6d. A new edition has recently been issued, or you could pick up a cheap second-hand copy in the booksellers' shops in Lothian Street and George IV. Bridge, Edinburgh, for about half this amount. Try, for instance. William Bryce, 54 Lothian Street. This book is eminently practical, and contains clear elementary instructions on the use of the microscope and mounting; so would serve admirably. If you were to work through such a book. mounting your slides as you went on, you would gain a knowledge of botany and provide yourself with an occupation that would never weary, but lead you on to further interests. It is a mistake, too, to suppose that botany cannot be pursued in winter. It is equally a mistake to idly examine many diverse things, with no aim or object in view.

V. T. (Chorley).—Of the specimens of pond life which you send me, No. 1 is a Turbellarian worm.

a Planarian. The green forms to which you allude are possibly *Vortex viridis*; No. 2 is a young leech; No. 3 is a larva of one of the Nemocera, probably *Tanypus maculatus*; No. 4 is a polychaete annelid in an immature stage. For Nos. 1, 2, and 4 the readiest book of reference would perhaps be the "Cambridge Natural History," vol. ii., and for No. 3 the same work, vol. vi., pp. 468–9, or Prof. Miall's "Natural History of Aquatic Insects." You could refer also for No. 1 to the "Encyclopædia Britannica" article "Planarians."

EXTRACTS FROM POSTAL MICROSCOPICAL SOCIETY'S NOTEBOOKS.

[These extracts were commenced in the September number at page 119. Beyond necessary editorial revision, they are printed as written by the various members. - ED. Microscopy, S.-G.]

NOTES BY JOHN T. NEEVE.

MARINE ALGAE.—*Odonthalia dentata* (fig. 1). with tetraspores: this species is found on the

FIG. 1. *Odonthalia dentata* with tetraspores.

Scottish coasts, and does not occur south of Berwick, except occasionally as a derelict cast up on the shore. I have not found it myself, but have received a few specimens from Arbroath,

FIG. 2. *Rhodomela subfusca* with tetraspores.

where it grows in abundance. The frond, as its name indicates, is tooth-branched and of a deep-red colour, from four to twelve inches high. *Rhodomela subfusca* (fig. 2), with tetraspores: this plant is abundant with us in the south, and is often cast up on shore after a storm. The fronds are filamentous, rather brown in colour, drying to a black and uninteresting specimen; but the reproductive organs are beautiful, the tetraspores, like the preceding species, bearing a hand-like resemblance. They are found thus only in winter. *Harveyilla mirabilis* (fig. 3): this tiny plant grows on the previous species in the form of a very minute globose body, looking to the naked eye

of a very interesting order of sea-weeds, which I have put in to show the structure of the so-called siphons. [These would be only shown in a cross-section of stem; see Mrs. Major's note.—ED. Microscopy, S.-G.] In *P. nigrescens* and *P. atro-rubescens* tetraspores are seen, whilst *P. brodiaei* has beautiful urn-shaped cystocarps. *Callithamnion corymbosum* does not occur on our coast at Deal, but is found

FIG. 3. *Harveyilla mirabilis.*

FIG. 5. *P. atro-rubescens.*

like a fruit of its host. I have observed it in various stages of growth, and have found all three reproductive organs. The larger plants prove only to be vegetative after having fulfilled their propagation. Certain sections in the slide show tetraspores. The larger cells are part of the filament of the host. *Bryopsis plumosa* belongs to the

on the Devon and Cornish coasts, and is a most beautiful microscopic object. *Delesseria sanguinea* (fig. 7): having found the reproductive organs this winter, I am sending them in as fresh a state as possible. I am of opinion that another mode of reproduction exists, as I have found old plants with young fronds as perfect as their parent fronds were when in their summer beauty.

REMARKS.—I must apologise for having kept the box past the allotted time; but these special series are of exceptional interest to me, and the system is well worthy of imitation. In the specimens referred to, will be remarked the special inclination of the marine algae for the tetraspore or vegetative form of reproduction. When one considers the vast volume of the surrounding water one cannot help marvelling at the immense numbers of

FIG. 4. *Polysiphonia nigrescens.*

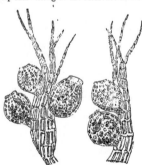

FIG. 6. *P. brodiaei.*

Chlorosperms, or green sea-weeds, and is an object of great beauty when seen in the tide pools, looking like minute green feathers. Reproduction is by the direct change of the endochrome into zoospores. *Polysiphonia nigrescens* (fig. 4), *P. atro-rubescens* (fig. 5), and *P. brodiaei* (fig. 6), are three species

antherozoids, or better pollinoids, which must be produced in order to keep up the intermittent sexual generation. Probably many millions of pollinoids are wasted for every one that fulfils its life's object, whilst every tetraspore that is detached has a good chance of becoming a perfect plant—provided, of course, that it escapes the clutches of the more highly-organised inhabitants of the ocean.—*Thomas S. Beardsmore.*

I wish I could be as successful in preserving the colour of my sea-weeds as Mr. Neeve has been. He has a great advantage in living near the sea, and thus being able to mount his specimens in fresh condition, whereas mine have always been kept some time before mounting. I cannot find *Harveyilla mirabilis* in Harvey's "Phycologia," and should like to know to which class it belongs. I suppose it has been added to the list of marine flora since that book was published. I would suggest that these beautiful slides deserve a better finish, and I trust Mr. Neeve will forgive me for making the following suggestions: i.e. the first coat of the ring should be of brown varnish or thick gold size, to secure the cover-glass, and the white zinc cement used is too thin. The mounting medium appears to be glycerine jelly.—*T. A. Skelton.*

The forms of fructification of marine algae are well exhibited in these slides, and are worthy of careful examination. I hope Mr. Neeve may contribute a further series upon this subject. Miss Phillips is to be congratulated upon her very successful illustrations. To draw microscopical preparations is excellent practice, as it helps, more than any other method, to fix in one's mind the characters and minute differences which exist between allied species. I have added a small

FIG. 7. *Delesseria* in winter.

drawing (fig. 7) from a well-known illustration of the naked-eye appearance of *Delesseria* in the winter. It may be of service to members not having such an illustration in helping them to understand from whence the leafy sporophylls which are shown in the slides are obtained. The branches are the midribs of fronds from which the membranous part has decayed or fallen away. These midribs are clothed with tufts of the sporophylls or leafy lobes containing the tetraspores. I should have liked to have added some illustrations of *Callithamnion*, but feared to detain the slides. This slide is a very pretty one, and shows well the branched feathery fronds composed of a single row of tubular cells. Tetraspores arranged in berry-like receptacles and having lateral attachments to the branches (ramuli) may be well made out with a ¼-inch or ⅛-inch objective. Antheridia may also be observed.—*J. R. L. Dixon, M.R.C.S., L.R.C.P.*

There are beautiful figures of sections of *Polysiphonia polynorpha* in Sowerby's "English Botany," showing the "siphons" alluded to.—*F. C. Major.*

[We do not think the foregoing notes call for any particular editorial comment. Figs. 1 to 6 are by Miss Florence Phillips ; fig. 7 is by Dr. J. R. L. Dixon.—ED. Microscopy, S.-G.]

NOTES & QUERIES

LARVA OF PIERIS RAPAE IN JANUARY.—On the 21st of January last I found a half-grown larva of this species at rest in our garden. Mr. Barrett mentions in his "British Lepidoptera" that "the larva has been found feeding even into December, but always passes the latter part of the winter in the pupa state." I do not recollect any record of a partly-fed larva so late in the winter as this.— *1. H. Mead-Briggs, Rock House, Lynmouth.*

WOODPECKER FEEDING ON GROUND.—Whilst watching on December 16th last the movements of a pair of woodpeckers (*Gecinus viridis* Lin.) on Wimbledon Common, near London, a friend and I were surprised to see one of the birds settle on the ground. There it searched for worms and insects, and we observed it consuming some of each. Presumably the bark of the trees did not yield sufficient food, so the bird was forced to thus deviate from its usual arboreal habits.—*John E. S. Dallas, 19 Ulverscroft Road, East Dulwich, S.E.*

SPARROWS IN FROST.—On a shady piece of gravel walk hard with frost I noticed half a dozen sparrows, both males and females, apparently trying hard to get a dust bath. They opened their wings, pressed their bodies close to the ground, and shuffled along like creeping mice for about six or eight inches at a time ; then, getting up, shook themselves, frequently repeating the process. I thought for a moment that it might be a sort of love-dance, but the cocks and hens were equally engaged, and I saw nothing to indicate that it was more than an attempt at a bath ; but as there was certainly no dust, one would think it must have been an unsatisfactory performance—*F. T. Mott, Birstal Hill, Leicester, Feb. 1901.*

NATURE PICTURES OF LEPIDOPTERA.—Referring to the last June number of SCIENCE-GOSSIP (p. 32), in which Mr. Verity asked about impressions of butterflies' wings, I recently came across the following extract from "The Home Naturalist": "Spread a thick solution of gum arabic over a sheet of fine paper, so that it readily adheres to the finger. Take off the wings of a dead butterfly and lay them carefully on the gummed paper, spread out as in the act of flying. Place another sheet of paper over the wings, and rub them gently with the finger, or the smooth handle of a knife. The wings" (? impressions) "will then be left on the paper. Afterwards draw the form of the body, with the head and antennae in the space between the wings." Another way I have seen is to gum some tissue paper all over, and also gum the page of the book, or sheet of paper, whichever one may use, placing the wings on the latter and putting the former over them. This has the disadvantage, however, of making the book bulky, as the wings are left in ; and also of not being able to see the colours as brightly as one could wish, because, however fine the tissue and clear the gum, there is always a deadening effect on the colours.—*J. H. Wright, 18 Chorley Old Road, Bolton.*

ASTRONOMY

CONDUCTED BY F. C. DENNETT.

	1901	*Rises.*	*Sets.*	*Position at Noon.* R.A.	*Dec.*
	Mar.	*h.m.*	*h.m.*	*h.m.*	° '
Sun	6	6.38 a.m.	5.46 p.m.	23.6	5.49 S.
	16	6.14 a.m.	6.4 p.m.	23.42	1.54 S.
	26	5.52 a.m.	6.20 p.m.	0.19	2.3 N.

		Rises.	*Souths.*	*Sets.*	*Age at Noon.*
	Mar.	*h.m.*	*h.m.*	*h.m.*	*d. h.m.*
Moon	6	7.31 p.m.	0.33 a.m.	6.31 a.m.	15 9.15
	16	3.46 a.m.	8.26 a.m.	1.13 p.m.	25 9.15
	26	9.36 a.m.	5.45 p.m.	1.0 a.m.	5 23.7

		Souths.	*Semi-*	*Position at Noon.* R.A.	*Dec.*
	Mar.	*h.m.*	*diameter.*	*h.m.*	° '
Mercury	6	0.15 p.m.	5·3''	23.9	1.30 S.
	16	11.7 a.m.	5·2''	22.40	6.12 S.
	26	10.32 a.m.	4·4''	22.45	8.18 S.
Venus	6	11.20 a.m.	5·1''	22.14	12.15 S.
	16	11.28 a.m.	5·0''	23.2	7.46 S.
	26	11.35 a.m.	5·0''	23.47	2.56 S.
Mars	6	11.11 p.m.	6·8''	10.8	16.0 N.
	16	10.20 p.m.	6·4''	9.56	16.44 N.
	26	9.34 p.m.	6 0''	9.48	16.58 N.
Jupiter	16	7.11 a.m.	16·9''	18.43	22.51 S.
Saturn	16	7.32 a.m.	7·4''	19.5	22.3 S.
Uranus	16	5.30 a.m.	1·8''	17.3	22.49 S.
Neptune	16	6.10 p.m.	1·2''	5.45	22.12 N.

MOON'S PHASES.

		h.m.			*h.m.*
Full	Mar. 5	8.4 a.m.	3rd Qr.	Mar. 13	1.6 p.m.
New	„ 20	0.53 p.m.	1st Qr.	„ 27	4.39 a.m.

In apogee March 9th at 10 a.m.; and in perigee on 21st at 10 a.m.

METEORS.

				h.m.		°
Mar. 1	(κ Persei)	Radiant R.A.	3.10	Dec.	45 N.
1–4	τ Leonids	„	„	11.4	„	5 N.
1–28	κ Cepheids	„	„	20.32	„	78 N.
11 to May 31	Draconids	„	„	17.32	„	50 N.
14	(ν Virginis)	„	„	11.40	„	10 N.
24	β Ursids	„	„	10.44	„	58 N.
27	(η Cor. Bor.)	„	„	15.16	„	32 N.
28	(26 Draconis)	„	„	17.32	„	62 N.

CONJUNCTIONS OF PLANETS WITH THE MOON.

						° '
Mar. 4	Mars*†	10 a.m.	Planet	9.53 N.
„ 14	Jupiter†	11 p.m.	„	3.25 S.
„ 15	Saturn*	9 a.m.	„	3.26 S.
„ 19	Mercury*	6 a.m.	„	3.47 S.
„ 19	Venus†	11 p.m.	„	6.31 S.
„ 31	Mars	3 a.m.	„	8.56 N.

* Daylight. † Below English horizon.

OCCULTATIONS, AND NEAR APPROACH.

				Angle		*Angle*
		Magni-	*Dis-*	*from*	*Re-*	*from*
Mar.	*Star.*	*tude.*	*appears.*	*Vertex.*	*appears.*	*Vertex.*
			h.m.	°	*h.m.*	°
2–3	κ Cancri	5·0	10.46 p.m.	116	0.3 a.m.	267
25	l Tauri	5·5	6.32 p.m.	58	7.42 p.m	245
26	71 Orionis	5·1	11.11 p.m.	150	Near approach.

THE SUN now appears in a very quiescent state, but observation should not be relaxed. Spring is said to commence at 7 a.m. on March 21st, when the Sun crosses the equator and enters sign Aries.

MERCURY is in inferior conjunction with the Sun at 3 p.m. on March 7th, after which it becomes a morning star, but is poorly placed for observation.

VENUS too near the Sun for observation.

MARS in Leo is well placed for observation all night, but its diameter is only 13·8'' at the beginning, decreasing to 11·6'' at the end of the month.

JUPITER and SATURN are near together, in Sagittarius, both morning stars rising in the southeast about three hours before sunrise.

URANUS is at a similar low declination, preceding Saturn by about two hours.

NEPTUNE must be looked for early in the evening, as it is on the meridian at sunset near the middle of the month.

COMET c 1900 was only visible with quite large telescopes on favourable occasions, its faint nucleus being less than twelfth magnitude. It must have passed its perihelion on December 3rd, 91,000,000 miles from the Sun. Its orbit is elliptical, and its period about seven years.

NEW MINOR PLANETS.—Already five have been discovered this century by Professor Max Wolf of Heidelberg, on January 9th, 16th, 17th, and 18th. Herr Carnera was assisting on the 17th at the time of the discovery. There is thought to be a mistake about those reported to have been discovered by Professor W. R. Brooks.

Two new variable stars have been discovered by Mr. R. T. A. Innes at the Cape Observatory, to be called 24,1900, Arãe, and 25,1900, Octantis. The former is remarkable for the shortness of its period, which amounts to only $0^{d.}\,3115$, or 7 h. 28 m. 34 s.; its magnitude changes from 8·9 to 9·75. The magnitude of the latter varies between 7·7 and 10·3.

THE GOLD MEDAL of the Royal Astronomical Society was this year awarded to Professor E. C. Pickering, of Harvard College Observatory, and presented to the American Ambassador, Mr. Choate, on behalf of the recipient, at the annual general meeting at Burlington House on February 8th.

MOON'S PATH.—There are some phenomena connected with the earth and moon which, as popularly understood, are not correct, and with the favour of your permission are therefore open to discussion. The moon does not describe a circular path in revolving round the earth, but an irregular zig-zag track; and, in consequence, this irregular movement is at variance with the law affecting the inertia of matter. The earth, in its annual circuit round the sun, has in round numbers a daily course of about 2,000,000 miles, and the moon has a daily course of about 56,000 miles round the earth. Now at full moon the moon is carried forward in space the sum of these two velocities, while during the last quarter it is propelled in front of the earth's path 2,000,000 miles a day, and has at the same time a lateral motion of its own of 56,000 miles a day. Again, at new moon its direction is changed, and has now a retrograde motion of 2,000,000 miles a day, less its own circuit motion of 56,000 miles. If we be traced showing the path of the moon in its orbit round the earth, it will be seen to be a zig-zag course. The only force of which we know connecting the moon to the earth is the force of gravity, which acts in a straight line between their centres. What explanation can be given of the fact of the moon being impelled at an enormous speed in front of the earth's path against the force of gravity?—*John Clark, Birmingham.*

THE real daily motions of the earth and moon are about 1,600,000 miles and 55,000 miles respectively. Mr. Clark's difficulty seems to me met by the fact that we must not regard the path round the sun as that of the earth, but as that of the earth and moon. The true orbit of the system is that of the common centre of gravity of the two bodies. Both the centre of the earth and the moon travel along zig-zag courses, being now accelerated, now retarded, in their motions as seen from the centre of the system, which, by the way, whilst it is within the sun, is not the centre of the sun. The sun's direct attraction on the moon is more than twice that of our earth. The lateral motion of the moon is so slight that she is always travelling along a path concave to the sun.—*F. C. D.*

SPECTROSCOPES.—These instruments are of two classes, which are alike in having a fine slit admitting the light it is sought to analyse, and a collimating lens to render the rays of light from the slit parallel. After being spread out into a spectrum it is usual, except in pocket instruments, to have a small telescope with an astronomical eye-piece to observe the lines, if any, which are shown. The difference of the two classes is the means used to effect the dispersion of the spectrum. In one class it is a prism, or prisms, employed, and the light rays of the various wave-lengths are refracted to a varying extent, the result being the beautiful spectrum with which most of us are familiar—if not in practice, at least in book illustrations. In the other class a diffraction plate takes the place of the prisms, having upon it a series of parallel lines, some thousands to the inch. An expensive

Rowland plate, ruled at the Johns Hopkins University, may be employed, or a copy of the same prepared by Mr. Thos. Thorp, of Whitefield, Manchester. These diffraction gratings give greater dispersion of the red rays than do prisms giving a spectrum of the same length. This is shown on the diagram, where A represents the position of the principal Fraunhofer lines in the prismatic spectrum, and B the same lines in the diffraction spectrum. The spectra given by the gratings are known as "Normal," because the exact wave-length of every line may be measured; whilst measures made with the prismatic spectrum are simply arbitrary, and differ with every variety of glass used. A prismatic spectroscope needs more prisms to produce greater dispersion, but with the diffraction grating a spectrum of a different order, obtained by a slight motion of the grating, gives the dispersion required. A Thorp transmission grating, 14,500 lines to the inch, in my possession, without any observing telescope separates quite widely the D lines in the solar spectrum. A plate of this description is valued at only a few shillings; whilst prisms, to produce equal dispersion, would cost pounds. A solar spectroscope on this plan, though lighter than

a prismatic one, costs less than a quarter of the money—a consideration to many people. A grating used without a slit, but with a cylindrical lens, will show the lines in the spectra of the brightest stars, even with an instrument so small as the 3-inch Wray SCIENCE-GOSSIP telescope. Mr. Thorp has a form of grating mounted on a prism which gives a very brilliant direct-vision spectrum of great dispersion, separating the D lines well. This latter is useful in spectroscopes of the miniature class where a maximum of dispersion is desirable combined with a minimum of bulk — a class of instrument we have found most useful at times, especially for studying the spectra of Aurorae, or showing the true character of a brightening in the northern sky.

CHAPTERS FOR YOUNG ASTRONOMERS.

BY FRANK C. DENNETT.

(*Continued from p.* 281.)

JUPITER.

SOMETIMES most rapid changes have been noticed on Jupiter. With the achromatic telescope at the Dearhorn observatory its greatest diameter has been found to be not always identical with the equator, and to vary in direction some five degrees in the course of a week.

The spectrum shows the presence of watery vapour in the atmosphere of Jupiter, which however does not appear to be quite identical with our own. This will be seen by the somewhat rough copy taken from Sir William Huggins' map from the "Philosophical Transactions of the Royal Society" for May 1864.

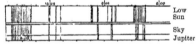

FIG. 6. SPECTRUM OF JUPITER.

Although Jupiter does not exhibit well-marked phases like Mars, if examined when near quadrature it shows a very evident shading-off of the limb farthest from the sun.

The first known observation of Jupiter was made at Alexandria about 6 a.m. on September 4th, B.C. 240, when it occulted the star δ Cancri. Castor was also occulted on November 22nd, 1716, as witnessed by Pound.

It is thought to be very probable that Jupiter, instead of being a dark body like the earth or moon, gives out a certain amount of light to its satellites.

Most useful work can be done in noting the exact time at which the various markings, bright spots, dark spots, belt ends, or broadenings, cross the central meridian. Calculate the longitude, then the exact relative motions on the planet's surface can be learnt when observation is compared with observation.

(*To be continued.*)

TOTAL ECLIPSE OF THE SUN in next May should be seen well in Sumatra, where Professor E. Barnard intends to go. An expedition is also to be sent from the University of California at the expense of Mr. W. H. Crocker of San Francisco.

GEOLOGY

CONDUCTED BY EDWARD A. MARTIN, F.G.S.

A LONDON GRAVEL SECTION.—A gravel section in London Wall, in the City of London, has been laid bare prior to building operations, by the demolition of a large number of warehouses. Beneath some eight feet of made ground was found about ten feet of dark carbonaceous clay containing freshwater shells—such as *Planorbis, Cyclas,* etc., and numerous traces of vivianite. Lower is a thin bed of broken bones, these resting on gravel to a thickness of about ten feet, below which a shaft had been sunk for some distance in London Clay. The layer of bones, which seem to have been deliberately broken out of all recognition, was extremely puzzling. To whatever cause they are to be attributed, they were laid there before the water flowed in, in which the clay was afterwards deposited. We have here, in fact, a natural deposit resting upon an artificial one. Stowe says that in Moorfields there were numerous brickfields, the brick-earth being excavated for brick-making. The water then accumulated in the low-lying spots which were thus made, these afterwards being favourite skating-grounds. We can scarcely think that the clay could have been formed in the comparatively short period since Stowe's time. I regard the layer of bones as of older date. There must have been some reason for the breaking of the bones. Perhaps the layer here exposed may represent the floor of Roman wild-beast dens, in which case there would have been ample time to allow of the deposition of the clay. Opinions on this interesting section would be valuable.—*E. A. Martin.*

INDURATED CHALK.—In reference to Mr. E. R. Sawer's remarks, *ante*, p. 201, about the induration of chalk, may I point out that beds of indurated chalk occur at Corfe, in Dorsetshire? The rock is so hard that it is used for mending the roads in some parts of the neighbourhood. The cavities which occur in this chalk contain crystals of calcite, and the same mineral may be found as veins traversing the mass. This is not a case of "contact-metamorphism." The alteration in the character of the rock was probably induced during the uplifting of the beds through heat generated by pressure. If water were present, hydrothermal action may have produced some changes; but I do not remember that there was sufficient visible evidence in support of the hydrothermal theory to the exclusion of others.—*Cecil Carus-Wilson.*

GEOLOGICAL SOCIETY OF LONDON.—At the annual meeting, held on February 15th, Mr. J. J. H. Teall was re-elected president, and Messrs. J. E. Marr, H. W. Monckton, Professor H. G. Seeley, and W. Whitaker vice-presidents of the Society. Messrs. R. S. Herries and Professor W. W. Watts remain as secretaries, and Mr. W. T. Blandford as treasurer. The report of the Council showed a total fellowship of 1,334 at the end of 1900. The Wollaston Medal was awarded to Professor Charles

Barrois, the Murchison Medal to Mr. A. J. Jukes-Browne, the Lyell Medal to Dr. R. H. Traquair, and the Bigsby Medal to Mr. G. W. Lamplugh.

DIRECTOR-GENERAL OF GEOLOGICAL SURVEY.—It is with considerable regret one hears of the forthcoming retirement on March 1st of Sir Archibald Geikie from the Director-Generalship of the Geological Survey. It is forty-six years since Sir Archibald's first official connection with the Survey, and he is still in the possession of a valuable store of energy and enthusiasm. We are not, therefore, surprised to hear that he has at present no intention of laying aside his unofficial work, and we trust he will continue to wield both pen and hammer for many years to come. Those who have heard him speak or lecture must have been struck by the originality of his descriptions, and by the intelligible manner in which he has placed his facts and opinions before the audience. It is intended, as mentioned in SCIENCE-GOSSIP last month, to entertain Sir Archibald early in March at a complimentary dinner.

INSECT WING FROM CARBONIFEROUS ROCKS.—Owing to the rarity with which insect remains have been found in our carboniferous rocks, special interest attaches to a short article in the "Geological Magazine," by Mr. H. A. Allen, describing portion of a wing, with a neuration apparently differing from any hitherto obtained. This has come to light from the top of the four-foot seam in the Lower Coal Measures of Llanbradach Colliery, Cardiff. The specimen was obtained by Mr. G. Robbings, and has been provisionally placed in the genus *Fouquea*, as *F. cambrensis.*

DEEP COLLIERY SHAFT.—A shaft is shortly to be sunk to a depth of 1,850 feet at Lyktrens in Pennsylvania, and from the bottom of the shaft a tunnel will be driven 1,200 feet long, in order to reach the anthracite seam. At present the deepest shaft in that region is at Wadesville, which is 1,600 feet deep.

UINTACRINUS IN CHALK.—The remarkable free-swimming crinoid, *Uintacrinus*, discovered twenty-five years ago in the Niobrara chalk of Kansas, and almost at the same time in the Lower Senonian of Westphalia, promises, after all, to be one of the commonest crinoidal forms in the Marsupites zone of our own English chalk. *Uintacrinus* is a stalk-less crinoid. A complete morphological description of *U. socialis* will be found in the "Proceedings of the Zoological Society for 1893" (pages 974-1004), by Dr. F. A. Bather. Since this description was written Mr. Bather has explored the cliffs east of Margate with complete success, and he finds that, next to columnals of *Bourgueticrinus*, cup-plates and brachials of *Uintacrinus* are the most common fossils. Other parts of Kent besides Margate have produced similar evidence. Mr. C. Griffith has found similar plates at Grately, near Andover. The Recklinghausen glauconiferous sandstone has been also searched by Dr. Bather, and he finds that the genus is as plentiful there as at Margate.

CHARNWOOD FOREST.—The Quarterly Report of the Leicester Literary and Philosophical Society for October 1900 contains in the geological section three interesting photographs of Charnwood Forest, showing a quarry between High Sharpley and Cademan, a band of grit in the Blackbrook series, and the Hanging Rocks, Woodhouse Eaves. The paper is by Professor W. W. Watts, M.A., F.G.S.

NOTICES OF SOCIETIES.

*Ordinary meetings are marked †. excursions * ; names of persons following excursions are of Conductors. Lantern Illustrations §.*

GEOLOGISTS' ASSOCIATION.

March 1.—† "The Post-Pliocene Non-Marine Mollusca of the South of England." A. S. Kennard and B. B. Woodward, F.L.S., &c.
„ 1.—‡ "The Pleistocene Fauna of West Wittering, Sussex." J. P. Johnson.
„ 2.—* Natural History Museum. Cromwell Road. Henry Woodward, F.R.S., F.G.S.
„ 23.—Royal College of Surgeons, Lincoln's Inn Fields. Professor Charles Stewart, F.R.S.

MANCHESTER MUSEUM, OWENS COLLEGE.

March 2.—§ "Monotremes." Professor S. J. Hickson.
„ 9.—† "Our Neolithic Ancestors." Professor W. Boyd Dawkins, F.R.S.
„ 10.—† "Hyaena-Dens in Britain." Professor W. Boyd Dawkins, F.R.S.

PRESTON SCIENTIFIC SOCIETY.

March 6.—† "Seeds and Germination." Miss Holden.
„ 13.—† "Architecture and the Evolution of the British Nation" E. H. Turner, A.C.A.
„ 20.—† "Nuts." H. Barbour.
„ 27.—† "Liquid Air." R. Wallace Stewart, D.Sc.

NORTH LONDON NATURAL HISTORY SOCIETY.

March 7.—† Short Papers on " Scientific Progress in 1900."
„ 21.—§ " Pictures of Bird-Life." O. G. Pike.
„ 30.—* Visit to Epping Forest Museum.

SOUTH LONDON ENTOMOLOGICAL AND NATURAL HISTORY SOCIETY.

March 14.—§ " The Life History of a Dragon-fly." W. J. Lucas, B.A., F.E.S.
„ 28.— " The Lepidoptera of the Guildford District." E. B. Bishop.

BIRKBECK NATURAL HISTORY SOCIETY.

March 9.—* Geological Ramble round Ilford. Martin A. C. Hinton.
„ 9.—† " The Palaeolithic Inhabitants of the Thames Valley." Martin A. C. Hinton.

HULL SCIENTIFIC AND FIELD NATURALISTS' CLUB.

March 6.—§ " The Grave-Mounds of the Yorkshire Wolds and their Contents." T. Sheppard, F.G.S.
„ 20.—§ " A Trip to Portugal to see the Eclipse of May 28, 1900." Rev. H. P. Slade, M.B.W.A.

SELBORNE SOCIETY.

March 7.—§ " The Wealden Formation and its Wonderful Contents." E. A. Martin, F.G.S.

LAMBETH FIELD CLUB AND SCIENTIFIC SOCIETY.

March 4.—§ " Early English Architecture." C. H. Dedman.
„ 16.—* Visit to Westminster Abbey. E. W. Harvey Piper.
„ 18.— " A Chat on the Mycetozoa." F. W. Evens.

STREATHAM SCIENCE SOCIETY.

March 2.—† " British Reptiles." M. G. Palmer.
„ 16.—† " Bombyces." P. Hamm.

NOTICES TO CORRESPONDENTS.

To CORRESPONDENTS AND EXCHANGERS.—SCIENCE-GOSSIP is published on the 25th of each month. All notes or short communications should reach us not later than the 18th of the month for insertion in the following number. No communications can be inserted or noticed without full name and address of writer. Notices of changes of address admitted free.

BUSINESS COMMUNICATIONS.—All business communications relating to SCIENCE-GOSSIP must be addressed to the Manager, SCIENCE-GOSSIP, 110 Strand, London.

SUBSCRIPTIONS.—The volumes of SCIENCE-GOSSIP begin with the June numbers, but Subscriptions may commence with any number, at the rate of 6s. 6d. for twelve months (including postage), and should be remitted to the Manager, SCIENCE-GOSSIP, 110 Strand. London, W.C.

EDITORIAL COMMUNICATIONS, articles, books for review, instruments for notice, specimens for identification, etc., to be addressed to JOHN T. CARRINGTON, 110 Strand, London, W.C.

NOTICE.—Contributors are requested to strictly observe the following rules. All contributions must be *clearly* written on one side of the paper only. Words intended to be printed in *italics* should be marked under with a single line. Generic names must be given in full, excepting where used immediately before. Capitals may only be used for generic and not specific names. Scientific names and names of places to be written in round hand.

ANSWERS TO CORRESPONDENTS.

J. R. (Beith).—The specimen enclosed is one of the small tortoiseshell butterfly (*Vanessa urticae*). It is not unusual to find them in winter, if disturbed, as they hibernate in the imago or perfect state.

C. E. S. (London).—We fear you will not obtain, at the price, all you require. Consult for trees, in the reading-room at the British Museum, Bloomsbury, London's " Arboretum et Fructicetum Britannicum ; or, Trees and Shrubs of Britain." It costs £4 to buy. To give a list of insects inhabiting each would constitute a great work ; so you had better consult books on the different orders of insects inhabiting Britain with regard to their arboreal habits.

EXCHANGES.

NOTICE.—Exchanges extending to thirty words (including name and address) admitted free, but additional words must be prepaid at the rate of threepence for every seven words or less.

WANTED to exchange, Kodak or ½-pl. Mahogany Stand Camera, for a 3- or 4-wick Oil Lantern.—A. Nicholson, 67 Greenbank Road, Darlington.

WANTED, a Miniature Spectroscope, with adjustable slit. Offered, a good ⅓-inch micro-objective.—F. Compton, care of SCIENCE-GOSSIP Office, 110 Strand, W.C.

OFFERED, good clutches of curlews', arctic terns', stock doves' (C. oenas), and other eggs. Wanted, woodlark, black guillemot, woodpeckers', and many others.—E. G. Potter, 14 Bootham Crescent, York.

"EUROPEAN Butterflies and Moths," W. F. Kirby, Parts 1-43, in 6d. parts : " European Ferns," James Britten, Parts 1-26, in 7d. parts. Exchange. What offer ? Hand-camera preferred.—E. S. Sugden, 48 Victoria Road N., Southsea.

WANTED, about fifty each of *Periplaneta americana* and *Blatta orientalis* ; also some larvae of cockchafers, alive or dead, preferably the former.—Frederick J. Bridgman, 4 Avenue Road, Highgate.

WANTED, London Catalogue, Ed. IX., Nos. 6, 89, 136, 181, 297, 342, etc. Offered, 264, 298, 331, 716, 826, 833, 870, 1186 ; many others. Lists exchanged.—A. Hosking, 48 Norwich Street, Cambridge.

OFFERED, magazine hand-camera, "Zeus" : good condition ; cost when new 46s. ; in exchange for good collection of foreign stamps.—R. Clapperton, 23 Albert Place, Galashiels.

FOSSILS.— Crag, London Clay, Upper and Lower Greensand, and Wealden wanted in exchange for Palaeozoic and Jurassic species.—Dr. Brendon Gubbin, 15 Redland Grove, Bristol.

OFFERED, Exotic Butterflies, Ornith. miranda M. dohertyi, croesus (both sexes of each) ; also fine specimens in papers from Mexico, Peru, Brazil, and Australia, in exchange for other exotic diurnals.—W. Danuatt, Donnington, Vanbrugh Park, Blackheath, S.E.

FOSSILS.—Permian and Jurassic duplicates offered in exchange for others. List.—Harold Tarbuck, Land and Mine Surveyor, The Chestnuts, Ryhope, *viâ* Sunderland.

OFFERED, Wray's 6¼"×5" R.R. Lens, with extra tube to convert into 10"×8". Wide angle, new condition, in part exchange for Watson's " Fram " or " Student's " microscope stand.—Thomas Peters, 45 Lord Street, Leigh, Lancs.

CONTENTS.

NEW OR TEMPORARY STARS.

By Frank C. Dennett.

FROM time to time mysterious stars have been noted in the heavens that after glowing brilliantly for a time have gradually dwindled away. The first of these of which we have any record appeared between β and ρ Scorpii so long ago as B.C. 134, whilst the last is still shining in

Ma-tuan-lin also tells of a star blazing out in our era A.D. 173 between α and β Centauri. In April A.D. 386 one appeared between λ and φ Sagittarii, and is said to have remained visible and stationary until July. Another, rivalling Venus, came into view in 389, near α Aquilae, and was seen for three

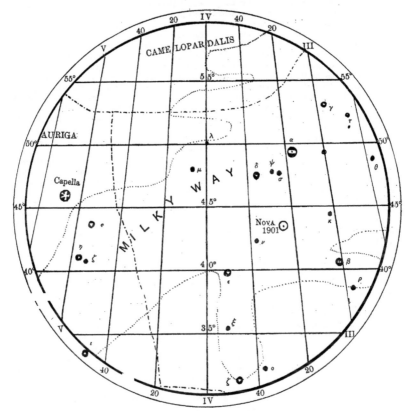

POSITION OF THE NEW STAR, NOVA PERSEI, 1901.

the constellation Perseus. The first rivalled Venus in lustre, and helped to stimulate Hipparchus of Nicaea, in Bithynia, to set about constructing a catalogue of 1,022 stars. Chinese records, too, tell of this star, as well as of another between α Herculis and α Ophiuchi, in A.D. 123. The Chinese

weeks. During the reign of the Emperor Otho, A.D. 945, a star surpassing Venus in brilliance is said to have appeared in the part of Cassiopeia bordering on Cepheus. In December 1230 another appeared between the constellations Serpens and Ophiuchus, remaining visible until March 1231.

"In the year 1264 an eminently bright star" burst out in Cassiopeia, also near Cepheus, "which kept itself in the same place, and had no proper motion." There has been doubt as to whether the stars of 945 and 1264 were not really comets, but the evidence seems rather in favour of their being stars.

We enter upon an era of greater exactitude when the great star of 1572 in Cassiopeia forced attention. Schuler of Wittemberg is said to have observed it upon August 6th, though Dr. Halley writes that Cornelius Gemma did not see it on November 8th, when he "considered that part of the heavens in a very serene sky, and saw it not; but that the next night, November 9th, it appeared with a splendour surpassing all the fixed stars." Tycho Brahe was at the time staying with an uncle at the monastery of Herritzwadt, working in the chemical laboratory until the evening. In coming home on November 11th his attention was attracted by the brilliant visitor in R.A. 0 h. 19·2 m., Dec. 63° 35' N. (1890), about equal to Sirius. It became brighter even than Venus, and was then visible in daylight. At this time its colour was white; but it quickly began to lose its brilliance, at the same time becoming yellow, and, as it diminished, reddish, and finally tinged with blue. Hind and Plummer found a small variable 10–11-magnitude star within 1' of its place. In 1604 the pupils of Kepler, John Bronowski and Möstlin, whilst observing Mars, Jupiter. and Saturn, in close proximity in the constellation Ophiuchus, were interrupted by bad weather for a day or two, but on resuming work on October 10th they were surprised to find a star in brightness between Jupiter and Venus situated R.A. 17 h. 25 m., Dec. 21° 23' S. (1890). Its scintillation was exceptionally great. but its actual colour does not seem to have changed like that of 1572. It soon faded. By January 1605 it had fallen below the brightness of Arcturus, and in March was described as of third magnitude. Between February and March 1606 it became invisible. At the present time there appears to be a variable star of about the twelfth magnitude at this place. On June 20th, 1670, a Carthusian monk named Anthelm found a third-magnitude star in the head of Vulpecula, not far from β Cygni, which by August 10th had fallen to the fifth magnitude. After three months it disappeared; but on March 17th, 1671, it reappeared as a star of the fourth magnitude, and a month later Dominique Cassini found it very variable in brightness. After at length reaching the third magnitude it faded and was missing during February 1672; but at the end of March it brightened to the sixth magnitude, though it was not afterwards seen. Its position was R.A. 19 h. 43·5 m., Dec. 27° 1' N., and close to this spot there is an 11·12 magnitude which is said to be decidedly variable, and somewhat hazy-looking.

After the lapse of 178 years J. Russell Hind found a reddish-yellow fifth-magnitude star in Ophiuchus, R.A. 16 h. 54 m., Dec. 12° 45' S. (1890) on April 28th. 1848, where he was certain no star exceeding ninth magnitude could have existed on April 5th. It remained steady until May 9th. By the 10th it had increased to fourth magnitude; but by the 18th had fallen to the sixth. from which it gradually receded to between the eleventh and twelfth magnitudes, and. still so remains. showing a very slight variation.

In 1866 the celebrated " Blaze Star." T Coronae Borealis, made its appearance near ε in the same constellation, R.A. 15 h. 55·3 m., Dec. 26° 12' N. (1890). Professor Julius Schmidt of Athens found no trace of it. at a little after 9 p.m., on May 12th, or of any star of more than fourth magnitude in its place, but a little before midnight Birmingham at Tuam in Ireland found the place occupied by a second-magnitude star. Through a telescope it appeared nebulous, and looked like a yellow star seen through a blue film. On May 14th it was seen in America, and observed by Baxendell of Manchester upon the 15th to have already fallen to the fourth magnitude. By May 24th it was only 8·5, and afterwards fell to ninth magnitude. After the 25th the blue tinge was lost, and the star passed through the many tints of orange and yellow. On June 26th it appeared as of tenth magnitude. and so remained until August 20th. then commencing to steadily increase to between the sixth and seventh magnitudes on September 15th, continuing steady until November 9th. when it began to decrease till the tenth magnitude was reached. Its colour in September was a pretty bright yellow, but at last assumed a dullish white. Before the "blaze" traces of variation had been shown, for Sir John Herschel found it 6·3 magnitude in 1842, whilst in 1855 Argelander found it 9·5 in magnitude.

Since the previous Nova a new instrument had come into use, the spectroscope, and Sir William Huggins and Dr. A. Miller on May 16 commenced to study the nature of the light. Like other stars it was converted into a rainbow band crossed by black absorption lines, but with this difference, the lines that showed the presence of hydrogen were reversed, bright instead of dark. By some means there was a large accession of heat, and a great atmosphere of hydrogen was rendered incandescent. This is the first step towards the knowledge of the nature of these phenomena.

Quite early in the evening of November 24th, 1876, Schmidt had his attention attracted by a third-magnitude star appearing in Cygnus, R.A. 21 h. 37·3 m., Dec. 42° 23' N. (1890). Only two nights previously he had been examining this very region, and was positive no bright star was then in the position. At midnight it was decidedly yellow. News was sent to Paris and Vienna. but until December 2nd the weather proved bad, and by then the magnitude had fallen to the fifth. Cornu, with the 15-inch Paris achromatic, found its spectrum faint, only containing several bright lines, eight of which he succeeded in measuring, and found to be due to incandescent sodium (or more probably helium), hydrogen, and magnesium, the atmosphere being almost identical with that of our own Sun. Hind found its light reduced to seventh magnitude by December 12th, and by January 10th, 1877, it had fallen another magnitude. At the end of December 1876 the tint was described as deep red in colour, but the object now appears to be a tiny nebula.

In 1885, on August 22nd, a star was simultaneously discovered by Mr. Isaac Ward, of Belfast, and a Hungarian lady, the Baroness de Podmaniczky. It was in, or in front of, the great nebula in Andromeda, which is believed to have exhibited unusual brightness for some days previously. The star seems to have reached its greatest brilliancy, sixth magnitude, about the 31st, but by the end of September had fallen to the tenth, and gradually vanished from view. The spectrum of this star appeared to be continuous.

An unsigned postcard to Dr. Copeland, at the Royal Observatory, Edinburgh, on February 1st, 1892, called attention to a new star of fifth magnitude in Auriga, not far from the star χ. The message came from Dr. Anderson as afterwards transpired, a gentleman of whom it was recently said that it was believed he knew the sky so well he could detect a new fifth-magnitude star in almost any part of the heavens. Professor E. C. Pickering, of Harvard College, U.S.A., had latterly begun setting his "policeman" to "patrol the heavens" on every fine night. This is a small photographic transit instrument which automatically sweeps the meridian, recording all visible stars. The instrument gave evidence to the effect that it had seen this new comer in R.A. 5 h. 25 m., Dec. 30° 21′ N., on thirteen occasions between December 10th, 1891, and January 20th, 1892, but it was certainly not so bright as the eighth magnitude on December 8th. It remained fourth or fifth magnitude through February, then rapidly fell to the twelfth by the end of March. On April 26th it was found to be sixteenth magnitude, in August brightening to the tenth, but afterwards falling to the twelfth magnitude. The spectrum was interesting, showing both bright bands and black lines close to them, on the violet side. The various substances seemed to each give two sets of lines, one bright, displaced towards the red, one dark, displaced towards the violet. The amount of displacement showed a relative velocity of about five hundred miles a second, as if a solid globe or a great mass of meteors moving from us had plunged into a nebulous mass moving towards us. It now appears as a nebulous star, and gives a similar spectrum to those objects. In 1893 Mrs. Fleming, on

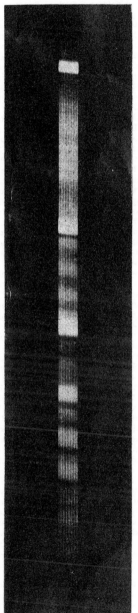

SPECTRUM OF NOVA PERSEI. *From a Photograph taken at Stonyhurst College Observatory, England.*

March 3rd, 8 P.M., 1901.

October 26th, found Nova Normae, of the seventh magnitude, which gave a spectrum that is practically the duplicate of Nova Aurigae, and now also appears to be nebula. The same lady discovered another in Carina in 1895, which in three months fell from the eighth to the eleventh magnitude, having a similar spectrum and finally a similar ending. In the same year, on July 8th, she discovered another of seventh magnitude in Centaurus, R.A. 13 h. 34·3 m., Dec. 31° 8′ S., which whilst it gave a somewhat different spectrum became nebula.

The next, and last Nova that demands our attention is that in Perseus, which was discovered by Dr. T. D. Anderson, of Edinburgh, the discoverer of Nova Auriga, on the early morning of February 22nd, when the star had a very low altitude. Its colour was then bluish-white, and its magnitude 2·7 ; but the same evening, just before seven o'clock, Dr. Copeland found it brighter than Aldebaran, and little more than an hour later considered it equal to Procyon, which it resembled in colour. Twenty-four hours afterwards Dr. Halm found it 0·2 magnitude brighter than Capella, or brighter than any temporary star since that of 1604. On this evening the spectroscope was brought to bear. At Potsdam it was described as like that of Rigel, the spectrum having no bright lines. At Edinburgh with a direct vision prism on the 6-inch refractor it appeared to have a perfectly continuous spectrum ; but with a larger spectroscope, on the 15-inch achromatic, Drs. Halm and Copeland found about half a dozen delicate dark lines from near D to F. The same evening, but presumably a few hours later, Professor Pickering, of Harvard College Observatory, recorded many bright lines, but on the 25th he telegraphed : " S. greatly changed, now resembles N. Aurigae." By this time its light had begun to decrease, but on that evening its position at Greenwich was found to be R.A. 3 h. 24 m. 28·21 s., Dec. 43° 33′ 54″·8 N.. a place where no star previously appears on the map. The district had been photographed at Harvard on February 19th, at which time it certainly could not have exceeded eleventh magnitude. So late as midnight on February 21st it is said that it certainly could not have been so bright as third magnitude, as an English observer was examining that region of the sky. On and after the 27th the colour of the star became a warm yellow or reddish tint, the spectroscope showing the reason, as the brilliant C line of hydrogen was most vivid.

On March 3 I turned our 3-inch " Wray Science-Gossip " achromatic telescope on the newcomer. It was an ordinary-looking, golden-yellow star, quite sharp, surrounded by diffraction rings, like any other. When, however, the cylindrical lens and prism were used its brilliant red line was most striking ; with the Thorp diffraction grating, the rest of the spectrum became visible, six broad bright lines, especially C (red) and F of hydrogen, the yellow helium and green magnesium lines

being readily seen. On the violet side of each bright line there was a black line visible. At this time the Nova was barely brighter than Algol (β Persei). On the same evening the Rev. Walter Sidgreaves, of Stonyhurst College Observatory, obtained a photograph of the spectrum, which he has kindly permitted us to reproduce. " The extreme brightest broad line," he writes, " is Hβ." There is first to be noted the broad bright lines of great width which Sir Norman Lockyer has found, from comparison with the spectra of other stars, to occupy about their normal position in the spectrum, indicating a condition of rest. Next there is a spectrum of dark lines violently displaced towards the violet, so much so as to indicate that the matter yielding it is rushing towards us with a velocity of about 700 miles a second. There is yet another point of interest in the spectrum. Nearly down the centre of each bright line of hydrogen and calcium, but probably too delicate for reproduction, there is a fine dark line—most likely a reversal—which Sir Norman considers will be of great service in accurately determining the wave-length of the other bright lines. The star is still fading ; as I write on March 11th it is between third and fourth magnitudes, but, although less vivid, its spectrum appears to be much as on the 3rd of the month.

The cause of the outburst would seem to be a mass or swarm of meteoric matter crashing into nebula, the force of the collision producing the great amount of heat necessary to raise both a quantity of the solid matter as well as of gas to a condition of incandescence. Already we see the cooling in progress, and probably after a few months it will relapse into the condition of a minute faint planetary nebula. Doubtless there will be many watchers to see if this really occurs.

Perhaps we may wonder when the collision took place. Many months must elapse before we can get any definite, or even indefinite, knowledge on this point, not until it can be known whether the object yields a sensible parallax. It is probable that at the least a century has passed since the outburst occurred, probably longer ; and seeing that all this time light has been travelling earthward at a rate of 11,179,800 miles per minute, the real distance is entirely beyond the human mind to grasp.

One most remarkable fact is to be noted, which is that nearly all these temporary stars make their appearance within the bounds, or upon the borders, of the Milky Way, the reason being not at present known. Perhaps our last visitor may help to throw some light on this interesting subject.

The new star appears to have also been discovered simultaneously by Herr Grimmler, in Bavaria, and a lady student at Vassar College. In conclusion I desire to express my thanks both to Sir William Huggins and to Father Sidgreaves for their kindly help.

London, March 1901.

SOME BRITISH DIVING BEETLES.

By E. J. Burgess Sopp, F.R.Met.Soc., F.E.S.

(Concluded from page 293.)

THE wings (fig. 7) of both sexes of *Dytiscus* are well developed and of ample proportions, which enable the insects to change their habitation when forty-five miles distant from land, and it not infrequently happens during the summer and early autumn evenings that a water-beetle, probably

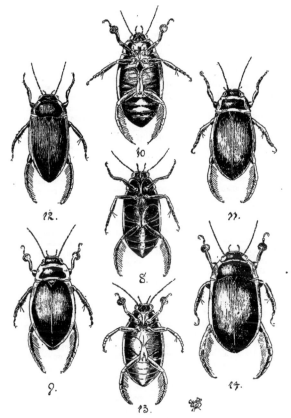

Fig. 2. *BRITISH DYTISCI. Drawn by E. J. Burgess Sopp.*

8. *Dytiscus punctulatus* Fab. Female, showing black under side. 9. *D. circumflexus* Fab. Male. 10. *D. circumflexus* Fab. Male, showing markings on under side. 11. *D. circumcinctus* Ahr. Female, showing narrow border round eyes. 12. *D. punctulatus* Fab. Female. 13. *D. lapponicus* Gyll. Male, under side with lateral markings. 14. *D. dimidiatus* Berg. Male.

at will. This they do towards dusk or after nightfall, in common with many other of the Dytiscidae. Darwin, in his "Origin of Species," makes mention of a *Colymbetes* flying on board H.M.S. "Beagle" when mistaking the light for the reflection of moonlight on water, flies into a house through an open door or window. The insects usually ascend a water-plant or other elevation from which to start on their aërial journeys, although in captivity they occasionally demonstrate that they are not entirely dependent upon such aids to the successful spreading of their wings. I have known both *Dytiscus punctulatus* and *D. marginalis* to escape during the evening from their glass jars and fly from an upstairs room down to the lighted hall below. They make a loud buzzing sound during flight, heard at a considerable distance.

In common with many other imagines and some few larvae they stridulate when at rest, generally towards nightfall, the notes produced partaking of two very distinct sounds—the one a harsh "shrill," and the other a lower humming or "buzzing" sound, which some authors have associated with the peculiar alulae or winglets situated under the base of the elytra, the use of which is not apparent. The mode of stridulation in the genus can, however, scarcely be said to be satisfactorily solved beyond question. The opinion has been expressed that the "shrills" are generated by the rubbing of the underside of the elytra against the end of the abdomen (Miall); but if these areas are examined stridulating organs are not easily perceptible. Mr. C. J. Gahan made no allusion to the Dytisci in his exhaustive paper "On the Stridulating Organs in the Coleoptera," read at a recent meeting of the Entomological Society of London, for the reason, he tells me, that up to that time he had been unable to detect with certainty the presence of stridulating organs in the genus. Since then Mr. Gahan has brought to my notice an article by Hermann Reeker, "Die Tonapparate der Dytiscida," in which it is stated that stridulation in *Pelobius*, *Dytiscus*, and others of the Dytiscidae is generated by rubbing one of the large wing nervures against the elytra. The transverse ridges on these nervures are certainly very noticeable, and according to Reeker are also more numerous in this family than amongst the generality of the coleoptera. Mr. Gahan, however, calls attention to the fact that many beetles that are not known to stridulate, as well as others which are well known to stridulate in a different way, "have the corresponding wing ner-vures just as prominent and as strongly ribbed across," and he is of Dr. Sharp's opinion ("Camb. Nat. Hist.") that Reeker's explanation is certainly erroneous as applied to *Pelobius*, although it may possibly be correct with regard to *Dytiscus*. Whether the Dytisci, like the Gyrini or "Whirli-gig Beetles," stridulate previous to flight I do not know, but am personally of opinion that they do, since the cries are almost invariably emitted during the evening, at which time the insects also fly. In captivity, when placed for a change in an ordinary 6-foot bath, transformed by the arrange-ment of water-plants into a natural pool, I have known the males to crawl up a protruding rush and stridulate during the twilight for eight or ten minutes at a time.

In *Dytiscus* the two terminal spiracles are con-siderably larger than in other members of the family Dytiscidae; this exceptional development being, in fact, one of the distinguishing features of the genus. They are of great service to the beetle in its peculiar method of breathing, for, poising itself, with posterior legs at right angles to the body and tail upwards at the surface of the water, it is able to bring them at once into direct contact with the atmosphere. In addition to pro-tecting the more delicate under-wings beneath, the elytra also render valuable service by forming a reservoir for air into which the spiracles lead, these being in this genus more dorsally situated than in the majority of the coleoptera (fig. 5). For, although fitting perfectly to a portion of the body, there is left towards the apex, between the flattened back of the insect and the concave wing-covers, a hollow water-tight compartment, which, when stored with fresh air, enables the diver to remain submerged for a considerable period. This reservoir renders the posterior por-tion of the beetle doubly buoyant, so that, if re-maining still and unanchored, being lighter than the water, it rises to the surface tail upwards, in the right position for aëration. This is effected by slightly raising the wing-covers and first ejecting any used air they may contain. The males are more active and breathe more frequently than the females, a fact easily perceived by keeping speci-mens of both sexes under observation. From experiments carried out by Dr. Sharp, our greatest authority on the carnivorous water-beetles, he found that in *Dytiscus marginalis* the average time passed under water bore to the time the beetles were exposed to the surface for aëration a ratio of about 12 to 1. In less highly developed species the period is very much greater, amounting in the case of *Pelobius hermanii*, one of the most primi-tive types of the family, to as much as 375 to 1.

Dr. Sharp's communication to the Linnean Society in 1876 "On the Respiratory Action of the Carnivorous Water-beetles (Dytiscidae)" contains so much interesting information that I cannot refrain from quoting a short summary of his observations bearing on the genus under considera-tion. After giving various details *in extenso*, he says:— 'The male of *Dytiscus marginalis* rose on an average once in about 8⅓ minutes for breathing, and remained on an average about 54 seconds at surface for each respiration. The longest interval it was observed to pass without breathing was 19 minutes. The duration of a respiration varied from 5 seconds to 300 seconds, and the time it was ex-posed bore to the time it was quiescent a ratio of 1 : 9⅓. . . . . The female *D. marginalis* rose on the average once in about 12⅘ minutes for breathing, and remained on an average about 55½ seconds at surface for each respiration. The longest interval it was observed to pass without breathing was 32½ minutes. The duration of a respiration was from

3 seconds to 280 seconds, and the time it was exposed bore to the time it was quiescent a ratio of 1 : 13⅘."

The female of *D. punctulatus* (fig. 12), although generally spending more time at the surface than *D. marginalis*, occasionally remains submerged for longer periods, sometimes refraining from aëration for considerably over 40 minutes. She also swims nearer the surface than our commonest species, spending a great deal of her time with a portion of the back slightly protruding from the water.

In very severe winters the Dytisci probably bury themselves in the mud; but retirement from an active piratical life is of short duration, and a slight rise in the temperature suffices to again bring the insects abroad in their favourite haunts. In mild winters it seems unlikely the beetles hibernate at all. Both at Ferndown, Dorset, in the south, and at Hoylake in the north-west of England, I have taken our two commoner Dytisci throughout the winter months, even when the weather has been decidedly cold.

Our beetles of the genus *Dytiscus* may be conveniently divided into two groups, in one of which the underside is black and in the other some shade of yellow or yellowish-red. To the first of these groups belongs one species.

Dytiscus punctulatus Fabr. (figs. 8 and 12) is a flattened, long-oval insect, measuring from an inch to an inch and one-eighth in length, by about half an inch in breadth (26–29 mm. by 12–14 mm.). The general form is narrower in proportion to its length than is that of our following and commoner species. The upper surface is black; with just the slightest tinge of brown, which is usually more apparent in the females, the lateral margins of the thorax and elytra being bordered with yellow. The clypeus is yellow, as in all our British species, and the antennae reddish. The underside is black and polished, the pitchy legs slender and furnished with brown hairs, and the coxal processes rounded at the apex. The upper surface of the males is smooth and shining, there being but slight punctuation, and that mainly towards the apical area of the elytra. The females are duller, and have their wing-cases deeply furrowed from near the base to beyond the middle, there being ten grooves on the basal portion of each elytron, the remainder of the upper surface being finely and somewhat closely punctured.

D. punctulatus is a native of Northern and Central Europe. In Britain it is a local insect in some districts, although widely distributed throughout the kingdom. It may be taken at all periods of the year, excepting during severe frost, in similar situations to *D. marginalis*, but is also stated by Dr. Hofmann to be "chiefly found in running water."

To our second group belong the five remaining British Dytisci. All of these are easily distin-

guished from the foregoing species in having the underside other than black and the thorax distinctly bordered on more than two of its margins.

Dytiscus marginalis Linn. (figs. 2, 3, and 4), although varying considerably in size, is on an average larger than the preceding beetle, ranging from an inch and one-eighth to an inch and three-eighths in length, and from five-eighths to three-quarters of an inch in breadth (25–34 mm. by 15–18 mm.). The upper surface is olivaceous or greenish-black, with the lateral margins of the elytra and the whole of the thorax bordered with yellow. The antennae and legs are red, and the underside of a uni-colorous yellow, with the coxal processes both shorter and more pointed than in the former species. The males are highly polished; but in this and the three following species is presented to us the peculiar and as yet unexplained circumstance of the females assuming two distinct and well-known forms, whilst very rarely specimens have been captured which may be said to occupy the position of "missing links" between them (¹). In the first and commonest type the insect is dull and deeply furrowed, as in the preceding species; whereas in the second it is smooth and shining, like the male, save that it has slight punctuation on the thorax and a rather more liberal allowance towards the apical area of the elytra than is to be met with in the opposite sex.

D. marginalis is widely distributed over the middle and northern portions of Europe, Asia, and America. It has also been recorded from Japan, and ranges from Geneva to well within the Arctic Circle. It is by far the commonest of the British *Dytisci*, being generally distributed in stagnant water throughout the kingdom.

Dytiscus circumflexus Fabr. (figs. 9 and 10) is on an average rather longer and more slender than our last beetle (27–35 mm. by 15–17 mm.). The upper surface is olivaceous or black, with a more or less greenish tinge, both the size and colour of this insect varying considerably. The antennae and legs are red, and the margins of the thorax and elytra are bordered with yellow, as in *D. marginalis*. This species is, however, easily separable from the latter beetle, on account of its yellow underside being boldly decorated by black markings (fig. 10). The coxal processes are moreover, longer, narrower, and sharper, and the scutellum often exhibits a reddish tint. The sexes differ as in the last beetle, the females being similarly dimorphic.

D. circumflexus is a native of Central and Southern Europe and Northern Africa. In its more northern range the sulcate form of female predominates, the insects exhibiting also a broader form and darker hue than those of Spain, Algeria,

(1) "In Dr. Power's collection there is a female *D. circumcinctus* that comes between the two forms, the sulci being only rudimentary but distinctly traceable."—Fowler, "Col. Brit. Is.," i. 205.

etc., where the smooth type of female is more generally abundant and of a brighter green coloration. In Britain the species is decidedly uncommon, and confined chiefly to the London district.

Dytiscus circumcinctus Ahr. (fig. 11) is less variable in size than is our commonest species, to which it bears a strong resemblance. It measures about one and a quarter inches in length and five-eighths of an inch in breadth (31–33 mm. by 15–16 mm.). The colour of the upper surface. is greenish-black, with the wing-cases and thorax bordered as in the last two species; antennae and legs red. From *D. marginalis* it may be distinguished by having the coxal processes more elongate and sharp. and from *D. circumflexus* by the absence of the black markings on the underside, whilst it differs from both of these species in having a narrow yellow border round the eyes. The sexes differ as in *D. marginalis*, the males being glabrous and the females dimorphous, either dull and sulcate, or smooth and polished, but with more punctuation towards the apex of the elytra than the males.

This insect is a native of North-East Europe and North America. In Britain it is decidedly rare, but the fact of its having been recorded from localities as widely separated as Eastbourne. the Cambridgeshire fens, and Askham Bog near York ("Col. Brit. Is."). should stimulate ardent coleopterists to endeavour to locate it in intermediate districts.

Dytiscus lapponicus Gyll. (fig. 13) is one of our smaller species. Although very variable, its general average size is scarcely so large as that of *D. punctulatus* (23–28 mm. by 12–15 mm.). In form it is slightly more oval than the latter beetle, and on account of the thorax being narrower in proportion, especially at the base, the general outline of the insect is less continuous than that of any of our other species. The colour of the upper surface is pitchy or blackish-brown ; the fact that the elytra are finely striped with longitudinal yellow lines also imparts to the beetles a generally lighter appearance than is presented by other members of the genus. The underside is yellow with black markings along the lateral edges (fig. 13). the coxal processes being long and sharp. This species may be distinguished from *D. marginalis* and *D. circumflexus* by the broad yellow border round the eyes, and from *D. circumcinctus* by the black markings on the abdomen. The males are smooth and shining. with the elytra punctured towards their apex. the females being. as a rule, dull and deeply furrowed. although occasionally resembling the stronger sex.

D. lapponicus extends from Central Russia and Germany throughout Northern Europe and Siberia. With us it is a very local insect, and confined for the most part to the Highlands of Scotland and Ireland. Yet, although usually a scarce insect. it has not infrequently occurred to collectors in considerable quantity. Mr. James J. F. X. King. of Glasgow, to whom I am indebted for my British specimens, obtained it in Mull in September 1899 in great number.

Dytiscus dimidiatus Berg. (fig. 14) is the last and largest of our indigenous species. It ranges in size from one and a quarter to one and a half inches in length, whilst attaining a breadth of about three-quarters of an inch at the broadest part of the elytra (32–37 mm. by 17–18 mm.), which is situated rather farther back towards the apex in this than in others of the genus. The upper side is pitchy-black; the antennae and legs red; and coxal processes blunt. The underside is somewhat darker, and exhibits a more ruddy appearance than in any of the preceding insects. *D. dimidiatus* differs from the rest of the group in having only the side margins of the thorax broadly bordered with yellow, a narrow streak being also traceable along the anterior edge. Like *D. punctulatus*, the females of this species exhibit but one form, being always deeply sulcate from near the base to just beyond the middle of the elytra ; the males, as in the other species, being smooth and polished.

D. dimidiatus occurs on the Continent and in Asia Minor, and was formerly plentiful in the fen districts of England ; but, notwithstanding that a recently published encyclopaedia brackets it with *D. marginalis* as one of our two commonest British species, there is little doubt that this beetle is now by far the rarest of our indigenous Dytisci, and bids fair to occupy a place ere long in the growing list of our extinct insect fauna.

The Dytisci are easily kept in captivity, and make interesting pets, becoming in time so tame that they do not appear to mind being lifted out of the water and examined in the hand. In their natural element their movements are varied and marked by extreme gracefulness ; whilst at times, particularly in some of the cat-like methods of cleaning themselves, their performances become amusingly grotesque. The naturalist who is possessed of an aquarium, or who is sufficiently handy with his tools to fabricate one after the manner described in an early number of SCIENCE-GOSSIP (January 1886), will find much to learn from a study of the habits and actions of these interesting insects. Especially will this be so if able to arrange a bath or other large receptacle for water, as previously described (*ante*, p. 326), so as to give the beetles ample room and opportunity to display themselves to advantage. Then they will be found to prove much more instructive and educating than goldfish, or the generality of the usual occupants of our private aquaria.

Saxholme, Hoylake.

"THE CAMBRIAN NATURAL OBSERVER" for January and February 1901 is to hand, and contains much that is interesting. including a history of the Astronomical Society of Wales, of which it is the organ. The Society appears to be making good progress.

MECHANICS OF CONDUCTION OF SAP.

By HAROLD A. HAIG.

(Concluded from page 306.)

LET us now inquire more closely into the nature of both cell-wall and parietal protoplasm; the former of these is, in the root-hairs, composed entirely of cellulose, the molecules of which are of a considerable size. It has been supposed that between certain groups of these molecules there exist spaces, the so-called "micellae" which are

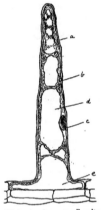

FIG. 1. A root-hair, showing :—*a*, cell-wall; *b*, protoplasm; *c*, nucleus; *d*, vacuoles; *e*, original cell from which hair arose.

of a size comparable with the molecular group. Between the constituent molecules of each group still smaller spaces are supposed to exist, these being the "tagmata" of certain botanists. It is quite reasonable to suppose that these spaces do exist; at any rate, the micellae are probably present, and the supposition affords a ready explanation of the absorption of salts and water; but some salts that are positively injurious to such a cell-wall or to the protoplasm, or substances which have a very large molecule (colloïd), do not appear to have the power of getting through a membrane of cellulose. It is, in fact, only salts with a molecule comparable in size with the micellar spaces or the tagmata that can get through, and so be submitted to the discriminative power of the primordial utricle.

As far as the intimate structure of this latter layer goes, but little is known, except that it is made up of two parts—one, the "ectoplasm," being that nearest the cell-wall, and which probably exercises the power of selection; the other, the "endoplasm," that is connected with the nutrition

of the cell. Whether, however, spaces comparable with those in the cell-wall really exist, or whether a real affinity is exerted in the case of the ectoplasm, is not known. It is conjectured, however, that a sort of sifting process also goes on here, but there has not been much recent research upon the subject. The actual process will not be elucidated except by the aid of extremely refined experiments. It is probably due to that "vital activity" which is supposed to exist in all plants.

We must now answer the question, What happens after this raw sap has been taken into the root-hairs and other epiblemal cells in their

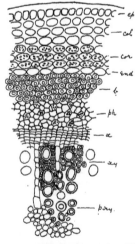

FIG. 2. Trans-section across young stem (cucurbita), to show arrangement of the vascular elements :—*ep.*, epidermis; *col.*, collenchyma; *cor.*, cortex (these three constitute the "bark)"; *b.*, bast-fibres; *end.*, endodermis; *ph.*, phloem; *x.* cambium; *xy.*, xylem (reticulate and pitted vessels); *p.xy.*, protoxylem (annular and spiral tracheides).

vicinity? The problem will be found to resolve itself into the task of determining the nature and structure of the other elements occurring both in root, stem, and leaves. These elements constitute what is known as the "vascular system" of the plant, and are contained in the fibro-vascular bundles which in Dicotyledons ([1]) are composed of the two

([1]) It does not matter whether we take Monocotyledons or Dicotyledons; the essential elements are of course present in both.

M 3

containing portions, "xylem" or wood, and phloem, with a layer of meristematic tissue, the cambium, between. Of these the xylem forms the tissue which conducts the raw sap upwards towards the leaves ; the phloem forming a tissue whose function is to distribute the elaborated sap from the leaves to the cortex and other parts of the plant.

In the ultimate ramifications of the roots, the raw sap passes from the root-hairs to the central cylinder of the root by a process of osmosis, and here gets through into the elements of the xylem. These are in young roots of a simplified nature, and consist for the most part of elongated elements with spiral or annular thickenings upon the walls (see figs. 2 and 3). Further up we get, on account of secondary thickening, additional xylem elements of many kinds, such as the forms with pits, reticulated walls, and others. All these fall into the category of "tracheides," and are designated as annular, spiral, or reticulated, as the case may be (see fig. 3.)

We must consider, How is the raw sap able to pass up this long line of xylem elements? The answer is that the xylem, with its system of tracheides, many of which possess numerous pits in their walls, affords a transmitting channel for the sap. It was formerly thought that the walls of the tracheides were traversed by the fluid, but this has been proved not to be the case. It passes by means of the spaces, and by them only; moreover, it has been shown(2) that the air in these spaces exists there under a negative pressure, so that by means of certain other forces that will be mentioned later (root-pressure and transpiration) the sap easily penetrates from one space to another, and so upwards.

If we take a short piece of hazel stem, cut both ends flat, and block out the pith at either end by means of some impermeable varnish, we can demonstrate the extreme facility with which water passes along the wood. Immerse one end in some water, and observe the other end closely. In less than one minute, if the piece is not too long (10 cm.), the cut surface will become wet and small drops of water will appear. If the piece is taken out and held upside down, the water remaining on the uppermost surface will sink down and appear on the lower surface, being, in fact, aided in this case by gravity. We thus see that the wood is highly permeable, and the fact that the spaces in the xylem elements contain bubbles of rarefied air, surrounded by a thin aqueous film, would of itself be sufficient to explain the rapid upward suction or attraction by means of the surface tension force called into play. Moreover, the existence of a large number of pits between adjacent elements forms a great factor in the upward conduction. These pits, it must be remembered, are not actual apertures, but are closed by the thin "middle lamella" that is present between

(2) Detmer and Moore, " Practical Plant Physiology."

adjacent cell-walls. This thin lamella is highly permeable to a watery solution of salts.

With regard to the physical forces brought into play in aiding the rapidity and continual upward flow of raw sap, two, namely "root-pressure" and "transpiration," must be especially mentioned. The first of these is dependent upon that pressure that is set up during osmosis, and known as "osmotic pressure," which in itself is dependent upon the partial semi-permeability of the cell-wall and ectoplasm. It has, moreover, been supposed that at certain points in the upward progress certain cells exist whose upper wall is more permeable than the lower wall to the same solution, and that these form fresh relays, whereby new force of pressure is attained.(3)

The best way of demonstrating root-pressure is to procure a rapidly growing plant in a pot and cut off its upper foliage. To the end of the cut stem affix by means of a short rubber tubing a manometer, the space between cut surface and mercury being filled with oil. In a few hours the level of the mercury in the two limbs will be found to be different, indicating an increase of pressure on that side connected with the cut stem. The experiment is easy to perform, and shows us what a considerable force may be obtained from this pressure, a force that is constantly employed in the forcing-up of sap.

Although root-pressure forms a powerful factor in sap transmission, "transpiration" is even more efficacious. To study this we must first of all examine the terminations of the fibro-vascular bundles in the leaves, which are the main organs of transpiration. A section across the lamina of a leaf near its junction with the petiole will cut across several of the leaf-bundles as they come up from the stem. It will be seen on microscopical examination they are of a simpler formation than those in the stem or root, the xylem consisting of merely a few scattered annular or spiral tracheides, situated towards the upper surface of the leaf, and a small mass of succulent phloem on the under surface (see fig. 4). A thin layer of cambium may be present if near the petiole. The ultimate terminations in the mesophyll of these bundles consist of a few spiral tracheides, surrounded by a layer of elements that resemble embryonic phloem-cells; and these tracheides stretch in amongst the succulent, turgid cells of the middle layer, or "spongy parenchyma" of the leaf. In order to fully understand the process that goes on during transpiration we may inspect fig. 6, in which A represents a cell of the spongy parenchyma ; B, C, D, and E are single elements of a fibro-vascular bundle, B being the terminal tracheide ; and F is the soil, with its dilute solution of salts.

Let us now start with an equally concentrated solution of salts in all these receptacles. A is ex-

(3) The fact that in the ultimate rootlets the cells become turgid also constitutes a great aid to root-pressure.

posed to the air, and as it presents a large surface evaporation takes place, and some of the water is withdrawn from the liquid in that part. In consequence more water will be taken in from B, the fluid in which will in turn become more concentrated; and the process extends through C, D, and E. The result of this is that the liquid at F, which remains at the same concentration, will be continuously drawn up towards A, and an upward flow of liquid is thus produced. The same process goes on in the plant, the water evaporated from the

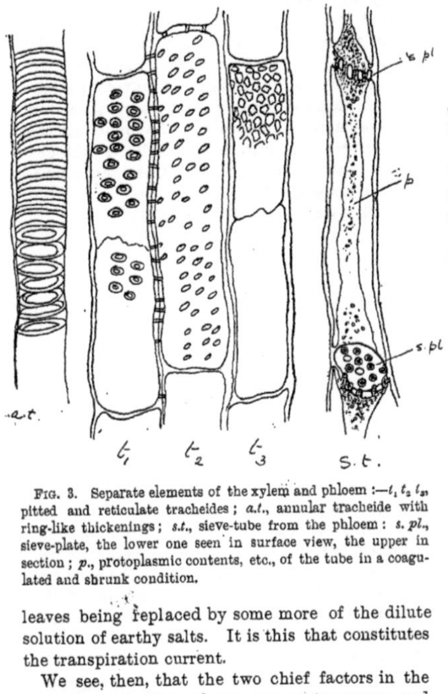

escapes. At the same time, it must not be supposed that the rate of transpiration remains constant for any one plant. It varies naturally with the hygrometric conditions of the air, more water

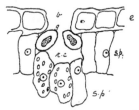

FIG. 5. Section across a stoma:—*v.*, the vestibule; *e.*, epidermis; *g.*, guard-cells; *r.c.*, respiratory-cavity; *sp.*, spongy parenchyma.

being evaporated into "dry" air than into "moist" air. In this way the transpiration current varies also in rate, and it will be found that plants will only flourish in air that exactly suits their hygrometric conditions, although they may to some extent become adapted to abnormal states of the atmosphere. It is found also that even root-

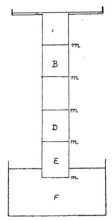

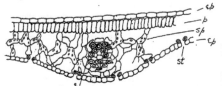

FIG. 3. Separate elements of the xylem and phloem:—*t₁ t₂ t₃* pitted and reticulate tracheides; *a.t.*, annular tracheide with ring-like thickenings; *s.t.*, sieve-tube from the phloem: *s. pl.*, sieve-plate, the lower one seen in surface view, the upper in section; *p.*, protoplasmic contents, etc., of the tube in a coagulated and shrunk condition.

leaves being replaced by some more of the dilute solution of earthy salts. It is this that constitutes the transpiration current.

We see, then, that the two chief factors in the upward conduction of sap are root-pressure and

FIG. 4. Trans-section across a blade of a leaf in region of midrib:—*ep.*, epidermis; *p.*, palisade parenchyma; *sp.*, spongy parenchyma; *b*, bundle, with xylem on upper phloem on under aspect; *st.*, stomata.

transpiration from the green parts of a plant. Transpiration occurs in any part of a plant that possesses stomata (see fig. 5) in its epidermis; for it is through these that the evaporated water

FIG. 6. Diagram to illustrate the mechanics of transpiration; for explanation see text. *m*, *m* are the walls of the tracheides that are adjacent to one another.

pressure varies periodically, the flow being more active shortly after midday, and least in the early hours of the morning.

The elaborated sap in the cells of the mesophyll of the leaf is conducted away from the leaf by means of the phloem elements on the under side of that organ, and, reaching the phloem of the stem, part of it is directly distributed by osmotic processes to the outlying cortex, another part being conducted and stored by some peculiar elements

M 4

of the phloem strand known as the "sieve-tubes" (see fig. 3). Each sieve-tube is an elongated cell, with its end walls thickened and perforated by a number of holes, through which at certain times of the year a direct protoplasmic and structural continuity is established. There is in each element a layer of parietal protoplasm, in which are numerous drops of mucilage; and adjacent to each adult sieve-tube is another elongated cell of smaller calibre, the "companion-cell," which is full of a granular protoplasm and has a well-defined nucleus. The sieve-tubes act chiefly in the distribution of the organic food-material of the plant, the mucilage containing much proteid that gives a brownish coloration on addition of iodine. At certain times in the year the sieve-plates are closed on either side by a thick cushion of callose, the "callus," and it is at these periods that the storing function of these elements comes into play. In the spring these cushions are dissolved, and the organic materials pass on and are further distributed. The downward progress is aided chiefly by the pressure of turgescent cells in their immediate vicinity.

There exist also sieve-plates on the side-walls of each sieve-tube, and by means of these organic material passes into the companion-cell, or elsewhere, where it is again dissolved, and passes, probably in some other form, by osmosis into the surrounding cortex or the cambium.

We have now studied some of the chief methods in which the nutrient fluids, the raw and elaborated sap, of a plant are conducted and distributed. During our observations we have seen that many physical forces or processes are brought into play, such as osmosis, evaporation, and capillarity. An exhaustive study of these various methods is required in order to fully understand the true relations between physical forces and vital activity, and it must not be here supposed that the former are the only agents. In fact, the two classes aid one another in the furtherance of all these phenomena.

53 *Ordnance Road,*
St. John's Wood, London, N.W.

COLOURING OF WATER BY MICRO-ORGANISMS.

By JAMES BURTON.

IT is well known, not alone to microscopists, that large or small bodies of water are sometimes coloured by the presence of various organisms, either animals or plants, often of microscopic size. Every roadside pond is liable to become of a thick soupy appearance and green colour from the multiplication in it of the very common *Euglena*, or some other of the unicellular algae, such as *Protococcus*. Frequently portions in similar localities appear pink or red, owing to the existence in them of immense numbers of some of the Daphniae or water-fleas. In the two cases now to be described, the colour, though extremely marked and characteristic, was the result of the presence of less common organisms.

Early in October the ornamental water in the Botanical Gardens, Regent's Park, appeared of an almost uniform pale green. On close examination this was seen to be due to some minute bodies diffused through the water; they were not merely floating on the surface, but seemed about equally distributed at all visible depths. Every twig and thread of water-weed, etc., at the margin was covered with what looked to the unassisted eye like tiny green balls, while in the quiet corners and backwaters towards which the breeze was blowing, the same bodies were collected in such quantities as to resemble thick light-green paint. Under the microscope it was found that the tiny balls were of irregular outline, and consisted of small algae in colonies of various sizes, formed of more or less spherical groups. These were made up of very numerous individuals, oval or pear-shaped, so minute that the green colour noticeable in the aggregations was not distinguishable in them. The groups were hollow and surrounded by a thin layer of jelly or mucilage. In many cases there seemed to be spines radiating from the individuals, but these have no real existence, and the appearance is probably due to the mucilage composed of the swollen outer cell-walls of the separate members not having yet entirely coalesced.

The colonies, I think, have no motion within themselves, but, being of nearly the same specific gravity as the water, are very readily moved about by any slight current, such as would be set up by wind, or by the sun shining on the surface and causing a difference of temperature between different layers. Owing to the disengagement of gas under the influence of light, there is a tendency in the organisms to rise to the surface, while the gelatinous envelopes make them cling to one another and to any object with which they come in contact. Thus are larger and more noticeable masses formed, which, however, have very little cohesion, and disperse again readily. My somewhat doubtful identification of *Coelosphaerium kutzingianum* was confirmed by an authority who kindly took the trouble to examine specimens. A figure is given in Dr. Cooke's "Introduction to Freshwater Algae," and the size of the individual cells is stated to be 2 to 5 μ, and that of the families 60 μ and more. The alga is probably not rare; but as it was not recognised by two or three microscopists to whom it was shown, it is most likely seldom noticed, and certainly does not commonly occur in such numbers as to give any tint to the water it inhabits.

Attempts to mount these algae in several preparations of glycerine were not successful, the groups breaking up. Chlor-zinc-iodine (Schulze's solution) gave better results. So did some other fluid media; but the distinctive characteristics are hardly likely to be enduring.

A somewhat more remarkable instance, both as to the colour and its cause, came under notice in January, 1898, in a farm pond at Hampstead. When first seen the water appeared of a rosy-pink tint, owing to a growth which had formed on dead leaves and debris of various kinds. About a week later, however, the pond presented a striking aspect. When some distance from it, the water seemed to be of a beautiful intense red-purple, so exactly resembling what might be reflected from the sky in a fine winter sunset that I involuntarily turned round as I approached, almost expecting to see the sun setting behind. On closer examination it was seen that every leaf and twig at the bottom was of this brilliant tint. Some floating patches of Confervae looked like masses of vivid purple, without a particle of their normal green being visible. The organisms producing this effect were spread in a thin layer over everything, and also formed delicate filaments lightly attached, which, however, were dissipated by the slightest movement. On agitating the Confervae or leaves, the colour-containing matter was at once diffused through the water.

Under the microscope it was found to consist of exceedingly minute bodies, so small that a definite outline could scarcely be made out with a power of 500 diameters. These were surrounded by a thin layer of mucilage, and mostly aggregated into hollow spheres; many were solitary, but some were gathered in masses. The filaments it was almost impossible to examine in their original form, but they were composed of the same minute bodies disposed more or less in line. A friend kindly brought the matter under the notice of a professor of botany, who at once identified the organism as a bacterium now named *Beggiatoa rosco-persicina*. He referred me to a paper by Dr. Lankester, published in "The Quarterly Journal of Microscopical Science" for 1873, N.S., vol. xiii. Dr. Lankester there describes, under the name of *Bacterium rubescens*, an organism he discovered in some jars containing putrescent remains of animals and plants which had been undisturbed for a short time. The point to which he pays most attention is the remarkable colour of the "plastids," which he considered characteristic of the species. There is little doubt that it is the same species mentioned by Dr. Cooke in his "British Freshwater Algae" as *Pleurococcus rosco-persicinus* Rabh., with the remark that it is "certainly not a good pleurococcus." He gives the size of the individual cells as ·0015 to ·004 m. It is not mentioned in the same author's "Introduction to Freshwater Algae." I do not see the reason for classing this bacterium with *Beggiatoa*, as to me it seems it would be more correctly considered as a *Micrococcus*.

Apart from the colour, the most interesting fact about these lower forms of life is that, while ordinarily present to a small extent, occasionally, owing to favourable conditions of environment and food supply, they multiply so enormously as to have the effect described. Thus giving a visible example of what must occur invisibly during epidemics of diseases, such as influenza and plague, which, according to modern science, are caused by micro-organisms distantly related to them.

20 *Fortune Green Road, West Hampstead, London, N.W.*

AN INTRODUCTION TO BRITISH SPIDERS.

BY FRANK PERCY SMITH.

(Continued from page 304.)

Tapinocyba incurvata Cb. (*Walckenaera incurvata* in "Spiders of Dorset.")

Length of male 1.7 mm.

Cephalo-thorax yellow-brown, with a green tinge. Legs pale yellow, the femora being somewhat darker. Abdomen black, marked with numerous impressed dots.

Tapinocyba dolosa Cb. (*Neriene dolosa* in "Spiders of Dorset.")

Length of male 2 mm.

The abdomen is of a globular form. Said to be an extremely rare spider.

Tapinocyba ingrata Cb. (*Walckenaera ingrata* in "Spiders of Dorset.")

Length of male 1.5 mm.

Cephalo-thorax dull yellowish-brown. Legs reddish-brown. Abdomen blackish-brown, tinged with green. A very rare spider.

Tapinocyba subitanea Cb. (*Walckenaera subitanea* in "Spiders of Dorset.")

Length of male 1.2 mm.

Cephalo-thorax brown. Abdomen blackish-brown, tinged with olive.

Tapinocyba praecox Cb. (*Walckenaera praecox* in "Spiders of Dorset.")

Length of male 1.7 mm.

Very similar to *T. subitanea*, but the curve of the posterior eyes is considerably stronger.

Cnephalocotes elegans Cb.

Cnephalocotes laesus L. Koch. (*Cnephalocotes interjectus* Cb.)

Cnephalocotes curtus Sim.

Cnephalocotes silus Cb.

Drawings of structural details will be given for the differentiation of these closely allied species.

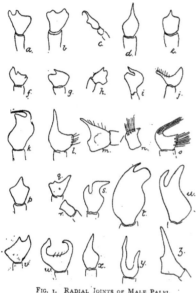

FIG. 1. RADIAL JOINTS OF MALE PALPI.

a. Tapinocyba pallens, from above. *b. T. subaequalis*, ditto. *c. T. incurvata*, side view. *d. T. subitanea*, from above. *e. T. praecox*, ditto. *f. Cnephalocotes obscurus*, ditto. *g. C. elegans*, ditto. *h. C. curtus*, side view. *i. C. laesus*, from above. *j. C. silus*, ditto. *k. Troxochrus ignobilis*, ditto. *l. Styloctetor broccha*, ditto. *m. S. broccha*, side view. *n. S. penicillatus*, ditto. *o. S. penicillatus*, from above. *p. Metapobactrus prominulus*, ditto. *q. Pocadicnemis pumilus*, ditto. *r. P. pumilus*, side view. *s. Araeoncus humilis*, from above. *t. A. crassiceps*, ditto. *u. A. vaporariorum*, ditto. *v. Peponocranium ludicrum*, ditto. *w. Lephocarenum nemorale*, ditto. *x. L. parallelum*, ditto. *y. L. blackwallii*, ditto. *z. L. mengei*, side view.

Troxochrus ignobilis Cb. (*Walckenaera ignobilis* in " Spiders of Dorset.")

Length of male 1.25 mm.

Cephalo-thorax dark brown. Legs reddish-brown, the femora being the darkest. Abdomen sooty black. A rare species.

Troxochrus hiemalis Bl.

Length of male 1.7 mm., female 1.75 mm.

The general colouring of this species is very similar to that of *T. ignobilis*. It is, however, considerably larger.

Araeoncus crassiceps Westr. (*Walckenaera affinitata* Cb.)

Length of male 1.7 mm.

The caput is raised and overhangs the clypeus.

Araeoncus vaporariorum Cb.

In this species the caput is far lower than in *A. crassiceps*.

The radial joint of the male palpus is extremely large, and projects greatly over the digital joint.

GENUS *PEPONOCRANIUM* SIM.

The posterior row of eyes is very strongly curved. The tibial spines are rather long, and are placed towards the end of the joint.

Peponocranium ludicrum Cb. (*Walckenaera ludicra* Bl.)

Length. Male 1.5 mm., female 1.6 mm.

In this species the cephalic eminence of the male is very large and conspicuous, sloping backward from the lower edge of the clypeus. Near the apex of this

FIG. 2. CEPHALO-THORACES OF MALE SPIDERS IN PROFILE.

a. Tapinocyba pallens. b. T. subaequalis. c. T. incurvata. d. T. subitanea. e. T. praecox. f. Cnephalocotes obscurus. g. C. elegans. h. C. laesus. i. C. silus. j. Troxochrus scabriculus. k. T. ignobilis. l. T. hiemalis. m. Metapobactrus prominulus. n. Pocadicnemis pumilus. o. Araeoncus humilis. p. A. vaporariorum. q. A. crassiceps.

prominence are situated the posterior central eyes, the remaining eyes being placed in a transverse group some distance lower down upon its front part.

(*To be continued.*)

THE ZEM-ZEM WATER OF MECCA.

By C. Ainsworth Mitchell, B.A. (Oxon), F.I.C.

THE Zem-zem well of Mecca is probably the most celebrated well in the world. To all Mohammedans it is an object of especial veneration, whilst among Europeans it has the reputation of having been the means of disseminating more than one outbreak of cholera. It is pointed to as the identical well from which Hagar filled her bottle of water when driven forth with her son into the wilderness, though the tradition appears to have a modern origin. Various derivations of the name "Zem-zem" have been given. According to some it is a word formed in imitation of the splashing of its waters; whilst others attribute its origin to "Zam! zam!" ("Fill! fill!"), Hagar's exclamation at the well. Others again connect it

the analysis of the Zem-zem water shows extreme pollution, or that the well has been the medium of spreading infection far and wide. The water is not only drunk in Mecca, but is exported to all parts of the world for the use of the faithful. and a great trade is done in the sale of bottles and vessels to the pilgrims.

The curious tin bottle shown in the accompanying sketch is one of several brought back from Mecca by Sir Richard Burton in 1853, and given to the present writer by Lady Burton shortly before her death. Readers of Burton's "Pilgrimage to Mecca" will remember how, disguising himself as a Mohammedan dervish, and risking his life, he succeeded in making his way to Mecca and taking

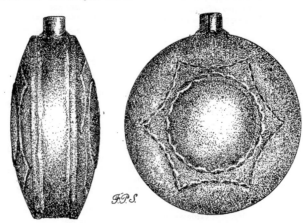

TIN BOTTLE OF WATER FROM THE HOLY WELL, MECCA.
ACTUAL SIZE 3½ IN. DIAMETER × 2 IN. WIDTH.

with the mythology of the old fire-worshippers. Be this as it may, the well has long been regarded as peculiarly sacred, and miraculous properties are attributed to its water.

Each pilgrim to Mecca makes a point of drinking and bathing in it; and as the supply is naturally limited, an Arab stands on the parapet of the well and draws up the water. A pilgrim then advances and receives the contents on his head. He drinks what he can, and the remainder of the water flows down over him and falls through a grating into the well, whence it is again drawn, to be poured over succeeding pilgrims.

When we consider that this practice has been going on year after year, it is not surprising that

part in all the ceremonies at the tomb of the prophet. One of the objects of his pilgrimage was to see the Zem-zem well and obtain some of its water, and his account of its properties is well worth quoting:—" The produce of Zem-zem is held in great esteem. It is used for drinking and religious ablution, but for no baser purposes, and the Meccans advise pilgrims always to break their fast with it. It is apt to cause boils, and I never saw a stranger drink it without a wry face. Sale is decidedly correct in his assertion: the flavour is a salt-bitter, much resembling an infusion of a teaspoonful of Epsom salts in a large tumbler of tepid water. Moreover, it is exceedingly 'heavy' to the digestion. For this reason Turks and other strangers prefer rain-water collected in a cistern, and sold for five farthings a gugglet. It was a favourite amusement with me to watch them whilst they drank the holy water, and to taunt their scant and irreverent potations."

It is interesting to compare with this account the results of the analysis made by the writer. On opening the bottles, which were hermetically soldered, they were found to be filled with water in which were suspended beautiful silken crystals, which proved to be a compound of tin, evidently derived from the action of the water upon the interior of the vessel. On filtering the water a clear and colourless filtrate was obtained, whilst the crystals and some earthy matter were left on the filter. The water had a slight odour, which was more perceptible on gently warming.

The amount of solid matter left on evaporation was in the proportion of 219·5 grains per gallon. If this be compared with Thames water, which contains about 29 grains per gallon, or with New River water, with 25 grains, some idea will be obtained of the large proportion of salts which this figure represents. It was typical of a strongly saline mineral water.

The chlorine, largely present in the form of common salt, was 69·3 grains per gallon, with which again compare Thames water with 1·4 grain and New River water with 1·8 grain. A large proportion of this chlorine was probably derived from the soil; but at the same time it is not going too far to attribute some of it to the pilgrims, seeing that salt is excreted by the human skin. A water containing as much as 9 grains of chlorine per gallon is looked upon with suspicion, unless it is satisfactorily accounted for by the known nature of the bearing strata.

The quantity of magnesium, which was probably all present in the form of magnesium sulphate or Epsom salts, was not so large as might have been expected from the descriptions which have been given of the taste of the water. It amounted to 6·6 grains per gallon. The degree of hardness was determined by the well-known Clarke's test. In this the water is shaken with a solution of soap of known strength, which is added in gradually increasing quantities until a permanent lather is obtained. This does not happen until the whole of the carbonate of lime or its equivalent in other salts has been neutralised by the soap. The amount of soap solution used is expressed in terms of carbonate of lime, each degree corresponding to 1 grain per gallon. The hardness of the Zem-zem water was 43°, as against 17° for Thames water and 15° for New River water. Hence it will be seen that it was excessively hard.

A very important test for judging the character of a water is the determination of the amount of "free" and organically combined ammonia. A large proportion of the ammonium was present in the form of salts, probably combined with part of the chlorine, as ammonium chloride or sal-ammoniac. The combined organic ammonia amounted to 2·2 parts per million. A good drinking water does not contain as much as 0·1 part per million of such ammonia, and the high proportion found here is what one would expect in a highly contaminated water. The amount of nitrates was also indicative of pollution.

The following table gives the quantity of the chief constituents of the water :—

	Per cent.		Per cent.
Aluminium	0·8	Potassium ..	24·3
Calcium	0·5	Ammonium ..	5·3
Silica	3·0	Chlorine	69·3
Magnesium	6·6	Sulphates ..	30·7
Sodium	38·3	Nitrates	19·9

It is interesting to note that on bacteriological examination the water was found to be absolutely sterile, though this is not surprising when we remember that it had been sealed up in an air-tight vessel and in total darkness for over forty years.

I have to thank Mr. Frank Percy Smith for having kindly made the excellent drawing from the original tin flask, illustrating this article.

57 Chancery Lane, London, March 1901.

BUTTERFLIES OF THE PALAE-ARCTIC REGION.

By HENRY CHARLES LANG, M.D., M.R.C.S., L.R.C.P. LOND., F.E.S.

(Continued from p. 300.)

Genus *COLIAS (continued).*

In the second section of Group I., about to be described, we have several species that are quite as brightly coloured as those of the first section; and, as in the case of *Colias regia* and *C. eogene*, some that reach the maximum of intensity in the coloration of the genus. They are all characterised, however, by the absence of the basal patch on the h.w. of the ♂.

10. **C. romanovi** Gr.-Gr. Rom. Mem. vol. iv., 1890.

46—50 mm.

♂ deep orange, with borders very much as in *Colias aurora*, but the nervures are not black. The border of the f.w. is not sharply defined at its inner edge, but shaded or streaked. It has a few yellow rays near the apex; there are also very frequently to be seen upon it some traces of yellow spots. In spite of this last character *C. romanovi* must come into this group; first, because the spotted border in the male does not seem to be the normal condition of the species, though four out of five males in my collection possess it. Indeed, the name *maculata* is sometimes given to this form. Secondly, because of its proximity to *C. aurora* and the allied species. Gr.-Gr. says of it, " Elle se distingue de celles-ci par un trait fort caractéristique, c'est l'absence des taches empesées." It is so like *C. aurora* that the absence of the costal patch is the best distinguishing character. ♀ greatly resembles the ♀ of *C. aurora*, but the light spots on the marginal border are not so large nor so distinct. There is not so much basal shading, and the neuration is not black. It is quite possible that previously to 1890 this insect was confounded with *C. aurora*, especially as the ♂ often has a patch of light-coloured scales at the base of h.w. beneath the subcostal nervure. This, however, on a merely casual examination, will be seen to differ entirely from the structure found in the last group. Fringes orange-red mixed with yellow.

Hab. Turkestan, Pamir, Transalai. VI.—VII. e. Flies rapidly in rocky places. (R. & H.)

C. romanovi. Male.

11. **C. pamiri** Alph. Rom. Mem. vol. iv., 1890. 49—52 mm..

♂ very like *C. myrmidone* above, both in colour and markings, but f.w. have the borders less sharply defined internally, and more deeply indented; there is some yellow striation near the apex between the nervules. H.w. marginal border more indented than in *C. myrmidone*, and the sex mark absent. ♀ closely resembling that of *C. myrmidone*, but the ground colour is deeper orange. H.w. clearer orange and not shaded towards the base as in that species. The yellow spots are smaller, deeper yellow, and more completely placed within the black border. Disc. spot in both sexes rounded. U.s. f.w. with a white centre to disc. spot. H.w. more velvety in appearance, and of a green tint. Disc. spot smaller and bordered with rosy red, somewhat as in *C. fieldii.*

Hab. Pamir. In marshy Alpine meadows, at considerable elevations.

12. **C. regia** Gr.-Gr. 45—50 mm.

♂ not unlike *C. myrmidone* in the general character of the markings, but the ground colour of the wings is deeper than in any other *Colias*, being of so deep an orange tint that it may almost be described as red, with violet reflection. ♀ varies greatly as regards the yellow spots on the black

border; there are at most four near the apex, but generally only very slight traces of spots. U.w. with the marginal border broadly black, the basal area of wings more shaded than in ♂, the disc. spot large and deep orange. Fringes of all the wings bright red. U.s. f.w. with central area reddish-orange, costa and ou. marg. light green ; disc. spot white centred. An ante-marginal row of black spots. H.w. uniformly green, not very dark. Disc. spot large and silvery.

Hab. Turkestan, Transalai, Alai. VII. Flight impetuous and jerky, only in precipitous rocky places where vegetation is absent. (R. & H.)

13. **C. thisoa** Mén. Cat. Rais. p. 244, *myrmidone* var. Ld. Ann. Soc. Belg. xiii. pp. 20, 21. Lg. B. E. p. 57, pl. xiii. fig. 3, ♂ and ♀. 40 – 42 mm.

♂ very greatly resembles *C. myrmidone* ♂, but besides the absence of the sex mark on the h.w. the marginal band of f.w. is of more even width throughout its whole length, and is more distinctly veined with yellow, especially towards the apex. ♀ much more distinct from *C. myrmidone,* the ground colour of the wings is much deeper orange, the marginal bands are broader and more even, the spots upon it are fewer, smaller, and of a deeper yellow. H.w. always more shaded than in *C. myrmidone*, but very variable ; some specimens resemble *C. romanovi* in coloration; others have the h.w. very dark, sometimes almost entirely black. The discoidal spot is conspicuous and of a reddish-orange colour. U.s. f.w. orange with a white-centred disc. spot. Ou. marg. greenish-yellow, with three black spots. H.w. greenish-yellow with a pearly disc. spot, and some reddish marks at the base, and parallel to ou. marg. Fringes not very conspicuously red, in some specimens being even light yellow, especially on h.w.

Hab. Caucasus, Ararat, Turkestan, Altai, Ala-Tau-Kuldja, Tarbagtai, and Tianschan Mountains. At elevations of 7,000 and 8,000 feet. VI.—VIII.

(*To be continued.*)

OBJECT LESSONS IN SCHOOLS.—The Board of Education has issued a circular giving special courses of object lessons on common things. These have been sent to the masters of schools in country districts. They are only suggestive, and much is left to the masters with regard to their adaptation. The object of these lessons is to introduce to rural children instruction upon common things with which they meet every day. Some of the subjects suggested are: " Birds and their Habits," " Life-History of Common Insects." " Growth and Habits of some Wild and Garden Flowers," " Trees and the Commoner Kinds of Timber," " Living Things in Still and Running Water," " Soils, Mud, Sand, Clay, and Gravel," " Air, Weather-Charts, Rainfall, Frost and Heat, Ventilation," " Breathing, the Heart and Blood," " Clothing and Warming," " Chemistry of Food of Man, Beast, and Plant," " Levers and Pulleys," " Natural History Calendars," " Outdoor Studies in Geography and Land-Measuring."

LAND AND FRESHWATER MOLLUSCA OF HAMPSHIRE.

BY LIONEL E. ADAMS, B.A., WITH KIND ASSISTANCE OF B. B. WOODWARD, F.L.S., F.G.S.

(*Continued from page* 303.)

Buliminus montanus Drap. Selborne (W. J.).

Buliminus obscurus Müll. Generally distributed. Var. *albina* Moq., Winchester (J. W. Taylor); Ventnor (A. L.).

Cochlicopa lubrica Müll. Common. Var. *lubricoides* Fér., Christchurch (C. A.); var. *hyalina* Jeff., Isle of Wight (A. L.).

Azeca tridens Pult. No type recorded. Var. *crystallina* Dup., Petersfield (C. A.).

Caecilianella acicula Müll. Above chalk-pit, Afton Down, Isle of Wight. Also from Holocene deposits in the Test Valley, at Ventnor and at Totlands Bay.

Pupa secale Drap. A colony on the face of the cliff near Blackgang Chine (A. L.).

Pupa cylindracea Da Costa. Common and very fine in woods near Winchester (L. E. A.); Christchurch (C. A.); Isle of Wight (A. L.). Var. *albina* Moq., Christchurch (C. E. W.); Isle of Wight (A. L.). See also Proc. Malac. Soc. i. 296.

Pupa muscorum Linn. North Hants, rare (H. P. F.); Christchurch (C. E. W.); Hengistbury (C. A.); round Havant (C. E. W.); Isle of Wight (A. L.).

Vertigo minutissima Hartm. Ventnor (C. A.); subfossil in calcareous loam, Isle of Wight (A. L.).

Vertigo antivertigo Drap. Christchurch (C. A.); Ventnor, rare (A. L.).

Vertigo pygmaea Drap. Highcliff and Hengistbury, a few (C. A.); Ventnor (A. L.).

Vertigo moulinsiana Dup. Near Bishopstoke.

Balea perversa Linn. Isle of Wight (C. A.); round Havant on flint walls (C. E. W.); frequent in New Forest (C. A.). Var. *simplex* Moq., two specimens, Havant (A. L.).

Clausilia laminata Mont. North Hants, common (H. P. F.); Winchester (L. E. A.); Havant (C. E. W.); Isle of Wight (A. L.).

Clausilia bidentata Ström. Common. Var. *albina* Moq., Christchurch (C. A.); Havant (A. L.).

Clausilia rolphii Gray. Common in woods near Winchester, Petersfield (L. E. A.); Racton, Finchdean, abundant (C. E. W.).

Succinea putris Linn. Common.

Succinea elegans Risso. Common.

Carychium minimum Müll. Common.

Leuconia bidentata Mont. Whitecliff Bay.

Ancylus fluviatilis Müll. North Hants, common (H. P. F.); Christchurch in Avon and Stour (C. A.). Var. *albida* Jeff., Somerford Brook (C. A.); var. *stricta* Morch.

Velletia lacustris Linn. Hayling (W. J.); Havant (C. E. W.). Var. *albida* Jeff., River Stour, from Christchurch, where only a white form is found (C. A.); Hayling (W. J.).

Limnaea auricularia Linn.; Christchurch, in both rivers, never fine (C. A.).

Limnaea pereger Müll. Common in North Hants.

Limnaea stagnalis Linn. North Hants, common (H. P. F.): Christchurch, in R. Stour (C.A.).

Limnaea palustris Müll. Widely distributed. Var. *elongata* Moq., Isle of Wight (W. J.); var. *lacunosa* Zgl. (J. W. Taylor).

Limnaea truncatula Müll. Widely distributed.

Limnaea glabra Müll. Holmsley (C. A.).

Planorbis corneus Linn. Christchurch (C. A.) Itchin at Winchester (F. J.).

Planorbis nautileus Linn. Type not recorded. Var. *crista* Linn.; moderately abundant in Somerford Brook (C. A.).

Planorbis albus Müll. Widely distributed throughout the county.

Planorbis carinatus Müll. North Hants, common (H. P. F.); common in south of county.

Planorbis marginatus Drap. Common in south of county, but not mentioned in list for north.

Planorbis vortex Linn. Much less local round Christchurch than *P. spirorbis* (C. A.); Langstone (C. E. W.). Not included in Mr. Fitzgerald's list for North Hants. Not noticed in Isle of Wight (A. L.).

Planorbis spirorbis Linn. Widely distributed throughout the mainland and also in the Isle of Wight.

Planorbis contortus Linn. Somerford (C. A.); Langstone (C. E. W.); North Hants, common (H. P. F.)

Planorbis fontanus Lightf. North Hants, rare (H. P. F.); Christchurch, rare (C. A.).

Physa fontinalis Linn. Sandown, Isle of Wight (A. L.); rare in North Hants (H. P. F.); Havant (C. E. W.); Christchurch and district, common (C. A.).

Physa hypnorum Linn. Only two examples at Christchurch (C. A.); Langstone (C. E. W.); Havant (A. L.).

Paludestrina stagnalis Bast. Southampton (J. C. Melvill's Coll. *fide* E. R. Sykes).

Paludestrina jenkinsi Smith. Specimens labelled " *Hydrobia ferrusina*, Hampshire, Sowerby," in the Jeffreys collection at Washington, U.S., prove to be this species.

Bithynia tentaculata Linn. Widely distributed throughout the county. Var. *albida* (Rimmer), North Hants, rare (H. P. F.); Var. *producta* Menke, Tuckton (C. A.); *M. decollatum* (Jeff.), Holmsley (C. A.)

(*To be concluded.*)

THE question of the preservation of the stones at Stonehenge has been placed in the hands of the Charities and Records Committee, to whom the subject of ancient monuments has been referred.

MR. FRED PULLAR, who in conjunction with Sir John Murray recently published an extensive survey of the depths of many Scottish freshwater lochs, died at the end of February while rescuing a young lady who had fallen through the ice on Airthrey Loch.

IT has been decided to erect a marble bust of Dr. Gerhard Armauer Hansen, the discoverer of the bacillus of leprosy, in the Lungegard Hospital, Bergen, on the occasion of his sixtieth birthday in July next. The place chosen is the institution where he discovered the bacillus.

WE have received from the University Correspondence College Press a copy of the Matriculation Directory for 1901. It is a useful guide to those intending to enter for the examination in June or later, as it not only gives a list of textbooks, but also clearly indicates the style of questions and answers required.

THE well-known botanist Dr. J. C. Agardh, of Lund, sometimes called the Nestor of European botanists, died on January 17th at the age of eighty-seven. He was chiefly known for his work in marine algae, and had been for some years a correspondent of the Section of Botany at the Paris Academy of Sciences.

WE have already referred to "Man," the new monthly magazine issued under the auspices of the Anthropological Institute. The first three numbers are now before us, but considering that each contains only sixteen pages of literary matter with one plate, the charge to the public of a shilling per number is decidedly high compared with other magazines.

MR. HARRY F. WITHERBY, who has lately made an expedition to the White Nile in search of birds, is commencing in "Knowledge" a series of illustrated articles descriptive of the country, its people, its wild animals, and its birds. In the first instalment the author deals with his journey by river and the Desert Railway from Cairo to Khartoum.

WE have received from Mr. Charles Morley, Museum Press, Lockwood, Huddersfield, twelve sheets of Nature notes for the months of the year. They are sold at one shilling the set. Opposite each date the compiler of these sheets places some Nature note. For instance, January 4th: "Redbreast whistles." January 16th: "House sparrows chirp." February 5th: "Don't dig under chestnuts." February 10th: "Jackdaws frequent church steeples" (presumably because it was a Sunday this year). May 10th: "Cockchafer flies," a habit we have also observed on their appearance in certain human beings.

HIS MAJESTY THE KING has signified to the President and Council of the Marine Biological Association his pleasure in becoming the Patron of the Association.

WE have received "Proceedings of the Twenty-first Annual Meeting of the Society for Promotion of Agricultural Science," held at St. Louis, U.S.A., when Mr. William Trelease reviewed the advantage of public Botanic Gardens, working in conjunction with school plots in rural districts, for educational purposes. It appears that much has already been done in this direction.

DR. HUGH ROBERT MILL, F.R.S.E., lectured before the Royal Meteorological Society at its last meeting upon climate and the effects of climate. He specially drew attention to the difference between weather and climate, and exhibited a number of lantern slides showing the effect of climate in creating land forms.

MESSRS. ALEXANDER and MARTIN HEYNE, sons of the well-known Leipsic naturalist, Ernst Heyne, have opened rooms at 110 Strand, London, W.C., as an agency for the supply of natural history specimens, apparatus, and museum accessories. We understand they have a large stock of insects, especially of Palaearctic butterflies. It will be remembered that the former gentleman was connected with the authorship of Rühl's book on this subject, and other works.

FOLLOWING the plan adopted by the Cornell University, which issues "Nature Study Leaflets," the Agricultural Education Committee, of 10 Queen Anne's Gate, London, is now publishing a series of Teacher's Leaflets entitled "Nature-Knowledge." They are edited for the Committee by Mr. Wilfred Mark Webb, F.L.S., and are intended for distribution in schools, especially in rural districts. They are illustrated, and will doubtless appeal to the scholars as well as to the teachers. Four of these have already been published; No. 2, "Lilies from Leaves," being the most interesting.

MESSRS. N. ANNANDALE and H. C. Robinson, Hon. Research Assistant in the Zoological Department of University College, Liverpool, are about to leave England for a year's residence in the Malay Peninsula. They intend to reside in the neighbourhoods of Patani and Biseret, with the object of making collections in all branches of natural history. Especial attention will be paid to records of pre-Malayan tribes of Negrito origin. On the outward voyage Mr. Robinson proposes to study the plankton of the Red Sea and Indian Ocean by pumping sea water through fine silk nets.

THE preparations for the British Antarctic Expedition are now practically completed. The vessel, named the "Discovery," has been launched in Dundee and is to be equipped in London. The ship has been built on whaler lines, but with considerably greater strength than is usual. to withstand ice pressure. The expedition will leave in July or August, arriving in Melbourne in November. Commander R. F. Scott, R.N., is naval officer in charge; the scientific department being under the direction of Professor Gregory, of Melbourne University, assisted by Mr. Hodgson as biologist, and Mr. Schakleton as physicist. Those of our readers who are interested in Antarctic expeditions will find on p. 341 of this number of SCIENCE-GOSSIP a review of the book by M. Borchgrevink on the enterprise financed by Sir George Newnes.

NOTICES BY JOHN T. CARRINGTON.

Who's Who in 1901. xx + 1,234 pp., 7½ in. × 5 in. (London: A. & C. Black. 1901.) 5s. net.

This indispensable book of reference becomes stouter and more important-looking as its years advance. With this edition is incorporated "Men and Women of the Time," and new biographies have been added to include all Companions of Orders not previously inserted. In addition to biographies there is much other general information, which is brought down to October 31st last.

Convalescents' Diet. By HELEN LUCY YATES. viii + 88 pp., 6½ in. × 4¼ in. (London: H. Virtue & Co. 1901.) 1s.

The author of this useful little work, who has written another book on cookery from the French point of view, gives especial hints with regard to a suitable dietary for patients suffering from dyspepsia, neuralgia, anemia, and other troublesome complaints. None the less important are the recipes for preparing refreshing and restorative drinks; in fact, there is hardly a page in the book on which something useful will not be found for others as well as convalescents.

Comparative Physiology of the Brain and Psychology. By Professor JACQUES LOEB. x + 309 pp., 8½ in. × 5½ in. Illustrated with 89 figures. (London: John Murray. 1901.) 6s.

The object of this work, which is one of Mr. Murray's Comparative Science Series, is to serve as a short introduction to the comparative physiology of the brain and the central nervous system, especially the latter, as observed in non-vertebrate animals. Hitherto physiologists have turned their attention chiefly to the nervous organisation of vertebrates, and psychologists have concerned themselves only with the interpretation of the brain functions as shown in the higher orders of mammals, without reference to the physiology producing such functions, and have, therefore, entirely ignored the consideration of comparative psychology. Dr. Loeb, who is professor of Physiology at the University of Chicago, in the volume before us endeavours to resolve into their elementary components the central nervous system. He first deals with the class of processes called reflexes, which in the opinion of many psychologists may be defined as "the mechanical effects of acts of volition of past generations." It is also a usually accepted opinion that the ganglion-cell is the only place where such mechanical effects could be stored. It has therefore been considered the essential element of reflex mechanism, nerve-fibres being regarded simply as conductors. Professor Loeb, however, does not agree with this view, and he points out that the identity of the reactions of animals and plants to light show that the phenomena cannot depend upon specific qualities of the central nervous system, assuming, of course, that plants do not possess nerves. He also adds that in the case of *Ciona intestinalis* he had successfully shown the

complicated reflexes would continue after the central nervous system had been removed. The author is of opinion that a study of comparative physiology brings out the fact that irritability and conductibility are the only qualities essential to reflexes. Some of the experiments recorded are exceedingly interesting—for instance, those on a number of *Nereis*, which, when placed in a square aquarium, instead of crawling contentedly round the glass after their brains had been removed, on arriving at a corner attempted to continue straight on through the glass, continuing their efforts until they died. The author investigates the theories generally held on instinct, the nervous system, heredity, and associative memory. In many of his conclusions he differs considerably from other psychologists, and his points are well worthy of consideration by the reader.   F. W.

Manual of Elementary Science. By R. A. GREGORY, F.R.A.S., and A. T. SIMMONS, B.Sc. viii + 429 pp., 7 in. × 5 in., with 260 figs. (London and New York: Macmillans. 1901.) 3s. 6d.

This is a carefully compiled series of chapters suitable for the guidance of pupil-teachers taking elementary science with a view to presenting themselves in that subject at the Queen's Scholarship Examination. Also it provides a series of experiments, complete in themselves, illustrative of the principles of physical science, capable of being performed with simple apparatus. Each chapter is divided into numbered sections, corresponding to definite ideas. Evidently every effort has been employed to make, as far as possible, the statements in the descriptive text justifiable by the experiments proceeding. As a whole, the book can be strongly recommended, not only to those for whom it is intended, but also to many readers with scientific tastes who desire some further instruction in common phenomena. As may be easily understood, the section devoted to the celestial sphere and its diurnal motions is well done, having doubtless had the especial attention of Professor Gregory. The book is well worth its published price.

First on the Antarctic Continent. By C. E. Borchgrevink, F.R.G.S. xv + 333 pp., 9½ in. × 5¾ in. With 190 illustrations and 3 charts. (London: George Newnes, Ltd. 1901.) 10s. 6d. net.

We can well imagine the feelings of Sir George Newnes when he had placed in his hands the first copy of this handsome and deeply interesting book. It is the popular record of what has been done through Sir George's munificence in providing the whole funds for equipping and maintaining the recent British Antarctic expedition under the author's command. Sir George is never behindhand in supporting useful public works, and one cannot help reflecting on the different results accruing from the same amount of money spent on this enterprise, or on some good-natured charity for maintaining useless people who ought to disappear in Nature's battle for the survival of the fittest. M. Borchgrevink seems to have proved himself a capable commander with sufficient resource, especially during the terrible E.S.E. gales of wind that sweep the fringe of the Antarctic cap on the Australian side. These appear to be the most dreaded dangers of a winter spent in those forlorn regions, in consequence of drifting snow and pebbles. It is estimated that these gales, raging at a pace of from eighty to one hundred miles an hour, reduced the explorers' opportunities by 20 per cent. The aspect of travel in the Antarctic is quite different from the Arctic. In the former the land attains mountainous altitudes estimated at 12.000 feet, whilst the broad valleys are occupied by immense and difficult glaciers. Unlike the Arctic, there are not any of the usual mammal land-fauna, such' as bears, foxes, musk-oxen, and reindeer; therefore land-travel depends wholly upon the food taken with the party. This cause, suggests the author, will confine Antarctic exploration to centres near the coast and local expeditions therefrom, rather than long journeys, as with sledges near the North Pole. The southern ice cap is so much larger and apparently wilder than the northern, that we may readily understand different forms of exploration will be necessary. This book does not attempt a description of the

Insect life is evidently very sparse, a few being found by Dr. Klovstad among mosses. There is no description, but they were probably apterous. The marine algae and the ocean fauna prove bipolarity in many cases, though the fauna is not so on land. That land animals were not discovered does not, as the commander points out, prove the non-existence of land animals of any kind. At any rate, around both polar caps certain lichens like the reindeer moss flourish in abundance. From the geological point of view, this first year of human residence should add greatly to knowledge. The author had once previously touched at Cape Adare in 1895, hence his choice of that point for the winter camp. Whether good or not, we must ask the reader to form independent judgment but considering the previous small know-

A CRYSTAL PALACE.

(From Borchgrevink's "First on the Antarctic Continent.")

scientific side of the expedition excepting in appendices; still its pages teem with nature-notes. Many of them are entertaining, as, for instance, the account of the arrival, love-making, nesting, habits of the young, and other features connected with the countless hordes of penguins that breed along shingle banks. It would be worth a winter's discomfort to see the fun so well described by the author. The land flora is limited to mosses and lichens, reindeer lichen being the largest plant. The marine life is evidently abundant in specimens and numerous in species. In summer there are immense jelly-fish weighing up to 100 lbs. Of crustacea and starfish there are abundance; several species of marine fish occur at all the seasons, with seals; whales and many sea birds come in summer.

ledge of South Victoria Land, the commander had but little data for choice of base. M. Borchgrevink has written an interesting account of the voyage; his style is good, and at times rises above the ordinary level of travel books. His pathetic yet manly description of Hanson's death—the only man he lost from the expedition—appeals to those who can appreciate his position. Another event, in which he nearly lost his own life. is modestly but effectively told; there he describes the birth of an iceberg and consequent great wave. The illustrations in the book are admirable and abundant. One of these we reproduce by permission of the publishers, showing a monarch among icebergs. We strongly recommend this latest addition to the works on Arctic travel.

CONDUCTED BY F. SHILLINGTON SCALES, F.R.M.S.

ROYAL MICROSCOPICAL SOCIETY, February 20.— Mr. A. D. Michael, Vice-President, in the chair, A photograph of *Amphipleura pellucida*, taken by Mr. Brewerton, was exhibited. Mr. Nelson said the photograph was interesting because it showed the transverse striae as thin in comparison with the spaces between them. Some optical theorists maintained that the striae and spaces must be of equal width, whereas he had affirmed that the striae were much finer than the spaces. In many photographs of this object they appeared to be of equal width, but that was because the object had been badly photographed. In the example before the meeting the photograph had been properly taken, and therefore exhibited the difference in the thickness of the line and the interspaces. Mr. Rogers brought to the meeting a contrivance for exhibiting a fly in the act of feeding. This differed in some respects from Mr. Macer's arrangement for a like purpose, being a brass plate 3″ × 1″, underneath which a brass cone was soldered to contain the fly, the plate lying on the stage of the microscope like an ordinary slide. Mr. E. M. Nelson read a paper on the tube-length of the microscope, explaining the difference between the mechanical and optical tube-length, illustrating the subject with drawings and formulae. The Chairman thought there was no subject connected with the technique of the microscope about which ideas were more vague than that of the tube-length. Many were of opinion that it was the length of the brass tube, although it had often been pointed out in the Society's room that what was really meant was the optical tube-length. The subject did not seem to be very well understood. Very little practical information had been published that would enable a person to ascertain the tube-length of his microscope, but Mr. Nelson had now given them a method by which this could be found. Mr. F. W. Millett's "Report on the Recent Foraminifera of the Malay Archipelago" was taken as read. The Chairman called attention to a set of slides of Bacteria and Blood Parasites exhibited by Mr. Conrad Beck. Some mounted rotifers which had been sent from Natal by the Hon. Thomas Kirkman were likewise exhibited.

THE QUEKETT MICROSCOPICAL CLUB.—We feel that we shall be doing a service to our readers, especially those resident in or near London, if we call attention to the assistance they may derive from membership in the Quekett Microscopical Club. In this magazine (S.-G., vol. vi., p. 313), quoting from the journal of the Club, we alluded to the intimate connection between the Club and SCIENCE-GOSSIP in its early days. It should be understood that the Club in no way trespasses upon the ground covered by, or endeavours to compete with, the Royal Microscopical Society, the centre of microscopy throughout the world. The Club was originated, and has always existed, mainly for the benefit of those who take an interest in microscopical research as a means of intelligent recreation ; in other words, for the encouragement of the study of microscopy amongst amateurs. Several of its Presidents have been eminent men of science, such as Lankester, Beale, Huxley, Cobbold, Carpenter, Dallinger, and others. Established in 1865, it has had its home since 1890 at 20 Hanover Square. The meetings are held on the first and third Fridays in each month. Those on the first Friday are called "Gossip" meetings, and are especially helpful to beginners, being for the exhibition of objects and for conversation. Those on the third Friday are mainly devoted to the reading and discussion of papers descriptive of new instruments and the ordinary business of the Club. On Saturday afternoons, during the spring, summer, and autumn, excursions are arranged to well-known collecting-grounds near London. These are of great value to those interested in pond life, and also help to promote the spirit of comradeship—a special feature of the Club. An excellent journal, which is reviewed in these columns, appears twice a year—in April and November. This is sent free to all members. The Club is in possession of a valuable microscopical library as well as of a cabinet comprising about 6,000 slides. The subscription is ten shillings per annum, and there is no entrance fee. The Secretary is Mr. G. C. Karop, 198 Holland Road, W.

SHEFFIELD MICROSCOPICAL SOCIETY.—The Transactions and Annual Report of this society for the session 1899–1900 have just reached us. The report shows a membership amounting to 158, being a slight increase on the previous year; also a satisfactory financial position. The Society is now twenty-two years old, so has more than attained its majority. During the past year there have been twelve fortnightly meetings, supplemented by discussion and conversation ; but we note, from remarks made by the President in his annual address, that the Society has apparently neglected that most valuable means of sustaining the interest of members and attracting new recruits to microscopy, *i.e.* the organising of field excursions. Amongst the papers read was one on "The Bryozoa, Recent and Fossil," by Mr. G. R. Vine, B.Sc.; another on "The Adulteration of Foodstuffs," by Mr. G. E. Scott-Smith, F.I.C., F.C.S. ; and an account by Dr. Sorby, F.R.S., of his methods of preparing and preserving marine animals. This is by means of an equal mixture of glycerine and sea-water, in which the animals are placed for ten minutes direct from the sea-water. They are then transferred to water, and kept there for another ten minutes to remove most of the glycerine. They thus become quite limp, and can be easily arranged on the slides, and allowed to dry at ordinary temperatures in the open air before being finally mounted in balsam. By this means the objects are not only permanently preserved, but are made comparatively thin, flat, and transparent, and even the natural red colour of the blood is preserved. This is a great improvement upon the ordinary method of preserving in alcohol or formalin, which, as we ourselves have too often found, not only renders the specimens opaque, but causes contraction and distortion. We alluded to this method in SCIENCE-GOSSIP (vol. vi., page 30). The report contains portraits of Dr. Sorby ; of the founder of the Society, Mr. William Jenkinson; of the first President, Mr. G. R. Vine ; and of the President for the year,

Mr. G. T. W. Newsholme, F.C.S. The Hon. Secretary is Mr. John Austen, of 27 High Street, Sheffield. During the past year the Society has been enriched by the presentation of a Powell and Lealand No. 1 microscope, with Zeiss' apochromatic objectives and eyepieces, together with lamp and other accessories. It is sad to have to record that the donor of this princely gift, Mr. Samuel Cocker, of Sheffield, who had himself been a member for only a few months, died about a month later, in January of last year, before his gift was ready for presentation.

MOUNTING IN GLYCERINE.—Mr. John H. Schaffner, in the American "Journal of Applied Microscopy," recommends a rather original method of mounting objects in glycerine, which has proved very satisfactory in his hands. Ordinarily speaking, glycerine is a most difficult and often treacherous medium in which to mount, if only on account of the difficulty of securely and permanently closing the cell. Mr. Schaffner's method is to transfer the objects from water to pure glycerine by adding the glycerine gradually and permitting the water to evaporate until absolutely pure glycerine alone is left. The objects are then placed in a small drop of glycerine jelly on a slide, and the whole of the remaining broad space to be subsequently covered by the cover glass is filled with Canada balsam. The two media will not mix, and the wide surrounding film of Canada balsam makes a secure mount, even without sealing.

METHOD OF PARAFFIN INFILTRATION.— Mr. C. M. Thurston in the "Journal of Applied Microscopy" states that he has found the following method of paraffin infiltration very successful. The essential feature consists in applying heat to the upper surface of the paraffin of such an intensity as to melt the paraffin only for a sufficient depth to submerge the tissues requiring infiltration. The object lies on the unmelted paraffin, and recedes from the heat if the heat increases, and the paraffin melts more deeply. Small glass cups, 4 c.m. in diameter by 5 c.m. deep, are filled with melted paraffin, which is allowed to cool. The cups are then placed under a copper plate suspended on a tripod or retort stand. The flame should be at such a distance and of sufficient intensity to melt the paraffin 1 or 2 cm. deep.

BAUSCH & LOMB'S "BB" MICROSCOPE.—Messrs. A. E. Staley & Co., of 35 Aldermanbury, E.C., have submitted to our notice several of the stands and accessories manufactured by the Bausch & Lomb Optical Company of America, to whose catalogue we recently briefly referred, and for whom Messrs. Staley are the English agents. As we have endeavoured to make these columns a record of microscopical progress, and to keep our readers informed of all new stands and accessories that are brought out from time to time by the various makers, we are glad of an opportunity to call attention to the stands of this firm. The cheapest was a non-inclinable horseshoe stand, fitted with coarse adjustment of sound workmanship, substantial stage and mirrors, and sold for twenty-one shillings, or with a divisible 1½- and ½-inch objective and one eyepiece for £1 18s. 6d. This is really a workmanlike instrument, and would serve many purposes of a beginner. Nearly all the instruments are on the Continental model, with pillar and horseshoe foot, and micrometer screw fine adjustment of the triangular bar form. We illustrate the "BB" stand

herewith. It is of brass throughout, with spiral rack and pinion coarse adjustment; fine adjustment as mentioned above, with graduated head ; graduated nickelled draw-tube, working in cloth-lined sleeve, which gives a very smooth movement ; and vulcanite stage 3 × 3¼ inches. The draw-tube, however, only extends to 190 millimetres (7½ inches), which, though sufficient for objectives corrected for the Continental length of tube, is not long enough for those corrected for the standard English length. Both main tube and draw-tube are fitted with the English "society" screw, and the eyepieces are of R.M.S. No. 2 gauge. In the stand illustrated the sub-stage, which is also of the "society" gauge, is adjustable by means of spiral screw; but an ingeniously arranged iris-diaphragm is fitted so as to work immediately beneath the opening of the stage itself for use with condenser removed, the condenser having also its independent iris-diaphragm. The price of this stand alone without eyepieces, objectives, or condenser is £5 ; in

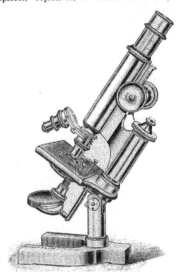

BAUSCH & LOMB "BB" MICROSCOPE.

polished wood case, with one eyepiece and ⅔-inch and ¼-inch objectives, £7 4s. 10d. ; and with Abbé condenser with iris-diaphragm, 15s. extra. We have tested both objectives and found them excellent, and particularly free from colour, whilst a ¹⁄₁₂-inch immersion objective, of 1·32 N.A., sold at £5 3s., proved to be one of the best lenses we have had through our hands. The same microscope is made without inclination and a plain sub-stage ring for condenser at a proportionately lower price, also with an elaborate sub-stage of the later Continental type, which we think, however, unnecessarily complex. The workmanship of all the stands is excellent.

For further articles on Microscopic subjects see pp. 329, 332, and 333.

MEETINGS OF MICROSCOPICAL SOCIETIES.

Royal Microscopical Society, 20 *Hanover Square, London.* April 17, 8 p.m.
Quekett Microscopical Club, 20 *Hanover Square, London.* April 5, 19, 8 p.m.

ANSWERS TO CORRESPONDENTS.

F. L. (Sunderland).—The mite to which you refer is *Cheyletus eruditus.* It is not a cheese-mite, but belongs to an allied family, the Trombidiidae. An account of it can be found in SCIENCE-GOSSIP for 1869, page 5. It is there stated to feed on cheese-mites and other acari. According to the late Mr. Richard Beck, by whom it was first discovered, parthenogenesis has been observed in this species.

W. W. (Clayton West).—I know nothing of the microscope to which you refer other than the advertisement. I recommend you to have nothing to do with it. The twelfth immersion objective alone, if good, would be worth £5. If you are about to buy a microscope, go to a good maker, and do not be persuaded into buying an inferior instrument, such as is unfortunately too often displayed for sale even in responsible opticians' windows. If I can advise you on the matter I shall be pleased to do so.

EXTRACTS FROM POSTAL MICROSCOPICAL SOCIETY'S NOTEBOOKS.

[These extracts were commenced in the September number, at page 119 of the present volume. Beyond necessary editorial revision, they are printed as written by the various members. Correspondence thereon will be welcomed.— ED. Microscopy, S.-G.]

DEVELOPMENT OF GNAT.—My object in these slides is to illustrate the transformation of an insect, a subject I wish we saw more illustrations of in our Society. Unfortunately, I am not able to send a slide of the eggs of the gnat, but in "Science for All" Mr. Hammond says that they are laid in small boat-shaped masses which float on the surface of the water. The eggs themselves are of an oval form with a kind of knot at one end, and are arranged side by side and closely packed together. In Duncan's "Transformation of Insects" it is thus written: "The male gnats have pretty hairy antennae, like little feathers, and the females have antennae which are almost plain. It is therefore not difficult to distinguish one from the other, and it is rather important, for the females are the blood-suckers. When about to lay their eggs they seek the water, and with the assistance of their long hind legs collect and agglutinate them together and place the little boat-shaped mass upon the surface of the water, and then leave it to its fate." The larvae are soon hatched, and grow with great rapidity. They are almost always seen with their heads downwards and their tails towards the surface of the water. After the larvae have grown to a certain size they undergo a change of skin and become nymphs or pupae, and it may be noticed that the nymphs come up to the surface of the water but do not present their tails like the larvae, so as to obtain air, but allow their backs to touch the surface, just where there are two respiratory tubes. When the perfect insect is about to emerge from the nymph stage it floats on the surface of the water, perfectly at rest, and the skin of the back, which is exposed to the air, dries and splits open. Then the perfectly-formed insect begins to come out: first it protrudes its head, then a portion of its body, and after a short time one leg after the other is disengaged from the nymph skin; after a little while it tries its wings and flies away. Fig. 1 shows the beautiful breathing and swimming organs of the larva. Fig. 2 shows the larva, and fig. 3 the pupa complete, taken from Duncan. The pupa only shows the respiratory tubes on its back. Fig. 4 represents the mouth organs of the female and fig. 5 of the male gnat. It will be noticed that the female gnat has no halteres.—*T. G. Jefferys.*

The best account of the gnat known to me is that given by Professor Miall in his "Natural History of Aquatic Insects," from which I make the following excerpts: "Small stagnant pools and ditches are the favourite haunts of the larvae and pupae of the gnat. A ditch in a wood choked with fallen leaves is one of the best hunting-grounds, and in the summer months they may be found by the thousand in such places. The larva, when at rest,

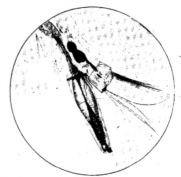

FIG. 1. BREATHING AND SWIMMING ORGANS OF LARVA OF GNAT.

floats at the surface of the water. Its head, which is provided with vibratile organs suitable for sweeping minute particles into the mouth, is directed downwards, and, when examined by a lens in a good light, appears to be bordered below by a gleaming band. There are no thoracic limbs; the hind limbs, which are long and hooked in the chironomous larvae, and reduced to a hook-bearing sucker in *Simulium*, now disappear altogether; a new and peculiar organ is developed from the eighth segment of the abdomen. This is a cylindrical respiratory syphon, traversed by two large air-holes, which are continued along the entire length of the body to supply every part with air. The larva ordinarily rests in such a position that the tip of the respiratory syphon is flush with the surface of the water, and thus suspended it feeds incessantly, breathing uninterruptedly at the same time." Professor Miall's explanation as to how it is possible for a larva heavier than water to remain floating at the surface without effort, as the larva of the gnat appears to do, is too long to give here. It deals with the surface film. "After three or four months the larvae are ready for pupation. By this time the organs of the future fly are almost completely formed, and the pupa assumes a strange shape, very unlike that of the larva. At the head end is a great rounded mass which encloses the wings and legs of the fly, besides the mouth parts and other organs of the head. Each appendage has its own sheath, part of the proper

FIG. 2. LARVA OF GNAT.

pupal skin, and the appendages are cemented together by some substance which is dissolved or softened by alcohol. At the tail end is a pair of flaps which form an efficient swimming fan. The body of the pupa, like that of the larva, is abundantly supplied with air-tubes, and a communication with the outer air is still maintained, though

FIG. 3. PUPA OF GNAT.

in an entirely different way. The air-tubes no longer open towards the head. Just behind the heart of the future fly is a pair of trumpets, so placed that in a position of rest the margins of the trumpets come flush with the surface of the water. Floating in this position the pupa remains so long as it is undisturbed : but if attacked by any of the

FIG. 4. FEMALE.   FIG. 5. MALE.

HEAD PARTS OF GNAT.

predatory animals which abound in the fresh water it is able to descend by the powerful swimming movements of its tail." Then follows an explanation, too long to quote, as to why the respiratory organs are changed from the tail end in the larva to the head end in the pupa. "But a time comes when the fly has to escape from the

pupa-case. The skin splits along the back of the thorax, and here the fly emerges, extricating its legs, wings, head, and abdomen from their closely-fitting envelope." Drawing from photograph (fig. 6) represents the manner of emergence. "The mouth of the female gnat is provided with a case of instruments for piercing the skin and drawing blood. The foremost of these is a tube split along its hinder side, which lies in front of the rest, and is used in suction. This, though long and slender, is stouter than the delicate parts behind it, and it serves to stiffen and protect them then come fine, long, and slender blades of great delicacy. Two pairs correspond to the mandibles and maxillae of other insects, though here they are so simplified and attenuated that it is not easy to make out the correspondence. The maxillae are furnished near their tips with a row of

FIG. 6. EMERGENCE OF GNAT.

extremely minute saw-teeth. There is also a fifth unpaired implement, which is an extraordinary development of a part of the insect's mouth, which is usually quite inconspicuous. Besides these piercing implements, the gnat is provided with a soft, flexible sheath which represents the labium. This takes the shape of a tube split along its fore side, which surrounds and protects the delicate parts within. The extremity is divided into two lobes."—*J. J. Wilkinson.*

[Mr. Wilkinson's quotations from Professor Miall need no further explanation. The gnat (*Culex pipiens*) would make an excellent study for microscopical beginners, perhaps even more so than the common cockroach. We may call attention in addition to the beautiful antennae of the male gnat, and to the scales upon the wings and body, which latter can be readily removed by means of a camel-hair brush, and so transferred to a slide. The larva in particular makes a most interesting microscopical object, owing to its transparency, which enables the tracheal tubes, the digestive tube, and contractile vessel that performs the duty of the heart to be readily made out. The gnat *C. pipiens* must not be confused with the allied genus *Chironomus,* or Midges.—ED. Microscopy, S.-G.]

ASTRONOMY

CONDUCTED BY F. C. DENNETT.

1901	Rises.	Sets.	Position at Noon.	
			R.A.	Dec.
	Apr.	h.m.	h.m.	h.m.	° ′
Sun ..	5 .. 5.30 a.m. .. 6.37 p.m. ..		0.55 .. 5.55 N.
	15 .. 5.7 a.m. .. 6.53 p.m. ..		1.32 .. 9.37 N.
	25 .. 4.47 a.m. .. 7.9 p.m. ..		2.0 .. 13.3 N.

	Rises.	Souths.	Sets.	Age at Noon
	Apr.	h.m.	h.m.	h.m.	d. h.m.
Moon ..	5 .. 8.34 p.m. .. 0.39 a.m. ..	5.39 a.m. .. 15 23.7
	15 .. 3.11 a.m. .. 8.50 a.m. ..	2.42 p.m. .. 25 23.7
	25 .. 10.46 a.m .. 6.15 p.m. ..	1.5 a.m. .. 6 14.23

	Souths.	Semi-	Position at Noon.	
		diameter.	R.A.	Dec.
	Apr.	h.m.	″	h.m.	° ′
Mercury ..	5 .. 10.23·9 a.m. .. 3·7″ .. 23.16 .. 6.47 S.
	15 .. 10.31·2 a.m. .. 3·2″ .. 0.3 .. 2.31 S.
	25 .. 10.48·7 a.m. .. 2·8″ .. 1.0 .. 3.48 N.
Venus ..	5 .. 11.40·6 a.m. .. 4·9″ .. 0.33 .. 2.2 N.
	15 .. 11.46·7 a.m. .. 4·9″ .. ′.19 .. 6.57 N.
	25 .. 11.53·7 a.m. .. 4·9″ .. -2.5 .. 11.37 N.
Mars ..	5 .. 8.52·1 p.m. .. 5·6″ .. 9.46 .. 16.44 N.
	15 .. 8.16·1 p.m. .. 5·1″ .. 9.48 .. 16.7 N.
	25 .. 7.42·8 p.m. .. 4·7″ .. 9.54 .. 15.10 N.
Jupiter ..	15 .. 5.24·3 a.m. .. 18·5″ .. 18.55 .. 22.40 S.
Saturn ..	15 .. 5.39·6 a.m. .. 7·8″ .. 19.10 .. 21.55 S.
Uranus ..	15 .. 3.31·6 a.m. .. 1·9″ .. 17.2 .. 22.47 S.
Neptune ..	15 .. 4.13·9 p.m. .. 1·2″ .. 5.46 .. 22.14 N.

MOON'S PHASES.

		h.m.			h.m.
Full	.. Apr. 4 .. 1.20 a.m.	3rd Qr. Apr. 12 .. 3.57 a.m.
New	.. „ 18 .. 9.37 p.m.	1st Qr. „ 25 .. 4.15 p.m.

In apogee April 5th at 6 a.m.; and in perigee at 9 p.m. on the 18th.

METEORS.

			h.m.	°
Mar. 11 to May 31	Draconids Radiant R.A.17.32 Dec. 50 N.
Apr. 5 „ 10	Lyrids* „ „ 18.8 „ 35 N.
„ 9 „ 12	(42 Herculis) „ „ 16.36 „ 51 N.
„ 12 to June 30	Coronids „ „ 15.40 „ 23 N.
„ 17 „ 25	β Serpentids „ „ 15.24 „ 17 N.
„ 19	(μ Serpentis) „ „ 15.16 „ 2 S.
„ 20	(ξ Herculis) „ „ 17.57 „ 32 N.
„ 29 to May 6	η Aquaridst „ „ 22.28 „ 2 S.

* Principally visible April 17 to 22.
† Just before sunrise.

CONJUNCTIONS OF PLANETS WITH THE MOON.

				° ′
April 11 ..	.. Jupiter*† .. Noon	.. Planet 3.50 S.
„ 11 ..	.. Saturn† .. 7 p.m. ..	„ 3.42 S.
„ 17 ..	.. Mercury* .. 9 a.m. ..	„ 7.5 S.
„ 18 ..	.. Venus† .. 7 p.m. ..	„ 3.47 S.
„ 27 ..	.. Mars* .. 2 p.m. ..	„ 7.57 N.

* Daylight. † Below English horizon.

OCCULTATIONS, AND NEAR APPROACH.

				Angle		Angle
		Magni-	Dis-	from	Re-	from
Apr.	Star.	tude.	appears.	Vertex.	appears.	Vertex.
			h.m.	°	h.m.	°
4 ..	ι Virginis	5·5 .. 7.39 p.m. .. 239 near approach.
8 ..	ω² Scorpii	4·6 .. 3.59 a.m. .. 71 .. 5.19 a.m. .. 267
8 ..	ω¹ „	4·1 .. 4.5 a.m. .. 5 .. 4.20 a.m. .. 342

THE SUN seldom has disturbances visible, but still should be watched. A small interesting group was on the disc early in March.

MERCURY reaches its greatest western elongation, 27° 48′, at 6 a.m. on April 4th, and two hours later attains the part of its orbit most distant from the Sun; yet from the position of the two bodies in the heavens, only 37 minutes elapse between their rising, so that the planet cannot be observed.

VENUS is too near the Sun for observation.

JUPITER and SATURN are both near each other in Sagittarius, and rise in the south-east about two and a half hours before midnight at the beginning of the month and about two hours earlier at the end. Saturn's rings are still open beyond his poles: on 15th his diameter is 15″·6, whilst the minor axis of his outer ring is 15″·96.

URANUS, still in the southern part of Ophiuchus, rises about two hours earlier than the two brighter planets.

NEPTUNE is getting too far to the north-west for very favourable observation, although he does not set until after midnight.

MARS is in Leo all the month, a little north-west of Regulus. Its apparent diameter is rapidly decreasing. Two drawings, made by aid of Wray's 3-inch SCIENCE-GOSSIP telescope and the 216 power, are appended to show how much a small good instrument will delineate of Mars, even when unfavourably placed, as he has been during this opposition. The long funnel-shaped dark marking on the second figure was called the Kaiser

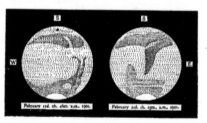

February 12d. 0h. 0m. a.m., 1901.   February 21d. 1h. 15m. a.m., 1901.

Sea by Proctor and Green, but is now known as Syrtis Major. The dark patch nearest the central meridian, in the southern part of the first figure, is now known as Sinus Sabaeus, but was formerly called Herschel II. Strait. The two long dark markings in the eastern part of this figure are the Indus and Hydaspes. The north Polar Cap is readily visible in both figures.

THE NEW STAR IN PERSEUS.—Particulars of this interesting object will be found on page 324 in this number of SCIENCE-GOSSIP.

THE LEONIDS.—An observer at York Factory, Hudson Bay, on November 15th, 1900, describes a "very general display of shooting stars. Some very big ones, north-west to south-east. Sky full in shoals. November 16, shooting stars seen until daylight." This communication has been sent to the President of the Toronto Astronomical Society through Mr. R. F. Stupart, Director of the Toronto Observatory and Superintendent of the Meteorological Service of Canada.

SPECTRA OF SUN AND JUPITER on p. 318. The double line in the diagram at 1000 should be marked D and the line at 800 C.

CONDUCTED BY C. AINSWORTH MITCHELL,
B.A.OXON., F.I.C., F.C.S.

INTERNATIONAL ATOMIC WEIGHTS.--The Atomic Weights Committee of the German Chemical Society. has just issued its report for 1901, in which it has drawn up two tables of atomic weights. In one of these the values are based upon Oxygen = 16, and Hydrogen = 1·008 ; and in the other upon H = 1, and O = 15·88. The Committee suggests that the former shall be termed "International Atomic Weights," and be used wherever absolute values are required, as in physico-chemical researches. For the second table (H = 1), which, on the whole, is more convenient, especially for teaching purposes, the title "Didactic Atomic Weights" is proposed. Argon, helium,. krypton, neon, thulium (= 171) and xenon find a place in these tables, which include in all 76 elements.

DISADVANTAGES OF ALUMINIUM AND THEIR REMEDY.—The great expectations which were formed of the prospects of aluminium, when first it was obtained at a relatively cheap rate, have, unfortunately, not been realised. As time went on it was discovered that, contrary to what was first believed, it readily acted upon by solutions of many salts, by acids, including vinegar, and especially by alkalies, though it excelled copper in the resistance which it offered to nitric acid. To obviate this drawback it is now frequently coated with a layer of silver ; but this in itself is no easy matter, for aluminium is so porous that it retains water and impurities obstinately, with the result that bubbles are formed in the electro deposit, or that this peels off in the polishing process. The chief remedy is to thoroughly cleanse the aluminium, and to give it a preliminary coating with another metal, preferably copper, before silvering. Several patents have recently been taken out for processes on these lines. Another great drawback to the general use of aluminium is that it has hitherto been very difficult to unite two separate pieces of the metal. This objection, however, will be probably met by the welding process of Heraeus, which has recently been protected in several countries. In this the welding is so complete that the juncture is practically invisible, and the aluminium can be rolled out to a thin sheet without separating into its component parts.

ALLEGED CONVERSION OF PHOSPHORUS INTO ARSENIC. - Some years ago, in an address to the British Association, Sir William Crookes brought forward an ingenious speculation to account for the remarkable relationship which exists between the different elements. He suggested that the whole of these might possibly have been derived from a single primeval element, which as it cooled condensed under recurring conditions (except as to temperature) to form the different substances now known to us as "elements,"

beginning with the lightest, hydrogen. and ending with the heaviest, uranium. This process might be compared to the beats of a pendulum in which the elements which condensed at definite points on the forward stroke were allied in their properties to those condensing at the same points in the return stroke. It was also suggested that this primeval element might be helium, which had not then been discovered, and the existence of which had only been inferred from its absorption lines in the solar spectrum. Now, in this speculation, for which of course there is no experimental proof, we have something closely akin to the germinal idea of the "philosopher's stone" of the mediæval alchemists. It is quite conceivable that it is only for the want of sufficiently powerful means, i.e. the "lapis philosopherum," that we are unable to decompose the bodies we now call "elements" into their hypothetical constituents or to convert them into one another. In fact, during the last few years more than one claim has been brought forward for such a transmutation. The American "argentaurum," which professed to be gold produced from silver, is a case in point, but the examination of the process by other chemists showed that any gold in the final product was there originally. A still more recent instance is the alleged conversion of phosphorus into arsenic, which, as is well known, show many remarkable relationships in their properties. Fittica claimed that when phosphorus was heated with nitre it was partially converted into arsenic. His experiments were repeated by C. Winkler, however, who came to the conclusion that any arsenic found by Fittica must have been present as an impurity in the phosphorus. This is confirmed by Noelting and Feuerstein, who have recently shown it is not an easy matter to obtain phosphorus absolutely free from arsenic, although the latter can be eliminated by distilling the phosphorus twice in a current of steam.

ALCOHOL FROM SAWDUST.—It is generally known, especially since the recent investigations into the cause of arsenic-poisoning in beer. that starch is converted into a fermentable sugar, glucose, when treated with dilute acid. Dr. Simonsen, of Christiania, has recently solved the difficult problem of obtaining a similar product from sawdust. Cellulose, which is one of the principal constituents of sawdust, is closely allied to starch, but all attempts to "invert" it with sulphuric or hydrochloric acid had previously proved unsuccessful. The first experiments were made with cellulose (paper). which was treated with dilute acid under pressure. The conditions for the best yield of sugar were thus obtained. Then, basing his process on the results of experiments on both a small and manufacturing scale, Simonsen succeeded in obtaining about 22 per cent. of fermentable sugar from sawdust. the inversion being made in a closed boiler under a pressure of about eight atmospheres. The sugar was apparently entirely derived from the cellulose, the allied compound, lignin, being unaffected by the treatment. The yield of sugar from pine sawdust was greater than from fir sawdust, but the largest quantity (about 31 per cent.) was obtained from birch sawdust. Of this sugar about 75 per cent. was fermentable. and the spirit distilled from the fermented liquid was remarkably pure and of good flavour. About 1½ gallon of absolute alcohol was obtained from 225 lbs. of sawdust.

CONDUCTED BY B. FOULKES-WINKS, M.R.P.S.

EXPOSURE TABLE FOR MARCH.

The figures in the following table are worked out for plates of about 100 Hurter & Driffield. For plates of lower speed number give more exposure in proportion. Thus plates of 50 H. & D. would require just double the exposure. In the same way, plates of a higher speed number will require proportionately less exposure.

Time, 10 A.M. to 2 P.M.

Between 9 and 10 A.M. and 2 and 3 P.M: double the required exposure. Between 8 and 9 A.M. and 3 and 4 P.M. multiply by 4.

SUBJECT	F.5·6	F.8	F.11	F.16	F.22	F.32	F.45	F.64
Sea and Sky..	$\frac{1}{161}$	$\frac{1}{110}$	$\frac{1}{110}$	$\frac{1}{40}$	$\frac{1}{21}$	$\frac{1}{12}$	$\frac{1}{6}$	$\frac{1}{4}$
Open Landscape and Shipping	$\frac{1}{116}$	$\frac{1}{60}$	$\frac{1}{31}$	$\frac{1}{16}$	$\frac{1}{8}$	$\frac{1}{4}$	$\frac{1}{2}$	1
Landscape, with dark fore-ground, Street Scenes, and Groups ..	$\frac{1}{32}$	$\frac{1}{10}$	$\frac{1}{8}$	$\frac{1}{4}$	$\frac{1}{2}$	1	2	4
Portraits in Rooms ..	2	4	8	16	32	—	—	—
Light Interiors	4	8	16	32	1	2	4	8
Dark Interiors	16	32	1	2	4	8	16	30

The small figures represent seconds, large figures minutes. The exposures are calculated for sunshine. If the weather is cloudy, increase the exposure by half as much again ; if gloomy, double the exposure.

ROYAL PHOTOGRAPHIC SOCIETY.—At the annual general meeting held on Tuesday, February 12th, 1901, in the Society's Rooms at 66 Russell Square, W.C., Mr. Thomas R. Dallmeyer, F.R.A.S., the President. in the chair, the Society's ' Silver Progress Medal ' was formally presented to Dr. R. L. Maddox. Mr. Bedding, Editor of the " British Journal of Photography," received the medal on behalf of Dr. Maddox, who was unfortunately unable to be present, owing, we are sorry to say, to age and infirmities preventing him travelling to London. The President expressed the pleasure it gave him at being privileged to make this presentation. Dr. Maddox has devoted much time and study to Photomicrography, of which he was a very brilliant exponent ; but the reason the Society had awarded him this Progress Medal was for the introduction of the gelatine-bromide of silver dry plate. Of photography it might truly be said that the introduction of the gelatino-bromide dry plate had done more for the art than had any single step of progress since the days of Daguerre. Of the many thousands who practise the art to-day, from the operator in the studio to the amateur who is satisfied with a little "snapshotting" on his or her holiday tour, all are more or less indebted to Dr. Maddox. The "Journal" of the Society is just to hand, and we see that as yet there are only two meetings down for April : they are April 2nd,

Lantern meeting, when Mr. Charles Reid will give an illustrated lecture entitled " Animals and Birds in their Native Haunt "; and April 9th, an ordinary meeting, when Mr. Wm. Webster will read a paper entitled " Notes from Five Years' Work with X-Rays."

TELEPHOTO-LENSES.—Messrs. Staley & Co. have submitted to us for inspection and trial a new tele-attachment, made by Bausch & Lomb, of Rochester, U.S.A. The leading feature of the attachment is that it can be fitted to any existing lens of a given focus. They are made in various sizes and negative power, so that all that is necessary is to send through a dealer the focal length of any lens that it is desired to convert into a telephoto-lens, and a suitable attachment can be supplied. These are so made that they may very readily be attached to a lens, and as easily detached when not in use. They are of moderate magnifying power, enlarging from two to four and a half times, and can be used on any ordinary camera with moderate amount of extension. These attachments are very low in price, varying from 40s. upwards, according to the focal length of lens to which they are to be attached. We understand that these lenses can also be supplied as a complete telephoto-lens and shutter ; that is to say, a rapid symmetrical lens, with Bausch & Lomb shutter and negative attachment, price complete for ¼-plate being 75s. At the usual distance the lenses commonly employed for photography are all that is necessary for satisfactory results. When, however, it is impossible to approach close enough to the object to secure an image of the desired size with the ordinary lens the telephoto-lens is the only resource. The amateur of experience knows how often this happens.

NOVELTIES.—Mr. Arthur Rayment, of Forest Gate, E., has introduced several novelties in connection with folding film cameras. The "Adapter Guénault " is a small metal dark slide that enables the possessor of a film camera to use ordinary glass dry plates or cut films. It is easily adapted, and requires no alteration to the camera, the dark slide being held in position merely by friction. We are convinced that this simple little arrangement will prove a real boon to users of folding Kodaks. who will now be enabled to make one or two exposures, and to develop them at once, without having to wait until a whole spool of film is exposed before development is possible. Mr. Rayment is also adapting to folding Kodaks the " Steinheil Anastigmat" lens working at F. 6·3, and fitted to Steinheil's time and instantaneous shutter. Thus the value of a folding Kodak can be greatly increased and its efficiency modernised. We must not overlook that useful little novelty of Guénault's, a pneumatic shutter release, which can be rapidly applied to any folding Kodak, and acts in conjunction with the shutter supplied with the camera. This will be found especially valuable to amateurs who have a difficulty in holding the camera steady during exposure. It will be useful also for time exposure, for it obviates the necessity of touching the camera. Whilst on the subject of folding Kodak cameras we would remind our readers that the Kodak Company are now supplying spools with four exposures in two separate pieces, permitting two exposures to be made. These two can be removed from the camera, and another two-exposure film remains in the camera for future use.

PHOTOGRAPHY FOR BEGINNERS.

By B. FOULKES-WINKS, M.R.P.S.

(*Continued from page* 310.)

SECTION I. CAMERAS (*continued*).

AUTOMATIC CHANGING CAMERAS.— Our next illustration is representative of a very large class of hand-cameras, all more or less alike, but varying in price according to finish and quality of lens and shutters. The "Midg" (see *ante*, p. 277) is a very cheap camera of this type; it is fitted with single lens, has a time and instantaneous shutter, and two Brilliant finders; capacity twelve ¼-plates and register for number of exposed plates. This camera is manufactured by Messrs. Butcher & Son, of Blackheath, the retail price being a guinea. It is also made in three other qualities, all fitted with rapid rectilinear lenses, ranging from £2 10s. to £4 4s., according to quality of lens and general finish.

The "Salex" hand-camera is another type of automatic changing hand-camera. It is well made, and has a very good rapid rectilinear lens fitted with Bausch & Lomb "Unicum" time and instantaneous shutter, iris diaphragms, rack focussing,

FIG. 1. THE "SALEX."

and two special Brilliant real image finders. This pattern is manufactured by the City Sale and Exchange Company, price £5 5s. No. 2 "Salex" is fitted with "Cooke" lens working at 6·3, and has rising front each way. Price £10 10s.

The "Bullard" camera is quite a new system of automatic changing, and is of American manufacture. It is made only for 5 × 4 plates, and has a

FIG. 2. THE 5 × 4 "BULLARD."

capacity of 18 plates of that size. The changing is very rapid and sure, and each plate is registered as it is exposed. The front of the camera, carrying

lens and shutter, is made to fold, and when closed up forms a neat leather-covered box. The lens is of the rapid rectilinear type, and the shutter is a Bausch & Lomb time and instantaneous, with pneumatic release. The camera is fitted with a reversing finder, thus enabling it to be used vertically or horizontally. Price £4 15s.

The "Tella" is an automatic changing flat film camera, and is of an exceedingly neat appearance. It is very small and compact. The changing is one of the most ingenious inventions that we have seen applied to cameras. The films are inserted in the chamber at the back of the camera in packets of twenty-five, the film chambers being made to hold any number up to fifty. There is a small celluloid separator between every film. Each serves to separate a film from the rest as it becomes the front one. This film is then isolated by means of a septum that is automatically inserted between the first film and the remainder, thus preventing

FIG. 3. THE "TELLA."

the light acting upon any but the one film exposed in front of the septum. After this film has been exposed, it is passed away into a chamber for exposed films, and the number registered, whilst by the same action another film is brought into position for exposing. In connection with this changing system, there is a very pretty device which shows the operator whenever there is a film ready for exposure. The films may be purchased in packets of twenty-five, and the parcel inserted bodily. They may be also purchased in packets of twelve of any manfacture, in which case they are inserted into the camera one at a time, and a small separator placed between each. Thus it will be seen that any make of film may be used.

The shutter is a very efficient one, and works in the diaphragm slot of the lens. It is a pneumatic regulation shutter, with a range of speed from $\frac{1}{100}$ to $\frac{1}{2}$ a second and time. The finders are of the "Real Image Brilliant" type, and are very carefully adjusted.

The ¼-plate "Tella," fitted with an ordinary F. 8 rapid rectilinear lens, costs £7 17s. 6d. The No. 3 "Tella," ¼-plate size, fitted with the "Cooke" anastigmat lens, working at F. 6·3, and having rising front for both vertical and horizontal pictures, is priced £14 14s. The No. 4 "Tella" for 5 × 4 films, and fitted with 6-inch Cooke lens, is £19 19s. All these instruments are beautifully finished, and for flat film cameras are undoubtedly the best on the market.

The "Frena" is another form of hand-camera for use with cut films. The films are inserted in the camera in packets of twenty, the capacity of each camera being forty, or two packets. In this system the films are notched along each side, and a piece of mill-board is packed between each film, which is also notched so as to alternate with the notching on the film, and it is upon this notching that the whole system of changing the film depends. This is an exceedingly clever device, and, if the instructions are followed, the changing is both rapid and sure. The "Frena" is made in box form, in several sizes and types; also as a folding camera, with detachable box.

FIG. 4. THE "FRENA."

The No. 00, memorandum size, viz. 3½ × 2⅝ in., is the smallest size made, and is very useful for travellers, where bulk is a consideration. Although the picture is small, it is a very pretty size, lending itself admirably for enlargement purposes. The lens is a "Beck" 4¼-in. achromatic, working at F. 11, F. 16, F. 22, and F. 64. The "Memo Frena" has a fixed focus; that is to say, every object over 14 ft. distance will be fairly in focus. It is fitted with time and instantaneous shutter, working from 1/50 to ¼ of a second and time. There is an indicator to record how many films have been exposed; it is also fitted with a swing back, and with two fenders, one vertical, the other horizontal. The price of camera fitted with forty films is £2 18s. 6d.

The No. 0 "Frena" is similar in every respect to the No. 00, except that it is fitted with Beck's "Autograph" Rapid Rectilinear lens, with a working aperture of F. 8. The price, with forty films, is £5 5s.

The No. 2 "Frena" is for 4¼ × 3¼ in. films, and is fitted with a 5½-in. single achromatic lens working from F. 11. The other details are as in No. 00. The price, with forty films, is £5 8s. 6d.

The No. 2 "Frena" is the same as No. 0, except in size of film, which is 3¼ × 4¼, or ordinary ¼-plate size; price £8 17s. 6d. when charged with 40 ¼-plate films.

The No. 3 "Frena" is for 5 × 4 in. films, and is fitted with a 6½-in. Beck Autograph Rapid Rectilinear lens, with a working aperture of F. 8.

Magnifiers may be fitted to all the types of "Frena" cameras. These are a series of single lenses which fit on to the front of the permanent lens, by which means objects near to the camera may be photographed. Thus, for example, the nearest object that can be sharply photographed with the No. 2 "Frena" would be 20 ft. away, whilst by the addition of suitable magnifiers objects may be taken as near to the camera as 4 ft. These magnifiers are generally sold in sets

of four, which will bring an object in focus at the following distances: No. 1, 27 to 9 ft.; No. 2, 10 to 6 ft.; No. 3, 6 to 4½ ft.; and No. 4, 4½ to 3½ ft.

These cameras may now be had fitted with the Beck-Steinheil orthostigmat lens for an additional cost of from £3 to £6.

The new folding "Frena" is an adaptation of the box form of camera, having precisely the same changing arrangement, but in which the front of the camera carrying the lens and shutter is made to fold, thus reducing the size of the instrument very considerably. This alteration also carries with it the additional advantages of allowing lenses of different foci to be used, and permits of focussing the image on a ground-glass screen or to scale, and allows the use of a rising front for both vertical and horizontal views. Another great convenience in this new form of "Frena" is that

¼-PLATE PICTURE TAKEN WITH NO. 2 "FRENA."

the whole film-changing portion of the camera may be removed, and leave a complete camera which can be used with double dark slides, thus enabling the possessor of such an instrument to use either films or glass plates. The folding "Frena" cameras are made in three classes—Nos. 6, 7, and 8—and range in price from £11 18s. 6d. to £3 17s. The No. 8, which is the best of this class, is fitted with a Beck-Steinheil 2-foci ortho-stigmat lens, 4¾-in. focus in combination, and 8½-in. focus when used as a single long-focus lens. The full aperture of this lens is F. 6·3.

(*To be continued.*)

NOTES & QUERIES

AN ABNORMAL NEWT.—When dipping in a pond about three years ago for creatures to stock my aquarium I caught a newt which seems to be different from any of those that I have seen described in books. It was found under the roots of an osmunda fern that was growing on the border of a small pond about twenty yards by ten yards in extent. This pond contains great numbers of the palmate newt, *Lophinus palmatus*, in the early

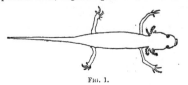

Fig. 1.

part of the summer, but I have never found any other species there. I thought that this newt was the larva of the great water-newt, *Triton cristatus*, and so took it home and kept it in an aquarium 2 ft. by 1 ft. in size, with about seven inches of water. The newt had well-developed gills, which I expected it would absorb in the autumn, on leaving the water and taking refuge on an island arranged in the centre of the aquarium. To my

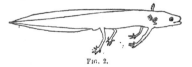

Fig. 2.

surprise, however, it never left the water at all, but remained in the larval condition with its gills nearly as well developed as when first caught. This interesting creature has hardly grown half an inch during these three years, and except for this small increase in size has not altered in any way. The food on which the newt feeds consists of the smaller inhabitants of the aquarium—such as water-mites, young leeches, hydras, and many

Fig. 3.

others—but when these are scarce I give it worms and flies. When I agitate the surface of the water it comes for the flies; but finds some difficulty in seizing them, as they are so buoyant on the surface. In warm weather the newt often stays on a rock with its nose just out of the water and the rest of the body submerged, and this seems to show that it has developed lungs, as does also the fact that

it is very fond of giving up bubbles of air. There are three gills on each side, one pair being much smaller than the others. They are capable of movement, as sometimes they are carried well forward and at others lying back against the sides. Another remarkable feature of this newt is the

Fig. 4.

great prominence of the under lip, which would lead one to suppose that it was the straight-lipped newt. The colour, generally, is dark brown on the dorsal surface, shading into silvery grey ventrically, while the whole surface is slightly mottled. The crest on the tail always remains the same height through summer and winter. Very often there are tears in its edge, made, I think, by the larvae of dragonflies which live in the same aquarium. At the end of the tail is a short filament about half the length of that on *L. palmatus*. The length of the newt is within an eighth of an inch of three inches. The figures of explanation are: Fig. 1. An outline drawing of the newt as seen when lying at the bottom of a dish; the legs are thus spread out in their natural positions. Fig. 2. A side view taken through a glass vessel. Fig. 3. This shows the protruberence of the under lip. Fig. 4. A shaded drawing to show the general character of the markings.—*A. T. Mundy, Cornwood Vicarage, Ivybridge, Devon, March* 1901.

WOODPECKER FEEDING ON GROUND.—It is no uncommon sight to see woodpeckers feeding on the ground, particularly when there are ants' nests about. It is curious to see them drive their beaks into the ground in search of the ants and their pupae.—*D. Wilson-Barker, Greenhithe.*

WOODPECKER FEEDING ON GROUND.—With regard to the note of Mr. Dallas (*ante*, p. 316), this is not such an uncommon circumstance as many persons imagine. Ant-hills are very favourite hunting-grounds for these birds. I have often watched them at work: they penetrate with their strong bills deeply into the ant-hill in order to obtain the pupae, or so-called ant-eggs, as well as ants themselves. I have seen specimens of these birds with the breast quite stained with red earth, among which they had been searching for ants.—*E. Wheeler, 71 Queen's Road, Clifton, Bristol.*

FORMALIN AS A PRESERVATIVE FOR PLANTS.— The use of formalin for the preservation of zoological specimens is now very general, and its advantages were mentioned in the last number of this journal (*ante*, p. 313). Its application to the preservation of plants and flowers, however, is quite new, and the experiments of Mr. J. W. Peck, which are described in the "Pharmaceutical Journal," are extremely interesting and suggestive. The most satisfactory results were obtained with a 5 per cent. solution of formaldehyde, *i.e.* an eighth of the strength of the commercial formalin, which contains 40 per cent. of formaldehyde. The flowers and portions of plants immersed in this and kept in the dark remained intact, whilst the tissues became more or less translucent, showing the structure. After seventeen months yellow calceolaria flowers had lost but little of their colour, whilst a tulip and hyacinth had lost about 30 per cent. A pansy exposed to diffused light in a 5 per cent. solution was rapidly bleached. with the exception of the lower yellow petal. A white tulip became translucent, but retained its external form perfectly. The odour of mignonette was still perceptible after four months, notwithstanding the penetrating odour of the formalin itself. Unfortunately the solution soon bleached blue colours. A blue hyacinth became opaque white in two days and translucent in six months. Green leaves became only slightly translucent, and were otherwise unchanged. In order to prevent the bleaching action of sunlight it was found essential to keep the specimens in as dark a place as possible. The preservative action of the formalin is due to its destroying all external microorganisms, and preventing the inter-action of the plant-cells by contracting their protoplasm.—*C. A. Mitchell, Chancery Lane, London.*

A SECTION IN WESTMINSTER.—The workmen employed in the excavations for the foundations of the new Government offices in Parliament Street, Westminster, have brought to light several finds of considerable interest. The section, which attained a depth of 30 ft., showed made ground containing old-fashioned tobacco pipes, pottery. etc., overlying alluvial peat with Pleistocene river drift below. In the peat, which contained numerous mollusca, several human remains were found—namely, a frontal bone, a portion of a tibia, and several vertebrae, ribs, and metatarsals. The peat also yielded a complete lower jaw and two horn-cores of *Bos primigenius,* a horn-core with a portion of the skull of *B. longifrons,* a complete skull of *Equus caballus,* two incisor teeth of *Sus scrofa,* and a skull of a horned sheep. The only mammalian remains obtained from the drift were a single molar of the *Elephas primigenius* and an antler of the reindeer (*Rangifer tarandus*). They were both in excellent preservation.—*Gilbert White, 31 North Side, Clapham Common, S.W.*

NOTICES OF SOCIETIES.

*Ordinary meetings are marked †, excursions * ; names of persons following excursions are or Conductors. Lantern Illustrations §.*

NORTH LONDON NATURAL HISTORY SOCIETY.

April 4.—§ " Some Notes on British Spiders." Frank P. Smith.
　　" 18.—† "The Tree in its Relation to Primitive Thought." Mrs. H. M. Halliday.
　　" 27.—° Visit to Royal Botanic Society's Gardens.

SOUTH LONDON ENTOMOLOGICAL AND NATURAL HISTORY SOCIETY.

April 25.—† " Birds and their Nests." R. Kearton, F.Z.S.

BIRKBECK NATURAL HISTORY SOCIETY.

April 13.—* Natural History Museum, South Kensington. A. B. Rendle, M.A., D.Sc., F.L.S.

LAMBETH FIELD CLUB AND SCIENTIFIC SOCIETY.

April 1.—§ " The Natural Beauties of Crohamhurst."
　　" 8.—* Effingham and Bookham. E. A. Martin, F.G.S.
　　" 20.—* Crohamhurst.

LONDON GEOLOGICAL FIELD CLASS.

April 27.—° Nutfield to Caterham. Anticlinal of the Weald. (Weald Clay to Chalk.) Professor H. G. Seeley, F.R.S.

CAMERA CLUB.

April 1.— Discussion : " The Development of Rollable and other Films."
　　" 18.—† " Man's Place in Nature." Percy Ames.

NOTTINGHAM NATURAL SCIENCE RAMBLING CLUB.

April 6.—† " The Teachings of Geology."
　　" 20.—† " Meaning of the Colours on a Geological Map."

PRESTON SCIENTIFIC SOCIETY.

April 10.— " Peculiarities of Animal Forms." James Harrison, A.R.C.S.
　　" 24.— " The Origin of the Higher Vertebrates : a Study in Evolution " James Marsden.

MANCHESTER MUSEUM, OWENS COLLEGE.

April 8.—† " Selection." W. E. Hoyle.

NOTICES TO CORRESPONDENTS.

TO CORRESPONDENTS AND EXCHANGERS.—SCIENCE-GOSSIP is published on the 25th of each month. All notes or short communications should reach us not later than the 18th of the month for insertion in the following number. No communications can be inserted or noticed without full name and address of writer. Notices of changes of address admitted free.

BUSINESS COMMUNICATIONS.—All business communications relating to SCIENCE-GOSSIP must be addressed to the Manager, SCIENCE-GOSSIP, 110 Strand, London,

EDITORIAL COMMUNICATIONS, articles, books for review, instruments for notice, specimens for identification, etc., to be addressed to JOHN T. CARRINGTON, 110 Strand, London, W.C.

SUBSCRIPTIONS.—The volumes of SCIENCE-GOSSIP begin with the June numbers, but Subscriptions may commence with any number, at the rate of 6s. 6d. for twelve months (including postage), and should be remitted to the Manager, SCIENCE-GOSSIP, 110 Strand, London, W.C.

NOTICE.—Contributors are requested to strictly observe the following rules. All contributions must be *clearly* written on one side of the paper only. Words intended to be printed in *italics* should be marked under with a single line. Generic names must be given in full, excepting where used immediately before. Capitals may only be used for generic, and not specific names. Scientific names and names of places to be written in round hand.

CHANGE OF ADDRESS.

J. Burton, from 39 Ingham Road to 20 Fortune Green Road, West Hampstead, N.W.

EXCHANGES.

NOTICE.—Exchanges extending to thirty words (including name and address) admitted free ; but additional words must be prepaid at the rate of threepence for every seven words or less.

ABOUT 300 specimens of Helix nemoralis, 12 varieties, and 100 H. aspersa for dried specimens of Foreign Plants.—F. T. Mott, Birstal Hill, Leicester.

WANTED, a bit of Sugar Cane, fresh or dry ; also, when in season, a few fresh stems of Equisetum hyemale. Botanical or insect slides offered in exchange.—W. White, Litcham, Swaffham.

WANTED to correspond with Collectors abroad for the exchange of land and marine shells, crustacea and echinodermata.—H. W. Parritt, 8 Whitehall Park, London, N.

EXCHANGE—Fifty-one micro-slides in pine box for well-set British Lepidoptera.—Edward Kitchen, 116 Eversleigh Road, Battersea, London, S.W.

WANTED, wild flowers (fresh or pressed) in exchange for microscopical material, &c.—A. Nicholson, 67 Greenbank Road, Darlington.

CONTENTS.

THE NATURE OF ANIMAL FAT.

By C. Ainsworth Mitchell, B.A. (Oxon.)

WE owe to the distinguished French chemist Chevreul our first insight into the chemistry of fats and oils. His classical work "Recherches sur les Corps Gras," published in 1823, was supplemented, and to some extent corrected, in numerous investigations—notably by Liebig and his pupils, and more recently by von Hübl, Hazura, Hehner and many others. Notwithstanding all this research, however, the subject remains obscure in many respects, and presents numerous problems which will only be solved when more accurate methods of analysis have been devised.

What is popularly known as the "fat" of an animal is really part of the connective tissue containing cells, in which the original protoplasm has been gradually displaced by the true fat leaving the cell walls intact. These cells are usually grouped together and supported by the fibrous

The nature of these fatty acids varies with the particular fat, but the mixture usually consists of palmitic, stearic, and oleic acids, with, at all events, in some cases linolic and linolenic acids.

Palmitic acid derives its name from being a chief constituent of palm oil. It is a white, soft, soapy solid, practically without taste or odour.

Stearic acid, from στέαρ, otherwise tallow, is, as its name suggests, an important constituent of tallow. In its general properties it resembles palmitic acid, but melts at a higher temperature, and is less soluble in alcohol and other solvents. A mixture of palmitic and stearic acids recalls the behaviour of an alloy of different metals, the mixture having a lower melting point than either of its constituents. Until recently there was no exact method of quantitatively separating these two acids.

FIG. 1. CRYSTALS FROM MIXED LARD AND BEEF FAT.

FIG. 2. CRYSTALS FROM BEEF FAT.

substance of the connective tissue, as is shown in figs. 3 and 4.

In order to separate the fat from the tissue the cell walls must be broken down, and in manufacturing processes, such as the rendering of lard, this is done by means of heat.

The isolated fat of most land animals consists essentially of compounds of various acids, known as fatty acids, with glycerin, these compounds being termed "glycerides."

If we boil a fat with a strong solution of soda or potash, decomposition takes place, the fatty acids combining with the alkali to form soap, whilst the glycerin is set free. On now treating the soap solution with hydrochloric acid the soap is decomposed, and a mixture of different fatty acids rises to the surface of the liquid.

Oleic acid, from *oleum*, otherwise oil, is a colourless liquid when pure. It also differs from the preceding acids in being unsaturated. Thus, on treatment with a solution of iodine it absorbs a large proportion of that element, yielding a definite compound.

Linolic acid and linolenic acid derived their name from linseed-oil—*oleum linium*—in which they were first discovered. They resemble oleic acid, but are still more unsaturated—linolenic acid much more so than linolic acid. The drying properties of linseed-oil are attributed to linolenic acid.

The manner in which the glycerin is combined with these various acids is not known with certainty. Glycerin has three available groups that enter into combination, so that there are numerous possible glycerides. Thus the fat may consist of tri-glycerides such as :—

Glycerin ⎧Stearic A.   G.⎧Palmitic A.   G.⎧Oleic A.
        ⎨Stearic A.   ⎨Palmitic A.   ⎨Oleic A.
        ⎩Stearic A.   ⎩Palmitic A.   ⎩Oleic A.

These compounds have been artificially prepared, and are known as stearin, palmitin, and olein.

Or there may be a series of mixed glycerides such as:—

G.⎧Stearic A.   G.⎧Stearic A.   G.⎧Palmitic A.
  ⎨Stearic A.   ⎨Palmitic A.   ⎨Oleic A.
  ⎩Palmitic A.   ⎩Oleic A.   ⎩Linolic A.

There is evidence that in some cases, at all events, such mixed glycerides do exist.

When we examine the mixed fatty acids obtained from any fat we are met by an initial difficulty. There is no exact method of separating the solid acids from the liquid or unsaturated acids, and we have in most cases to be content with drawing inferences from the percentage of iodine absorbed by the mixture, or its incompletely separated fractions. In the case of stearic acid, however, Mr. Hebner and the present writer have devised the following means of effecting a quantitative separation. Strong alcohol is saturated with pure stearic acid, and the flask containing the solution left all night in a mixture of ice and water, the clear liquid being filtered off next morning from any deposit. On treating the mixed fatty acids with this saturated solvent, and cooling the flask overnight in ice-water, the alcohol, having already been saturated with pure stearic acid, is unable to dissolve any more, although it can take up all the other constituents.

The liquid is filtered off by means of the apparatus shown in fig. 5, and the deposit in the small flask weighed. The filter is the thistle-funnel, the mouth of which is covered with linen and immersed in the flask, and a suction pump is attached to the indiarubber tube fixed to the side tubulure of the large flask.

By this method light has been thrown on many obscure points, as, for instance, the nature of fat crystals. If we dissolve a fat in ether and allow it to crystallise, characteristic microscopic crystals are frequently obtained. Thus lard yields straight crystals with chisel-shaped ends, whilst if beef-fat be taken, the crystals have the form shown in fig. 2. A mixture of lard with about fifteen per cent. of beef-fat gives crystals of intermediate character (fig. 1), the grouping resembling those of fig. 2, and the chisel-shaped ends those of ordinary lard.

In certain cases lard from the flare of the pig yields crystals closely resembling those in fig. 1, and the explanation of this was found in the following experiment: A lard containing sixteen per cent. of stearic acid was dissolved in ether, and yielded an abundant deposit of crystals with the characteristic chisel-shaped ends, which were found to contain thirty-two per cent of stearic acid. This deposit was again crystallised from ether. The crystals were then needle-shaped, but had distinct chisel-shaped ends, as in fig. 1. They contained forty-seven per cent. of stearic acid. For a third time the deposit was recrystallised. The percentage of stearic acid had now risen to fifty-nine per cent., and the crystals were hardly distinguishable in form from those of beef-fat (fig. 2).

It thus appears that the difference in form between the two kinds of crystals is solely due to a larger proportion of stearic acid, and that in exceptional cases the fat from certain parts of the pig contains more of that acid, and thus yields pointed crystals in one crystallisation.

Experiments were next made with fat from different parts of the same pig, and the following amounts of stearic acid were found: head, 9 ; ham, 9 ; breast, 11 ; flare, 15 ; and back, 9 per cent.

In a similar series of determinations with sheep-fat the results were: back, 25 ; neck, 16·4 ; breast, 1 ; ham, none ; and kidney, 27 per cent. of stearic acid. The ham fat, which contained no stearic

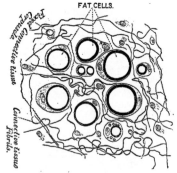

FAT CELLS.

FIG. 3. FAT-CELLS FROM RABBIT.

acid, was liquid at the ordinary temperature, and that of the breast nearly liquid.

A specimen of beef-fat, which was very hard, was found to contain fifty per cent. of stearic acid.

These results show that there is a relation between the consistency of the fat and the proportion of stearic acid.

Some interesting observations on the variations in the consistency of the fat from different parts of several animals have recently been made by Henriques and Hansen. They have found that the fat gradually varies in composition from the exterior to the interior parts of the body. The fat immediately below the skin has the lowest melting-point, and that in the centre of the body the highest, whilst the melting-point of the fat in the intermediate parts varies with its distance from the interior.

It is curious that horse-fat derived from any part of the animal is soft and of about the consistency of butter. Even that from the kidneys is semi-fluid and quite different from the kidney fat of the sheep, which is hard and tallow-like.

The food of an animal appears to have a considerable influence on the consistency and composition of its fat. It is well known that the fat of American pigs is much more fluid and capable of combining with more iodine than that of European pigs. This is possibly due to the presence of a larger proportion of the more unsaturated linolic acid derived from the oil in the maize on which American pigs are frequently fed.

The effect of captivity on the fat of wild animals is shown in a series of interesting results obtained by Amthor and Zink.

The amount of liquid fatty acids (oleic, linolic etc.), as measured by the proportion of iodine with which they could combine, was found to be lower in the fat of domestic animals than in that of the corresponding wild animals, and the more fluid character of the fat was also indicated by its higher melting-point, as shown in the following table:—

Fat.	Melting-point ° C.	Per cent. of Iodine absorbed.
Domestic Cat	39—40 ..	54.5
Wild Cat	37—38 ..	57.8
Tame Rabbit	40—42 ..	69.6
Wild Rabbit	35—38 ..	99.8
Goose	32—34 ..	67.6
Wild Goose	— ..	99.6
Wild Goose (2 years' captivity) ..	— ..	67.0
Duck	36—39 ..	58.5
Wild Duck	— ..	84.6

In the case of birds, the fat of the domestic goose, duck and hen resembled lard, whilst that of the related wild birds was oily.

The fat of the wild boar was found to differ from

Fig. 4. Fat-cells, some showing Nucleus, and one Fat-crystals.

ordinary lard in possessing more liquid fatty acids, and especially in having drying properties.

This remarkable property of drying, which has never been previously recorded of any animal fat, was also possessed by the fat of the hare and wild rabbit, and to a lesser extent by that of the blackcock.

The fat of the polecat was quite liquid, whilst that of the dog and the cat resembled lard in appearance and general characteristics. Fox-fat was like that of dog, but more liquid.

The fat of marine animals and of fishes has been less studied than that of land animals, and many of the results which have been published require confirmation.

Speaking generally, they are oils that contain very unsaturated liquid fatty acids, and some

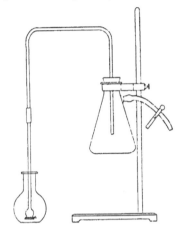

Fig. 5. Apparatus for Determining Stearic Acid.

of which resemble linseed-oil, though without possessing its drying properties.

I have to express my thanks to Messrs. Chas. Griffin & Co. for permission to use the blocks in figs. 3 and 4, and to Dr. Sykes, editor of the "Analyst," for the loan of the block of fig. 5.

57 *Chancery Lane, London, W.C.*

ALGOL.—Professor A. A. Nijland, of Utrecht, finds that the light curve of this star during decrease, is not so regular as is usually described, but in reality shows a marked break.

FLYING FISH.—From time to time a question interesting to naturalists is raised as to the mode of flight of the "flying fish." Does it really fly as do birds, or is its passage through the air simply a blowing along by the wind in conjunction with the impetus given by its leap from the water? I believe that close and constant observation would prove that the pectoral fins of the fish do not vibrate, and that they are incapable of flapping as the wings of birds flap; also that the fish does not "fly" in calm weather. Further, that it generally "flies" up the wind, and that the fact of its rising above the waves in a forward movement is due to the upward direction of the wind from the waves. The flight of the fish resembles in many particulars the soaring flight of oceanic birds; a rapid vibratory motion of the anal fins is sometimes very noticeable. It would be interesting to hear other opinions on this point, and I should personally be much obliged for any information or for photographs showing the fish in the act of flight.—*D. Wilson Barker, Greenhithe.*

N 2

FRUITING OF LESSER CELANDINE.

By Charles E. Britton.

IT has been a long-accepted belief concerning this plant, not in this country alone but also upon the continent of Europe, that there is a vital defect in its sexual organisation of such a character that reproduction in the normal manner of flowering plants, by formation of seed, is rare. This sterility, partial or more complete, is accompanied and compensated for by the formation of tubers, a non-sexual means of reproduction which occurs at the underground part of the plant, appearing as thick, fleshy, root-like swellings, or are formed above the soil in the axils of the stem-leaves. The function of the subterranean tubers is essentially to serve as a means of conveying the life of the plant over to the next period of growth ; besides which, however, by the annual increase of tubers, the species is propagated. It is chiefly owing to these root-tubers that the individual plants of *Ranunculus ficaria* grow in clumps, which are again associated in larger communities. The stem-tubers, from their place of formation in the axils of the leaves, are capable, according to the influence of various agents, of spreading the species into areas in which it may be unrepresented. At the conclusion of the period of flowering, leaves, stems, and roots disappear, and early in the following year the tubers give origin to new plants.

In recent years at least two series of observations have been made upon the lesser celandine—one by Mr. I. H. Burkill, who published in the "Journal of Botany" for 1897 some studies in the "Fertilisation of Spring Flowers on the Yorkshire Coast." As regards the species under consideration, this investigator paid special attention to the number and variety of the insects visiting its flowers, and came to the conclusion that "*R. ficaria*, which so extremely rarely sets seed, is an enigma. This failure in seed-production cannot be due to want of fertilisation, for the flowers are visited by a considerable variety of insects, though not very freely." Professor Federico Delpino also studied the lesser celandine, and came to the rather startling conclusion that our familiar plant is the dwarf functionally female form of the stouter, larger-flowered *R. caltheafolius*, a Continental plant with a rather restricted·range, in this manner accounting for the sterility of the pollen, and the readiness in which the species is propagated asexually. It is no new theory that the South European plant just named is, indeed, closely allied to the lesser celandine, and at one time it was expected that it would reward search directed for it in this country.

Having occasionally encountered *R. ficaria* bearing fruit, I was inclined to regard with suspicion so sweeping a statement that seed-production is a rare phenomenon in this species. In 1898 the opportunity occurred of paying some attention to this question, as on a certain occasion in May abundant fruit was observed in North Middlesex and South Hertfordshire. The amount of seed borne by plants of the lesser celandine in the neighbourhood of Potter's Bar, South Mimms, North Mimms, and Shenley was sufficient to dispose of the question of the rarity of seed-production. During the same month of May the species was observed fruiting on Great Bookham Common and near Ockham, Surrey. On this occasion the conclusion arrived at was that fruit, if not freely produced, was by no means rare. In the extreme south-east corner of Surrey abundance of fruit was observed in one locality. From one clump were gathered twelve heads of well-developed fruit and nine heads with undeveloped carpels occurring with the almost mature achenes. After May I failed to observe other instances of seed-production, but during the next year followed up the subject. In late April and early May of 1899 I was at Midhurst, in Sussex, and during my walks in the neighbourhood noted twelve localities in which *R. ficaria* was bearing fruit either plentifully or abundantly. In June fruit was being produced at the foot of the downs near Brook, in Kent. During the first fortnight in May, 1900, the species was noticed fruiting well near Westhumble, Westcott on roadside banks, New Oxted on dry banks, West Horsley on roadside bank, and abundantly at Chipstead Bottom, all these localities being in the county of Surrey. During the same period the production of fruit was observed in various places in the neighbourhood of Fawkham, Kent. The enumeration of these localities does not exhaust the list of places where the lesser celandine was observed in fruit. It may also be as well to mention that these observations were made during the course of general botanising.

Seeing that a certain amount of fruit is produced, the question arises as to its origin, bearing in view the alleged sterility of the pollen. Obviously examination of the pollen of *R. ficaria* is quite as important as observations directed to the frequency of seed-production. The pollen-grains are yellow, spherical, or tetrahedral in a younger state, with densely granular protoplasm. Sometimes a proportion consists of shrivelled, imperfect grains, the percentage of which varies, occasionally being considerable, and at other times less so.

During the late spring of 1898 I prepared about eight cultures of pollen, in water and in sugar-and-water, with the result that no examples of emission of pollen-tubes were noticed. Strangely enough, not even in the case of pollen selected from plants bearing fruit was germination observed. This negative evidence did not, however, lead me to assume that seed is produced partheno-genetically without the influence of the male element. The following spring gave very satis-factory results as to the capability of the pollen to germinate. Usually the grains germinated more readily in the water cultures than in the sugar-and-water preparations. In the case of pollen taken from a cultivated plant, which the previous year was found bearing fruit at Shenley, more germinating grains were observed in the sugar-and-water preparations than in the water cultures. The observations of 1899 were made on pollen obtained from plants flowering from March to June, and some of the pollen was obtained from plants bearing fruit. Apart from germination, the chief noticeable feature of most cultures of pollen was that the proportion of imperfect pollen-grains was much smaller than in the pollen-cultures of the previous year; in fact, the shrivelled pollen-grains were scarcely present. An exception was in the case of pollen taken from plants growing in a damp situation by a stream near Theydon Bois in Essex. The proportion of imperfect grains was there as much as about six to every eight regularly formed grains. Continued observations during the spring of 1900 were as successful as those of the previous year. In one instance only did the grains fail to germinate, and in these preparations most of the pollen-grains absorbed water to such an extent as to burst the membranes.

In all preparations of pollen, it is but a minority of the grains which emit pollen-tubes. Some of the pollen-tubes reach a great length in comparison with the size of the grains, being twenty or more times as long as the diameter of the grains from which they originate. A portion only germinate, and I think these alone possess the power of emitting pollen-tubes. It is well, however, to retain in view that the conditions presented by water and sugar-and-water cultures are vastly different from those occurring in nature. It may perhaps not be out of place to mention here that seed of the lesser celandine procured at Shenley and in Surrey germinated readily in the year following gathering.

The imperfect character of these observations upon *R. ficaria* is very apparent, and from them it would be extremely unwise to draw con-clusions as to the general behaviour of the species. Further, the results of observations, however ex-tensive, of one observer only can never be taken as characteristic of the species under investigation. To achieve this end it is necessary that observa-tions of many investigators should be pieced together into one continuous whole. Personally, I entertain the opinion that fruit of the lesser celandine, if not generally produced, is by no means rare, and can, when looked for, usually be found. To the habit of the plant is due the fact that fruit, even when produced, is apparently absent. Usually after flowering, the peduncle of a plant is strengthened in order to support the maturing fruit. This is not so in *R. ficaria*, in which species the fruit-stalks are flaccid and lie low among the leaves. Occasionally, the stalks become curved as much as is the habit among the water crowfoots, and the. heads of achenes seem almost to burrow into the soil. This depressed habit of the fruit-bearing stalks, together with the large achenes and the formation of tubers, are probably the causes of the plants being gregarious.

That there is an undoubted connection between the formation of tubers by which asexual repro-duction is effected and the production of seed is evident. Usually in seed-bearing plants of the lesser celandine the production of tubers is not marked on the parts above the soil. That a plentiful formation of axillary tubers. is preju-dicial to the production of seed is suggested by the circumstance that in one locality near South Mimms, in 1898, where the plants were abundantly producing axillary tubers, one or several together, only two plants could be found bearing seed, and this so small in quantity that it was limited to two stalks, each with five well grown carpels among the undeveloped ovaries. If, as is likely to be the case, the two methods of reproduction cannot very well exist side by side, it may be that, in localities where the chief mode of propagation is by means of tubers, there is this inability, not from any defect in the sexual consti-tution, but by reason of the plastic food materials being diverted from the perhaps fertilised ovaries towards the tubers. It may be thought that dryness of situation, in a measure, affects seed-production, though my observations do not lead to such a conclusion. It is true fruiting plants have been met with in dry situations; but, on the other hand, such plants have been found in damp places, such as at the base of hedge-banks and by ditches. In 1898 a locality where fruit was produced abundantly was in a wet, clayey field; and the instance mentioned of tuber-forma-tion near South Mimms occurred on a very dry, sandy bank. Especially wet situations, as by watercourses and streams, may favour tuber forma-tion, to the detriment of the sexuality of the plants, for in such places the chances of the flowers perfecting seed may be considerably lessened. It may have been due to the influence of the surroundings that failure attended my observations directed to ascertain the capability of the pollen to germinate, when pollen was procured from plants growing in situations as above indicated. One cause that may contribute towards the conclusion as to the rarity of seed-production is that situations such as hedge-banks, where *R. ficaria* is most evident, support a much stronger general vegetation towards the close than at the prime of the flowering period of our plant, which, like other low-growing herbs, suffers the risk of being obscured by its taller neighbours.

Flowering at a time when few other plants are in bloom; having few competitors possessing flowers of a similar organisation; making a great display of colour; affording abundant pollen and honey to its insect-visitors, which are fairly numerous in kind and in number, and some, as I have observed, are very assiduous in their attentions to the flowers; possessing a peculiar sensitiveness of the floral envelopes, which open or close according to variations in the temperature of the atmosphere; if, with all these advantages, the flowers nevertheless extremely rarely set seed, then, indeed, as Mr. I. H. Burkill says, *R. ficaria* is an enigma. If more attention is given by observers to the question of seed-production in the lesser celandine, it will be found, I anticipate, that the species is, after all, not such an enigma, and that fruit is more generally produced than is thought.

35 *Dugdale Street, Camberwell, London, S.E.*
February 27th, 1901.

AN INTRODUCTION TO BRITISH SPIDERS.

By Frank Percy Smith.

(Continued from page 334.)

GENUS *LOPHOCARENUM* MENGE.

The posterior row of eyes is strongly curved, its convexity being directed backwards. The tibial spines are very small, and those of the fourth pair of legs are situated near to the middle of the joint. The abdomen has upon its upper side a scutum, or shield, which is furnished with numerous distinct punctures.

Lophocarenum nemorale Bl. (*Walckenaera nemoralis* Bl.)
Length. Male 1.5 mm.
The caput of the male is furnished with a large prominence upon which the posterior central eyes are placed, the remaining eyes being placed in a transverse group near the junction of this lobe with the cephalo-thorax.

Lophocarenum parallelum Bl. (*Walckenaera parallela* Bl.)
Length. Male 1.5 mm., female 1.75 mm.
This species is very similar to the last, but the cephalic lobe of the male is not nearly so prominent, it being low and rounded. As in *L. nemorale* the posterior central eyes are placed upon the front surface of this lobe.

Lophocarenum blackwallii Cb.
The radial joint of the male palpus when viewed from above is seen to consist of two long unequal branches, almost parallel to each other.

Lophocarenum mengei Sim.
The caput is distinctly elevated, forming a somewhat rounded lobe.
Walckenaera turgida Bl. is probably referable to this genus.

GENUS *SAVIGNIA* Bl.

The small spider upon which this genus is founded was, when first described by Blackwall, stated to possess but six eyes. The two small eyes upon the front part of the cephalic eminence were soon discovered, and the spider removed to the genus *Walckenaera*. It has since been included in the genera *Prosoponcus* Sim. and *Diplocephalus* Bertkau; but as it appears to possess characteristics sufficiently distinct to warrant the formation of a separate genus for its reception, the above name is here adopted.

Savignia frontata Bl. (*Walckenaera frontata* in "Spiders of Dorset.") Fig. 1, page 303.
Length of male, 2 mm.
The curious form of the caput, which will also be figured in profile, will at once distinguish this species, which is not rare.

GENUS *ENTELECARA* SIM.

The anterior row of eyes is curved, the convexity being directed backwards. The tibiae of the first pair of legs have each two fine spines, and are longer than the metatarsi, which latter are distinctly longer than the tarsi, in the proportion of about 4 to 3.

Entelecara acuminata Wid. (*Walckenaera altifrons* Cb.)
Length of male, 1.7 mm.
The caput is very high, and when viewed from in front has the appearance of a cone surmounted by a sphere.

Entelecara flavipes Bl. (*Walckenaera implana* Cb.)
Length. Male 1.5 mm., female 1.7 mm.

Cephalo-thorax blackish brown, minutely punctured. Legs pale yellow. Abdomen black. A rare
species.

Entelecara erythropus Westr. (*Walckenaera
erythropa* in "Spiders of Dorset.")

Length of male, 2 mm.

The radial joint of the male palpus is of a curious
form, consisting of two branches, which when viewed
from above have the appearance of the letter F.

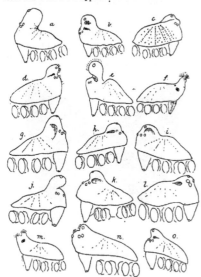

FIG. 1. CEPHALO-THORACES OF MALE SPIDERS IN PROFILE.

*a. Peponocranium ludicrum. b. Lophocarenum nemorale.
c. L. parallelum. d. L. blackwallii. e. L. mengei. f.
Savignia frontata. g. Entelecara acuminata. h. E. flavipes. i. E. erythropus. j. E. thorellii. k. E. trifrons.
l. Evansia merens. m. Diplocephalus cristatus. n. D. permixtus. o. D. fuscipes.*

Entelecara thorellii Westr. (*Walckenaera
fastigata* Bl.)

Length of male, 2 mm.

Cephalo-thorax dark brown. Legs yellowish red.
Abdomen brownish black. The cephalic lobe is
very prominent and distinct.

Entelecara trifrons Cb. (*Walckenaera trifrons* in "Spiders of Dorset.")

Length of male, 2.2 mm.

Very similar in general form to *E. thorellii*, but
easily distinguishable by the form of the radial joint
of the male palpus, which will be figured.

GENUS *EVANSIA* CAMBR.

The anterior row of eyes is slightly curved, its
convexity being directed backwards. The posterior
row is strongly curved, its convexity also being

directed backwards. The eyes of this row are equidistant, and almost equal in size. Legs with coarse
hairs, strongest upon the femora. Tarsi of fourth
pair of legs much longer than the metatarsi.

Evansia merens Cb.

Length. Male 2.25 mm.

Cephalo-thorax pale brown. Abdomen dark brown,
with a darker central stripe and several pale markings
towards its posterior part. Legs yellow, tinged with
orange.

A single specimen of this spider was discovered by
Mr. W. Evans in 1899.

GENUS *DIPLOCEPHALUS* BERTKAU.

Posterior eyes in a straight line, or in a slightly
curved row, having its convexity directed backwards;
equidistant, or the space between the centrals slightly
less than that between one of them, and the adjacent
lateral. Anterior eyes in a straight or nearly straight

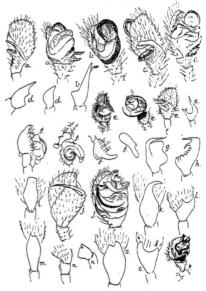

FIG. 2. PARTS OF PALPI OF MALE SPIDERS.

*a. Entelecara acuminata. b. E. flavipes. c. E. erythro-
pus. d. E. thorellii. e. E. trifrons. f. Evansia merens.
g. Diplocephalus cristatus. h. D. permixtus. i. D. fuscipes.
j. D. latifrons. k. D. picinus. l. D. beckii. m. Thyreo-
sthenius biovatus. n. Dismodicus bifrons. o. Typhocrestus
dorsuosus. p. Savignia frontata.*

line. Tibias with long erect bristles. The cephalo-
thorax of the male in many species is of a most curious
conical form, the apex of the projection bearing the
anterior central eyes, and having, upon its upper
surface, a projection bearing the posterior central
eyes.

Diplocephalus cristatus Bl.   (*Walckenaera cristata* in "Spiders of Dorset.")

Length.   Male 2 mm., female 2.2 mm.

The caput when viewed from above reminds one somewhat of *Savignia frontata* Bl.   When seen in profile, however, the bifid form of the caput is at once evident.   This species is the type of the genus.

Diplocephalus permixtus Cb.   (*Walckenaera permixta* in "Spiders of Dorset.")

Length of male, 1.7 mm.

The form of the cephalic region resembles that of *D. cristatus* Bl., but the projecting portions are less separated.   The radial joint of the male palpus will, however, separate the two species without difficulty.

Diplocephalus fuscipes Bl.   (*Walckenaera fuscipes* in "Spiders of Dorset.")

Length of male, 2 mm.

In this species the caput is not nearly so deeply cleft as in *D. cristatus* Bl.   The form of the radial joint is the most certain clue to its identity.

I have received specimens of this species, as well as of *D. cristatus* Bl. quite recently, from Mr. W. Falconer, of Slaithwaite, Huddersfield.

Diplocephalus latifrons Cb.   (*Walckenaera latifrons* in "Spiders of Dorset.")

Length of male, 1.7 mm.

The caput of this species, viewed in profile, bears a close resemblance to that of *D. fuscipes* Bl.   The width of the cephalic eminence, as seen from in front, however, is greater, and the radial joint of the palpus is totally different, approaching in form that of *D. cristatus* Bl.

Diplocephalus picinus Bl.   (*Walckenaera picina* in "Spiders of Dorset.")

Length of male, 1.8 mm.

The form of the caput is very similar to that of *D. fuscipes* Bl.   The radial joint of the male palpus, however, is quite different, being produced into a curved prolongation which projects greatly over the digital joint.

Diplocephalus beckii Cb.   (*Walckenaera beckii* in "Spiders of Dorset.")

Length of male, 1.6 mm.

Diplocephalus alpinus (*Plaesiocraerus alpinus* of various authors.)

Diplocephalus speciosus Cb.   (*Plaesiocraerus speciosus* Cb. 1895.)

Structural details of the last three extremely rare species will be seen on reference to the figures.

(To be continued.)

GENERA INSECTORUM.—Under the direction of M. P. Wytsman, of 108 Boulevard du Nord, Brussels, it is proposed to publish a list of all the groups of insects.   This appears to be a very large undertaking and expensive.   Full particulars may be obtained from M. Wytsman.

BUTTERFLIES OF THE PALAE-ARCTIC REGION.

BY HENRY CHARLES LANG, M.D., M.R.C.S., L.R.C.P. LOND., F.E.S.

(Continued from p. 337.)

Genus *COLIAS* (*continued*).

14. **C. eogene** Feld.   Reise Novara, p. 196, pl. 27, fig. 7.

41—45 mm.

♂ differs from *C. thisoa* in the very deep reddish-orange of the ground colour, with stronger purple reflection; in the breadth of the marginal borders, which are often as wide as in *C. edusa*;

C. romanovi.   Female.   (*Ante, p.* 337.)

also in the neuration, which is blackish, as in *C. aurora*.   Disc. spot f.w. not rounded, but narrow or semi-lunar ; that of h.w. large and bright orange.   H.w. deeply shaded towards base and in. marg.   ♀ differs from *C. thisoa* in the darker and less brilliant orange of the ground-colour, the marginal borders of both wings are less strongly defined on their inner edges.   The spots are more numerous, more regular, and of a more orange-yellow tint. nearly approaching the ground colour.   The disc. spot of h.w. is more pyriform than circular in outline, and of a brighter orange than the rest of the wing.   As in the case of *C. thisoa* ♀, the h.w. are liable to be very deeply shaded, so that the ground colour in some specimens is almost black, obliterating the marginal band and spots.   The fringes in both sexes are brilliant rosy red.   U.s. differs considerably from that of *C. thisoa.*   F.w. bright orange, marginal border and costa being light green, disc. spot small and white-centred, a trace of a row of sub-marginal spots.   H.w. rich yellowish-green, with a trace of a sub-marginal row of dark red spots.   Disc. spot small but brilliantly silvery-white, edged with reddish-purple.   Neuration whitish, especially in ♀, fringes bright rose colour.

HAB.   Turkestan, the Himalayas, and Southern Thibet ; always at considerable elevations, from 11,000 feet and upwards.   VII.—VIII.

a. var. *theia* Stgr.   MSS. Cat. 1882.   Somewhat smaller than the type, and much paler in the ground-colour. The disc. spot of f.w. is smaller, and in some specimens nearly obsolete. U.s. paler as regards the green coloration of the h.w. Disc. spot h.w. very small. This form seems to be intermediate between the type and the next. HAB. The mountains of Osch Turkestan, and extending to the north and west (Elwes).

b. var. *stoliczkana* Moore. 35—40 mm. ♂ much smaller than the type, paler in colour, and with a distinct row of sub-marginal spots on u.s., h.w. Nervures black. ♀ with f.w. greenish-yellow.

C. eogene. Male.

C. eogene. Female.

C. eogene. Var. *stoliczkana.*

The disc. spot is much smaller on h.w. than in typical *C. eogene.*

HAB. N.E. Thibet, Lobnoor, Ladak, Sikkim. At great elevations—from 16,000 to 17,000 feet.

c. var. *erythias* Grum. ♀ rather brighter than type. H.w. brighter. The disc. spot larger. The marginal band narrower, more obtusely angled and lighter, and the spots are more numerous, besides being larger and more definite. HAB. Hinducoosh, Kaschmir. (R. and H.)

d. var. *elissa* Grum. Smaller than type, lighter in both sexes, especially as regards the h.w. The yellow marginal spots are more distinct in ♀, and have a tendency to coalesce. Disc. spot of h.w. disproportionately developed. Neuration of u.s.

h.w. of a greenish colour; darker than that of the ground-colour. HAB. Northern Pamir. (R. and H.)

e. ab. *cana* Gr.-Gr. The white form of ♀. HAB. Cashmere.

f. var. *arida* Alph. ♂ has the marginal bands narrower than in type. Ground-colour orange, somewhat lighter. ♀ white, like *C. diva* ♀, or blackish, with yellowish-green spots. Disc. spot h.w. orange or red. Very variable. HAB. Kokonoor and Lob-noor districts. This is probably only a local form of *C. eogene,* peculiar to the arid deserts of the above-mentioned localities. There is a fine series of this form in the collection of the late Mr. H. J. Leech, from which the above note was made.

15. **C. staudingeri** Alph. Hor. Ent. Ross. xvi. p. 35, pl. xiv. 4 (1883). 43—51 mm.

Rather larger than *C. eogene,* not so deeply coloured. Borders narrower. Disc. spot f.w. long and narrow. Fringe of ou. marg. more distinctly red. U.s. f.w. greenish, except along inner margin, which is orange mixed with grey. H.w. greenish, disc. spot small and pink.

HAB. Kuldja. At higher altitudes than *C. thisoa.* 7,000—12,000 feet. Flight very swift, like that of *C. eogene.*

16. **C. viluiensis** Mén. Schrk., p. 18, T.I. 7. 45—50 mm.

♂ ground colour orange, but paler than in the preceding species, and with less reddish tinge.

C. viluiensis. Male.

C. viluiensis. Female.

F.w., border much as in *C. thisoa,* but with only a slight trace of yellow veining near apex. Disc. spot much as in that species. H.w. with a very pale

N 4

disc. spot hardly distinguishable from the ground-colour. ♀ something like *C. aurora* ♀, but smaller and much duller in colour. The nervures are black. U.s. somewhat as in *C. thisoa*, but the dark green of that species is replaced by a light yellowish-green. The sub-marginal spots on f.w. are either entirely wanting or very slight. Disc. spot h.w. small and with a very faint edging of red. Antennae very dull red; fringes of all the wings narrow and inconspicuous. The above description is from a pair sent to me by the late Dr. Staudinger from the "Transbaical."

HAB. Transbaical. Northern Siberia on the Jenisei and Vilui Rivers (Elw.), Pokrofka Amur (R. and H.).

17. C. felderi Gr.-Gr.

22—25 mm.

♂ somewhat paler than *C. hecla*. F.w. disc. spot rounded, marginal border narrow with yellow rays at apex f.w. H.w. with a very narrow border and a row of lightish spots internal to the border, as in *C. myrmidone*, etc. Disc. spot faint. U.s. h.w. the disc. spot is very small rose-red, and surrounded by a broad edge of reddish-brown. Described from specimens in Leech's Collection, in which only the ♂ is represented.

HAB. Sinin Alps, Central Asia.

NOTE.—On p. 261, *ante*, third line from bottom of second column; for "It seems to occur," read "The ♀ seems to occur."

(*To be continued.*)

LAND AND FRESHWATER MOLLUSCA OF HAMPSHIRE.

BY LIONEL E. ADAMS, B.A.,
AND B. B. WOODWARD, F.L.S., F.G.S.

(*Concluded from page 303.*)

Bithynia leachii Shepp. Sparingly, Christchurch, Winkton, Ringwood, Tuckton (C. A.), Itchin at Winchester (C. F. A.).

Vivipara vivipara Linn. Itchin at Winchester (F. J.).

Vivipara contecta Millet. Itchin at Winchester (F. J.).

Valvata piscinalis Müll. Fairly abundant throughout the county.

Valvata cristata Müll. North Hants, common (H. P. F.); rivers Avon and Stour (C. A.).

Pomatias reflexus Linn. North Hants, common (H. P. F.); generally distributed in Isle of Wight (A. L.). Var. *pallida* Moq.; somewhat rare in Isle of Wight (A. L.).

Neritina fluviatilis Linn. Itchin at Winchester (F. J.); abundant in rivers Avon and Stour with great variety of markings (C. A.). Var.

cerina Colb., about 4 per cent. of the whole in the above-mentioned district (C. A.); var. *trifasciata* Colb., very common (C. A.); var. *undulata* Colb., 2 specimens (C. A.).

Unio tumidus Retz. Basingstoke Canal (coll. H. C. Leslie, in Hartley Inst.). This species has evidently made its way into the county along the canal. There is no record of any example of the genus occurring in the district, either living or in the post-pliocene deposits, if we except fragments recorded by Bristow from the Tot-lands Bay deposit. These fragments, we are inclined to think, were more likely portions of *Anodonta*.

Anodonta cygnea Linn. River Stour at Tuckton Bridge, generally undersized, occasionally attaining five inches; at Beaulieu, six inches (C. A.); near Havant (C. E. W.). There are very large specimens in Alton Museum from Woolmer Forest (J. T. C.).

Sphaerium rivicola Leach. Itchin at Winchester (F. J.).

Sphaerium corneum Linn. Round Christchurch in all suitable places (C. A.); Sandown, Isle of Wight (A. L.).

Sphaerium lacustre Müll. Havant (C. E. W.) and also (A. L.).

Pisidium amnicum Müll. North Hants Common (H. P. F.); Stour at Tuckton, length 8 mm., breadth 10 mm.; Ringwood, Brockenhurst, abundant and fine (C. A.).

Pisidium pusillum Gmel. River Stour, Christchurch to Tuckton and district (C. A.); North Hants, common (H. P. F.); Isle of Wight (A. L.).

Pisidium fontinale Drap. Rivers Avon and Stour (C. A.); Isle of Wight (A. L.). Var. *cinerea* Ald., Sandown (A. L.); var. *henslowana* Shepp., Tuckton (C. A.); var. *pulchella* Jen., Avon and Stour at Christchurch (C. A.).

Pisidium nitidum Jenyns. Three specimens from Avon and Stour at Christchurch (C. A.); North Hants (H. P. F.); colony in a pond at Hambledon (C. S. Coles).

Pisidium milium Hud. Hayling, rare (C. E. W.).

In addition to the foregoing the following species, in fossil or sub-fossil condition, have been met with in alluvial deposits:—

Vertigo substriata Jeff. Test Valley.
Vertigo pusilla Müll. Test and Itchin Valleys.
Succinea oblonga Drap. Totlands and Freshwater Bays.
Paludestrina confusa Frauenf. Stone.
Paludestrina ventrosa Mont. Southampton Dock. Stone.

N.B.—The nomenclature has been revised and brought up to recent knowledge by Mr. B. B. Woodward.

January 30th, 1901.

CLASSIFICATION OF BRITISH TICKS.

By Edward G. Wheler.

THE Ixodidae have received so little considera- tion at the hands of British naturalists that there does not exist amongst our literature any classification of the family having pretension to accuracy or completeness. This may be sufficient to account for the fact that when inviting correspondence last year, through the columns of Science-Gossip, I did not receive any reply from a fellow-countryman who had made a serious attempt to study the British ticks; though I have been favoured with much kindly assistance from correspondents who had turned their atten- tion to foreign species.

There seems no doubt that the best classification of the genera, giving descriptions of the known species, is that con- tained in a very care- fully compiled series of articles in the "Mé- moires de la Société Zoologique de France" for the years 1896–97– 99 (vols. ix., x. and xii.). These articles were written by M. G. Neumann, Professeur à l'École |vétérinaire de Toulouse, and are en- titled "Revision de la famille des Ixodidés." These papers are out of print, and are not likely to fall readily into the hands of an English reader.

It is with the object of popularising the sys- tematic study of the

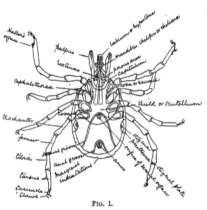

FIG. 1.

NOMENCLATURE OF EXTERNAL PARTS OF SHEEP-TICK.

ticks that I venture to print the following résumé of M. Neumann's classification, giving copies of such of his figures as may assist in explaining the letterpress ([1]). To condense as much as possible, I have confined my descriptions, which are in great part taken from those of M. Neumann, to the more salient characteristics that may pro- bably suffice for identifying the sub-family and genus to which a specimen may belong. To these are added remarks on the number of known species in each genus, a description of those which have been identified in this country, a list of synonyms, and other points of interest. Most of the characteristics referred to are such as may be examined readily without having recourse

(1) The illustrations copied are figs. 5, 7, 10, 18, 19, 21 and 24.

to any more powerful magnifier than a pocket lens.

A short description, with a diagram (fig. 1), of the various parts of a tick referred to in this paper may assist readers in following Professor Neumann's classification.

Ticks pass through four stages in their exist- ence: the egg, the larva, the pupa or nymph and the adult. In the larval, pupal and adult female stages the body consists of a highly distendable cuticle. This, in the sub-family Ixodinae, is partly covered by a hard scutellum, or shield, on the back, and is provided with a false head, or capitulum, that carries the palpi, and the mouth organs, con- sisting of a hard chitinous labium or hypostome provided with a tube for the suction of blood, and armed with rows of barbs for clinging on to the flesh of the host. On each side are situ- ated the mandibles, also called chelifers, or cheli- cerae. They are re- tractile, and doubtless serve to cut a slot in the skin to make a pas- sage for the insertion of the labium, and afterwards to force it into the flesh of the host. For these pur- poses the chelifers are furnished with a series of teeth or hooks. Col- lectively these organs are called the rostrum.

The adult male is similar, but he has a shield that, with the excep- tion in many cases of a narrow margin, covers the whole of the body. The latter is incapable of being much distended by the suction of the host's blood. In the sub-family Argasinae these shields are altogether absent. In the larval stage ticks have but six legs, but in all other stages eight legs.

In adults the sexual organ is situated far forward between the haunches of the legs; behind it is the anus, usually surrounded in part by a groove, and on each side, near the fourth pair of legs, is placed a stigmal plate or peritreme for respiration, in the centre of which is the stigma. The plates are absent in the larval state. There is reason for believing that the sexual organ of the male above referred to is either immature or obsolete in certain, if not all, of the species of the genus *Ixodes*, and that in such cases the sexual functions are performed by the mouth organs, all of which are inserted with the exception of the palpi (2). This curious anomaly has hitherto only been observed in two species, *Ioxdes reduvius* in England and *I. pilosus* in Cape Colony. The sexual orifice is absent in the larval and pupal stages. The tarsus, or last joint, of the first pair of legs is furnished with a peculiar organ, known as "Haller's organ," which is probably one of touch, hearing, or smell · but the function is not understood. The second pair of legs are the shortest, and the fourth pair the longest.

The life history of a tick is sharply divided between a free and a parasitical existence. In the first state it lives absolutely without food of any sort for prolonged periods, and passes its time either in a semi-torpid condition, or else is actively occupied in searching for a host on which to establish itself. A headless female of *Ixodes reduvius* (3), lacking all the mouth organs by which feeding would be possible, survived over a year in captivity, and was eventually lost. *Argas persicus* is similarly stated to have lived without food for three-and-a-half years in captivity (4). At such times all growth is suspended, and the tick is debarred from making any advance towards metamorphosis from one stage of its existence to another.

In the parasitical states life is supported by sucking the blood of the host until the body of the tick has, unless it is a male, become distended to a considerable extent. and it is in this condition that these pests are generally noticed owing to their increased size. When replete, certain species fall to the ground, and there remain while development is proceeding inside the distended cuticle. After a time the skin is split open, and the creature emerges with its rostrum, shield, legs and other external parts, increased in size and fully developed. The body is proportionately diminished, so that the animal's entire dimensions are practically about the same as before ; but the new body, being formed of a similarly distensible cuticle, is again ready for repletion so soon as another host is attacked.

Some species never leave the host they have first found, but pass all their metamorphoses upon its person. In this respect the habits of different Ixodidae vary considerably. Mr. Lounsbury, Government Entomologist at the Cape of Good Hope, informs me that the "red tick," *Rhipicephalus evertsi*, passes the first moult on, and the second moult off, the host *R. decoloratus*, the "blue tick," never leaves the host that it has once found, after being hatched out of the egg, until, if

a female, it is ready in its turn for oviposition. *Argas reflexus* only attacks by night, after the manner of the bed bug, a practice which may enable it to escape destruction from the beaks of the fowls or pigeons which are its usual prey, as at such times they would be asleep, or at least in a drowsy condition. Mr. Lounsbury says "it has the peculiarity of undergoing an additional moult. and, what is more, when adult, alternates egglaying with feeding, the interval being about the same as between the moults." *Ixodes reduvius* seeks a fresh host after each moult, but as yet little is known of the habits of any other British species. Referring to the South African "bonttick," Mr. Lounsbury writes: "Females do not appear to complete their engorgement until they have mated." If this be so with *I. reduvius*, and perhaps other species, it may account for the numerous dead and half-distended specimens that may generally be found on the host.

The length of life depends mainly on the climatic conditions, and whereas Messrs. Dixon & Spreull state that *Rhipicephalus decoloratus*, the Texas cattle-tick, is only sixty days in passing the whole period of its existence, it is probable that our British species average about a year and a half, varying largely according to circumstances.

The damage done to stockowners by these pests in other countries is enormous. Mr. Cooper Curtice says : (5) "Cattle-ticks cause the quarantine of eighty-one counties in North Carolina. The cattle traffic in thirteen States and the Indian Territory is seriously interfered with on account of the ticks.

Mr. P. R. Gordon, Chief Inspector of Stock for the Government of Queensland, states in his Annual Report for 1898 that previous to that year no less than £44,000 had been spent in that colony in connection with the investigations and experiments made in combating "tick," or "Texas fever." The searching character of these investigations has probably proved the salvation of stock raising in Australia, as they resulted in the discovery by Mr. C. J. Pound. Director of the Stock Institute, that inoculation by the blood of immune beasts would produce immunity in previously susceptible stock.

From Cape Colony Mr. Lounsbury writes in his Report for 1899: "Heartwater," another tick-inoculated disease, "seems to have gained fresh impetus of late years, and is spreading by leaps and bounds into the Midlands." "The market value of these properties is depreciated by the infection from 30 per cent. to 60 per cent., I am reliably informed." This disease attacks sheep and goats, and is carried by a tick named *Amblyomma hebraeum*.

To the ravages caused by ticks abroad may be added the damage done to sheep at home by

(2) See R. T. Lewis, in the "Quekett Microscopical Society Journal" of October 1900.

(3) Mentioned by me in "Louping-ill and the Grass-tick" in the "Royal Agricultural Journal" of December 1899 (vol. x., part iv.).

(4) Referred to by C. Fuller in his "Bovine-tick Fever," 1896, p. 8.

(5) Regulations for the Control of Contagious Diseases of Live Stock, &c., May 1st, 1900, North Carolina Department of Agriculture.

louping-ill, and doubtless other diseases will eventually be traced to these parasitical pests.

As the study of ticks is of so much economic importance, a few hints as to the methods of collecting and preserving them may not be out of place. The object in view is usually identification of species, or investigation of the life-history of the parasite. For the former purpose the large distended females, which are generally those first noticed on the host, are of comparatively little use. The great distension of body obliterates some material characteristics and obscures others. Where these large females are observed careful search should be made for the much smaller specimens in the other stages, by which identification is facilitated.

Ticks of a uniform brownish colour may generally be preserved without damage in spirits of wine, but those having variety of colour should be immersed in 3 per cent. formalin. For examination and future reference I find it convenient to mount them dry in cells as microscopic objects. This keeps them clean and free from dust. Those that have been soaked in formalin must be very thoroughly washed and dried. Even then they will be found to deposit an oily dew on the slide and cover-glass. This can only be removed by remounting, which may have to be done more than once. Treated in this manner they have so far retained their colours excellently. They may also be mounted in Canada balsam as transparent objects, but so mounted they are far harder to identify, and I have found difficulty in clearing the body of its contents when preparing them in this manner after they have been much distended.

For the purpose of studying their life-history ticks may be kept alive for long periods in tightly corked glass bottles, but many species require to be supplied with a little very slightly damped sand and fresh moss. Provided there is enough moisture to keep the moss alive, and no more, lest the ticks become mouldy, they will survive many months. Air does not seem necessary to them. If collected in any immature stage when fully distended they will undergo metamorphosis, or when adult the females will lay eggs, in confinement.

It is scarcely necessary to emphasise the importance of keeping notes of the date and place where specimens are found, together with any circumstances attending their capture, especially the prevalence of disease amongst hosts infested by them.

(To be continued.)

NEW ARCTIC EXPEDITION.—It is reported from Vancouver, British Columbia, that Mr. Bernier proposes to exploit a new Arctic expedition. He has made arrangements for building a specially designed ship, which will be arranged so as to store provisions that will last for seven years. The contract for the ship is said to be £16,000.

IRISH PLANT NAMES.

By JOHN H. BARBOUR, M.B.

(Continued from page 306.)

COMPOSITAE (continued).

GALLAN GREANCAIR. gallan, "branch"; grean, "gravel." refers to the roots branching in gravelly soil. *Tussilago farfara.* colt's-foot.

GALLAN MOR. POBAL or PIOBAL PUBAL. "names." *Petasites officinalis.* pestilent wort, butter-bur.

BOGLUS. "soft plant." *Senecio paludosus.* bird's tongue.

BUAFANAN NA HEASGARAN. easgar, "plague"; eascar, "enemy"; buafanan, "toad." BUIDE BAILSCEAN. yellow rain. BUADALLAN or BUAD-GALLAN. Buad, "food"; gallan, "branch," "rock." *Senecio jacobaea.* ragweed. seggrum. St. James' wort.

GRON LUS. gron, "a stain." CRANLUS. "tree herb." *vulgaris.* simson. groundsel.

COCOIL. COPOGTUAITIL. tuaite, "rural"; copog, "a large leaf"; "a large rural leaf." LEADAN LIOSTA. liosda, "lingering," "heavy"; "the hair which sticks." MEACAN DOGA. MEACAN TOBAC. MEACAN TUAN. MEACAN TUATAIL. meacan, "a root," therefore plant; doga, "mischievous." tobac. "elect"; tuan, "hedge." tuatail, "awkward"; hence such names as "mischievous plant," "hedge plant," "awkward plant." *Arctium lappa.* burdock, clotbur. cucculeys in King's Co.

DEILGNEAC. deilgne, "thorns," "prickles." *Cnicus lanceolatus.* spear thistle.

CURRAC or CURAC NA CUAIG. currac, "a cap." cuaig, possibly cnac, "cup." GORMAN. gorm. "blue." *Centaurea cyanus.* bluebottle. hurtsickle.

MULLAC DUB. mullac, "the top"; dub, "great" or "black." "Blackheads." NIANSGOT. *Centaurea nigra.* matfellon. horse-knops.

CASERBAN or CASTEARBAN. purely "bitter." LUS AN TSUICAIR. suice, "sooty," but it may be related to Lat. succurrere, "to run under." Whichever is most correct, I think this name refers to the root. *Cichorium intybus.* succory.

DUILLEOG BRIGIDE, DUILLEOG MAIT, DUILLEOG MIN are respectively "St. Bridget's leaflets." "excellent leaflet," "tender leaflet." *Lapsana communis.* nipplewort, dock cresses.

CLUAS LIAT or CLUAS LUC. liat, "hoary"; luc. "mouse"; cluas, "ear"; "hoary or mouse ears." *Hieracium pilosella.* mouse-ear. hawkweed.

BAINE MUC or MUIC. "pig's milk." BLIOCT FOCADAN. blioct, "milk"; focan, "a plant." FOFANAN MIN. "a little thistle." *Sonchus oleraceus.* sow-thistle.

FEOCADAN. FOTANAN. *Sonchus arvensis.* corn sow-thistle.

BEARNAN BEARNAC. both words refer to "gaps," "cuts." CEISEARBAN. "bitter." SEARBAN. "sour."

FIACAL LEOGAIN. fiacal, "a tooth": leogain, "relating to lion." CASTEARBAN NA MUIC. "too bitter to please a pig." *Taraxacum officinale* and *T. dens-leonis.* "dandelion."

VACCINIACEAE.

BRAIGLEOG. BRAOILLEOG. FRACOG. frac, "bleakness." FRAOCAIN. CRAN FRAOCAIN. *Vaccinium myrtillus.* bilberry.

ERICACEAE.

FRAOC. rage. *Calluna. Erica.* grig-ling.

PRIMULACEAE.

BAINE-BO-BLEACT. all words refer to "cow's milk." SEICGEIRGIN, SEICEARBAN. "sour sorrel." SEICEIRGIN. MUISEAN or LUS NA MUISEAN. muisean, "primrose." *Primula vulgaris.* primrose.

BAINE-BO-BUIDE. "yellow cow's milk." *Primula veris.* cowslip.

FALCAIRE FIODAIN or FALCAIRE FUAIR. fiodain, "a witness"; fuair, "found"; falcaire, "scoffer," hence "scoffing witness." RUIN RUISE, "wayside grass." *Anagallis arvensis.* male pimpernel.

LUS COLUIMCILLE. "St. Columcille's herb." LUS SIOTCAINT. siotcaint, "peace"; "peace plant." SEAMAR MUIRE. "lady's trefoil." *Lysimachia nemorum.* yellow loosestrife.

OLEACEAE.

CRAN FUINSEAN. CRAN FUINSEOG. CRAN FUINSE, or these words without cran stand for ash. FUINSEOG COILLE. coille, "a wood." NIUN or NION. letter N. UINSEAN. *Cf.* fuinsean really same word. *Fraximus excelsior.* ash.

GENTIANEAE.

DREIMIRE MUIRE. "virgin's ladder." *Erythraea centaurium.* centaury.

LUS AN CRUBAIN. cruban, "crabfish"; crubaim, "I bend." CEADARLAC. *Gentiana campestris.* gentian.

DEAGA BUIDE. deaga, "bug," "chafer," "yellow chafer." DREIMIRE BUIDE, "yellow-ladder." *Chlora perfoliata.* yellow centaury, or yellow wort.

PACARAN CAPUIL. pacaire, "pedlar." PONAIRE CAPUIL. PONAIRE CURRAIG. Ponaire, "beans," Heb. pol; capull, "a horse"; corrac and curraig, "marsh"; therefore "horse or marsh beans." PONRA CURRAIG. PARA CAPUIL. *Cf.* ante. *Menyanthes trifoliata.* buckbean.

BORAGINEAE.

BARRAISTE. possibly derived from borrac, "branches of trees," or borrad, "a swelling," both being suggested. BOG-GLAS. *Borago officinalis.* borage.

BOGLAS. BAGLUS. BOLG-LUS. bolg, "cow." bog, "soft." lus, "plant." *Lycopsis arvensis.* bugloss.

TEANGA CON. teanga, "tongue"; con, gen. of cu, "hound." *Cynoglossum officinale.* hound's-tongue.

LUS MIDE or LUS MIOLA. mide, "a neck," Co. Meath.; miol, any low animal; therefore we get "the neck plant" or "scorpion herb." GOTARAE. *Myosotis arvensis.* scorpion grass, forget-me-not.

LUS NA CNAM BRISTE, or LUS NA CANAB BRISTE. cnam, "a bone"; briste, "broken"; "the plant for a broken bone." *Symphytum officinale.* comfrey.

SOLANACEAE.

CAOC-NA-CEARC. caco, "blasted," cearc, "hen"; "hen blast." GAFFAN. CLOC, "cloak," "pall," "pupil of eye," and henbane itself. DEODA. deoc, "drink." *Hyoscyamus niger.* henbane.

DREIMIRE GORM. FUAT GORM. SLATGORM. dreimire, "ladder"; gorm, "blue"; fuat, "hatred"; slat, "a switch"; hence, "blue ladder or switch," "blue hatred." MIOG BUIDE. MIOTOG BUIDE. miog, "smile"; miotog, "bite"; "yellow smile, or bite." *Solanum dulcamara.* woody nightshade.

CONVOLVULACEAE.

CUNAC. *Cuscuta epilinum.* dodder.

SCROPHULARINAE.

BIOLAB UISQE. BIOLAR MUIRE. "river cresses," "lady's cresses." LOCAL MOTAIR. loc, "a lake"; motair, "tuft." *Veronica beccabunga.* brooklime.

FUALACTAR. *Veronica serpyllifolia.* Paul's betony. thyme-leaved speedwell.

AN ULAC. EALCAH ANOLURAC. *Veronica chamaedrys.* wild germander.

LUSCRE. cre, "keel," "earth." SEAMAR CRE. keeled trefoil. *Veronica officinalis.* fluellin.

CUINEAL MUIRE. cuinead, "mourning," with mourning candles are associated, hence "women's candles." *Verbascum thapsus.* hag's taper. crow's lungwort.

DUN LUS. "fortress or hill plant." FOTRUM. FARAC DUB. farac, "a mallet," "beetle." LUS NA ECNAPAN. dim. of cnap, "knot"; "plant of little knots." *Scrophularia nodosa.* kernelwort. rose-noble in Co. Tipperary.

BOLOAN BEIC. bolgan, "a little bag"; beic, "bee" "little bee-bag." MEARAEAN SIOTAIN or MEARAAN NA MINA SIGE. "fairy hillocks" or "fairy fingerings." SIOTAIN SLEIBE. SIAN SLEIBE. siotain, "fairy," "fairy mounts." *Digitalis purpurea.* foxglove.

CAOIMIN. LIN RADARC. lin, "eye"; radarc, "sight." LUSNA LEAC. leaó, "a flag-stone." RAIN AN RUISG. rain, "point of anything"; ruisg, gen. of rorg, "eye"; "eye-spark." RUISNIN RAIDRE, LIN RAIDRE, GLAN RUISE, all mean much the same—"little path, or eye ray," "wayside smile." RAEMIN RAIDRE. "little field ray." *Euphrasia officinalis.* eyebright.

(To be continued.)

BOOKS TO READ

NOTICES BY JOHN T. CARRINGTON.

We have received the following books, but in consequence of pressure on our space the notices must be deferred : " Stalk-eyed Crustacea of British Guiana, West Indies, and Bermuda," by Charles G. Young, M.A., M.D. ; " Annual Report of Smithsonian Institution, 1898," 2 vols. : " Total Solar Eclipse of May, 1900," by E. Walter Maunder, F.R.A.S. ; " Disease in Plants," by H. Marshall Ward, Sc.D., F.R.S. ; " Missouri Botanical Garden," Eleventh Annual Report : " Earth's Atmosphere," by Dr. T. L. Phipson : " Technics of the Hand-Camera," by W. B. Coventry, M.I.C.E. : " First Aid to the Injured," by H. Drinkwater, M.D. ; " Electricity Simplified," by A. T. Stuart, A.I.E.E. ; " Practical Electrician's Pocket-book for 1901," by H. T. Crewe ; " Scientific Roll," by Alexander Ramsay : " Laws Regulating Sale of Game," by Dr. C. Hart Merriam ; " Principles of Magnetism and Electricity," by P. L. Gray, B.Sc. ; " Colour Photography," by A, R. Smith ; " Chemistry," by James Knight, M.A., B.Sc. ; " Alga Flora of Yorkshire," by W. West, F.L.S. ; " Practical Enlarging," by J. A. Hodges, F.R.P.S. ; " Experimental Farms Re orts " ; " Maryland Geological Survey," with Atlas ; " Fruit Trees," by Wm. Saunders, LL.D., F.R.S C., F.L.S. ; Reports of the " Entomological Society of Ontario " ; " Scottish Arboricultural Society " ; " Society for Psychical Research " ; " Hull Scientific Club " ; " Botanical Exchange Club " ; " Croydon, New and Old " ; " Journal Quekett Microscopical Club " ; " Marine Biological Association " ; " Millport Marine Biological Station " ; " Annals of Andersonian Society " ; " Leicester Literary Society " ; " Vitality," by L. S. Beale, F.R.S., F.R.C.P. ; " Matriculation Directory " ; " Price List of Exotic Butterflies and Moths " ; " Report Sarawak Museum " ; " Report Bristol Museum Reference Library " ; " Report Wellington College Science Society " ; " Transactions English Arboricultural Society," by J. Davidson ; " Laws of Industrial Property " ; " Handbooks of Essex Field Club " ; " Proceedings of Association of Economic Entomologists " ; " Sunny Days at Hastings and St. Leonards," by H. W. Saunders and P. Row ; " Godalming and its Surroundings," by T. F. W. Hamilton ; " Journal of Board of Agriculture " ; " Report of Canadian Department of Agriculture, Central Experimental Farm " ; " Description of Apparently New Species of Goat," by D. G. Elliot, F.R.S.E. ; " Genus Eupomotis," by D. G. Elliot, F.R.S.

The Englishwoman's Year-Book and Directory for 1901. Edited by EMILY JANES. xxxv + 378 pp., 7½ in. × 5 in. (London : A. & C. Black. 1901.) 2s. 6d. net.

This is the twenty-first year of publication of this handbook of women, giving particulars of some women who work and those who have become celebrated. This book contains a large amount of information, and there seems to be little left out of its pages that could be of use to women. The particulars are arranged under sectional headings. We should have liked to have observed more space devoted under the heading of " Science," there being eight pages only, still it is satisfactory to notice that the number of educated women devoting their time to scientific pursuits and investigations is increasing. Perhaps, however, in their modesty, particulars are in many instances withheld, as we do not observe several names of those who are doing considerable work of this kind.

The Story of Art in the British Isles. By J. ERNEST PHYTHIAN. 216 pp., 6 in. × 3¾ in., with 28 illustrations. (London : George Newnes, Ltd. 1901.) 1s.

This, being one of Messrs. Newnes' Library of Useful Stories, gives an outline of Art in the British Islands. Commencing in the prehistoric period with a sketch of the earliest known example in these islands of imitation of animal form, we have reproduced the well-known sketch of a horse

from the Creswell Crags found in Robin Hood's cave, on the borders of Nottinghamshire and Derbyshire. Thence it carries us forward, step by step, to the Art of the present period. In this case the author very properly includes in Art, Archaeology, Architecture, Drawing and Painting. The chapter on " Celtic Christian Art " is of especial interest, as the drawings and descriptions are largely of Celtic monuments not much known by the general public. More examples, however, might have been given of Art in furniture, pottery, and some other familiar household surroundings.

Alpine Plants. By W. A. Clark, F.R.H.S. vii + 108 pp., 7½ in. × 5 in., with 9 plates. (London : L. Upcott Gill. New York : Scribner's Sons. 1901.) 3s. 6d.

Considering the ease with which Alpine plants can be grown by persons having even small gardens, it is a source of surprise that we so seldom meet with this delightful group of flowers under cultivation. Such a collection, however, is one for the botanist rather than for the horticulturist. Want of knowledge of their habits and requirements is, perhaps, the true cause of so many persons shirking their growth, for there are few places where some of them cannot be planted and grown with success. The author of the book before us is one of the most experienced authorities in this country upon the Alpine garden, having had charge for some time past of one that is celebrated for its success, in the nurseries of Messrs. Backhouse, near York. Mr. Clark does not attempt in his pages anything further than plain instructions for growing the more difficult and rarer plants, with which most people fail. If his instructions be followed, many an unsatisfactory garden will be changed to a source of pleasure. The illustrations are generally good, as is the whole production of the book in its neat cloth cover.

Imitation. By RICHARD STEEL. xii + 197 pp., 7½ in. × 5 in. (London : Simpkin, Marshall.) 1900. 3s.

In this book the author draws attention to " the mimetic force in Nature and Human Nature." The earlier pages are occupied by a series of papers read before the Literary and Philosophical Society of Liverpool, which have received revision and considerable addition in the form of several chapters, bringing the subject to further development. The first part of the work deals with imitation in Economics, Psychology, Ethics, Religion, Politics, Law, Custom, Fashion, Language, Poetry and Arts. With Chapter VI. commences the consideration of imitation in Habit and Instinct, in Animal and Vegetable Life, also in Heredity. Thence Mr. Steel proceeds to the consideration of imitation in the inorganic world, and finally is a chapter on retrospective considerations and conclusion. There is an appendix on " Imitation in Reasoning." We must refer our readers to the author's pages for the plan and arguments of the theory of the influence of imitation. They are well worth consideration, and the book should be widely read. In a new edition, we imagine, the author will somewhat rearrange his plan, as it has, perhaps, in its present form the fault of taking for granted that the author and reader have previously discussed the subject ; so the latter is sometimes left too much to his own judgment as to the conclusions intended by the writer. The addition of a special chapter elaborating the last paragraph, under the heading Conclusion, would be of service.

Birds of Siberia. By HENRY SEEBOHM, F.L.S., F.Z.S., F.R.G.S. xix + 512 pp., 9½ in. × 6½ in. with map and 112 illustrations. (London : John Murray, 1901.) 12s. net.

Many of our readers will recollect that the late Mr. Seebohm spent two years in Siberia, chiefly with the object of ascertaining the breeding places of certain birds which nest within the Arctic Circle. In 1875, in company with another well-known ornithologist, Mr. J. A. Harvie-Brown, he visited the Valley of Petchora, and two years later accomplished a more lengthy and adventurous journey to Yenesei. The outcome of these travels was two occupied by Nature notes, but the second has the additional interest of commercial-geographical exploration in company with the well-known Captain Wiggins, who in modern times has made such strenuous efforts to develop the little-understood trade which awaits with its riches the enterprise of those who have the energy and capital. to bring the North Asiatic commerce into direct touch with Britain. The illustrations are, as a whole, excellent, and by the courtesy of the publisher we give a reproduction of one as an example. Among them we must not omit reference to the drawings, which appear as tail-pieces to the

LITTLE STINT'S NEST. EGGS AND YOUNG.
(From " Birds of Siberia.")

fascinating books with the respective titles of "Siberian Europe" and "Siberian Asia." Both these works were soon out of print, and before his death the author decided to combine them in one volume. This is now before us, published with ample and elegant illustrations by Mr. John Murray. Those who read the two previous volumes will, we feel sure, return to this re-issue with pleasant memories. Notwithstanding its present title it must not be imagined that the pages are devoted entirely to bird life. They sparkle with quaint incident of travel and curious lore relating to the native Samoyede and other indigenous tribes occupying Northern Siberia and its tundras. The work is divided into two parts: 1. The Petchora Valley, and 2. The Yenesei. The first portion is largely chapters, of a remarkable series of old Russian Greek church crosses ; a collection of which works of art form some of the most valued portions of the gatherings made by Mr. Seebohm in the remote and only partially civilised regions of Northern Siberia visited by him.

Practical Electrician's Pocket Book. Edited by H. T. CREWE, E.I.M.E. lxxiv + 198 pp., 5½ in. × 3¾ in. Illustrated. (London : S. Rentell & Co., Ltd. 1901.) 1s. 6d.

The Electrician's Pocket Book for 1901 has been much improved in this, its third year's issue. Some additional matter appears which increases the usefulness of the book. The general production of this useful little work is also more satisfactory.

CONDUCTED BY C. AINSWORTH MITCHELL,
B.A.OXON., F.I.C., F.C.S.

SIDE-CHAIN THEORY OF IMMUNITY.—It has been demonstrated in various researches that when bacteria or the toxines which they produce are introduced into the living body of an animal specific antagonistic substances are formed, which afford protection against the same poison when transferred to another animal. To account for this phenomenon Ehrlich has proposed what is known as the side-chain theory. He assumes that a toxine molecule has two components, one of which, the "haptophore" group, is unsaturated and stable, whilst the other, the "toxophore" group, is readily decomposed. The haptophore group attaches itself to the attacked cell and thus enables the toxophore group to exert its specific action. The living molecule of protoplasm is also regarded as consisting of two parts, a central group with attached unsaturated side-chains with which the haptophore groups of toxines combine. Side-chains are then generated to replace those fixed by the toxine, and these being formed in excess of the requirement, circulate in the blood forming the specific immune substances or anti-toxines. Now, on introducing such immune sera into the blood of another animal, these free side-chains attach themselves to the haptophore group of any toxine subsequently gaining admittance, and at the same time combine with an active principle the "addiment." This is normally present in the blood, but in too dilute a form to have much effect upon a toxine; but when thus concentrated by means of the side-chains it is enabled to destroy the intruding substance by means of its "zymophore" group. In support of this view Ehrlich points out that when serum containing the immune body or side-chains is heated to 55° C. it loses its power of destroying bacterial poison, but regains it after the addition of ordinary serum containing no immune substance, but only the hypothetical "addiment."

IMMUNISATION OF MILK TO RENNET.—The discovery of the specific ferment which coagulates milk was made by Heintz in 1872, and it was subsequently isolated by Hammarsten from the stomachs of numerous animals. It becomes inactive when heated to 63° C. in a neutral solution, but is uninjured by cooling to 0° C. Its activity is measured by the time required by a given quantity of rennet to coagulate a definite quantity of milk. This time is nearly inversely proportional to the amount of ferment. The most active preparations obtained by Hammarsten coagulated 4,800,000 times their weight of casein from the milk; but beyond a certain point a further addition of the ferment caused no acceleration. Morgenroth has recently made the remarkable discovery that on injecting rennet in gradually increasing minute amounts into animals an *anti-rennet* is apparently formed in the serum, and on adding the latter to

ordinary milk, coagulation by rennet is prevented or retarded. The strongest immune serum obtained by Morgenroth prevented coagulation, when added in the proportion of 2 per cent. to milk into which 1 part of rennet in 20,000 was subsequently introduced. In ordinary milk coagulation occurred when rennet was added in the proportion of 1 in 3,000,000. If we apply Ehrlich's side-chain theory to this phenomenon, we must regard the rennet introduced into the serum as attaching itself by its "haptophore" group to the side-chains of the protoplasmic cells. Side-chains are then generated in excess and diffused through the serum, forming the immune substance or anti-rennet. Ordinary milk, according to this theory, contains an "addiment" which is too dilute to have much action upon rennet; but on treating the milk with the immune serum, the anti-rennet attaches itself to this "addiment," and thus enables it to concentrate its "zymophore" group upon the intruding rennet, and destroy it.

ARSENIC ON MALT.—Some interesting particulars on this subject were given by Mr. W. Thomson, Public Analyst for Stockport, in a recent lecture before the Society of Arts. He stated that owing to the method in which malt was dried, arsenic must have been present in beer for more than a century. The germinated barley was dried by means of anthracite or coke, and some of the arsenic which was volatised from the fuel, even when there was no smoke, condensed upon the malt. In one case he found ordinary soot to contain 1·5 grain of arsenic to the pound, or 22½ times the maximum dose, its origin being probably the pyrites in the coal. In anthracite coal he detected from $\frac{1}{100}$ to $\frac{1}{200}$ grain per pound, whilst coke contained from $\frac{1}{8}$ to $\frac{1}{100}$ grain. Of sixty-two samples of malt which he examined, only seven were quite free from arsenic, the others containing from $\frac{1}{1000}$ to $\frac{1}{7}$ grain. In the latter case, if the whole of the arsenic present passed into the beer, about ½ grain per gallon would be present—a serious amount. It has been found, however, by Mr. A. C. Chapman, Secretary of the Society of Public Analysts, that a considerable proportion of the arsenic on malt is taken up by the yeast during the fermentation, whilst another part is possibly precipitated on boiling the wort. A member of a firm of malsters informed the writer that a sample of malt which had been found to contain arsenic was mechanically brushed and again submitted to the analyst, who now certified it to be free from contamination. The evidence, however, on this point is conflicting.

THE BLOOD OF DIFFERENT ANIMALS.—In the March issue of the "Bulletin de la Société Chimique de Paris" M. S. Cotton describes some interesting observations which he has made, on the well-known property possessed by blood, of liberating oxygen from hydrogen peroxide. On treating 1 cubic centimetre of blood with 250 c.c. of hydrogen peroxide the following quantities of oxygen in c.c. were collected:—

	Minimum.	Maximum.
Man	580	610
Horse	320	350
Pig	320	350
Ox	165	170
Guinea-pig	115	125
Sheep	60	65

The blood of female and young animals gives somewhat higher results than old male animals.

AN ASTRONOMICAL SOCIETY FOR THE MIDLANDS, with Birmingham as its centre, seems likely to be formed as a branch of the British Astronomical Association.

As we go to press we observe newspaper reports that a large comet has appeared visible in South Africa and Australia. It is said to be seen shortly before sunrise, and to have a trifurcate tail, reaching over ten degrees.

IT has been found necessary to change the name of the "Streatham Science Society," which is now entitled the Norwood Natural Science Society. The hon. secretary is Mr. Ben. H. Winslow, of 31 South Croxted Road, West Dulwich, S.E.

MESSRS. SANDHURST & CROWHURST, of 71 Shaftesbury Avenue, W., are making an important feature of the testing of eyesight, and for this purpose have designed and fitted up a special room for this purpose.

WE have received a copy of "The Student's Friend," an educational journal published in Bombay, which appears to be the only monthly magazine of this kind issued in India. It contains amongst other interesting matter two illustrated articles on astronomical subjects.

THE vessel which has been built for the German Antarctic Expedition and recently launched at Kiel, has by order of the Emperor been christened "Gauss," in memory of the late Professor Karl Friedrich Gauss, who did so much to stimulate Antarctic research.

MR. T. E. FRESHWATER, a well-known amateur microscopist and photographer, who has spent the whole of his business life in the service of Newton & Co., scientific instrument makers, at Temple Bar, London, has been offered, and accepted, a partnership in that firm.

THE "Public School Magazine," now in its seventh volume, has been acquired by Messrs. A. & C. Black, under whose auspices it will in future be published. The magazine retains its present position as the only publication devoted exclusively to matters concerning the Public Schools and those who are, or have been, connected with them.

MR. EDWARD RICHARD HENRY, C.S.I., the new Chief of the Criminal Investigation Department, Scotland Yard, was for many years Inspector-General of Police in Bengal. It will be remembered that it was he who introduced into India the Bertillon-Galton anthropometric system for the identification and scientific study of criminals.

DR. J. P. LOTSY WAGENINGEN, Holland, has issued a circular to botanists, inviting them to join an International Association of Botanists. He has already secured several of the leaders of the science towards the foundation. A first general meeting will take place at Geneva on the 7th August next, at 10 A.M., in the botanical laboratory of the University.

THE great success attending the meetings of the International Association of Academics, held in Paris last month, is largely to be attributed to the friendly reception accorded to each other by the various delegates.

DR. E. VON OPPOLZER, MM. F. Rossard and Ch. André have independently sent the announcement that Eros is variable. M. Rossard, of Toulouse, says from 9·3 to 11·0 magnitude, with a period of 2h. 22m. It has been suggested that there are really two tiny planets revolving round each other, having diameters as 3 : 2, the orbital plane passing through the Earth.

ACETYLENE burners with incandescence mantles were exhibited in the Paris Exhibition, but these were liable to "strike back" and to smoke. The Carbide and Acetylene Company of Berlin announce that it has overcome these drawbacks, due to the explosive nature of a mixture of acetylene and air, and to the richness of the acetylene in carbon. That firm is now|manufacturing serviceable burners.

DR. J. A. VOELCKER, the new President of the Society of Public Analysts, is well known as an authority on agricultural chemistry. Some two or three years ago the University of Oxford conferred upon him the honorary degree of M.A. for his scientific work in this direction. The retiring president is Mr. W. W. Fisher, M.A., F.I.C., Aldrichian Demonstrator of Chemistry at Oxford.

THE mussel is beginning to be recognised as an equal source of danger with the oyster in disseminating epidemics of typhoid fever, and an agitation is on foot to place the beds from which they are obtained under the direction of the sanitary authorities, giving them power to register various beds as wholesome and condemn others. To these molluscs should also be added cockles and periwinkles, which are equally sources of danger.

THE well-known collector of Lepidoptera, Mr H. McArthur, is arranging to work for that order of insects during the coming season in either the Outer Hebrides, or Orkneys and Shetlands, as may be desired by the majority of those for whom he will collect. His terms are eight subscriptions of £10 each, and among the subscribers the proceeds of his season's work will be divided. His address is 35 Aveil Street, Fulham Palace Road, London, W.

THE London Geological Field Class have now commenced their Saturday afternoon excursions. Visits have been arranged to places of geological interest both north and south of London, and students will have an excellent opportunity of investigating the geology and physical geography of the Thames basin. Particulars may be obtained from Mr. H. R. Bentley, hon. sec., 43 Gloucester Road, Brownswood Park, N.

WE hear that Mr. Thomas Southwell, F.Z.S., M.B.O.U., is engaged upon the preparation for publication of "Letters and Notes on the Natural History of Norfolk," being from the MSS. of Sir Thomas Browne, M.D. These MSS., which are in the Sloane Collection in the Library of the British Museum, cover observations made during a large proportion of the seventeenth century, and have already been published in Bond's Antiquarian Library; but Mr. Southwell's edition will be accompanied by critical notes of value to naturalists. The work will be issued by Jarrold & Sons, of London.

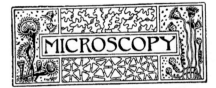

CONDUCTED BY F. SHILLINGTON SCALES, F.R.M.S.

ROYAL MICROSCOPICAL SOCIETY, March 20th. Mr. A. D. Michael, vice-president, in the chair.— The secretary called attention to an excellent portrait of Peter Dollond, presented to the Society by Mr. C. L. Curties. Mr. Nelson referred to two old microscopes which had recently come into the possession of the Society. The first—a non-achromatic microscope—has the name of Carpenter, 24 Regent Street, engraved upon it, and its date may be assigned to about 1825. It was especially interesting from the fact that the late Hugh Powell, before he began to make microscopes on his own account, made them for the trade, and in this instrument they doubtless had an early specimen of Hugh Powell's work. The other microscope, of the Culpeper and Scarlet type, was signed Dollond, and its date was probably not later than 1761. Messrs. Staley & Co. sent for exhibition a Bausch and Lomb Camera Lucida, described in the Journal last December. It was intended for reproducing natural size an object diagramatically. Mr. E. M. Nelson read a paper, " On the Working Aperture of Objectives for the Microscope," in which he showed that in recording delicate observations it was advisable to state the precise ratio of the utilised diameter of the objective to the full available aperture. He then proceeded to explain the different methods by which this ratio—which he termed the working ratio, or W.R.—could be measured. Dr. Tatham confirmed Mr. Nelson's views in regard to the necessity for recording the working aperture of objectives, and expressed his appreciation of the value of the methods proposed by the author for obtaining this measurement. A paper by Mr. H. G. Madan, F.C.S., " On a Method of Increasing the Stability of Quinidine as a Mounting Material," was read by Mr. Nelson in the absence of the author. Mr. Madan found that by keeping quinidine heated to a certain temperature for a considerable time it was converted into colloid quinidine, which condition it had retained for a year, but whether the tendency to revert to the crystalline form was entirely overcome time alone could show. Mr. Karop said of all media, quinidine, on the whole, was the best yet discovered for mounting diatoms, but was very troublesome on account of its tendency to crystallisation. He hoped the material, prepared as suggested by Mr. Madan, would be offered for sale, when he would give it a trial. Mr. Rousselet read a paper " On some of the Rotifera of Natal," by the Hon. Thomas Kirkman, illustrated by mounted specimens shown under microscopes. Mr. Rousselet had appended a technical description of *Pterodina trilobata*, one of the rotifers mentioned in the paper, a mounted specimen of which was among those exhibited. An excellent drawing of this rotifer by Mr. Dixon Nuttall was also shown. Mr. W. H. Merrett read a paper " On the Metallography of Iron and Steel," demonstrating the subject by the exhibition of a large number of lantern slides of sections of different classes of these metals, under various conditions of hardness, stress, etc. The methods by which these sections had been prepared and polished were also explained.

INSTANTANEOUS PHOTOMICROGRAPHY.—Mr. A. C. Scott, of Rhode Island College, has devised an arrangement by which he has been able to obtain instantaneous photographs of microscopic living organisms. A powerful light is, of course, necessary, and in his own work he has used an arc light of 2,200 volts, giving about 4,000 candle-power. This light is placed at a distance slightly greater than the focal length of the condensing lens to obviate such concentration of heat as would be detrimental to the microscope objective. The camera is of the usual vertical type, but the important essential is a combined shutter and view-tube, which is clamped by means of three thumb-screws to the draw-tube of the microscope : this apparatus is fastened above the ocular, and after the latter has been inserted in the draw-tube. The mechanism of this apparatus is described by Mr. Scott as follows: " Upon a movable brass plate inside a light-tight box is a 90° prism, mounted in such a way that all the light which passes through the microscope is projected upon a piece of ground glass at the end of a cone, which may be lengthened or shortened in order to give correct focus to the object, when it is properly focussed upon the ground glass of the camera directly above the microscope. Next to the prism is a hole in the brass plate for allowing light to pass from the microscope directly to the photographic plate, when the prism is moved by a spring and pneumatic release, and finally a sufficient area of the brass plate to cover the opening when exposure has been made. To take a photograph, the microscopic animal is placed in a drop of water upon a suitable glass plate, the light is turned on and the shutter so set that the object may be focussed upon the ground glass of the cone. The plate-holder is inserted and the dark slide drawn, leaving the plate exposed inside the camera bellows. The movements of the animals are easily seen upon the ground glass, and when the desired position is obtained the shutter is released, the prism moves out of the way and the light passes to the plate." The apparatus is not yet perfected to its inventor's complete satisfaction, but he states that exposures as short as one-fortieth of a second have been very satisfactory, and considers that thoroughly satisfactory negatives can be obtained with low-power objectives in one-hundredth of a second. The magnification has, however, ranged up to 200 diameters. Mr. Charles Baker, of High Holborn, in his last catalogue, mentions a somewhat similar arrangement for instantaneous photomicrography in which a pneumatic shutter with a prism attachment enables the object to be viewed on a ground-glass screen at right angles to the optic axis up to the moment of exposure. We have not, however, seen this apparatus. Mr. Andrew Pringle, in his well-known book on practical photomicrography, describes a vertical camera for the same purpose, but of different construction. This camera is fitted with a pair of " goggles " and a velvet bag for the head. An instantaneous shutter, made of thin sheet aluminium, lies almost in the plane of the sensitive plate and bears white discs upon which the focussing is done, and the image watched until the time for exposure.

FLOATING FORAMINIFERA.—In reference to Dr.
G. H. Bryan's interesting article entitled "Experiences in Floating Foraminifera" (ante, p 296), it
may be useful to those who are working in the
same direction to know that calcareous and other
light mineral substances can be separated from
heavier in sands by making use of the blast of an
ordinary blowpipe. There is a note in the "Geographical Journal," May 1897, on "Drifting Sands"
that fully explains my meaning. A little practice
soon enables the student to determine the strength
of blast necessary for the separation of particles
varying in size and density in any given sand-mass.
The sand should be allowed to fall through a small
paper funnel, and the descending stream "played"
upon with a constant blast of uniform pressure.
The method I employed for separating sand constituents by means of a vibrating inclined plane
(vide "Nature," vol. xxxix. 1889, p. 591) is useful
when it is desired to collect samples of the denser
minerals only. The inclined plane of "frosted"
glass which I made use of in my researches on
"musical" sands (see "Musical Sand," 1888) is
the best method that can be adopted for bringing
about the separation of rounded grains from those
of the angular type. I shall be pleased to send
any of your readers who are interested in the study
of sands a copy of my note on "Drifting Sands.—
Cecil Carus-Wilson, Royal Societies Club, S.W.,
March 7th, 1901.

ANSWERS TO CORRESPONDENTS

G. B. (Darlington).—I am obliged to you for
your letter, and am glad that the article referred
to interested your Society. The specimen you
send is, as you suggest, a dendrite on limestone.
Such dendrites are frequently met with, and you
will find them illustrated in text-books on geology.
They are due to thin films of mineral matter that
have formed between fissures or joints in the rock,
and in so doing have branched out into this moss-
or fern-like appearance. The fissure is necessarily
a very fine one—so fine as to be almost invisible
without the aid of a pocket lens. You will find an
interesting illustrated article on dendrites by Mr.
Carrington in SCIENCE-GOSSIP, vol. i., N.S., p. 267.

G. E. H. (Hornsey).—I have never tried to
make permanent mounts of amoebae, and cannot
therefore speak from my own experience, but
I would suggest fixing with 2 per cent. chromic
acid added to the water containing the amoebae
in a watch-glass. This is one of the methods recommended by Lee. As dilute stains you might try
methyl-green, Bismarck brown, or haematoxylin,
and as mounting medium, glycerine or Farrant's
solution. I am sorry I cannot give you more
definite information. Will you let me know what
results you get, for the benefit of other microscopists? Minute crustacea, such as Cyclops and
Daphnia, are generally mounted unstained in
Canada balsam, after soaking in turpentine; and
the same applies to Hydra. You can fix with
alcohol and so get rid of the water. Weak osmic
acid will both kill and often differentiate the
tissues; but be very careful with this reagent, as its
vapour is most irritating to both eyes and throat.
Perhaps the best stain is picro-carmine. All the
above reagents etc. can be obtained from Chas.
Baker, or Watson & Sons, both of High Holborn,
W.C. I am afraid there is no other book dealing
with the mounting of the above, other than Lee's

"Microtomist's Vade-mecum," which deals with
advanced methods only.

MEETINGS OF MICROSCOPICAL SOCIETIES.

Royal Microscopical Society, 20 Hanover Square,
London. May 15, 8 p.m.
Quekett Microscopical Club, 20 Hanover Square,
London. May 3, 17, 8 p.m.

For further articles on Microscopic subjects see
pp. 353 and 358.

EXTRACTS FROM POSTAL MICROSCOPICAL
SOCIETY'S NOTEBOOKS.

[Beyond necessary editorial revision, these
extracts are printed as written by the various
members. Correspondence thereon will be welcomed. - ED. Microscopy, S.-G.]

SECTIONS THROUGH STEMS OF BRAZILIAN
LIANAS.—The term Lianas was first used in the
French colonies and afterwards adopted by English,
German and other travellers to designate the
woody, climbing and twining plants which abound
in tropical forests and constitute a remarkable and
ever-varying feature of the scene. They over-
top the tallest trees, descend again to the ground
in vast festoons, pass from one tree to another, and
bind the whole together in a maze of living network. Many lianas become tree-like in the thickness of their stems, and often kill by constriction
the trees which originally supported them.
Botanically considered, lianas belong to orders
which are often quite different.—John Terry.

The mounter of these sections states that the
peculiar distribution of the wood is due to the
cohesion of other stems. I have looked in the
"Encyclopaedia Britannica" and other large works
of reference, but can find no information concerning these plants. Travellers in tropical regions
refer to the wonderful growths of the forest, but
they rarely give any of those minute details which
are essential in such cases as the one before us.
Yet, though I do not know anything of these
plants, I have no hesitation in saying that I do not
believe this cohesion theory. Take, for instance,
fig. 1. The central portion starts on its journey
through life; it comes across two companions, who
metaphorically link arms and agree to unite and
travel together. By-and-by they come across two
more, and they likewise agree to accompany them,
also one on each side; soon they find two more, -
little ones this time, and persuade them to
join the company. Now, it will be noticed
that they always seem to join on in pairs, one
on each side, and that one side numerically
balances the other, whilst each woody area on
the right exhibits nearly the same development
as its counterpart on the left, even down to the two
little ones at the extremities; further, that each
woody area exhibits the same structure as its
neighbour. Chance acquaintances do not usually
exhibit such a singleness of purpose and formation.
Secondly, what has become of the organic centres
or axes of these coherent members? How, on this
theory, account for the unbroken sclerenchyma
(the hard bast) encircling the entire section, or
its absence in the interior, and where are the
remains of the cortices of the last arrivals? To my
way of thinking, this theory gives no answer. In
the hot and humid atmosphere of these tropical

forests plant life seems in a manner to lose command over itself and exhibit phenomena with which we in temperate climes are unacquainted. For instance, notice the central portion. The wood at first is developed with great regularity and strongly lignified. Then on the entire periphery a great change becomes visible, a more open formation taking the place of the previously regular tissue. Zones of thick and thin, lignified and unlignified tissues alternate, which in a temperate clime would be taken to denote successive periods of growth and rest. Beyond this we come to a zone of cambium, or wood-producing tissue, on the outside of which is the soft bast or phloem, and then wood again, and this order continues till we come to the bark. In the plants with which we are familiar the wood cambium is invariably found occupying one definite position, and I think few are familiar with any plants which have more than one regular. system of wood cambium. Fascicular cambium and cork cambium do not here concern us. This cambium is derived from the thin-walled cells of the ground tissue, and under great stimulation it is conceivable that any portion of thin-walled elementary tissue with active contents might develop a cambium. Now as regards the object before us, my views are that the outermost cells of the primary soft phloem or else the inner cells of the cortex developed a second cambium which proceeded to form new wood and phloem tissue, and to develop on exactly the same lines and at the same time as the original cambium. So that now we have two sets of wood-forming cambium simultaneously at work. When this new growth had developed the same thing occurred again until the present state of development had been arrived at. As the greatest increase in thickness has always taken place in one direction, it naturally follows that the cortex and sclerenchyma will be folded back laterally and that we shall always find this hard tissue in its usual place, *i.e.* at the periphery. In this section we see four bands of

FIG. 1. TRANSVERSE SECTION OF STEM OF BRAZILIAN LIANAS.

cambium and four bands of wood and four bands of phloem-tissue on either side of the centre, and all appear to have been in an active condition at the time of gathering. The cambium at the outer edge of the latest woody increase is very clear. Of course my theory requires practical proof such as could only be obtained by observation on the living plant and by tracing its development. These remarks are made about fig. 1, but they apply also to fig. 3, and I do not think it would be very difficult to account for the complicated arrangement shown in this section. Perhaps some of our members will give their views on the subject. The only section which would satisfactorily prove the cohering theory would be one which showed component areas with a totally dissimilar structure. In a case like fig. 3 one could better understand their division into nine separate twigs rather than their cohesion.—*Thos. S. Beardsmore.*

I think if Mr. Beardsmore regards fig. 1 as a series of climbers, the central being the original stem, the others clinging to it one upon another, and the section cut through a plane where there were three stems on either side, the cohesion difficulty will disappear. The inner xylem of the younger climbers is not formed owing to lateral pressure due to cohesion, but the xylem has been formed outwardly. We have a familiar example of younger climbers from the same shrub surrounding and clinging to the older stems in the honeysuckle. Fig. 3 is another example of tropical climbers, some of which produce trunks resembling cables.—*F. C. Fuller.*

FIG. 2. TRANSVERSE SECTION OF STEM OF BRAZILIAN LIANAS.

The type of fig. 1 is figured and described in Kerner & Oliver's "Natural History of Plants," and the explanation there given agrees with Mr. Beardsmore's theory. We may therefore accept it without hesitation as the true one. A structure somewhat analogous is found in the Cycads, but here there are several complete rings of cambium, each ring forming xylem internally and phloem externally. In fig. 2 certain parts of the cambium ring appear to give up the formation of xylem and devote themselves entirely to the manufacture of phloem. This seems to take place in a remarkably orderly manner, the demand for an increased supply of phloem being made by the plant at fairly definite intervals. This arrangement seems capable of an interesting physiological explanation, which may be something like the following: The humid atmosphere of a Brazilian forest would perhaps tend to check rapid transpiration, while the absorbed liquid being fairly rich in inorganic food materials a very active transpiration current would not be necessary. Consequently xylem, which is the essentially water-conducting tissue of the plant, would not need to be so extensively developed as in a plant growing under other conditions. But assimilation is rapid, and large quantities of organic material are manufactured. This needs phloem for its conduction from place to place. Hence the somewhat abnormal development of this tissue in the stem. Another interesting point about this slide is the presence of tüllen or tyloses in the vessels. In this connection the following paragraph, taken from Vine's "Text-book of Botany," may be of interest: "When a tracheal cell with a fitted wall abuts upon cells containing living protoplasm, it not infrequently happens that the thin pit-membranes begin to bulge, in consequence of the pressure upon them of the contents of the living cells, into the cavity of the tracheal cell, and actually grow. Cell division may take place in these ingrowths, so that a mass of cellular tissue is found in the cavity of the tracheal cell. These ingrowths are termed tyloses; they are constantly to be found in some kinds of wood—*e.g.* Robinia—occasionally in many others." With regard to fig. 3, though I am unwilling to admit the cohesion theory, it is hard to suggest any other explanation. If the case were really one of cohesion, would there not be some line of demarcation between the respective stems? In fig. 4 the regular groups of lignified cells towards the periphery seem to be strands of sclerenchyma developed in the phloem.

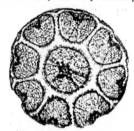

Fig. 3. Transverse Section of Stem of Brazilian Lianas.

That it is not of the same nature as the true xylem towards the centre is indicated by the fact that the peripheral lignified tissue contains no vessels. If this view be taken the structure of this stem is comparatively normal.—*C. J. Wilkinson.*

Mr. Beardsmore states that he can find no information in botanical works concerning Lianas, or, as I have generally seen it spelt, Lianes. I find on referring to Henfrey's "Botany" and Balfour's "Class-book of Botany" that both contain information on the subject, although perhaps not all the information one might wish. Henfrey, on p. 233, and Balfour, on p. 85, give an illustration of the stem in fig. 1. The former describes it as a peculiar fasciculated stem of a Malpighiaceous plant of South America. The stem presents an example of an anomalous exogen. It consists of numerous woody masses, having each distinct pith, and surrounded by cellular tissue resembling that of the outer bark. Such a stem, he says, looks as if it were formed of several united together. He offers no explanation of how this is produced, but refers one for an account of anomalous exogens to the works of De Jussieu and others. The only book that I have at hand that offers any explanation of how these appearances may be produced is Goebel's "Classification and Special Morphology of Plants," and the information appears to be taken for the most part from De Bary. A few short extracts may be interesting. In describing the histology of the stem in dicotyledons he says: "Very striking deviations from normal structure are to be found in the Sapindaceae. Some species of the order are formed in the usual manner, but in others the transverse section of the stems shows, in addition to the usual ring of wood, a number of smaller closed rings of different sizes in the secondary phloem, each of which increases in thickness, like the ordinary ring, by means of a layer of cambium" (fig. 3). Nägeli supposes the principal cause of this to be that the primary vascular bundles of the stem do not lie in a circle or transverse section, but in groups more towards the

outside or inside. When the interfascicular cambium is formed in the fundamental tissue the isolated bundles are connected together, according to their grouping on the transverse section, into one closed ring in *Paullinia*, or into several in *Serjana* (*Serjana*, Balfour, p. 780). Goebel groups these abnormal arrangements under several headings: under C, "renewed thickening rings," there are some remarks that apply to fig. 1. Growth in thickness begins in the normal manner and then ceases, but is afterwards continued by a new cambium zone formed in the parenchyma, outside the first one. The process may be repeated and a number of concentric zones be formed. This mode of proceeding, which has already been described in *Cycas* and *Gnetum* among the Gynosperms, is found in Dicotyledons in the Menispermaceae, and in the stem of *Avicennia*. In the latter cases all the zones of increase which succeed to the normal one are formed in the secondary phloem. In the stem of some lianes (*Bauhinia*), in *Wisteria* and others the new zones are formed in the secondary phloem (not in the primary phloem, as Mr. Beardsmore suggests). Referring to the mode previously described under *Cycas* and *Gnetum*, I find (p. 344): "In *Gnetum*, as in the Cycadeae and many Dicotyledons, the growth in thickness from the first cambium ring ceases after a time, and a new zone of meristem (*i.e.* actively dividing tissue) is found in the secondary cortex outside the ring. In this zone xylem-strands are formed on the inside and phloem-strands on the outside, alternating with medullary rays. As this process is repeated more than once, a transverse section of an older stem or branch of *Gnetum scandens*, for example, shows several concentric rings of growth, each consisting of a xylem-ring and a phloem-ring." In fig. 1 these zones are found laterally and not concentrically. Mr. Wilkinson's interesting account of fig. 2 leaves

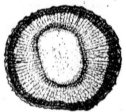

Fig. 4. Transverse Section of Stem of Brazilian Lianas.

very little to add. There are many other interesting points contained in Goebel's book, but I think it unnecessary to quote further, as anyone interested in the subject can read the entire account for himself. Nothing is said, however, about cohesion. As Mr. Beardsmore says, it needs no elaborate botanical knowledge to settle this question. An appeal to common sense decides at once against the cohesion theory. I have endeavoured to record the unusual appearances which these slides present by photographs. The task has been rather a difficult one for me, as I have never before photographed with such very low powers, and I found some trouble in getting an even illumination of the object and in judging the length of time necessary for exposure.—*J. R. L. Dixon.*

ASTRONOMY.

CONDUCTED BY F. C. DENNETT.

	1901	Rises.	Sets.	Position at Noon. R.A.	Dec.
	May	h.m.	h.m.	h.m. s.	° ' ''
Sun	.. 5	.. 4.28 a.m.	.. 7.26 p.m.	.. 2.47.16	.. 16. 7.54 N.
	15	.. 4.10 a.m.	.. 7.42 p.m.	.. 3.26.14	.. 18.45.59 N.
	25	.. 3.58 a.m.	.. 7.56 p.m.	.. 4. 6.10	.. 20.52.30 N.

		Rises.	Souths.	Sets.	Age at Noon
	May	h.m.	h.m.	h.m.	d. h.m.
Moon	.. 5	.. 9.26 p.m.	.. 0.53 a.m.	.. 5.15 a.m.	.. 16 14.23
	15	.. 2.27 a.m.	.. 9.16 a.m.	.. 4.21 p.m.	.. 25 14.23
	25	.. 11.56 a.m	.. 6.28 p.m.	.. 0.26 a.m.	.. 7 6.22

		Souths.	Semi-	Position at Noon. R.A.	Dec.
	May	h.m.	diameter.	h.m.s.	° ' ''
Mercury	.. 5	.11.17·3 a.m.	.. 2·6''	.. 2· 8. 0	..11.30.33 N.
	15	..11.59·7 a.m.	.. 2·5''	.. 3.29.48	..19.17.55 N.
	25	.. 0.49·8 p.m.	.. 2·7''	.. 4.59. 9	..24.29.15 N.
Venus	.. 5	.. 0. 2·1 p.m.	.. 4·9''	.. 2.52.45	..15.49.22 N.
	15	.. 0.12·2 p.m.	.. 4·9''	.. 3.42.17	..19.23. 6 N.
	25	.. 0.24·2 p.m.	.. 4·9''	.. 4.33.40	.. 2r. 3.16 N.
Mars	.. 5	.. 7.12·7 p.m.	.. 4·4''	..10. 4.10	..13.57.19 N.
	15	.. 6.45·6 p.m.	.. 4·0''	..10.16.26	..12.30.15 N.
Jupiter	.. 15	.. 3.26·7 a.m.	..20·2''	..18.55.15	..22.42.26 S.
Saturn	.. 15	.. 3.40·9 a.m.	.. 8·2''	..19, 9.34	..21.57.44 S.
Uranus	.. 15	.. 1.29·0 a.m.	.. 1·9''	..16.58. 7	.. 22.42.13 S.
Neptune	.. 15	.. 2.19·4 p.m.	.. 1·2''	.. 5.49.52	..22.16.22 N.

MOON'S PHASES.

			h.m.				h.m.
Full	..	May 3	.. 6.19 p.m.	3rd Qr.	May 11	..	2.38 p.m.
New	..	„ 18	.. 5.38 a.m.	1st Qr.	„ 25	..	5.40 a.m.

In apogee May 2nd at 8 a.m.; in perigee 17th at 7 a.m.; and in apogee again on 29th at 5 p.m.

METEORS.

				h.m.	°
Mar. 11 to May 31	Draconids	Radiant	R.A.17.32	Dec. 60 N.
Apr. 5 „ 10	Lyrids	„	„	18.8	„ 35 N.
„ 12 to June 30	Coronids	„	„	15.40	„ 23 N.
May 6	η Aquarids*	„	„	22.28	„ 2 S.
„ 1	(ν Herculis)	„	„	15.56	„ 46 N.
„ 3 to 9	α Serpentids	„	„	15.36	„ 10 N.
„ 11	(α Cor. Bor.)	„	„	15.24	„ 27 N.
„ 15	η Aquilids	„	„	19.36	„ 0
„ 29 to June 4	η Pegasids	„	„	22.12	„ 27 N.
* Just before sunrise.

CONJUNCTIONS OF PLANETS WITH THE MOON.

					° '
May 8	..	.. Jupiter*†	.. 7 p.m.	.. Planet	3.50 S.
„ 9	..	.. Saturn*	.. 1 a.m.	.. „	3.48 S.
„ 18	..	.. Mercury*	.. 1 p.m.	.. „	1.39 N.
„ 18	..	.. Venus*	.. 1 p.m.	.. „	0.38 N.
„ 25	..	.. Mars*	.. 3 p.m.	.. „	6.59 N.

* Daylight. † Below English horizon.

OCCULTATIONS AND NEAR APPROACHES.

		Magni-	Dis-	Angle from Vertex.	Re-	Angle from Vertex.
May	Star.	tude.	appears.	°	appears.	°
			h.m.		h.m.	
2	.. ι Virginis	5·5	.. 3.18 a.m.	.. 50 below horizon.	
5	.. λ Libræ	5·0	.. 4.16 a.m.	.. 334 near approach.	
8	.. 21 Sagittarii	4·9	.. 0.39 a.m.	.. 86	.. 1.51 a.m.	.. 305
9	.. d „	4·9	.. 0.26 a.m.	.. 153	.. 1.23 a.m.	.. 247
14	.. λ Piscium	4·7	.. below horizon	.. 2.48 a.m.	... 266
31–1..	B.A.C. 5109	5·4	..11.49 p.m.	.. 34	.. 0.36 a.m.	.. 319

THE SUN continues very free from disturbances either bright or dark, but should be watched.

There is no real night from May 21st to July 23rd, twilight lasting all the time the Sun is below the horizon.

MERCURY is in superior conjunction with the Sun at 6 p.m. on May 14th, after which it becomes an evening star, and at 6 p.m. on 18th is in conjunction with and 1° 4' north of Venus. During the last week of the month Mercury does not set until more than one hour after the Sun, and will be within the view of the observer in the constellation Taurus, near Gemini.

VENUS is in superior conjunction with the Sun at 1 a.m. on May 1st, but is too near the Sun for observation.

MARS passes the meridian in daylight all the month, and so must be looked for as soon as it is sufficiently dark. His apparent diameter decreases apace, but at this time the details of his surface often seem better defined, and he bears high magnifying powers well.

JUPITER AND SATURN remain near each other in Sagittarius, rising just after midnight at the beginning of the month, and about an hour and a half earlier at the end. Their low altitude is not helpful to good observation.

URANUS, in the southern part of Ophiuchus, precedes Jupiter by nearly two hours, and so is nearly as well placed for observation as he will be this year.

NEPTUNE is too close to the Sun to be seen.

ECLIPSE OF THE MOON.—When the Moon rises, at 7.28 p.m., on May 3rd, she will be wholly covered by the penumbra of the earth's shadow, the last contact with the same taking place at 8.55 p.m., 38° west of the north point. It will be barely noticeable.

ECLIPSE OF THE SUN.—This will be total across the Indian Ocean, Sumatra, Borneo, and the southern part of New Guinea, in the early morning of May 18th, but will be quite invisible in England. or, indeed, in Europe.

A NEW VARIABLE STAR, near Nova Persei, known as B.D. + 43° 726, has had its character defined by Mr. A. Stanley Williams, who has favoured us with a rough diagram, which is reproduced, to aid in identification. "Its visual

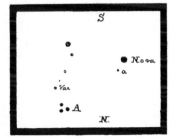

DIAGRAM OF NEW VARIABLE STAR NEAR NOVA PERSEI.

magnitude at present is about 8½. The period of variation is probably long; possibly a year or more. It is probably nearly at a maximum now." It is catalogued as 8·9 magnitude. Photographically its magnitude is less than when observed visually. On January 11th it was photographically 11·47 magnitude, but by February 28th had increased to 10·53 magnitude. It had previously been catalogued as having a peculiar spectrum.

THE NEW STAR IN PERSEUS continues to decrease in brightness, but not with regularity, having shown remarkable fluctuations. By March 18th it was 3·71 magnitude, and on 19th 5·25; by 20th it had again reached 3·7, then recommenced falling. On 21st it was 4·0 magnitude, but by 22nd again fell to 5·25; on 24th it was 4·0, on 25th 5·38, and on 26th 3·9 magnitude. It decreased to a little less than 4·0 magnitude on 28th, and seems to have become a trifle brighter on 30th. On March 31st it was a little below 4·0 magnitude, but by April 1st had fallen below 5th magnitude. On April 5th it had nearly reached 4th magnitude, and was still about the same brightness on the 8th, but by the 10th had fallen to an equality with 30 Persei, which is about 5th magnitude. A photograph of the region showing stars to below 11th magnitude, taken only 28 hours before Dr. Anderson's discovery, by Mr. A. Stanley Williams, of Hove, shows no trace of the Nova.

ANOTHER NEW VARIABLE STAR in Cygnus has been discovered by Dr. T. D. Anderson, known as 2.1901, in R.A. 19h. 12·2m., Dec. N. 29° 55'. Variation was observed from 9·5 magnitude on December 26th, to 10·4 on 16th of February.

ANSWERS TO CORRESPONDENTS.

JOHN CLARK (Birmingham).—We cannot spare space to enter into the discussion. The moon's path described on p. 317 is not theory, but ascertained fact. The mass of the moon is to that of the earth as 1 : 88, nearly; so that the centre of gravity of the system is well within the globe of the earth.

CHAPTERS FOR YOUNG ASTRONOMERS.

BY FRANK C. DENNETT.

(Continued from p. 318.)

JUPITER'S SATELLITES.

SINCE Galileo first discovered three of these little bodies on January 7th, 1610, and the other a few days later, they have been constant objects of interest to the observer. In size, the four which are within the reach of the ordinary telescopes do not differ much from that of our own moon. They are usually known as I., II., III., and IV. in the order of their distance from the planet. Their real and apparent diameters, and distances from Jupiter, are :—

I.	....	2,390 miles	.... 1″.048	....	267,000 miles
II.	....	2,120 ,,	.... 0″.911	.,..	425,000 ,,
III.	....	3,980 ,,	.... 1″.513	....	628,000 ,,
IV.	....	2,970 ,,	.... 1″.278	....	1,192,000 ,,

From time to time they have been seen with the naked eye, especially when two or more of them have been near together, and when the brilliance of the planet has been modified by twilight. The missionary Stoddart at Oroomiah, in Persia, could often see them under the latter conditions, as also could Mrs. McCance at Putney Hill, London, when II., III. and IV. were near together on October 12th, 1878. About 20 minutes before the auroral display on April 21st, 1859, Levander and others at Devizes were able to see two of the satellites in twilight. Dr. R. F. Hutchinson, of Mussoorie, India, has twice seen two of the moons with the

naked eye. Jacob, Banks, Webb, Mason, Boyd, Buffham and others also add their testimony of seeing these tiny bodies without artificial aid. The slightest optical aid will bring them in view. Their apparent stellar magnitudes being about 7, 7, 6 and 7. The tiny fifth satellite discovered by the great Lick telescope is too small to be considered in this matter, being hopelessly invisible, except with the very largest telescope in existence.

If it were only for their motions around Jupiter, the satellites would be very interesting; but their eclipses, occultations, transits and shadow transits add much for observation and study, as also do their variations in brightness. M. Camille Flammarion from observation in 1874-5 found that IV. was dark and sombre compared with the others, but its brightness is very variable, sometimes being equal to the 6th, and at other periods falling to the 10th magnitude. As regards dimensions he places the satellites in the decreasing order III., IV., I., II. For the intrinsic light reflected from equal areas of surface he places them in the order, I., II., III., IV., and found the decreasing order of their variability to be IV., I., II., III. The writer, as a result of observations from 1877 to 1881, found, in decreasing order, the intrinsic surface brightness, II., I., III., IV. Relative variability, IV., I., III., II. I. was usually brightest in the half of its orbit nearest to the earth; II. in the eastern half; and III. also in the eastern half; whilst IV. was dullest in the quadrant nearest the earth on the western side, and a little the brightest in the other quadrant on the same side. These observations lead to the conclusion that the Jovian satellites, like our own moon, turn once on their own axis during one revolution of their orbit; in other words, always present the same face to the planet. This result has been confirmed more or less completely by several observers at different times during the past century. The study is one which does not need great instrumental power; but it does necessitate persistent observation, month in and month out, on every available opportunity. It is necessary to note the brilliance of each satellite and also the position of each in its orbit east or west of the planet, and also whether in the superior—or farther, or inferior—or nearer, half of its path, which may be found by reference to the "Nautical Almanac." The time of each record should be noted. There must also be meteorological changes on these moons, for no revolution will alone explain many of the changes to be observed.

Notwithstanding that the moons, with the exception of V., are all more distant from Jupiter than our own moon is from the earth, their motions, owing to the much greater attractive power of their primary, are much quicker than that of the moon. I. has a month which lasts only 42 h. 29 m.; II., 3 d., 13 h. 13 m.; III., 7 d. 4 h.; and even IV. only 16 d., 18 h. 5 m. Barnard's tiny V. only takes 11 h. 49 m. to travel the circuit of its orbit. There is a peculiar relationship between the motions of the first three satellites which makes it impossible for them all three to be in conjunction together. If II. and III. are in conjunction, I. is always at the part of its orbit on the opposite side of the planet.

The colours of the moons are not identical. I. is usually found to be white or bluish, II. and III. yellow, and IV. dusky red.

(To be continued.)

CONDUCTED BY B. FOULKES-WINKS, M.R.P.S.

EXPOSURE TABLE FOR MAY.

The figures in the following table are worked out for plates of about 100 Hurter & Driffield. For plates of lower speed number give more exposure in proportion. Thus plates of 50 H. & D. would require just double the exposure. In the same way, plates of a higher speed number will require proportionately less exposure.

Time, 8 A.M. to 4 P.M.

Between 7 and 8 A.M. and 4 and 5 P.M. double the required exposure. Between 6 and 7 A.M. and 5 and 6 P.M. multiply by 4.

SUBJECT	F.5·6	F.8	F.11	F.16	.22	F.32	F.45	F.64
Sea and Sky ..	1/100	1/80	1/60	1/30	1/15	1/12	1/8	1/4
Open Landscape and Shipping	1/100	1/60	1/30	1/16	1/8	1/4	1/2	1
Landscape, with dark foreground, Street Scenes, and Groups ..	1/32	1/16	1/8	1/4	1/2	1	2	4
Portraits in Rooms ..	2	4	8	16	32	—	—	—
Light Interiors	4	8	16	32	1	2	4	8
Dark Interiors	16	32	1	2	4	8	16	30

The small figures represent seconds, large figures minutes. The exposures are calculated for sunshine. If the weather is cloudy, increase the exposure by half as much again ; if gloomy, double the exposure.

CLOUD NEGATIVES.—Owing to the sky in landscape pictures receiving, as a rule, some eight or ten times the normal exposure, it is seldom that we get any cloud effects in landscape negatives ; it therefore follows that we must look to some artificial means of inserting clouds in the picture. The course generally adopted is to print clouds into the picture by means of a cloud negative, which may either be purchased ready-made, or the photographer may prefer to make his own. We adopt this latter method, and would recommend the amateur to take every opportunity of securing cloud negatives on the same size plates that he generally uses for landscapes. These should be indexed and full data attached, so that one can be selected to suit any picture into which he may wish to print clouds. We would lay great stress upon this power of selection, as there is nothing so objectionable in a picture as to see, perhaps, heavy storm clouds printed into a picture of a bright sunlight landscape. A very common error with amateur photographers is to purchase one or two cloud negatives, and to print these into every picture they take. We need hardly point out the absurdity of this practice. Another method, and we think the most satisfactory, is to use an isochromatic screen in conjunction with an isochromatic plate, and if to this can be added a foreground shutter, fitted on the lens, the best possible relation of densities can be obtained, the different colours in the landscape are more correctly rendered, and the most pleasing amount of cloud effect will be secured. The intensity of the isochromatic screen we prefer is one that requires about three times the normal exposure, and we may say that we have found the B. J. Edwards Isochromatic Instantaneous Anti-Halo plates the most suitable for all-round purposes. If these plates are used, great care should be exercised during development, as they are sensitive to the yellow rays, and slightly so to the red. It is therefore necessary that all manipulations should be carried out in a deep ruby light, and we advise that the plate should be covered up as much as possible during development, only uncovering it occasionally to examine its progress. Where a more pronounced cloud effect is desired than will be obtained under the above conditions we recommend that the sky in the negative should be locally treated with a five per cent. solution of bromide of potassium. This is most conveniently applied by means of a small camel's-hair brush. Select one for this purpose that is made in a quill and wound with string ; do not on any account use a metal one. The clouds in a negative may sometimes be improved by a little judicious local reduction. For this purpose we advise a weak solution of persulphate of ammonium, which may be applied by means of a tuft of cotton-wool. There is another method of securing the true relation of clouds in landscape negatives that has distinct advantages over any other process where it is necessary to make very rapid exposures, and the isochromatic screen in consequence cannot be used. This method consists of making two negatives of the same view from the same position, one being exposed for the landscape and one for the clouds, the probable relation of exposure being one to eight—viz. the cloud negative will require about one-eighth the exposure required for the landscape negative. These negatives are developed in the usual way, and the print is then made from the landscape negative. As a rule, the sky in the landscape negative will be sufficiently dense to give a perfectly white sky on which the clouds in the cloud negative can then be printed. If the sky in the landscape negative should print through, so as to discolour the sky portion of the picture, it is advisable to paint out the sky with some opaque colour. This is best applied to the back or glass side of negative, which will then give a landscape picture with a pure white sky. The print is then adjusted over the cloud negative in precisely the same position, and the clouds allowed to print in, care being taken to shield the landscape portion of the print from the action of light by means of a card, or brown paper mask, cut to the shape of the horizon in the landscape.

THORNTON-PICKARD CATALOGUE.—We have just received a copy of the new Catalogue of the Thornton-Pickard Manufacturing Company. We notice this firm is placing on the market a "new patent iris shutter" which will work between the lenses. This should prove of great value to hand-camera makers, and amateurs generally. There is no doubt it can be relied upon as efficient, since all goods of this manufacture by this firm are of the most perfect workmanship and finish. The "Ruby" and "Amber" cameras still hold a front place.

PHOTOGRAPHY FOR BEGINNERS.

By B. Foulkes-Winks, M.R.P.S.

(*Continued from page* 350.)

Section I. Cameras (*continued*).

Bag-changing Cameras.—There are two principal makes of bag-changing cameras on the market at the present time. They are the "Newman & Guardia" and the "Adams." There are several others of minor importance, but as we are dealing more with types of cameras than with each in particular, we shall confine our remarks to these two which cover the whole principle of bag-changing. In the "Newman & Guardia" changing-box, the plates are inserted in metal sheaths, and then placed in the box, which is constructed to hold twelve plates, or twenty-four cut films. When the box is thus charged, the back is replaced in position, and the box is ready for use. The accompanying sketch shows the box in this condition ready to be inserted into the camera. Before making an exposure, it is necessary to draw out the black ebonite shutter, and this can remain withdrawn during the exposure and changing of the whole twelve plates, provided the box is being used in a camera fitted with a self-capping shutter such as is used in the N. & G. cameras. After an exposure has been made, the plate is changed by means of a lifter (see metal lifter on the back of the box, fig. 1). This lifter raises the plate into the bag. It is then taken hold of by the fingers and drawn right up into the bag and passed over to the

Fig. 1. Newman & Guardia Changing-box.

front of the box, when it is pushed down in front of the exposed plate. Fig. 2 shows the plate being changed, and just about to be pushed down in front of the exposed plate. There is a travelling partition supplied with each box: this consists of a sheath of aluminium with two springs attached. In filling the box with plates this dividing piece is inserted behind the first plate and in front of the second. This partition travels backwards as the plates are inserted in the front of box, in the process of changing, until it becomes the last sheath in the box. When in this position it indicates that all the plates have been changed; it also serves the purpose of keeping the front plate pressed up to the register of the box. There is an indicator fitted to each box which automatically registers the number of plates exposed. The changing-box is supplied with all the cameras made by this firm. The pattern "Special B" is the most popular form, and is made for two and three foci. It is made in three sizes—namely, $\frac{1}{4}$-plate, 5 × 4, and $\frac{1}{2}$-plate, and fitted with Zeiss-Satz lenses. The other manufactures are Pattern A

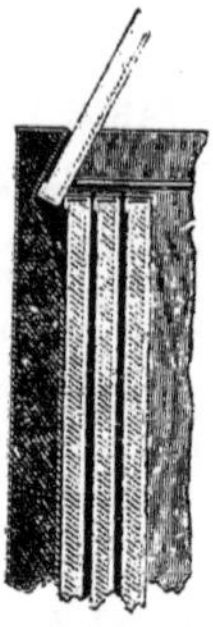

Fig. 2. N. & G. Box, showing Position of Plate when Changing.

for $\frac{1}{4}$-plate only and single focus, fitted either with Wray's $5\frac{1}{2}$-in. Rapid Rectilinear lens, working at F. 8, or Zeiss's Series II. $5\frac{1}{2}$-in. Anastigmat, working aperture F. 6·3. Pattern B is similar in every respect to Pattern A, except that it is made in 5 × 4 size. The "Special B" is made in three sizes, and is fitted only with the Zeiss-Satz lens, and is so constructed that the single lenses may be used at will; thus the $\frac{1}{4}$-plate size will allow the three different lenses of the Zeiss-Satz Anastigmat being used and give three lenses of the following foci: $5\frac{5}{8}$ in., 9 in., and $11\frac{1}{4}$ in., or the $\frac{1}{4}$-plate two foci will give two combinations of 5 in. and 9 in. equivalent foci. The 5 × 4 size will give the following combinations of the No. 7 Zeiss lens: $6\frac{1}{2}$ in. and $11\frac{1}{4}$ in. in the two foci, and 7 in., $11\frac{1}{4}$ in., and $13\frac{3}{4}$ in. in the three foci. The Special B in the $\frac{1}{2}$-plate size will give lenses of the following focal lengths: two foci No. 10, $7\frac{7}{8}$ in. and $13\frac{3}{4}$ in., and the three foci, $8\frac{1}{2}$ in., $13\frac{3}{4}$ in., and $16\frac{1}{4}$ in. Pattern H.S. is a camera specially adopted for high speed work, and is fitted with the Zeiss "Planar" lens, with a working aperture of F. 3·8. In addition to the ordinary shutter this camera is fitted with a focal plain shutter giving exposures from one-half to one-thousandth of a second. In addition to the above patterns Messrs. Newman & Guardia make a special camera for $\frac{1}{2}$-plate, a twin lens camera in $\frac{1}{4}$-plate and 5 × 4 size, and a stereoscopic camera, fitted with Goerz lenses.

The following specification is applicable to all the cameras of this manufacture. The body and front are built separately, and effectually conceal and protect the working parts. There are no projections of any kind. All the parts are arranged in the most convenient position for rapid work, whether used in the hand or on a tripod. Any lens or different lenses can be used, and all the patterns (except one) provide double or even triple extension. There is at all times easy access to the working parts, for the purpose of cleaning etc.

Every adjustment can be made from the outside, so that the camera need never be opened when in work, except for changing plates All the friction slides (rises, extension) are of metal, and capable of adjustment for wear. All the parts are detachable without removing the leather covering, and interchangeable. All scales are accurately measured and engraved. The front is fitted to the body by a slide, which allows it to be detached for cleaning, and gives the vertical rise. The bottom of the front conceals the index plate, protects the lever for setting the shutter and the iris diaphragm, the hand and pneumatic releases, etc.

Any lens, or several lenses, can be used. They are fitted with the utmost care, and the centring

FIG. 3. "N. & G." SPECIAL B.

and optical adjustments are rigorously tested. The Iris diaphragm is specially constructed: it cannot get out of order or wear shiny. The scale of apertures is measured and engraved for each lens. The shutter is of high efficiency, yet simple in construction. It works between the combinations, and is made entirely of metal. It is provided with hermetically closed pneumatic regulation, which ensures accuracy for years. It gives automatic exposures from $\frac{1}{2}$ to $\frac{1}{100}$th second and "time" at will. Its action can be seen from outside.

A self-capping device works automatically in front of the lens, and protects it from dust, spray, or other injuries. It also shuts off the light while the shutter is being set. Focussing is effected by central rackwork which acts with the utmost smoothness and precision. The knob is completely

FIG. 4. N. & G. SPECIAL B, SHOWING MOVEMENTS.

sunk in the side of the camera. The scales are divided by actual trial for each lens. Special facilities are provided for tripod work. The back of the camera forms a dark focussing chamber, and contains a strong focussing screen the frame of which is utilised for a variety of purposes. The instruments are complete in every respect, and provided with special finders and T-levels for horizontal and vertical pictures.

(*To be continued.*)

CORRESPONDENCE.

WE have pleasure in inviting any readers who desire to raise discussions on scientific subjects to address their letters to the Editor at 110 Strand, London, W.C. Our only restriction will be in case the correspondence exceeds the bounds of courtesy, which we trust is a matter of great improbability. These letters may be anonymous. In that case they must be accompanied by the full name and address of the writer, not for publication, but as an earnest of good faith. The Editor does not hold himself responsible for the opinions of the correspondents.—*Ed. S.-G.*

FORAMINIFERA.

To the Editor of SCIENCE-GOSSIP.

SIR,—Allow me to plead for further workers in the most interesting and little-worked branch of scientific research, the Foraminifera. The British Isles can boast of probably not more than some twenty serious students of this subject. Yet the study of Foraminifera, an order of "Rhizopoda," a class of the sub-kingdom "Protozoa," is so engrossing and full of interest that anyone with the least taste for research, or with even a modified tendency towards collecting, would only have to take a cursory glance through a microscope at some of these minute wonders of the marine floor to be at once seized with the desire to know and see more of them.

Two volumes of the "*Challenger* Report" are occupied by this one order of "Rhizopoda," which of itself will show what a large subject it embraces. One of the chief claims, in my opinion, that Foraminifera have to induce a beginner to study them is that one never knows at what moment some species, altogether new to science, may not be found in the field of the microscope. I need not expatiate on the pleasure of such discoveries. Again, new forms of known species are often found, and, if not quite in the same category as a new species, are sufficiently inspiriting to goad one on to further work.

If these minute shells are studied only with a view to collecting, I can conceive no more wonderful or beautiful collection when neatly arranged. If investigated with the object of advancing the world's knowledge, a tremendous scope for research and thought is open to all. Our field of operations, too, is so vast—viz. the bottom of all the oceans, seas and bays of the world—that some new knowledge is bound to be the result of energy in their investigation.

Another great advantage to anyone endeavouring to decide on some way of filling leisure time is that the cost of the apparatus and material necessary to study Foraminifera is not great. From £12 to £15, or less, will obtain practically all that is necessary for commencement. The requirements are: a microscope, binocular for choice, with inch and $\frac{3}{4}$-inch powers; a condenser, some microscope slides, a fine camel's-hair brush, and some marine ooze, mud, or sand. Other small requisites can be obtained for a mere trifle.

The sea bottom from almost any place is generally found to be teeming with minute animal life or their tests, otherwise shells. There are very few of us without the means of obtaining a sample from our own coasts, which have not yet been adequately searched, as is proved by the continual additions to the list of British Foraminifera. Therefore, without going further, plenty of interesting home research remains to be done by earnest workers. Fossil Foraminifera, if not so numerous, present another, perhaps more scientific, sphere of action, and possibly one of great geological interest. For example, suppose one has obtained from the Arctic regions a sample of some marine deposit of bygone days. Search among the material is practically sure of reward by finding specimens of Foraminifera, of course in a fossil condition, now only represented in its living or, as it is termed, "recent state" in the Equatorial regions of the earth. The inference to be drawn is that where the specimen on the slide was living it occurred in a climate not far removed in temperature from that of our Equatorial regions, where the identical species is now to be obtained in a living form, unless our material was carried by some prehistoric Gulf Stream, to the Arctic regions, and so deposited. Is not an enigma such as this enough to induce anyone to take up the study of those wonderful microscopic shells, and try to prove by analogy, circumstantial evidence, or some means why a species should be found fossil and living in such widely differing temperatures?

A similar class of scientific conundrums, if I may so call them, meets one quite as frequently in the study of recent animals. A species may be plentiful in the Adriatic, but not known to exist anywhere else except at Valparaiso, where the form is identical. Why should this be? Perhaps if sufficient search were made, which cannot be with the present limited number of workers, the form might, given the same conditions, be found at intermediate places, unless indeed environment and evolution have so altered the appearance of the species in the intermediate space that it is unrecognisable. It is improbable that a species would occur only in these two far-apart localities.

In conclusion, I would say that if this appeal for more workers should induce anyone to take up this study, the student will never regret its commencement. The pity is that life is too short, and that the days follow too quickly to enable one to work enough for satisfaction. I shall be most happy at any time to help, so far as I can, by advice or any other means in my power, anyone who may care to look even into the fringe of this far too little known study.

Yours faithfully,
W. B. THORNHILL.

Castle Cosey,
Castle Bellingham, Ireland.

NOTICES OF SOCIETIES.

*Ordinary meetings are marked †, excursions * ; names of persons following excursions are of Conductors. Lantern Illustrations §.*

NORTH LONDON NATURAL HISTORY SOCIETY.

May 2.—† "Icework in Britain." Miss H. K. Brown.
„ 16.—* Evening at Winchmore Hill. O. G. Pike.
„ 24.—* New Forest. R. W. Robbins.
„ 27.—* Cuxton. J. A. Simes.

LAMBETH FIELD CLUB AND SCIENTIFIC SOCIETY.

May 6.—§ "Wild Flowers at Home." Edward Step, F.L.S.
„ 11.—* Epping Forest. C. H. Cooper.
„ 18.—* Box Hill and Leatherhead (Geological).
„ 27.—* Oxted and Limpsfield. W. Rivers.

NOTTINGHAM NATURAL SCIENCE RAMBLING CLUB.

May 4.—* Stapleford.

MANCHESTER MUSEUM, OWENS COLLEGE.

May 27.—† "Evolution." W. E. Hoyle.

GEOLOGISTS' ASSOCIATION.

May 4.—* Swanscomb. A. E. Salter, B.Sc., F.G.S.
„ 11.—*Leighton Buzzard. A. M. Davies, B.Sc., F.G.S.
„ 18.—* Grays. M. A. C. Hinton and A. S. Kennard.
„ 25.—* The New (G.W.R.) line from Wootton Bassett to Patchway. Rev. H. H. Winwood, M.A., F.G.S.
„ 29.—* Bristol and Dundry Hill. Rev. H. H. Winwood.

PRESTON SCIENTIFIC SOCIETY.

May 1.—† "Flora." W. Clitheroe.

LONDON GEOLOGICAL FIELD CLASS.

May 4.—* Arlesey. Professor H. G. Seeley, F.R.S.
„ 11.—* Gomshall. Professor H. G. Seeley, F.R.S.
„ 18.—* Aylesford to Maidstone. Professor H. G. Seeley.

NOTICES TO CORRESPONDENTS.

TO CORRESPONDENTS AND EXCHANGERS.—SCIENCE-GOSSIP is published on the 25th of each month. All notes or short communications should reach us not later than the 18th of the month for insertion in the following number. No communications can be inserted or noticed without full name and address of writer. Notices of changes of address admitted free.

BUSINESS COMMUNICATIONS.—All business communications relating to SCIENCE-GOSSIP must be addressed to the Manager, SCIENCE-GOSSIP, 110 Strand, London.

EDITORIAL COMMUNICATIONS, articles, books for review, instruments for notice, specimens for identification, etc., to be addressed to JOHN T. CARRINGTON, 110 Strand, London, W.C.

SUBSCRIPTIONS.—The volumes of SCIENCE-GOSSIP begin with the June numbers, but Subscriptions may commence with any number, at the rate of 6s. 6d. for twelve months (including postage), and should be remitted to the Manager, SCIENCE-GOSSIP, 110 Strand, London, W.C.

EXCHANGES.

NOTICE.—Exchanges extending to thirty words (including name and address) admitted free; but additional words must be prepaid at the rate of threepence for every seven words or less.

OFFERED.—Collection of English wild plants, mounted on cartridge paper, also small collection of West Australian flowers. Several "Hobbies." What offers?—Miss Graham, 42 Tisbury Road, Hove, Brighton.

OFFERED.—Clau. dubia, Sph. lacustre, Cy. elegans, Hel. ericetorum, H. rupestris, Cardium echinatum, etc. Wanted, Kalmia and Prunus cuttings, young greenhouse plants and bulbs. No shells.—A. Whitworth, St. John's, Isle of Man.

WANTED.—"Science-Gossip," Nos. 13 to 24, 205 to 216, 229 to 240 (O.S.), and Nos. 21, 22, 24 to 28, 38, 41, 47, and 75 (N.S.). Full value given in British L., F.W., and Marine Shells.—W. Hy. Heathcote, F.L.S., Preston.

WANTED.—2 or 2½ inch Telescope with Astronomical Eyepiece. Offered, Heads or Horns of South African Antelopes.—J. G. Brown, Pell Street, Port Elizabeth, South Africa.

WANTED.—Geological material for Microscopic mounting in exchange for sawn sections or mounted sections, or would saw sections of material to share mutually.—A. Stott, Five Ways, Aldridge, Walsall.

CONTENTS.

Lightning Source UK Ltd.
Milton Keynes UK
UKHW020744251118
332796UK00002B/177/P

9 780332 512396